4TH EDITION

THE
ANATOMY
AND PHYSIOLOGY
Learning System

Edith Applegate, MS
Professor of Science and Mathematics
Kettering College of Medical Arts
Kettering, Ohio

SAUNDERS

ELSEVIER

SAUNDERS
ELSEVIER

3251 Riverport Lane
St. Louis, Missouri 63043

Library of Congress Cataloging-in-Publication Data

Applegate, Edith J.
  The anatomy and physiology learning system / Edith Applegate. – 4th ed.
    p. ; cm.
  Includes index.
  ISBN 978-1-4377-0393-1 (pbk. : alk. paper) 1. Human physiology. 2. Human anatomy. I. Title.
    [DNLM: 1. Anatomy. 2. Physiology. QS 4 A648a 2011]
  QP34.5.A67 2011
  612–dc22
                                    2010002996

Executive Editor: Suzi Epstein
Developmental Editor: Lynda Huenefeld
Publishing Services Manager: Deborah L. Vogel
Project Manager: Pat Costigan
Design Direction: Karen Pauls

Printed in India

Last digit is the print number: 11

Preface

The Anatomy and Physiology Learning System includes a textbook, a student study guide, and an Evolve site to help instructors present a quality educational experience for students preparing to enter one of the health-related professions. It is designed for students in college, technical school, adult education, and career training, either private or public. Used together, the textbook and student study guide are effective as a self-study A&P review for licensure examinations. The textbook presents fundamental information and concepts in a manner that encourages learning and understanding. The topics are consistent with those needed by students in the health-related professions. Students with minimal science background, as well as those with greater preparation, will appreciate the clear and concise writing style that makes the book easy to read and comprehend. Numerous clinical applications add interest and provide a springboard for discussions or projects. The colorful illustrations in the textbook make it visually attractive and facilitate learning.

The new edition includes a combined vocabulary section called **Building Your Medical Vocabulary** that will help students learn etymology of terms, giving them the building blocks to use and recognize new terms. A new **Vocabulary Quiz** at the end of each chapter stresses the importance of these terms while helping students learn them. One hundred **new colorful illustrations** have been designed exclusively for this new edition and will help drive home concepts while keeping students visually stimulated.

The textbook component is comprehensive and can be used as a stand-alone item. The study guide, which follows the same outline as the textbook, provides an added dimension to the learning process in the form of written exercises and quizzes that reinforce facts and concepts. Used together, the textbook and study guide provide a package that facilitates learning, enhances understanding, and relates information to medical applications. It offers a strong foundation that enables students to apply their knowledge to further study within their chosen health-related professions.

Edith Applegate

Acknowledgments

This book, like all others, reflects the efforts of many people. Foremost is my husband, Stan Applegate, who did a major portion of the data input, helped with proofreading, and cautiously critiqued my work. He was truly a partner in this entire project and patiently encouraged me throughout the process. From this project's inception to its completion, the editorial staff at Elsevier has been a source of support and encouragement. The production staff kept the project on schedule even if I missed deadlines. Designers and artists met the challenge of designing a book that is new, fresh, unique, and exciting in appearance. The copy editors and proofreaders were gracious and kind when correcting my errors. I greatly appreciate the work of these and many others I have not mentioned. Thank you to everyone who has had a part in the completion of *The Anatomy and Physiology Learning System*, Fourth Edition.

Dedicated to

FAMILY
who have loved me;
sons, **David** *and* **Douglas,**
who bring pride and joy to my life;
my best friend and husband for over fifty years, **Stan,**
who makes life a joyous experience to be treasured.

FRIENDS
editors who have stimulated me;
students who have challenged me;
colleagues who have encouraged me;
others who have enriched my life.

Note to Students

Dear Student,

You are starting on an adventure that is interesting, rewarding, exciting, and fascinating. It will also be challenging and require a significant portion of your study time. The human body is intricate, complex, and marvelous. Your adventure through the body, although challenging, will be rewarding because you will learn more about yourself and it will help prepare you to serve in the health care professions. As you study, relate what you learn to your own body. It will make the study more relevant to you.

Anatomy and physiology forms the foundation for course work in the health related professions. Nearly every day, students tell me that what they learned in anatomy and physiology has helped them in some other course. Learn the subject well. You'll be glad you did.

I have written *The Anatomy and Physiology Learning System* with you in mind. It contains the topics that are most beneficial to students in health-related professions. I have included features and study aids that will help you learn. The **chapter outline and objectives** help to keep you organized. Be sure to learn the **key terms.** Answer the **quick check** questions as you progress through the chapter and test yourself with the **chapter quizzes.** Practice your use of scientific and medical terms with the **vocabulary quizzes.** Review the **functional relationships** page to understand how all body systems work together.

The *Learning System* includes a study guide to complement the textbook. The organization of the study guide matches that in the textbook. It includes **learning exercises** and **labeling/coloring exercises** that reinforce what you learn from class and the textbook. The **review questions** and **vocabulary practice** provide additional ways to master the material. The **testing comprehension** pages give you an opportunity to demonstrate what you have learned. Finally, you can relax and have a little fun with the **fun and games** pages. These study tools offer multiple ways of going over the material so you can maintain concentration and they have been used successfully by thousands of students.

If computers are your "thing," you should try the **Evolve** website that is maintained by Elsevier, Inc. to complement *The Anatomy and Physiology Learning System.* The site contains additional information and activities that will enhance your learning experience.

Study methods are an individual thing. Not all methods work for all people, and you have to select those that work best for you. I will share some hints that have worked for my students.

- Read the appropriate material in the textbook <u>before</u> class. By doing this you will gain more from the class because you are familiar with the terminology and ideas.
- Attend class faithfully, sit near the front if possible, pay attention to the teacher, and maintain a notebook with good notes.
- Use all the study aids available to you because variety adds interest and improves concentration. Repetition, repetition, repetition helps to reinforce the material.
- Study **every day**. This keeps the material fresh in your mind.
- Keep current. Thoroughly understand one day's material before your next class appointment.
- Study for understanding rather than rote memory.
- If there is something you don't understand, make an appointment with your teacher for further explanation. Make your teacher earn his or her paycheck!

It looks like my brief note to you has turned into a long letter and I know you are anxious to get started on your adventure in anatomy and physiology. I hope you enjoy your study of the human body. In closing, I offer my best wishes for success in this course, in your selected curriculum, and in your chosen career.

Sincerely,

Edith Applegate
Kettering College of Medical Arts
3737 Southern Boulevard
Kettering, OH 45429-1299

Contents

CHAPTER 1
Introduction to the Human Body   1

Anatomy and Physiology   2
Levels of Organization   2
Organ Systems in the Body   3
Life Processes   6
Environmental Requirements for Life   7
Homeostasis   7
  Negative Feedback   8
  Positive Feedback   8
Anatomical Terms   9
  Anatomical Position   9
  Directions in the Body   9
  Planes and Sections of the Body   10
  Body Cavities   10
  Regions of the Body   11
Chapter Summary   14
Building Your Medical Vocabulary   16

CHAPTER 2
Chemistry, Matter, and Life   21

Elements   22
Structure of Atoms   23
  Isotopes   23
Chemical Bonds   24
  Ionic Bonds   24
  Covalent Bonds   25
  Hydrogen Bonds   27
Compounds and Molecules   27
  Nature of Compounds   27
  Formulas   28
Chemical Reactions   28
  Chemical Equations   28
  Types of Chemical Reactions   29
  Reaction Rates   29
  Reversible Reactions   30
Mixtures, Solutions, and Suspensions   31
  Mixtures   31
  Solutions   31
  Suspensions   31
Electrolytes, Acids, Bases, and Buffers   31
  Electrolytes   31
  Acids   31

Bases   32
The pH Scale   32
Neutralization Reactions   33
Buffers   33
Organic Compounds   33
  Carbohydrates   33
  Proteins   35
  Lipids   36
  Nucleic Acids   37
  Adenosine Triphosphate   38
Chapter Summary   39
Building Your Medical Vocabulary   42

CHAPTER 3
Cell Structure and Function   45

Structure of the Generalized Cell   46
  Plasma Membrane   46
  Cytoplasm   48
  Nucleus   48
  Cytoplasmic Organelles   49
  Filamentous Protein Organelles   50
Cell Functions   51
  Movement of Substances Across the Cell Membrane   51
  Filtration   54
  Cell Division   56
  DNA Replication and Protein Synthesis   58
Chapter Summary   63
Building Your Medical Vocabulary   65

CHAPTER 4
Tissues and Membranes   69

Body Tissues   70
  Epithelial Tissue   70
  Connective Tissue   74
  Muscle Tissue   77
  Nervous Tissue   78
Inflammation and Tissue Repair   79
  Inflammation   79
  Tissue Repair   79
Body Membranes   80
  Mucous Membranes   80
  Serous Membranes   80

Synovial Membranes   81
Meninges   81
Chapter Summary   82
Building Your Medical Vocabulary   84

CHAPTER 5
Integumentary System   87

Structure of the Skin   89
Epidermis   89
Dermis   90
Subcutaneous Layer   91
Skin Color   91
Epidermal Derivatives   91
Hair and Hair Follicles   91
Nails   92
Glands   92
Functions of the Skin   93
Protection   93
Sensory Reception   93
Regulation of Body Temperature   94
Synthesis of Vitamin D   94
Burns   94
Chapter Summary   97
Building Your Medical Vocabulary   99

CHAPTER 6
Skeletal System   101

**Overview of the
Skeletal System   103**
Functions of the Skeletal System   103
Structure of Bone Tissue   103
Classification of Bones   104
General Features of a Long Bone   104
Bone Development and Growth   105
Divisions of the Skeleton   107
Bones of the Axial Skeleton   107
Skull   107
Hyoid Bone   113
Vertebral Column   113
Thoracic Cage   116
Bones of the Appendicular Skeleton   117
Pectoral Girdle   117
Upper Extremity   118
Pelvic Girdle   121
Lower Extremity   123

Fractures and Fracture Repair   126
Articulations   127
Synarthroses   127
Amphiarthroses   127
Diarthroses   128
Chapter Summary   132
Building Your Medical Vocabulary   135

CHAPTER 7
Muscular System   139

**Characteristics and Functions of the Muscular
System   141**
Structure of Skeletal Muscle   141
Whole Skeletal Muscle   141
Skeletal Muscle Attachments   142
Skeletal Muscle Fibers   142
Nerve and Blood Supply   143
Contraction of Skeletal Muscle   143
Stimulus for Contraction   143
Sarcomere Contraction   144
Contraction of a Whole Muscle   145
Energy Sources and Oxygen Debt   146
Movements Produced by Skeletal Muscles   149
Skeletal Muscle Groups   150
Naming Muscles   150
Muscles of the Head and Neck   152
Muscles of the Trunk   153
Muscles of the Upper Extremity   156
Muscles of the Lower Extremity   159
Chapter Summary   164
Building Your Medical Vocabulary   167

CHAPTER 8
Nervous System   169

Functions of the Nervous System   171
Organization of the Nervous System   171
Nerve Tissue   171
Nerve Impulses   174
Resting Membrane   174
Stimulation of a Neuron   174
Conduction Along a Neuron   175
Conduction Across a Synapse   176
Reflex Arcs   177
Central Nervous System   178
Meninges   178
Brain   179

Ventricles and Cerebrospinal Fluid   184
Spinal Cord   184
Peripheral Nervous System   187
Structure of a Nerve   187
Cranial Nerves   187
Spinal Nerves   188
Autonomic Nervous System   188
Chapter Summary   195
Building Your Medical Vocabulary   198

CHAPTER 9
The Senses   201

Receptors and Sensations   202
General Senses   202
Touch and Pressure   202
Proprioception   202
Temperature   202
Pain   203
Gustatory Sense   203
Olfactory Sense   204
Visual Sense   204
Protective Features and Accessory Structures
of the Eye   205
Structures of the Eyeball   206
Pathway of Light and Refraction   208
Photoreceptors   208
Visual Pathway   209
Auditory Sense   210
Structure of the Ear   210
Physiology of Hearing   212
Pitch and Loudness   213
Sense of Equilibrium   213
Static Equilibrium   214
Dynamic Equilibrium   214
Chapter Summary   217
Building Your Medical Vocabulary   220

CHAPTER 10
Endocrine System   223

Introduction to the Endocrine System   225
Comparison of the Endocrine and Nervous
Systems   225
Comparison of the Exocrine and Endocrine
Glands   225
Characteristics of Hormones   225
Chemical Nature of Hormones   225
Mechanism of Hormone Action   225
Control of Hormone Action   226

Endocrine Glands and Their Hormones   226
Pituitary Gland   226
Thyroid Gland   230
Parathyroid Glands   231
Adrenal (Suprarenal) Glands   232
Pancreas—Islets of Langerhans   235
Gonads (Testes and Ovaries)   236
Pineal Gland   237
Other Endocrine Glands   237
Prostaglandins   237
Chapter Summary   240
Building Your Medical Vocabulary   242

CHAPTER 11
Cardiovascular System: The Heart   245

Overview of the Heart   247
Form, Size, and Location of the Heart   247
Coverings of the Heart   247
Structure of the Heart   248
Layers of the Heart Wall   248
Chambers of the Heart   248
Valves of the Heart   249
Pathway of Blood Through the Heart   251
Blood Supply to the Myocardium   251
Physiology of the Heart   252
Conduction System   252
Cardiac Cycle   254
Heart Sounds   255
Cardiac Output   255
Chapter Summary   257
Building Your Medical Vocabulary   259

CHAPTER 12
Cardiovascular System: Blood Vessels   263

Classification and Structure of Blood Vessels   264
Arteries   264
Capillaries   265
Veins   265
Physiology of Circulation   266
Role of the Capillaries   266
Blood Flow   267
Pulse and Blood Pressure   269
Circulatory Pathways   272
Pulmonary Circuit   272
Systemic Circuit   272
Fetal Circulation   283

Chapter Summary   285
Building Your Medical Vocabulary   287

CHAPTER 13
Cardiovascular System: Blood   289

Functions and Characteristics of the Blood   290
Composition of the Blood   290
   Plasma   290
   Formed Elements   291
Hemostasis   295
   Vascular Constriction   296
   Platelet Plug Formation   296
Blood Typing and Transfusions   297
   Agglutinogens and Agglutinins   297
   ABO Blood Groups   297
   Rh Blood Groups   299
Chapter Summary   301
Building Your Medical Vocabulary   304

CHAPTER 14
Lymphatic System and Body Defense   307

Functions of the Lymphatic System   309
Components of the Lymphatic System   309
   Lymph   310
   Lymphatic Vessels   310
   Lymphatic Organs   310
Resistance to Disease   312
   Nonspecific Defense Mechanisms   312
   Specific Defense Mechanisms   315
   Acquired Immunity   319
Chapter Summary   323
Building Your Medical Vocabulary   325

CHAPTER 15
Respiratory System   327

Functions and Overview of Respiration   329
Ventilation   329
   Conducting Passages   329
   Mechanics of Ventilation   333
   Pressures in Pulmonary Ventilation   333
   Respiratory Volumes and Capacities   335
Basic Gas Laws and Respiration   335
   Properties of Gases   335
   External Respiration   337
   Internal Respirations   337

Transport of Gases   338
   Oxygen Transport   338
Regulation of Respiration   340
   Respiratory Center   340
   Factors That Influence Breathing   340
   Nonrespiratory Air Movements   342
Chapter Summary   345
Building Your Medical Vocabulary   348

CHAPTER 16
Digestive System   351

Overview and Functions of the Digestive System   353
General Structure of the Digestive Tract   353
Components of the Digestive Tract   354
   Mouth   354
   Pharynx   357
   Esophagus   357
   Stomach   357
   Small Intestine   359
   Large Intestine   362
Accessory Organs of Digestion   363
   Liver   363
   Gallbladder   365
   Pancreas   365
Chemical Digestion   367
   Carbohydrate Digestion   367
   Protein Digestion   367
   Lipid Digestion   368
Absorption   368
Chapter Summary   372
Building Your Medical Vocabulary   375

CHAPTER 17
Metabolism and Nutrition   377

Metabolism of Absorbed Nutrients   378
   Anabolism   378
   Catabolism   378
   Energy from Foods   378
   Heat Production   384
   Heat Loss   384
   Temperature Regulation   384
Concepts of Nutrition   384
   Carbohydrates   385
   Proteins   385
   Lipids   386

Vitamins   387
Minerals   388
Water   388
Chapter Summary   390
Building Your Medical Vocabulary   392

CHAPTER 18
Urinary System and Body Fluids   395

Functions of the Urinary System   397
Components of the Urinary System   397
Kidneys   397
Ureters   400
Urinary Bladder   401
Urethra   401
Urine Formation   402
Glomerular Filtration   402
Tubular Reabsorption   402
Tubular Secretion   404
Regulation of Urine Concentration and Volume   404
Micturition   405
Characteristics of Urine   406
Physical Characteristics   406
Chemical Composition   406
Abnormal Constituents   406
Body Fluids   407
Fluid Compartments   407
Intake and Output of Fluid   407
Electrolyte Balance   408
Acid-Base Balance   409
Chapter Summary   413
Building Your Medical Vocabulary   416

CHAPTER 19
Reproductive System   419

Overview of the Reproductive System   421
Male Reproductive System   421
Testes   421
Duct System   423
Accessory Glands   424
Penis   424
Male Sexual Response   425
Hormonal Control   425
Female Reproductive System   426
Ovaries   426
Genital Tract   429

External Genitalia   430
Female Sexual Response   430
Hormonal Control   430
Mammary Glands   433
Chapter Summary   436
Building Your Medical Vocabulary   439

CHAPTER 20
Development and Heredity   441

Fertilization   442
Preembryonic Period   442
Cleavage   442
Implantation   444
Formation of Primary Germ Layers   445
Embryonic Development   445
Formation of Extraembryonic Membranes   446
Formation of the Placenta   446
Organogenesis   447
Fetal Development   447
Parturition and Lactation   448
Labor and Delivery   448
Adjustments of the Infant at Birth   450
Physiology of Lactation   451
Postnatal Development   451
Neonatal Period   451
Infancy   452
Childhood   452
Adolescence   452
Adulthood   452
Senescence   452
Heredity   452
Introduction   452
Chromosomes and Genes   452
Gene Expression   453
Genetic Mutations   455
Hereditary Disorders   455
Gene Therapy   457
Chapter Summary   458
Building Your Medical Vocabulary   461
Answers for QuickCheck Questions   463
General Glossary   467

Introduction to the Human Body

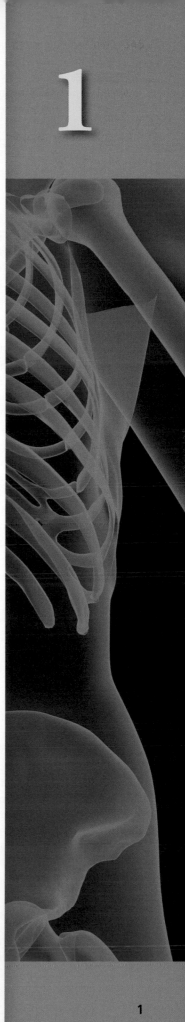

1

CHAPTER OBJECTIVES

Anatomy and Physiology
- Define the terms *anatomy* and *physiology* and discuss the relationship between these two areas of study.

Levels of Organization
- List the six levels of organization within the human body.

Organ Systems in the Body
- Name the 11 organ systems of the body and briefly describe the major role of each system.

Life Processes
- List and define 10 life processes in the human body.

Environmental Requirements for Life
- List five physical environmental factors necessary for survival of the individual.

Homeostasis
- Discuss the concept of homeostasis.
- Distinguish between negative feedback mechanisms and positive feedback mechanisms.

Anatomical Terms
- Describe the four criteria that are used to describe the anatomical position.
- Use anatomical terms to describe body planes, body regions, and relative positions.
- Distinguish between the dorsal body cavity and the ventral body cavity, and list the subdivisions of each cavity.

KEY TERMS

Anabolism (ah-NAB-oh-lizm) Building up, or synthesis of, reactions that require energy and make complex molecules out of two or more smaller ones

Anatomical position (an-ah-TOM-ih-kul poh-ZIH-shun) Standard reference position for the body

Anatomy (ah-NAT-o-mee) Study of body structure and the relationships of its parts

Catabolism (kah-TAB-oh-lizm) Reactions that break down complex molecules into two or more smaller ones with the release of energy

Differentiation (dif-er-en-she-AY-shun) Process by which cells become structurally and functionally specialized

Homeostasis (hoh-mee-oh-STAY-sis) Normal stable condition in which the body's internal environment remains the same; constant internal environment

Metabolism (meh-TAB-oh-lizm) Total of all biochemical reactions that take place in the body; includes anabolism and catabolism

Negative feedback (NEG-ah-tiv FEED-bak) Mechanism of response in which a stimulus initiates reactions that reduce the stimulus

Physiology (fiz-ee-AHL-oh-jee) Study of the functions of living organisms and their parts

Positive feedback (POS-ih-tiv FEED-back) Mechanism of response in which a stimulus initiates reactions that increase the stimulus and the reaction keeps building until a culminating event occurs that halts the process

The **human body** is an awesome masterpiece. Imagine billions of microscopic parts, each with its own identity, working together in an organized manner for the benefit of the total being. The human body is more complex than the greatest computer, yet it is very personal and efficient. The study of the human body is as old as history itself because people have always had an interest in how the body is put together, how it works, why it becomes defective (illness), and why it wears out (aging).

The study of the human body is essential for those planning a career in health sciences just as knowledge about automobiles is necessary for those planning to repair them. How can you fix an automobile if you do not know how it is put together or how it works? How can you help fix a human body if you do not know how it is put together or how it works?

Knowledge of the human body is also beneficial to the non–health care professional. Using this knowledge will help you keep your body healthy. It will help you rate your activities as being beneficial or detrimental to your body. Knowledge of anatomy and physiology makes you better able to communicate with medical personnel, understand treatments that may be prescribed, and critically evaluate advertisements and reports in magazines. In addition to all this, the study of the human body is appealing. It lets you learn more about yourself.

Anatomy and physiology are dynamic, applied sciences. New technology leads to new discoveries that may change the way certain aspects are viewed. For the student to keep up with the new discoveries and changes, it is necessary to have a strong foundation on which to build new knowledge. This book is an introduction that provides a basis for the continuing study of anatomy and physiology, whether you are a health care provider or a health care consumer.

Anatomy and Physiology

Anatomy (ah-NAT-oh-mee) is the scientific study of the structure or morphology of organisms and their parts. Human anatomy is the study of the shape and structure of the human body and its parts. It encompasses a wide range of study, including the development and microscopic organization of structures, the relationship between structures, and the interrelationship between structure and function. The broad field of anatomy has numerous subdivisions. Gross human anatomy deals with the large structures of the human body that can be seen through normal dissection. It includes topics in surface anatomy, systemic anatomy, regional anatomy, and surgical anatomy. Microscopic anatomy deals with the smaller structures and fine detail that can be seen only with the aid of a microscope. Cytology and histology are subdivisions of microscopic anatomy.

Physiology (fiz-ee-AHL-oh-jee) is the scientific study of the functions or processes of living things. It answers how, what, and why anatomical parts work. Some physiologists specialize in the study of a particular system, such as the digestive system. This is called **systemic physiology**. Many people currently studying in the field of physiology deal with individual cells and how they work. This is called **cellular physiology**. Another aspect of physiology is **immunology**, which is the study of the body's defense mechanisms. **Pharmacology** is the study of drug action in the body. This is also under the broad heading of physiology.

Anatomy and physiology are interrelated because structure and function are always closely associated. The function of an organ, or how it works, depends on how it is put together. Conversely, the anatomy or structure provides clues to understanding how it works. The structure of the hand, with its long, jointed fingers, is related to its function of grasping things. The heart is designed as a muscular pump that can contract to force blood into the blood vessels. By contrast, the lungs, whose function is the exchange of oxygen and carbon dioxide between the outside environment and the blood, are made of a very thin tissue. Imagine what would happen if the heart were made of thin tissue and the lungs were made of thick muscle. Structure and function are always related. **Diagnostic imaging** and **pathology** are two areas of medicine that encompass both anatomy and physiology. In diagnostic imaging, the structures and functions within the body are observed by using various imaging techniques. The pathologist studies structural and functional changes associated with disease.

QUICK **APPLICATIONS**
Often congenital anatomical abnormalities must be surgically repaired so that disruptions in physiology are corrected. For example, a cleft palate (anatomy) is repaired so that food will enter (physiology) the pharynx instead of the nasal cavity. Broken bones (anatomy) are reset so that function (physiology) is restored.

Levels of Organization

One of the most outstanding features of the complex human body is its order and organization—how all the parts, from tiny atoms to visible organs, work together to make a functioning whole. There are six levels to the organizational scheme of the body (Figure 1-1).

Starting with the simplest and proceeding to the most complex, the six levels of organization are chemical, cellular, tissue, organ, body system, and total organism. The structural and functional characteristics of all organisms are determined by their chemical makeup. The **chemical** level, discussed in Chapter 2, deals with the interactions of atoms (such as hydrogen and oxygen) and their combinations into molecules (such as water). Molecules contribute to the makeup of a cell, which is the basic unit of life. **Cells**, discussed in Chapter 3, are the basic living units of all organisms. Estimates indicate that there are about 100 trillion dynamic, living cells in the human body. These cells represent a variety of sizes, shapes, and structures and provide a vast array of functions. Cells with similar structure and function are grouped together as **tissues**. All of the tissues of the body are grouped into four main types: epithelial, connective, muscle, and nervous. Chapter 4 addresses the tissue level of organization. Two or more tissue types that combine to form a more complex structure and work together to perform one or more

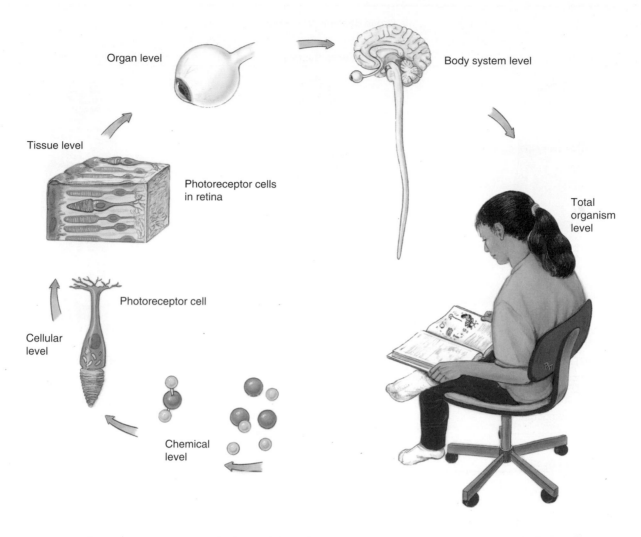

Figure 1-1 *Organizational scheme of the body. From simple to complex, the levels are chemical, cellular, tissue, organ, body system, and total organism.*

functions make up **organs**, the next higher level of organization. Examples of organs include the skin, heart, ear, stomach, and liver. A **body system** consists of several organs that work together to accomplish a set of functions. Some examples of body systems include the nervous system, the digestive system, and the respiratory system. Finally, the most complex of all the levels is the **total organism**, which is made up of several systems that work together to maintain life. Beginning with Chapter 5, this book deals primarily with the organs and organ systems that make up the total organism, the human body.

QUICK CHECK

1.1 Would an immunologist be considered an anatomist or a physiologist? Why?

1.2 Would an anatomist be more likely to study the structure or the function of the human body?

1.3 On an organizational scale, are organs more complex or less complex than tissues?

Organ Systems in the Body

There are 11 major organ systems in the human body, each with specific functions, yet all are interrelated and work together to sustain life. Each system is described briefly here and then in more detail in later chapters. Figure 1-2 illustrates and summarizes the organ systems.

Integumentary System

Integument means "skin." The **integumentary** (in-teg-yoo-MEN-tar-ee) system consists of the skin and the various accessory organs associated with it. These accessories include hair, nails, sweat glands, and sebaceous (oil) glands. The components of the integumentary system protect the underlying tissues from injury, protect against water loss, contain sense receptors, assist in temperature regulation, and synthesize chemicals to be used in other parts of the body.

Skeletal System

The **skeletal** (SKEL-eh-tull) system forms the framework of the body and protects underlying organs such as the brain, lungs,

Organ Systems of the Body

Integumentary System	Skeletal System	Muscular System

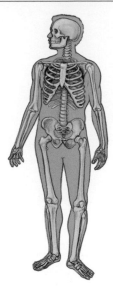
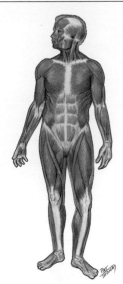

COMPONENTS: Skin, hair, nails, sweat and sebaceous glands

FUNCTIONS: Covers and protects body; regulates temperature

COMPONENTS: Bones, cartilage, ligaments

FUNCTIONS: Provides body framework and support; protects; attaches muscles to bones; provides calcium storage

COMPONENTS: Muscles

FUNCTIONS: Produces movement; maintains posture; provides heat

Nervous System	Endocrine System	Cardiovascular System

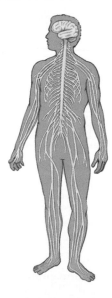
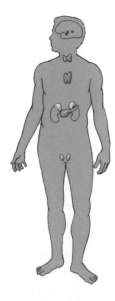
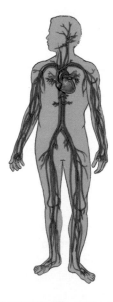

COMPONENTS: Brain, spinal cord, nerves, sense receptors

FUNCTIONS: Coordinates body activities; receives and transmits stimuli

COMPONENTS: Pituitary, adrenal, thyroid, other ductless glands

FUNCTIONS: Regulates metabolic activities and body chemistry

COMPONENTS: Heart, blood vessels, blood

FUNCTIONS:Transports material from one part of the body to another; defends against disease

Figure 1-2 *Organ systems of the body.*

Organ Systems of the Body—cont'd

Lymphatic System

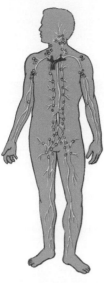

COMPONENTS: Lymph, lymph vessels, lymphoid organs

FUNCTIONS: Returns tissue fluid to the blood; defends against disease

Digestive System

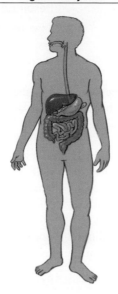

COMPONENTS: Mouth, esophagus, stomach, intestines, liver, pancreas

FUNCTIONS: Ingests and digests food; absorbs nutrients into blood

Respiratory System

COMPONENTS:Air passageways, lungs

FUNCTIONS: Exchanges gases between blood and external environment

Urinary System

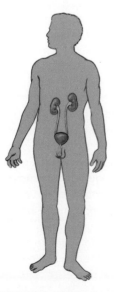

COMPONENTS: Kidneys, ureters, urinary bladder, urethra

FUNCTIONS: Excretes metabolic wastes; regulates fluid balance and acid-base balance

Reproductive System

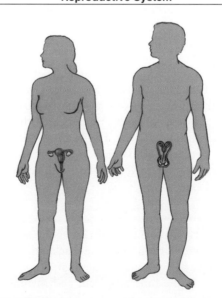

COMPONENTS:Testes, ovaries, accessory structures

FUNCTIONS: Forms new individuals to provide continuation of the human species

Figure 1-2, Cont'd.

and heart. It consists of the bones and joints along with ligaments and cartilage that bind the bones together. Bones serve as attachments for muscles and act with the muscles to produce movement. Tissues within bones produce blood cells and store inorganic salts containing calcium and phosphorus.

Muscular System

Muscles are the organs of the **muscular** (MUS-kyoo-lar) system. As muscles contract, they create the forces that produce movement and maintain posture. Muscles can store energy in the form of glycogen and are the primary source of heat within the body.

Nervous System

The **nervous** (NER-vus) system consists of the brain, spinal cord, and associated nerves. These organs work together to coordinate body activities. Nerve cells, or neurons, are specialized to transmit impulses from one point to another. In this way, body parts can communicate with each other and with the outside environment. Some nerve cells have special endings called sense receptors that detect changes in the environment.

Endocrine System

The **endocrine** (EN-doh-krin) system includes all the glands that secrete chemicals called hormones. These hormones travel through the blood and act as messengers to regulate cellular activities. The endocrine and nervous systems work together to coordinate and regulate body activities to maintain a proper balance. The nervous system typically acts quickly, whereas the endocrine system acts slowly but with a more sustained effect. The endocrine system also regulates reproductive functions in both males and females.

Cardiovascular System

The **cardiovascular** (kar-dee-oh-VAS-kyoo-lar) system consists of the blood, heart, and blood vessels. The blood transports nutrients, hormones, and oxygen to tissue cells and removes waste products such as carbon dioxide. Certain cells within the blood defend the body against disease. The heart acts as a pump to create the forces necessary to maintain blood pressure and to circulate the blood. The blood vessels serve as pipes or channels for the flow of blood.

Lymphatic System

The **lymphatic** (lim-FAT-ik) system consists of a series of vessels that transport a fluid called lymph from the tissues back into the blood. In addition to lymph, the system includes lymph nodes and lymphoid organs such as the tonsils, spleen, and thymus that filter the lymph to remove foreign particles as a protection against disease. Lymphoid organs also function in the body's defense mechanism by enhancing the activities of cells that inactivate specific pathogenic agents. The lymphatic system is sometimes considered to be a part of the cardiovascular system.

Digestive System

The organs of the **digestive** (dye-JES-tiv) system include the mouth, pharynx, esophagus, stomach, small intestine, and large intestine (colon), which make up the digestive tract. Accessory organs of this system include the teeth, tongue, salivary glands,

liver, gallbladder, and pancreas. The functions of this system are to ingest food, process it into molecules that can be used by the body, and then eliminate the residue.

Respiratory System

The **respiratory** (reh-SPY-rah-tor-ee or res-per-ah-TOR-ee) system brings oxygen, in the form of air, into the lungs, removes the carbon dioxide that is produced in metabolism, and provides a membrane for the exchange of these gases between the blood and lungs. The system consists of the nasal cavities, pharynx, larynx, trachea, bronchi, and lungs.

Urinary System

The kidneys, ureters, urinary bladder, and urethra make up the **urinary** (YOO-rin-air-ee) system. The kidneys remove various waste materials, especially nitrogenous wastes, from the blood and help to regulate the fluid level and chemical content of the body. The product of kidney function is urine, which is transported through the ureters and urethra. The urinary bladder serves as a reservoir or storage area for the urine.

Reproductive System

The purpose of the **reproductive** (ree-pro-DUK-tiv) system is the production of new individuals. The primary organs of the system are the gonads that produce the reproductive cells. These are the ovaries in females and testes in males. In addition to gonads, there are accessory glands, supporting structures, and duct systems for the transport of reproductive cells. In the female, the reproductive system produces ova or eggs, receives sperm from the male, and provides for the support and development of the embryo and fetus. The male reproductive system is concerned with the production and maintenance of sperm and the transfer of these cells to the female.

Life Processes

All living organisms have certain characteristics that distinguish them from nonliving forms. The basic processes of life include organization, metabolism, responsiveness, movement, and reproduction. In humans, who represent the most complex form of life, there are additional requirements such as growth, differentiation, respiration, digestion, and excretion. All of these processes are interrelated. No part of the body, from the smallest cell to a complete body system, works in isolation. All function together, in fine-tuned balance, for the well-being of the individual and to maintain life. Disease and death represent a disruption of the balance in these processes. The following is a brief description of the life processes.

Organization

At all levels of the organizational scheme, there is a division of labor. Each component has its own job to perform in cooperation with others. Even a single cell will die if it loses its integrity or organization.

Metabolism

Metabolism (meh-TAB-oh-lizm) includes all the chemical reactions that occur in the body. One phase of metabolism is

catabolism (kah-TAB-oh-lizm), in which complex substances are broken down into simpler building blocks and energy is released. **Anabolism** (ah-NAB-oh-lizm) is a building-up process in which complex substances are synthesized from simpler ones. This usually requires energy.

Responsiveness

Responsiveness or irritability is concerned with detecting changes in the internal or external environments and reacting to that change. It is the act of sensing a stimulus and responding to it. Conductivity is the ability to transmit a stimulus or information from one point to another.

Movement

There are many different types of movement within the body. On the cellular level, molecules move from one place to another. Blood moves from one part of the body to another. The diaphragm moves with every breath. The ability of muscle fibers to shorten and thus to produce movement is called contractility.

Reproduction

For most people, reproduction refers to the formation of a new person, the birth of a baby. In this way, life is transmitted from one generation to the next through reproduction of the organism. In a broader sense, reproduction also refers to the formation of new cells for the replacement and repair of old cells as well as for growth. This is cellular reproduction. Both are essential to the survival of the human race.

Growth

Growth refers to an increase in size either through an increase in the number of cells or through an increase in the size of each individual cell. For growth to occur, anabolic processes must occur at a faster rate than catabolic processes.

Differentiation

Differentiation (dif-er-en-she-AY-shun) is a developmental process by which unspecialized cells change into specialized cells that have distinctive structural and functional characteristics. Through differentiation, cells develop into tissues and organs.

Respiration

Respiration refers to all the processes involved in the exchange of oxygen and carbon dioxide between the cells and the external environment. It includes ventilation, the diffusion of oxygen and carbon dioxide, and the transport of the gases in the blood. Cellular respiration includes the cell's use of oxygen and release of carbon dioxide in metabolism.

Digestion

Digestion is the process of breaking down complex ingested foods into simple molecules that can be absorbed into the blood and used by the body.

Excretion

Excretion is the process that removes the waste products of digestion and metabolism from the body. It gets rid of byprod-

ucts that the body is unable to use, many of which are toxic and incompatible with life.

Environmental Requirements for Life

The 10 life processes just described are not sufficient to ensure the survival of the individual. In addition to these processes, life depends on certain physical factors from the environment.

Water

Water is the most abundant substance in the body. About 60% of the body weight is attributed to water. It provides a medium in which chemical reactions occur, transports substances from one place to another within the body, and helps to regulate body temperature.

Oxygen

Oxygen is necessary for the metabolic reactions that provide energy in the body.

Nutrients

Nutrients, such as carbohydrates, proteins, fats, vitamins, and minerals, come from the foods that we eat. They supply the chemicals that the body needs for energy and serve as raw materials for making new tissues for growth, replacement, and repair.

Heat

Heat, a form of energy, is necessary to keep the chemical reactions in the body proceeding at an appropriate rate. In general, the more heat there is, the faster the reaction, up to an optimum point. After that, the reaction rate decreases. Consequently, the amount of heat in the body must be regulated. Temperature is a measure of the amount of heat that is present.

Pressure

Pressure is the application of a force. Atmospheric pressure is the force of the air acting on our bodies. This pressure plays an important role in breathing and in the exchange of gases between the lungs and the outside air. Hydrostatic pressure is the force applied by fluids. Blood pressure, a form of hydrostatic pressure, is essential for the circulation of blood.

QUICK CHECK

1.4 Which body system is responsible for the exchange of oxygen and carbon dioxide between the blood and the external environment?
1.5 Which body system transports oxygen, carbon dioxide, and nutrients from one part of the body to another?
1.6 What factors from the environment are necessary to sustain human life processes?

Homeostasis

Homeostasis (hoh-mee-oh-STAY-sis) refers to the constant internal environment that must be maintained for the cells of the body. The word is derived from two Greek words: *homeo,* which means "alike" or "the same," and *stasis*, which means "always" or "staying." Putting these together, the word *homeostasis* means "staying the same." When the body is healthy, the internal environment stays the same. It remains stable within very limited normal ranges.

Everyone is familiar with aspects of the external environment—whether it is cold or hot, humid or dry, smoggy or clear. The internal environment is not quite as obvious. It refers to the tissue fluid that surrounds and bathes every cell of the body. Normal activities of the cell depend on the internal environment being maintained within very limited normal ranges. The chemical content, volume, temperature, and pressure of the fluid must stay the same (homeostasis), regardless of external conditions, so that the cell can function properly. If the conditions in the tissue fluid deviate from normal, mechanisms respond that try to restore conditions to normal. If the mechanisms are unsuccessful, the cell malfunctions and dies. Cell death leads to illness or disease. If homeostasis is not restored, the eventual result is death of the individual.

Negative Feedback

Any condition or stimulus that disrupts the homeostatic balance in the body is a **stressor**. When a stressor causes internal conditions to deviate from normal, all the body systems work to bring conditions back to the normal range. This is usually accomplished by a **negative feedback** mechanism; this mechanism works similar to a thermostat connected to a furnace and an air conditioner, as illustrated in Figure 1-3. When the temperature in the room decreases (stressor) below the thermostat setting (normal), the sensing device in the thermostat detects the change and causes the furnace to add heat to the room. When the room becomes too warm, the furnace stops and the

air conditioner begins to cool the room. Negative feedback mechanisms do not prevent variation, but they keep variation within a normal range.

An example of a physiologic negative feedback mechanism is illustrated in Figure 1-4. When blood pressure decreases below normal, body sensors detect the deviation and initiate changes that bring the pressure back within the normal range. When the pressure increases above normal, changes occur to decrease the pressure to normal. Variations in blood pressure occur, but homeostatic mechanisms keep them within the limits of a normal range. There are numerous examples of negative feedback in the body. In fact, negative feedback is the primary mechanism for maintaining homeostasis.

The nervous and endocrine systems work together to control homeostasis, but all the organ systems in the body help maintain the normal conditions of the internal environment. The brain contains centers that monitor temperature, pressure, volume, and the chemical conditions of body fluids. Endocrine glands secrete hormones in response to deviations from normal conditions, and these hormones affect other organs. The changes required to bring conditions back to the normal range are mediated by the various organ systems. Good health depends on homeostasis. Illness results when the negative feedback mechanisms that maintain homeostasis are disrupted. Medical therapy attempts to assist the negative feedback process to restore balance, or homeostasis.

Positive Feedback

Negative feedback mechanisms inhibit a change or deviation from normal. They create a response that is opposite to the deviation and that restores homeostasis. **Positive feedback** mechanisms stimulate or amplify changes. In the example of

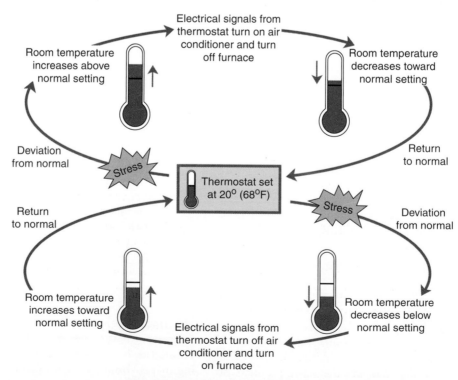

Figure 1-3 *Negative feedback mechanism.*

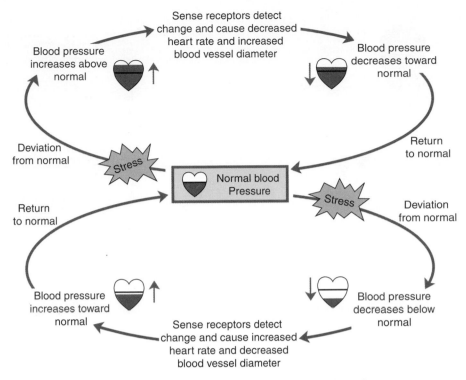

Figure 1-4 *Physiologic example of a negative feedback mechanism.*

a thermostat, if there is an increase in temperature, a positive feedback mechanism brings about an even greater increase. Such responses tend toward instability because the variable, in this case the temperature, deviates further and further from the normal range. The process or change continues faster and faster until it comes to a quick conclusion. You may wonder why there is a need for positive feedback in the body. There are times when positive feedback is beneficial because it quickly terminates a process that, if allowed to continue for a prolonged period, might be detrimental. The events in the birth of a baby and the formation of a blood clot are examples of physiological positive feedback mechanisms.

QUICK APPLICATIONS

At the end of pregnancy, the pressure of the fetal head on the cervix of the uterus stimulates the release of the hormone oxytocin, which stimulates increased uterine contractions. The contractions promote the release of more oxytocin, which stimulates more contractions, and so on, until there is a culminating event, the birth of the baby. This is an example of positive feedback.

Anatomical Terms

You need to understand certain basic terms to communicate effectively in the health care profession. In other words, you have to speak the language. This section explains some basic terms that relate to the anatomy of the body. These terms are used to describe directions and regions of the body. This is the beginning of a new vocabulary that will expand as you progress through the book.

Anatomical Position

If directional terms are to be meaningful, there must be some knowledge of the beginning position. If you give a person directions to go somewhere, you must have a starting reference point. When using directions in anatomy and physiology, we assume that the body is in anatomical position. In this position, the body is standing erect, the face is forward, and the arms are at the sides with the palms and toes directed forward. Figure 1-5 illustrates the body in anatomical position.

Directions in the Body

Directional terms are used to describe the relative position of one part to another. Note that in the following list of directional terms, the two items in each pair of terms are opposites.

Superior (soo-PEER-ee-or) means that a part is above another part, or closer to the head. The nose is superior to the mouth. **Inferior** (in-FEER-ee-or) means that a part is below another part, or closer to the feet. The heart is inferior to the neck.

Anterior (an-TEER-ee-or) (or ventral) means toward the front surface. The heart is anterior to the vertebral column. **Posterior** (pos-TEER-ee-or) (or dorsal) means that a part is toward the back. The heart is posterior to the sternum.

Medial (MEE-dee-al) means toward, or nearer, the midline of the body. The nose is medial to the ears. **Lateral** (LAT-er-al) means toward, or nearer, the side, away from the midline. The ears are lateral to the eyes.

Proximal (PRAHK-sih-mal) means that a part is closer to a point of attachment, or closer to the trunk of the body, than another part. The elbow is proximal to the wrist. The opposite of proximal is **distal** (DIS-tal), which means that a part is

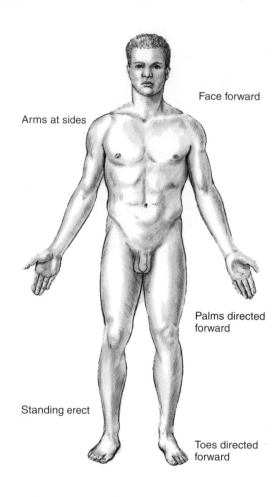

Figure 1-5 *The body in anatomical position.*

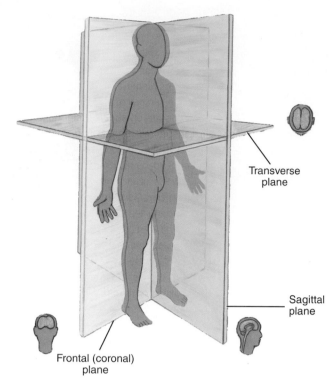

Figure 1-6 *Transverse, sagittal, and frontal planes of the body.*

farther away from a point of attachment than is another part. The fingers are distal to the wrist.

Superficial (soo-per-FISH-al) means that a part is located on or near the surface. The superficial (or outermost) layer of the skin is the epidermis. The opposite of superficial is **deep**, which means that a part is away from the surface. Muscles are deep to the skin.

Visceral (VIS-er-al) pertains to internal organs or the covering of the organs. The visceral pericardium covers the heart. **Parietal** (pah-RYE-ih-tal) refers to the wall of a body cavity. The parietal peritoneum lines the wall of the abdominal cavity.

Planes and Sections of the Body

To aid in visualizing the spatial relationships of internal body parts, anatomists use three imaginary planes, each of which is cut through the body in a different direction. Figure 1-6 illustrates these three planes.

A **sagittal** (SAJ-ih-tal) **plane** refers to a lengthwise cut that divides the body into right and left portions. This is sometimes called a longitudinal section. If the cut passes through the midline of the body, it is called a **midsagittal plane**, and it divides the body into right and left halves.

A **transverse plane** or horizontal plane is perpendicular to the sagittal plane and cuts across the body horizontally to

divide it into superior and inferior portions. Sections cut this way are sometimes called cross sections.

A **frontal plane** divides the body into anterior and posterior portions. It is perpendicular to both the sagittal plane and the transverse plane. This is sometimes called a **coronal** (ko-ROH-nal) **plane**.

Body Cavities

The spaces within the body that contain the internal organs or viscera are called body cavities. The two main cavities are the **dorsal cavity** and the larger **ventral cavity**, which are illustrated in Figure 1-7. The dorsal cavity is divided into the **cranial** (KRAY-nee-al) **cavity**, which contains the brain, and the **spinal** (SPY-nal) **cavity**, which contains the spinal cord. The cranial and spinal cavities join with each other to form a continuous space.

The ventral cavity is much larger than the dorsal cavity and is subdivided into the **thoracic** (tho-RAS-ik) **cavity** and the **abdominopelvic** (ab-dahm-ih-noh-PEL-vik) **cavity**. The thoracic cavity is superior to the abdominopelvic cavity and contains the heart, lungs, esophagus, and trachea. It is separated from the abdominopelvic cavity by the muscular diaphragm. Although there is no clear-cut partition to divide it, the abdominopelvic cavity is separated into the superior **abdominal** (ab-DAHM-ih-nal) **cavity** and the inferior **pelvic** (PEL-vik) **cavity**. The stomach, liver, gallbladder, spleen, and most of the intestines are in the abdominal cavity. The pelvic cavity contains portions of the small and large intestines, the rectum, the urinary bladder, and the internal reproductive organs.

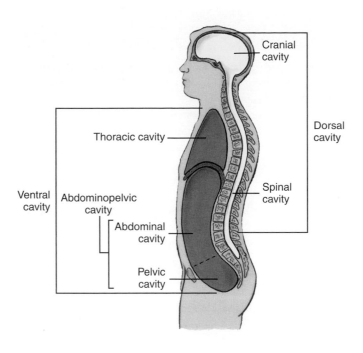

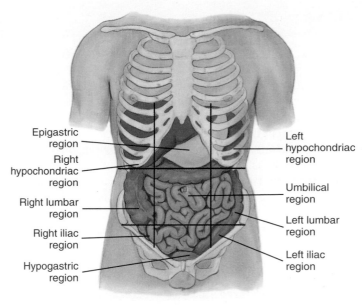

Figure 1-9 *Nine abdominopelvic regions formed by two sagittal planes and two transverse planes.*

Figure 1-7 *The two major cavities in the body and their subdivisions.*

To help describe the location of body organs or pain, health care professionals frequently divide the abdominopelvic cavity into regions using imaginary lines. One such method uses the midsagittal plane and a transverse plane that passes through the umbilicus. This divides the abdominopelvic area into four quadrants, illustrated in Figure 1-8. Another system uses two sagittal planes and two transverse planes to divide the abdominopelvic area into the nine regions illustrated in Figure 1-9. The three central regions are, from superior to inferior, the **epigastric** (ep-ih-GAS-trik), **umbilical** (um-BIL-ih-kal), and **hypogastric** (hye-poh-GAS-trik) regions. Lateral to these, from superior to inferior, are the right and left **hypochondriac** (hye-poh-KAHN-dree-ak), right and left **lumbar**, and right and left **iliac** (ILL-ee-ak) or **inguinal** (IN-gwih-nal) regions.

Regions of the Body

The body may be divided into the **axial** (AK-see-al) portion, which consists of the head, neck, and trunk, and the **appendicular** (ap-pen-DIK-yoo-lar) portion, which consists of the limbs. The trunk, or **torso**, includes the thorax, abdomen, and pelvis. In addition to these terms and the nine abdominopelvic regions identified in the previous section, there are numerous other terms that apply to specific body areas. Some of these are listed in Table 1-1 and are identified in Figure 1-10.

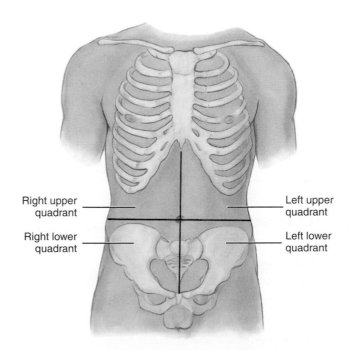

Figure 1-8 *Abdominopelvic quadrants that are formed by a midsagittal plane and a transverse plane through the umbilicus.*

QUICK **CHECK**

1.7 If the blood glucose level increases above normal, which is more likely to return it to normal: a positive feedback mechanism or a negative feedback mechanism?

1.8 Which of the body planes would not intersect the hypochondriac and epigastric regions at the same time?

1.9 Where would you expect the popliteal fossa to be located?

FROM THE PHARMACY

Pharmacology is the study of drugs. The word is derived from two Greek words, *pharmakos*, which means "medicine" or "drug," and *logos*, which means "study." The definitions for five common words derived from the Greek word *pharmakos* are given here.

Pharmacologist Specialist in the study of drugs

Pharmaceutical Medicinal drug

Pharmacopeia Authoritative treatise describing drugs, chemicals, and medicinal preparations

Pharmacy Place where medicines are compounded or dispensed; also the act or practice of preparing, preserving, and dispensing drugs

Pharmacist Specialist in the practice of preparing, preserving, and dispensing drugs

Throughout history, people have searched for ways to treat and cure disease. The earliest known medical documents, written on Egyptian clay tablets and papyri, date from as early as 3000 BC. These documents include detailed anatomical observations and descriptions of the treatment for numerous ailments. The earliest prescriptions are from this same period.

The first pharmacy in the Western world was founded in 1241 in Trier, Germany, and is still operating today. During the early history of the United States, the physician diagnosed the illness and also prepared and dispensed the medications for its treatment. In the eighteenth century, the practice of pharmacy became viewed as a separate vocational field and pharmacies began to be established in the United States. As the practice of medicine advanced, the need for educated and skilled pharmacists became evident. The first school of pharmacy, the Philadelphia College of Pharmacy and Science, was founded in 1821. Thirty years later, the first professional association for pharmacists was established. Today, all pharmacists must be licensed and adhere to a strict code of ethics and follow the standards set by the United States Pharmacopeia.

TABLE 1-1 Body Area Terms

Abdominal (ab-DAHM-ih-nal)	Portion of the trunk between the thorax and pelvis; celiac region
Antebrachial (an-te-BRAY-kee-al)	Region between the elbow and wrist; forearm; cubital region
Antecubital (an-te-KYOO-bih-tal)	Space in front of elbow
Axillary (AK-sih-lair-ee)	Armpit area
Brachial (BRAY-kee-al)	Proximal portion of upper limb: arm
Buccal (BUK-al)	Region of cheek
Buttock (BUT-tuck)	Posterior aspect of lower trunk; gluteal region
Carpal (KAR-pal)	Region between forearm and hand; wrist
Celiac (SEE-lee-ak)	Portion of the trunk between the thorax and pelvis; abdomen
Cephalic (seh-FAL-ik)	Head
Cervical (SER-vih-kal)	Neck region
Costal (KAHS-tal)	Ribs
Crural (KROO-rhal)	Portion of lower extremity between knee and foot; leg
Cubital (KYOO-bih-tal)	Region between elbow and wrist; forearm; antebrachial
Cutaneous (kyoo-TAY-nee-us)	Skin
Femoral (FEM-or-al)	Part of lower extremity between hip and knee; thigh region
Frontal (FRUN-tal)	Forehead
Gluteal (GLOO-tee-al)	Posterior aspect of lower trunk; buttock region
Groin (GROYN)	Depressed region between abdomen and thigh; inguinal
Inguinal (IN-gwih-nal)	Depressed region between abdomen and thigh; groin
Leg (LEG)	Portion of lower extremity between knee and foot; crural region
Lumbar (LUM-bar)	Region of lower back and side between lowest rib and pelvis
Mammary (MAM-ah-ree)	Pertaining to the breast
Navel (NAY-vel)	Middle region of abdomen; umbilical region

TABLE 1-1 Body Area Terms—*Cont'd*

Occipital (ahk-SIP-ih-tal)	Lower portion of the back of the head
Ophthalmic (off-THAL-mik)	Pertaining to the eyes
Oral (OH-ral or AW-ral)	Pertaining to the mouth
Otic (OH-tik)	Pertaining to the ears
Palmar (PAWL-mar)	Palm of hand
Pectoral (PEK-toh-ral)	Part of trunk inferior to neck and superior to diaphragm; thoracic or chest region
Pedal (PED-al)	Pertaining to the foot
Pelvic (PEL-vik)	Inferior region of abdominopelvic cavity
Perineal (pair-ih-NEE-al)	Region between anus and pubic symphysis; includes region of external reproductive organs
Plantar (PLAN-tar)	Sole of foot
Popliteal (pop-LIT-ee-al)	Area behind knee
Sacral (SAY-kral)	Posterior region between hip bones
Sternal (STIR-nal)	Anterior midline of the thorax
Tarsal (TAHR-sal)	Ankle and instep of foot
Thigh (THIGH)	Part of lower extremity between hip and knee; femoral region
Thoracic (tho-RAS-ik)	Chest; part of trunk inferior to neck and superior to diaphragm; pectoral region
Umbilical (um-BIL-ih-kal)	Navel; middle region of abdomen
Vertebral (ver-TEE-bral or VER-teh-bral)	Pertaining to spinal column; backbone

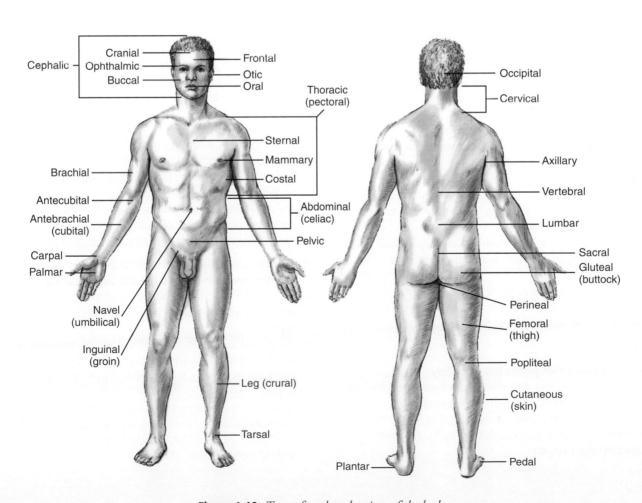

Figure 1-10 *Terms for selected regions of the body.*

CHAPTER SUMMARY

Anatomy and Physiology

■ **Define the terms** *anatomy* **and** *physiology* **and discuss the relationship between the two areas of study**

- Anatomy is the scientific study of structure.
- Physiology is the scientific study of function.
- Anatomy and physiology are interrelated because structure has an effect on function and function influences structure.

Levels of Organization

■ **List the six levels of organization within the human body**

- From the simplest to the most complex, the six levels of organization within the human body are chemical, cellular, tissue, organ, body system, and total organism.

Organ Systems in the Body

■ **Name the 11 organ systems of the body and briefly describe the major role of each one**

- Integumentary system consists of the skin with its derivatives; it covers and protects the body.
- Skeletal system includes the bones, cartilages, and ligaments; it forms the framework of the body.
- Muscular system consists of all the muscles in the body; it produces movement and heat.
- Nervous system includes the brain, spinal cord, and nerves; it receives/transmits stimuli and coordinates body activities.
- Endocrine system consists of the ductless glands; it regulates metabolic activities.
- Cardiovascular system includes the blood, heart, and blood vessels; it transports substances throughout the body.
- Lymphatic system includes lymph, lymphatic vessels, and lymphoid organs; it is a major defense against disease.
- Digestive system consists of the gastrointestinal tract and accessory organs; it is responsible for the ingestion, digestion, and absorption of food.
- Respiratory system includes the air passageways and lungs; it is responsible for the exchange of gases between the external environment and the blood.
- Urinary system consists of the kidneys, urinary bladder, and ducts; it functions to eliminate metabolic wastes from the body.
- Reproductive system consists of the ovaries and testes with the associated accessory organs; its function is to form new individuals for the continuation of the species.

Life Processes

■ **List 10 life processes in the human body**

- The basic characteristics that distinguish living from nonliving forms are organization, metabolism, responsiveness, movement, and reproduction. Human life has additional characteristics such as growth, development, digestion, respiration, and excretion.

Environmental Requirements for Life

■ **List five physical environmental factors necessary for survival of the individual**

- Physical factors from the environment that are necessary for human life include water, oxygen, nutrients, heat, and pressure.

Homeostasis

■ **Discuss the concept of homeostasis**

- Homeostasis refers to a constant internal environment.
- A lack of homeostasis leads to illness or disease.
- All organ systems of the body, under direction from the nervous and endocrine system, work together to maintain homeostasis.

■ **Distinguish between negative feedback mechanisms and positive feedback mechanisms**

- Homeostasis is usually maintained by negative feedback mechanisms, which inhibit changes.
- Positive feedback mechanisms are stimulating and cause a process or change to occur at faster and faster rates, leading to a culminating event.

Anatomical Terms

■ **Give the four criteria that are used to describe the anatomical position**

- The body is erect with feet flat on the floor and facing forward, arms are at the sides, and palms and toes are directed forward.

■ **Use anatomical terms to describe body planes**

- Six pairs of opposite terms are used to describe the relative position of one body part to another—superior/inferior, anterior/posterior, medial/lateral, proximal/distal, superficial/deep, and visceral/parietal.
- A sagittal plane divides the body into right and left parts; a transverse, or horizontal, plane divides it into upper and lower regions; and a frontal, or coronal, plane divides it into front and back portions.
- The axial portion of the body consists of the head, neck, and trunk; the appendicular portion consists of the limbs or appendages; and specific anatomic terms are used to designate body regions.

■ **Distinguish between the dorsal body cavity and the ventral body cavity, and list the subdivisions of each one**

- The dorsal body cavity consists of the cranial cavity, which contains the brain, and the spinal cavity, which contains the spinal cord.
- The ventral body cavity is subdivided into the thoracic cavity, which contains the heart and lungs, and the abdominopelvic cavity, which contains the digestive, urinary, and reproductive organs.
- A convenient and commonly used method divides the abdominopelvic cavity into nine regions: epigastric, umbilical, hypogastric, right and left hypochondriac, right and left lumbar, and right and left iliac.

CHAPTER QUIZ

Recall

Match the definitions on the left with the appropriate term on the right.

_____ 1. One part is closer to a point of attachment than another part

_____ 2. Stable internal environment

_____ 3. Divides body or part into anterior and posterior portions

_____ 4. Refers to skin

_____ 5. Complex substances are broken down into simpler building blocks and energy is released

_____ 6. Process by which unspecialized cells change into cells with distinctive structural and functional characteristics

_____ 7. Study of function and functional relationships

_____ 8. Contains the brain and spinal cord

_____ 9. Midline superior abdominal region

_____ 10. Space in front of the elbow

A. Antecubital
B. Catabolism
C. Cutaneous
D. Differentiation
E. Dorsal cavity
F. Epigastric
G. Frontal plane
H. Homeostasis
I. Physiology
J. Proximal

Thought

1. Which of the following is more complex than tissues on the organizational scale?
   A. Molecules
   B. Organs
   C. Cells
   D. Organelles

2. Two body systems function to regulate and coordinate body activities. These systems are the:
   A. Cardiovascular and nervous
   B. Lymphatic and endocrine
   C. Nervous and reproductive
   D. Endocrine and nervous

3. A student wants to separate a brain specimen into right and left halves. Along what plane should the student make the cut?
   A. Transverse
   B. Sagittal
   C. Frontal
   D. Midsagittal

4. Which of the following groups of organs are found in the abdominopelvic cavity?
   A. Spinal cord, lungs, liver
   B. Liver, stomach, heart
   C. Brain, spinal cord, liver
   D. Gallbladder, stomach, small intestine

5. Which body regions are inferior to the costal region?
   A. Sternal and umbilical
   B. Inguinal and pelvic
   C. Inguinal and axillary
   D. Buccal and brachial

Application

Ima Dock, MD, is able to visualize organ A of her patient in midsagittal, frontal, and transverse body planes. Which of the following is organ A?
A. Kidney
B. Spleen
C. Gallbladder
D. Urinary bladder

Building Your Medical Vocabulary

Human anatomy and physiology, like any technical subjects, have their own vocabulary. It is necessary to learn the language to understand the concepts. The sciences of medicine, anatomy, and physiology were born in the eighteenth-century universities in which Greek and Latin were the languages used in lectures and in writing. As a result, the special vocabulary used in these sciences consists of roots, prefixes, and suffixes based on Greek and Latin. These three principal parts are arranged in different combinations and sequences to make words. Once you have learned the basic parts, you can recombine them to form or analyze new words. It is like having a mix-and-match wardrobe of words! As you progress through this course, you will discover that the terminology makes a lot of sense and is easy to understand. Practice it on your friends and family. You will find that once you learn the tricks of the trade, it is easy and fun!

You need to be able to recognize, spell, pronounce, and use the roots, prefixes, and suffixes of these words to be an effective student in anatomy and physiology and in the medical sciences. In everyday life, these terms will help you read articles and advertisements with more understanding. They will allow you to communicate more effectively with the medical personnel with whom you come in contact. Now it is time to start building the mix-and-match wardrobe of word parts.

The **root** is the main part or subject of a word. It is the word's foundation and conveys the central meaning of the word. For example, the root *hemat* means "blood," *gastr* means "stomach," and *path* means "disease."

A **prefix** is a syllable, or group of syllables, placed before a root to alter or modify the meaning of the root. *Hyper* is a prefix that means "over" or "above." *Hypersensitive* means "overly sensitive." The root has been modified by the prefix before it. *Intra* is a prefix that means "within" or "inside." *Intracellular* means "within the cell." Another commonly used prefix is *hypo*, which means "deficient" or "below." Putting this prefix with the root *gastr* forms *hypogastric*, which means "below the stomach."

A **suffix** is a syllable, or group of syllables, attached to the end of a root to modify the meaning of the root. The suffix *itis* means "inflammation of" and *gastr* is a root that refers to the "stomach." Thus, *gastritis* means "inflammation of the stomach." The suffix *oma* means "tumor," "mass," or "swelling." If you put this with the root *hemat*, you have *hematoma*, which is a "swelling or mass filled with blood."

Sometimes a root has both a prefix and a suffix. For example, *epi* is a prefix that means "above" or "upon." *Cardi* is a root meaning "heart," and *itis* is a suffix that means "inflammation." Putting these together makes *epicarditis*, which literally means "inflammation upon the heart."

To make pronunciation easier, words parts are often linked together by a **combining vowel**. In the word *cardiopulmonary*, the vowel "o" is used to connect the two roots, *cardi* and *pulmon*. The vowel "o" is also used to link the root *crani* with the suffix *tomy* to make the word *craniotomy*. The letter "o" is the most commonly used combining vowel.

The following sections list some commonly used roots, combining forms, prefixes, and suffixes with their definitions.

You should learn these word parts and practice putting them together to make words that are used in anatomy, physiology, and clinical practice. There are written exercises in the Student's Study Guide that will help you to learn this terminology.

Word Roots and Combining Forms with Definitions and Examples

ROOT/COMBINING FORM	DEFINITION	EXAMPLE
aden/o	gland	adenitis: inflammation of a gland
ana-	apart	anatomy: to cut apart
carcin/o	cancer	carcinoma: cancerous tumor
cardi/o	heart	cardiology: study of the heart
cephal/o	head	cephalad: toward the head
cis/o	to cut	incision: to cut into
dors/o	back	dorsal: toward the back
electr/o	electricity	electrocardiogram: electrical recording of the heart
gastr/o	stomach	gastritis: inflammation of the stomach
home/o	same	homeostasis: staying the same
integ-	skin; covering	integument: the covering of the body
path/o	disease	pathology: study of disease
pelv/o	basin	pelvis: basin-shaped region of the body cavity
physi/o	nature, function	physiology: study of the function
proxim/o	nearest	proximal: nearest the point of attachment
radi/o	x-ray	radiograph: recording of an x-ray examination
ren/o	kidney	renal: pertaining to the kidney
skelet/o	dried hard body	skeletal: pertaining to bones (dried hard bodies)
vas/o	vessel	vascular: pertaining to blood vessels
viscer/o	internal organs	visceral: pertaining to the internal organs

Prefixes with Definitions and Examples

PREFIX	DEFINITION	EXAMPLE
a-, an-	no, not, without	anorexia: without appetite
ab-	away from	abduct: move away from the midline or axis of the body

PREFIX	DEFINITION	EXAMPLE
ad-	toward	adduct: move toward the midline or axis of the body
ante-	before, forward	anteversion: tilted forward
end-	within	endotracheal: within the trachea
epi-	upon, above	epigastric: above the stomach
ex-	outside, outward	expiration: breathing out
hyper-	excessive, above	hypersensitive: excessively sensitive
hypo-	deficient, below	hypodermis: below the skin
inter-	between, among	intercostal: between the ribs
intra-	within, inside	intracellular: within the cell
macro-	large	macrocephaly: enlarged brain
micro-	small	microglia: small cells in the nervous system
peri-	surrounding, around	pericardium: membrane around the heart
sub-	under, less, below	subcutaneous: below the skin
supra-	above, excessive	suprarenal: above the kidney

Suffixes with Definitions and Examples

SUFFIX	DEFINITION	EXAMPLE
-ac	pertaining to	cardiac: pertaining to the heart
-al	pertaining to	pericardial: pertaining to the membrane around the heart
-algia	pain	arthralgia: joint pain
-gram	a record or picture	electrocardiogram: recording of the electrical activity of the heart
-graphy	process of recording	radiography: process of recording x-ray examinations a picture or record
-ic	pertaining to	gastric: pertaining to the stomach
-ism	process of, condition	hyperthyroidism: condition of excessive thyroid activity
-ist	specialist	anatomist: specialist in anatomy
-itis	inflammation	dermatitis: inflammation of the skin
-logy	study of, science of	physiology: study of function

SUFFIX	DEFINITION	EXAMPLE
-osis	condition of	leukocytosis: condition of excessive white blood cells
-opsy	process of viewing	biopsy: process of viewing specimens under a microscope
-scopy	visual examination	microscopy: visualizing small objects with a microscope
-tomy	to cut into	tracheotomy: to cut into the trachea
-y	process, condition	neuropathy: condition of having a disease of the nervous system

The **plural form** of most English words is derived by adding "s" or "es" to the end of the word. In Greek and Latin, and consequently in medical terminology, the ending may be completely changed to designate the plural form. This is illustrated in the list of plural forms given below.

Commonly Used Plural Endings

SINGULAR ENDING	PLURAL ENDING
a as in aort**a**	**ae** as in aort**ae**
en as in foram**en**	**ina** as in foram**ina**
is as in test**is**	**es** as in test**es**
is as in ir**is**	**ides** as in ir**ides**
nx as in phala**nx**	**ges** as in phalan**ges**
on as in spermatozo**on**	**a** as in spermatozo**a**
um as in ov**um**	**a** as in ov**a**
us as in bronch**us**	**i** as in bronch**i**
x as in thora**x**	**ces** as in thora**ces**
y as in arter**y**	**ies** as in arter**ies**

Correct spelling is extremely important in medical terminology. When in doubt, look it up in the dictionary. Guessing has no place in medicine because one wrong letter may give the word an entirely new meaning. This is illustrated by the examples given below of similar words that have different meanings.

Examples of Similar Words That Have Different Meanings

TERM WITH LETTER DIFFERENCE	MEANING OF TERM
A**b**duct	To lead **away** from midline
A**d**duct	To lead **toward** midline
Art**e**ritis	Inflammation of an **artery**
Art**h**ritis	Inflammation of a **joint**
Il**e**um	Portion of **small intestine**
Il**i**um	A **pelvic bone**

Scientific and medical terms may look like they are hard to pronounce, especially if you have only read them and have not heard them. This text uses a phonetically spelled pronunciation guide, and you should practice saying each new word aloud to reinforce the pronunciation in your mind. Try pronouncing the following words by using the phonetic guide that is provided. The syllable that is capitalized is the one that should be stressed or emphasized.

Anabolism	(ah-NAB-oh-lizm)
Anatomical	(an-ah-TOM-ih-kul)
Catabolism	(kah-TAB-oh-lizm)
Differentiation	(dif-er-en-she-AY-shun)
Homeostasis	(hoh-mee-oh-STAY-sis)

Many of the words used in scientific terminology are long and easily misspelled. Be careful! Practitioners use many short-cut abbreviations, but there are some difficulties in using some of these. At a quick glance, q.o.d., q.i.d., and q.d. may look alike, especially considering the differences in penmanship. However, q.o.d. means "every other day," q.i.d. means "four times a day," and q.d. means "every day." You can see that mis-reading these abbreviations could have disastrous results. The use of abbreviations differs from country to country and region to region. An abbreviation that is standard in the United States may not be understood in Mexico or Germany, for example. For these reasons, each hospital, clinic, and other medical facility has a list of "Do Not Use" abbreviations.

The abbreviations presented in each chapter of this book represent standard, common usage. Be aware, however, that some of these may be on your facility's "Do Not Use" list and should be avoided. Some common abbreviations that are used in clinical practice are given below.

Clinical Abbreviations

ABBREVIATION	MEANING
Abd	abdomen
ant	anterior
b.i.d.	twice per day
BP	blood pressure
C	centigrade, Celsius
cap	capsule
cm	centimeter

ABBREVIATION	MEANING
Dx	diagnosis
ENT	ear, nose, throat
F	Fahrenheit
gm	gram
gtt	drops
H&P	history and physical
kg	kilogram
mcg	microgram
mg	milligram
ml	milliliter
PRN	as needed
soln	solution
susp	suspension
tab	tablet
t.i.d.	three times a day

Clinical Terms

Acronym (ACK-roh-nym) Word formed from the initial letters of the successive parts of a compound term

Acute disease (ah-CUTE dih-ZEEZ) Disease that has a sudden onset, severe symptoms, and a short duration

Chronic disease (KRAHN-ik dih-ZEEZ) Disease that continues over a long time, showing little change in symptoms

Diagnosis (dye-ag-NO-sis) Identification of a disease; determination of the cause and nature of a disease

Eponym (EP-oh-nym) Name of a disease, structure, operation, or procedure that is based on the name of an individual, usually the person who discovered it or described it first

Prognosis (prahg-NO-sis) Prediction of the course of a disease and the recovery rate

Remission (re-MISH-uhn) Partial or complete disappearance of the symptoms of a disease without achieving a cure

Sign (SYN) Evidence of disease, such as fever, that can be observed, measured, or evaluated by someone other than the patient

Symptom (SIMP-tum) Evidence of disease, such as pain or headache, that can only be observed or evaluated by the patient

Syndrome (SIN-drohm) Combination of signs and symptoms occurring together that characterize a specific disease

VOCABULARY QUIZ

1. What are the three parts of word in scientific and medical terminology?

2. Where does the prefix occur in a word?

3. What is the most commonly used combining vowel?

4. What is the root in the word *visceral*?

5. What is the root that means "stomach"?

6. A physician who specializes in the study of diseases is a (a) physiologist; (b) cardiologist; (c) pathologist.

7. Susan went to her physician because she had a fever and was coughing. The fever and cough are (a) signs; (b) symptoms; (c) syndromes.

8. Toby was writing a paper about disorders of the foot. He knew that one of the bones in the toes is a phalanx. He needed to use the plural form of the word. The word he should use is (a) phalanxes; (b) phalanges; (c) phalances.

9. Julie was examining the chart of her patient, Ms. Jones. She learned that Ms. Jones was to receive a given medication twice a day. The abbreviation the physician likely used when writing the orders was (a) TID; (b) PRN; (c) BID.

10. Dr. N. Dokrin diagnosed his patient as having an adenocarcinoma. When he discussed this condition with his patient, he told him that he had (a) an inflammation of a gland; (b) a cancerous tumor of a gland: (c) a cancerous tumor on a bone.

Using the definitions of word parts given in this chapter, match each of the following definitions with the correct word.

_____ 11. Process of cutting apart

_____ 12. Skin or covering of the body

_____ 13. Study of function

_____ 14. Structure shaped like a basin

_____ 15. Pertaining to the heart and blood vessels

A. Physiology
B. Pelvis
C. Cardiovascular
D. Anatomy
E. Integument

Use word parts given in the chapter to form words that have the following definitions.

16. Pertaining to the back _____

17. Inflammation of the stomach _____

18. Study of the heart _____

19. Staying the same _____

20. Visual examination of the stomach _____

Chemistry, Matter, and Life

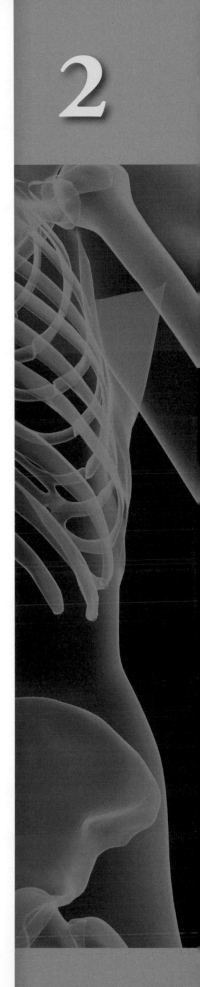

2

CHAPTER OBJECTIVES

Elements
- Define *matter, element,* and *atom.*
- Use chemical symbols to identify elements.

Structure of Atoms
- Illustrate the structure of an atom with a simple diagram showing the protons, neutrons, and electrons.
- Distinguish between the atomic number and mass number of an element.
- Describe the electron arrangement that makes an atom most stable.

Isotopes
- Distinguish between isotopes and other atoms.
- Describe a radioactive isotope.

Chemical Bonds
- Describe the difference between ionic bonds, covalent bonds, and hydrogen bonds.

Compounds and Molecules
- Describe the relationship between atoms, molecules, and compounds and interpret molecular formulas for compounds.

Chemical Reactions
- Describe and write chemical equations for four types of chemical reactions and identify the reactants and products in each.
- Discuss five factors that influence the rate of chemical reactions.

Mixtures, Solutions, and Suspensions
- Distinguish between mixtures, solutions, and suspensions.

Electrolytes, Acids, Bases, and Buffers
- Differentiate between acids and bases and discuss how they relate to pH and buffers.

Organic Compounds
- Describe the five major groups of organic compounds that are important to the human body.

KEY TERMS

Acid (AS-id) Substance that ionizes in water to release hydrogen ions; a proton donor; a substance with a pH less than 7.0

Atom (AT-tum) Smallest unit of a chemical element that retains the properties of that element

Base (BASE)Substance that ionizes in water to release hydroxyl (OH^-) ions or other ions that combine with hydrogen ions; a proton acceptor; a substance with a pH greater than 7.0; alkaline

Buffer (BUFF-fur) Substance that prevents, or reduces, changes in pH when either an acid or a base is added

Carbohydrate (kar-boh-HYE-drayt) Organic compound that contains carbon, hydrogen, and oxygen with the hydrogen and oxygen present in a 2:1 ratio; sugar, starch, cellulose

Compound (KAHM-pownd) Substance formed from two or more elements joined by chemical bonds in a definite, or fixed, ratio; smallest unit of a compound is a molecule

Covalent bond (koe-VAY-lent BOND) Chemical bond formed by two atoms sharing one or more pairs of electrons

Element (EL-eh-ment) Simplest form of matter that cannot be broken down by ordinary chemical means

Inorganic Compound (in-or-GAN-ik KAHM-pownd) Chemical components that do not contain both carbon and hydrogen.

Ionic bond (eye-ON-ik BOND) Chemical bond that is formed when one or more electrons are transferred from one atom to another

Isotope (EYE-so-tohp) Atoms of a given element that have different numbers of neutrons and consequently different atomic weights

Lipid (LIP-id) Class of organic compounds that includes oils, fats, and related substances

Molecule (MAHL-eh-kyool) Particle composed of two or more atoms that are chemically bound together; smallest unit of a compound

Organic Compound (or-GAN-ik KAHM-pownd) Chemical components that contain carbon and hydrogen atoms covalently bonded together.

Protein (PRO-teen) Organic compound that contains nitrogen and consists of chains of amino acids linked together by peptide bonds

Radioactive isotope (ray-dee-oh-ACK-tiv EYE-so-tohp) Isotope with an unstable atomic nucleus that decomposes, releasing energy or atomic particles

Solute (SOL-yoot) Substance that is dissolved in a solution

Solvent (SOL-vent) Fluid in which substances dissolve

Chemistry is the science that deals with the composition of matter and the changes that may occur in that composition. *Matter* can be defined as anything that has mass and takes up space. For all practical purposes, mass is the same as weight, and thus these terms will be used interchangeably in this book. Technically, however, this is not correct. The mass of something remains the same throughout the universe. Weight varies with gravitational pull. An object that weighs 10 pounds on earth weighs less in space because there is less gravitational pull; however, its mass is the same wherever it is. Matter includes the solids, liquids, and gases that are in our bodies and in the environment around us. Because our bodies are composed of matter, an understanding of chemistry or the composition of matter is important to understanding human body structure and function.

Elements

All matter, living and nonliving, is composed of elements. An element is the simplest form of matter; it cannot be broken down into a simpler form by ordinary chemical means.

Examples of elements include gold, silver, oxygen, and hydrogen. There are about 90 naturally occurring elements and another 16 that are man-made in the laboratory. Of the 90 elements that occur naturally, only 21 are needed by living organisms. Four of the elements—carbon (C), oxygen (O), hydrogen (H), and nitrogen (N)—make up a little over 95% of the human body by weight. Table 2-1 lists the elements that are essential for human life. You will notice from the table that of the 21 elements listed, 10 are found in trace amounts. Even though they are present in minute amounts, they are essential for maintaining good health. This is why you eat your vegetables!

Instead of writing out the complete name of an element, scientists use a shorthand method or chemical symbol to identify an element. The chemical symbol is usually the first one or two letters of either the English or Latin name of the element. The chemical symbol for hydrogen is H, carbon is C, oxygen is O, Cl is chlorine, and Na is sodium (from *natrium*, the Latin word for "sodium"). The symbols for the elements in the human body are given in Table 2-1.

TABLE 2-1 **Elements Essential for Human Life**

Element	Symbol	Atomic Number	Approximate Percent of Human Body by Weight	Importance
Oxygen	O	8	65.0	Required for cellular metabolism; component of water; present in most organic compounds
Carbon	C	6	18.5	Component of all organic compounds
Hydrogen	H	1	9.5	Present in most organic compounds; component of water
Nitrogen	N	7	3.3	Component of all proteins, phospholipids, and nucleic acids
Calcium	Ca	20	1.5	Component of bones and teeth; important in muscle contraction, nerve impulse conduction, and blood clotting
Phosphorus	P	15	1.0	Component of phospholipids, nucleic acids, and adenosine triphosphate; important in energy transfer
Potassium	K	19	0.4	Principal cation inside cells; important in muscle contraction and nerve impulse conduction
Sulfur	S	16	0.3	Component of most proteins
Sodium	Na	11	0.2	Principal cation in fluid outside cells; important in muscle contraction and nerve impulse conduction
Chlorine	Cl	17	0.2	Principal anion in fluid outside cells; important in fluid balance
Magnesium	Mg	12	0.1	Part of many important enzymes; necessary for bones and teeth
Fluorine	F	9	Trace	Aids in development of teeth and bones
Chromium	Cr	24	Trace	Aids in carbohydrate metabolism
Manganese	Mg	25	Trace	Necessary for carbohydrate metabolism and bone formation
Iron	Fe	26	Trace	Component of hemoglobin; part of many enzymes
Cobalt	Co	27	Trace	Component of vitamin B_{12}
Copper	Cu	29	Trace	Necessary to maintain blood chemistry
Zinc	Zn	30	Trace	Necessary for growth, healing, and overall health
Selenium	Se	34	Trace	Aids vitamin E action and fat metabolism
Molybdenum	Mo	42	Trace	Component of enzymes necessary for metabolism
Iodine	I	53	Trace	Component of thyroid hormones

Structure of Atoms

An element is composed of atoms that are all of the same kind. The element gold is made up entirely of gold atoms, and the element silver is made up entirely of silver atoms. An **atom** is the smallest particle of an element that still retains the properties of that element, and it is almost unbelievably small. It takes over 100 million average-sized atoms lined up side by side to make 1 inch, or 2.54 centimeters.

Even though it is extremely small, an atom is made up of still smaller subunits or subatomic particles called **protons, neutrons,** and **electrons.** A dense region, called the nucleus, contains the protons and neutrons. Electrons are outside the nucleus. The number and nature of the subatomic particles in the atoms of an element determine the physical and chemical characteristics of the element. Protons, located in the nucleus, have a positive electrical charge, and each has a mass of 1 atomic mass unit (amu). The number of protons in the nucleus is called the **atomic number.** All the atoms in an element have the same number of protons in the nucleus so they have the same atomic number. Elements are arranged by their atomic number in the periodic table. Figure 2-1 is a simplified and abbreviated periodic table that shows the first 54 elements. The atomic number is above the chemical symbol in the periodic table.

Neutrons, also found in the nucleus, have the same mass as protons but have no charge. Protons and neutrons together account for the mass of the atom, and their number, collectively, is called the **mass number** of an atom. The element sodium is made up of sodium atoms, which have 11 protons and 12 neutrons in the nucleus. The atomic number of sodium is 11 (the number of protons), and the atomic mass number is 23 (the number of protons plus neutrons).

Electrons are minute, negatively charged particles with almost no mass. Their number and arrangement determine how an atom reacts. Electrons are located in the space surrounding the nucleus. The number of negatively charged electrons in an atom is always equal to the number of positively charged protons so that the atom is electrically neutral. The sodium atom, described in the previous paragraph, has 11 protons and 12 neutrons in the nucleus. Because the number of electrons equals the number of protons, there will be 11 electrons in the space surrounding the nucleus. It is impossible to know where a given electron will be at any given time, but it is possible to predict the region in which it will be located. Electrons are located in **energy levels,** or **shells,** around the nucleus. In general, electrons with higher energy levels are located in shells farther away from the nucleus than electrons with lower energy levels. The shell closet to the nucleus has the lowest energy level and can hold two electrons. The next higher energy level can hold eight electrons. Higher energy levels can hold more than eight electrons, but an atom is most stable when there are eight electrons in the outermost shell, which has the highest energy level. Simplified diagrams of the atomic structure of some biologically important elements are shown in Figure 2-2.

Isotopes

The number of neutrons in the nucleus may vary for different atoms of a given element, which changes the atomic weight. For example, most hydrogen atoms have one proton and one electron, which gives an atomic weight of 1 amu. A small number of hydrogen atoms have a neutron in the nucleus with the proton, and this gives them an atomic weight of 2 amu. This is called *deuterium*. It still has the characteristics of hydrogen because it has one proton. Figure 2-3 illustrates the structure of an atom of deuterium. Atoms of a given element that have different numbers of neutrons, and consequently different atomic weights, are called **isotopes.** Isotopes are included in the calculations of an element's atomic weight. For example, the periodic table in Figure 2-1 gives the atomic weight of hydrogen as 1.01. This value includes the amount of the isotope deuterium that occurs with normal hydrogen.

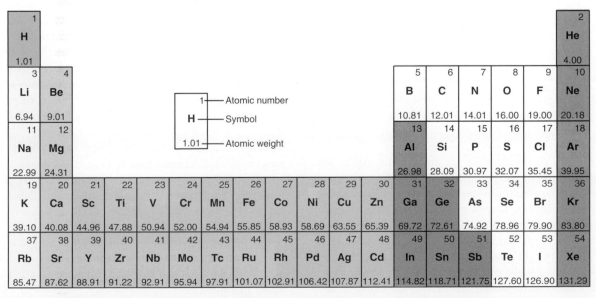

Figure 2-1 *An abbreviated and simplified periodic table of the elements from hydrogen to xenon, atomic numbers 1 through 54.*

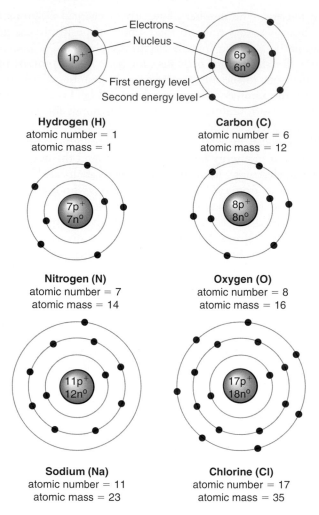

Figure 2-2 *Diagrams of the atomic structure of some biologically important elements—hydrogen, carbon, nitrogen, oxygen, sodium, and chlorine.*

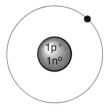

Figure 2-3 *Diagram of the atomic structure of deuterium, an isotope of hydrogen with one proton and one neutron in the nucleus.*

Some isotopes of an element are stable, but others have unstable atomic nuclei that decompose, releasing energy or atomic particles. Unstable isotopes are called **radioactive isotopes,** and the energy or atomic particles they emit are called **atomic radiations.** Because all isotopes of a given element, whether radioactive or nonradioactive, have the same number of electrons, they all react in the same way chemically. For example, oxygen has eight different isotopes, and they all will function identically in metabolic reactions. This feature enables the medical profession to use radioactive isotopes in diagnosis and therapy.

QUICK APPLICATIONS
Radiation therapy uses radioactive isotopes for the treatment of certain types of cancer. All cells are susceptible to the effects of the radiation, but cancer cells are more readily damaged than normal cells. The goal is to destroy as many cancer cells as possible and to leave most of the normal cells intact.

QUICK CHECK
2.1 The most abundant element in the body is abbreviated with the letter "O." What is this element?
2.2 How many electrons, protons, and neutrons are in an element with an atomic number equal to 17 and a mass number equal to 35?
2.3 How many energy levels or shells contain electrons in the element referred to in Question 2.2?
2.4 What are isotopes?

Chemical Bonds

An atom's chemical behavior, or reactivity, is determined largely by the electrons in the outermost energy shell. This is the shell with the highest energy level. Atoms have a tendency to transfer or share electrons to achieve a stable configuration in the outer shell. When electrons from the outermost energy level are transferred or shared between atoms, attractive forces called *chemical bonds* develop that "hold" together the atoms to form a molecule. There are two types of intramolecular chemical bonds: ionic bonds and covalent bonds. A third type of bond, the hydrogen bond, is intermolecular. It provides a weak bond between molecules.

Ionic Bonds

Ionic (eye-ON-ik) **bonds** are formed when one or more electrons are transferred from one atom to another. Because electrons have a negative charge, atoms are no longer neutral when they lose or gain electrons. The charged particles that result when atoms lose or gain electrons are called *ions.* The symbols for ions are the symbols for the atoms from which they were derived with a superscripted plus (+) or minus (–) to indicate the charge. If more than one electron is lost or gained, then a number is used with the plus or minus sign. Table 2-2 lists some of the important ions in the human body.

When an atom loses one or more electrons, it then has more protons than electrons, and it becomes a positively charged ion. Positively charged ions are called **cations** (KAT-eye-onz). A sodium (Na) atom has 11 protons in the nucleus and 11 electrons in the energy shells. It is electrically neutral. When it loses an electron from its outermost shell, it still has 11 positively charged protons but only 10 negatively charged electrons. It has one more positive charge than it has negative charges, so it becomes a positively charged sodium ion (Na^+). This is illustrated in Figure 2-4.

Atoms that gain or pick up one or more electrons and then have more electrons than they do protons become negatively

TABLE 2-2 Important Ions in the Body

Ion	Symbol	Importance
Calcium	Ca^{2+}	Component of bones and teeth; necessary for blood clotting and muscle contraction
Sodium	Na^+	Principal cation in fluid outside cells; important in muscle contraction and nerve impulse conduction
Potassium	K^+	Principal cation in fluid inside cells; important in muscle contraction and nerve impulse conduction
Hydrogen	H^+	Important in acid-base balance
Hydroxide	OH^-	Important in acid-base balance
Chloride	Cl^-	Principal anion in fluid outside cells
Bicarbonate	HCO_3^-	Important in acid-base balance
Ammonium	NH_4^+	Important in acid-base balance; removes toxic ammonia from body
Phosphate	PO_4^{3-}	Component of bones, teeth, and high-energy molecules; important in acid-base balance
Iron	Fe^{2+}	Important component of hemoglobin for oxygen transport

charged ions. A negatively charged ion is called an **anion** (AN-eye-on). A chlorine atom has 17 protons and 17 electrons to make it electrically neutral. When it gains or picks up an electron, it still has 17 positively charged protons but now has 18 negatively charged electrons to make it a negatively charged chloride ion (Cl^-) (see Figure 2-4).

When an atom such as sodium loses or gives up an electron to another atom such as chlorine, two charged particles called *ions* are formed. One is a positively charged *cation;* the other is a negatively charged *anion.* Because of the opposite charges, the two ions are attracted to each other. The force of attraction between two oppositely charged ions is an ionic bond.

Oppositely charged ions that are held together by ionic bonds are called **ionic compounds.** Sodium chloride (table salt), formed from sodium ions and chloride ions, is an example of an ionic compound.

Covalent Bonds

Covalent (koh-VAY-lent) **bonds** are formed when two atoms share a pair of electrons. Two hydrogen atoms, for example, can share their electrons to form a molecule of hydrogen gas. Because only one pair of electrons is shared, a single covalent bond is formed (Figure 2-5). Carbon has four electrons in its outer shell that it can share with other atoms to form covalent bonds. If it shares these electrons with four hydrogen atoms, then four single covalent bonds are formed and a molecule of methane gas results (Figure 2-6).

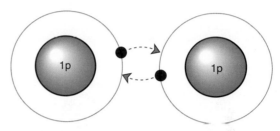

Two hydrogen atoms each with one proton in the nucleus and one electron in the energy shell

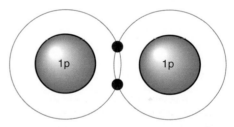

A molecule of hydrogen gas, H–H (H_2)
The two hydrogen atoms share an electron pair to form a single covalent bond

Figure 2-5 *Single covalent bond in hydrogen gas.*

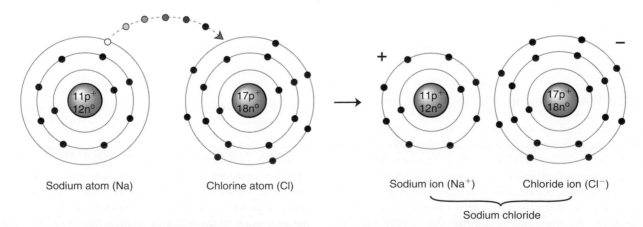

Sodium atom (Na)   Chlorine atom (Cl)   Sodium ion (Na^+)   Chloride ion (Cl^-)

Sodium chloride

Figure 2-4 *Formation of ionic bonds. An electron is transferred from the outer shell of the sodium atom to the outer shell of the chlorine atom. This transfer results in a positively charged sodium ion and a negatively charged chloride ion. The opposite charges attract to form the ionic bond of sodium chloride.*

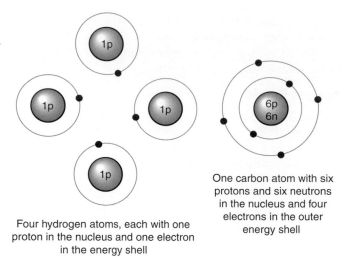

Four hydrogen atoms, each with one proton in the nucleus and one electron in the energy shell

One carbon atom with six protons and six neutrons in the nucleus and four electrons in the outer energy shell

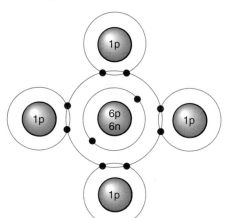

A molecule of methane gas (CH_4)
Each hydrogen atom shares an electron with the carbon atom and the carbon atom shares an electron with each of the hydrogen atoms to form four single covalent bonds

Figure 2-6 *Single covalent bonds in methane gas.*

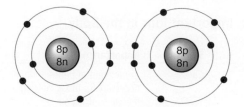

Two atoms of oxygen, each with eight protons and eight neutrons in the nucleus and six electrons in the outer energy level.

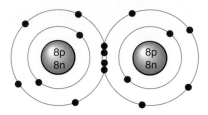

A molecule of oxygen gas.
Each oxygen atom shares two electrons to form a double covalent bond.

Figure 2-7 *Double covalent bond in oxygen gas.*

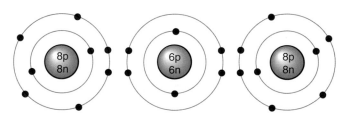

Two oxygen atoms and one carbon atom. Each oxygen atom has six electrons in the outer energy level and the carbon atom has four electrons in the outer energy level.

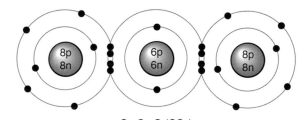

O=C=O (CO_2)
A molecule of carbon dioxide. Each oxygen atom shares two electrons with the carbon atom and the carbon atom shares two electrons with each oxygen to form a double covalent bond between the carbon and each oxygen.

Figure 2-8 *Double covalent bonds in carbon dioxide.*

Some atoms may share more than one pair of electrons with another atom. For example, an oxygen atom shares two pairs of electrons with another oxygen atom to form a molecule of oxygen gas (Figure 2-7). This forms a double covalent bond. Carbon may share two electron pairs with each of two atoms of oxygen to form two double covalent bonds as in carbon dioxide (Figure 2-8). A few atoms, such as nitrogen, may share three pairs of electrons to form triple covalent bonds (Figure 2-9).

A covalent bond formed by sharing a pair of electrons is sometimes indicated by drawing a straight line between the symbols for the two atoms. Two lines indicate a double covalent bond, and three lines designate a triple covalent bond.

There are times when the two electrons of a covalent bond are not shared equally between the two atoms. Instead, the electrons tend to spend more time around the nucleus of one atom than the other. Another way of stating this is that the electrons are closer to the nucleus of one atom than the other. This unequal sharing of electrons is called a **polar covalent bond.** Water is a good example of this (Figure 2-10). The shared pair of electrons spends more time with, or is closer to, the nucleus

of the oxygen atom than the hydrogen atom. The result is a partial negative charge at the oxygen end of the molecule and a partial positive charge on the hydrogen end to form a polar molecule. The molecule as a whole is neutral; but because it has opposite partial charges, it is polar, and the bonds are polar covalent bonds. Molecules with polar covalent bonds have a weak attraction for ions or other covalent molecules. Polar covalent bonds are stronger than ionic bonds.

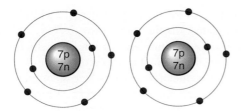

Two nitrogen atoms each with 7 protons and 7 neutrons in the nucleus and 5 electrons in the outer energy shell.

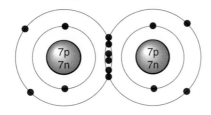

N≡N (N_2)

A molecule of nitrogen gas. The two nitrogen atoms share three electron pairs to make a triple covalent bond.

Figure 2-9 *Triple covalent bond in nitrogen gas.*

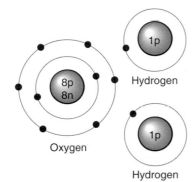

Hydrogen

Oxygen

Hydrogen

One oxygen and two hydrogen atoms

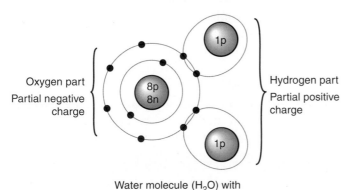

Oxygen part
Partial negative charge

Hydrogen part
Partial positive charge

Water molecule (H_2O) with polar covalent bonds

Figure 2-10 *Polar covalent bonds between oxygen and hydrogen.*

Hydrogen Bonds

The ionic and covalent bonds just described hold together atoms within molecules. They are **intramolecular** bonds. The **hydrogen bond** is an **intermolecular** bond. It is the attraction between two molecules. The electropositive hydrogen end of a polar covalent

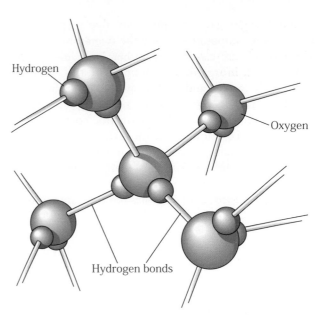

Hydrogen

Oxygen

Hydrogen bonds

Figure 2-11 *Intermolecular hydrogen bonds in water. The electropositive hydrogen end of a water molecule has a weak attraction for the electronegative oxygen end of other water molecules to form hydrogen bonds.*

molecule has a weak attraction for the negative portion of other polar covalent molecules or for negative ions. This weak attraction is called a hydrogen bond. Figure 2-11 illustrates how water molecules are held together by hydrogen bonds.

QUICK CHECK

2.5 What type of chemical bond is formed when two elements share one or more pairs of electrons?

2.6 Hydrogen bonds between covalent molecules of water are an example of what type of chemical bond?

Compounds and Molecules

Nature of Compounds

A **molecule** is formed when two or more atoms chemically bond together. The atoms may be different, as they are in water when oxygen and hydrogen bond together, or they may be alike, as they are in oxygen gas when two atoms of oxygen combine. If the atoms are different, the new substance is a compound. The atoms in a molecule, whether they are alike or different, are held together by either ionic or covalent bonds.

A **compound** is formed when two or more different types of atoms chemically combine in a definite, or fixed, ratio to form a new substance that is different from any of the original atoms. Water is an example of a compound. It is formed from two atoms of hydrogen and one atom of oxygen that chemically combine. The ratio is always two hydrogen atoms to one oxygen atom, and the result of the chemical combination, water, is different from either hydrogen or oxygen alone. Carbon dioxide is another example. It is formed from one atom of carbon and two atoms of oxygen that chemically combine. Oxygen gas is formed from two atoms of oxygen, but it is not a compound because the two atoms are alike.

Compounds are classified as either inorganic or organic. Briefly defined, **inorganic compounds** are those that do not contain both carbon and hydrogen. Water, carbon dioxide, and sodium chloride are examples of inorganic compounds. There are approximately 50,000 inorganic compounds in the world. **Organic compounds** are those that contain carbon and hydrogen atoms that are covalently bonded together. These are sometimes called the compounds of living organisms, although with modern technology many are created synthetically in the laboratory. There are about 20 times more organic compounds than inorganic compounds.

A molecule is the smallest unit of a compound formed from two or more different atoms that still retains the properties of the compound. Molecules can be broken down into the atoms from which they were made, but then they will no longer have the properties of the compound. The principal particles of matter are summarized in Table 2-3.

TABLE 2-3 Principal Particles of Matter

Particle	Description
Proton (p⁺)	Relatively large particle; has a positive charge; found within nucleus of atoms; number of protons in nucleus equals atomic number
Neutron (n⁰)	Relatively large particle; carries no charge; found within nucleus of atoms
Electron (e⁻)	Extremely small particle; has a negative charge; in constant motion around nucleus
Atom	Smallest particle of an element that has properties of that element; made up of protons, neutrons, and electrons
Ion	Atom that has lost or gained one or more electrons so that it is electrically charged
Molecule	Particle formed by chemical union of two or more atoms; smallest particle of a compound

QUICK **APPLICATIONS**

Water is the most abundant compound in living organisms. It accounts for about two thirds of the weight of the adult human body. Water is physiologically important because it is a good solvent, it breaks down molecules by hydrolysis, it remains liquid over a broad temperature range, and it can absorb and transport relatively large quantities of heat without dramatic changes in its own temperature.

Formulas

The **molecular formula** is a shorthand way to indicate the type and number of atoms in a molecule. The chemical symbols of the elements are used to indicate the types of atoms in the molecule, and a subscript number follows each symbol to show the number of atoms. If there is no subscript, the number 1 is implied. The molecular formula for water is H_2O. H is the

symbol for hydrogen; the subscript 2 after the H indicates two atoms of hydrogen; O is the symbol for oxygen; and because there is no subscript the number 1 is implied, which indicates there is one atom of oxygen. The molecular formula for water, H_2O, indicates that one molecule of water is composed of two atoms of hydrogen and one atom of oxygen. A molecule of carbon dioxide, CO_2, is composed of one atom of carbon and two atoms of oxygen. The molecular formula for glucose, an important sugar in the body, is $C_6H_{12}O_6$, which indicates that there are 6 atoms of carbon, 12 atoms of hydrogen, and 6 atoms of oxygen in the molecule.

As mentioned earlier, covalent bonds are sometimes represented by straight lines between the symbols for the elements in the molecule. These representations show how atoms are arranged and joined together in a molecule. Illustrations of these bonding arrangements are called **structural formulas** (Figure 2-12).

$$H-H \qquad O=O \qquad H_{\diagdown O \diagup} H \qquad O=C=O$$
$$H_2 \qquad\qquad O_2 \qquad\qquad H_2O \qquad\qquad CO_2$$

Figure 2-12 *Structural formulas for hydrogen gas, oxygen gas, water, and carbon dioxide.*

QUICK **CHECK**

2.7 Identify the elements in a molecule of sodium bicarbonate: $NaHCO_3$.
2.8 An atom is the smallest unit of an element. What is the smallest unit of a compound?

Chemical Reactions

A **chemical reaction** is the process by which atoms or molecules interact to form new chemical combinations. Bonds break and new ones form to create different molecules. The atoms and molecules present before the chemical reaction occurs are called the **reactants.** The new atoms and molecules that are created as a result of the reaction are called the **products.**

Chemical Equations

Reactions between molecules are represented by **chemical equations** that indicate the number and type of molecules involved in the reactions. The molecular formulas for the reactants are written on the left side of the equation, and the formulas for the products are written on the right. They are connected by an arrow that indicates the direction of the reaction, and the numeral before a molecular formula denotes the number of molecules of that particular substance involved. The following equation for the reaction between methane gas and oxygen provides a simple example:

$$\underset{\text{Methane}}{CH_4} + \underset{\text{Oxygen}}{2O_2} \rightarrow \underset{\text{Carbon dioxide}}{CO_2} + \underset{\text{Water}}{2H_2O}$$

This chemical equation indicates that 1 molecule of methane reacts with 2 molecules of oxygen gas to form 1 molecule of carbon dioxide and 2 molecules of water. The number of each *type of atom* must be the same in both the reactants and the products; that is, the equations must be balanced. In the above example, there is 1 carbon atom in the reactants and 1 carbon atom in the products. There are 4 hydrogen atoms and 4 oxygen atoms in the reactants and the same number in the products. Because changing a subscript in a formula indicates a different type of molecule, equations can be balanced only by adjusting the numerals before the molecular formulas.

The reaction between glucose and oxygen is an important one in the body. It provides much of the energy needed for body processes and daily activities:

$$C_6H_{12}O_6 \quad + \quad 6O_2 \quad \rightarrow \quad 6CO_2 \quad + \quad 6H_2O$$
Glucose   Oxygen   Carbon dioxide   Water

This balanced chemical equation indicates that 1 molecule of glucose reacts with 6 molecules of oxygen to yield 6 molecules of carbon dioxide and 6 molecules of water. It is balanced because there are 6 carbon atoms, 12 hydrogen atoms, and 18 oxygen atoms on both the reactant and the product side of the equation.

Types of Chemical Reactions

Synthesis Reactions

When two or more simple reactants combine to form a new, more complex product, the reaction is called **synthesis, combination,** or **composition.** These are the anabolic reactions in the body. This is represented symbolically by the following equation:

$$A + B \rightarrow AB$$

For example, oxygen and hydrogen combine to form water in a synthesis reaction. Two simple molecules combine to form a more complex molecule as indicated in this equation:

$$2H_2 + O_2 \rightarrow 2H_2O$$

When two simple molecules combine to form a more complex molecule by the removal of water, the reaction is called **dehydration synthesis.** Many anabolic reactions in the body, for example, the conversion of glucose to glycogen for storage, are of this type.

Decomposition Reactions

When the bonds in a complex reactant break to form new, simpler products, the reaction is **decomposition.** In the body, these are the catabolic reactions of metabolism. When water is used to break the bonds, the reaction is called **hydrolysis.** The digestion of food involves hydrolysis reactions. Decomposition reactions, represented symbolically by the following equation, are the reverse of synthesis reactions:

$$AB \rightarrow A + B$$

Single Replacement Reactions

Single replacement reactions, also called **single displacement reactions,** occur when one element in a compound is replaced by another. The general pattern for this type of reaction is represented symbolically as follows:

$$A + BC \rightarrow AC + B$$

Double Replacement Reactions

Double replacement reactions (also called **double displacement** or **exchange reactions**) occur when substances in two different compounds replace each other. These reactions are partially decomposition and partially synthesis. The bonds in the original reactants must break (decomposition) before the new products can be formed (synthesis). The general equation pattern for a double replacement reaction is the following:

$$AB + CD \rightarrow AD + CB$$

where A has replaced C and C has replaced A from the original reactants. A and C have exchanged places.

Exergonic and Endergonic Reactions

Chemical reactions are important in the body because this is the way in which molecules are produced when they are needed. The reactions are also important for the energy changes that occur when bonds break and new bonds form. Energy is stored in the chemical bonds of molecules. In **exergonic** (eks-er-GAHN-ik) reactions, there is more energy stored in the reactants than in the products. The extra energy is released. In other words, Reactants → Products + Released Energy. Some of the energy is released in the form of heat, which helps maintain body temperature. A common exergonic reaction that occurs in the body involves adenosine triphosphate (ATP), which breaks down to adenosine diphosphate (ADP) and a phosphate group, with the release of energy:

$$ATP \rightarrow ADP + phosphate + energy$$

Endergonic (en-der-GAHN-ik) reactions have more energy stored in the products than in the reactants. An input of energy from exergonic reactions is needed to drive these reactions. The products of endergonic reactions store energy in their chemical bonds. For these reactions, Reactants + Energy → Products. In the human body, the large carbohydrate, lipid, and protein molecules are synthesized by endergonic reactions.

Reaction Rates

Chemical reactions occur at different rates. Some are very slow, like the rusting of iron or the tarnishing of silver. Other reactions occur much faster, such as the setting of epoxy cement or the burning of paper. Some reactions occur so fast that they become explosive, like dynamite or the gasoline in a car. The rate at which chemical reactions occur is influenced by the nature of the reacting substances, temperature, concentration, catalysts, and surface area.

Certain substances are more reactive than others, depending on how readily bonds are broken and formed. Reactions involving ions are extremely fast because there are no bonds to break. Reactions in which covalently bonded molecules are involved require that bonds be broken and that new ones be formed. These occur more slowly. When hydrogen gas is mixed with oxygen gas, the reaction to produce water proceeds very slowly because the covalent bonds between the hydrogen atoms in the molecules of hydrogen gas and the covalent bonds between the oxygen atoms in the oxygen gas must first be broken. If a spark is introduced into this mixture of hydrogen and oxygen gas, the reaction occurs very rapidly because the spark supplies sufficient energy to break the covalent bonds.

As temperature increases, the speed of most chemical reactions also increases. In general, for every 10° C increase in temperature, the reaction rate nearly doubles. The opposite is also true. If temperature decreases, reaction rates decrease.

Within limits, the greater the concentration of the reactants, the faster will be the speed of the reaction. Because there are more molecules to react, the rate of reaction is increased. Under normal conditions, the concentration of oxygen molecules in body cells is sufficient to sustain the reactions that are necessary for life. If the concentration of oxygen decreases, the rate of the reactions decreases. This interferes with cell function and ultimately causes death.

A **catalyst** (KAT-ah-list) is a substance that changes the rate of a reaction without itself being chemically altered in the process. It increases the rate of a chemical reaction. In the body, organic catalysts called **enzymes** speed up the metabolic processes to a rate that is compatible with life. At normal body temperature, 37 ° C, most of the chemical reactions in the body would take place too slowly to sustain life if it were not for the enzymes. Enzymes are protein molecules that are very specific for the reactions they control. Each enzyme controls only one chemical reaction. For example, the enzyme lactase speeds up the hydrolysis of lactose into smaller molecules of glucose and galactose in the process of digestion. Because enzymes are very specific, no other enzymes are able to catalyze this reaction and lactase in not effective in any other reactions.

The rate of a chemical reaction also depends on the surface area of the reactants. Lumps of coal or charcoal briquettes burn quite slowly; however, coal dust is explosive and dangerous to coal miners. When a piece of coal is ground into coal dust, it greatly increases the total surface area of the particles. Many medications, such as antacids, are given as finely ground particles in a liquid so that they can react more rapidly.

 QUICK **APPLICATIONS**
When a person has a fever, reaction rates in the body increase. Some of this is manifested in an increased pulse rate and an increased respiratory rate. In certain types of surgery, the body is cooled so that metabolism is decreased during the procedure.

Reversible Reactions

Many chemical reactions are **reversible.** This means that they can proceed from reactants into products or reverse the direction and proceed from products back into reactants. This is shown symbolically by using a double arrow with the points in opposite directions, as follows:

$$A + B \rightleftharpoons AB$$

Under certain conditions, it is a synthesis reaction and proceeds from left to right. At other times, it is a decomposition reaction and proceeds from right to left. Whether a reversible reaction proceeds in one direction or the other depends on the relative proportions of the reactants and products, the energy necessary for the reaction, and the presence or absence of catalysts. At **equilibrium,** the reaction proceeds in both directions

at the same rate. In other words, the rate of product formation is the same as the rate of reactant formation, and the amounts of products and reactants remain relatively constant.

The reaction between carbon dioxide and water to form hydrogen ions and bicarbonate ions is reversible:

$$CO_2 + H_2O \rightleftharpoons H^+ + HCO_3^-$$

If carbon dioxide is added to the system, the reaction proceeds to the right to produce more hydrogen and bicarbonate ions until equilibrium is reestablished. When hydrogen ions are added, the reaction proceeds to the left to produce more carbon dioxide and water until equilibrium is again reached. This reversible system is one mechanism that regulates the amount of hydrogen ions in the body, which is critical to normal functioning of metabolic processes.

 QUICK **APPLICATIONS**
The body adjusts the hydrogen ion concentration in the blood by altering the breathing rate. If more hydrogen ions are needed in the body, the breathing rate decreases, which increases the amount of carbon dioxide in body fluids. This causes the reaction to proceed to the right and increases the number of hydrogen ions. If the hydrogen ion concentration is too high, the breathing rate increases, more carbon dioxide is exhaled, and the reaction proceeds to the left to decrease the number of hydrogen ions.

 QUICK **CHECK**
2.9 In the chemical equation $CH_4 + 2O_2 \rightarrow CO_2 + 2H_2O$, how many atoms of oxygen are present in the reactants? How many oxygen atoms are in the products?
2.10 During digestion, a molecule of sucrose forms the simpler molecules of glucose and fructose. This is an example of what type of reaction?
2.11 Charcoal briquettes burn rather slowly, but coal dust can be explosive. Why?

 FROM THE PHARMACY

The health care profession relies on the use of pharmaceutical preparations, and the health care professional needs to recognize the different nomenclatures used for these medications. The **chemical name** is the nomenclature that indicates the nature of the active ingredient. It is often long, complex, and difficult to pronounce. For example, the chemical name for Advil, a common analgesic, is α-methyl-4-(2-methylpropyl)benzene. That is quite a mouthful, and it is one of the shorter chemical names! Generic and brand names are more frequently used in health care practice. The **generic name** is a nonproprietary name that is assigned by the United

Mixtures, Solutions, and Suspensions

Mixtures

A **mixture** is a combination of two or more substances, in varying proportions, that can be separated by ordinary physical means. The substances retain their original properties after they have been combined in a mixture. The components of a mixture may be elements (such as iron and sulfur), compounds (such as sugar and water), or elements and compounds (such as iodine and alcohol).

Solutions

Solutions are mixtures in which the component particles remain evenly distributed. All solutions consist of two parts: the **solute** and the **solvent.** The solute is the substance that is present in the smaller amount and that is being dissolved. It may be a gas, liquid, or solid. The solvent, usually a gas or liquid, is the component that is present in the larger amount and that does the dissolving. In a sugar solution, the sugar is the solute and the water is the solvent. Water is the most common solvent and is called the universal solvent. Alcohol and carbon tetrachloride are also commonly used solvents. When alcohol is the solvent, the solution is called a **tincture.** For example, when iodine dissolves in alcohol, the solution is called tincture of iodine. The composition of a solution is variable; that is, it may be weak or concentrated. In a sugar solution, whether there is a small amount of sugar or a large amount, it is still a solution. Although a solution is always clear and the solute does not settle, the components may be separated by physical means such as evaporation. For example, when the water evaporates from a sugar and water solution, rock sugar forms.

Suspensions

Some mixtures involving a liquid settle unless they are continually shaken. If sand is mixed with water, shaken, and then allowed to settle, the particles of sand will fall to the bottom. This is a type of mixture called a **suspension.** A suspension is cloudy and its particles settle. Blood cells form a suspension in the plasma.

One type of mixture that is particularly important in the body is the **colloidal suspension.** The particles in a colloidal suspension are so small that they remain suspended in the liquid but they do not dissolve. Mayonnaise, although not of particular importance in the body, is a colloidal suspension. In this case, the vinegar is the suspending medium and the beaten egg provides the colloidal particles. More relevant, perhaps, is the fact that the fluid that fills the cells of the body, the cytoplasm, is a colloidal suspension.

Electrolytes, Acids, Bases, and Buffers

Acids, bases, and salts belong to a large group of compounds called electrolytes. There are numerous electrolytes in the body, and their concentrations are an important aspect of health care.

Electrolytes

Electrolytes (ee-LEK-troh-lites) are substances that break up, or **dissociate,** in solution to form charged particles, or ions. These compounds are called *electrolytes* because the ions can conduct an electrical current. When an ionic compound such as sodium chloride, NaCl, is placed in water, the positively charged sodium ion is attracted to the negatively charged oxygen end of the water molecule. The negatively charged chloride ion is attracted to the hydrogens of the water molecule. Because the polar covalent bonds of the water are stronger than the ionic bonds of the sodium chloride, the sodium chloride breaks apart, or dissociates, into cations and anions in the water. Figure 2-13 illustrates the dissociation of sodium chloride in water. Refer to Table 2-2 for some of the cations and anions in the body.

QUICK APPLICATIONS

The electrocardiogram and electroencephalogram are graphic tracings of the electrical currents created by the movement of electrolytes in the heart and brain, respectively.

Acids

Everyone is familiar with acids of various types. Orange juice, lemon juice, vinegar, coffee, and aspirin all contain acids. Acids have a sour taste. An **acid** is defined as a **proton donor.** Think about the structure of a hydrogen atom: it has one proton in the nucleus and one electron in the electron shell. When the hydrogen atom loses its one electron and becomes a hydrogen ion, its

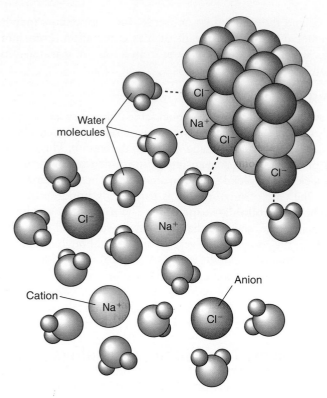

Figure 2-13 *Dissociation of sodium chloride in water. The sodium becomes positively charged cations and the chloride becomes negatively charged anions.*

structure consists of the single proton in the nucleus. A hydrogen ion is a proton. Any substance that releases hydrogen ions in water is a proton donor and is an acid. Hydrochloric acid, the acid in the stomach, forms hydrogen ions (H^+) and chloride ions (Cl^-). Because it forms hydrogen ions, it is a proton donor:

$$HCl \rightarrow H^+ + Cl^-$$
Hydrochloric acid   Hydrogen ion   Chloride ion

Carbonic acid also plays an important role in human physiology. When it dissociates or ionizes in water, it forms hydrogen ions and bicarbonate ions:

$$H_2CO_3 \rightarrow H^+ + HCO_3^-$$
Carbonic acid   Hydrogen ion   Bicarbonate ion

The strength of an acid depends on the degree to which it dissociates in water. Hydrochloric acid is a strong acid because it dissociates readily to produce an abundance of hydrogen ions. There are very few hydrogen chloride molecules in the solution. On the other hand, carbonic acid is a relatively weak acid because most of the molecules remain intact in water and only a few dissociate into hydrogen ions and bicarbonate ions.

Bases

Household ammonia, milk of magnesia, and egg white are just a few of the bases with which everyone is familiar. Bases feel slippery and taste bitter. A **base** is defined as a **proton acceptor.** A base accepts the protons that an acid donates. Sodium hydroxide (NaOH) is a common example of a base. When placed in water, it dissociates into sodium ions (Na^+) and hydroxide ions (OH^-):

$$NaOH \rightarrow Na^+ + OH^-$$
Sodium hydroxide   Sodium ion   Hydroxide ion

It is a base because the hydroxide ions accept the protons or hydrogen ions from an acid to form water:

$$H^+ + OH^- \rightarrow HOH(H_2O)$$
Hydrogen ion   Hydroxide ion   Water
Proton donor   Proton acceptor

The degree of ionization or dissociation of a base determines its strength. A strong base is highly ionized in solution and at equilibrium will have more ions than molecules. Weak bases are poorly ionized, and at equilibrium their solutions have more molecules than ions.

The pH Scale

The term **pH** is used to indicate the exact strength of an acid or a base. The pH scale, illustrated in Figure 2-14, ranges from 0 to 14 and measures the hydrogen ion concentration of a solution. The greater the number of hydrogen ions, the more **acidic** is the solution, and the lower is the pH. Fewer hydrogen ions result in basic or **alkaline** solutions with a higher pH. A **neutral** solution has the same number of proton donors (hydrogen ions) as

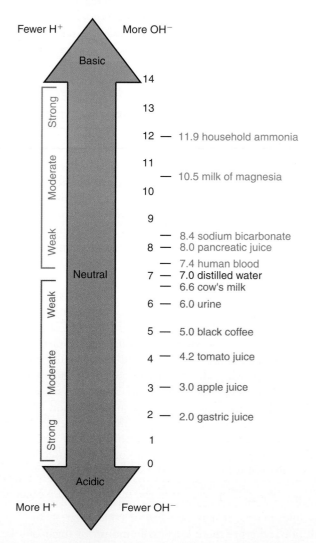

Figure 2-14 *pH scale and pH values of some common substances.*

proton acceptors (hydroxide ions), and the pH is 7. Pure water is a neutral solution with a pH of 7. Solutions that are acidic have a pH of less than 7, and the lower is the pH, the stronger is the acid. Basic or alkaline solutions have a pH greater than 7, and the higher the pH, the stronger it is. Because the scale is logarithmic, each unit on the pH scale represents a 10-fold difference in the hydrogen ion concentration. For example, a solution with a pH of 6 has 10 times the hydrogen ion concentration as a solution with a pH of 7, and a solution with a pH of 5 has 10 times more than a solution with a pH of 6. A solution with a pH of 5 has 100 times the hydrogen ion concentration of a pH 7 solution. Figure 2-14 compares the pH values for some familiar acids and bases.

QUICK **APPLICATIONS**

The pH of human milk ranges from 7.0 to 7.4. The initial breast secretion, colostrum, has a pH of 7.45. This drops to a low of 7.0 for milk during the second week of lactation. The pH remains near this level for about 3 months, then increases gradually to 7.4 at about 10 months.

Neutralization Reactions

Neutralization is the reaction between an acid and a base. The hydrogen ion (H^+) from the acid reacts with the hydroxide ion (OH^-) from the base to form water. The acid removes or neutralizes the effect of the base and vice versa. The other product is called a **salt**. Salts are ionic compounds produced from neutralization reactions and consist of cations other than H^+ and anions other than OH^-. They are formed from the positive ion of the base and the negative ion of the acid:

$$HCl + NaOH \rightarrow NaCl + H_2O$$
Acid — Base — Salt — Water

Neutralization reactions help maintain the proper pH of the blood. They also have many practical applications. One such application involves antacid medications. Antacids are basic and are often used to neutralize excess acid in the stomach.

Buffers

If you add a few drops of a strong acid to some distilled water (pH = 7), the pH of the solution will decrease because hydrogen ions have been added. If a few drops of acid are placed in a buffer solution, the pH will change only slightly. A **buffer** is a solution that resists change in pH when either an acid or a base is added.

A buffer solution contains a weak acid and a salt of that same acid, the salt functions as a weak base. One of the most common buffering systems and one that is important in human physiology involves carbonic acid (H_2CO_3) and its salt, sodium bicarbonate ($NaHCO_3$). When hydrogen ions from a strong acid are added, they react with the salt to form a weaker acid and a neutral salt:

$$HCl + NaHCO_3 \rightarrow H_2CO_3 + NaCl$$
Strong acid — Buffer salt — Weak acid — Neutral salt

In this way the hydrogen ions from the strong acid are incorporated into the weak acid and will have less effect on the pH. If a base such as NaOH is added to the buffer system, it will be neutralized by the acid to form water and a salt:

$$NaOH + H_2CO_3 \rightarrow NaHCO_3 + H_2O$$
Base — Buffer acid — Buffer salt — Water

When either an acid or a base is added to a buffered solution, something neutral (water or salt) and a component of the buffer (weak acid or its salt) are formed. The pH does not change. Buffers are very important physiologically. Even a slight deviation from the normal pH range can cause pronounced changes in the rate of cellular chemical reactions and thus threaten survival. Buffers provide one of the homeostatic control mechanisms that maintain normal pH.

QUICK **APPLICATIONS**

In the human body, acid-base balance is regulated by chemical buffer systems, the lungs, and the kidneys.

QUICK **CHECK**

2.12 Black coffee is an acid. Is its pH greater or less than 7.0?
2.13 Many of the foods we eat are acids, yet the pH of body fluids remains relatively constant. Why?

Organic Compounds

The term **organic chemistry** is inherited from the day when the science of chemistry was comparatively primitive. Compounds formed by living processes were termed organic. In the nineteenth century, it was discovered that organic compounds can be created artificially in the laboratory from substances that do not arise from life processes. Since then, the field of organic chemistry has grown until today more than 1 million different organic compounds have been identified.

Organic compounds are important because they are associated with all living matter in both plants and animals. Nearly all compounds related to living substances contain carbon; thus, organic compounds are carbon compounds. Carbohydrates, fats, proteins, hormones, vitamins, enzymes, and many drugs are organic compounds. Wool, silk, linen, cotton, nylon, and rayon all contain organic compounds. Add to this the perfumes, dyes, flavorings, soaps, gasoline, and oils and you see that the study of organic compounds is extensive.

The most important groups of organic compounds in the body are the carbohydrates, proteins, lipids, nucleic acids, and adenosine triphosphate.

Carbohydrates

Carbohydrate (kar-boh-HYE-drayt) molecules are composed of carbon, hydrogen, and oxygen. They are the product of photosynthesis, a process by which plants convert solar energy into chemical energy. Carbohydrate molecules range in size from small to very large. They are an important energy source in the body, contribute to the structure of some cellular components, and form a reserve supply of stored energy.

The simplest carbohydrate molecules are the **monosaccharides** (mahn-oh-SAK-ah-rides) or simple sugars. The most important simple sugar is **glucose,** which has 6 carbon atoms, 12 hydrogen atoms, and 6 oxygen atoms ($C_6H_{12}O_6$). Because it has six carbon atoms, it is called a hexose. Glucose is also known as dextrose or grape sugar. It occurs normally in the blood and abnormally in the urine. Glucose requires no digestion; therefore, it can be given intravenously. Figure 2-15 illustrates two methods of representing the structural formula for glucose. Two other important hexoses are **fructose** and **galactose.** Even though they have the same molecular formula as glucose, $C_6H_{12}O_6$, their atoms are arranged differently, which gives them different structural formulas. Fructose is found in fruit juices and honey. It is called fruit sugar and is the sweetest of all sugars. Galactose is not found free in nature but is one of the components of lactose or milk sugar. When ingested, fructose and galactose are often converted to glucose in the liver. Figure 2-16 compares the structural formulas for glucose, fructose, and galactose. Notice that fructose and galactose have 6 carbon atoms, 12 hydrogen atoms, and 6 oxygen atoms, the same as glucose, but each arrangement of atoms is slightly different.

Not all monosaccharides are hexoses. Some are **pentoses,** which contain five carbon atoms. **Ribose** and **deoxyribose** are important pentose monosaccharides in the body. They are components of nucleic acids, which are discussed later in this chapter. Figure 2-17 depicts structural formulas for ribose and deoxyribose. Note that there is one less oxygen atom in deoxyribose.

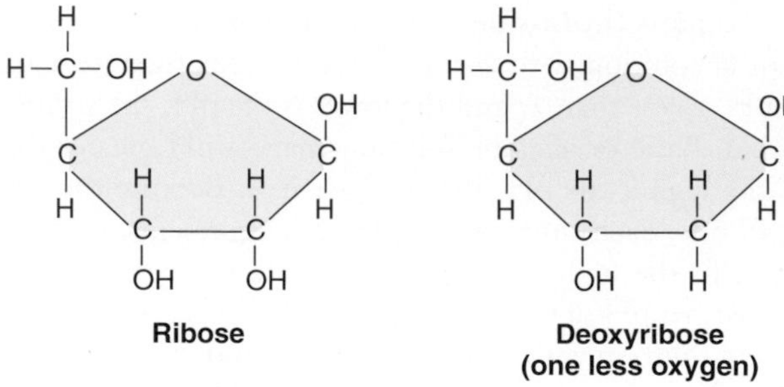

Figure 2-17 *Structural formulas for two pentoses.*

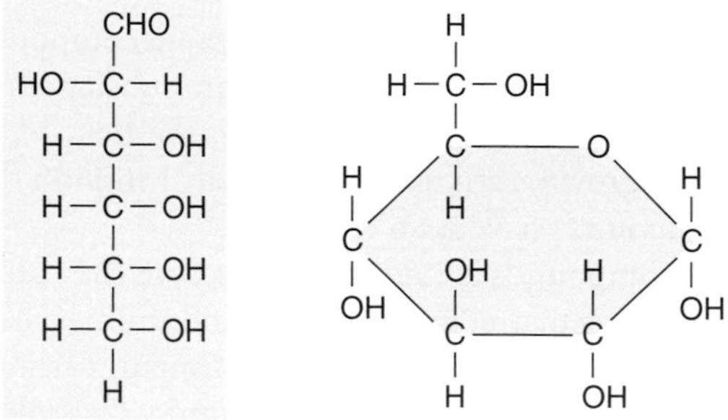

When two hexose monosaccharides are linked together by a dehydration synthesis reaction, a **disaccharide** (die-SAK-ah-ride), or double sugar, is formed. Disaccharides have the molecular formula $C_{12}H_{22}O_{11}$. Three common disaccharides are sucrose, maltose, and lactose. Sucrose is common, ordinary table sugar. It is formed from one glucose and one fructose. Maltose is malt sugar, found in germinating grain and in malt. The building blocks of maltose are two molecules of glucose. Lactose is milk sugar, the sugar found in milk. When lactose is digested or broken down by hydrolysis, it produces glucose and galactose. Certain bacteria produce enzymes that convert lactose to lactic acid, which sours the milk.

Polysaccharides (pahl-ee-SAK-ah-rides) are long chains of monosaccharides linked together. This is illustrated in Figure 2-18. Three important polysaccharides, all composed of glucose units, are starch, cellulose, and glycogen. Starch is the storage food of plants. It exists as small granules that are insoluble in water, but if heated, the granules rupture and form a colloidal gel. Cellulose is the supporting tissue of plants. It is not affected by any enzymes in the human digestive system, so it is not digestible and contributes to the "roughage" in the diet. Glycogen, composed of many glucose units, is animal starch. It is the storage form of carbohydrates in the body and is found particularly in the liver and in muscle. Table 2-4 summarizes the physiologically important carbohydrates.

Figure 2-15 *Structural formulas for glucose, $C_6H_{12}O_6$.*

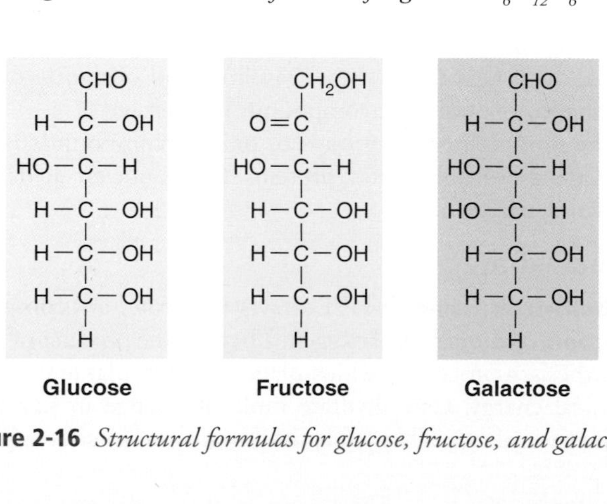

Glucose   Fructose   Galactose

Figure 2-16 *Structural formulas for glucose, fructose, and galactose.*

 QUICK APPLICATIONS

When blood sugar (glucose) levels decline, stored glycogen is broken down by hydrolysis into its component glucose units to restore blood glucose levels to normal.

QUICK CHECK

2.14 What elements are present in a carbohydrate?
2.15 Name three hexose molecules.
2.16 Name three disaccharides and the hexose components of each.

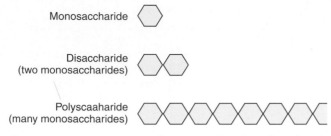

Figure 2-18 *Representation of a monosaccharide, a disaccharide, and a polysaccharide.*

TABLE 2-4 Physiologically Important Carbohydrates

Carbohydrate	Characteristics
Monosaccharides	Simplest carbohydrate molecules
Hexoses	Six carbon atoms
Glucose	Occurs normally in the blood
Fructose	Found in fruit juice and honey
Galactose	Component of milk sugar
Pentoses	Five carbon atoms
Ribose	Component of nucleic acids (RNA)
Deoxyribose	Component of nucleic acids (DNA)
Disaccharides	Double sugars: two monosaccharides linked together
Sucrose	Table sugar; glucose plus fructose
Maltose	Malt sugar; glucose plus glucose
Lactose	Milk sugar; glucose plus galactose
Polysaccharides	Long chains of monosaccharides
Starch	Carbohydrate storage form in plants; primary source of calories in food
Glycogen	Carbohydrate storage form in animals (animal starch)
Cellulose	Supporting tissue of plants; provides fiber and/or roughage in the diet

Proteins

All **proteins** (PRO-teens) contain the elements carbon, hydrogen, oxygen, and nitrogen. Most of them also contain sulfur and some include phosphorus. The building blocks of proteins are **amino acids,** molecules characterized by an amino group (—NH$_2$) and a carboxylic acid group (—COOH). The structural formulas for leucine and valine, two essential amino acids, are illustrated in Figure 2-19. There are about 20 different amino acids that occur commonly in proteins. Of these, humans can synthesize 11 from simple organic molecules, but the other 9 are "essential amino acids" and must be provided in the diet (Table 2-5). Amino acids are linked together by **peptide bonds,** produced by dehydration synthesis, to form chains that vary in

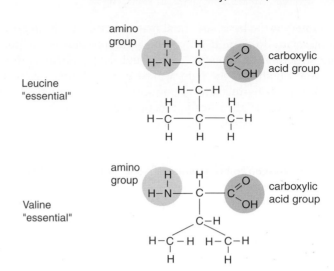

Figure 2-19 *Structural formulas for leucine and valine, two essential amino acids. Note the amino group and the carboxylic acid group on each one to make it an amino acid.*

TABLE 2-5 Common Amino Acids

Name	Abbreviation
Alanine	Ala
Arginine	Arg
Asparagine	Asn
Aspartic acid	Asp
Cysteine	Cys
Glutamic acid	Glu
Glutamine	Gln
Glycine	Gly
Histidine*	His
Isoleucine	Ile
Leucine	Leu
Lysine	Lys
Methionine	Met
Phenylalanine	Phe
Proline	Pro
Serine	Ser
Threonine	The
Tryptophan	Trp
Tyrosine	Tyr
Valine	Val

*Essential only in infants.
Names in bold indicate essential amino acids

length from less than 100 to more than 50,000 amino acids. The chains are then folded to form a three-dimensional protein molecule. The number, types, and sequence of amino acids, as well as the folding pattern, are unique for each specific protein.

Protein is the basic structural material of the body and performs many important functions. Some proteins are important structural components of cells and tissues whereas others act as antibodies in the fight against disease. Muscles contain specific proteins that are responsible for contraction. Other proteins provide identification marks and receptor sites on cell membranes. Enzymes and hormones have a critical role in all metabolic processes. Hemoglobin is a large protein molecule found inside red blood cells that has a specialized function to transport oxygen to body tissues. In addition to all of this, proteins provide a source of energy.

QUICK CHECK

2.17 List at least four functions of proteins in the body.

2.18 What are the building blocks of proteins and what elements do they contain?

Lipids

Lipids represent a group of organic compounds that, similar to carbohydrates, are composed of carbon, hydrogen, and oxygen. They differ from carbohydrates in that they have a lower oxygen content. This is illustrated by the molecular formula for the fat glycerol tristearate, $C_{57}H_{110}O_6$. For the number of carbon and hydrogen atoms, there is a much smaller ration of oxygen than in carbohydrates. Lipids are insoluble in water but are generally soluble in nonpolar solvents such as alcohol, acetone, ether, and chloroform. Lipids include a variety of compounds such as fats, phospholipids, and steroids that have important functions in the body. These are summarized in Table 2-6.

The most common members of the lipid group are the fats, or **triglycerides** (try-GLIS-ser-ides). They are the body's most highly concentrated source of energy and energy storage. They

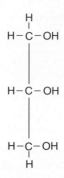

Figure 2-20 *Structural formula for glycerol. Note the three carbon atoms and the hydroxyl (OH) group on each carbon.*

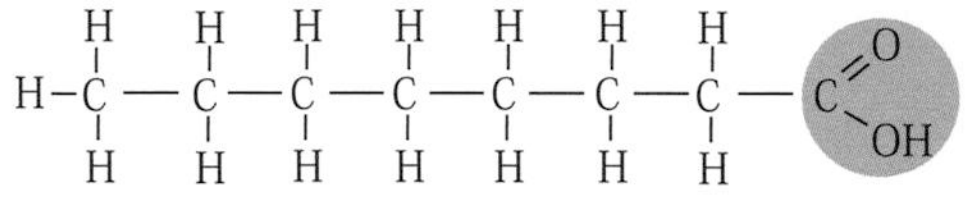

Figure 2-21 *Structural formula for capryllic acid, a representative fatty acid.*

also provide protection, padding, and insulation. The building blocks of fats are **glycerol** and **fatty acids.** Glycerol is a three-carbon molecule that has a hydroxyl group (—OH) on each carbon (Figure 2-20). Fatty acids are carbon chains with a carboxylic acid group (—COOH) at one end (Figure 2-21). The carbon chains in fatty acids vary in length but have an even number of carbon atoms. The carboxylic acid group gives it its acidic properties. If all the carbons in a fatty acid are joined together with single covalent bonds, then it is a **saturated fatty acid** (Figure 2-22). It is saturated with hydrogen. If, however, there are double covalent bonds between some of the carbons in the chain, then it is an **unsaturated fatty acid** (Figure 2-23). Table 2-7 lists some of the common fatty acids.

Three fatty acids combine with one glycerol, by dehydration synthesis, to form a triglyceride or fat (Figure 2-24). In natural fats, the fatty acids in the triglyceride are different. Triglycerides that contain only saturated fatty acids are called **saturated fats,** but if some of the fatty acids are unsaturated, they are **unsaturated fats.** Saturated fats are typically animal fats, and unsaturated fats are vegetable oils. Structurally, animal fats and vegetable oils are similar except that the oils are synthesized from unsaturated fatty acids. **Hydrogenated fats** are produced by artificially adding hydrogen to unsaturated oils. This process breaks the double bonds between the carbon atoms and "saturates" the molecule with hydrogen. The configuration of the molecule changes and the liquid oil becomes solid at room temperature.

TABLE 2-6 Important Lipid Groups

Group	Components	Characteristics
Triglycerides	Glycerol and fatty acids	Most common lipid in body, efficient energy storage; insulation
Phospholipids	Glycerol, fatty acids, phosphate, and nitrogen groups	Component of cell membranes; also found in nerve tissue
Steroids	Complex carbon rings derived from lipids	Most common steroid is cholesterol; includes sex hormones, adrenocortical hormones, vitamin D

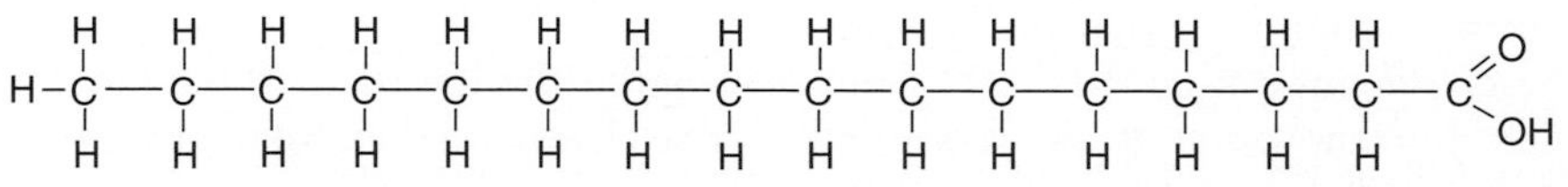

Figure 2-22 *Structural formula for palmit acid, a representative saturated fatty acid. Note the even number of carbon atoms (16), the carboxylic acid group at the end, and the single bonds between the carbon atoms.*

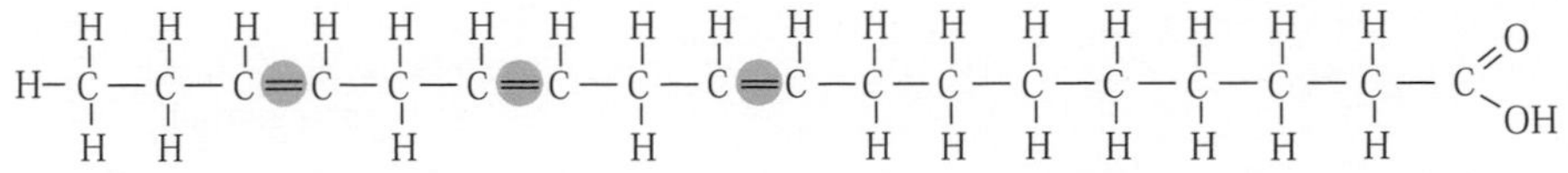

Figure 2-23 *Structural formula for linolenic acid, a representative unsaturated fatty acid. Note the even number of carbon atoms (18), the carboxylic acid group at one end of the chain, and the double bonds between some of the carbon atoms.*

TABLE 2-7 Some Common Fatty Acids

Name	Number of Carbon Atoms	Formula	Saturated or Unsaturated
Butyric	4	C_3H_7COOH	Saturated
Caproic	6	$C_5H_{11}COOH$	Saturated
Caprylic	8	$C_7H_{15}COOH$	Saturated
Capric	10	$C_9H_{19}COOH$	Saturated
Lauric	12	$C_{11}H_{23}COOH$	Saturated
Myristic	14	$C_{13}H_{27}COOH$	Saturated
Palmitic	16	$C_{15}H_{31}COOH$	Saturated
Stearic	18	$C_{17}H_{35}COOH$	Saturated
Oleic	18	$C_{17}H_{33}COOH$	Unsaturated (1)*
Linoleic	18	$C_{17}H_31COOH$	Unsaturated (2)
Linolenic	18	$C_{17}H_{29}COOH$	Unsaturated (3)
Arachidic	20	$C_{19}H_{39}COOH$	Saturated
Arachidonic	20	$C_{19}H_{31}COOH$	Unsaturated (4)

*The number in parentheses by the unsaturated fatty acids indicates the number of double bonds.

Phospholipids contain a phosphate group and a nitrogenous group in addition to the glycerol and fatty acids. They are an important component of cell membranes in the body and are particularly abundant in nerve and muscle cells.

Steroids are compounds that are derivatives of lipids and have four interconnected rings of carbon atoms. The most common steroid in the body is cholesterol, which is particularly abundant in the brain and nerve tissue. It is the chief component of gallstones. Sex hormones, steroid hormones from the cortex of the adrenal gland, and vitamin D are all steroids and are derivatives of cholesterol.

QUICK CHECK
2.19 Carbohydrates and lipids contain carbon, hydrogen, and oxygen. What is the difference in chemical content between these two classes of organic compounds?

Nucleic Acids

Nucleic acids are large, complex, organic compounds that contain carbon, hydrogen, oxygen, nitrogen, and phosphorus. The building blocks of nucleic acids are **nucleotides** (NOO-klee-oh-tides). A nucleotide consists of a five-carbon sugar or pentose, an organic nitrogenous base, and a phosphate group (Figure 2-25). Nucleotides are joined together by dehydration synthesis into long chains to form the nucleic acid (Figure 2-26). There are two classes of nucleic acids: deoxyribonucleic acid (DNA) and ribonucleic acid (RNA).

Deoxyribonucleic acid (dee-ahk-see-rye-boh-noo-KLEE-ik AS-id), or **DNA,** is the genetic material of the cell. The sugar in DNA is deoxyribose, and the nitrogenous bases are adenine, thymine, cytosine, and guanine. The specific sequence of the nitrogenous bases constitutes the genetic code, which is all the genetic information of the cell in a code form. There are two chains of nucleotides in DNA that are loosely joined together by hydrogen bonds and then these two chains are twisted into a double helix (Figure 2-27).

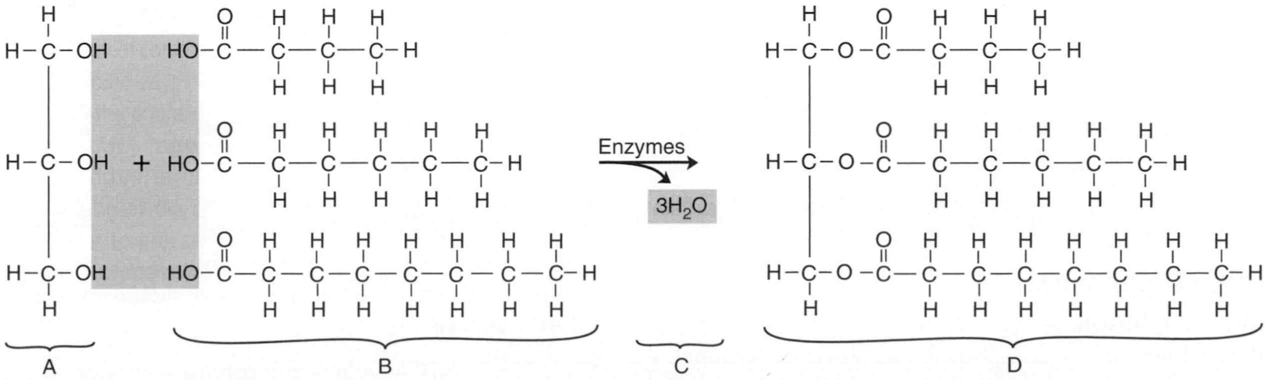

Figure 2-24 *Formation of a triglyceride. Glycerol (A) reacts with 3 fatty acids (B) and by dehydration synthesis (C) forms a triglyceride (D). Three water molecules are removed by dehydration synthesis. In this example all of the fatty acids are saturated and the resulting triglyceride is a saturated fat.*

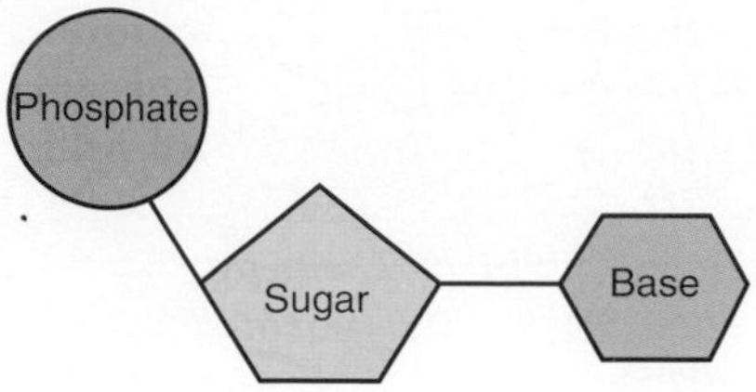

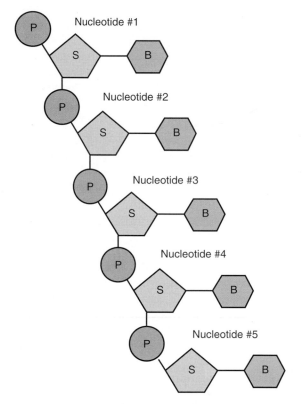

Figure 2-25 *Components of a nucleotide: phosphate, pentose sugar, and nitrogenous base.*

Figure 2-26 *Structure of a nucleic acid. Nucleic acids are long chains of nucleotides joined by dehydration synthesis.*

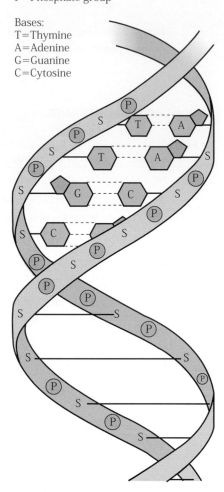

S=Deoxyribose sugar
P=Phosphate group

Bases:
T=Thymine
A=Adenine
G=Guanine
C=Cytosine

Figure 2-27 *Structure of deoxyribonucleic acid. Note the two chains of nucleotides twisted into a double helix. The dotted lines represent hydrogen bonds. There are two hydrogen bonds between thymine and adenine, and three hydrogen bonds between cytosine and guanine.*

Ribonucleic acid (rye-boh-noo-KLEE-ik AS-id), or **RNA,** has several functions in the synthesis of proteins within the cell. RNA is a single chain of nucleotides in which the sugar is ribose and the nitrogenous bases are adenine, uracil, cytosine, and guanine. DNA, the genetic material, contains the instructions for making proteins, but various types of RNA carry out the processes. DNA and RNA are compared in Table 2-8.

Adenosine Triphosphate

Adenosine triphosphate (ah-DEN-oh-sin trye-FOS-fate), or **ATP,** is a high-energy compound composed of adenine (a nitrogenous base), ribose (a five-carbon sugar), and three phosphate groups (Figure 2-28). The phosphate groups are linked together by high-energy chemical bonds that release chemical energy when they are broken. This chemical energy is usable

TABLE 2-8 Comparison of DNA and RNA

Feature	DNA	RNA
Sugar	Deoxyribose	Ribose
Bases	Adenine, guanine, thymine, cytosine	Adenine, guanine, uracil, cytosine
Strands	Two strands in double helix	Single strand

by the cells of the body. When the phosphate group on the end is split off by hydrolysis, a molecule of adenosine diphosphate (ADP) remains:

ATP → ADP + phosphate + energy

Another phosphate can be split from ADP to release more energy and adenosine monophosphate (AMP). When energy is available, ATP can be resynthesized from ADP. The breakdown

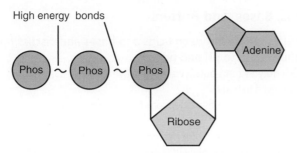

Figure 2-28 *Components of adenosine triphosphate (ATP). Note the three phosphate groups with high-energy bonds, ribose (a pentose sugar), and adenine (a nitrogenous base).*

QUICK **CHECK**

2.20 **What is the difference between saturated and unsaturated fatty acids?**

2.21 **Analysis of a given nucleic acid reveals cytosine, uracil, adenine, and guanine. Is this nucleic acid DNA or RNA?**

2.22 **Analysis of an organic compound shows it contains only C, H, and a relatively small amount of O. Is this compound likely a carbohydrate, protein, lipid, or nucleic acid?**

of nutrients releases energy that is used to synthesize ATP. In this way, the nutrient energy is converted to a form that is usable by the body through the high-energy bonds of ATP.

CHAPTER SUMMARY

Elements

■ **Define matter, element, and atom.**
- Matter is anything that takes up space and has weight.
- An element is the simplest form of matter.
- An atom is the smallest unit of an element.

■ **Use chemical symbols to identify elements.**
- Chemical symbols are abbreviations used to identify elements.

Structure of Atoms

■ **Illustrate the structure of an atom with a simple diagram showing the protons, neutrons, and electrons.**
- Protons are positively charged particles in the nucleus of an atom and have a mass of 1 amu.
- Neutrons, also in the nucleus, have the same mass as protons but have no charge.
- Electrons are negatively charged particles with negligible mass that are in constant motion in orbits outside the nucleus.

■ **Distinguish between atomic number and mass number of an element.**
- The atomic number of an element is the number of protons in the nucleus of an atom.
- The mass number of an element is the number of protons plus the number of neutrons in the nucleus of an atom

■ **Describe the electron arrangement that makes an atom most stable.**
- The most stable atoms have eight electrons in their highest energy level or shell.

Isotopes

■ **Distinguish between isotopes and other atoms.**
- An atom of a given element that has a different number of neutrons in the nucleus, and consequently a different atomic weight, is an isotope.

■ **Describe a radioactive isotope.**
- A radioactive isotope has an unstable nucleus that decomposes and releases energy or atomic particles.

Chemical Bonds

■ **Describe the difference between ionic bonds, covalent bonds, and hydrogen bonds.**
- Chemical bonds are forces that hold together atoms.
- An ion is an atom that has lost or gained one or more electrons; a positively charged ion is a cation; a negatively charged ion is an anion.
- Ionic bonds are the attraction forces between cations and anions to form ionic compounds.
- Covalent bonds result when atoms share electrons; atoms may share more than one pair of electrons, which results in double or triple covalent bonds; an unequal sharing of electrons results in polar covalent bonds.
- Hydrogen bonds are intermolecular bonds, or attractions between molecules, that are formed by the attraction between the electropositive hydrogen end of a polar covalent compound and the negative charges of other molecules or ions.

Compounds and Molecules

■ **Describe the relationship between atoms, molecules, and compounds and interpret molecular formulas for compounds.**

- Atoms combine in definite ratios to form molecules, which are the smallest units of compounds; the atoms in the molecule are held together by chemical bonds.
- Molecular formulas use chemical symbols to indicate the types of atoms in a molecule or a compound and numerical subscripts show how many of each atom are present.

Chemical Reactions

■ **Describe and write chemical equations for four types of chemical reactions and identify the reactants and products in each.**

- Chemical equations are an abbreviated method of showing the reactants and products in a chemical reaction; the reactants are written on the left side of the equation and the products are written on the right side.
- Synthesis reactions form a complex molecule from two or more simple molecules.
- Decomposition reactions break down large molecules into simpler ones.
- In single replacement reactions, an atom in a reactant is replaced by a different atom.
- Double replacement reactions involve the exchange of two or more elements to form new compounds.
- Exergonic reactions release energy. Endergonic reactions require energy, which is then stored in the chemical bonds.

■ **Discuss five factors that influence the rate of chemical reactions.**

- The nature of the reacting substances affects the reaction rate.
- Reaction rates increase as temperature increases.
- Increasing the concentration of the reactants to an optimum increases the rate of the reaction.
- Enzymes and other catalysts increase the rate of a reaction.
- Breaking the reactants into small particles increase the total surface area of the reactants and increases the reaction rate.

Mixture, Solutions, and Suspensions

■ **Distinguish between mixtures, solutions, and suspensions.**

- A mixture consists of two or more substances that can be physically separated.
- Solutions consist of a solute that is being dissolved and a solvent that does the dissolving.
- In most suspensions, the particles settle if left undisturbed; in colloidal suspensions, the particles are so small they remain suspended but do not dissolve.

Acids, Bases, and Buffers

■ **Differentiate between acids and bases and discuss how they relate to pH and buffers.**

- Electrolytes form positive and negative ions when they are dissolved in water.
- Acids are proton (hydrogen ion) donors; bases accept protons.
- A pH value indicates the hydrogen ion concentration of a solution; a pH of 7.0 is neutral; acids have a pH less than 7.0; bases have a pH greater then 7.0
- Neutralization reactions occur between acids and bases to produce salts and water.
- Buffers, which contain a weak acid and a salt of that same acid, resist pH changes by neutralizing the effects of stronger acids and bases.

Organic Compounds

■ **Describe the five major groups of organic compounds that are important to the human body.**

- Carbohydrates, an important energy source, contain carbon, hydrogen, and oxygen and include monosaccharides, disaccharides, and polysaccharides. Glucose, fructose, and galactose are monosaccharides, or simple sugars, with six carbon atoms, while ribose and deoxyribose have five carbons. Sucrose, maltose, and lactose are disaccharides or double sugars consisting of two hexose monosaccharides linked together. Starch, cellulose, and glycogen are polysaccharides, which consist of long chains of glucose molecules.
- Proteins are formed from amino acids linked together by peptide bonds; they contain carbon, hydrogen, oxygen, nitrogen, usually sulfur, and often phosphorus.
- Lipids contain carbon, hydrogen, and oxygen and are insoluble in water but will dissolve in solvents such as alcohol and ether. The building blocks of triglycerides, commonly known as fats, are glycerol and fatty acids. Saturated fats contain only fatty acids that have single bonds between the carbon atoms. Phospholipids, which contain phosphates and nitrogen, are important components of cell membranes. Steroids are derivates of lipids and include cholesterol, certain hormones, and vitamin D.
- Nucleic acids are chains of nucleotides; they contain carbon, hydrogen, oxygen, nitrogen, and phosphorus. DNA is the genetic material of the cell and RNA functions in the synthesis of proteins within the cell.
- Adenosine triphosphate (ATP) is a high-energy compound that supplies energy in a form that is usable by body cells.

Recall

Match the definitions on the left with the appropriate term on the right.

_____ 1. Positively charged particle in the nucleus of an atom

_____ 2. Atom that has lost or gained an electron

_____ 3. Substance being dissolved

_____ 4. Alcohol is the solvent

_____ 5. Positively charged ion

_____ 6. Has a pH less than 7

_____ 7. Bond between amino acids

_____ 8. Adenosine triphosphate

_____ 9. Building block of DNA

_____ 10. Smallest unit of a compound

A. Acid

B. ATP

C. Cation

D. Ion

E. Molecule

F. Nucleotide

G. Peptide

H. Proton

I. Solute

J. Tincture

Thought

1. The symbol for sodium is
   A. Na
   B. So
   C. S
   D. N

2. The molecular formula for sodium bicarbonate is $NaHCO_3$. What is the total number of atoms in one molecule of sodium bicarbonate?
   A. 3
   B. 6
   C. 4
   D. 12

3. What type of reaction is represented by the following equation?

   $$HgCl_2 + H_2S \rightarrow HgS + 2HCl$$

   A. Synthesis
   B. Decomposition
   C. Single replacement
   D. Double replacement

4. What type of organic molecule is represented by the formula $C_6H_{12}O_6$?
   A. Carbohydrate
   B. Protein
   C. Lipid
   D. Steroid

5. A nucleotide molecule consists of
   A. Thymine, adenine, and guanine
   B. Phosphate, adenine, and a nitrogenous base
   C. Pentose, a phosphate group, and a nitrogenous base
   D. Phosphate, adenosine, and ribose

Application

1. A chemical reaction is proceeding very slowly. Suggest four things you might try to speed up the reaction.

2. The following reaction is reversible:

   $$CO_2 + H_2O \rightleftarrows H^+ + HCO_3^-$$

   As you breathe, you exhale CO_2 and remove it from the body. What effect does this have on the above reaction? Does the reaction proceed to the right or to the left as a result of exhaling CO_2?

Building Your Medical Vocabulary

There are approximately 50,000 medical words currently in use. That number is beyond the scope of this book. Here the emphasis is only on the most commonly used roots, prefixes, suffixes, abbreviations, and terminology. Fortunately, many complex medical terms can be broken down into their parts and then understood. Careful attention to the spelling of the word parts will help in spelling new complex words.

Word Parts and Combining Forms with Definitions and Examples

PART/ COMBINING FORM	DEFINITION	EXAMPLE
aer/o	air, gas	aerobic: requires air or oxygen
alkal-	basic	alkaline: basic, pH less than 7
bar/o	pressure, weight	hyperbaric: characterized by greater than normal pressure
carb/o	charcoal, coal,	carbonuria: excretion of urine carbon containing carbon dioxide or other carbon compounds
calor-	heat	calorie: a unit of measurement of heat
di-	two	diatomic: containing two atoms
end/o	within, inner	endergonic: energy is added to reactants and is stored within the product molecule
erg/o	work, energy	endergonic: energy is added to reactants and is stored within the product molecule
-esis	state, condition	diuresis: state or condition of increased kidney function
ex/o	out of, away from	exergonic: energy is released or taken out of a molecule in a reaction
-genesis	to form, produce	glycogenesis: to form or produce glycogen
gluc/o	sweetness, sugar, glucose	glucogenesis: to form or produce glucose sugar
hex-	six	hexagonal: having six sides
hydr/o	water	hydrotherapy: using water in treatment of disease
-ide	pertaining to	chloride: pertaining to chlorine

PART/ COMBINING FORM	DEFINITION	EXAMPLE
lact/o	milk	lactogenic: capable or producing milk
lip/o	fat	lipogenesis: production of fat
-lys	to take apart	hydrolysis: using water to take apart a molecule
-meter	measure, instrument used to measure	calorimeter: an instrument used to measure heat energy
mono-	one	mononuclear: having only one nucleus
-ose	sugar	pentose: a five-carbon sugar
oxy-	oxygen	oxyhemoglobin: oxygen combined with hemoglobin
pent/o	five	pentolysis: the taking apart or splitting of a five-carbon sugar
poly-	many	polyplegia: paralysis of many muscles
-ptosis	downward displacement,	nephroptosis: downward displacement of a kidney fall, sag
sacchar/o	sugar, sweet	saccharolytic: capable of taking apart a sugar
tetra-	four	tetrachloride: a compound with four chlorine atoms
tri-	three	triceps: a muscle with three heads
uni-	one	unicellular: having one cell

Clinical Abbreviations

ABBREVIATION	MEANING
$\bar{a}$	before
ad lib	as desired
ADL	activities of daily living
BRP	bathroom privileges
CHO	carbohydrate
DNR	do not resuscitate
ECF	extended care facility
ICU	intensive care unit
NA	not applicable; not available

ABBREVIATION	MEANING
oz	ounce
Rx	prescription, treatment, therapy
stat	immediately
T.O.	telephone order
VS	vital signs
WA	while awake

Clinical Terms

Endemic (en-DEM-ick) Presence of a disease within a given population at all times

Epidemic (ep-ih-DEM-ick) Sudden and widespread outbreak of a disease within a given population

Etiology (ee-tee-AHL-oh-jee) Study of the causes of disease

Functional disorder (FUNK-shun-al dis-OR-der) Disorder in which there are no detectable physical changes to explain the symptoms

Iatrogenic illness (eye-at-roh-JEN-ick IHL-nehs) Unintended adverse condition in a patient resulting from medical treatment

Idiopathic disorder (id-ee-oh-PATH-ick dis-OR-der) Illness that occurs without any known cause

Infectious disease (in-FECK-shus dih-ZEEZ) Illness caused by a pathogenic organism

Nosocomial infection (nos-oh-KOH-mee-ahl in-FECK-shun) Infection acquired from the place of treatment

Organic disorder (or-GAN-ick dis-OR-der) Disease accompanied by pathologic physical changes that explain the symptoms

Pandemic (pan-DEM-ick) Occurring over a large geographic area; a widespread epidemic

VOCABULARY QUIZ

Use word parts given in this chapter to form words that have the following definitions.

1. Many sugars _____

2. Produce fat _____

3. Molecule with two phosphates _____

4. Taking energy away from _____

5. Pertaining to two oxygens _____

Match each of the following definitions with the correct word.

_____ **6.** Forming glycogen

_____ **7.** Less oxygen than ribose

_____ **8.** Basic, pH greater than 7.0

_____ **9.** Pertaining to carbon and water

_____ **10.** Breaking down fat

A. Alkaline
B. Carbohydrate
C. Deoxyribose
D. Glycogenesis
E. Lipolysis

Answer the following questions.

11. What clinical term is used to denote the presence of a disease within a given population at all times?

12. What is the clinical term for an illness that occurs without any known cause?

13. What is the work of an etiologist?

14. What is the meaning of the abbreviation T.O.?

15. What is the meaning of the abbreviation DNR?

You should be familiar with the underlined words or word parts in the following questions. For each one, select the response that provides the best meaning.

16. John acquired a <u>nosocomial</u> infection. This is (a) an illness without any known cause; (b) a disorder in which there are no detectable physical changes to explain the symptoms; (c) a condition that is acquired from the place of treatment.

17. Susan was scheduled for a hyper<u>baric</u> treatment. The practitioner explained that she would (a) receive a treatment with oxygen at higher than normal pressure; (b) receive a treatment in a pool of warm water; (c) receive an injection of barium.

18. Henry's chemistry teacher was explaining the differences between <u>deoxy</u>ribose and ribose. From the word parts in the terms, Henry knew that (a) deoxyribose had fewer oxygen atoms than ribose; (b) deoxyribose was an isotope of ribose; (c) deoxyribose had more hydrogen atoms than ribose.

19. Glucose is a <u>hex</u>ose sugar. This means that it has (a) 3 carbon atoms; (b) 5 carbon atoms; (c) 6 carbon atoms.

20. A <u>calor</u>imeter is an instrument used to (a) measure temperature; (b) measure pressure; (c) measure heat energy.

Cell Structure and Function

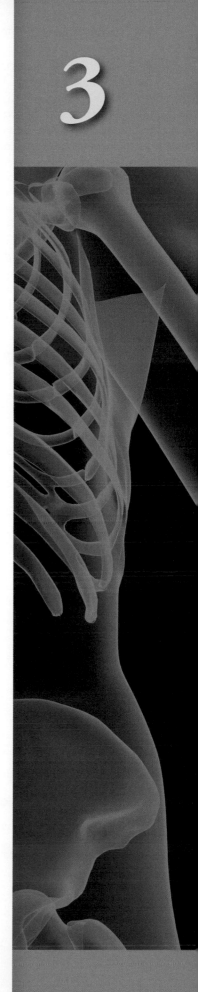

3

CHAPTER OBJECTIVES

Structure of the Generalized Cell
- Describe the cell membrane and list five functions of the proteins in the membrane.
- Describe the composition of the cytoplasm.
- Describe the components of the nucleus and state the function of each component.
- Identify and describe each of the cytoplasmic organelles and state the function of each organelle: mitochondria, ribosomes, endoplasmic reticulum, Golgi apparatus, and lysosomes.
- Identify and describe each of the filamentous protein organelles and state the function of each one: cytoskeleton, centrioles, cilia, or flagella.

Cell Functions
- Explain how the cell membrane regulates the composition of the cytoplasm.

- Describe the various mechanisms that result in the transport of substances across the cell membrane: simple diffusion, facilitated diffusion, osmosis, filtration, active transport, pumps, endocytosis, and exocytosis.
- Name the phases of a typical cell cycle and describe the events that occur in each phase.
- Explain the difference between mitosis and meiosis.
- Describe the process of DNA replication.
- Summarize the process of protein synthesis, including the roles of DNA and RNA.

KEY TERMS

Active transport pump (AK-tiv TRANS-port PUMP) Membrane transport process that moves substances against a concentration gradient and requires cellular energy

Cytokinesis (sye-toh-kih-NEE-sis) Division of the cytoplasm at the end of mitosis to form two separate daughter cells

Diffusion (dif-YOO-zhun) Movement of atoms, ions, or molecules from a region of high concentration to a region of low concentration

Endocytosis (en-doh-sye-TOH-sis) Formation of vesicles to transfer substances from outside the cell to inside the cell

Exocytosis (eks-oh-sye-TOH-sis) Formation of vesicles to transfer substances form inside the cell to outside the cell

Meiosis (mye-OH-sis) Type of nuclear division in which the number of chromosomes is reduced to one half the number found in a body cell; results in the formation of an egg or sperm

Mitosis (mye-TOH-sis) Process by which the nucleus of a body cell divides to form two new cells, each identical to the parent cell

Osmosis (os-MOH-sis) Diffusion of water through a selectively permeable membrane

Passive transport (PASS-iv TRANS-port) Membrane transport process that does not require cellular energy

Phagocytosis (fag-oh-sye-TOH-sis) Cell eating; a form of endocytosis in which solid particles are taken into the cell

Pinocytosis (pin-oh-sye-TOH-sis) Cell drinking; a form of endocytosis in which fluid droplets are taken into the cell

45

Every individual begins life as a single cell, a fertilized egg. This single cell divides into 2 cells, then 4, 8, 16, and on and on, until the adult human body has an estimated 75 trillion cells. Cells are the structural and functional units of the human body. Homeostasis depends on the interaction between the cell and its environment.

Structure of the Generalized Cell

During development, cells become specialized in size, shape, characteristics, and function, resulting in a large variety of cells in the body. This is called *differentiation.* It is impossible, in the scope of this book, to describe each different type of cell in detail. For descriptive purposes, it is convenient to imagine a typical, generalized cell that contains the components of all the different cell types. Not all the components of a "generalized" cell are present in every cell type, but each component is present in some cells and has its particular function to maintain life. A generalized cell is illustrated in Figure 3-1, and the structure and functions of the cellular components are summarized in Table 3-1.

Plasma Membrane

Every cell in the body is enclosed by a **plasma (cell) membrane.** The plasma membrane separates the material outside the cell (extracellular) from the material inside the cell (intracellular). It maintains the integrity of the cell. If the membrane ruptures or is broken, the cell dies. The nature of the membrane determines what can go into, or out of, the cell. It is selectively permeable, which means that some substances can pass through the membrane but others cannot.

The main structural components of the plasma membrane are **phospholipids** and **proteins.** Some carbohydrate and cholesterol molecules are present, too. The phospholipid molecules are arranged in a double layer (bilayer) with the polar phosphate portions in contact with the outside and inside of the cell and the nonpolar lipid portions sandwiched between the phosphate layers (Figure 3-2). You can visualize this as a cheese sandwich. The two slices of bread represent the phosphate layers, and the cheese is the lipid portion. The polar phosphate layers are **hydrophilic,** which means that they attract water and other polar molecules. These are the layers that are in contact with the intracellular and extracellular fluids. The middle, nonpolar lipid portion is **hydrophobic,** which means that water will not mix with it. Nonpolar organic molecules such as ether and chloroform dissolve readily in the lipid layer and pass through the membrane.

❀ *QUICK* **APPLICATIONS**

Trichloromethane, commonly known as chloroform, is a nonpolar organic solvent. Chloroform vapor is an anesthetic. Because it is nonpolar, it can diffuse through the cell membrane. James Young Simpson was the first to use chloroform as an anesthetic in 1846 and it was widely used in surgery until the early twentieth century. It was found to be carcinogenic and toxic to the liver and its use was discontinued.

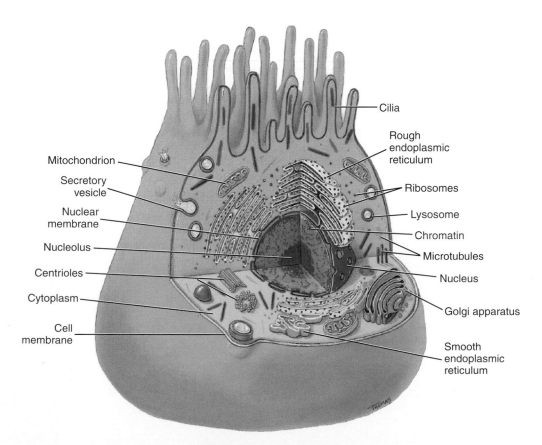

Figure 3-1 *Generalized cell.*

TABLE 3-1 Structure and Function of Cellular Components

Component	Structure	Function
Plasma membrane	Bilayer of phospholipid and protein molecules	Maintains integrity of cell; controls passage of materials into and out of cell
Cytoplasm	Water; dissolved ions and nutrients; suspended colloids	Medium for chemical reactions; suspending medium for organelles
Nucleus	Spherical body near center of cell; enclosed in a membrane	Contains genetic material; regulates activities of cell
Nuclear membrane	Double-layered membrane around nucleus; has pores	Separates cytoplasm from nucleoplasm; pores allow passage of material as necessary
Chromatin	Strand of DNA in nucleus	Genetic material of cell; becomes chromosomes during cell division
Nucleolus	Dense, nonmembranous body in nucleus; composed of RNA and protein molecules	From ribosomes
Mitochondria	Rod-shaped bodies enclosed by a double-layered membrane in cytoplasm; folds of inner membrane from cristae	Major site of ATP synthesis; converts energy from nutrients into a form that is usable by body
Ribosomes	Granules of RNA in cytoplasm	Protein synthesis
Endoplasmic reticulum	Interconnected membranous channels and sacs in cytoplasm	Transports material through cytoplasm; rough endoplasmic reticulum aids in synthesis of protein; smooth endoplasmic reticulum is involved in lipid synthesis
Golgi apparatus	Group of flattened membranous sacs usually near nucleus	Packages products for secretion; forms lysosomes
Lysosomes	Membranous sacs of digestive enzymes in cytoplasm	Digest material taken into cell, debris from damaged cells, and worn-out cell components
Cytoskeleton	Protein microfilaments and microtubules in cytoplasm	Provides support for cytoplasm; helps in movement of organelles
Centrioles	Pair of rod-shaped bodies composed of microtubules; located near nucleus at right angles to each other	Distribute chromosomes to daughter cells during cell division
Cilia	Membrane-enclosed bundles of microtubules that extend outward from cell membrane; short and numerous	Move substance across surface of cell
Flagella	Similar to cilia except usually long and single	Cell locomotion

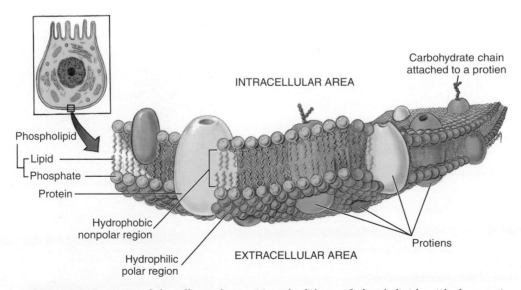

Figure 3-2 *Structure of the cell membrane. Note the bilayer of phospholipids with the proteins scattered throughout.*

Protein molecules are scattered throughout the phospholipid molecules. Some of the proteins contribute to the structural support of the membrane, whereas others form channels that allow water and water-soluble substances to pass through the membrane. Although the lipid in the membrane is a barrier to these ions and molecules, some of the proteins selectively permit their passage. Other proteins in the membrane act as receptor sites for hormones that affect the metabolic activity of the cell. Proteins also act as carrier molecules that combine with ions and molecules to transport them across the membrane.

Proteins on the surface of the cell act as markers for identification. They are the "fingerprints" of the cell. These proteins enable the body to recognize and accept its own cells but reject those that are nonself or foreign. This is the basis of tissue and organ transplant rejection and the defense mechanisms against disease. Autoimmune diseases result when the self-recognition process becomes faulty.

QUICK APPLICATIONS

Cystic fibrosis is an inherited disorder that afflicts 1 in every 2000 live births in whites. In cystic fibrosis, the cell membrane proteins that function as channels for transporting chloride ions out of the cell are defective. Because chloride ion transport is altered, secretions such as mucus, sweat, and pancreatic juice are very salty and thick. Thick mucus in the lungs leads to impaired breathing and increased infections. The ducts in the pancreas become plugged, which stops the flow of digestive enzymes. Life expectancy, with therapy, is about 27 years.

Cytoplasm

The **cytoplasm** (SYE-toh-plazm) is the gel-like fluid inside the cell. When viewed with an ordinary light microscope, it appears homogeneous and empty. Electron microscopy reveals that the cytoplasm is highly organized, with numerous small structures, called **organelles** (or-guh-NELZ), suspended in it. These organelles, or "little organs," are the functional machinery of the cell, and each organelle type has a specific role in the metabolic reactions that take place in the cytoplasm.

Chemically, the cytoplasm is primarily water, the **intracellular fluid.** About two thirds of the water in the body is in the cytoplasm of cells. The fluid contains dissolved electrolytes, metabolic waste products, and nutrients such as amino acids and simple sugars. Proteins suspended in the fluid give it colloidal properties and a gel-like consistency. Various **inclusions** may be suspended in the cytoplasm. These are bodies that are temporarily in the cell but that are not a part of the permanent metabolic machinery of the cell. Examples of inclusion bodies are membrane-enclosed fluid vacuoles, secretory products, glycogen granules, pigment granules, and lipid droplets.

QUICK CHECK

3.1 What are the two main structural components of the plasma membrane?

3.2 What is the water in the cytoplasm called?

Nucleus

The **nucleus** (NOO-klee-us) is the control center that directs the metabolic activities of the cell. All cells have at least one nucleus at some time during their existence; some, however, such as red blood cells, lose their nucleus as they mature. Other cells, such as skeletal muscle cells, have multiple nuclei.

The nucleus is a relatively large, spherical body that is usually located near the center of the cell (see Figure 3-1). It is enclosed by a double-layered **nuclear membrane** that separates the cytoplasm of the cell from the **nucleoplasm,** the fluid portion inside the nucleus. At numerous points on the nuclear surface, the membrane becomes thin and is interrupted by pores that allow large molecules, such as ribonucleic acid (RNA), to pass from the nucleus into the cytoplasm. The nuclear membrane is more permeable than the cell membrane.

The nucleus contains the genetic material of the cell. In the nondividing cell, the genetic material, deoxyribonucleic acid (DNA), is present as long, slender, filamentous threads called **chromatin** (see Figure 3-1). When the cell starts to divide or replicate, the chromatin condenses and becomes tightly coiled to form short, rodlike **chromosomes.** Each chromosome, composed of DNA with some protein, contains several hundred genes arranged in a specific linear order. Human cells have 23 pairs of chromosomes that together contain all the information necessary to direct the synthesis of more than 100,000 different proteins.

The **nucleolus** (noo-KLEE-oh-lus) ("little nucleus") appears as a dark-staining, discrete, dense body within the nucleus (see Figures 3-1 and 3-3). It has no enclosing membrane, and the number of nucleoli may vary from one to four in any given cell. The nucleolus has a high concentration of RNA and is the region

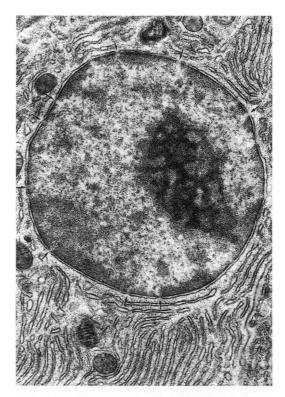

Figure 3-3 *Cell nucleus. Electron micrograph (original magnification, ×16,762). Note the double-layered nuclear membrane that is interrupted by nuclear pores (arrows). There is a large dense nucleolus and the granular substance is chromatin. Adapted from Gartner and Hiatt: Color Textbook of Histology. 3rd ed. Philadelphia, Elsevier/Saunders, 2007. Originally from Fawcett DW: The Cell. Philadelphia, WB Saunders, 1981.*

of ribosome formation. In growing cells and other cells that are making large amounts of protein, the nucleoli are very large and distinct, reflecting the function of RNA in protein synthesis.

Cytoplasmic Organelles

Cytoplasmic organelles are "little organs" that are suspended in the cytoplasm of the cell. Each type of organelle has a definite structure and a specific role in the function of the cell.

Mitochondria

Mitochondria (mye-toh-KON-dree-ah) are elongated, oval, fluid-filled sacs in the cytoplasm that contain their own DNA and can reproduce themselves (see Figure 3-1). The membrane around a mitochondrion consists of two layers with a small space between the layers. The outer layer is smooth, but the inner layer has many invaginations that project like partitions into the interior matrix of the mitochondrion. These invaginations are called **cristae** (KRIS-tee) (Figure 3-4). Enzymes necessary for the production of adenosine triphosphate (ATP) are located along the cristae. Mitochondria could be called the "power plant" of the cell because it is here that energy from nutrients is converted into a form that is usable by the cell. The enzymes in the mitochondria catalyze the reactions that form the high-energy bonds of ATP.

Ribosomes

Ribosomes (RYE-boh-sohmz) are small granules of RNA in the cytoplasm. The RNA in the ribosomes is from the nucleolus, and when fully assembled, ribosomes function in protein synthesis. Some ribosomes are found free in the cytoplasm. These function in the synthesis of proteins for use within that same cell. Other ribosomes are attached to the membranes of the endoplasmic reticulum and function in the synthesis of proteins that are exported from the cell and used elsewhere.

Ribosomes can attach to, or detach from, the endoplasmic reticulum, depending on the type of protein that is being produced.

Endoplasmic Reticulum

The **endoplasmic reticulum** (ER) (end-oh-PLAZ-mik reh-TICK-yoo-lum) is a complex series of membranous channels that extend throughout the cytoplasm. The interconnected membranes form fluid-filled flattened sacs and tubular canals. The membranes are connected to the outer layer of the nuclear membrane, to the inner layer of the cell membrane, and to certain other organelles. The endoplasmic reticulum provides a path to transport materials from one part of the cell to another.

Some of the membranes of the endoplasmic reticulum have granular ribosomes attached to the outer surface (see Figure 3-1). This is called **rough endoplasmic reticulum** (RER) and, because of the ribosomes, it functions in the synthesis and transport of protein molecules. Other portions of the endoplasmic reticulum lack the ribosomes and appear smooth. This is the **smooth endoplasmic reticulum** (SER), which functions in the synthesis of certain lipid molecules such as steroids. There is also evidence that indicates that the SER is involved in the detoxification of drugs.

Golgi Apparatus

The **Golgi apparatus** (GOL-jee ap-ah-RAT-us) is a series of four to six flattened membranous sacs, usually located near the nucleus, and connected to the endoplasmic reticulum (see Figure 3-1). It is the "packaging and shipping plant" of the cell.

Proteins from the rough endoplasmic reticulum and lipids from the smooth endoplasmic reticulum are carried through the channels of the endoplasmic reticulum to the Golgi apparatus. Within the Golgi apparatus, the proteins may be modified or concentrated or have a carbohydrate component added. Then they are surrounded by a piece of the Golgi membrane and are pinched off the end of the apparatus to become a **secretory vesicle,** a temporary inclusion in the cytoplasm (Figure 3-5).

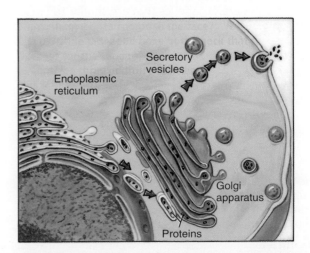

Figure 3-5 *Golgi apparatus. Proteins from the rough endoplasmic reticulum are transferred to the Golgi apparatus, where they are modified. They are surrounded by a piece of the Golgi membrane and pinched off the end and become secretory vesicles.*

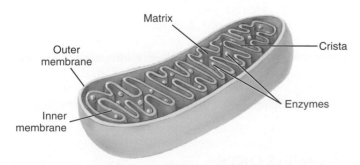

Figure 3-4 *Structure of a mitochondrion. Note the smooth outer membrane and the invaginations (cristae) of the inner membrane. Enzymes necessary for the production of ATP are located along the cristae.*

The secretory vesicles move to the cell membrane and release their contents to the exterior of the cell.

The Golgi apparatus is especially abundant and well developed in glandular cells that secrete a product, but they also function in nonsecretory cells. In these cells they appear to package intracellular enzymes in the form of lysosomes. Because of the secretory vesicles pinching off the ends of the flattened membranous sacs, the Golgi apparatus is sometimes described as looking like a stack of pancakes with syrup dripping off the edge.

Lysosomes

Lysosomes (LYE-soh-sohmz) are membrane-enclosed sacs of various digestive enzymes that have been packaged by the Golgi apparatus. When cells are damaged, these enzymes destroy the cellular debris. They also function in the destruction of worn-out cell parts. The enzymes break down particles such as bacteria that have been taken into the cell. When a white blood cell phagocytizes or engulfs bacteria, the enzymes from the lysosomes destroy it. Lysosomal activity also seems to be responsible for decreasing the size of some body organs at certain periods. Atrophy of muscle because of lack of use, reduction in breast size after breast-feeding, and decrease in the size of the uterus after parturition all seem to be caused by lysosomal function.

> **QUICK CHECK**
> 3.5 In which cellular organelle are proteins concentrated and prepared for secretion?
> 3.6 Which cellular organelle contains the enzymes that are necessary for the production of ATP?
> 3.7 Which cellular organelles are granules of RNA?

> **QUICK APPLICATIONS**
> Peroxisomes are organelles that are similar to lysosomes. They contain the enzymes peroxidase and catalase, which are important in the breakdown of hydrogen peroxide, a substance that is toxic to cells. Peroxisomes are often found in liver and kidney cells, where they function in the detoxification of harmful substances.

> **QUICK APPLICATIONS**
> Normally, the membrane around the lysosome is impermeable to the powerful digestive enzymes it contains. If the cell is injured or deprived of oxygen, the membrane becomes fragile and the contents escape. This results in self-digestion of the cell, a process called autolysis. This also happens normally when cells are destroyed as part of a reconfiguration process during embryologic development. It also occurs in white blood cells during inflammation.

Filamentous Protein Organelles

Several types of protein filaments are considered to be cellular organelles. The cytoskeleton and centrioles are in the cytoplasm, but the cilia and flagella project outward, away from the cell surface.

Cytoskeleton

The **cytoskeleton** helps to maintain the shape of the cell. At times it anchors certain organelles in position, but it may also move organelles from one position to another. Some parts of the cytoskeleton may move a portion of the cell membrane, whereas others may move the entire cell. The cytoskeleton also plays a role in muscle contraction.

The cytoskeleton is made up of protein **microfilaments** and **microtubules.** Microfilaments are long, slender rods of protein that support small projections of the cell membrane called **microvilli.** Microtubules are thin cylinders, larger than the microfilaments, and are composed of the protein **tubulin.** In addition to their role as part of the cytoskeleton, microtubules are also found in centrioles, cilia, and flagella.

Centrioles

A dense area called the **centrosome** (SEN-troh-sohm), located near the nucleus, contains a pair of **centrioles** (SEN-tree-ohlz) (see Figure 3-1). Each centriole is a nonmembranous rod-shaped structure composed of microtubules. The two members of the pair are at right angles to each other. Centrioles function in cell reproduction by aiding in the distribution of chromosomes to the new daughter cells.

Cilia

Cilia (SIL-ee-ah) are short, cylindrical, hairlike processes that project outward from the cell membrane. Each cilium consists of specialized microtubules surrounded by a membrane and anchored under the cell membrane. Cilia have an organized pattern of movement that creates a wavelike motion to move substances across the surface of the cell. They are found in large quantities on the surfaces of cells that line the respiratory tract, where their motion moves mucus, in which particles of dust are embedded, upward and away from the lungs.

Flagella

Similar in structure to cilia, **flagella** (fluh-JELL-ah) are much longer and fewer. In contrast to cilia, which move substances across the surface of the cell, flagella beat with a whiplike motion to move the cell itself. In the human, the tail of the spermatozoon, or sperm cell, is a single flagellum that causes the swimming motion of the cell.

> **QUICK CHECK**
> 3.8 What are the numerous short, hairlike filamentous organelles that move mucus along the respiratory tract?
> 3.9 What is the function of centrioles?

Cell Functions

The structural and functional characteristics of different types of cells are determined by the nature of the proteins present. Cells of various types have different functions because cell structure and function are closely related. It is apparent that a very thin cell is not well suited for a protective function. Bone cells do not have an appropriate structure for nerve impulse conduction. Just as there are many cell types, there are varied cell functions. The specific functions of cells will become more apparent as the tissues, organs, and systems are studied. This section deals with the more generalized cell functions—the functions that relate to the sustained viability and continuation of the cell itself. These functions include movement of substances across the cell membrane, cell division to make new cells, and protein synthesis.

Movement of Substances Across the Cell Membrane

The cell membrane provides a surface through which substances enter and leave the cell. It controls the composition of the cell's cytoplasm by regulating the passage of substances through the membrane. If the membrane breaks, this control is removed and the cell dies. The survival of the cell depends on maintaining the difference between extracellular and intracellular material. Mechanisms of movement across the cell membrane include both passive and active processes. Passive transport mechanisms move substances down a concentration gradient and do not require cellular energy. These include simple diffusion, facilitated diffusion, osmosis, and filtration. Active mechanisms move substances against a concentration gradient and require cellular energy in the form of ATP. These include active transport pumps, phagocytosis, pinocycosis, and exocytosis. Membrane transport mechanisms are summarized in Table 3-2 and in Figure 3-6.

TABLE 3-2 Summary of Membrane Transport Mechanisms

Mechanism	Description
Passive	
Simple diffusion	Molecular movement down a concentration gradient
Facilitated diffusion	Carrier molecules transport down a concentration gradient; requires membrane
Osmosis	Movement of solvent toward high solute (low solvent) concentration; requires membrane
Filtration	Movement of solvent using hydrostatic pressure; requires membrane filter
Active	
Active transport	Movement of ions/molecules against a concentration gradient; requires carrier molecule and ATP
Phagocytosis	Ingestion of solid particles by creating vesicles; requires ATP
Pinocytosis	Ingestion of fluid by creating vesicles; requires ATP
Exocytosis	Secretion of cellular products by creating vesicles, then liberating contents to outside of cell; requires ATP

ATP, adenosine triphosphate.

QUICK CHECK

3.10 What cellular component regulates the composition of the cytoplasm?
3.11 What are two fundamental differences between passive and active transport mechanisms?

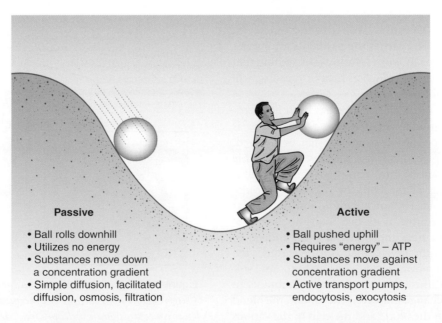

Passive

- Ball rolls downhill
- Utilizes no energy
- Substances move down a concentration gradient
- Simple diffusion, facilitated diffusion, osmosis, filtration

Active

- Ball pushed uphill
- Requires "energy" – ATP
- Substances move against concentration gradient
- Active transport pumps, endocytosis, exocytosis

Figure 3-6 *Comparison of passive and active transport mechanisms.*

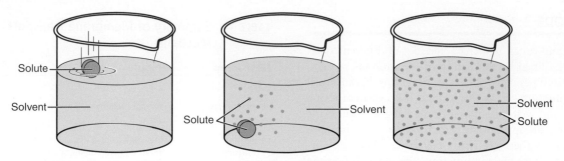

Figure 3-7 *Simple diffusion. Molecules of solute diffuse throughout the solvent until equilibrium exists.*

Simple Diffusion

Simple diffusion (dif-YOO-zhun) is the movement of atoms, ions, or molecules from a region of high concentration to a region of low concentration. Odors permeate a room because the aromatic molecules diffuse through the air. A crystal of dye will color a whole beaker of water because the dye particles diffuse from the region of high concentration in the dye crystal to regions of low concentration in the water (Figure 3-7).

Atoms, ions, and molecules are constantly moving at high speed. Each particle moves in a straight line until it collides with another particle or the edge of the container; then it changes directions until it hits another particle. Such random motion accounts for the mixing that occurs when two different substances are put together. Eventually the particles will be evenly distributed and **equilibrium** will exist. This does not mean that the particles cease their motion but as soon as a particle moves in one direction, others move in the opposite direction so that the concentration does not change.

If two solutions have different concentrations, then a **concentration gradient** exists between them. The gradient is the difference in the solute concentrations. When particles move from an area of high concentration to an area of low concentration, they move down, or with, a concentration gradient. Generally, diffusion rates are faster if the gradients are steeper (there is a greater difference in concentrations). Anything that increases the speed of movement, such as heat or pressure, will also increase the rate of diffusion.

In the examples of diffusion cited, there has been no membrane involved. Diffusion can occur across a membrane as long as the membrane is permeable to the substances involved. Most physiologic examples involve a selectively permeable membrane. Many substances move through the extracellular and intracellular fluids of the body by diffusion. Because movement is down a concentration gradient and no cellular energy is involved, it is a form of passive transport or movement. Lipid-soluble substances and particles small enough to pass through protein membrane channels diffuse through the cell membrane. Oxygen and carbon dioxide are both lipid soluble so they are able to diffuse through the cell membrane. In this way, the gases are exchanged between the air and the blood in the lungs, and between the blood and the cells of the various tissues (Figure 3-8).

Facilitated Diffusion

Facilitated diffusion is a special type of diffusion that involves a carrier molecule. The movement of glucose from the blood into cells is an example. Most sugar molecules, including glucose, are not soluble in lipids and are too large to pass through the membrane pores or channels. Yet they are able to diffuse across the membrane. A glucose molecule combines with a special protein carrier molecule in the cell

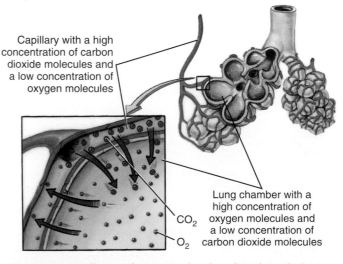

Capillary with a high concentration of carbon dioxide molecules and a low concentration of oxygen molecules

Lung chamber with a high concentration of oxygen molecules and a low concentration of carbon dioxide molecules

CO_2

O_2

Figure 3-8 *Diffusion of oxygen and carbon dioxide in the lungs. Oxygen moves from the higher concentration in the lung into the lower concentration in the capillary. Carbon dioxide moves in the opposite direction.*

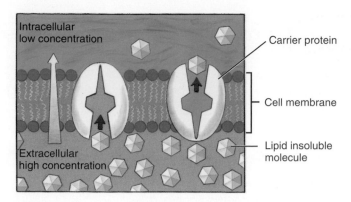

Intracellular low concentration

Carrier protein

Cell membrane

Lipid insoluble molecule

Extracellular high concentration

Figure 3-9 *Facilitated diffusion. The carrier protein "picks up" an insoluble molecule from the area of high concentration and releases it in the area of low concentration.*

membrane. This combination is soluble in lipid, so it diffuses across the membrane. When it reaches the inside of the cell, the glucose portion is released and the protein carrier is free to pick up another glucose molecule and "carry" it across the membrane (Figure 3-9). Because the solute particles move down a concentration gradient and there is no expenditure of cellular energy, it is a passive process of diffusion, although it is "facilitated" by a carrier molecule. Facilitated diffusion is limited by the number of carrier molecules that are present.

QUICK CHECK

3.12 By what transport mechanism do oxygen and carbon dioxide move between the capillaries and the lungs?

3.13 How does facilitated diffusion differ from simple diffusion?

Osmosis

In the process of diffusion just discussed, any type of particle can move. Gases move through gases, gases move through liquids, solids dissolve and move through liquids, and liquids move through liquids. A membrane may or may not be involved in simple diffusion.

Osmosis (os-MOH-sis) involves the movement of **solvent** (water) molecules through a **selectively permeable** membrane from a region of higher concentration of water molecules (where the **solute** concentration is lower) to a region of lower concentration of water molecules (where the solute concentration is higher). Figure 3-10 illustrates osmosis. When equilibrium is reached, the solutions on both sides of the membrane have the same concentration but the solution that was more concentrated at the start (had more solute) will now have a greater volume. Water molecules continue to pass through the membrane after equilibrium, but because they move in both directions at the same rate there is no change in concentration or volume.

If a red blood cell, which contains 5% glucose, is placed in a container of 5% glucose solution, water will move in both directions at the same rate because the glucose concentrations inside and outside the cell are the same. Solutions that have the same solute concentration are **isotonic** (Figure 3-11, *A*).

When a red blood cell is placed in a 10% glucose solution, water will leave the cell (where there are more water molecules) and enter the surrounding fluid (where there are fewer water molecules). When fluid leaves the cells, they will shrink or **crenate.** The 10% glucose solution is **hypertonic** (greater solute concentration) to the cell (Figure 3-11, *B*).

When a red blood cell is placed in distilled water, water will enter the cell because there are more water molecules outside the cell than there are inside. The distilled water is **hypotonic** (lower solute concentration) to the cell. As water enters the cell, it will swell, because of the increased volume. If enough water

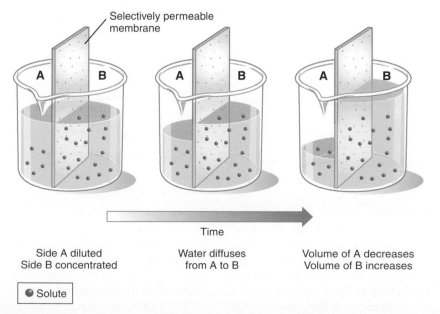

Selectively permeable membrane

A B

A B

A B

Time

Side A diluted
Side B concentrated

Water diffuses from A to B

Volume of A decreases
Volume of B increases

● Solute

Figure 3-10 *Osmosis. Solvent molecules move across the membrane but solute molecules do not because the membrane is selectively permeable.*

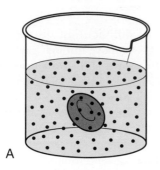
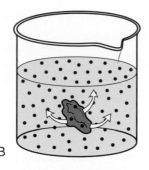
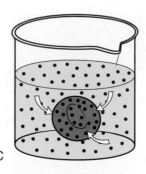

Figure 3-11 **A,** *Isotonic solution. The extracellular concentration equals the intracellular concentration and there is no net movement of solvent.* **B,** *Hypertonic solution. The extracellular concentration is greater than the intracellular concentration and solvent moves from the cell to the surrounding fluid. The cell shrinks (crenates).* **C,** *Hypotonic solution. The extracellular concentration is less than the intracellular concentration and solvent moves into the cell. The cell expands.*

goes into the cell, it may rupture. This is called **lysis.** When this happens to a red blood cell, it is called **hemolysis** (hee-MAHL-ih-sis) (Figure 3-11,*C*).

The terms *isotonic, hypotonic,* and *hypertonic* are relative. They are used to compare two solutions. A 5% glucose solution is hypertonic to distilled water but hypotonic to a 10% glucose solution.

> **QUICK CHECK**
> 3.14 What type of membrane is involved in osmosis?
> 3.15 What happens to cells when they are placed in a hypertonic solution?

Filtration

In diffusion and osmosis, particles, whether solute, solvent, or both, pass through a membrane by virtue of their own random movement, which is directed by concentration gradients. In **filtration,** however, pressure pushes the particles through a membrane. It is pressure gradients, rather than concentration gradients, that direct the movement. Drip coffee makers, for example, use this principle. Water drips first through the coffee and then through the water, and small particles pass through a filter. The large granules of coffee are too big to go through the pores in the filter. The size of the pores determines the size of the particles that can pass through the filter. The pressure is created by the weight of the water on the paper filter. In the laboratory, filtration is often used to separate small solid particles from a liquid.

Contraction of the heart creates pressure in the blood. This fluid pressure or **hydrostatic pressure,** which is greater inside the blood vessels, pushes fluid, dissolved nutrients, and small ions through the capillary walls to form tissue fluid. The large protein molecules and blood cells are unable to pass through the pores in the capillary membrane. Blood is filtered through specialized membranes in the kidney as the initial step in urine formation. Water and small molecules and ions pass through the filtration membrane while blood cells and protein molecules remain in the blood.

> **QUICK CHECK**
> 3.16 What causes the movement of particles through the membrane in filtration?
> 3.17 What limits the size of the particles that can pass through a filtration membrane?

FROM THE PHARMACY

There are three routes by which medications are administered. In the **enteral** route, the medications are administered directly into the gastrointestinal (GI) tract. This includes oral preparations, nasogastric (NG) tubes, gastrostomy tubes, and intestinal tubes. Oral preparations are the most common forms in this group. The **parenteral** route refers to injections of various forms and includes intradermal (ID), subcutaneous (subQ), intramuscular (IM), and intravenous (IV) injections. In the **percutaneous** route, the medications are absorbed through the mucous membranes or skin. Methods of percutaneous administration include putting solutions onto the mucous membranes of the ear, eye, nose, mouth, or vagina; applying topical creams, powders, ointments, or lotions to the skin; and inhaling aerosolized liquids or gases to carry medication to the nasal passages, sinuses, and lungs.

Active Transport Pumps

In the transport mechanisms discussed thus far, no cellular energy has been involved and the molecules and/or ions moved from a region of high concentration or pressure to one of low concentration or pressure. **Active transport** differs from these processes in that it moves molecules and ions "uphill" against a concentration gradient and uses cellular energy in the form of ATP in the process (see Figure 3-6). If ATP is not available, active transport ceases immediately. Active transport also uses a carrier molecule. Amino acids and glucose are transported from the small intestine into the blood by active transport, and all living cells depend on the active transport of electrolytes.

As a result of active transport pumps, some substances accumulate in significantly higher concentrations on one side of the cell membrane than on the other. In a resting cell, sodium ions are more concentrated outside the cell membrane than inside the cell. Potassium is just the opposite; its concentration is higher inside the cell. Normal passive transport, such as diffusion and osmosis, tends to equalize the concentrations on the two sides of the membrane. Active transport, in this case known as the sodium/potassium pump, moves sodium and potassium ions against concentration gradients so that sodium ions are pumped out of the cell and potassium ions are pumped into the cell. For every pump cycle, three sodium ions are transported out of the cell and two potassium ions are transported into the cell. This maintains a high extracellular sodium concentration and a high intracellular potassium concentration. The process requires ATP and a protein carrier molecule. Figure 3-12 illustrates the sodium/potassium active transport pump.

Endocytosis

Endocytosis (en-doh-sye-TOH-sis) refers to the formation of vesicles to transfer particles and droplets from outside to inside the cell. In this case, the material is too large to enter the cell by diffusion or active transport pump. The cell membrane surrounds the particle or droplet, and then that portion of the

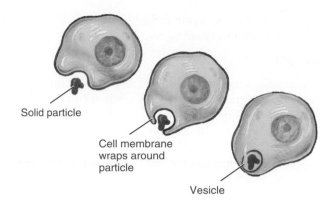

Figure 3-13 *Phagocytosis. The cell membrane engulfs a particle to form a vesicle in the cytoplasm.*

membrane pinches off to form a vesicle in the cytoplasm. The process requires energy in the form of ATP. **Phagocytosis** (fag-oh-sye-TOH-sis), which means "cell eating," involves solid material (Figure 3-13). The cell membrane engulfs a particle to form a vesicle in the cytoplasm. Lysosomes fuse with the vesicle and the enzymes digest the particle. Certain white blood cells are called phagocytes because they engulf and destroy bacteria in this manner. Another form of endocytosis is **pinocytosis** (pin-oh-sye-TOH-sis), or "cell drinking." It differs from phagocytosis in that the vesicles that are formed are much

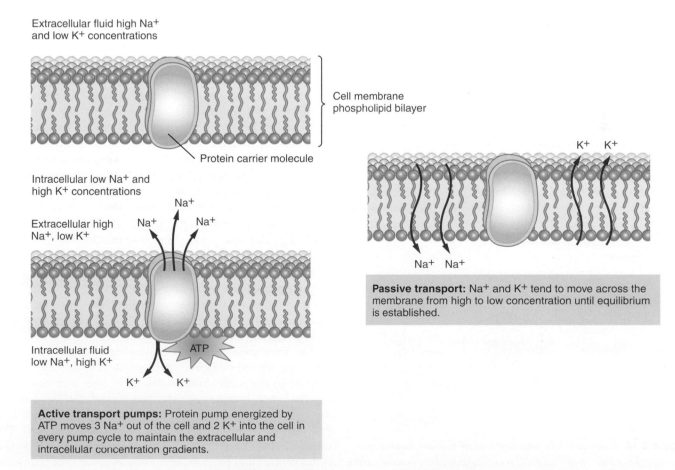

Figure 3-12 *Sodium/potassium active transport pump.*

smaller and their contents are fluids. Pinocytosis is important in cells that function in absorption.

Exocytosis

In certain cells, secretory products are packaged into vesicles by the Golgi apparatus and are then released from the cell by a process called **exocytosis** (eck-soh-sye-TOH-sis). The secretory vesicle moves to the cell membrane, where the vesicle membrane fuses with the cell membrane and the contents are discharged to the outside of the cell (Figure 3-14). Secretion of digestive enzymes from the pancreas and secretion of milk from the mammary glands are examples of exocytosis. Exocytosis and endocytosis are similar except they work in opposite directions. They are both active processes that require cellular energy (ATP). Exocytosis releases substances to the outside of the cell, and endocytosis transports substances to the inside of the cell.

QUICK CHECK

3.18 Active transport moves substances against a concentration gradient. What two types of molecules are necessary to support this process?

3.19 Is phagocytosis an example of endocytosis or exocytosis?

3.20 What is the difference between phagocytosis and pinocytosis?

Cell Division

Cell division is the process by which new cells are formed for growth, repair, and replacement in the body. This process includes division of the nuclear material and division of the cytoplasm. Periods of growth and repair are special periods in the life of an individual when it is obvious that new cells are needed either to increase the number of cells or to repair tissues after an injury. General maintenance and replacement needs of the body may not be quite as obvious. More than 2 million red blood cells are worn out and replaced in the body every second of every day. Skin cells are continually sloughed off the body's surface and must be replaced. The lining of the stomach is replaced every few days. All cells in the body (somatic cells), except those that give rise to the eggs and sperm (gametes),

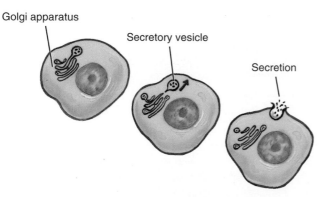

Figure 3-14 *Exocytosis. The secretory vesicles that are formed by the Golgi apparatus move to the cell membrane and the contents are discharged to the outside of the cell.*

Golgi apparatus

Secretory vesicle

Secretion

reproduce by **mitosis** (mye-TOH-sis). Egg and sperm cells are produced by a special type of nuclear division called **meiosis** (mye-OH-sis) in which the number of chromosomes is halved. Division of the cytoplasm is called **cytokinesis.**

Mitosis

All somatic cells reproduce by mitosis, in which a single cell divides to form two new "daughter cells," each identical to the parent cell. Humans have 23 pairs of chromosomes (or 46 chromosomes) in their cells. Each new cell that forms must also have 23 pairs or 46 chromosomes. For this to occur, events must proceed in an organized manner. Chromosome material must replicate exactly and then the chromosomes must separate precisely so that each new cell receives a set of chromosomes that is a carbon copy of the parent cells. For descriptive purposes, it is convenient to divide the events of mitosis into stages, as illustrated in Figure 3-15. It is important to remember that the process is a continuous one and that there are no starting and stopping points along the way.

The period between active cell divisions is called **interphase.** This is a time of growth and metabolism and is usually the longest period of the cell cycle. In cells that are rapidly dividing it may last for as little as a few hours, but in other cells it may take days, weeks, or even months. Some highly specialized cells, such as nerve and muscle cells, may never divide and spend their whole life in interphase.

During interphase, the cell increases in size and synthesizes an exact copy of the DNA in its nucleus so that when the cell begins to divide it has identical sets of genetic information. Also just before division, the cell synthesizes an additional pair of centrioles and some new mitochondria. In addition to these synthetic activities that are a preparation for division, normal cellular function takes place during interphase.

After interphase, the cell begins mitosis. The first stage of mitosis is **prophase.** During prophase, the chromatin shortens, thickens, and becomes tightly coiled to form chromosomes. As a result of the replication in interphase, each chromosome has two identical parts, called **chromatids,** that are joined by a special region on each called the **centromere.** The two pairs of centrioles separate and go to opposite ends of the cytoplasm. Microtubules called **spindle fibers** form and extend from the centromeres to the centrioles. The nucleolus and nuclear membrane disappear during the latter part of prophase.

Prophase ends when the nuclear membrane disintegrates, and this signals the beginning of the next stage, **metaphase.** The chromosomes align themselves along the center of the cell during metaphase. This is the time when the chromosomes are most clearly visible and distinguishable.

The third stage of mitosis is **anaphase.** After the chromosomes are aligned along the center of the cell, the centromeres separate so that each chromatid now becomes a chromosome. At this time, there are actually two sets of chromosomes in the cell. The two chromatids (now chromosomes) from each pair migrate to the centrioles at opposite ends of the cell. The microtubules that are attached to the centrioles and centromeres shorten and pull the chromosomes toward the centrioles. At the end of anaphase, the cytoplasm begins to divide.

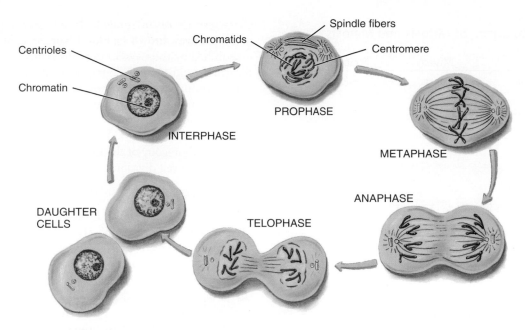

Figure 3-15 *Mitosis. Interphase is the period between active cell divisions. Prophase is the first stage of mitosis. The process continues through metaphase, anaphase, and telophase. Cytokinesis occurs in telophase. With the division of nuclear material and cytoplasm, two exact copies of the parent cell are produced.*

The final stage of mitosis is **telophase.** This stage is almost the reverse of prophase. After the chromosomes reach the centrioles at the ends of the cell, a new nuclear membrane forms around them. The spindle fibers disappear. The chromosomes start to uncoil to become long, slender strands of chromatin, and nucleoli appear in the newly formed nucleus. During this time, the cell membrane constricts in the middle to divide the cytoplasm and organelles into two parts that are approximately equal. Division of the cytoplasm is called **cytokinesis** (sye-toh-kih-NEE-sis). Except for size, the two newly formed daughter cells are exact copies of the parent cell. The two daughter cells now become interphase cells to carry out designated cellular functions and to undergo mitosis as necessary. The events of mitosis are summarized in Table 3-3.

Normally, body cells divide at a rate required to replace the dying ones. Normal cells are subject to control mechanisms that prevent overpopulation and competition for nutrients and space. Occasionally, a series of events occurs that alters some cells so they lack the control mechanisms that tell them when to stop dividing. When the cells do not stop their mitotic activity, they form an abnormal growth called a tumor, or neoplasm.

QUICK APPLICATIONS

A benign neoplasm consists of highly organized cells that closely resemble normal tissue. In contrast, a malignant neoplasm, or cancer, consists of unorganized and immature cells that are incapable of normal function. These cells may detach from the tumor site and travel in the blood or lymph to another site and establish a new tumor. This is called metastasis and is probably the most devastating property of malignant cells.

TABLE 3-3 Summary of Mitotic Events

Stage	Events
Interphase	DNA, mitochondria, and centrioles replicate
Prophase	Chromatin shortens and thickens to become chromosomes; centrioles move to opposite ends of cell; spindle fibers form; nucleolus and nuclear membrane disappear
Metaphase	Chromosomes align along center of cell
Anaphase	Centromeres separate and spindle fibers shorten to pull chromatids (chromosomes) toward centrioles at opposite ends of cell
Telophase	Chromosomes uncoil to become long filaments of chromatin; nuclear membrane and nucleolus reappear; cytokinesis occurs; daughter cells form and enter interphase

Meiosis

Meiosis is a special type of cell division that occurs in the production of the gametes, or eggs and sperm. These cells have only 23 chromosomes, one half the number found in somatic cells, so that when fertilization takes place the resulting cell will again have 46 chromosomes, 23 from the egg and 23 from the sperm. Meiosis is discussed in greater detail in Chapter 19. In brief, meiosis consists of two divisions, but DNA is replicated only once. The result is 4 cells, but each one has only 23 chromosomes. Table 3-4 and Figure 3-16 compare mitosis and meiosis.

TABLE 3-4 Comparison of Mitosis and Meiosis

Feature	Mitosis	Meiosis
Type of cell where it occurs	Somatic cells	Reproductive cells
Chromosomes in parent cell	46 (23 pairs)	46 (23 pairs)
Chromosome replication	Yes—once	Yes—once
Number of cytoplasmic divisions	1	2
Number of cells formed	2	4
Number of chromosomes in each new cell formed	46 (23 pairs)	23

QUICK CHECK

3.21 During which stage of mitosis does cytokinesis occur?

3.22 In cell division, when is the genetic material replicated?

3.23 How is the number of chromosomes reduced in the formation of eggs and sperm?

3.24 How many cells are formed by mitosis of one somatic cell? How many are formed by the meiosis of one cell?

DNA Replication and Protein Synthesis

Proteins that are synthesized in the cytoplasm function as structural materials, enzymes that regulate chemical reactions, hormones, and other vital substances. Because DNA in the nucleus directs the synthesis of the proteins in the cytoplasm, it ultimately determines the structural and functional characteristics of an individual. Whether a person has blue or brown eyes, brown or blond hair, or light or dark skin is determined by the types of proteins synthesized in response to the genetic information contained in the DNA in the nucleus. The portion of a DNA molecule that contains the genetic information for making one particular protein molecule is called a **gene.** If a cell produced for replacement or repair is to function exactly as its predecessor, then it must have the same genes, a carbon copy of the DNA. This is the purpose of DNA replication in cell division.

DNA Replication

As described in Chapter 2, DNA consists of two long chains of nucleotides that are loosely held together by hydrogen bonds and then are twisted to form a double helix. Each nucleotide in the chains has a phosphate, a sugar called deoxyribose, and a nitrogenous base. The bases are adenine, thymine, cytosine, and guanine. Uncoiled, the DNA looks something like a ladder. The sugar and phosphate alternate to form the uprights of the ladder. The bases from each chain project toward each other and are held together by hydrogen bonds to form the crossbars or rungs of the ladder (Figure 3-17). There is a specific pattern to the combination of bases. Adenine is always opposite thymine, and cytosine always pairs with guanine.

When DNA replicates during interphase before starting mitosis, the DNA molecule uncoils, the hydrogen bonds between the complementary base pairs break, and the two strands separate. Enzymes then catalyze the formation of new complementary strands from nucleotides that are present in the nucleoplasm. Because adenine always pairs with thymine, and guanine always pairs with cytosine, the new complementary strands are identical to the previous

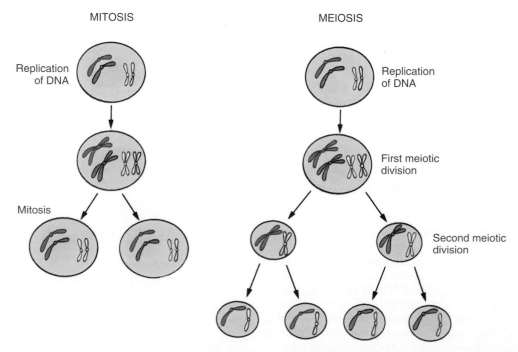

Figure 3-16 *Comparison of mitosis and meiosis. The result of mitosis in humans is two cells, each with 46 (23 pairs) chromosomes. Meiosis results in four cells, each with 23 chromosomes.*

S=Deoxyribose sugar
P=Phosphate group

Bases:
T=Thymine
A=Adenine
G=Guanine
C=Cytosine

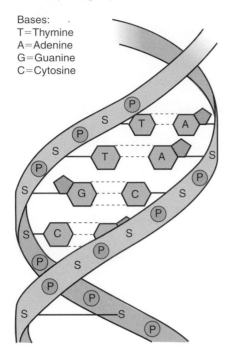

Figure 3-17 *DNA molecule before replication. A DNA molecule consists of two nucleotide chains that are twisted into a double helix. Each nucleotide in the chain consists of a phosphate group, deoxyribose sugar, and a nitrogenous base. The bases are thymine, adenine, guanine, and cytosine.*

ones (Figure 3-18). As a result of this process, two identical molecules of DNA are produced, each with one strand from the old molecule and one new strand. The cell is now ready to begin mitosis.

QUICK CHECK

3.25 What is a gene?
3.26 In DNA replication, what nitrogenous base always pairs with thymine?

QUICK APPLICATIONS

Genetic disorders are pathologic conditions caused by mistakes, or mutations, in a cell's genetic code. Mutations may occur naturally, or they may be induced by mutagens such as radiation and certain chemicals. If the mutations occur in the gametes, the faulty code is passed from one generation to the next. Errors in the genes (DNA) cause the production of abnormal proteins, which result in abnormal cellular function. For example, in sickle cell anemia, a genetic blood disorder, red blood cells have abnormal hemoglobin because there is an "error" in the gene that directs hemoglobin synthesis.

Role of DNA and RNA in Protein Synthesis

The genetic information, contained in the DNA, that controls protein synthesis is located in the nucleus. The process of protein synthesis takes place in the cytoplasm outside the nucleus. Because the DNA is unable to leave the nucleus, there must be a "messenger" molecule that can take the information from the DNA to the cytoplasm. This molecule is **messenger RNA** (mRNA). The basic structure of RNA is discussed in Chapter 2. Briefly, RNA is a single chain of nucleotides, the sugar is ribose, and uracil is substituted for thymine in the nitrogenous bases.

The first step in protein synthesis is to transfer the genetic information from DNA to mRNA. This process is called **transcription** and is illustrated in Figure 3-19. First, the DNA strands uncoil and separate; then the RNA nucleotides pair with the complementary DNA nucleotides on the coding strand of DNA. Because there is no thymine in RNA, adenine on DNA will pair with uracil on RNA. Enzymes catalyze the formation of chemical bonds between the RNA nucleotides to form a molecule of mRNA. The sequence of bases on the mRNA is determined by the sequence on DNA because of the complementary base-pairing. If the sequence on the coding strand of DNA is CTTACCCGT (where C = cytosine, G = guanine, A = adenine, T = thymine, and U = uracil), then the sequence on mRNA will be GAAUGGGCA. After they are formed, the mRNA molecules leave the nucleus through the pores in the nuclear membrane to become associated with ribosomes in the cytoplasm and to act as templates for protein synthesis. The genetic information is carried in groups of three nucleotides called **codons.** Each codon, a sequence of three bases on mRNA, codes for a specific amino acid. The three codons in the above example are GAA, UGG, and GCA. These code for glutamic acid, tryptophan, and alanine, respectively. There are also instruction codons that indicate where protein synthesis is to start and where it is to stop.

QUICK CHECK

3.27 In transcription, what nitrogenous base pairs with adenine?
3.28 What is a codon?

There are two additional types of RNA in the cytoplasm, **ribosomal RNA** (rRNA) and **transfer RNA** (tRNA). These are also produced in the nucleus and then move to the cytoplasm. The rRNA, as its name implies, is part of the ribosomes. One portion of tRNA consists of three nucleotides, called an **anticodon.** Another portion attaches to an amino acid. There are specific tRNA nucleotide base sequences for each of the 20 amino acids.

In the cytoplasm, a ribosome, which coordinates the activities of the codons and anticodons, attaches to mRNA near a codon. A tRNA with an anticodon complementary to the mRNA codon brings its specific amino acid into place. The ribosome moves along the mRNA, codon by codon, which allows complementary tRNA anticodons to put their respective

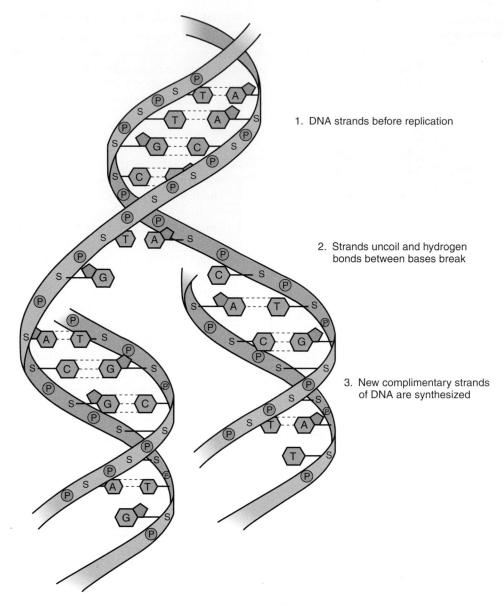

1. DNA strands before replication

2. Strands uncoil and hydrogen bonds between bases break

3. New complimentary strands of DNA are synthesized

Figure 3-18 *DNA replication. In the replication of DNA, cytosine always pairs with guanine and adenine always pairs with thymine.*

amino acids in place. Enzymes catalyze the formation of peptide bonds between the amino acids. Once a tRNA has released its amino acid, it is free to pick up another amino acid of the same kind and "transfer" it to the amino acid chain. The process is repeated again and again, until the amino acid sequence for the protein is completed. The process of creating a protein in response to the codons on mRNA is called **translation,** which is illustrated in Figure 3-20.

QUICK CHECK

3.29 What are the three types of RNA?

3.30 If a coding strand of DNA has the base sequence of CAT, what will be the base sequence of the codon? What will be the base sequence of the anticodon?

FOCUS ON AGING

Many of the cellular effects of aging are attributed to damage to DNA. Normal cells have built-in mechanisms to repair minor DNA damage, but this ability appears to diminish in aging cells. Because DNA directs protein synthesis, DNA damage is reflected in changes in the membranes and enzymes that are made by the cell. The cell membrane exhibits changes in its transport of ions and nutrients. Membrane-bound organelles, such as mitochondria and lysosomes, are present in reduced numbers. In addition, they are less effective, presumably because of changes in their membranes and in the enzymes that regulate their reactions.

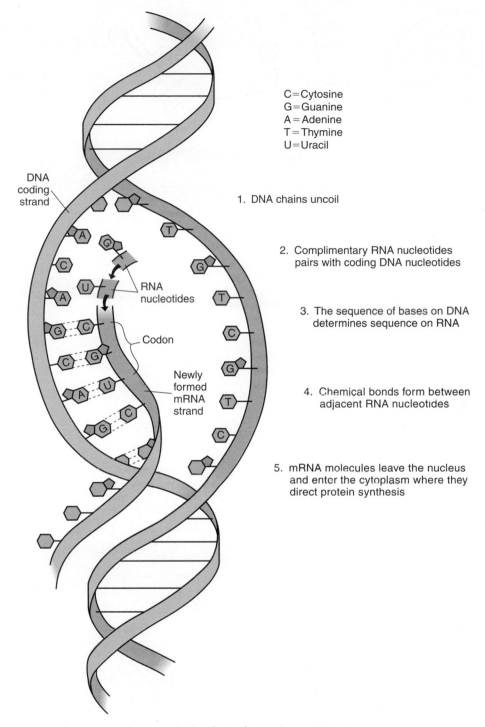

C = Cytosine
G = Guanine
A = Adenine
T = Thymine
U = Uracil

DNA coding strand

RNA nucleotides

Codon

Newly formed mRNA strand

1. DNA chains uncoil

2. Complimentary RNA nucleotides pairs with coding DNA nucleotides

3. The sequence of bases on DNA determines sequence on RNA

4. Chemical bonds form between adjacent RNA nucleotides

5. mRNA molecules leave the nucleus and enter the cytoplasm where they direct protein synthesis

Figure 3-19 *Synthesis of mRNA—transcription.*

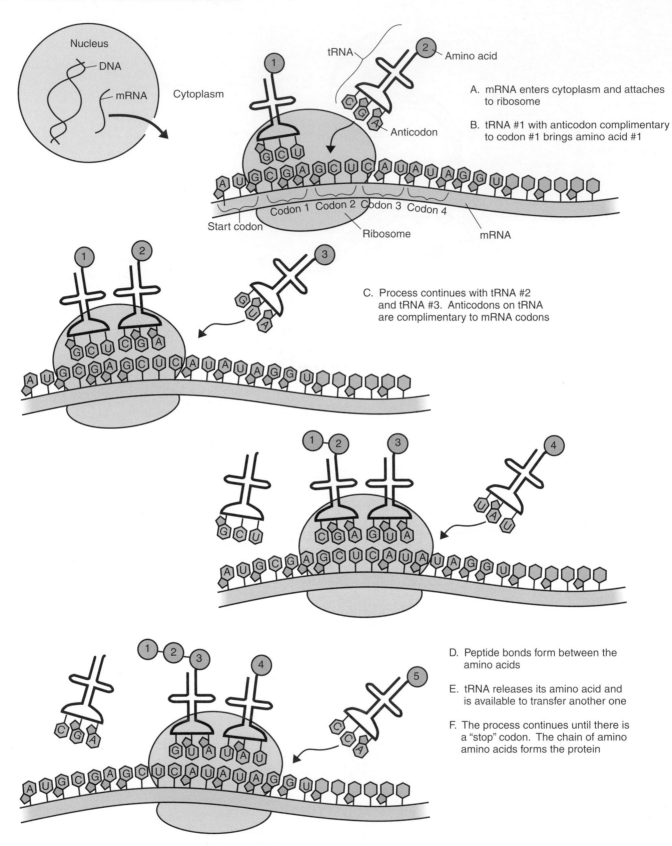

Figure 3-20 *Protein synthesis—translation.*

CHAPTER SUMMARY

Structure of the Generalized Cell

■ **Describe the composition of the cell membrane and list five functions of the proteins in the membrane.**

- A selectively permeable cell membrane separates the extracellular material from the intracellular material.
- The cell membrane is a double layer of phospholipid molecules with proteins scattered throughout.
- Proteins in the cell membrane provide structural support, form channels for passage of materials, act as receptor sites, function as carrier molecules, and provide identification markers

■ **Describe the composition of the cytoplasm.**

- Cytoplasm, the gel-like fluid inside the cell, is largely water with a variety of solutes and has organelles suspended in it.

■ **Describe the components of the nucleus and state the function of each one.**

- The nucleus, formed by a nuclear membrane around a fluid nucleoplasm, is the control center of the cell.
- Threads of chromatin in the nucleus contain DNA, the genetic material of the cell.
- The nucleolus is a dense region of RNA in the nucleus and is the site of ribosome formation.

■ **Identify and describe each of the cytoplasmic organelles and state the function of each organelle: mitochondria, ribosomes, endoplasmic reticulum, Golgi apparatus, and lysosomes.**

- Mitochondria are enclosed by a double membrane and function in the production of ATP.
- Ribosomes are granules of RNA that function in protein synthesis.
- Endoplasmic reticulum is a series of membranous channels that function in the transport of molecules; rough endoplasmic reticulum has ribosomes associated with it and transports proteins; smooth endoplasmic reticulum does not have ribosomes and it transports certain lipids.
- Golgi apparatus modifies substances that are produced in other parts of the cell and prepares these products for secretion.
- Lysosomes contain enzymes that break down substances taken in at the cell membrane; they also destroy cellular debris.

■ **Identify and describe each of the filamentous protein organelles and state the function of each one: cytoskeleton, centrioles, cilia, and flagella.**

- Cytoskeleton is formed from microfilaments and microtubules and helps to maintain the shape of the cell.
- Centrioles are located in the centrosome, a dense region near the nucleus, and function in cell division.
- Cilia are short, hairlike projections that move substances across the surface of a cell.
- Flagella are long, threadlike, projections that move the cell.

Cell Functions

■ **Explain how the cell membrane regulates the composition of the cytoplasm.**

- The cell membrane controls the composition of the cytoplasm by regulating movement of substances through the membrane.

■ **Describe the various mechanisms that result in the transport of substances across the cell membrane: simple diffusion, facilitated diffusion, osmosis, filtration, active transport pumps, endocytosis, and exocytosis.**

- Simple diffusion is the movement of particles from a region of higher concentration to a region of lower concentration; it may take place through a permeable membrane.
- Facilitated diffusion requires a special carrier molecule but still moves particles down a concentration gradient.
- Osmosis is the diffusion of solvent or water molecules through a selectively permeable membrane; cells placed in a hypotonic solution will take in water through osmosis and will swell due to the increased intracellular volume; cells placed in a hypertonic solution will lose fluid due to osmosis and will shrink or crenate.
- Filtration uses pressure to push substances through a membrane; pores in the membrane filter determine the size of particles that will pass through it.
- Active transport pumps move substances against a concentration gradient, from a region of lower concentration to a region of higher concentration, and require a carrier molecule and cellular energy.
- Endocytosis is a process by which solid particles (phagocytosis) and liquid droplets (pinocytosis) are taken into the cell.
- Exocytosis moves secretory vesicles from inside the cell to the outside of the cell.

■ **Name the phases of a typical cell cycle and describe the events that occur in each phase.**

- Somatic cells reproduce by mitosis, which results in two cells identical to the one parent cell.
- Interphase is the period between successive cell divisions; it is the longest part of the cell cycle.
- Successive stages of mitosis are prophase, metaphase, anaphase, and telophase; cytokinesis, division of the cytoplasm, occurs during telophase.

■ **Explain the difference between mitosis and meiosis.**

- Reproductive cells divide by meiosis.
- In meiosis, a single parent cell produces four cells, each with one half the number of chromosomes as the parent cell.

■ **Describe the process of DNA replication.**

- The DNA molecule uncoils, the hydrogen bonds between the complementary base pairs break, and the two strands separate.
- Enzymes catalyze the formation of new complementary strands, resulting in two molecules of DNA identical to the original.

■ **Summarize the process of protein synthesis, including the roles of DNA and RNA.**
- DNA in the nucleus directs protein synthesis in the cytoplasm. A gene is the portion of a DNA molecule that controls the synthesis of one specific protein molecule.
- Protein synthesis utilizes mRNA and tRNA in transcription and translation.
- During the process of transcription, the genetic code is transferred from DNA to mRNA, which carries the information to the sites of protein synthesis in the cytoplasm.

- A sequence of three nucleotide bases on mRNA represents a codon that codes for a specific amino acid.
- Translation is the process of creating a protein in response to mRNA codons.
- Anticodons on tRNA have bases that are complementary to the codons of mRNA; each tRNA carries its specific amino acid to the developing molecule, pairs with the complementary codon on mRNA to determine the amino acid sequence, and releases the amino acid to the developing protein.

CHAPTER QUIZ

Recall

Match the definitions on the left with the appropriate term on the right.

_____ **1.** Attracts water
_____ **2.** Inside the cell
_____ **3.** Small granules of RNA in cytoplasm
_____ **4.** Cellular ingestion of solid particles
_____ **5.** Diffusion of water through a selectively permeable membrane
_____ **6.** Greater solute concentration
_____ **7.** Division of cytoplasm
_____ **8.** Transfer of genetic information from DNA to mRNA
_____ **9.** Sequence of three bases on mRNA
_____ **10.** Results in eggs and sperm

A. Codon
B. Cytokinesis
C. Hydrophilic
D. Hypertonic
E. Intracellular
F. Meiosis
G. Osmosis
H. Phagocytosis
I. Ribosomes
J. Transcription

Thought

1. The plasma membrane is composed primarily of:

   A. Phospholipid and cholesterol
   B. Cholesterol and glycoprotein
   C. Phospholipid and protein
   D. Protein and carbohydrate

2. Which of the following organelles functions in the production of ATP?

   A. Golgi apparatus
   B. Endoplasmic reticulum
   C. Centrioles
   D. Mitochondria

3. A cell is placed in a hypotonic solution. What will happen?

   A. Water will enter the cell.
   B. The cell will shrink.
   C. Nothing, because the plasma membrane is impermeable to water.
   D. Solute particles will leave the cell.

4. The stage of mitosis in which cytokinesis occurs is:

   A. Prophase
   B. Telophase
   C. Anaphase
   D. Metaphase

5. In DNA replication:

   A. Adenine pairs with guanine.
   B. Cytosine pairs with thymine.
   C. Cytosine pairs with guanine.
   D. Adenine pairs with uracil.

Application

A given cell is active in the synthesis of secretory proteins. Name the three organelles that are likely to be abundant in this cell.

Building Your Medical Vocabulary

Building a vocabulary is a cumulative process. As you progress through this book, you will use word parts, abbreviations, and clinical terms from previous chapters. Each chapter will present new word parts, abbreviations, and terms to add to your expanding vocabulary.

Word Parts and Combining Forms with Definitions and Examples

PART/ COMBINING FORM	DEFINITION	EXAMPLE
ab-	away from	abduct: to take away from
ana-	apart	anaphase: the part of the cell cycle when chromatids pull apart
bio-	life	biology: study of life
-cele	swelling, tumor	hydorcele: a tumor or swelling filled with fluid
cyt/o	cell	cytology: study of cells
dist-	far	distal: a directional term for far from a given part
-elle	little, small	organelle: little organs
eti/o	cause	etiology: study of causes of diseases
extra-	outside, beyond	extracellular: outside the cell
hiat-	gap, cleft,	aortic hiatus: an opening the in diaphragm for the opening aorta
hyper-	excessive, above	hypertoxicity: an excessive toxic quality
hypo-	beneath, below	hypodermis: below the skin
-ic-	pertaining to	aortic: pertaining to the aorta
intra-	within, inside	intracellular: within the cell
iso-	equal, same	isometric: same measure
-osis	condition of	leukocytosis: condition of having too many white blood cells
-ostomy	an opening that is surgically created	colostomy: an artificial opening in the abdomen forthe purpose of evacuating the colon
-phag-	to eat, devour	dysphagia: difficulty in eating or swallowing

PART/ COMBINING FORM	DEFINITION	EXAMPLE
pharmac/o	drug	pharmacologist: a specialist in the study of drug action
-phil-	to love, have affinity for	hydrophilic: pertaining to the affinity for water
-phob-	to hate, dislike	claustrophobic: pertaining to the dislike or fear of closed places
pin/o	to drink	pinocytosis: condition of the cell taking in fluid droplets or drinking
-plasia	development,	hyperplasia: increase in size because of an increase formation, growth in the formation of cells
-plasm-	matter	cytoplasm: the matter of the cell
quadr-	four	quadriceps: a group of four muscles
-reti-	network, lattice	reticulocyte: an immature blood cell that shows a diffuse network of fibrils when strained
-som-	body	somatic: pertaining to the body
sub-	below, less, under	subcutaneous: below the skin
ton-	solute strength	hypertonic: pertaining to excessive solute strength
-ul-, -ule	small, tiny	venule: a small or tiny vein
-um	structure, tissue	pericardium: a membrane or tissue around the heart

Clinical Abbreviations

ABBREVIATION	MEANING
ant	anterior
CA, Ca	cancer; carcinoma
chol	cholesterol
DNA	deoxyribonucleic acid
ea	each

ABBREVIATION	MEANING
ECG, EKG	electrocardiogram
ED	emergency department
FH	family history
HR	heart rate
LBW	low birth weight
MRI	magnetic resonance imaging
NB	newborn
NKA	no known allergies
NPO	nothing by mouth
PET	positron emission tomography

Clinical Terms

Anaplasia (an-ah-PLAY-zee-ah) Loss of differentiation of cells; reversion to a more primitive cell type; characteristic of cancer

Anomaly (ah-NAHM-ah-lee) Deviation from normal

Atrophy (AT-roh-fee) Wasting away; a decrease in the size of a cell, tissue, organ, or part

Benign (bee-NYNE) Not malignant, not recurring

Carcinogen (kar-SIN-oh-jen) Agent that causes cancer; known carcinogens include chemicals and drugs, radiation, and viruses

Congenital disorder (kahn-JEN-ih-tahl dis-OR-der) Abnormal condition that is present at birth and continues to exist from the time of birth

Cytology (sye-TAHL-oh-jee) Study of cells including their origin, structure, function, and pathology

Dysplasia (dis-PLAY-zee-ah) Abnormality in development; alteration in size, shape, and organization of cells

Genetic disorder (jeh-NET-ick dis-OR-der) Condition or disease that is caused by a defective gene and may appear at any time in life; also called hereditary disorder

Hyperplasia (hye-per-PLAY-zee-ah) Abnormal increase in the number of cells resulting from an increase in the frequency of cell division

Hypertrophy (hye-PER-troh-fee) Enlargement of an organ attributable to an increase in the size of the individual constituent cells

Malignant (mah-LIG-nant) Tending to become worse and result in death; refers to tumors having the characteristics of invasiveness, anaplasia, and metastasis

Metaplasia (meh-tah-PLAY-zee-ah) Transformation of one cell type into another cell type

Metastasis (meh-TASS-tah-sis) Spread of a tumor to a secondary site

Necrosis (neh-KROH-sis) Death of cells or groups of cells

Neoplasm (NEE-oh-plazm) Any new and abnormal growth; a tumor

VOCABULARY QUIZ

Use word parts given in this chapter or in previous chapters to form words that have the following definitions.

1. Condition of cell eating _____

2. Same solute strength _____

3. Condition of attracting water _____

4. Matter of the cell _____

5. Within the cell _____

Using the definitions of word parts given in this chapter or in previous chapters, define the following words.

6. Pinocytosis _____

7. Somatic _____

8. Cytology _____

9. Hydrophobic _____

10. Endoplasmic _____

Match each of the following definitions with the correct word.

_____ 11. Appears to have layers but

_____ 12. Tumor of epithelial tissue

_____ 13. Fat tissue

_____ 14. Osseous tissue

_____ 15. Tumor of neuroglia cells

A. Adipose does not
B. Bone
C. Carcinoma
D. Glioma
E. Pseudostratified

16. What is the clinical term for a "benign tumor formed of muscle tissue"?

17. A patient has a benign tumor derived from fat cells. What is the clinical term for this condition?

18. What is the work of a histologist?

19. What is the meaning of the abbreviation pc?

20. What is the meaning of the abbreviation ASA?

Tissues and Membranes

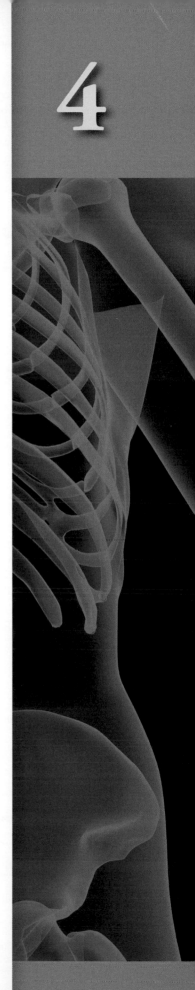

4

CHAPTER OBJECTIVES

Body Tissues
- Define the term *tissue*.
- List the four main types of tissues found in the body.
- Describe the general characteristics of epithelial tissue.
- Describe the various types of epithelial tissues in terms of structure, location, and function.
- Describe the classification of glandular epithelium according to structure and method of secretion; give an example of each type.
- Describe the general characteristics of connective tissue.
- Name three types of connective tissue cells and state the function of each one.
- Describe the features and location of the various types of connective tissue.

- Describe the general characteristics of muscle tissue.
- Distinguish between skeletal muscle, smooth muscle, and cardiac muscle in terms of structure, location, and control.
- Name two categories of cells in nerve tissue and state their general functions.

Inflammation and Tissue Repair
- Describe the four manifestations of inflammation and how they develop.
- Differentiate between regeneration and fibrosis by describing each process.

Body Membranes
- Differentiate between epithelial membranes and connective tissue membranes.
- Describe four types of membranes and specify the location and function of each.

KEY TERMS

Chondrocyte (KON-droh-syte) Cartilage cell

Collagenous fibers (koh-LAJ-eh-nuss FYE-burs) Strong and flexible connective tissue fibers that contain the protein collagen

Elastic fibers (ee-LAS-tick FYE-burs) Yellow connective tissue fibers that are not particularly strong but can be stretched and will return to their normal shape when released

Erythrocyte (ee-RITH-roh-syte) Red blood cell

Fibroblast (FYE-broh-blast) Connective tissue cell that produces fibers

Histology (hiss-TAHL-oh-jee) Branch of microscopic anatomy that studies tissues

Leukocyte (LOO-koh-syte) White blood cell

Macrophage (MAK-roh-fahj) Large phagocytic connective tissue cell that functions in immune responses

Mast cell (MAST SELL) Connective tissue cell that produces heparin and histamine

Neuroglia (noo-ROG-lee-ah) Supporting cells of nervous tissue; cells in nervous tissue that do not conduct impulses

Neuron (NOO-ron) Nerve cell, including its processes; conducting cell of nervous tissue

Osteocyte (AH-stee-oh-syte) Mature bone cell

Tissue (TISH-yoo) Group of similar cells specialized to perform a certain function

Thrombocyte (THROM-boh-syte) Blood platelet

A tissue is a group of cells that have a similar structure and that function together as a unit. The microscopic study of tissues is called **histology** (hiss-TAHL-oh-jee). Nonliving material, called the *intercellular matrix,* fills the spaces between the cells. This may be abundant in some tissues and scarce in others. The intercellular matrix may contain special substances such as salts and fibers that are unique to a specific tissue and give that tissue distinctive characteristics.

BODY TISSUES

There are four main tissue types in the body: epithelial, connective, muscle, and nervous. Each is designed for specific functions.

Epithelial Tissue

Epithelial (ep-ih-THEE-lee-al) **tissues** are widespread throughout the body. They form the covering of all body surfaces, line body cavities and hollow organs, and are the major tissue in glands. They perform a variety of functions that include protection, secretion, absorption, excretion, filtration, diffusion, and sensory reception.

The cells in epithelial tissue are tightly packed with very little intercellular matrix. Because the tissues form coverings and linings, the cells have one free surface that is not in contact with other cells. Opposite the free surface, the cells are attached to underlying connective tissue by a noncellular **basement membrane.** This membrane is a mixture of carbohydrates and proteins secreted by the epithelial and connective tissue cells. Because epithelial tissues are typically **avascular,** they must receive their nutrients and oxygen supply by diffusion from the blood vessels in the underlying tissues. Another characteristic of epithelial tissues is that they regenerate, or reproduce, quickly. For example, the cells of the skin and stomach are continually damaged and replaced, and skin abrasions heal quite rapidly.

Epithelia are classified according to cell shape and the number of layers in the tissue (Figure 4-1). Classified according to shape, the cells are squamous, cuboidal, or columnar, and the shape of the nucleus corresponds to the cell shape. **Squamous** cells are flat and the nuclei are usually broad and thin. **Cuboidal** cells are cubelike, as tall as they are wide, and the nuclei are spherical and centrally located. **Columnar** cells are tall and narrow, resembling columns, and the nuclei are usually in the lower portion of the cell near the basement membrane. According to the number of layers, epithelia are **simple** if they have only one layer of cells and **stratified** if they have multiple layers. Stratified epithelia are named according to the type of cells at the free surface of the tissue.

QUICK APPLICATIONS

Carcinomas are solid cancerous tumors that are derived from epithelial tissue. Approximately 85% of all malignant neoplasms are carcinomas.

Simple Squamous Epithelium

Simple squamous epithelium (Figure 4-2) consists of a single layer of thin, flat cells that fit closely together with very little intercellular matrix. Because it is so thin, simple squamous epithelium is well suited for areas in which diffusion and filtration take place. The alveoli or air sacs of the lungs, where diffusion of oxygen and carbon dioxide gases occurs, are made of simple squamous epithelium. This tissue is also found in the kidney, where the blood is filtered. Capillary walls, where oxygen and carbon dioxide diffuse between the blood and tissues, are made of simple squamous epithelium. Because it is so thin and delicate, this tissue is damaged easily and offers very little protective function.

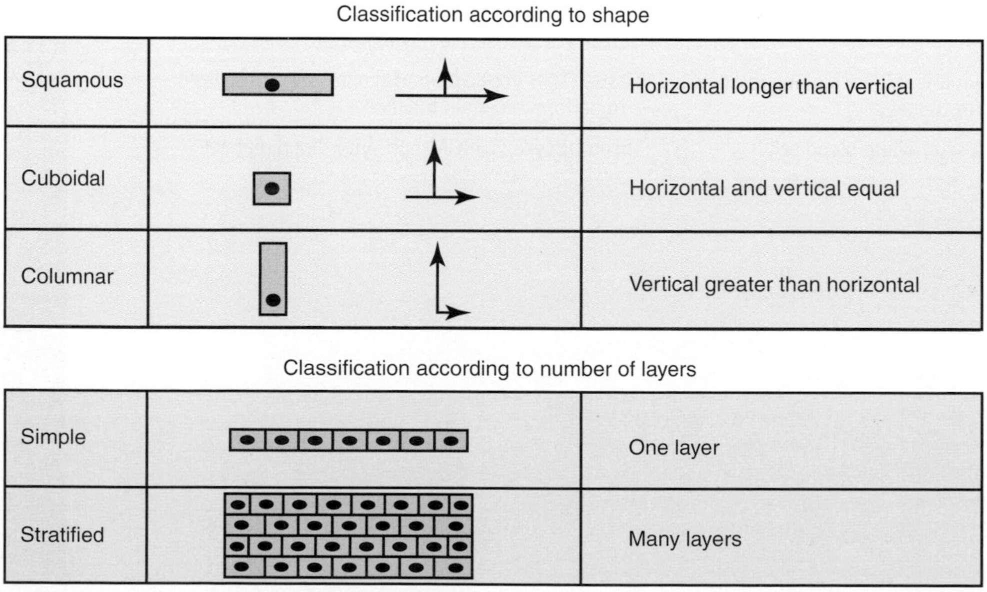

Classification according to shape

Squamous			Horizontal longer than vertical
Cuboidal			Horizontal and vertical equal
Columnar			Vertical greater than horizontal

Classification according to number of layers

Simple		One layer
Stratified		Many layers

Figure 4-1 *Classification of epithelium according to shape and according to the number of layers.*

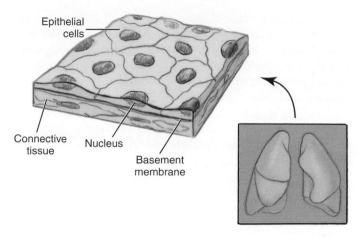

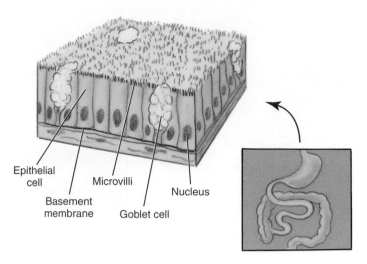

Figure 4-2 *Simple squamous epithelium in the alveoli of the lungs. It is also found in capillary walls and in the renal corpuscles of the kidney.*

Figure 4-4 *Simple columnar epithelium in the lining of the stomach and intestines.*

Simple Cuboidal Epithelium

Simple cuboidal epithelium (Figure 4-3) consists of a single layer of cube-shaped cells. These cells have more volume than squamous cells and also have more organelles. Simple cuboidal epithelium is found as a covering of the ovary, as a lining of kidney tubules, and in many glands such as the thyroid, pancreas, and salivary glands. In the kidney tubules, the tissue functions in absorption and secretion. In glands, simple cuboidal cells form the secretory portions and the ducts that deliver the products to their destination.

Simple Columnar Epithelium

A single layer of cells that are taller than they are wide makes up **simple columnar epithelium** (Figure 4-4). The nuclei are in the bottom portion of the cell near the basement membrane. Simple columnar epithelium is found lining the stomach and intestines, where it secretes digestive enzymes and absorbs nutrients. Because the cells are taller (or thicker) than either squamous or cuboidal cells, this tissue offers some protection to underlying tissues.

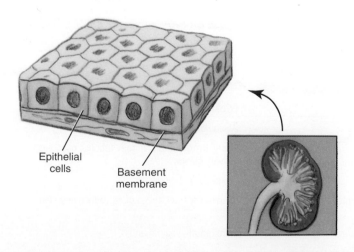

Figure 4-3 *Simple cuboidal epithelium in the kidney tubules. It is also found in many glands and as a covering of the ovary.*

In regions where absorption is of primary importance, such as in parts of the digestive tract, the cell membrane on the free surface has numerous small projections called **microvilli.** Microvilli increase the surface area that is available for absorption of nutrients. **Goblet cells** are frequently interspersed among the simple columnar cells. Goblet cells are flask- or goblet-shaped cells that secrete mucus onto the free surface of the tissue. **Cilia** may be present to move secretions along the surface.

QUICK CHECK

4.1 What term is used to denote an epithelial tissue that has multiple layers of cells?

4.2 What type of epithelial tissue is found in the alveoli (air sacs) of the lungs?

4.3 What type of epithelium consists of a single layer of cells that are taller than they are wide?

Pseudostratified Columnar Epithelium

Pseudostratified columnar epithelium (Figure 4-5) appears to have multiple layers (stratified) but it really does not. This is because the cells are not all the same height. Some cells are short and some are tall, and the nuclei are at different levels. Close examination reveals that all the cells are attached to the basement membrane but that not all cells reach the free surface of the tissue. Cilia and goblet cells are often associated with pseudostratified columnar epithelium. This tissue lines portions of the respiratory tract in which the mucus, produced by the goblet cells, traps dust particles and is then moved upward by the cilia. Pseudostratified columnar epithelium also lines some of the tubes of the male reproductive system. Here the cilia help propel the sperm from one region to another.

Stratified Squamous Epithelium

Stratified squamous epithelium, the most widespread stratified epithelium, is thick because it consists of many layers of cells (Figure 4-6). Because stratified epithelia are named

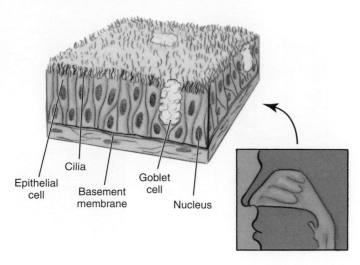

Epithelial cell
Cilia
Basement membrane
Goblet cell
Nucleus

Figure 4-5 *Pseudostratified columnar epithelium in the respiratory tract. It also lines some of the male reproductive system.*

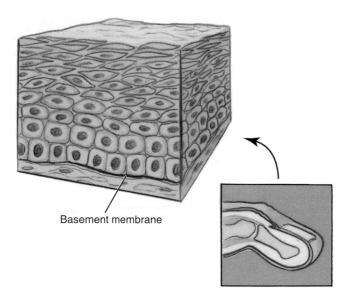

Basement membrane

Figure 4-6 *Stratified squamous epithelium from the outer layer of the skin. Note the numerous cell layers and the flattened cells at the surface.*

according to the shape of the surface cells, it follows that the cells on the surface are flat. The cells on the bottom layer, next to the basement membrane, are usually cuboidal or columnar, and these are the cells that undergo mitosis. As the cells are pushed toward the surface, they become thinner, so the surface cells are squamous. As the cells are pushed farther away from the basement membrane, it is more difficult for them to receive oxygen and nutrients from underlying connective tissue and the cells die. As cells on the surface are damaged and die, they are sloughed off and replaced by cells from the deeper layers. Because this tissue is thick, it is found in areas in which protection is a primary function. Stratified squamous epithelium forms the outer layer of the skin and extends a short distance into every body opening that is continuous with the skin.

Transitional Epithelium

Transitional epithelium (Figure 4-7) is a specialized type of tissue that has several layers but can be stretched in response to tension. The lining of the urinary bladder is a good example of this type of tissue. When the bladder is empty and contracted, the epithelial lining has several layers of cuboidal cells. As the bladder fills and is distended or stretched, the cells become thinner and the number of layers decreases. Table 4-1 summarizes the different types of epithelial tissue.

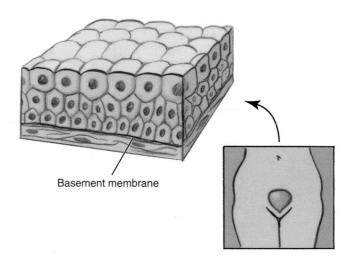

Basement membrane

Figure 4-7 *Transitional epithelium from the urinary bladder.*

TABLE 4-1 Summary of Epithelial Tissues

Type	Description	Location
Simple squamous	Single layer of thin, flat cells	Alveoli of lungs, capillary walls, kidneys
Simple cuboidal	Single layer of cuboidal cells	Ovary, thyroid gland, kidney tubules, pancreas, salivary glands
Simple columnar	Single layer of tall cells; often contains goblet cells	Stomach, intestines
Pseudostratified columnar	Single layer of uneven columnar cells; often contains cilia and goblet cells	Respiratory tract, tubes of reproductive system
Stratified squamous	Several layers with flat cells at the free surface	Skin, mouth, vagina, anus
Transitional	Specialized for stretching; several layers that decrease in number and cells that become thinner when distended	Urinary bladder

Glandular Epithelium

Glandular epithelium consists of cells that are specialized to produce and secrete substances. It normally lies deep to the epithelia that cover and line parts of the body. If the gland secretes its product onto a free surface via a duct, it is called an **exocrine gland.** If the gland secretes its product directly into the blood, it is a ductless gland, or **endocrine gland.** Endocrine glands are discussed in Chapter 10.

Exocrine glands that consist of only one cell are called **unicellular** glands. Goblet cells, which produce mucus in the lining of the digestive, respiratory, urinary, and reproductive tracts, are examples of unicellular glands. Most glands are **multicellular** because they consist of many cells. These glands have a secretory portion and a duct derived from epithelium.

Multicellular glands are classified according to their structure and to the type of secretion they produce.

A gland is **simple** if its duct has no branches. If the duct branches, then the gland is **compound.** The glands are **tubular** if the gland and duct merge with no change in diameter. Glands in which the distal part of the duct expands to form a saclike structure are called **acinar** (AS-ih-nur) or **alveolar** (al-VEE-oh-lar). The structural classification of glands is illustrated in Figure 4-8.

Glands are also classified according to their mode of secretion (Figure 4-9). **Merocrine** (MER-oh-krin) glands secrete a fluid that is released through the cell membrane by exocytosis with no loss of cytoplasm. Most glandular cells are of this type. Salivary glands, pancreatic glands, and certain sweat glands are merocrine glands. The cells of merocrine glands may be either **serous** cells or **mucous** cells. Serous cells secrete a thin, watery serous fluid that often contains enzymes. Mucous cells secrete a thick fluid called *mucus* that contains the glycoprotein mucin. In **apocrine** (AP-oh-krin) glands, the secretory product accumulates in one region of the cell and then that portion pinches off so that a small portion of the cell is lost with the secretion. The cell repairs itself and repeats the process. Examples of apocrine glands include certain sweat glands and mammary

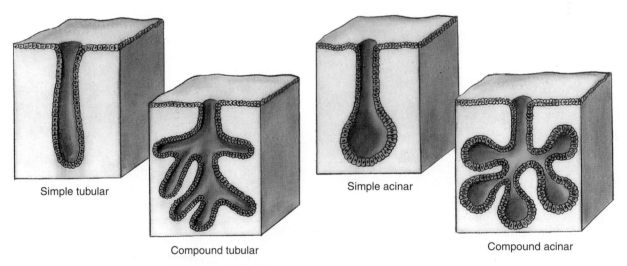

Simple tubular

Compound tubular

Simple acinar

Compound acinar

Figure 4-8 *Classification of glands according to structure.*

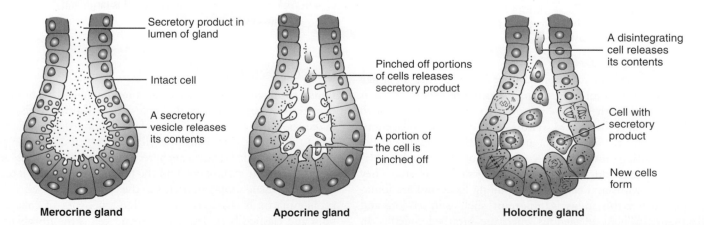

Secretory product in lumen of gland

Intact cell

A secretory vesicle releases its contents

Merocrine gland

Pinched off portions of cells releases secretory product

A portion of the cell is pinched off

Apocrine gland

A disintegrating cell releases its contents

Cell with secretory product

New cells form

Holocrine gland

Figure 4-9 *Classification of glands according to mode of secretion.*

TABLE 4-2 Summary of Glandular Modes of Secretion

Secretion Mode	Description	Examples
Merocrine	Fluid released through cell membrane; no cytoplasm lost	Salivary glands, pancreatic glands, certain sweat glands
Apocrine	Portion of cell pinched off with secretion	Mammary glands and certain sweat glands
Holocrine	Entire cell discharged with secretion	Sebaceous glands

glands. In **holocrine** (HOH-loh-krin) glands, the cells become filled with the secretory product and then rupture, releasing their products. The cells have to be replaced. Sebaceous (oil) glands are holocrine glands. Table 4-2 summarizes the modes of glandular secretion.

QUICK APPLICATIONS
A benign tumor of glandular epithelial cells is called an adenoma. An adenocarcinoma is a cancerous tumor arising from glandular cells.

QUICK CHECK
4.6 What is an example of a unicellular gland?
4.7 In what class of glands do secretory vesicles release the product from the cell into the lumen of the gland and the cell remains intact?

Connective Tissue

Connective tissues bind together structures, form the framework and support for organs and the body as a whole, store fat, transport substances, protect against disease, and help repair tissue damage. They occur throughout the body. Connective tissues are characterized by an abundance of intercellular matrix with relatively few cells. Connective tissue cells are able to reproduce but not as rapidly as epithelial cells. Most connective tissues have a good blood supply, but some do not.

QUICK APPLICATIONS
Sarcomas are cancerous tumors that are derived from connective tissue. These account for approximately 10% of all malignant neoplasms.

The intercellular matrix in connective tissue has a gel-like base of water, nonfibrous protein, and other molecules. Various mineral salts in the matrix of some connective tissues, such as bone, make them hard. Two types of fibers, collagenous (koh-LAJ-eh-nus) and elastic, are frequently embedded in the matrix. **Collagenous fibers,** composed of the protein collagen, are strong and flexible but are only slightly elastic. They are able to withstand considerable pulling force and are found in areas where this ability is important, such as in tendons and ligaments. When collagenous fibers are grouped together in parallel bundles, the tissue appears white, so they are sometimes

called *white fibers*. **Elastic fibers,** composed of the protein elastin, are not very strong, but they are elastic. They can be stretched and will return to their original shape and length when released. Elastic fibers, also called *yellow fibers*, are located where structures are stretched and released, such as the vocal cords.

QUICK APPLICATIONS
Collagen turns into a soft gelatin when it is boiled. Meat that has a lot of collagenous fibers is tough, but it becomes more tender when it is boiled or simmered because the collagen softens.

Numerous cell types are found in connective tissue. Three of the most common are the **fibroblast, macrophage,** and **mast cell.** As the name implies, fibroblasts produce the fibers that are in the intercellular matrix. Macrophages are large phagocytic cells that are able to move about and clean up cellular debris and foreign particles from the tissues. Mast cells contain heparin, which is an anticoagulant. They also contain histamine, a substance that promotes inflammation and is active in allergies.

There are several types of connective tissue. The characteristics and functions of each type depend on the mature of the matrix and the cells and fibers embedded in the matrix. A description of each type of connective tissue follows, and they are summarized in Table 4-3.

QUICK CHECK
4.8 What type of connective tissue fiber is strong and flexible to withstand a pulling force?
4.9 What type of connective tissue cell is large and phagocytic?

Loose Connective Tissue

Loose connective tissue, also called **areolar** (ah-REE-oh-lar) **connective tissue,** is one of the most widely distributed tissues in the body. It is the packing material in the body. It attaches the skin to the underlying tissues and fills the spaces between muscles. Most epithelial tissue is anchored to loose connective tissue by the basement membrane, and the blood vessels in the loose connective tissue supply nutrients to the epithelium above. The gel-like matrix is characterized by a loose network of collagenous and elastic fibers. The predominant cell is the fibroblast, but other connective tissue cells are also present (Figure 4-10).

TABLE 4-3 Summary of Connective Tissues

Type	Description	Functions/Locations
Loose (areolar)	Collagenous and elastic fibers produced by fibroblasts are embedded in gel-like matrix	Binds organs together/beneath skin, between muscles
Adipose	Cells filled with fat droplets so that nucleus and cytoplasm pushed to periphery; little intercellular matrix	Cushions, insulates, stores energy/beneath skin, around kidneys, heart, eyeballs
Dense fibrous connective tissue	Matrix filled with parallel bundles of collagenous fibers	Bind structures together/tendons and ligaments
Elastic connective tissue	Matrix filled with yellow elastic fibers	Elasticity/vocal cords and ligaments between adjacent vertebrae
Hyaline cartilage	Solid matrix with fibers and scattered cells; chondrocytes located in lacunae	Supports, protects, provides framework/ends of long bones, connects ribs to sternum, tracheal rings, fetal skeleton
Fibrocartilage	Numerous collagenous fibers in matrix	Cushions and protects/intervertebral disks, pads in knee joint, pad between two pubic bones
Elastic Cartilage	Numerous elastic fibers in matrix	Supports and provides framework/external ear, epiglottis auditory tubes
Osseous (bone)	Hard matrix with mineral salts; matrix arranged in lamellae around haversian canal; osteocytes in lacunae	Protects, supports, provides framework/bones of skeleton
Blood	Liquid matrix called plasma with erythrocytes, leukocytes, and platelets suspended in it	Transports oxygen, protects against disease, functions in clotting mechanism/blood vessels and heart

Figure 4-10 *Loose (areolar) connective tissue. Note the fibroblasts and two types of fibers embedded in a gel-like matrix.*

Figure 4-11 *Adipose tissue. Note the closely packed fat-filled cells.*

Adipose Tissue

Commonly called *fat,* **adipose** (ADD-ih-pose) **tissue** is a specialized form of loose connective tissue in which there is very little intercellular matrix. Some of the cells accumulate liquid fat, droplets, which push the cytoplasm and nucleus to one side. As a result, the cells swell and become closely packed together (Figure 4-11). Fat cells have the ability to take up fat and then release it at a later time. Adipose tissue forms a protective cushion around the kidneys, heart, eyeballs, and various joints. It also accumulates under the skin, where it provides insulation for heat. Adipose tissue is an efficient energy storage material for excess calories.

 QUICK APPLICATIONS
Suction lipectomy is a cosmetic surgical procedure in which adipose tissue is suctioned from fatty areas of the body.

Dense Fibrous Connective Tissue

Dense fibrous connective tissue is characterized by closely packed parallel bundles of collagenous fibers in the intercellular matrix (Figure 4-12). There are relatively few cells, and the ones that are present are fibroblasts to produce the collagenous fibers. This is the tissue in **tendons,** which connect muscles to

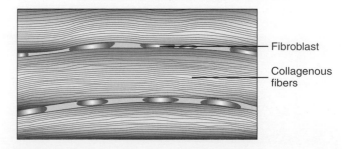

Figure 4-12 *Dense fibrous connective tissue. Note the fibroblast cells and parallel bundles of collagenous fibers.*

bones, and **ligaments,** which connect bones to bones. Dense fibrous connective tissue has a poor blood supply, and this, along with the relatively few cells, accounts for the slow healing of this tissue.

Elastic Connective Tissue

Elastic connective tissue has closely packed elastic fibers in the intercellular matrix. This type of tissue yields easily to a pulling force and then returns to its original length as soon as the force is released. The vocal cords and the ligaments that connect adjacent vertebrae are composed of elastic connective tissue.

> ### QUICK CHECK
> **4.10** What is the most widely distributed connective tissue in the body?
> **4.11** What type of connective tissue has cells that are filled with liquid fat droplets?
> **4.12** What type of connective tissue forms tendons and ligaments?

Cartilage

Cartilage has an abundant matrix that is solid, yet flexible, with fibers embedded in it. The matrix contains the protein **chondrin** (KON-drihn). Cartilage cells, or **chondrocytes** (KON-droh-sytes), are located in spaces called **lacunae** (lah-KOO-nee) that are scattered throughout the matrix. Typically, cartilage is surrounded by a dense fibrous connective tissue covering called the **perichondrium.** There are blood vessels in the perichondrium, but they do not penetrate the cartilage itself, and the cells obtain their nutrients by diffusion through the solid matrix. Cartilage heals slowly because there is no direct blood supply, and this also contributes to slow cellular reproduction. Cartilage protects underlying tissues, supports other structures, and provides a framework for attachments.

Hyaline cartilage (Figure 4-13) is the most common type of cartilage. It has fine collagenous fibers in the matrix and a shiny, white, opaque appearance. It is found at the ends of long bones, in the costal cartilage that connects the ribs to the sternum, and in the supporting rings of the trachea. Most of the fetal skeleton is formed of hyaline cartilage before it is replaced by bone.

Fibrocartilage has an abundance of strong collagenous fibers embedded in the matrix. This allows it to withstand compression, act as a shock absorber, and resist pulling forces. It is found in the intervertebral disks, or pads between the vertebrae; in the symphysis pubis, or pad between the two pubic bones; and between the bones in the knee joint.

Elastic cartilage has numerous yellow elastic fibers embedded in the matrix, which makes it more flexible than hyaline cartilage or fibrocartilage. It is found in the framework of the external ear, the epiglottis, and the auditory tubes.

> ### QUICK CHECK
> **4.13** What is the name of a cartilage cell?
> **4.14** What type of cartilage is found at the ends of long bones in the fetal skeleton?

Bone

Osseous tissue, or **bone,** is the most rigid of all the connective tissues. Collagenous fibers in the matrix give strength to bone, and its hardness is derived from the mineral salts, particularly calcium, that are deposited around the fibers. Bones form the framework for the body and help protect underlying tissues. They serve as attachments for muscles and act as mechanical levers in producing movement. Bone also contributes to the formation of blood cells and functions as a storage area for mineral salts.

Cylindrical structural units, called **osteons** or **haversian** (hah-VER-shun) **systems,** are packed together to form the substance of compact bone (Figure 4-14). The center or hub of the osteon is a tubular **osteonic** or **haversian** canal that contains a blood vessel. The matrix is deposited in concentric rings called **lamellae** (lah-MEL-ee) around the canal. **Osteocytes,** or bone cells, are located in lacunae between the lamellae so that they are also arranged in concentric rings. Slender processes from the bone cells extend, through tiny tubes in the matrix called **canaliculi** (kan-ah-LIK-yoo-lye), to other cells or to the osteonic canals. This provides a readily available blood supply for the bone cells, which allows a faster repair process for bone than for cartilage.

Blood

Blood is a unique connective tissue because it is the only one that has a liquid matrix. It is a vehicle for transport of substances throughout the body. **Erythrocytes,** or red blood cells,

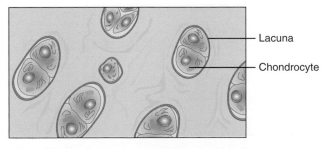

Figure 4-13 *Hyaline cartilage. Note the chondrocytes within the lacunae.*

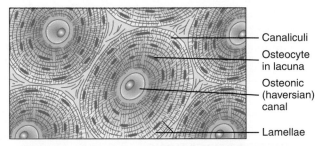

Figure 4-14 *Compact bone (osseous tissue). Note the osteons with a central haversian canal and concentric lamellae of matrix. Canaliculi extend from the osteocytes, which are located within lacunae.*

Figure 4-15 *Blood. Note the blood cells and platelets that are suspended in a liquid plasma.*

and **leukocytes,** or white blood cells, are suspended in a liquid matrix called **plasma** (Figure 4-15). The red blood cells transport oxygen from the lungs to the tissues. White blood cells are important in fighting disease. Another formed element in the blood is the **platelet,** or **thrombocyte,** which is not actually a cell but a fragment of a giant cell in the bone marrow. Platelets are important in initiating the blood clotting process. Blood is discussed in more detail in Chapter 11.

QUICK CHECK

4.15 What is the only connective tissue with a liquid matrix?
4.16 In osseous tissue, what are the concentric rings of matrix called?

Muscle Tissue

Muscle tissue is composed of cells that have the special ability to shorten or contract, to produce movement of body parts. The tissue is highly cellular and is well supplied with blood vessels. The cells are long and slender so they are sometimes called *muscle fibers,* and these are usually arranged in bundles or layers that are surrounded by connective tissue. The contractile proteins **actin** and **myosin** form microfilaments in the cytoplasm and are responsible for contraction. In muscle tissue, the cell membrane is called the sarcolemma and the cytoplasm is called sarcoplasm. There are three types of muscle tissue: skeletal muscle, smooth muscle, and cardiac muscle. A description of each type follows, and their features are summarized in Table 4-4.

Skeletal Muscle

Skeletal muscle tissue (Figure 4-16) is what is commonly thought of as "muscle." It is the meat of animals, and it constitutes about 40% of an individual's body weight. Skeletal muscle cells (fibers) are long and cylindrical with many nuclei (multi-nucleated) peripherally located next to the cell membrane. The cells have alternating light and dark bands that are perpendicular to the long axis of the cell. These bands are a result of the organized arrangement of the contractile proteins in the cytoplasm and give the cell a **striated** appearance. Skeletal muscle fibers are collected into bundles and wrapped in connective tissue to form the muscles, which are attached to the skeleton and which cause body movements when they contract in response to nerve stimulation. Skeletal muscle action is under conscious or voluntary control. Chapter 7 describes skeletal muscles in more detail.

Smooth Muscle

Smooth muscle tissue (Figure 4-17) is found in the walls of hollow body organs such as the stomach, intestines, urinary bladder, uterus, and blood vessels. It normally acts to propel substances through the organ by contracting and relaxing. It is called smooth muscle because it lacks the striations evident in skeletal muscle. Because it is found in the viscera or body organs, it is sometimes called **visceral muscle.** Smooth muscle cells are shorter than skeletal muscle cells, are spindle shaped and tapered at the ends, and have a single, centrally located nucleus. Smooth muscle usually cannot be stimulated to contract by conscious or voluntary effort, so it is called *involuntary muscle.*

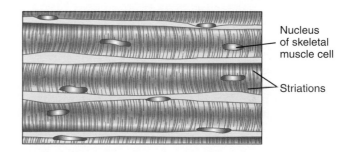

Figure 4-16 *Skeletal muscle. Note the long cylindrical fibers with striations.*

TABLE 4-4 Summary of Muscle Tissues

Feature	Skeletal	Smooth	Cardiac
Location	Attached to bones	Walls on internal organs and blood vessels	Heart
Function	Produces body movement	Contracts viscera and blood vessels	Pumps blood through heart and blood vessels
Cell shape	Cylindrical	Spindle shaped, tapered ends	Cylindrical, branching, intercalated disks join cells together end to end
Number of nuclei	Many; peripherally located	One; centrally located	One; centrally located
Striations	Present	Absent	Present
Speed of contraction	Fastest	Slowest	Intermediate
Length of contraction	Least	Greatest	Intermediate
Type of control	Voluntary	Involuntary	Involuntary

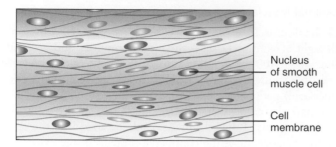

Figure 4-17 *Smooth muscle. Note the spindle-shaped cells with tapered ends.*

Cardiac Muscle

Cardiac muscle tissue (Figure 4-18) is found only in the wall of the heart. The cardiac muscle cells are cylindrical and appear striated, similar to skeletal muscle cells. Cardiac muscle cells are shorter than skeletal muscle cells and have only one nucleus per cell. The cells branch and interconnect to form complex networks. At the point where one cell attaches to another, there is a specialized intercellular connection called an intercalated (in-TER-kuh-lay-ted) disk. Cardiac muscle appears striated like skeletal muscle but its contraction is involuntary. It is responsible for pumping the blood through the heart and into the blood vessels.

 QUICK APPLICATIONS

All three muscle types—skeletal, smooth, and cardiac—produce movement. Contraction of skeletal muscle fibers moves body parts, such as arms and legs. You are usually aware of these movements. Cardiac muscle in the heart moves blood through the body. Smooth muscle contractions in organs like the stomach and intestine moves food and food residue along the digestive tract. The action of actin and myosin produces contractions of muscle fibers, which results in movement.

 QUICK CHECK

4.17 One type of muscle tissue has branching fibers, striations, and intercalated disks. Where in the body is this muscle tissue located?

4.18 What type of muscle tissue is considered voluntary?

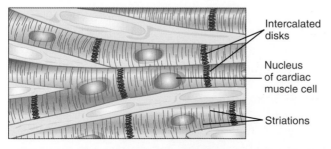

Figure 4-18 *Cardiac muscle. Note the branching striated cells and intercalated disks.*

Nervous Tissue

Nervous tissue is found in the brain, spinal cord, and nerves. It is responsible for coordinating and controlling many body activities. It stimulates muscle contraction, creates an awareness of the environment, and plays a major role in emotions, memory, and reasoning. To do all of these things, cells in nervous tissue need to be able to communicate with each other by way of electrical nerve impulses.

The cells in nervous tissue that generate and conduct impulses are called **neurons** (NOO-rons) or **nerve cells.** These cells have three principal parts: the dendrites, the cell body, and one axon (Figure 4-19). The main part of the cell, the part that carries on the general functions, is the **cell body. Dendrites** are extensions, or processes, of the cytoplasm that carry impulses to the cell body. An extension or process called an **axon** carries impulses away from the cell body.

Nervous tissue also includes cells that do not transmit impulses but instead support the activities of the neurons. These are the **glial** (GLEE-al) **cells** (or neuroglial cells), which combined are termed the **neuroglia** (noo-ROG-lee-ah). Supporting, or glial, cells bind together neurons and insulate the neurons. Some are phagocytic and protect against bacterial invasion, whereas others provide nutrients by binding blood vessels to the neurons. Further details on nerve tissue are presented in Chapter 8.

 QUICK CHECK

4.19 Which extension or process carries impulses away from the cell body?

4.20 Some cells in nervous tissue do not conduct impulses. What are these nonconducting supporting cells called?

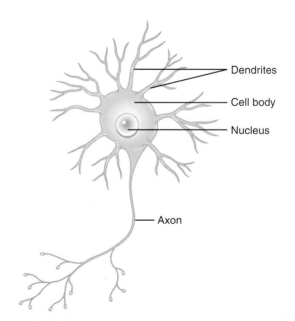

Figure 4-19 *Neuron (nervous tissue). Note the dendrites, axon, and cell body.*

Inflammation and Tissue Repair

Inflammation

Inflammation is a nonspecific defense mechanism that attempts to localize or contain tissue injury and to prepare the area for healing. It is a normal mechanism that is manifested by redness, swelling, heat, and pain.

Damaged tissues release chemicals that cause blood vessels in the area to dilate and become more permeable. Blood vessel dilation increases blood flow to the area, which accounts for the redness and the increased temperature. Increased blood vessel permeability permits fluid and phagocytic white blood cells to leave the blood vessels and to infiltrate the surrounding tissue spaces. Phagocytic white blood cells engulf invading bacteria and debris from the damaged cells. The increased fluid in the tissue spaces accounts for the swelling associated with inflammation, and swelling puts pressure on the nerves, causing pain. The fluid dilutes toxins and cleanses the area. It also contains clotting proteins that construct a clot to stop the loss of blood, to hold together the edges of the wound, and to isolate, or "wall off," the area to prevent bacteria and toxins from spreading to surrounding tissues. If the wound is in the skin, the portion of the clot that is exposed to air dehydrates and hardens to become a scab that forms a temporary covering over the break in the skin. Even though inflammation may be painful, it is a beneficial process that cleanses the damaged area and sets the stage for repair and restoration.

Tissue Repair

There are two types of **tissue repair:** regeneration and fibrosis, or scar formation. The type that occurs depends on the characteristics of the tissue and the severity of the damage. Usually both types are involved in the repair process. Because nearly everyone has experienced a skin wound, it is used to illustrate the process of repair.

Regeneration

Regeneration is the replacement of destroyed tissue by the proliferation of cells that are identical to the original cells. This occurs in superficial skin abrasions when mitosis of healthy and undamaged cells restores the skin to its original thickness. Regeneration occurs only in tissues that have mitotic capability. Skeletal muscle, cardiac muscle, and nerve tissue in the brain and spinal cord cannot regenerate. These damaged tissues have to be replaced by fibrous scar tissue, which may interfere with the normal function of the organ involved.

Fibrosis

Fibrosis is the replacement of destroyed tissue by the generation of fibrous connective tissue. This is the formation of scar tissue. Fibroblasts proliferate in the region of the wound and produce collagen fibers that form the basis of the scar. These fibers help draw together the edges of the wound and strengthen the area. This new, immature scar tissue is called **granulation tissue.** While the collagen fibers are forming, undamaged capillaries in the region develop buds that grow into the area of the wound to form new capillaries. The abundance of new capillaries, necessary to bring nutrients to the active cells, makes the immature scar tissue appear pink.

While the granulation tissue is forming, the surface epithelium regenerates and grows over the surface between the scab and the granulation tissue. When the epithelium is complete, the scab detaches. As the granulation tissue matures, it contracts, and the capillaries regress when the extra blood supply is no longer needed. The regenerating epithelium thickens until it matches the surrounding epithelium. The result is a fully regenerated layer of epithelium over an underlying area of fibrosis or scar tissue. The scar may or may not be evident, depending on the severity of the wound. Figure 4-20 illustrates tissue repair in a skin laceration.

 FROM THE PHARMACY

The study of pharmacology includes pharmacokinetics, pharmacodynamics, and pharmacotherapeutics. **Pharmacokinetics** involves the absorption, distribution, metabolism, and excretion of a medication. Absorption is what happens to a drug from the time it enters the body until it reaches the circulating fluid. Distribution is how the circulating fluids transport the drug to its action site. Metabolism is the mechanism by which the body breaks down, or detoxifies, the drug. This occurs primarily in the liver. Excretion is the elimination of the drug and its byproducts from the body.

Pharmacodynamics is the study of how the drug works, or the drug's action. Drugs generally react with receptors on the cells to produce a biological effect, either to alter the environment of the cell or to modify the function of the cell. These responses may be localized at the site of application, or the drug may be absorbed to produce systemic effects. A second method of response is for the drug to interact with a cellular enzyme to modify the action of the enzyme.

Pharmacotherapeutics is the study of drug effects, interactions, and client response. The primary effect is the desired intended reaction. Secondary effects are unintentional side effects or adverse reactions. Side effects usually are mild. Adverse effects imply more serious reactions. Drug interactions occur when the drug reacts with some other substance that is present, which alters the response. Some drugs may interact with food and must be administered when there is no food in the stomach. Cigarettes, caffeine, and alcohol may interact with drugs to modify their effect. Drugs may also interact with other drugs to produce undesirable effects. Client response includes dosage calculation and how the client reacts to a given medication. Two clients may respond differently to a given medication because of disease, immunological factors, psychological concerns, and genetics.

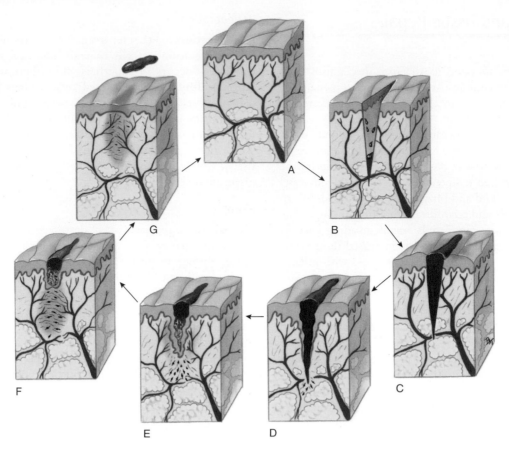

Figure 4-20 *Steps in tissue regeneration and repair.* **A,** *Normal skin.* **B,** *Wound with bleeding.* **C,** *Clot forms.* **D,** *Fibroblasts migrate to the area.* **E,** *Fibroblasts proliferate and begin forming fibers of collagen.* **F,** *Surface epithelium regenerates and grows between scab and granulation tissue.* **G,** *Formation of granulation tissue (scar) is complete and scab detaches.*

QUICK CHECK

4.21 What are the four manifestations of inflammation?

4.22 What type of tissue repair results in new cells that are identical in structure and function to the damaged cells?

Body Membranes

Body membranes are thin sheets of tissue that cover the body, line body cavities, cover organs within the cavities, and line the cavities in hollow organs. By this definition, the skin is a membrane because it covers the body, and indeed, the skin, or integument, is sometimes called the **cutaneous membrane.** This membrane is discussed in Chapter 5. This section examines two epithelial membranes and two connective tissue membranes. Epithelial membranes consist of epithelial tissue and the connective tissue to which it is attached. The two main types of epithelial membranes are the mucous membranes and serous membranes. Connective tissue membranes contain only connective tissue. Synovial membranes and meninges belong to this category.

Mucous Membranes

Mucous membranes are epithelial membranes that consist of epithelial tissue that is attached to underlying loose connective tissue. These membranes, sometimes called **mucosae,** line the body cavities that open to the outside. The entire digestive tract is lined with mucous membranes. Other examples include the respiratory, excretory, and reproductive tracts. The type of epithelium varies depending on its function. In the mouth, the epithelium is stratified squamous for its protection function, but the stomach and intestines are lined with simple columnar epithelium for absorption and secretion. The mucosa of the urinary bladder is transitional epithelium so that it can expand. Mucous membranes get their name from the fact that the epithelial cells secrete mucus for lubrication and protection.

Serous Membranes

Serous membranes line body cavities that do not open directly to the outside, and they cover the organs located in those cavities. A serous membrane, or **serosa,** consists of a thin layer of loose connective tissue covered by a layer of simple squamous epithelium called **mesothelium.** These membranes always have two parts. The part that lines a cavity wall is the **parietal** layer, and the part that covers the organs in the cavity is the **visceral** layer (Figure 4-21). Serous membranes are covered by

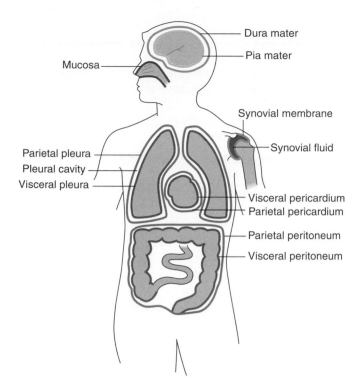

Figure 4-21 *Body membranes.*

a thin layer of **serous fluid** that is secreted by the epithelium. Serous fluid lubricates the membrane and reduces friction and abrasion when organs in the thoracic or abdominopelvic cavity move against each other or the cavity wall.

Serous membranes have special names given according to their location. The serous membrane that lines the thoracic cavity and covers the lungs is the **pleura,** with the parietal pleura lining the cavity and the visceral pleura covering the lungs. The **pericardium** (pair-ih-KAR-deeum) lines the pericardial cavity and covers the heart. The serous membrane in the abdominopelvic cavity is the **peritoneum** (pair-ih-toh-NEE-um).

QUICK APPLICATIONS

An inflammation of the serous membranes in the abdominal cavity is called peritonitis. This is sometimes a serious complication of an infected appendix.

Synovial Membranes

Synovial (sih-NOH-vee-al) **membranes** are connective tissue membranes that line the cavities of the freely movable joints such as the shoulder, elbow, and knee. Similar to serous membranes, they line cavities that do not open to the outside. Unlike serous membranes, they do not have a layer of epithelium. Synovial membranes secrete **synovial fluid** into the joint cavity, and this lubricates the cartilage on the ends of the bones so that they can move freely and without friction. In certain types of arthritis, these membranes become inflamed and the fluid becomes viscous. This reduces lubrication and increases friction, and movement becomes difficult and painful.

Meninges

The connective tissue coverings around the brain and spinal cord, within the dorsal cavity, are called **meninges** (meh-NIN-jeez). They provide protection for these vital structures. The outermost layer of the meninges is the toughest and is called the **dura mater** (DOO-rah MAY-ter). The middle layer, the **arachnoid** (ah-RAK-noyd), is quite fragile. The **pia mater** (PEE-ah MAY-ter), the innermost layer, is very delicate and closely adherent to the surface of the brain and spinal cord. Inflammation of the meninges is **meningitis.** Further discussion of the meninges appears in Chapter 8.

QUICK CHECK

4.23 The pericardium and peritoneum are examples of which type of epithelial membrane?
4.24 What is the name of the connective tissue membrane that lines joint cavities?

 FOCUS ON AGING

Because tissues consist of cells, cellular aging alters the tissues formed by those cells. There is a general loss of fluid in the tissues. Most of this dehydration is from intracellular fluid, but there is also a reduction in extracellular fluid. Elastic tissues lose some of their elasticity. Changes that occur in connective tissues exaggerate the normal curvatures of the spinal column and cause changes in the joints, in the arches of the foot, and in the intervertebral disks. These changes are manifested by a loss of height. Most organs show a loss of mass with aging because of tissue atrophy.

CHAPTER SUMMARY

Body Tissues

■ **Define the term *tissue*.**
- A tissue is a group of similar cells collected together by an intercellular matrix. Histology is the study of tissues.

■ **List the four types of tissues found in the body**
- The four main types of tissues in the body are epithelial, connective, muscle, and nerve tissue.

■ **Describe the general characteristics of epithelial tissue.**
- Epithelial tissues consist of tightly packed cells with little intercellular matrix; they have one free surface, are avascular, and reproduce readily. They cover the body, line body cavities, and cover organs within body cavities. The cells may be squamous, cubital, or columnar in shape, and they may be arranged in single or multiple layers.

■ **Describe the various types of epithelial tissues in terms of structure, location, and function.**
- Simple squamous epithelium consists of a single layer of flat cells. It is found in the alveoli of the lungs and the glomerulus of the kidneys, where it is well suited for diffusion and filtration.
- Simple cuboidal epithelium consists of a single layer of cells shaped like a cube; it is found in the kidney tubules, where it functions in absorption, and in glandular tissue, where it functions in secretion.
- Simple columnar epithelium consists of a single layer of cells that are taller than they are wide; it lines the stomach and intestines, where it functions in secretion and absorption; microvilli and goblet cells are frequently found in this tissue.
- Pseudostratified columnar epithelium appears to be stratified but is not; all cells are attached to the basement membrane but not all reach the surface; cilia and goblet cells are often associated with this tissue; it lines portions of the respiratory tract and some of the tubes of the reproductive tract.
- Stratified squamous epithelium consists of several layers of cells, and the ones at the surface are flat, squamous cells; protection is the primary function; it forms the outer layer of the skin and the linings of the mouth, anus, and vagina.
- Transitional epithelium has several layers but can be stretched in response to tension; it is found in the lining of the urinary bladder.

■ **Describe the classification of glandular epithelium according to structure and method of secretion; give an example of each type.**
- Exocrine glands secrete their product onto a free surface through a duct; endocrine glands are ductless glands and secrete their products into the blood.
- Goblet cells are unicellular exocrine glands; other exocrine glands are multicellular.
- The ducts of simple glands have no branches; compound glands have branched ducts; tubular glands have a constant diameter; acinar glands have a saccular distal end.
- Merocrine glands lose no cytoplasm with their secretion; this includes salivary glands, pancreatic glands, and most sweet glands.
- In holocrine glands, the entire cell is discharged with the secretory product; includes sebaceous (oil) glands.

■ **Describe the general characteristics of connective tissue.**
- Connective tissue has an abundance of intercellular matrix with relatively few cells.
- Strong and flexible collagenous fibers and yellow elastic fibers are frequently found in the matrix.

■ **Name three types of connective tissue cells and state the function of each one.**
- Fibroblasts produce the fibers found in connective tissue.
- Macrophages are phagocytic connective tissue cells that clean up cellular debris and foreign particles.
- Mast cells contain heparin, anticoagulant, and histamine, a substance that promotes inflammation.

■ **Describe the features and location of the various types of connective tissue.**
- Loose connective tissue is characterized by a loose network of collagenous and elastic fibers and a variety of connective tissue cells; the predominant cell is the fibroblast; it fills spaces in the body and binds together structures.
- Adipose tissue is commonly called fat; it forms a protective cushion around certain organs, provides insulation, and is an efficient energy storage material.
- Dense fibrous connective tissue is characterized by densely packed collagenous fibers in a matrix; it has a poor vascular supply and forms tendons and ligaments.
- Cartilage has a matrix that contains the protein chondrin and cells called chondrocytes that are located in spaces called lacunae; it is typically surrounded by a fibrous connective tissue membrane called perichondrium; blood vessels do not penetrate cartilage so cellular reproduction and healing occur slowly; most common type is hyaline cartilage, bound at the ends of long bones, trachea, costal cartilages, and fetal skeleton; fibrocartilage has an abundance of collagenous fibers in the matrix and is found in the intervertebral disks; elastic cartilage has an abundance of elastic fibers in the matrix and is found in the framework of the external ear.
- Bone, or osseous tissue, is a rigid connective tissue with mineral salts in the matrix to give strength and hardness; its structural unit is the osteon, or haversian system.
- Blood is a connective tissue that has a liquid matrix called plasma; its cells are the erythrocytes, which transport oxygen; leukocytes, which fight disease; and thrombocytes, which function in blood clotting.

■ **Describe the general characteristics of muscle tissue.**
- Muscle tissue has an abundance of cells and is highly vascular.
- Muscle cells are long and slender and are arranged in bundles.
- The cell membrane is called the sarcolemma, and the cytoplasm is called the sarcoplasm.
- Actin and myosin are contractile protein microfilaments in the sarcoplasm (cytoplasm).

■ **Distinguish between skeletal muscle, smooth muscle and cardiac muscle in terms of structure, location, and control.**
- Skeletal muscle fibers are cylindrical, multinucleated, striated, and under voluntary control; skeletal muscles are attached to the skeleton.

- Smooth muscle cells are spindle shaped, have a single centrally located nucleus, lack striations, and are involuntary. Smooth muscle is found in the walls of blood vessels and internal organs.
- Cardiac muscle has branching fibers, with one nucleus per cell, striations, and intercalated disks, and are involuntary. Cardiac muscle is in the wall of the heart.

■ **Name two categories of cells in nerve tissue and state their general functions.**
- Neurons and neuroglia are the cells found in nerve tissue.
- Neurons are the conducting cells of nerve tissue; they have a cell body with efferent processes called axons and afferent processes called dendrites.
- Neuroglia are the supporting cells of nerve tissue; they do not conduct impulses.

Inflammation and Tissue Repair

■ **Describe the four manifestations of inflammation and how they develop.**
- The four manifestations of inflammation are redness, swelling, heat, and pain.
- Blood vessel dilation increases blood flow to the area which causes the redness and heat.
- Increased vascular permeability results in an accumulation of fluid in the tissue spaces, which accounts for the swelling. The swelling puts pressure on the nerves to cause pain.

■ **Differentiate between regeneration and fibrosis by describing each process.**
- Regeneration is the replacement of destroyed tissue by cells that are identical to the original tissue cells.

- Fibrosis is the replacement of destroyed tissue by formation of fibrous connective tissue (scar tissue).
- Most tissue repair is a combination of regeneration and fibrosis.

Body Membranes

■ **Differentiate between epithelial membranes and connective tissue membranes.**
- Epithelial membranes are formed from epithelial tissue and connective tissue to which it is attached. Connective tissue membranes contain only connective tissue.

■ **Describe four types of membranes and specify the location and function of each.**
- Mucous membranes are epithelial membranes that line body cavities that open to the outside, such as the mouth, stomach, intestines, urinary bladder, and respiratory tract; they secrete mucus for protection and lubrication.
- Serous membranes are epithelial membranes that line body cavities that do not open to the outside and also cover the organs within these cavities. Serous membranes always consist of two layers, the visceral layer around organs and the parietal layer, which lines cavities. Serous fluid is secreted between the two layers. The pleura is the serous membrane around the lungs, pericardium is around the heart, and peritoneum is in the abdomen.
- Synovial membranes are connective tissue membranes that line joint cavities and secrete a synovial fluid into the joint cavity for lubrication.
- Meninges are connective tissue membranes around the brain and spinal cord. Dura mater is the outer layer of the meninges, arachnoid is the middle layer, and pia mater is the innermost layer.

CHAPTER QUIZ

Recall

. . . .

Match the definitions on the left with the appropriate term on the right.

_____	**1.** Flat cells	**A.** Adipose
_____	**2.** More than one layer of cells	**B.** Collagenous
_____	**3.** Unicellular glands that produce mucus	**C.** Goblet cell
_____	**4.** Entire cell discharged with secretion	**D.** Holocrine
_____	**5.** Strong, flexible connective tissue fibers	**E.** Macrophage
_____	**6.** Large, phagocytic cell	**F.** Myosin
_____	**7.** Fat tissue	**G.** Neuron
_____	**8.** Bone cell	**H.** Osteocyte
_____	**9.** Nerve cell	**I.** Squamous
_____	**10.** Protein in muscle	**J.** Stratified

Thought

. . . .

1. A tissue is composed of a single layer of flat cells, packed closely together. This tissue is likely to be found:

    A. Covering the outside of the body

    B. Forming the alveoli or air sacs of the lungs

    C. Lining the intestines

    D. As packing between neurons in the brain

2. A gland in which a portion of the secretory cell is pinched off and released with the secretion is a(n):

    A. Merocrine gland

    B. Sebaceous gland

    C. Apocrine gland

    D. Goblet cell

3. Tendons and ligaments are formed from:

   A. Dense fibrous connective tissue

   B. Osseous tissue

   C. Hyaline cartilage

   D. Extensions of muscle tissue

4. Muscle tissue that is spindle shaped with tapered ends and no apparent striations is:

   A. Found in the heart wall

   B. Called skeletal muscle

   C. Found in the wall of the stomach

   D. Usually attached to bones

5. Meninges are connective tissue membranes that:

   A. Secrete a serous fluid

   B. Line joint cavities

   C. Line the respiratory tract

   D. Cover the brain and spinal cord

Application

Approximately 85% of all cancers are carcinomas, meaning that they arise from epithelial tissue. Why do you think there are so many more carcinomas than there are sarcomas that arise from connective tissue?

BUILDING YOUR MEDICAL VOCABULARY

Building a vocabulary is a cumulative process. As you progress through this book, you will use word parts, abbreviations, and clinical terms from previous chapters. Each chapter will present new word parts, abbreviations, and terms to add to your expanding vocabulary.

Word Parts and Combining Forms with Definitions and Examples

PART/ COMBINING FORM	DEFINITION	EXAMPLE
a-	without, lacking	avascular: without blood vessels
aden/o	gland	adenoma: tumor of a gland
adip/o	fat	adipose: fat tissue
ambi-	both sides	ambilateral: pertaining to or affecting both sides
auscult-	listen	auscultation: listening to sounds produced with the body
-blast	to form, sprout	fibroblast: a connective tissue cell that forms or produces fibers
carcin/o	cancer	carcinogen: a substance that causes cancer
chrondr-	cartilage	perichondrium: membrane or tissue around cartilage
cohes-	stick together	cohesion: the sticking together of separate particles into a single mass
erythr/o	red	erythrocyte: red blood cell
fibr/o	fiber	fibroadenia: fibrous degeneration of glandular tissue
-glia-	glue	neuroglia: nerve glue; neural connective tissue
hist/o	tissue	histology: study of tissues
inocul-	implant, introduce	inoculation: introduction of pathogens into the body to produce immunity
leuk/o-	white	leukocyte: white blood cell
macro-	large	macrophage: a large phagocytic cell
multi-	many	multilobed: having many lobes
neur/o	nerve	neurolemma: membranous sheath around nerve fibers
-oma	tumor, swelling,	adenoma: tumor of a gland mass
os-	bone	os coxae: hip bone
os-	mouth, opening	ostium: an opening into a tubular organ
oste/o	bone	osteocyte: bone cell
pre-	before, in front of	preauricular: in front of the ear

PART/ COMBINING FORM	DEFINITION	EXAMPLE
pseud/o	false	pseudoplegia: false paralysis, hysterical paralysis
squam-	flattened, scale	squamous cells: flattened cells of epithelial tissue
strat-	layer	stratified: arranged in layers
stric-	narrowing	stricture: narrowing of a duct or passage
thromb/o	clot	thrombus: a blood clot
tox-	poison	toxin: a poisonous substance
vacu-	empty	vacuole: a space or cavity within the cytoplasm of the cell
vas/o	vessel	vascular: pertaining to or full of vessels

Clinical Abbreviations

ABBREVIATION	MEANING
ac	before meals
ASA	aspirin
cath	catheter, catheterization
ht	height
IM	intramuscular
LPN	licensed practical nurse
NS	normal saline
OD	overdose
OR	operating room
pc	after meals
PE	physical exam (with appropriate context)

ABBREVIATION	MEANING
PE	pulmonary embolus (with appropriate context)
PMH	past medical history
PO	by mouth
PWB	partial weight bearing

Clinical Terms

Adhesion (add-HEE-shun) Abnormal joining of tissues by fibrous scar tissue

Biopsy (BYE-ahp-see) Removal and microscopic examination of body tissue

Carcinoma (kar-sih-NOH-mah) Malignant growth derived from epithelial cells

Histology (hiss-TAHL-oh-jee) Branch of microscopic anatomy that studies tissues

Lipoma (lih-POH-mah) Benign tumor derived from fat cells

Marfan syndrome (mahr-FAHN SIN-drohm) Congenital disorder of connective tissue characterized by abnormal length of the extremities and cardiovascular abnormalities

Myoma (mye-OH-mah) Benign tumor formed of muscle tissue

Papilloma (pap-ih-LOH-mah) Benign epithelial tumor; may occur on any epithelial surface or lining

Pathology (pah-THAHL-oh-jee) Branch of medicine that studies the essential nature of disease, especially the structural and functional changes in tissues

Sarcoma (sar-KOH-mah) Malignant growth derived from connective tissue cells

Scurvy (SKUR-vee) Condition caused by a deficiency of vitamin C in the diet, which results in abnormal collagen synthesis

Systemic lupus erythematosus (sih-STEM-ik LOO-pus air-ith-eh-mah-TOH-sis) Chronic autoimmune connective tissue disease that is characterized by injury to the skin, joints, kidneys, nervous system, and mucous membranes but can affect any organ of the body

VOCABULARY QUIZ

Use word parts you have learned to form words that have the following definitions.

1. Cartilage cell _____

2. Tumor of a gland _____

3. Cell that forms clots _____

4. Large phagocytic cell _____

5. Nerve glue _____

Using the definitions of word parts you have learned, define the following words.

6. Avascular _____

7. Fibroblast _____

8. Erythrocyte _____

9. Stratified _____

10. Squamous _____

Match each of the following definitions with the correct word.

_____ **11.** Appears to have layers but

_____ **12.** Tumor of epithelial tissue

_____ **13.** Fat tissue

_____ **14.** Osseous tissue

_____ **15.** Tumor of neuroglia cells

A. Adipose does not
B. Bone
C. Carcinoma
D. Glioma
E. Pseudostratified

16. What is the clinical term for a "benign tumor formed of muscle tissue"? _____

17. A patient has a benign tumor derived from fat cells. What is the clinical term for this condition? _____

18. What is the work of a histologist? _____

19. What is the meaning of the abbreviation pc? _____

20. What is the meaning of the abbreviation ASA? _____

Integumentary System

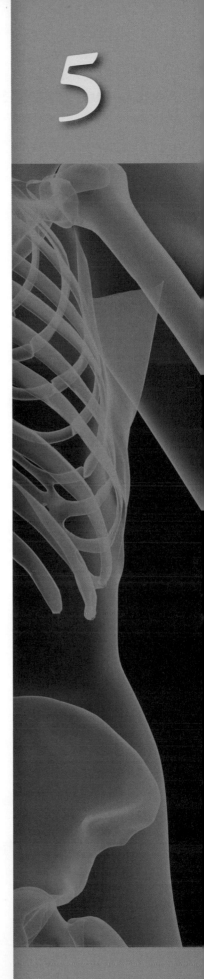

CHAPTER OBJECTIVES

Structure of the Skin
- Describe the structure of the epidermis.
- Describe the structure of the dermis.
- Name the supporting layer of the skin and describe its structure.

Skin Color
- Discuss three factors that influence skin color.

Epidermal Derivatives
- Describe the structure of hair and hair follicles and their relationship to the skin.

- Describe the structure of mails and how they are related to the skin.
- Discuss the characteristics and functions of the glands that are associated with the skin.

Functions of the Skin
- Discuss four functions of the integumentary system.

Burns
- Discuss the characteristics of first-, second-, and third-degree burns.
- Explain the rule of nines.

KEY TERMS

Arrector pili (ah-REK-tor PY lee) Muscle associated with hair follicles

Ceruminous gland (see-ROOM-in-us GLAND) Gland in the ear canal that produces cerumen or ear wax

Dermis (DER-mis) Inner layer of the skin that contains the blood vessels, nerves, glands, and hair follicles; also called stratum corium

Epidermis (ep-ih-DER-mis) Outermost layer of the skin

Keratinization (ker-ah-tin-ih-ZAY-shun) Process by which the cells of the epidermis become filled with keratin and move to the surface, where they are sloughed off

Melanin (MEL-ah-nin) A dark brown or black pigment found in parts of the body, especially skin and hair

Sebaceous gland (see-BAY-shus GLAND) Oil gland of the skin that produces sebum or body oil

Subcutaneous layer (sub-kyoo-TAY-nee-us LAY-ER) Below the skin; a sheet of areolar connective tissue and adipose tissue beneath the dermis of the skin; also called hypodermis or superficial fascia

Sudoriferous gland (soo-door-IF-er-us GLAND) Gland in the skin that produces perspiration; also called sweat gland

Functional Relationships of the
Integumentary System

Provides a barrier against hazardous materials and pathogens.

Reproductive
Gonads provide hormones that promote growth, maturation, and maintenance of skin.

Skin forms scrotum that protects testes; tactile receptors in skin provide sensations associated with sexual behaviors.

Urinary
Eliminates metabolic wastes and maintains normal body fluid composition.

Alternative excretory route for some salts and nitrogenous wastes; limits fluid loss.

Digestive
Provides nutrients needed for skin growth, maintenance, and repair.

Provides vitamin D for intestinal absorption of calcium.

Respiratory
Furnishes oxygen and removes carbon dioxide by gaseous exchange with blood.

Hairs of nasal cavity filter particles that may damage the upper respiratory tract.

Lymphatic/Immune
Prevents loss of interstitial fluid from skin, protects against skin infection, and promotes tissue repair.

Prevents pathogen entry; connective tissue cells in the skin activate the immune response.

Skeletal
Provides structural support.

Synthesizes vitamin D needed for calcium absorption and metabolism for bone growth and maintenance.

Muscular
Generates heat to warm the skin, muscle contraction pulls on skin to produce facial expressions.

Synthesizes vitamin D needed for absorption and metabolism of calcium essential for muscle contraction.

Nervous
Controls diameter of cutaneous blood vessels and sweat gland activity for temperature regulation.

Dermis contains receptors that detect stimuli related to touch, pressure, pain, and temperature.

Endocrine
Sex hormones influence hair growth, sebaceous gland activity, and distribution of subcutaneous adipose.

Synthesizes vitamin D needed for the absorption and metabolism of calcium, which acts as a messenger in some hormone actions.

Cardiovascular
Transports gases, nutrients, wastes, and hormones to and from the skin; hemoglobin provides color.

Prevents fluid loss from the blood; vasoconstriction of dermal vessels diverts blood flow to other organs.

← ─── Gives to Integumentary System
→ ─── Receives from Integumentary System

The skin and the glands, hair, nails, and other structures that are derived from it make up the **integumentary** (in-teg-yoo-MEN-tar-ee) **system.** Because it is on the outside of the body, this is the organ system with which we are most familiar. This is our contact with the external environment. It is the part that is, at least partially, exposed so others can see it. There is probably no other system that receives as much personal attention. We spend millions of dollars each year in attempts to "beautify" the integumentary system through creams, lotions, oils, color, rinses, conditioners, permanents, manicures, pedicures, polishes, and more. The list goes on and on. Yet because it is on the outside of the body, this system is subjected to continual abuse in the form of bumps, abrasions, cuts, scrapes, toxic chemicals, pollutants, wind, and sun. Fortunately, the skin is resilient and versatile. Generally, it quickly repairs itself and continues to perform its many functions year after year.

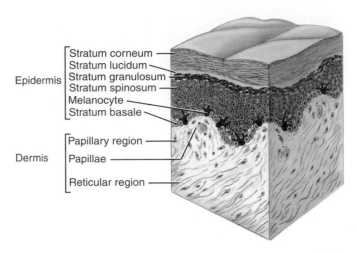

Figure 5-2 *Subdivisions of the epidermis and dermis. Note the five strata of the epidermis and the two layers in the dermis.*

 QUICK **APPLICATIONS**

For an "average" person, the skin weighs about 5 kilograms, has a surface area of approximately 2 square meters, and varies in thickness from 0.05 to 0.4 centimeter.

Structure of the Skin

The skin, sometimes called the **cutaneous** (kyoo-TAY-nee-us) **membrane,** consists of two distinct layers of tissues. The outer layer is the **epidermis,** and the inner layer is the **dermis.** These are anchored to underlying structures by a third layer, the **subcutaneous tissue.**

Figures 5-1 and 5-2 illustrate the structure of the skin.

Epidermis

The outer layer of the skin is the **epidermis.** This layer is stratified squamous epithelium (see Figure 5-1). Because the epidermis is epithelium, there are no blood vessels present and the cells receive their nutrients, by diffusion, from vessels in the underlying tissue. The cells on the bottom, near the basement membrane, receive adequate nutrients and actively grow and divide. As cells are pushed upward toward the surface by the growing cells next to the basement membrane, they receive fewer nutrients. They also undergo a process called **keratinization** (ker-ah-tin-ih-ZAY-shun). During keratinization, a protein called *keratin* is deposited in the cell, the chemical composition of the cell changes, and the cell changes shape. By the time the cells reach the surface, they are flat or squamous. The surface cells die from lack of nutrients and are sloughed off. They are replaced by other cells that are pushed upward from below. As cells are pushed upward, away from the nutrient supply, and become keratinized, they take on different appearances and characteristics to form distinct regions. In thick skin, such as that on the soles of the feet and palms of the hand, the epidermis consists of five regions, or strata, of cells (see Figure 5-2). In the skin that covers the rest of the body, the regions are thinner and there are only four strata. The stratum lucidum is present only in thick skin.

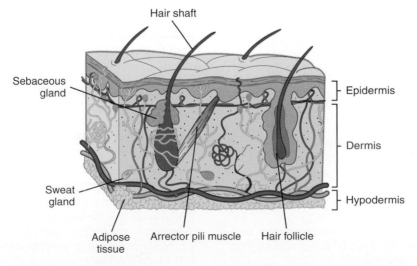

Figure 5-1 *Structure of the skin. Note the epidermis, dermis, and hypodermis.*

The bottom row of cells in the stratified squamous epithelium that makes up the epidermis consists of actively dividing (mitotic) columnar cells. This layer is the **stratum basale** (BAY-sah-lee), the layer next to the basement membrane and closest to the blood supply (see Figure 5-2). About one fourth of the cells in the stratum basale are **melanocytes** (meh-LAN-oh-sytes). These are specialized epithelial cells that produce a dark pigment called **melanin** (MEL-ah-nin). All individuals have the same number of melanocytes. However, melanocyte activity (the amount of melanin produced) differs according to genetic and environmental factors.

The **stratum spinosum** (spy-NOH-sum) consists of several layers of cells immediately above the stratum basale (see Figure 5-2). These cells have slender projections, or spiny processes, that connect them with other cells. When a cell in the stratum basale undergoes mitosis and divides into two cells, one of the daughter cells remains in the stratum basale and the other one is pushed upward into the stratum spinosum. Because the cells in the stratum spinosum show limited mitotic ability, this layer combined with the stratum basale is called the **stratum germinativum** (JER-mih-nah-tiv-um).

The **stratum granulosum** (gran-yoo-LOH-sum) is a very thin region consisting of two or three layers of flattened cells (see Figure 5-2). Keratinization begins in this layer. The cells appear granular, which gives this layer its name.

The **stratum lucidum** (LOO-sih-dum) appears as a translucent band just above the stratum granulosum (see Figure 5-2). It consists of a few layers of flattened, anucleate cells. The stratum lucidum is present only in thick skin.

The **stratum corneum** (KOR-nee-um) is the outermost or surface region of the epidermis and constitutes about three fourths of the epidermal thickness. It consists of 20 to 30 layers of flattened, dead, completely keratinized cells (see Figure 5-2). The cells in the stratum corneum are continually shed and replaced. About 5 weeks after a cell has been produced in the stratum basale, it is sloughed off the surface of the stratum corneum. The keratin that is present is a tough, water-repellent protein, and its inclusion in the stratum corneum provides protection against water loss from the body.

QUICK CHECK

5.1 The epidermis is constantly sloughed off and replaced. What layer of the epidermis is the source of new cells for replacement?

5.2 Which layer of the epidermis is found only in the thick skin of the soles of the feet and palms of the hands? It is not found in thin skin.

Dermis

The **dermis,** or **stratum corium** (KOR-ee-um), is dense connective tissue that is deeper and usually thicker than the epidermis (see Figure 5-1). Hair, nails, and certain glands, although derived from the stratum basale of the epidermis, are embedded in the dermis. The dermis contains both collagenous and elastic fibers to give it strength and elasticity. If the skin is overstretched, the dermis may be damaged, leaving white scars called **striae** (STRY-ee), commonly called "stretch marks." Fibers also form a framework for the numerous blood vessels and nerves that are present in the dermis but generally absent in the epidermis. Many of the nerves in the dermis have specialized endings called **sensory receptors** that detect changes in the environment such as heat, cold, pain, pressure, and touch. Because there are no receptors or nerves in the epidermis, these receptors in the dermis are the body's contact with the environment.

The dermis is divided into two indistinct layers. The upper **papillary** (PAP-ih-lair-ee) **layer** derives its name from the numerous **papillae,** or projections, that extend into the epidermis (see Figure 5-2). Blood vessels, nerve endings, and sensory receptors extend into the papillae to bring them into closer proximity to the epidermis and the surface. On the palms, fingertips, and the soles of the feet, the papillae form distinct patterns or ridges that provide friction for grasping objects. The patterns are genetically determined and are unique for each individual. These are the basis of fingerprints and footprints.

The **reticular** (reh-TICK-yoo-lar) **layer** of the dermis is deeper and thicker than the papillary layer (see Figure 5-2). Bundles of connective tissue fibers run in many different directions to provide the strength and resilience needed to stretch in many planes. More bundles usually run in one direction than in the others, and this produces **cleavage lines.** Incisions across cleavage lines tend to gape more and produce more scar tissue than those that are parallel to the cleavage lines.

QUICK APPLICATIONS

The dermis is the portion of an animal's skin that is used to make leather, because the collagen in the dermis becomes very tough when treated with tannic acid.

QUICK APPLICATIONS

A blister is a fluid-filled pocket between the dermis and the epidermis. When the skin is burned or irritated, some plasma escapes from the blood vessels in the dermis and accumulates between the two layers, where it forms the blister.

QUICK APPLICATIONS

Prolonged exposure to the harmful radiation in sunlight is associated with three types of skin cancer. Basal cell carcinoma results when ultraviolet (UV) exposure damages cells in the stratum basal and the cells multiply out of control. It usually develops slowly and does not metastasize. Squamous cell carcinoma may be due to exposure to UV or to carcinogens such as tar and oil chemicals. This type appears as an ulcer-like sore and can metastasize. Malignant melanoma is caused by radiation damage to melanocytes and the cells begin to proliferate uncontrollably. The cancer may develop from an existing mole or as a separate entity. Malignant melanoma is fast growing and readily metastasizes to the lung, liver, and brain. All three of these skin cancers require prompt medical attention.

QUICK APPLICATIONS

A pressure ulcer, also called a decubitus ulcer or bedsore, develops when you remain in one position for too long. The constant pressure reduces the blood supply to the skin and the affected tissue in that area dies. The most common sites for pressure ulcers is over bones that are close to the skin such as the elbow, hip, knee, shoulders, and back of the head.

Subcutaneous Layer

The **subcutaneous layer** (see Figure 5-1) is not actually a part of the skin, but it loosely anchors the skin to underlying organs. Because it is beneath the dermis, it is sometimes called the **hypodermis.** It is also referred to as **superficial fascia.** The subcutaneous layer consists largely of loose connective tissue and adipose tissue. The fibers in the loose connective tissue are continuous with those in the dermis and, as a result, there is no distinct boundary between the dermis and the subcutaneous tissue.

The adipose tissue in the subcutaneous layer cushions the underlying organs from mechanical shock and acts as a heat insulator in temperature regulation. Fat in the adipose tissue can be mobilized and used for energy when necessary. The distribution of subcutaneous adipose tissue is largely responsible for the differences in body contours between men and women.

Skin Color

Skin color is a result of many factors: some genetic, some physiologic, and some environmental. Basic skin color is caused by the dark pigment **melanin** produced by the melanocytes in the stratum basale of the epidermis. Melanocytes have long slender processes by which they transfer the melanin to surrounding cells in the skin and in hair. Everyone has about the same number of melanocytes. The activity of the melanocytes, however, is genetically controlled. A large number of melanin granules results in dark skin; fewer granules result in lighter skin. Although many genes are responsible for skin color, a single mutation can result in an inability to produce melanin. This results in a condition called **albinism** (AL-bih-nizm) in which individuals have very light skin, white hair, and unpigmented irises in the eyes.

Some people have the yellowish pigment **carotene** (KAIR-oh-teen) in addition to melanin. This gives a yellow tint to the skin. A pinkish tint in the skin is attributable to the blood vessels in the dermis. Ultraviolet light increases melanocyte activity so that more melanin is produced and the skin becomes darker or tanned.

QUICK APPLICATIONS

In people with light skin, when dermal blood vessels dilate and blood flow increases (e.g., during blushing and increased temperature), the skin may be quite red. If the vessels constrict and blood flow decreases, the individual is pale or "white as a sheet."

QUICK CHECK

5.3 Sensory receptors and blood vessels are embedded in which layer of the skin?
5.4 Melanin is the basic dark pigment responsible for skin color. Where in the skin is this produced?

Epidermal Derivatives

Accessory structures of the skin include hair, nails, sweat glands, and sebaceous glands. They are derived from the stratum basale of the epidermis and are embedded in the reticular layer of the dermis. Figure 5-1 illustrates some of the accessory structures associated with the skin.

Hair and Hair Follicles

Hair is found on nearly all body surfaces. All hair has essentially the same structure. It consists of a shaft and a root that are composed of dead, keratinized epithelial cells. The root is enclosed in a hair follicle that extends through the epidermis and is embedded in the dermis. The structure of hair and hair follicles is illustrated in Figure 5-3.

The **shaft** of a hair is that portion that extends beyond the surface of the epidermis. It is the part that you can see. Because it contains no nerves, it can be cut with no sensation of pain. The **root** is the portion of the hair that is below the surface of the skin. It is surrounded by a hair follicle. The shaft and root are continuous and together make up the hair, which is produced by the hair follicle. The central core of a hair is the **medulla.** This is surrounded by several layers of cells called the **cortex.** The outermost covering is a single layer of overlapping, keratinized cells called the **cuticle.** On the shaft of the hair, the

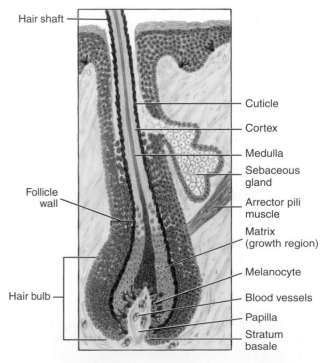

Figure 5-3 *Structure of hair and hair follicles.*

cuticle is exposed to the environment and subjected to abrasion. It tends to wear away at the tip of the shaft. When this happens, keratin fibers in the cells of the cortex and medulla project from the tip of the shaft, resulting in "split ends."

The root of a hair is enclosed in a tubular **hair follicle.** At the end deep in the dermis, the follicle expands to form a **hair bulb.** A **papilla** of dermis with a cluster of blood vessels projects into the center of the bulb. This provides the blood supply for the epithelial cells. As the follicle nears the hair bulb, the follicle becomes thinner so that only a single layer of stratum basale cells covers the papilla. The stratum basale cells, like those in the skin, provide the mitotic cells that divide and undergo keratinization to produce the hair. The stratum basale cells of the hair follicle can, if necessary, grow onto the surface of the skin to repair wounds there.

Hair color is determined by the type of melanin produced by the melanocytes in the stratum basale. Yellow, brown, and black pigments are present in varying proportions to produce different hair colors. With age, the melanocytes become less active. Hair in which melanin is replaced with air bubbles is white.

A bundle of smooth muscle cells, called the **arrector pili** (ah-REK-tor PY-lee) **muscle,** is associated with each hair follicle. Most hair follicles are at a slight angle to the surface of the skin. The arrector pili muscles are attached to the hair follicles in such a way that contraction pulls the hair follicles into an upright position or causes the hair to "stand on end." Contraction of the arrector pili muscles also causes raised areas on the skin, or "goose bumps." Action of the arrector pili muscles is controlled by the nervous system in response to cold and fright.

QUICK APPLICATIONS

The shape of the hair shaft determines whether hair is straight or curly. If the shaft is round, the hair is straight. If it is oval, the hair is wavy. If it is flat, the hair is curly or kinky. A permanent flattens the hair to make it curly.

QUICK CHECK

5.5 What portion of a hair extends beyond the surface of the epithelium and is visible?

5.6 What is the name of the band of smooth muscle fibers associated with a hair follicle?

Nails

Nails are thin plates of dead stratum corneum that contain a very hard type of keratin and cover the dorsal surfaces of the distal ends of the fingers and toes (Figure 5-4). Each nail has a **free edge;** a **nail body,** which is the visible portion; and a **nail root,** which is covered with skin. The **eponychium** (eh-poh-NICK-ee-um), or **cuticle,** is a fold of stratum corneum that grows onto the proximal portion of the nail body. Stratum basale from the epidermis grows under the nail body to form the **nail bed.** This is thickened at the proximal end to form the **nail matrix,** which is responsible for nail growth. As

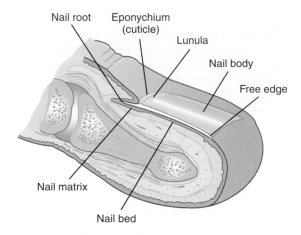

Figure 5-4 *Structure of a nail.*

cells are produced by the matrix, they become keratinized and slide over the nail bed. The portion of the body over the matrix appears as a whitish, crescent-shaped area called the **lunula** (LOO-nyoo-lah). Nails appear pink because of the rich supply of blood vessels in the underlying dermis.

Glands

The two major glands associated with the skin are the **sebaceous** (see-BAY-shus) **glands** and the **sweat glands.** A third type, the **ceruminous** (see-ROOM-ih-nus) **glands,** are modified sweat glands.

Sebaceous Glands

Most sebaceous glands are associated with hair follicles and are found in all areas of the body that have hair (see Figures 5-1 and 5-3). Those not associated with hair follicles open directly onto the surface of the skin. Functionally, sebaceous glands are holocrine glands. This means that the glandular cells are destroyed and released along with the secretory product. The oily secretion, called **sebum,** is transported by a duct into a hair follicle, and from there it reaches the surface of the skin. Sebum functions to keep hair and skin soft and pliable. It also inhibits growth of bacteria on the skin and helps to prevent water loss. Secretory activity of the sebaceous glands is stimulated by sex hormones; consequently, the glands are relatively inactive in childhood, become highly active during puberty, and decrease in activity during old age. Decreased sebum, in part, accounts for the dry skin and brittle hair that are common in older people.

QUICK APPLICATIONS

Acne is a problem that plagues many teenagers. Increased hormone activity at puberty causes an increase in sebaceous gland activity. Sebum and dead cells may block the hair follicle and form blackheads. Bacteria infect the blocked follicle, and the sebum/dead cell mixture accumulates until the follicle ruptures. This initiates an inflammatory response that soon appears on the surface as a pus-filled pimple.

Sweat (Sudoriferous) Glands

Sweat glands, also called **sudoriferous** (soo-door-IF-er-us) **glands,** are widely distributed over the body. They are most numerous in the palms and soles. It is estimated that an individual has more than 2.5 million sweat glands. There are two different types of sweat glands.

Merocrine sweat glands are the more numerous and more widely distributed of the two types. The glandular portion is a coiled tube that is embedded in the dermis of the skin, and the duct opens onto the surface of the skin through a **sweat pore** (see Figure 5-1). The secretion of these glands is primarily water with a few salts. When the body's temperature increases, the glands are stimulated to produce sweat, which evaporates and has a cooling effect. Sweat, or **perspiration,** is also produced in response to nerve stimulation as a result of emotional stress.

Apocrine sweat glands are larger than the merocrine glands, and their distribution is limited to the axillae and external genitalia. Their ducts open into the hair follicles in these regions. Their secretion consists of water and salts and also contains organic compounds such as fatty acids and proteins. These glands become active at puberty and are stimulated by the nervous system in response to pain, emotional stress, and sexual arousal. The secretion is odorless when released but is quickly broken down by bacteria to cause what is known as *body odor.*

Ceruminous Glands

Ceruminous glands are modified sweat glands that are found in the external auditory (ear) canal. They secrete an oily, sticky substance called **cerumen** (see-ROOM-men), or earwax, that is thought to repel insects and trap foreign material.

QUICK CHECK

5.7 Nails are derived from which layer of the epidermis?
5.8 Why do nails appear pink?
5.9 Which type of sweat gland is most widely distributed?

Functions of the Skin

Protection

The skin forms a protective covering over the entire body. The keratin in the cells waterproofs the skin and helps prevent fluid loss from the body. This waterproofing also prevents too much water from entering the body during swimming and bathing. Unbroken skin forms the first line of defense against bacteria and other invading organisms. The oily secretions of the sebaceous glands are acidic and inhibit bacterial growth on the skin. Melanin pigment absorbs light and helps protect underlying tissues from the damaging effects of ultraviolet light. Skin also protects underlying tissues from mechanical, chemical, and thermal injury.

Sensory Reception

The dermis contains numerous sensory receptors for heat, cold, pain, touch, and pressure. Even though hair itself has no sensory receptors, the movement of hair can be detected by receptors clustered around a hair follicle. The sensory receptors in the dermis relay information about the environment to the brain so that changes can be made to prevent or minimize injury. The sensory receptors are also a means of communication between individuals.

FROM THE PHARMACY

The "From the Pharmacy" boxes in Chapters 1 through 4 discussed some of the basic concepts of pharmacology. Beginning with this chapter, a few of the common preparations pertaining to the body system discussed in the chapter will be presented. Specific drugs are identified by generic name when appropriate. The brand name is given in parentheses.

Skin is in direct contact with the environment and, as a result, skin disorders are common. Because the skin is visible to ourselves and to others, skin lesions often have a psychological impact in addition to the medical concern. Numerous preparations are available to treat and, in some cases, prevent skin reactions. Many of these are available over the counter.

Prophylactic agents include sunscreen preparations. Extended exposure to ultraviolet rays from the sun may lead to premature aging of the skin. The skin damage may progress from minor irritation to precancerous lesions to some form of skin cancer. Sunscreen preparations are applied to either absorb or reflect the ultraviolet radiation. Those that reflect the rays are opaque and have to be applied as a thick paste, which makes them quite noticeable. People more commonly use preparations that absorb the rays. Sun protection factor (SPF) is the ratio of skin damage with and

without sunscreen. In general, a sunscreen with an SPF of 10 allows 10 times the exposure with the same skin damage that would occur without using sunscreen. The higher the SPF, the more protection the sunscreen provides.

Therapeutic agents are used to treat diseases of the skin such as infections, inflammations, dermatitis, and acne vulgaris. These agents include the anti-infectives, anti-inflammatory corticosteroids, topical anesthetics, and acne products. **Adapalene** (Differin) is a topical preparation that is an effective first-line medication for acne therapy. It is a vitamin A derivative and is available as a cream or gel. **Benzoyl peroxide–erythromycin** (Benzamycin) is a combination of the two drugs that may be used as a topical anti-infective agent to treat acne vulgaris. Benzoyl peroxide acts by releasing active oxygen, which inhibits anaerobic bacteria. It is a common agent in over-the-counter antiacne medications. Erythromycin acts by inhibiting protein synthesis in susceptible organisms. Erythromycin is also used orally in more severe acne or in cases that are resistant to therapy. **Hydrocortisone** (Hycort, Cortril) is a topical corticosteroid that may be used to treat the inflammation and itching of certain skin conditions. It is one of the few steroids that can be used on the face, but it should be used with caution. Preparations containing 1% hydrocortisone are sold over the counter.

Regulation of Body Temperature

Normally, body temperature is maintained at 37° C (98.6° F). It is important that body temperature be regulated because changes in temperature alter the speed of chemical reactions in the body. About 40% of the energy in a glucose molecule is stored in the bonds of adenosine triphosphate (ATP) to be used by the cells. The remaining 60% of the energy is in the form of heat to maintain body temperature. Often this amount is more than is necessary to maintain body temperature, and the excess heat must be removed from the body. The skin has two ways that it helps to regulate body temperature: by dilation and constriction of blood vessels and by activity or inactivity of the sweat glands. Both of these mechanisms are examples of negative feedback in maintaining homeostasis. The adipose tissue in the subcutaneous layer also helps by acting as an insulator.

When there is excess heat in the body, and body temperature rises above normal, the small arteries in the dermis dilate to increase blood flow through the skin. This brings the heat from the deeper tissues to the surface so that it can escape into the surrounding air. Sweat glands become active in response to increased body temperature. Moisture, in the form of sweat, accumulates on the surface of the skin and then evaporates to provide cooling.

If body temperature falls below normal, the sweat glands are inactive and the blood vessels in the skin constrict to reduce blood flow. This constriction of the blood vessels reduces the amount of heat that is transferred from the deeper tissues to the surface. However, if the skin becomes too cold, below 15° C (59° F), cutaneous blood vessels will begin to dilate to bring warm blood to the region. This process ensures that the tissues are not damaged by the cold. The role of the skin in temperature regulation is summarized in Figure 5-5.

QUICK APPLICATIONS

People who lose weight rapidly may feel cold because they have reduced their adipose insulation.

Synthesis of Vitamin D

Vitamin D is required for calcium and phosphorus absorption in the small intestine. The calcium and phosphorus are essential for normal bone metabolism and muscle function. Skin cells contain a precursor molecule that is converted to vitamin D when the precursor is exposed to ultraviolet rays in sunlight. It takes only a small amount of ultraviolet light to stimulate vitamin D production, so this should not be used as an excuse to expose the skin to sun unnecessarily and to risk the damage that may result.

QUICK CHECK

5.10 List four types of functions attributed to the integumentary system.
5.11 How does melanin in the stratum basale provide protection for the body?
5.12 Jake is playing a vigorous game of volleyball on a warm summer day. Describe two ways the integument helps maintain internal body temperature despite the heat and exercise.

Burns

A burn is tissue damage that results from heat, certain chemicals, radiation, or electricity. The seriousness of burns is a result of their effect on the skin. The most serious threat to survival after a severe burn is fluid loss because the waterproof protective

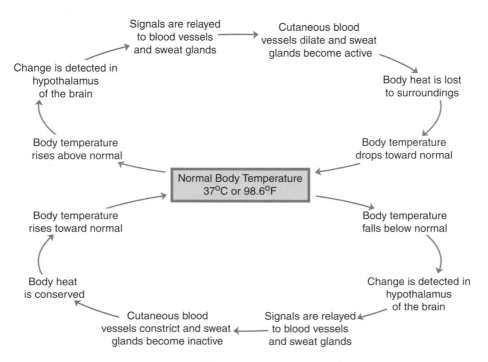

Figure 5-5 *Role of skin in temperature regulation. Changes in body temperature are detected in the hypothalamus and signals are relayed to blood vessels and sweat glands in the skin. The vessels and glands react to bring temperature closer to normal.*

Burn Type	Layers Involved	Appearance	Sensation	Prognosis
First Degree Superficial	Epidermis; stratum basale intact	Red and there may be some swelling	Painful	Regeneration occurs within a few days; no scarring
Second Degree Partial-thickness	Epidermis and a portion of the dermis	Redness, swelling, blisters	Painful	Heals within 2–4 weeks; deep second-degree burns produce scarring
Third Degree Full-thickness	Epidermis and dermis destroyed; burn extends into subcutaneous tissue or below	White or charred, no swelling	No pain because nerve endings are destroyed	Require skilled medical treatment because of problems associated with fluid loss and infection; skin grafts are necessary to protect area during healing; severe scarring

Figure 5-6 *Burn classification according to depth and tissues that are compromised.*

covering, the skin, is destroyed. As fluid seeps from the burned surfaces, electrolytes and proteins are also lost. The loss of fluids, electrolytes, and proteins leads to osmotic imbalances, renal failure, and circulatory shock. Another problem is the imminent danger of massive infection. A large, severely burned region is bacteria "heaven" because there is easy access, ideal growing conditions, and no means of attack by the immune system. Bacteria have easy access to tissues because the protective barrier provided by the skin is absent. The protein-rich fluid that seeps from burned areas is an ideal growth medium for bacteria, fungi, and other pathogens. Finally, the body's immune system, which normally fights off threats of disease, becomes exhausted within 2 or 3 days after a severe burn injury.

Burns are classified as first, second, or third degree according to their severity or depth. Figure 5-6 describes the characteristics of each type.

The treatment of burns depends on their severity and on the amount of surface area that is damaged. For this reason, it is helpful to be able to estimate quickly the amount of surface area that is burned. One easy method for making a quick estimate is called the **rule of nines.** In this method (Figure 5-7), each body region constitutes a percentage of the total that is some multiple of 9. Each upper limb is 9% of the total body surface area; each lower limb is 18%; the anterior and posterior trunk regions are each 18%; the head and neck together make up 9%; and the perineum is the final 1%.

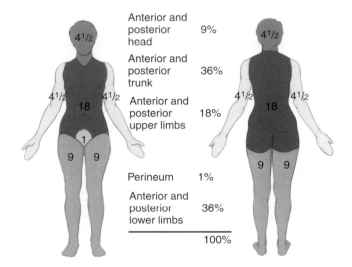

Anterior and posterior head	9%
Anterior and posterior trunk	36%
Anterior and posterior upper limbs	18%
Perineum	1%
Anterior and posterior lower limbs	36%
	100%

Figure 5-7 *Illustration of the rule of nines used to estimate the surface area that is burned.*

QUICK CHECK

5.13 What type of burn is painful and exhibits blisters?

5.14 If you receive a flash burn on your entire face, what percentage of the body has been burned?

✦ FOCUS ON AGING

As the skin ages, the number of elastic fibers decreases and adipose tissue is lost from the dermis and subcutaneous layer. This causes the skin to wrinkle and sag. Loss of collagen fibers in the dermis makes the skin more fragile and makes it heal more slowly. Mitotic activity in the stratum basale slows so that the skin becomes thinner and appears more transparent. Reduced sebaceous gland activity causes dry, itchy skin. Loss of adipose tissue in the subcutaneous layer and reduced sweat gland activity lead to an intolerance of cold and susceptibility to heat. The ability of the skin to regulate temperature is reduced. There is a general reduction in melanocyte

activity, which decreases protection from ultraviolet light, resulting in increased susceptibility to sunburn and skin cancer. Some melanocytes, however, may increase melanin production, resulting in "age spots."

Despite all the creams and "miracle" lotions, there is no known way to prevent skin from aging. Good nutrition and cleanliness may slow the aging process. Because skin that is exposed to sunlight ages more rapidly than unexposed skin, one of the best ways to slow the aging process is to avoid exposure by wearing protective clothing and by using sunblock whenever possible.

Representative Disorders of the
Integumentary System

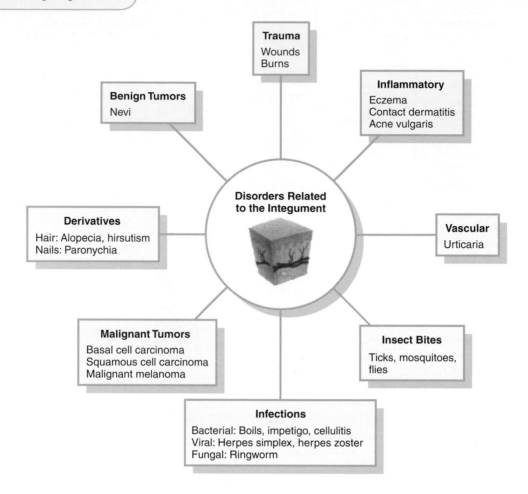

Trauma
Wounds
Burns

Benign Tumors
Nevi

Inflammatory
Eczema
Contact dermatitis
Acne vulgaris

Derivatives
Hair: Alopecia, hirsutism
Nails: Paronychia

**Disorders Related
to the Integument**

Vascular
Urticaria

Malignant Tumors
Basal cell carcinoma
Squamous cell carcinoma
Malignant melanoma

Insect Bites
Ticks, mosquitoes,
flies

Infections
Bacterial: Boils, impetigo, cellulitis
Viral: Herpes simplex, herpes zoster
Fungal: Ringworm

Acne vulgaris (ACK-nee vul-GAIR-is) Skin condition, usually occurring in adolescence, with blackheads, papules, nodules, and pustules on the face and upper trunk

Alopecia (al-oh-PEE-shee-ah) Absence of hair from skin areas where it normally grows; baldness; may be hereditary or due to disease, injury, or chemotherapy or may occur as part of aging

Basal cell carcinoma (BAY-sal SELL kar-sih-NOH-mah) Malignant tumor of the basal cell layer of the epidermis; most common form of skin cancer and usually grows slowly

Boils (BOYLZ) Local infection of the skin containing pus; a type of skin abscess; also called a furuncle

Burns (BURNZ) Damage to the skin caused by heat, chemicals, electricity, lightening, or ultraviolet rays (sun)

Cellulitis (sell-yoo-LYE-tis) Infection of connective tissue with severe inflammation of the dermis and subcutaneous layers of the skin

Contact Dermatitis (KAHN-takt der-mah-TYE-tis) Inflammation of the skin caused by direct contract between the skin and a substance to which an individual is sensitive; characterized by a rash, itching, swelling, blistering, oozing, and scaling; examples include poison ivy and poison oak

Eczema (ECK-zeh-mah) Inflammatory skin disease with red, itching, vesicular lesions that may crust over; common allergic reaction but may occur without any obvious cause

Herpes simplex (HER-peez SIM-plecks) Acute infectious viral disease characterized by watery blisters; cold sores, fever blisters

Herpes zoster (HER-peez ZAH-stir) Acute inflammation of spinal ganglia and characterized by vesicular eruptions on the skin along the route of sensory nerves; caused by the chickenpox virus; also called shingles

Hirsutism (her-SOO-tizm) Abnormal growth of hair

Impetigo (im-peh-TYE-go) Superficial skin infection caused by staphylococcal or streptococcal bacteria and characterized by vesicles, pustules, and crusted-over lesions; most common in children

Insect bites (IN-sekt BYTES) Injuries caused by the mouth parts and venom of insects (mosquitoes, flies, bees, etc.) and arachnids (spiders, ticks, scorpions)

Malignant melanoma (mah-LIG-nant mel-ah-NOH-mah) Cancerous growth composed of melanocytes; often arises in preexisting mole; an alarming increase in the prevalence of malignant melanoma is attributed to excessive exposure to sunlight

Nevus (NEE-vus) Elevated, pigmented lesion on the skin; commonly called a mole; a dysplastic nevus is a mole that does not form properly and may progress to a type of skin cancer; plural, nevi

Paronychia (pair-oh-NICK-ee-ah) Bacterial or fungal infection involving the folds of soft tissue around the fingernails

Ringworm (RING-werm) Fungal infection of the skin in which the fungus feeds on dead skin and perspiration; one type is athlete's foot

Squamous cell carcinoma (SKWAY-mas SELL kar-sih-NOH-mah) Type of skin cancer usually caused by exposure to ultraviolet rays or to carcinogens such as tar and oil chemicals, characterized by ulcer-like sores and metastasis

Urticaria (ur-tih-KAIR-ee-ah) Allergic transient skin eruptions characterized by elevated lesions, called wheals, and often accompanied by severe itching and burning; also called hives

Wart (WORT) Epidermal growth on the skin caused by a virus; plantar warts occur on the soles of the feet, juvenile warts occur on the hands and face of children, and venereal warts occur in the genital area

Wounds (WOONDZ) Any interruption of the continuity of the skin caused by physical means

CHAPTER SUMMARY

Structure of the Skin

■ Describe the structure of the epidermis.

- The integumentary system includes the skin with its glands, hair, and nails. The skin has an outer epidermis and an inner dermis. These are anchored to underlying tissues by the hypodermis and subcutaneous tissue.
- The epidermis is stratified squamous epithelium. In thick skin, it has five distinct regions.
- The bottom layer of the epidermis is the stratum basale. It is closest to the blood supply, is actively mitotic and contains melanocytes.
- The other layers are the stratum spinosum, stratum granulosum, stratum lucidum (present only in thick skin), and stratum corneum.
- The outermost layer, the stratum corneum, is continually sloughed off and replaced by cells from deeper layers.

■ Describe the structure of the dermis.

- The dermis, also called stratum corium, is composed of connective tissue with blood vessels, nerves, and accessory structures embedded in it. It consists of two layers: the upper papillary layer and the deeper reticular layer.

■ Name the supporting layer of the skin and describe its structure.

- The hypodermis, also called subcutaneous tissue, anchors the skin to the underlying muscles. The hypodermis is connective tissue with an abundance of adipose, which acts as a cushion, as a heat insulator, and can be used as an energy source.

Skin Color

■ Discuss three factors that influence skin color.

- Basic skin color is due to the amount of melanin produced by the melanocytes in the stratus basale.
- Carotene, a yellow pigment, gives a yellow tint to the skin.
- Blood in the dermal blood vessels gives a pink color to the skin.

Epidermal Derivates

■ Describe the structure of hair and hair follicles and their relationship to the skin.

- The central core of hair is the medulla, which is surrounded by the cortex and the cuticle. Hair is divided into the visible shaft and the root, which is embedded in the skin and surrounded by the follicle. The distal end of the hair follicle expands to form a bulb around a central papilla.

- Stratum basale cells in the bulb undergo mitosis to form hair, which increases the length of the hair. Hair color is determined by melanocytes in the stratum basale.
- Arrector pili muscles contract in response to cold and fear to make the hair stand on end.

■ Describe the structure of nails and how they are related to the skin.

- Nails are thin plates of keratinized stratum corneum. Each nail has a free edge, nail body, and a nail root. Other structures associated with the mail are the eponychium, nail bed, nail matrix, and lunula.
- Nails are derived from the stratum basale in the nail bed.

■ Discuss the characteristics and functions of the glands that are associated with the skin.

- Sebaceous glands are oil glands and are associated with hair follicles. These glands secrete sebum, which helps keep hair and skin soft and pliable and helps prevent water loss.
- Sweat glands are called suboriferous glands. Merocrine sweat glands open to the surface of the skin through sweat pores and secrete perspiration in response to nerve stimulation and in response to heat.
- Apocrine sweat glands are larger than merocrine glands and their distribution is limited to the axillae and external genitalia where they open into hair follicles. These glands become active at puberty and are stimulated in response to pain, emotional stress, and sexual arousal.
- Ceruminous glands are modified sweat glands found only in the external auditory canal where they secrete cerumen, or earwax.

Functions of the Skin

■ Discuss four functions of the integumentary system.

- Protection: The skin protects against water loss, invading organisms, ultraviolet light, and other injuries.
- Sensory reception: Sense receptors in the skin detect information about the environment and also serve as a means of communication between individuals.
- Regulation of body temperature: Constriction and dilation of blood vessels affect the amount of heat that escapes from the skin into the surrounding air. Sweat glands are stimulated in response to heat and are inactive in cold temperatures. Adipose in the subcutaneous tissue helps to insulate the body.
- Synthesis of vitamin D: Precursors for vitamin D are found in the skin. When the skin is exposed to ultraviolet light, these precursors are converted into active vitamin D.

Burns

■ **Discuss the characteristics of first-, second-, and third-degree burns.**

- First-degree burns are superficial, involve only the epidermis, are red and painful, and heal by regeneration.
- Second-degree burns are partial thickness burns, involve a portion of the dermis, form blister, are painful, and may produce scarring.
- Third-degree burns are full-thickness burns, which extend into the subcutaneous tissue or below. They are white or charred, cause no pain, present problems with fluid loss and infection, and produce severe scarring.

■ **Explain the rule of nines.**

- The rule of nines may be used to estimate the amount of body surface area that is burned.
- Each boy region constitutes a percentage of the total that is a multiple of 9.
- Head and neck total 9%, each upper extremity is 9%, each lower limb is 18%, the entire trunk is 36%, and the perineum is 1%.

CHAPTER QUIZ

Recall

Match the definitions on the left with the appropriate term on the right.

_____ **1.** Cells that produce a dark pigment
_____ **2.** Outermost layer of the skin
_____ **3.** Actively mitotic layer of the epidermis
_____ **4.** Subcutaneous layer
_____ **5.** Visible portion of hair
_____ **6.** Cuticle of fingernail
_____ **7.** Glands that open into hair follicles
_____ **8.** Muscle attached to hair follicles
_____ **9.** Tough, water-repellant protein in epidermis
_____ **10.** Upper layer of the dermis

A. Apocrine sweat glands
B. Arrector pili
C. Eponychium
D. Hair shaft
E. Hypodermis
F. Keratin
G. Melanocytes
H. Papillary layer
I. Stratum basale
J. Stratum corneum

Thought

1. Sandy Shore was walking barefoot along the beach when she stepped on a broken shell and cut her foot. The sequence in which the shell penetrated her foot was:
   A. Stratum corneum, stratum lucidum, stratum granulosum, stratum spinosum, stratum basale, stratum corium
   B. Stratum corium, stratum corneum, stratum basale, stratum granulosum, stratum spinosum, stratum lucidum
   C. Stratum corium, stratum granulosum, stratum lucidum, stratum spinosum, stratum basale, stratum corneum
   D. Stratum corneum, stratum granulosum, stratum spinosum, stratum lucidum, stratum basale, stratum corium

2. The layer of skin that is responsible for fingerprints and footprints is the:
   A. Stratum corneum   **C.** Reticular layer of dermis
   B. Stratum granulosum   **D.** Papillary layer of dermis

3. Receptors for the sensation of touch are located in the:
   A. Stratum corneum   **C.** Dermis
   B. Stratum basale   **D.** Hypodermis

4. In response to warm temperature:
   A. Cutaneous blood vessels dilate and sweat glands become active.
   B. Cutaneous blood vessels dilate and sweat glands become inactive.
   C. Cutaneous blood vessels constrict and sweat glands become active.
   D. Cutaneous blood vessels constrict and sweat glands become inactive.

5. Why are infants and the elderly more susceptible to temperature changes than other people?
   A. They have fewer sweat glands to carry away heat.
   B. They have less adipose tissue in the subcutaneous layer for insulation.
   C. They have fewer sebaceous glands to form sebum for insulation.
   D. Their blood vessels do not dilate and constrict as readily.

Application

Joyce is scheduled for surgery and is concerned about the scar that may result. Her surgeon explains that scarring will be minimal because the incision will be parallel to cleavage lines. Explain this so she will understand.

Building Your Medical Vocabulary

Building a vocabulary is a cumulative process. As you progress through this book, you will use word parts, abbreviations, and clinical terms from previous chapters. Each chapter will present new word parts, abbreviations, and terms to add to your expanding vocabulary.

Word Parts and Combining Forms with Definitions and Examples

PART/ COMBINING FORM	DEFINITION	EXAMPLE
albino/o	white	albinism: lack of pigment in skin, hair, and eyes, making them appear white
cer-	wax	cerumen: earwax produced by ceruminous glands
-cide	killing	bactericide: an agent that kills bacteria
cry/o	cold	cryotherapy: therapeutic use of cold
cutane/o	skin	subcutaneous: below the skin
cyan-	blue	cyanosis: a bluish discoloration of the skin caused by lack of oxygen
derm/o	skin	dermatitis: inflammation of the skin
-ectomy	surgical excision	appendectomy: surgical removal of the appendix
erythem-	redness	erythema: widespread redness of the skin
hidr/o	sweat	anhidrosis: a condition of producing no sweat
ichthy/o	scaly, dry	ichthyosis: rough, dry, and scaly skin
kerat/o	hard, horny tissue	keratosis: a condition of thickened and hard epidermal tissue usually due to aging or skin damage
-lucid-	clear, light	stratum lucidum: clear layer of the epidermis
melan/o	black	melanocyte: a cell that produces the dark or black pigment melanin
myc/o	fungus	mycosis: a fungus condition, such as ringworm or athlete's foot
necr/o	death	necrosis: death or destruction of tissue
onych/o	nail	onychomycosis: a fungal infection of the nails
pachy-	thick	pachyderma: abnormal thickening of the skin

PART/ COMBINING FORM	DEFINITION	EXAMPLE
pedicul/o	louse, lice	pediculosis: infestation with lice
pil/o	hair	pilose: covered with hair, hairy
-plasty	surgical repair	angioplasty: surgical repair of blood vessels or lymphatic channels
rhytid/o	wrinkles	rhytidectomy: plastic surgery to remove wrinkles in the skin
seb/o	oil	sebum: oily secretion from sebaceous glands
sud-	sweat	sudoriferous gland: sweat gland
trich/o	hair	trichomycosis: fungal infection of the hair
ungu/o	nail	subungual: beneath a nail
xer/o	dry	xeroderma: dry skin

Clinical Abbreviations

ABBREVIATION	MEANING
bx	biopsy
Decub.	decubitus
Derm.	dermatology
FUO	fever of unknown origin
HSV	herpes simplex virus
HCDPA	health care durable power of attorney
PPD	purified protein derivative (test for tuberculosis)
PUVA	psoralen ultraviolet A (treatment for psoriasis)
SLE	systemic lupus erythematosus
UV	ultraviolet

Clinical Terms

Albinism (AL-bih-nizm) Lack of pigment in the skin, hair, and eyes

Callus (KAL-us) Acquired localized thickening of the epidermis caused by repeated pressure or friction

Cicatrix (SICK-ah-tricks) Normal scar that remains after wound repair

Dermatitis (der-mah-TYE-tis) Inflammation of the skin

Ecchymosis (eck-ih-MOH-sis) Bluish-black mark on the skin; bruise

Eschar (ESS-kar) Slough produced by a burn or gangrene

Gangrene (GANG-green) Death of tissue due to loss of blood supply; may be the result of injury, frostbite, disease, or infection

Keloid (KEE-loyd) Thickened scarlike growth that rises above the skin surface and that occurs after trauma or surgical incision; caused by excessive collagen formation during repair

Petechia (peh-TEE-kee-ah) Small, pinpoint purplish red spots on the skin caused by intradermal hemorrhages

Pruritus (proo-RYE-tus) Severe itching

Psoriasis (so-RYE-ah-sis) Chronic, recurrent skin condition characterized by bright red patches covered by silvery gray scales, often on the elbows and knees

Pustule (PUST-yool) Small elevation of the skin containing pus

Purpura (PURR-pah-rah) Multiple hemorrhages and accumulation of blood under the skin; includes ecchymosis and petechia

Scabies (SKAY-beez) Contagious skin disease caused by the itch mite *Sarcoptes scabiei*

Wheal (WHEEL) Smooth swollen area on the skin, often accompanied by itching; characteristic of hives and insect bites

VOCABULARY QUIZ

Use word parts you have learned to form words that have the following definitions.

1. Black tumor _____

2. Below the nail _____

3. Surgical excision of a nail _____

4. Below the skin _____

5. Black pigment cell _____

Using the definitions of word parts you have learned, define the following words.

6. Pachyderma _____

7. Mycosis _____

8. Dermoplasty _____

9. Dermatology _____

10. Anhidrosis _____

Match each of the following definitions with the correct word.

_____ 11. Clear layer of the skin

_____ 12. Plastic surgery for removal

_____ 13. Condition of dry, scaly skin

_____ 14. Earwax

_____ 15. Condition of thick nails

A. Cerumen
B. Ichthyosis of wrinkles
C. Lucidum
D. Pachyonychia
E. Rhytidectomy

16. What is the clinical term for a "slough produced by a burn or gangrene"? _____

17. What is an ecchymosis? _____

18. What is the term for "inflammation of the skin"? _____

19. What is the meaning of the abbreviation bx? _____

20. What is the meaning of the abbreviation PPD? _____

Skeletal System

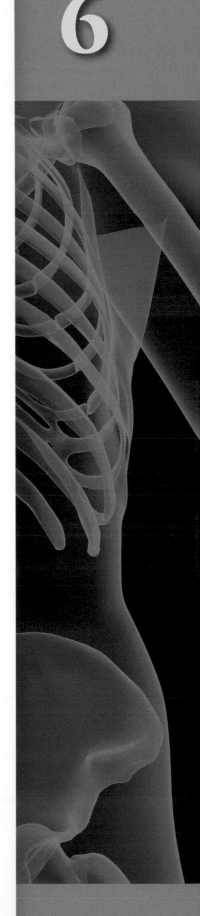

CHAPTER OBJECTIVES

Overview of the Skeletal System
- Discuss five functions of the skeletal system.
- Distinguish between compact and spongy bone on the basis of structural features.
- Classify bones according to size and shape.
- Identify the general features of a long bone.
- Name and define three types of cells in bone.
- Discuss intramembranous and endochondral ossification.
- Describe how bones increase in length and in diameter.
- Distinguish between the axial and appendicular skeletons, and state the number of bones in each.

Bones of the Axial Skeleton
- Identify the bones of the skull and their important surface markings.
- Identify the general structural features of vertebrae and compare cervical, thoracic, lumbar, sacral, and coccygeal vertebrae; state the number of each type.
- Identify the structural features of the ribs and sternum.

Bones of the Appendicular Skeleton
- Identify the bones of the pectoral girdle and their surface features.
- Identify the bones of the upper extremity and their surface features.
- Identify the bones of the pelvic girdle and their surface features.
- Identify the bones of the lower extremity and their surface features.

Fractures and Fracture Repair
- Identify these fractures by description and on diagrams: complete, incomplete, open, closed, transverse, spiral, comminuted, and displaced.
- Describe the process of fracture repair by using the terms *fracture hematoma, procallus, fibrocartilaginous callus, bony callus,* and *remodeling.*

Articulations
- Compare the structure and function of the synarthroses, amphiarthroses, and diarthroses. Give examples of each of these articulations.

KEY TERMS

Amphiarthrosis (am-fee-ahr-THROH-sis) Slightly movable joint; plural, amphiarthroses

Appositional growth (app-oh-ZISH-un-al GROHTH) Growth resulting from material deposited on the surface, such as the growth in diameter of long bones

Diaphysis (dye-AF-ih-sis) Long straight shaft of a long bone

Diarthrosis (dye-ahr-THROH-sis) Freely movable joint characterized by a joint cavity; also called a synovial joint; plural, diarthroses

Endochondral ossification (en-doh-KON-dral ah-sih-fih-KAY-shun) Method of bone formation in which cartilage is replaced by bone

Endosteum (end-AH-stee-um) Membrane that lines the medullary cavity of bones

Epiphyseal plate (ep-ih-FIZ-ee-al PLATE) Cartilaginous plate between the epiphysis and diaphysis of a bone; responsible for the lengthwise growth of a long bone

Epiphysis (ee-PIF-ih-sis) End of a long bone

Hematopoieses (hee-mat-oh-pay-EE-sis) Blood cell formation; also called hemopoiesis (hee-moh-pay-EE-sis)

Intramembranous ossification (in-trah-MEM-bran-us ah-sih-fih-KAY-shun) Method of bone formation in which the bone is formed directly in a membrane

Nutrient foramen (NOO-tree-ent for-A-men) Small opening in the diaphysis of bone for passage of blood vessels

Osteoblast (AH-stee-oh-blast) Bone-forming cell

Osteoclast (AH-stee-oh-clast) Cell that destroys or resorbs bone tissue

Osteocyte (AH-stee-oh-syte) Mature bone cell

Osteon (OH-stee-ahn) Structural unit of bone; haversian system

Periosteum (pair-ee-AH-stee-um) Tough, white outer membrane that covers a bone and is essential for bone growth, repair, and nutrition.

Synarthrosis (sin-ahr-THROH-sis) Immovable joint; plural, synarthroses

Functional Relationships of the
Skeletal System

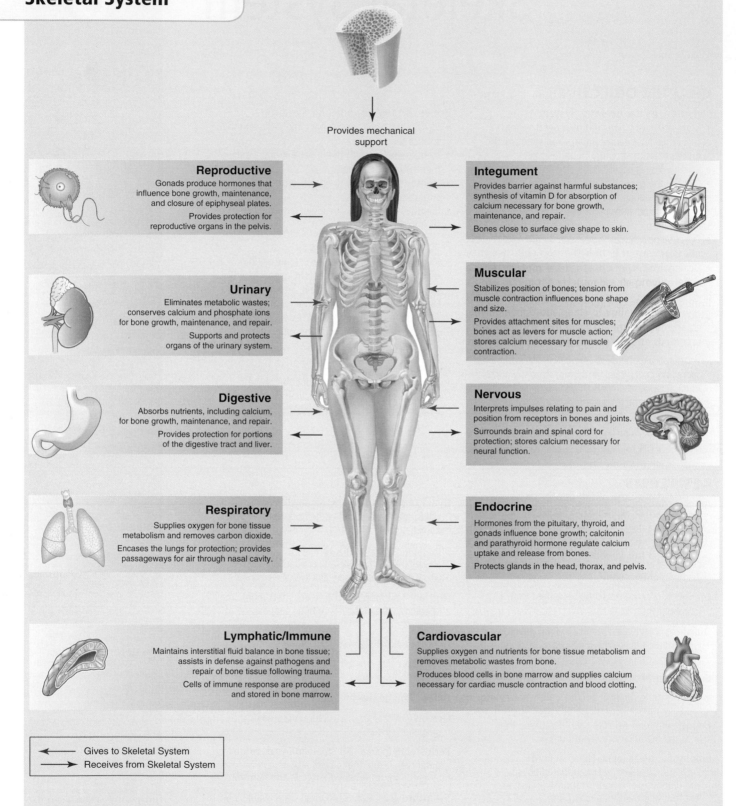

Provides mechanical support

Reproductive

Gonads produce hormones that influence bone growth, maintenance, and closure of epiphyseal plates.

Provides protection for reproductive organs in the pelvis.

Integument

Provides barrier against harmful substances; synthesis of vitamin D for absorption of calcium necessary for bone growth, maintenance, and repair.

Bones close to surface give shape to skin.

Urinary

Eliminates metabolic wastes; conserves calcium and phosphate ions for bone growth, maintenance, and repair.

Supports and protects organs of the urinary system.

Muscular

Stabilizes position of bones; tension from muscle contraction influences bone shape and size.

Provides attachment sites for muscles; bones act as levers for muscle action; stores calcium necessary for muscle contraction.

Digestive

Absorbs nutrients, including calcium, for bone growth, maintenance, and repair.

Provides protection for portions of the digestive tract and liver.

Nervous

Interprets impulses relating to pain and position from receptors in bones and joints.

Surrounds brain and spinal cord for protection; stores calcium necessary for neural function.

Respiratory

Supplies oxygen for bone tissue metabolism and removes carbon dioxide.

Encases the lungs for protection; provides passageways for air through nasal cavity.

Endocrine

Hormones from the pituitary, thyroid, and gonads influence bone growth; calcitonin and parathyroid hormone regulate calcium uptake and release from bones.

Protects glands in the head, thorax, and pelvis.

Lymphatic/Immune

Maintains interstitial fluid balance in bone tissue; assists in defense against pathogens and repair of bone tissue following trauma.

Cells of immune response are produced and stored in bone marrow.

Cardiovascular

Supplies oxygen and nutrients for bone tissue metabolism and removes metabolic wastes from bone.

Produces blood cells in bone marrow and supplies calcium necessary for cardiac muscle contraction and blood clotting.

←——— Gives to Skeletal System
———→ Receives from Skeletal System

The **skeletal system** consists of the bones and the cartilage, ligaments, and tendons associated with the bones. It accounts for about 20% of the body weight. Bones are rigid structures that form the framework for the body. People often think of bones as dead, dry, inert pipes and plates because that is how they are seen in the laboratory. In reality, the living bones in our bodies contain active tissues that consume nutrients, require a blood supply, use oxygen and discharge waste products in metabolism, and change shape or remodel in response to variations in mechanical stress. The skeletal system is strong but lightweight. It is well adapted for the functions it must perform. It is a masterpiece of design.

Overview of the Skeletal System

Functions of the Skeletal System

The skeletal system gives form and shape to the body. Without the skeletal components, we would appear as big "blobs" inefficiently "oozing" around on the ground. Besides contributing to shape and form, our bones perform several other functions and play an important role in homeostasis.

Support

Bones provide a rigid framework that supports the soft organs of the body. Bones support the body against the pull of gravity, and the large bones of the lower limbs support the trunk when standing.

Protection

The skeleton protects the soft body parts. The fused bones of the cranium surround the brain to make it less vulnerable to injury. The vertebrae surround and protect the spinal cord. The bones of the rib cage help protect the heart and lungs in the thorax.

Movement

Bones provide sites for muscle attachment. Bones and muscles work together as simple mechanical lever systems to produce body movement. A mechanical lever system has four components: (1) a rigid bar, (2) a pivot or fulcrum, (3) an object or weight that is moved, and (4) a force that supplies the mechanical energy for the movement. In the body, bones are the rigid bars, the joints between the bones are the pivots, the body or a part of it is the weight that is moved, and the muscles supply the force.

Storage

The intercellular matrix of bone contains large amounts of calcium salts, the most important being calcium phosphate. Calcium is necessary for vital metabolic processes. When blood calcium levels decrease below normal, calcium is released from the bones so that there will be an adequate supply for metabolic needs. When blood calcium levels are increased, the excess calcium is stored in the bone matrix. Storage and release are dynamic processes that go on almost continually.

Bone tissue contains lesser amounts of other inorganic ions such as sodium, magnesium, potassium, and carbonate. Fat is stored in the yellow bone marrow.

Blood Cell Formation

Blood cell formation, called **hematopoiesis** (hee-mat-oh-poy-EE-sis), takes place mostly in the red marrow of bones. Red marrow is found in the cavities of most bones in an infant. With age, it is largely replaced by yellow marrow for fat storage. In the adult, red marrow is limited to the spongy bone in the skull, ribs, sternum, clavicles, vertebrae, and pelvis. Red marrow functions in the formation of red blood cells, white blood cells, and blood platelets.

Structure of Bone Tissue

There are two types of bone tissue: compact and spongy. The names imply that the two types differ in density, or how tightly the tissue is packed together. There are three types of cells that contribute to bone homeostasis. Osteoblasts are bone-forming cells, osteoclasts resorb or break down bone, and osteocytes are mature bone cells. An equilibrium between osteoblasts and osteoclasts maintains bone tissue.

Compact Bone

The microscopic unit of compact bone, the osteon (haversian system), was described in Chapter 4. Briefly, the osteon consists of a central canal called the osteonic (haversian) canal that is surrounded by concentric rings (lamellae) of matrix. Between the rings of matrix, the bone cells (osteocytes) are located in spaces called lacunae. Small channels (canaliculi) radiate from the lacunae to the osteonic (haversian) canal to provide passageways through the hard matrix. In compact bone, the haversian systems are packed tightly together to form what appears to be a solid mass. The osteonic canals contain blood vessels that are parallel to the long axis of the bone. These blood vessels interconnect, by way of perforating (Volkmann's) canals, with vessels on the surface of the bone. The microscopic structure of compact bone is illustrated in Figure 6-1, and Figure 6-2 shows a photomicrograph of an osteon or haversian system.

Spongy (Cancellous) Bone

Spongy (cancellous) bone is lighter and less dense than compact bone (see Figure 6-1). Spongy bone consists of plates and bars of bone adjacent to small, irregular cavities that contain red bone marrow. The plates of bone are called **trabeculae** (trah-BEK-yoo-lee). The canaliculi, instead of connecting to a central haversian canal, connect to the adjacent cavities to receive their blood supply. It may appear that the trabeculae are arranged in a haphazard manner, but they are organized to provide maximum strength in the same way that braces are used to support a building. The trabeculae of spongy bone follow the lines of stress and can realign if the direction of stress changes.

QUICK CHECK

6.1 List five functions of the skeletal system.
6.2 What is the microscopic unit of compact bone?

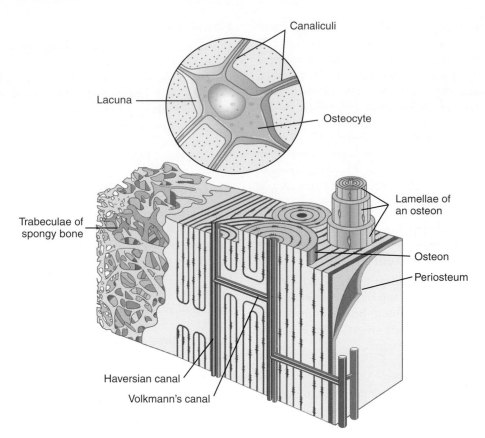

Figure 6-1 *Structure of compact and spongy bone. Note the osteons packed together for compact bone and the trabeculae of spongy bone.*

Figure 6-2 *Photomicrograph of an osteon. Note the central canal (C), the concentric lamellae (L), and lacunae that contain osteocytes. The* arrows *point to canaliculi. (From Gartner and Hiatt: Color Textbook of Histology. 3rd ed. Philadelphia, Elsevier/Saunders, 2007.)*

Classification of Bones

Bones come in a variety of sizes and shapes. Bones that are longer than they are wide are called long bones. They consist of a long shaft with two bulky ends or extremities. They are primarily compact bone but may have a large amount of spongy bone at the ends. Examples of long bones are those in the thigh, leg, arm, and forearm.

Short bones are roughly cube shaped with vertical and horizontal dimensions approximately equal. They consist primarily of spongy bone, which is covered by a thin layer of compact bone. Examples of short bones include the bones of the wrist and ankle.

Flat bones are thin, flattened, and often curved. They are usually arranged similar to a sandwich with a middle layer of spongy bone called the **diploë** (DIP-loh-ee). The diploë is covered on each side by a layer of compact bone; these layers are called the **inner** and **outer tables.** Most of the bones of the cranium are flat bones.

Bones that are not in any of the above three categories are classified as irregular bones. They are primarily spongy bone that is covered with a thin layer of compact bone. The vertebrae and some of the bones in the skull are irregular bones.

General Features of a Long Bone

Most long bones have the same general features, which are illustrated in Figure 6-3.

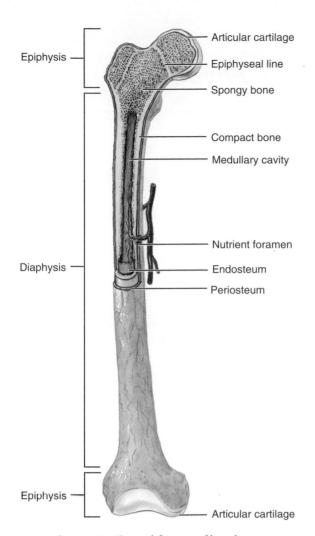

Epiphysis

Diaphysis

Epiphysis

Articular cartilage
Epiphyseal line
Spongy bone
Compact bone
Medullary cavity
Nutrient foramen
Endosteum
Periosteum
Articular cartilage

Figure 6-3 *General features of long bones.*

The shaft of a long bone is called the **diaphysis** (dye-AF-ih-sis). It is formed from relatively thick compact bone that surrounds a hollow space called the **medullary** (MED-yoo-lair-ee) **cavity.** In adults, the medullary cavity contains yellow bone marrow, so it is sometimes called the yellow marrow cavity. At each end of the diaphysis, there is an expanded portion called the **epiphysis** (ee-PIF-ih-sis). The epiphysis is spongy bone covered by a thin layer of compact bone. The end of the epiphysis, where it meets another bone, is covered by hyaline cartilage, called the **articular cartilage.** This provides smooth surfaces for movement in the joints. In growing bones, there is an **epiphyseal** (ep-ih-FIZ-ee-al) **plate** of hyaline cartilage between the diaphysis and epiphysis. Bones grow in length at the epiphyseal plate. Growth ceases when the cartilaginous epiphyseal plate is replaced by a bony **epiphyseal line.** Except in the region of the articular cartilage, the outer surface of long bones is covered by a tough, fibrous connective tissue called the **periosteum.** The periosteum is richly supplied with nerve fibers, lymphatic vessels, blood vessels, and osteoblasts. Blood vessels enter the diaphysis of the bone through small openings called **nutrient foramina.** The surface of the medullary cavity is lined with a thinner connective tissue membrane, the **endosteum,** which contains osteoclasts.

In addition to the general features that are present in most long bones, all bones have surface markings and characteristics that make a specific bone unique. There are holes, depressions, smooth facets, lines, projections, and other markings. These usually represent passageways for vessels and nerves, points of articulation with other bones, or points of attachment for tendons and ligaments. Some of the bony markings are described in Table 6-1.

QUICK CHECK

6.3 In flat bones, there are two layers of compact bone with a layer of spongy bone in the middle. What is this spongy layer called?
6.4 What is the shaft of a long bone called?
6.5 What covers the end of an epiphysis?

Bone Development and Growth

The terms **osteogenesis** and **ossification** are often used synonymously to indicate the process of bone formation. Parts of the skeleton form during the first few weeks after conception. By the end of the eighth week after conception, the skeletal pattern is formed in cartilage and connective tissue membranes, and ossification begins. Bone development continues throughout adulthood. Even after adult stature is attained, bone development continues for repair of fractures and for remodeling to meet changing lifestyles. Three types of cells are involved in the development, growth, and remodeling of bones. **Osteoblasts** are bone-forming cells; **osteocytes** are mature bone cells; and **osteoclasts** break down and reabsorb bone.

Intramembranous Ossification

Intramembranous ossification (in-tra-MEM-bran-us ah-sih-fih-KAY-shun) involves the replacement of sheetlike connective tissue membranes with bony tissue. Bones formed in this manner are called **intramembranous bones.** They include certain flat bones of the skull and some of the irregular bones. The future bones are first formed as connective tissue membranes. Osteoblasts migrate to the membranes and deposit bony matrix around themselves. When the osteoblasts are surrounded by matrix, they are called osteocytes.

Endochondral Ossification

Endochondral ossification (en-doh-KON-dral ah-sih-fih-KAY-shun) involves the replacement of hyaline cartilage with bony tissue. Most of the bones of the skeleton are formed in this manner and are called **endochondral bones.** In this process, illustrated in Figure 6-4, the future bones are first formed as hyaline cartilage models. During the third month after conception, the perichondrium that surrounds the hyaline cartilage "models" becomes infiltrated with blood vessels and osteoblasts and changes into a periosteum. The osteoblasts form a collar of compact bone around the diaphysis. At the same time, the cartilage in the center of the diaphysis begins to disintegrate. Osteoblasts penetrate the disintegrating cartilage and replace it with spongy bone. This forms a **primary ossification center.** Ossification

TABLE 6-1 Terms Related to Bone Markings

Term	Description	Examples
Projections for Articulation		
Condyle (KON-dial)	Smooth, rounded articular surface	Occipital condyle on occipital bone; lateral and medial condyles on femur
Facet (FASS-et)	Smooth, nearly flat articular surface	Facets on thoracic vertebrae for articulation with ribs
Head (HED)	Enlarged, often rounded, end of bone	Head of humerus; head of femur
Projections for Muscle Attachment		
Crest (KREST)	Narrow ridge of bone	Iliac crest or ilium
Epicondyle (ep-ih-KON-dial)	Bony bulge adjacent to or above a condyle	Lateral and medial epicondyles of femur
Process (PRAH-sess)	Any projection on a bone; often pointed and sharp	Syloid process on temporal bone
Spine (SPYN)	Sharp, slender projection	Spine of scapula
Trochanter (tro-KAN-turr)	Large, blunt, irregularly shaped projection	Greater and lesser trochanters on femur
Tubercle (TOO-burr-kul)	Small, rounded, knoblike projection	Greater tubercle of humerus
Tuberosity (too-burr-AHS-ih-tee)	Similar to a tubercle, but usually larger	Tibial tuberosity on tibia
Depressions, Openings, and Cavities		
Fissure (FISH-ur)	Narrow cleft or slit; usually for passage of blood vessels and nerves	Superior orbital fissure
Foramen (foh-RAY-men)	Opening through a bone; usually for passage of blood vessels and nerves	Foramen magnum in occipital bone
Fossa (FAW-sah)	A smooth, shallow depression	Mandibular fossa on temporal bone; olecranon fossa on humerus
Fovea (FOH-vee-ah)	A small pit or depression	Fovea capitis femoris on head of femur
Meatus (canal) (mee-ATE-us)	A tubelike passageway; tunnel	External auditory meatus in temporal bone
Sinus (SYE-nus)	A cavity or hollow space in a bone	Frontal sinus in frontal bone

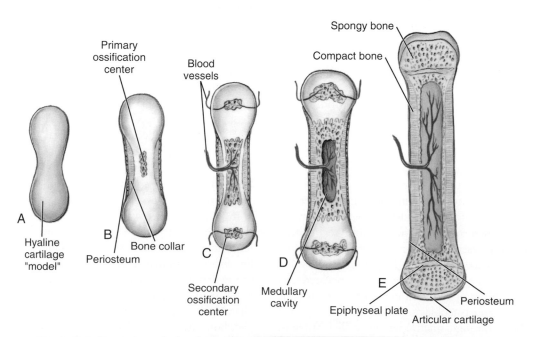

Figure 6-4 *Events in endochondral ossification.* **A,** *Hyaline cartilage "model."* **B,** *Periosteum and bone collar form.* **C,** *Blood vessels and osteoblasts infiltrate primary ossification centers.* **D,** *Osteoclasts form medullary cavity.* **E,** *Ossification is complete. Hyaline cartilage remains as articular cartilage and in the epiphyseal plate.*

continues from this center toward the ends of the bones. After spongy bone is formed in the diaphysis, osteoclasts break down the newly formed bone to open up the medullary cavity.

The cartilage in the epiphyses continues to grow so the developing bone increases in length. Later, usually after birth, **secondary ossification centers** form in the epiphyses. Ossification in the epiphysis is similar to that in the diaphysis except that the spongy bone is retained instead of being broken down to form a medullary cavity. When secondary ossification is complete, the hyaline cartilage is totally replaced by bone except in two areas. A region of hyaline cartilage remains over the surface of the epiphysis as the **articular cartilage.** Another area of cartilage remains between the epiphysis and diaphysis. This is the **epiphyseal plate** or growth region.

Bone Growth

Bones grow in length at the epiphyseal plate by a process that is similar, in many ways, to endochondral ossification. The cartilage in the region of the epiphyseal plate next to the epiphysis continues to grow through mitosis. The chondrocytes in the region next to the diaphysis age and degenerate. Osteoblasts move in and ossify the matrix to form bone. This process continues throughout childhood and adolescence until the cartilage growth slows and finally stops. When cartilage growth ceases, usually in the early 20s, the epiphyseal plate completely ossifies so that only a thin **epiphyseal line** remains and the bones can no longer grow in length. Bone growth is under the influence of growth hormone from the anterior pituitary gland and sex hormones from the ovaries and testes.

Even though bones stop growing in length in early adulthood, they can continue to increase in thickness or diameter throughout life in response to stress from increased muscle activity or to weight gain. The increase in diameter is called **appositional** (ap-poh-ZISH-un-al) **growth.** Osteoblasts in the periosteum form compact bone around the external bone surface. At the same time, osteoclasts in the endosteum break down bone on the internal bone surface, around the medullary cavity. These two processes together increase the diameter of the bone and, at the same time, keep the bone from becoming excessively heavy and bulky.

QUICK APPLICATIONS

The epiphyseal plates of specific long bones ossify at predictable times. Radiologists frequently can determine a young person's age by examining the epiphyseal plates to see whether they have ossified. A difference between bone age and chronologic age may indicate some type of metabolic dysfunction.

QUICK CHECK

6.6 What is the method of formation for the flat bones of the skull?

6.7 In a long bone, where is the primary ossification center located?

6.8 What is appositional growth?

Divisions of the Skeleton

The typical adult human skeleton consists of 206 named bones. In addition to the named bones, there are two other types that vary in number from one individual to another and do not have specific names. **Wormian** (WER-mee-an) **bones** or **sutural** (SOO-cher-ahl) **bones** are small bones in the joints between certain cranial bones. **Sesamoid** (SEH-sah-moyd) **bones** are small bones that grow in certain tendons in which there is considerable pressure. The **patella,** or kneecap, is an example of a sesamoid bone that is named, but other sesamoid bones are not named.

For convenience, the bones of the skeleton are grouped in two divisions, as illustrated in Figure 6-5. The 80 bones of the **axial skeleton** form the vertical axis of the body. They include the bones of the head, vertebral column, ribs, and breastbone or sternum. The **appendicular skeleton** consists of 126 bones and includes the free appendages and their attachments to the axial skeleton. The free appendages are the upper and lower extremities, or limbs, and their attachments are called girdles. The named bones of the body are listed by category in Table 6-2.

Bones of the Axial Skeleton

The axial skeleton, with 80 bones, is divided into the skull, hyoid, vertebral column, and rib cage.

Skull

There are 28 bones in the skull, illustrated in Figures 6-6 through 6-10. Eight of these form the cranium, which houses the brain. The anterior aspect of the skull, the face, consists of 14 bones. The remaining six bones are the auditory ossicles, which are tiny bones in the middle ear cavity. With the exception of the lower jaw, or mandible, and the auditory ossicles, the bones in the skull are tightly interlocked along irregular lines called **sutures.** Some of the bones in the skull contain **sinuses,** which are air-filled cavities that are lined with mucous membranes. The sinuses help to reduce the weight of the skull. The paranasal sinuses are arranged around the nasal cavity and drain into it.

There are numerous openings, or **foramina,** in the bones of the skull to allow for passage of blood vessels and nerves. Major foramina of the skull are listed in Table 6-3 and illustrated in Figures 6-6 through 6-10.

QUICK APPLICATIONS

The bones with paranasal sinuses are the frontal, sphenoid, ethmoid, and the two maxillae. The sinuses are lined with mucous membranes that are continuous with the nasal cavity. Allergies and infections cause inflammation of the membranes, which results in sinusitis. The swollen membranes may reduce drainage from the sinuses so that pressure within the cavities increases, resulting in sinus headaches.

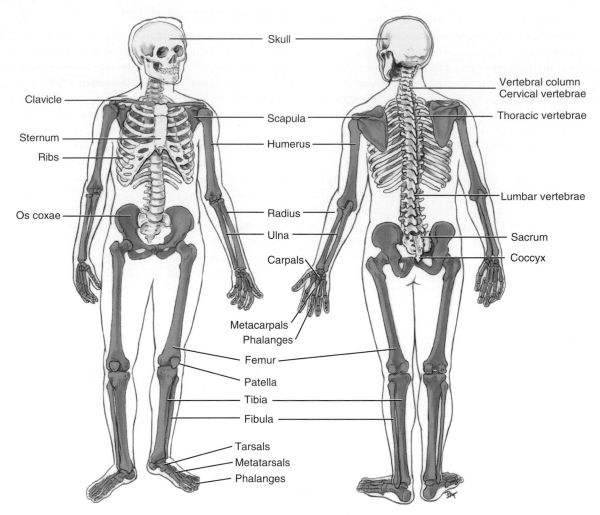

Figure 6-5 *Divisions of the skeleton with major bones identified.* Yellow = *axial skeleton.* Blue = *appendicular skeleton.*

Cranium

The eight bones of the cranium are interlocked to enclose the brain (see Figures 6-6 through 6-10).

Frontal Bone

The frontal bone forms the anterior portion of the skull above the eyes (forehead), a portion of the nose, and the superior portion of the orbit (eye socket). On the superior margin of each orbit, there is a **supraorbital foramen** (or supraorbital notch in some skulls) through which blood vessels and nerves pass to the tissues of the forehead. On either side of the midline, just above the eyes, there is a cavity in the frontal bone. These are the paranasal **frontal sinuses.**

Parietal Bones

The two parietal (pah-RYE-eh-tal) bones form most of the superolateral aspect of the skull. They are joined to each other in the midline by the **sagittal suture** and to the frontal bone by the **coronal suture.**

Occipital Bone

The single occipital (ahk-SIP-ih-tal) bone forms most of the dorsal part of the skull and the base of the cranium. It is joined to the parietal bones by the **lambdoid** (lamm-DOYD) **suture.** Wormian bones are frequently found in the lambdoid suture. The **foramen magnum** is a large opening on the lower surface of the occipital bone. The spinal cord passes through this opening. **Occipital condyles** are rounded processes on each side of the foramen magnum. They articulate with the first cervical vertebra.

Temporal Bones

The two temporal bones, one on each side of the head, form parts of the sides and base of the cranium. On each side, a temporal bone meets the parietal bone at the **squamous** (SKWAY-mus) **suture.** Near the inferior margin of the temporal bone, there is an opening, the **external auditory meatus,** which is a canal that leads to the middle ear. Just anterior to the external auditory meatus, there is a shallow depression, the **mandibular** (man-DIB-yoo-lar) **fossa,** that articulates with the mandible. Posterior and inferior to each external auditory meatus, there is a rough protuberance, the **mastoid process.** The mastoid process contains air cells that drain into the middle ear cavity. The **styloid process** is a long, pointed projection inferior to the external auditory meatus. A **zygomatic** (zye-goh-MAT-ik) **process** projects anteriorly from the temporal bone and helps form the prominence of the cheek.

TABLE 6-2 Names of Bones of the Body Listed by Category

Bones	Number
Axial Skeleton (80 Bones)	
Skull (28 bones)	
Cranial bones	8
Parietal (2)	
Temporal (2)	
Frontal (1)	
Occipital (1)	
Ethmoid (1)	
Sphenoid (1)	
Facial bones	14
Maxilla (2)	
Zygomatic (2)	
Mandible (1)	
Nasal (2)	
Palatine (2)	
Inferior nasal concha (2)	
Lacrimal (2)	
Vomer (1)	
Auditory ossicles	6
Malleus (2)	
Incus (2)	
Stapes (2)	
Hyoid	1
Vertebral column	26
Cervical vertebrae (7)	
Thoracic vertebrae (12)	
Lumbar vertebrae (5)	
Sacrum (1)	
Coccyx (1)	
Thoracic cage	25
Sternum (1)	
Ribs (24)	
Appendicular Skeleton (126 Bones)	
Pectoral girdles	4
Clavicle (2)	
Scapula (2)	
Upper extremity	60
Humerus (2)	
Radius (2)	
Ulna (2)	
Carpals (16)	
Metacarpals (10)	
Phalanges (28)	
Pelvic girdle	2
Coxal, innominate, or hip bones (2)	
Lower extremity	60
Femur (2)	
Tibia (2)	
Fibula (2)	
Patella (2)	
Tarsals (14)	
Metatarsals (10)	
Phalanges (28)	

QUICK APPLICATIONS

The mastoid air cells are separated from the cranial cavity by only a thin partition of bone. A middle ear infection that spreads to the mastoid air cells (mastoiditis) is serious because there is danger that the infection will spread from the air cells to the membranes around the brain.

Sphenoid Bone

The sphenoid (SFEE-noyd) bone is an irregularly shaped bone that spans the entire width of the cranial floor. It is wedged between other bones in the anterior portion of the cranium. This bone helps form the sides of the skull, the base of the cranium, and the lateral and inferior portions of each orbit. Within the cranial cavity, there is a saddle-shaped central portion, called the **sella turcica** (SELL-ah TUR-sih-kah), with a depression for the pituitary gland. Anterior to the sella turcica are two openings, one on each side, called the **optic foramina,** for the passage of the optic nerve. The **greater wings** extend laterally from the region of the sella turcica and are the portions of the sphenoid seen in the orbits and external wall of the skull. The sphenoid bone also contains paranasal **sphenoid sinuses.**

Ethmoid Bone

The ethmoid (ETH-moyd) bone is located anterior to the sphenoid bone and forms most of the bony area between the nasal cavity and the orbits. In the anterior portion of the cranial cavity, the **crista galli** (KRIS-tah GAL-lee) is seen as a triangular process that projects upward. It is an attachment for the membranes that surround the brain. On each side of the crista galli, there is a small, flat **cribriform** (KRIB-rih-form) **plate** that is full of tiny holes. The holes are the **olfactory foramina.** Nerve fibers from the sense receptors for smell in the nasal cavity pass through the olfactory foramina. The **perpendicular plate** of the ethmoid projects downward in the middle of the nasal cavity to form the superior part of the nasal septum. Delicate, scroll-like projections, called the **superior and middle nasal conchae** (KONG-kee) or **turbinates,** form ledges along the lateral walls of the nasal cavity. The conchae are lined with mucous membranes to warm and moisten inhaled air. The ethmoid bone contains many small, paranasal **ethmoidal sinuses.**

QUICK APPLICATIONS

The bones in the skull of a newborn are not completely joined together but are separated by fibrous membranes. There are six large areas of membranes called **fontanels,** or soft spots. The anterior fontanel is on the top of the head, at the junction of the frontal and parietal bones. The posterior fontanel is at the junction of the occipital and parietal bones. On each side of the head there is a mastoid (posterolateral) fontanel near the mastoid region of the temporal bone and a sphenoid (anterolateral) fontanel just superior to the sphenoid bone.

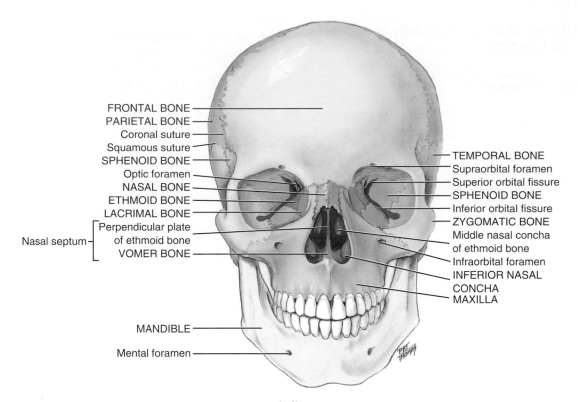

FRONTAL BONE
PARIETAL BONE
Coronal suture
Squamous suture
SPHENOID BONE
Optic foramen
NASAL BONE
ETHMOID BONE
LACRIMAL BONE
Nasal septum —
Perpendicular plate of ethmoid bone
VOMER BONE

TEMPORAL BONE
Supraorbital foramen
Superior orbital fissure
SPHENOID BONE
Inferior orbital fissure
ZYGOMATIC BONE
Middle nasal concha of ethmoid bone
Infraorbital foramen
INFERIOR NASAL CONCHA
MAXILLA

MANDIBLE
Mental foramen

Figure 6-6 *Skull, anterior view.*

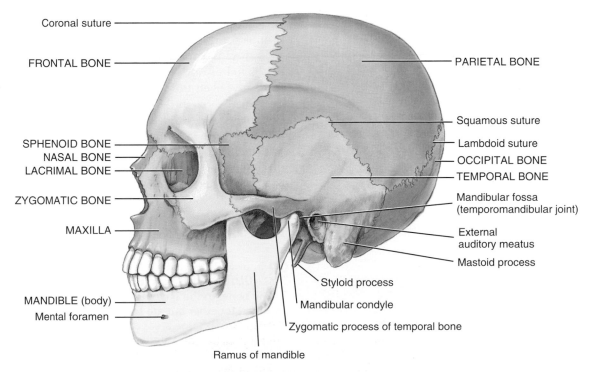

Coronal suture
FRONTAL BONE
SPHENOID BONE
NASAL BONE
LACRIMAL BONE
ZYGOMATIC BONE
MAXILLA
MANDIBLE (body)
Mental foramen

PARIETAL BONE
Squamous suture
Lambdoid suture
OCCIPITAL BONE
TEMPORAL BONE
Mandibular fossa (temporomandibular joint)
External auditory meatus
Mastoid process

Styloid process
Mandibular condyle
Zygomatic process of temporal bone
Ramus of mandible

Figure 6-7 *Skull, lateral view.*

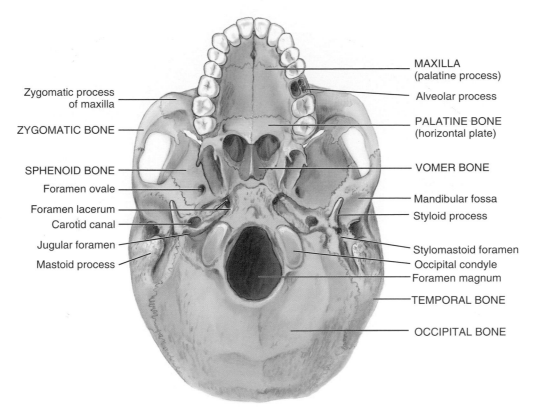

Zygomatic process of maxilla

ZYGOMATIC BONE

SPHENOID BONE

Foramen ovale

Foramen lacerum

Carotid canal

Jugular foramen

Mastoid process

MAXILLA (palatine process)

Alveolar process

PALATINE BONE (horizontal plate)

VOMER BONE

Mandibular fossa

Styloid process

Stylomastoid foramen

Occipital condyle

Foramen magnum

TEMPORAL BONE

OCCIPITAL BONE

Figure 6-8 *Skull, inferior view.*

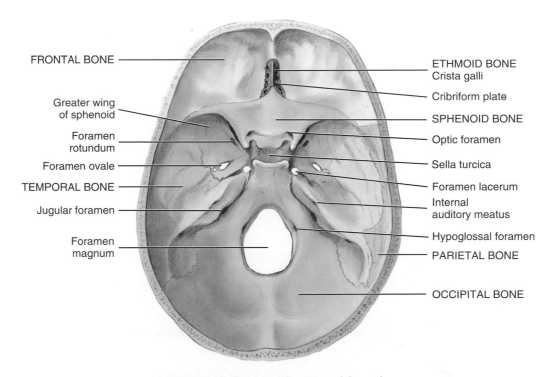

FRONTAL BONE

Greater wing of sphenoid

Foramen rotundum

Foramen ovale

TEMPORAL BONE

Jugular foramen

Foramen magnum

ETHMOID BONE
Crista galli

Cribriform plate

SPHENOID BONE

Optic foramen

Sella turcica

Foramen lacerum

Internal auditory meatus

Hypoglossal foramen

PARIETAL BONE

OCCIPITAL BONE

Figure 6-9 *Skull, cranial floor viewed from above.*

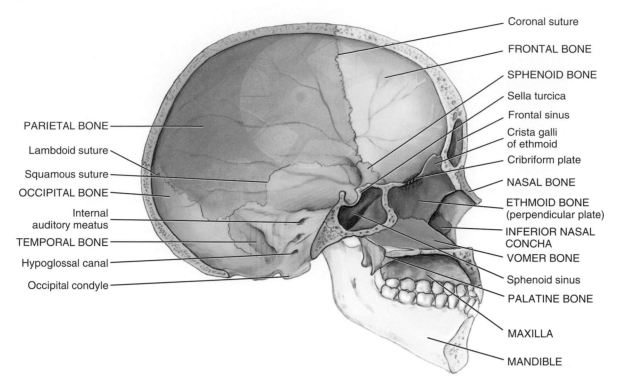

Figure 6-10 *Skull, midsagittal section.*

TABLE 6-3 Major Foramina of the Skull

Foramen	Location	Structures Transmitted
Carotid canal	Temporal bone	Internal carotid artery
Hypoglossal	Occipital bone	Hypoglossal nerve
Inferior orbital (fissure)	Floor of orbit	Maxillary branch of trigeminal nerve and infraorbital vessels
Infraorbital	Maxilla bone	Infraorbital vessels and nerves
Internal auditory (meatus)	Temporal bone	Vestibulocochlear nerve
Jugular	Temporal bone	Internal jugular vein; vagus, glossopharyngeal, and spinal accessory nerves
Magnum	Occipital bone	Medulla oblongata/spinal cord; accessory nerves; vertebral and spinal arteries
Mental	Mandible	Mental nerve and vessels
Nasolacrimal canal	Lacrimal bone	Nasolacrimal (tear) duct
Olfactory	Ethmoid bone	Olfactory nerves
Optic	Sphenoid bone	Optic nerve
Ovale	Sphenoid bone	Mandibular branch of trigeminal nerve
Rotundum	Sphenoid bone	Maxillary branch of trigeminal nerve
Stylomastoid	Temporal bone	Facial nerve
Superior orbital (fissure)	Orbit of eye	Oculomotor (III), trochlear (IV), ophthalmic branch of trigeminal (V), and abducens (VI) nerves
Supraorbital	Frontal bone	Supraorbital artery and nerve

QUICK CHECK

6.9 How many bones comprise the cranium?
6.10 How are the bones of the cranium held together?
6.11 What bone contains the foramen magnum?

Facial Bones

The facial skeleton consists of 14 bones. Thirteen of these are interlocked, and there is one movable mandible, the lower jawbone. These bones form the basic framework and shape of the face. They also provide attachments for the muscles that control facial expression and move the jaw for chewing. All the bones except the vomer and mandible are paired. Facial bones are illustrated in Figures 6-6 through 6-10.

Maxillary Bones

The maxillary bones, or **maxillae** (maks-ILL-ee), form the upper jaw. Additionally, they form the lateral walls of the nose, the floor of the orbits, and the anterior part of the roof of the mouth. The portion in the hard palate, or roof of the mouth, is the **palatine process.** The inferior border of each maxilla projects downward to form the **alveolar** (al-VEE-oh-lar) **process,** which contains the teeth. Each maxilla has a large paranasal **maxillary sinus.** These are the largest of all the paranasal sinuses.

Palatine Bones

The palatine (PAL-ah-tyne) bones are behind, or posterior to, the maxillae. Each one is roughly L-shaped. The **horizontal plates** form the posterior portion of the hard palate. The vertical portions help form the lateral walls of the nasal cavity.

Nasal Bones

The two nasal bones are small rectangular bones that form the bridge of the nose.

Lacrimal Bones

The small, thin lacrimal (LACK-rih-mal) bones are located in the medial walls of the orbits, between the ethmoid bone and the maxilla. Each one has a small **lacrimal groove** that is a pathway for a tube that carries tears from the eyes to the nasal cavity.

Zygomatic Bones

The zygomatic (zye-goh-MAT-ik) bones, also called **malar** bones, form the prominences of the cheeks and a portion of the lateral walls of the orbits. Each one has a **temporal process** that projects toward the zygomatic process of the temporal bone to form the **zygomatic arch.**

Inferior Nasal Conchae

The inferior nasal conchae (KONG-kee) are thin, curved bones that are attached to the lateral walls of the nasal cavity. They project into the nasal cavity just below the middle conchae of the ethmoid bone.

Vomer

The thin, flat vomer (VOH-mer) is in the inferior portion of the midline in the nasal cavity. It joins with the perpendicular plate of the ethmoid bone to form the **nasal septum.**

Mandible

The mandible (MAN-dih-bul) is the lower jaw. It has a horseshoe-shaped **body** that forms the chin and a flat portion, the **ramus,** that projects upward at each end. On the superior portion of the ramus, there is a knoblike process, the **mandibular condyle,** that fits into the mandibular fossa of the temporal bone to form the **temporomandibular** (tem-por-oh-man-DIB-yoo-lar) joint. The superior border of the mandible projects upward to form the **alveolar process** that contains the teeth. The reconstructed computed tomography (CT) image and the radiograph in Figure 6-11 show some of the bones of the cranium and face.

Auditory Ossicles

There are three tiny bones that form a chain in each middle ear cavity in the temporal bone. These are the **malleus, incus,** and **stapes.** These bones transmit sound waves from the tympanic membrane, or eardrum, to the inner ear where the sound receptors are located.

Hyoid Bone

The hyoid bone is not really part of the skull, so it is listed separately. It is a U-shaped bone in the neck, between the mandible and the larynx, or voice box. Figure 6-12 illustrates the position and shape of the hyoid bone. It is unique because it is the only bone in the body that does not articulate directly with another bone. Instead it is suspended under the mandible and anchored by ligaments to the styloid processes of the temporal bones. It functions as a base for the tongue and as an attachment for several muscles associated with swallowing.

QUICK CHECK

6.12 How many bones comprise the face?
6.13 What is the U-shaped bone in the neck that functions as a base for the tongue?
6.14 Name the three auditory ossicles.

Vertebral Column

The vertebral column extends from the skull to the pelvis and contains 26 bones called **vertebrae** (singular, vertebra). The bones are separated by pads of fibrocartilage called **intervertebral disks.** The disks act as shock absorbers and allow the column to bend. Normally there are four curvatures, illustrated in Figure 6-13, that increase the strength and resilience of the column. They are named according to the region in which they are located. The **thoracic** and **sacral curvatures** are concave anteriorly and are present at birth. The **cervical curvature** develops when an infant begins to hold its head erect. The **lumbar curvature** develops when an infant begins to stand and walk. Both the cervical and lumbar curvatures are convex anteriorly.

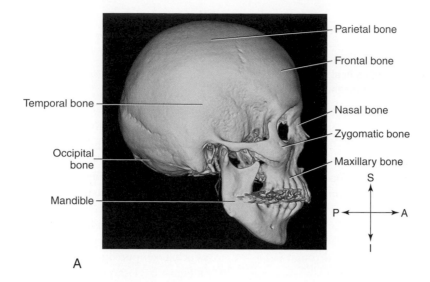

Parietal bone
Frontal bone
Temporal bone
Nasal bone
Zygomatic bone
Occipital bone
Maxillary bone
Mandible

A

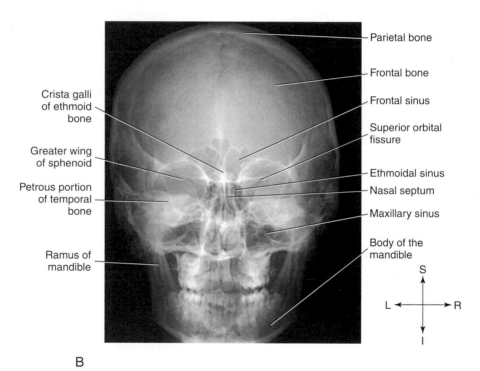

Parietal bone
Frontal bone
Frontal sinus
Crista galli of ethmoid bone
Superior orbital fissure
Greater wing of sphenoid
Ethmoidal sinus
Nasal septum
Petrous portion of temporal bone
Maxillary sinus
Ramus of mandible
Body of the mandible

B

Figure 6-11 *A, Reconstructed CT image of the lateral surface of the skull. **B**, Radiograph of the skull, posteroanterior projection, frontal view. (Adapted from Applegate E: The Sectional Anatomy Learning System, Concepts, ed 3. St. Louis, Elsevier/Saunders, 2010.)*

QUICK APPLICATIONS

An abnormally exaggerated lumbar curvature is called lordosis, or swayback. This is often seen in pregnant women as they adjust to their changing center of gravity. An increased roundness of the thoracic curvature is kyphosis, or hunchback. This is frequently seen in elderly people. Abnormal side-to-side curvature is scoliosis. Abnormal curvatures may interfere with breathing and other vital functions.

General Structure of Vertebrae

All vertebrae have a common structural pattern, illustrated in Figure 6-14, although there are variations among them. The thick anterior, weight-bearing portion is the **body** or **centrum.** The posterior curved portion is the **vertebral arch.** The vertebral arch and body surround a central large opening, the **vertebral foramen.** When all the vertebrae are stacked together in a column, the vertebral foramina make a canal that contains the spinal cord. **Transverse processes**

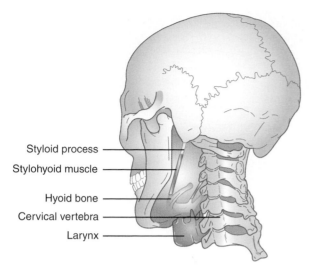

Figure 6-12 *Position and shape of the hyoid bone.*

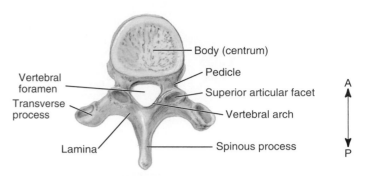

Figure 6-14 *General features of vertebrae, viewed from above.*

project laterally from the vertebral arch, and in the posterior midline there is a **spinous process.** These processes are places for muscle attachment. The spinous processes can be felt as bony projections along the midline of the back. The **pedicle** connects the transverse process to the body and the lamina is between the spinous process and the transverse process. Superior and inferior articular facets provide surfaces for adjacent vertebrae to articulate.

Although all vertebrae have similar general features, there are regional variations that distinguish one type of vertebra from another. These are illustrated in Figure 6-15.

Cervical Vertebrae

There are seven **cervical vertebrae,** designated C1 through C7. In general, the cervical vertebrae, shown in Figure 6-15, *C,* can be distinguished from other vertebrae because the cervical vertebrae have **transverse foramina** in the transverse processes and the spinous processes are forked, or **bifid.** The first two cervical vertebrae are greatly modified and there is no disk between them. The **atlas** (C1), illustrated in Figure 6-15, *A,* has no body, no spinous process, and short transverse processes. It is essentially a ring with large facets that articulate with the occipital condyles on the occipital bone. The **axis** (C2), shown in Figure 6-15, *B,* has a **dens,** or **odontoid** (oh-DON-toyd) **process** that projects upward from the vertebral body like a tooth. The odontoid process acts as a pivot for rotation of the atlas.

> **⚸ QUICK APPLICATIONS**
> The atlas holds up the skull and permits you to nod "yes." The axis allows you to rotate your head from side to side to indicate "no."

Thoracic Vertebrae

There are 12 **thoracic vertebrae,** designated T1 through T12. These can be distinguished from other vertebrae by the facets, located on the bodies and transverse processes, for articulation with the ribs. They also have long, pointed spinous processes. These features are illustrated in Figure 6-15, *D.*

Lumbar Vertebrae

There are five **lumbar vertebrae,** designated L1 through L5, that make up the part of the vertebral column in the small

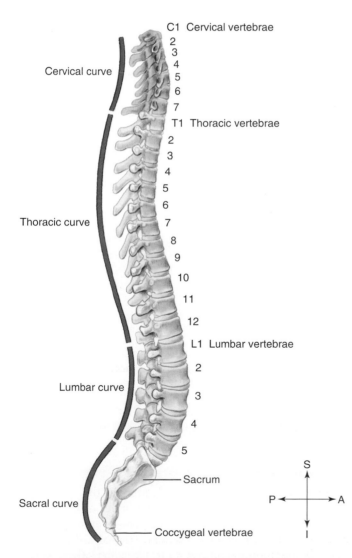

Figure 6-13 *Curvatures of the vertebral column. The thoracic and sacral curvatures are concave anteriorly, and the cervical and lumbar curvatures are convex anteriorly.*

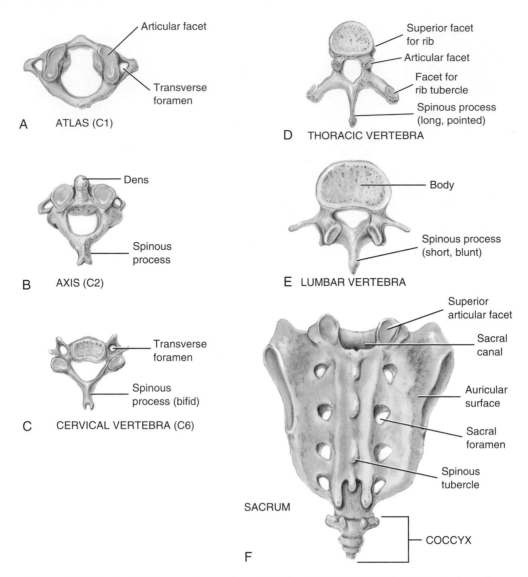

Figure 6-15 *Regional differences in vertebrae.* **A,** *Atlas (C1).* **B,** *Axis (C2).* **C,** *Cervical vertebrae (C3-C7).* **D,** *Thoracic vertebrae (T1-T12).* **E,** *Lumbar (L1-L5).* **F,** *Sacrum and coccyx.*

of the back. The lumbar vertebrae, shown in Figure 6-15, *E,* have large, heavy bodies because they support most of the body weight and have many back muscles attached to them. They also have short, blunt spinous processes.

Sacrum

The **sacrum,** shown in Figure 6-15, *F,* is a triangular bone just below the lumbar vertebrae. In the child there are five separate bones, but these fuse to form a single bone in the adult. The sacrum articulates with the pelvic girdle laterally, at the **sacro-iliac** (say-kro-ILL-ee-ak) **joint,** and forms the posterior wall of the pelvic cavity.

Coccyx

The **coccyx** (KOK-siks), or tailbone, is the last part of the vertebral column (see Figures 6-13 and 6-15, *F*). There are four (the number varies from three to five) separate small bones in the child, but these fuse to form a single bone in the adult. Several muscles have some point of attachment on the coccyx.

> **QUICK CHECK**
> 6.15 What type of vertebra has a transverse foramen?
> 6.16 Thoracic vertebrae have articular facets on the transverse processes. What is the purpose of these facets?
> 6.17 What is the most inferior portion of the vertebral column?

Thoracic Cage

The thoracic cage, or bony thorax, protects the heart, lungs, and great vessels. It also supports the bones of the shoulder girdle and plays a role in breathing. The components of the thoracic cage are the thoracic vertebrae dorsally, the ribs laterally, and the sternum and costal cartilage anteriorly.

Sternum

The **sternum,** or breastbone, is in the anterior midline (Figure 6-16). It consists of three parts: the superior, triangular

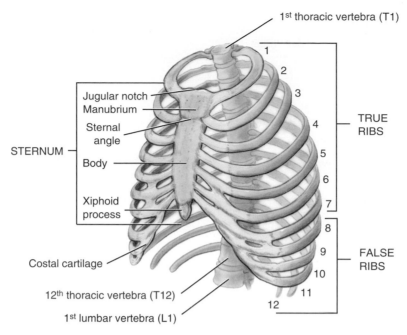

Figure 6-16 *Thoracic cage.*

manubrium (mah-NOO-bree-um); the middle, slender **body;** and the inferior, small **xiphoid** (ZYE-foyd) **process.** The xiphoid process, for muscle attachment, is composed of hyaline cartilage in the child, but it ossifies in the adult. An important anatomical landmark, the **jugular (suprasternal) notch** is an easily palpable, central indentation in the superior margin of the manubrium. The manubrium articulates with the clavicles and the first two pairs of ribs. The manubrium and body of the sternum meet at a slight angle that can be felt as a horizontal ridge below the jugular notch. This is the **sternal angle,** and it is a convenient way to locate the second rib (see Figure 6-16). The body of the sternum has notches along the sides where it attaches to the cartilage of the third through seventh ribs.

QUICK APPLICATIONS

The sternum is frequently used for a red marrow biopsy because it is accessible. The sample for biopsy is obtained by performing a sternal puncture, in which a large needle is inserted into the sternum to remove a sample of red bone marrow.

Ribs

Twelve pairs of **ribs,** illustrated in Figure 6-16, form the curved, lateral margins of the thoracic cage. One pair is attached to each of the 12 thoracic vertebrae. The upper seven pairs of ribs are called **true,** or **vertebrosternal** (ver-TEE-broh-stir-nal), **ribs** because they attach to the sternum directly by their individual **costal cartilage.** The lower five pairs of ribs are called **false ribs** because their costal cartilage does not reach the sternum directly. The first three pairs of false ribs reach the sternum indirectly by joining with the cartilage of the ribs above. These

are called **vertebrochondral** (ver-TEE-broh-kahn-dral) **ribs.** The bottom two rib pairs have no anterior attachment and are called **vertebral ribs,** or **floating ribs.**

QUICK CHECK

6.18 What is the posterior attachment of all ribs?
6.19 What are the three regions of the sternum?

Bones of the Appendicular Skeleton

The 126 bones of the appendicular skeleton are suspended from two yokes or girdles that are anchored to the axial skeleton. They are additions or appendages to the axis of the body. The appendicular skeleton is designed for movement. If a portion is immobilized for a period of time, we realize how awkward life can be without appendicular movement.

Pectoral Girdle

Each half of the **pectoral girdle,** or **shoulder girdle,** consists of two bones: an anterior **clavicle** (KLAV-ih-kul) and a posterior **scapula** (SKAP-yoo-lah). The term "girdle" implies something that encircles or a complete ring. The pectoral girdle, however, is an incomplete ring. Anteriorly, the bones are separated by the sternum. Posteriorly, there is a gap between the two scapulae because they do not articulate with each other or with the vertebral column. The bones of the pectoral girdle, illustrated in Figure 6-17, form the connection between the upper extremities and the axial skeleton. The clavicles and scapulae, with their associated muscles, also form the shoulder.

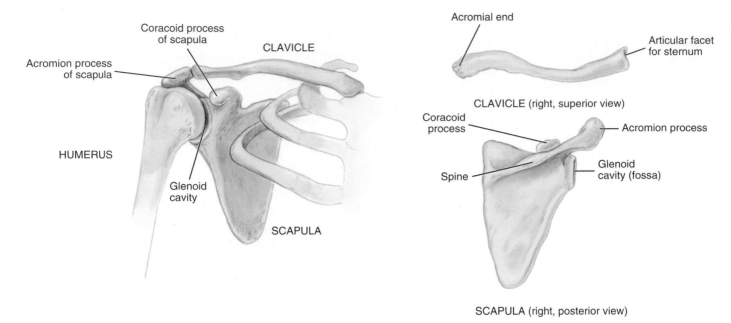

Figure 6-17 *Components of the pectoral girdle, clavicle and scapula.*

The **clavicle** is commonly called the **collarbone.** It is an elongated S-shaped bone that articulates proximally with the manubrium of the sternum. The distal end articulates with the scapula.

QUICK APPLICATIONS

The clavicle is the most frequently fractured bone in the body because it transmits forces from the arm to the trunk. The force from falling on the shoulder or outstretched arm is often sufficient to fracture the clavicle.

The **scapula,** commonly called the **shoulder blade,** is a thin, flat triangular bone on the posterior surface of the thoracic wall. It articulates with the clavicle and the humerus. The scapula has points of attachment for numerous muscles that are involved in movement of the shoulder and arm. A bony ridge called the **spine** divides the dorsal surface into two unequal portions. The lateral end of the spine broadens to form the **acromion** (ah-KRO-mee-on) **process,** which articulates with the clavicle and is a site of muscle attachment. It forms the point of the shoulder. Another process, the **coracoid** (KOR-ah-koyd) **process,** forms a hook that projects forward, under the clavicle. It also serves as a place for muscle attachments. Between the two processes there is a shallow depression, the **glenoid cavity** (fossa), where the head of the humerus connects to the scapula.

QUICK CHECK

6.20 What bone forms the posterior portion of the pectoral girdle?

6.21 What is the medial attachment of the clavicle?

Upper Extremity

The upper extremity (limb) consists of the bones of the arm, forearm, and hand.

Arm

The arm, or **brachium,** is the region between the shoulder and the elbow. It contains a single long bone, the **humerus,** illustrated in Figure 6-18. The **head** is the large, smooth, rounded end that fits into the scapula. Lateral to the head, there are

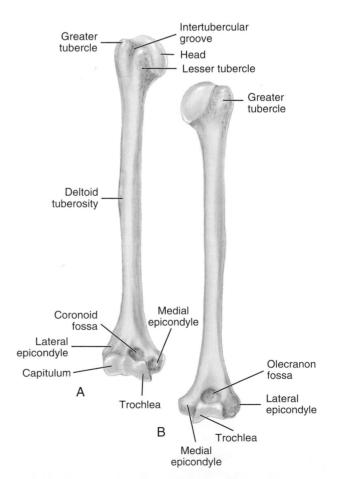

Figure 6-18 *Humerus. A, Anterior view. B, Posterior view.*

two blunt projections for muscle attachment. These are the **greater** and **lesser tubercles,** and the shallow groove between them is the **intertubercular groove.** The **deltoid tuberosity** is an elongated rough area along the shaft. The deltoid muscle attaches to the humerus along this region. The **lateral** and **medial epicondyles,** for the attachment of forearm muscles, project from the sides of the humerus at the distal end, near the elbow region. On the posterior surface, between the two epicondyles, there is a depression, the **olecranon fossa,** where the ulna fits with the humerus to form the hinged elbow joint. On the anterior surface is a shallow depression, the **coronoid fossa,** also for the ulna. Two smooth, rounded projections are evident on the distal end of the humerus. The **capitulum** is on the lateral side and articulates with the radius of the forearm. The **trochlea** is on the medial side and articulates with the ulna of the forearm. Table 6-4 describes important anatomical features on the humerus.

QUICK APPLICATIONS

Tennis elbow is an inflammation of the tissues surrounding the lateral epicondyle of the humerus. Six muscles that control movement of the hand attach in this region, and repeated contraction of these muscles irritates the attachments. The medical term for tennis elbow is *lateral epicondylitis.*

Forearm

The forearm, or **antebrachium,** is the region between the elbow and wrist. It is formed by the **radius** on the lateral side and the **ulna** on the medial side when the forearm is in anatomical position. When the hand is turned so the palm faces backward, the radius crosses over the ulna. The radius and ulna are illustrated in Figure 6-19.

The radius has a circular disklike **head** on the proximal end. This articulates with the capitulum of the humerus. Just inferior to the head and on the medial side of the bone, there is a small rough region called the **radial tuberosity,** which is an

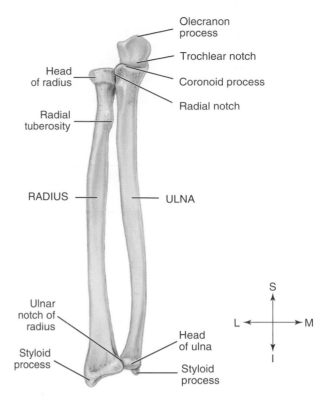

Figure 6-19 *Radius and ulna, anterior view. The radius is on the lateral side and the ulna is the medial bone.*

attachment for the biceps brachii muscle. On the distal end, the prominent markings of the radius are the **styloid process,** a pointed projection on the lateral side, and the **ulnar notch,** a smooth region on the medial side.

The proximal end of the ulna has a wrenchlike shape, with the opening of the wrench being the **trochlear notch,** or semilunar notch. The projection at the upper end of the notch is the **olecranon process,** which fits into the olecranon fossa of the humerus and forms the bony point of the elbow. The projection at the lower end of the notch is the **coronoid process,** which fits into the coronoid fossa of the humerus. The **radial**

TABLE 6-4 Important Marking on the Humerus

Marking	Description	Purpose
Head	Large, smooth, rounded surface on proximal end	Fits into glenoid cavity of scapula
Greater and lesser tubercles	Blunt projections, lateral to head, on proximal end	Attachment for muscles that move shoulder and arm
Intertubercular groove	Shallow groove between two tubercles	Holds tendon for biceps brachii muscle
Deltoid tuberosity	Rough area along shaft	Attachment for deltoid muscle
Lateral and medial epicondyles	Lateral projections from distal end	Attachment for muscles of forearm
Olecranon fossa	Depression on posterior distal end	Space for olecranon process of ulna when elbow is extended
Coronoid fossa	Shallow depression on anterior distal end	Space for coronoid process of ulna when elbow is flexed
Capitulum	Smooth, rounded condyle on lateral side of distal end	Articulates with head of radius of forearm
Trochlea	Smooth, pulley-shaped condyle on medial side of distal end	Articulates with ulna of forearm

TABLE 6-5 Important Markings on the Radius and Ulna

Marking	Description	Purpose
Radius		
Head	Circular, disklike proximal end	Articulates with capitulum of humerus and fits into radial notch of ulna
Radial tuberosity	Small, rough process on medial side, near head	Attachment for biceps brachii muscle
Syloid process	Pointed process on lateral side of distal end	Attachment for ligaments of wrist
Ulnar notch	Small, smooth facet on medial side of distal end	Articulates with head of ulna
Ulna		
Trochlear notch (semilunar notch)	Wrench-shaped opening or cavity at proximal end	Articulates with trochlea of humerus
Olecranon process	Projection at upper end of trochlear notch	Articulates with olecranon fossa of humerus when elbow is extended; provides attachment for muscles
Coronoid process	Projection at lower end of trochlear notch	Articulates with coronoid fossa of humerus when elbow is flexed; provides attachment for muscles
Radial notch	Smooth facet on lateral side of coronoid process	Articulates with head of radius
Head	Knoblike process at distal end	Articulates with ulnar notch of radius
Styloid process	Pointed process on medial side of distal end	Attachments for ligaments of wrist

notch, a smooth region on the lateral side of the coronoid process, is where the head of the radius fits. The **head** is at the distal end, and on the medial side of the head the pointed **styloid process** serves as an attachment point for ligaments of the wrist. The important features of the radius and ulna are described in Table 6-5. The radiograph in Figure 6-20 shows the relationships among the humerus, radius, and ulna.

Hand

The hand, illustrated in Figure 6-21, is composed of the wrist, palm, and five fingers. The wrist, or **carpus,** contains eight small **carpal bones,** tightly bound by ligaments and arranged in two rows of four bones each. The proximal row of bones, adjacent to the radius and ulna, contains the scaphoid, lunate, triquetral, and pisiform. The distal row of bones, from lateral to medial contains the trapezium, trapezoid, capitate, and hamate. The carpal bones are shown in Figure 6-21. The palm of the hand, or **metacarpus,** contains five **metacarpal bones,** one in line with each finger. These bones are not named but are numbered 1 to 5 starting on the thumb side. The 14 bones of the fingers are called **phalanges** (fah-LAN-jeez). Some people refer to these as digits. There are three phalanges in each finger (a proximal, middle, and distal phalanx) except the thumb, or pollex, which has two. The thumb lacks a middle phalanx. The proximal phalanges articulate with the metacarpals. Figure 6-22 shows a CT image of the carpals and metacarpals.

QUICK CHECK

6.22 **What bone is on the medial side of the forearm?**
6.23 **What bones form the palm of the hand?**
6.24 **What portion of the distal humerus articulates with the ulna?**

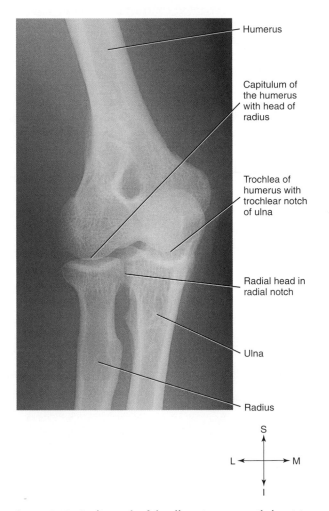

Figure 6-20 *Radiograph of the elbow in an extended position showing the relationships among the humerus, the radius, and the ulna. (Adapted from Applegate E: The Sectional Anatomy Learning System, Concepts, ed 3. St. Louis, Elsevier/Saunders, 2010.)*

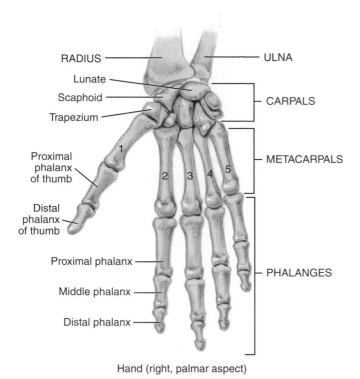

RADIUS — ULNA

Lunate

Scaphoid

Trapezium — CARPALS

1

Proximal
phalanx
of thumb

2   3   4   5

METACARPALS

Distal
phalanx
of thumb

Proximal phalanx

Middle phalanx — PHALANGES

Distal phalanx

Hand (right, palmar aspect)

Figure 6-21 *Hand. The carpals form the wrist, the metacarpals form the palm, and the phalanges form the fingers.*

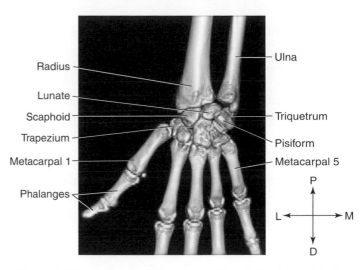

Radius

Lunate

Scaphoid — Triquetrum

Trapezium — Pisiform

Metacarpal 1 — Metacarpal 5

Phalanges

Ulna

P

L ←→ M

D

Figure 6-22 *CT image of the wrist and its relationships with the radius, ulna, and metacarpals. (Adapted from Applegate E: The Sectional Anatomy Learning System, Concepts, ed 3. St. Louis, Elsevier/Saunders, 2010.)*

Pelvic Girdle

The **pelvic girdle,** or **hip girdle,** attaches the lower extremities to the axial skeleton and provides a strong support for the weight of the body. It also provides support and protection for the urinary bladder, a portion of the large intestine, and the internal reproductive organs, which are located in the pelvic cavity.

The pelvic girdle consists of two **coxal** (hip) **bones,** illustrated in Figure 6-23, and their important features are described in Table 6-6. The coxal bones are also called the **ossa coxae,** or **innominate bones.** Anteriorly, the two bones articulate with each other at the **symphysis pubis;** posteriorly, they articulate with the sacrum at the **iliosacral joints.** During childhood, each coxal bone consists of three separate parts: the **ilium,** the **ischium,** and the **pubis.** In the adult, these bones are firmly fused to form a single bone. Where the three bones meet, there is a large depression, the **acetabulum** (as-seh-TAB-yoo-lum), which holds the head of the femur. The **obturator foramen** is a large opening between the pubis and ischium that functions as a passageway for blood vessels, nerves, and muscle tendons.

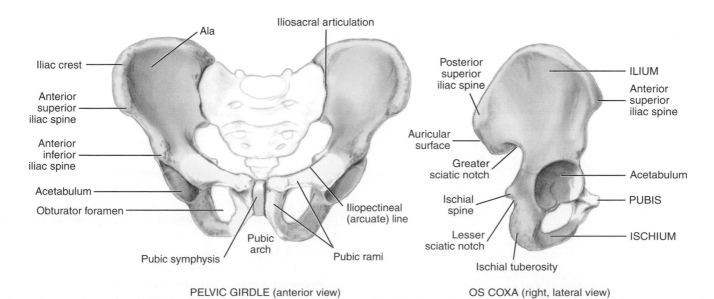

Ala

Iliosacral articulation

Iliac crest

Anterior
superior
iliac spine

Anterior
inferior
iliac spine

Acetabulum

Obturator foramen

Pubic symphysis

Pubic
arch

Pubic rami

Iliopectineal
(arcuate) line

PELVIC GIRDLE (anterior view)

Posterior
superior
iliac spine

Auricular
surface

Greater
sciatic notch

Ischial
spine

Lesser
sciatic notch

Ischial tuberosity

ILIUM

Anterior
superior
iliac spine

Acetabulum

PUBIS

ISCHIUM

OS COXA (right, lateral view)

Figure 6-23 *Bones of the pelvic girdle. The right and left ossa coxae form the pelvic girdle. Posteriorly, the two bones are separated by the sacrum. Anteriorly, they meet at the symphysis pubis.*

TABLE 6-6 Important Markings on the Coxal Bones

Marking	Description	Purpose
Acetabulum	Deep depression on lateral surface of coxal bone	Socket for articulation with head of femur (thigh bone)
Obturator foramen	Large opening between pubis and ischium	Passageway for blood vessels, nerves, muscle tendons; largest foramen in body
Ilium	Large flaring region that forms major portion of coxal bone	
Alae (wings)	Large flared portions of ilium	Large area for numerous muscle attachments; form false pelvis
Iliac crest	Thickened superior margin of ilium	Muscle attachment; forms prominence of hips
Anterior superior iliac spine	Blunt projection at anterior end of iliac crest	Attachment for muscles of trunk, hip, thigh; can be easily palpated
Posterior superior iliac spine	Blunt projection at posterior end of iliac crest	Attachment for muscles of trunk, hip, thigh
Anterior inferior iliac spine	Projection on ilium inferior to posterior superior iliac spine	Attachment for muscles, hip, thigh
Posterior inferior iliac spine	Projection on ilium inferior to posterior superior iliac spine	Attachment for muscles, hip, thigh
Greater sciatic notch	Deep indentation inferior to posterior inferior iliac spine	Passageway for sciatic nerve and some muscle tendons
Iliac fossa	Slight concavity on medial surface of alae	Attachment for iliacus muscle
Auricular surface	Large, rough region at posterior margin of iliac fossa	Articulates with sacrum to form iliosacral (sacroiliac) joint
Iliopectineal (arcuate) line	Sharp curved line at inferior margin of iliac fossa	Attachment for muscles; marks pelvic brim
Ischium	Lower, posterior portion of coxal bone	
Ischial spine	Projection near junction of ilium and ischium; projects into pelvic cavity	Attachment for a major ligament; distance between two spines indicates size of pelvic activity
Ischial tuberosity	Large, rough inferior portion of ischium	Muscle attachment; portion on which we sit; strongest part of coxal bones
Lesser sciatic notch	Indentation below ischial spine	Passageway for blood vessels and nerves
Pubis	Most anterior part of coxal bone	
Pubic symphysis	Anterior midline where two pubic bones meet	Form margins of obturator foramen
Pubic rami	Armlike portions that project from pubic symphysis	
Pubic arch	V-shaped arch inferior to pubic symphysis and formed by inferior pubic rami	Broadens or narrows dimensions of true pelvis

The major portion of the coxal bone is formed by the ilium, which has a large, flared region called the **ala,** or wing. The concavity on the medial surface of the ala is the **iliac fossa.** The superior margin of the ala is the **iliac crest,** which ends anteriorly as a blunt projection, the **anterior superior iliac spine.** Just inferior to this is another blunt projection, the **anterior inferior iliac spine.** Posteriorly, the crest ends in the **posterior superior iliac spine** and just inferior to this is a **posterior inferior iliac spine.** The **greater sciatic notch** forms a deep indentation in the posterior region. The sciatic nerve passes through this notch. The **auricular surface** is the rough region on the posterior part of the ilium, where it meets the sacrum to form the iliosacral (sacroiliac) joint. The **iliopectineal line,** a sharp line at the inferior margin of the iliac fossa, marks the pelvic brim.

The lower, posterior portion of the coxal bone is the ischium. The **ischial tuberosity** is the large, rough, inferior portion of the ischium. Near the junction of the ilium and ischium, the **ischial spine** forms a pointed projection. The indentation inferior to the spine is the **lesser sciatic notch.**

The most anterior portion of the coxal bone is the pubis. The two pubic bones meet at the pubic symphysis and extend laterally and inferiorly from this point. The armlike extensions are the **pubic rami.** Inferiorly, the pubic rami form a V-shaped arch called the **pubic arch.**

Together, the sacrum, coccyx, and pelvic girdle form the basin-shaped pelvis. The **false pelvis** (greater pelvis) is surrounded by the flared portions of the ilium bones and the lumbar vertebrae. The **true pelvis** (lesser pelvis) is smaller and inferior to the false pelvis. It is the region below the **pelvic**

TABLE 6-7 Differences Between Male and Female Pelvis

Characteristic	Description of Difference
Bone thickness	Female bones lighter, thinner, and smoother than male bones; markings more prominent in male
Pelvic cavity	Female pelvic cavity broad, oval, and shallow; male pelvic cavity narrow, deep, and funnel shaped
Pubic arch/angle	Less than 90 degrees in male; 90 degrees or greater in female
Ilium	Bones more flared in female than in male, giving female broader hips
Ischial spines	Farther apart in female
Ischial tuberosities	Farther apart in female
Acetabulum	Smaller and farther apart in female
Pelvic inlet	Wider and more oval in female; narrow, almost heart shaped in male

brim, or **pelvic inlet,** and it is encircled by bone. The large opening at the bottom of this region is the **pelvic outlet.** The dimensions of the true pelvis are especially important in childbirth. Differences between the male and female pelvis, described in Table 6-7, reflect the modifications of the female pelvis for childbearing.

QUICK APPLICATIONS

The female pelvis is shaped to accommodate childbearing. Because the fetus must pass through the pelvic outlet, the physician carefully measures this opening to make sure there is enough room. The distance between the two ischial spines is a good indication of the size of the pelvic outlet. If the opening is too small, a cesarean delivery is indicated.

QUICK CHECK

6.25 What is the purpose of the acetabulum?
6.26 What bone forms the largest portion of an os coxa?

Lower Extremity

The lower extremity (limb) consists of the bones of the thigh, leg, foot, and patella, or kneecap. The lower extremities support the entire weight of the body when we are erect and they are exposed to tremendous forces when we walk, run, and jump. With this in mind, it is not surprising that the bones of the lower extremity are larger and stronger than those in the upper extremity.

Thigh

The **thigh** is the region from the hip to the knee. It contains a single long bone, the **femur,** illustrated in Figure 6-24. It is the largest, longest, and strongest bone in the body.

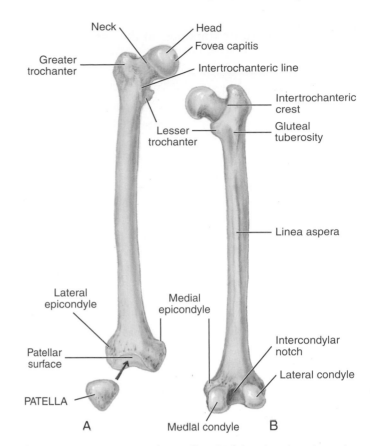

Figure 6-24 *Femur and patella (right).* ***A,*** *Anterior view.* ***B,*** *Posterior view.*

The large, smooth, ball-like **head** of the femur has a small depression called the **fovea capitis.** A ligament attaches here. Prominent projections at the proximal end, the **greater** and **lesser trochanters,** are major sites for muscle attachment. The **intertrochanteric crest** on the posterior side and the **intertrochanteric line** on the anterior side are between the trochanters and are also for muscle attachment. The **neck** is between the head and the trochanters. On the posterior surface, a rough area, the **gluteal tuberosity,** continues inferiorly as the **linea aspera.** Both of these are regions of muscle attachment. The distal end is marked by two large, rounded surfaces, the **lateral** and **medial condyles.** These form joints with the bones of the leg. The **intercondylar notch** is a depression between the condyles that contains ligaments associated with the knee joint. Two small projections superior to the condyles are the **epicondyles.** On the anterior surface, between the condyles, a smooth **patellar surface** marks the area for the kneecap. Table 6-8 describes important anatomical features on the femur.

QUICK APPLICATIONS

Elderly people, particularly those with osteoporosis, are susceptible to "breaking a hip." The femur is a weight-bearing bone, and when it is weakened, it cannot support the weight of the body and the neck of the femur fractures under the stress. Instead of saying, "Grandma fell and broke her hip," often it is more appropriate to say, "Grandma broke her hip, then fell."

TABLE 6-8 Important Markings on the Femur

Marking	Description	Purpose
Head	Large, ball-like proximal end	Fits into acetabulum of coxal bone
Fovea capitis	Small depression on head of femur	Attachment for ligamentum teres femoris
Greater trochanters	Prominent projection from proximal part of shaft	Attachment for gluteus maximus muscle
Lesser trochanters	Smaller projection inferior and medial to greater trochanter	Site of attachment for buttock and hip muscles
Intertrochanteric crest	Between trochanters on posterior side	Muscle attachment
Neck	Between head and trochanters	Muscle attachment; offsets thigh from hip joint for ease in movement
Intertrochanteric line	Between trochanters of anterior side	Muscle attachment
Gluteal tuberosity	Rough area below trochanters on posterior surface	Attachment for gluteal muscles
Linea aspera	Sharp ridge that is continuation of gluteal tuberosity	Muscle attachment
Lateral and medial condyles	Large, rounded surfaces on distal end of femur	Articulate with tibia
Lateral and medial epicondyles	Small projections just above condyles	Muscle attachment
Patellar surface	Smooth area between condyles on anterior surface	Articulates with patella
Intercondylar notch	Large, U-shaped depression between condyles on posterior surface	Contains ligaments associated with knee joint

Leg

The **leg** is the region between the knee and the ankle. It is formed by the slender **fibula** (FIB-yoo-lah) on the lateral side and the larger, weight-bearing **tibia** (TIB-ee-ah), or shin bone, on the medial side. The tibia articulates with the femur to form the knee joint and with the **talus** (one of the foot bones) to allow flexion and extension at the ankle.

The proximal end of the fibula is the **head** and the projection at the distal end is the **lateral malleolus,** which forms the lateral bulge of the ankle. The superior surface of the tibia is flattened and smooth, with two slightly concave regions called the **lateral** and **medial condyles.** The condyles of the femur fit into these regions. Just below the condyles, the **tibial tuberosity** forms a rough area for the attachment of ligaments associated with the knee. The **anterior crest** is a sharp ridge on the anterior surface and forms the shin. On the medial side of the distal end, the **medial malleolus** forms the medial bulge of the ankle. Figure 6-25 illustrates the tibia and fibula, and their important features are summarized in Table 6-9. Figure 6-26 is a radiograph that shows some of the relationships of the femur, tibia, and fibula.

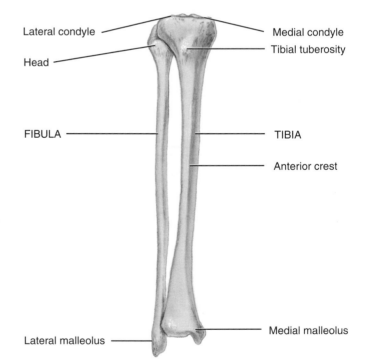

Lateral condyle — Medial condyle
Head — Tibial tuberosity
FIBULA — TIBIA
Anterior crest
Medial malleolus
Lateral malleolus —

Figure 6-25 *Tibia and fibula, anterior view* (right). *The fibula is on the lateral side of the leg and the tibia is on the medial side.*

QUICK APPLICATIONS

Skiers frequently fracture the distal part of the fibula as a result of a twisting, or shearing, force near the ankle. This is called Pott's fracture. Sometimes, the force is sufficient to fracture the medial malleolus at the same time.

Foot

The **foot,** illustrated in Figure 6-27, is composed of the ankle, instep, and five toes. The ankle, or **tarsus,** contains seven **tarsal bones.** These correspond to the carpals in the wrist. The larg-

TABLE 6-9 Important Marking on the Tibia and Fibula

Marking	Description	Purpose
Tibia		
Medial and lateral condyles	Slightly concave, smooth surfaces on proximal end of tibia	Articulate with condyle of femur
Tibial tuberosity	Large, rough area on anterior surface just below condyles	Attachment of patellar ligament
Anterior crest	Sharp ridge on anterior surface of shaft	Forms shin
Medial malleolus	Medial, rounded process at distal end	Forms medial bulge of ankle; attachment for ligaments
Fibula		
Head	Proximal end	Articulates with tibia
Lateral malleolus	Projection at distal end	Forms lateral bulge of ankle; attachment for ligaments

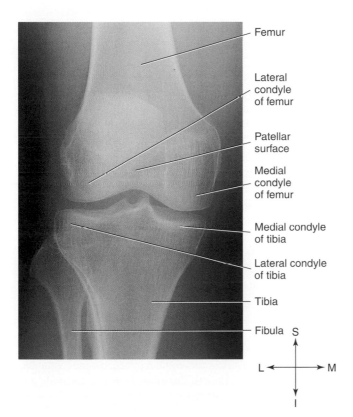

Femur

Lateral condyle of femur

Patellar surface

Medial condyle of femur

Medial condyle of tibia

Lateral condyle of tibia

Tibia

Fibula

Figure 6-26 *Radiograph of the knee showing some of the relationships of the femur, tibia, and fibula. (Adapted from Applegate E: The Sectional Anatomy Learning System, Concepts, ed 3. St. Louis, Elsevier/Saunders, 2010.)*

est tarsal bone is the **calcaneus** (kal-KAY-nee-us), or heel bone. The **talus,** another tarsal bone, rests on top of the calcaneus and articulates with the tibia. The tarsal bones are identified in Figure 6-27. The instep of the foot, or **metatarsus,** contains five **metatarsal bones,** one in line with each toe. The distal

ends of these bones form the ball of the foot. These bones are not named, but are numbered one through five starting on the medial side. The tarsals and metatarsals, together with strong tendons and ligaments, form the arches of the foot. The 14 bones of the toes are called **phalanges.** There are three phalanges in each toe (a proximal, middle, and distal phalanx), except in the great (or big) toe, or hallux, which has only two. The great toe lacks a middle phalanx. The proximal phalanges articulate with the metatarsals. The radiograph in Figure 6-28 shows the bones of the foot.

QUICK APPLICATIONS
Poorly fitted shoes may compress the toes so that there is a lateral deviation of the big toe toward the second toe. When this occurs, a bursa and callus form at the joint between the first metatarsal and proximal phalanx. This creates a bunion.

Patella

The **patella,** or **kneecap,** is a flat, triangular sesamoid bone enclosed within the major tendon that anchors the anterior thigh muscle to the tibia. It provides a smooth surface for the tendon as it turns the corner between the thigh and leg when the knee is flexed. It also protects the knee joint anteriorly.

QUICK CHECK
6.27 Sally developed some bone spurs on her heel. What bone was involved?
6.28 The bone of the lateral side of the leg is frequently fractured in skiing accidents. What bone is this?
6.29 What is the long bone in the thigh?

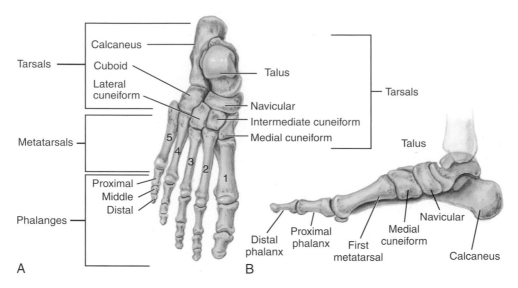

Figure 6-27 *Bones of the foot.* **A,** *Superior view.* **B,** *Lateral view.*

 FROM THE PHARMACY

When people consider disorders of the skeletal system, they often think of pain. This pain may be from fractures, inflammation, arthritis, or some other cause. *Analgesic* is the term used to describe a drug that relieves pain. Some analgesics, but not all, also have anti-inflammatory and antipyretic effects. For minor pain, many people obtain relief with over-the-counter medications. There are three major categories of these analgesics: aspirin, acetaminophen, and nonsteroidal anti-inflammatory drugs (NSAIDs).

Aspirin, or acetylsalicylic acid (ASA), is a widely used analgesic that also has anti-inflammatory and antipyretic effects. It is available with numerous brand names, including Bayer, St. Joseph, Ecotrin, and many others. Aspirin acts by inhibiting the synthesis of prostaglandins in both the central nervous system and the peripheral nervous system. Common adverse effects include stomach irritation, indigestion, nausea, and bleeding. Taking aspirin with a full glass of water helps reduce these effects.

Acetaminophen is available under numerous brand names including Tylenol, Anacin-3, and Liquiprin. It has analgesic and antipyretic effects similar to aspirin, but it does not have anti-inflammatory effects. Like aspirin, the mechanism of action is to inhibit prostaglandin synthesis. An advantage of acetaminophen products is that they may be used by people who are allergic to aspirin and they rarely cause gastrointestinal (GI) distress or bleeding.

Nonsteroidal anti-inflammatory drugs all started out as prescription drugs. Now some of these, in relatively low doses, have been approved by the FDA for change to over-the-counter status. Higher doses still require a prescription. Over-the-counter NSAIDs include **ibuprofen** (Motrin, Advil, and others) and **naproxen** (Alleve). These medications have analgesic, anti-inflammatory, and antipyretic effects similar to aspirin and they also act by inhibiting prostaglandin synthesis. They are rapidly metabolized and eliminated in the urine. Use of these drugs should be avoided in patients with renal insufficiency and they should be used with caution in the elderly. GI side effects can be serious but these may be minimized by taking the medication with food.

Women, especially, think of osteoporosis when considering disorders of the skeletal system. **Alendronate** (Fosamax) is a biphosphonate that is incorporated into bone and is used in the treatment of osteoporosis. It inhibits osteoclast activity to decrease bone destruction and it does this without affecting bone formation. Alendronate can have serious GI side effects and must be used carefully to prevent erosive esophagitis.

Fractures and Fracture Repair

A bone fracture is any break in the continuity of a bone. Traumatic injury is the most common cause of fractures, although metabolic disorders and aging may weaken bones to the point at which they can withstand very little stress and they fracture spontaneously. Fractures occur more frequently in children than in adults because children have slender bones and are more active. Fortunately for the children, their bones tend to heal more quickly than adults because of greater osteoblast activity in young people.

Figure 6-29 illustrates some of the more commonly used terms for describing fractures.

Usually, the pieces of bone in a displaced fracture can be brought into normal position by physical manipulation without surgery. This is called a **closed reduction.** In some cases, surgery is necessary to expose the fractured bone fragments and bring them into normal alignment. This process is called **open reduction.**

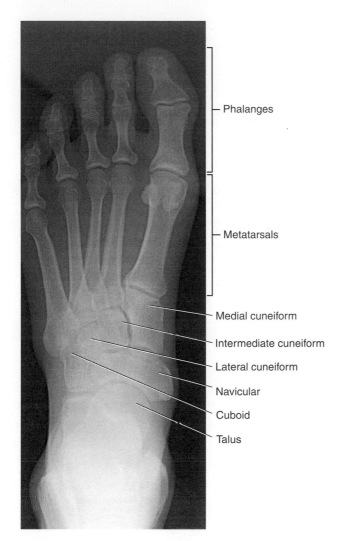

Figure 6-28 *Radiograph of the foot. (Adapted from Applegate E: The Sectional Anatomy Learning System, Concepts, ed 3. St. Louis, Elsevier/Saunders, 2010.)*

After the reduction of a fracture, a complex series of events occurs that usually results in satisfactory healing. Anyone who has ever had a fracture knows firsthand that the healing process is slow. Several factors contribute to this. The blood vessels to the bone are damaged by the break, and whenever vascular supply is reduced, there is a corresponding reduction in cellular metabolism and mitosis. The physical trauma of the break tears and destroys tissues in the region, and a lack of blood supply further damages tissues. This dead and damaged tissue inhibits repair. A third factor that makes bone repair slow is the fact that the bone cells reproduce slowly. An infection slows the process even further.

When a bone fracture occurs, the periosteum and numerous blood vessels that cross the fracture line are torn. The blood from the torn vessels forms a clot, or **fracture hematoma,** which plugs the gap between the ends of the bones. The fracture hematoma, usually formed within a few hours after the injury, stops blood circulation in the region so that cells in the area, including periosteal and bone cells, die. This dead tissue and reduced vascular supply in the traumatized area seem to initiate the healing process.

After the fracture hematoma forms, blood capillaries start to grow into the hematoma and organize it into a **procallus.** These new vessels bring phagocytic cells, such as neutrophils and macrophages, to start cleaning up the dead debris. The cleanup is an ongoing process that takes several weeks. While cleanup is in progress, fibroblasts from neighboring healthy tissue migrate to the fracture area and produce collagen fibers within the procallus. The fibers help tie the ends of the bones together. Chondroblasts, which develop from bone cells that do not receive enough blood, produce fibrocartilage that transforms the procallus into a **fibrocartilaginous callus.** This lasts about 3 weeks.

Osteoblasts from neighboring healthy bone tissue start to produce trabeculae of spongy bone. As the trabeculae grow, they infiltrate the fibrocartilaginous callus to make a **bony callus.** This bony callus, consisting of spongy bone, lasts 3 to 4 months before it is remodeled into bone that is very similar to the original. **Remodeling** is the final step in the repair process. Osteoblasts lay down new compact bone around the periphery whereas osteoclasts reabsorb spongy bone from the inside and form a new medullary cavity. Successful healing results in a repaired bone that is so similar in structure to the original that the fracture line may not be visible on a radiograph, although the bone may be slightly thicker in that region. If healing is complete, original bone strength is also restored.

QUICK CHECK

6.30 What is another name for a compound fracture?
6.31 What is the first event in the healing process after a fracture?

Articulations

An **articulation** (ahr-tik-yoo-LAY-shun), or joint, is where two bones come together. In terms of the amount of movement they allow, there are three types of joints: immovable, slightly movable, and freely movable.

Synarthroses

Synarthroses (sin-ahr-THROH-seez) are immovable joints. The singular form is synarthrosis. In these joints, the bones come in very close contact and are separated only by a thin layer of fibrous connective tissue. The **sutures** in the skull are examples of immovable joints.

Amphiarthroses

Slightly movable joints are called **amphiarthroses** (am-fee-ahr-THROH-seez). The singular form is amphiarthrosis. In this type of joint, the bones are connected by hyaline cartilage or fibrocartilage. The ribs connected to the sternum by costal cartilage are slightly movable joints connected by hyaline cartilage. The symphysis pubis is a slightly movable joint in which there is a fibrocartilage pad between the two bones. The joints between the vertebrae, the intervertebral disks, are also of this type.

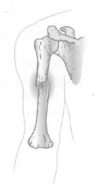

Complete—The break extends across the entire section of bone.

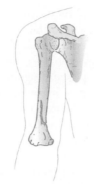

Incomplete— The fracture still has pieces of the bone partially joined together.

Open—Fracture in which the broken end of the bone protrudes through the surrounding tissues and skin, which presents an open pathway for infection; also called a compound fracture.

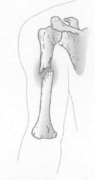

Closed—Fracture in which the bone does not extend through the skin so that there is less chance of bacterial invasion; also called a simple fracture.

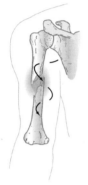

Spiral—The bone is broken by twisting.

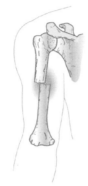

Transverse—The bone is broken at right angles to the long axis of the bone.

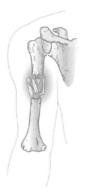

Comminuted—The bone is crushed into small pieces.

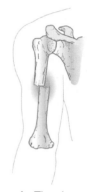

Displaced—The pieces of bone are not in correct alignment.

Figure 6-29 *Classification and description of fractures.*

Diarthroses

Most joints in the adult body are **diarthroses** (dye-ahr-THROH-seez), or freely movable joints. The singular form is diarthrosis. In this type of joint, the ends of the opposing bones are covered with hyaline cartilage, the **articular cartilage,** and they are separated by a space called the **joint cavity.** The components of the joints are enclosed in a dense fibrous **joint capsule** (Figure 6-30). The outer layer of the capsule consists of the ligaments that hold the bones together. The inner layer is the **synovial membrane** that secretes **synovial fluid** into the joint cavity for lubrication. Because all of these joints have a synovial membrane, they are sometimes called **synovial joints.**

Some diarthroses have pads and cushions associated with them. The knee has fibrocartilaginous pads, called **semilunar cartilages** or the **lateral meniscus** (meh-NIS-kus) and **medial meniscus,** which rest on the lateral and medial condyles of the tibia. The pads help stabilize the joint and act as shock absorbers. **Bursae** are fluid-filled sacs that act as cushions and help reduce

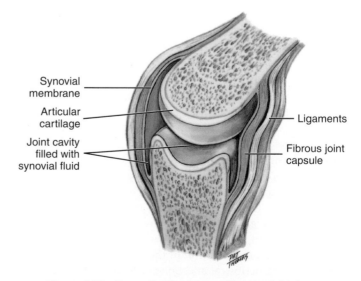

Synovial membrane

Articular cartilage

Joint cavity filled with synovial fluid

Ligaments

Fibrous joint capsule

Figure 6-30 *Generalized structure of a synovial joint.*

friction. Bursae are lined with a synovial membrane that secretes synovial fluid into the sac. They are commonly located between the skin and underlying bone or between tendons and ligaments. Inflammation of a bursa is called **bursitis.** Figure 6-31 illustrates some menisci and bursae associated with the knee joint.

 QUICK **APPLICATIONS**

The term *torn cartilage* refers to a damaged meniscus, usually the medial, in the knee. Frequently this can be repaired with relatively minor arthroscopic surgery. A torn ligament in the knee usually involves one of the cruciate ligaments. The surgical procedure to repair this damage is quite involved, and recovery of function may require months of rehabilitative therapy.

 QUICK **CHECK**

6.32 An athlete is diagnosed as having a torn meniscus. What joint is affected?

6.33 Which type of joint is the only one with a joint cavity?

6.34 The knee and elbow are classified as what type of freely movable joint?

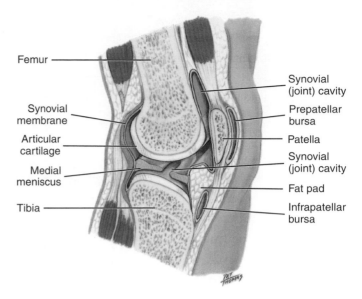

Figure 6-31 *Sagittal section of the knee joint illustrating bursae and menisci.*

There are six types of diarthrotic or freely movable joints based on the shapes of their parts and the types of movement they allow. These are illustrated and described in Figure 6-32.

 FOCUS ON AGING

The major age-related change in the skeletal system is the loss of calcium from the bones. Calcium loss occurs in both men and women, but it starts at an earlier age and is more severe in women. The exact reasons for the loss are unknown and possibly involve a combination of several factors. These may include an imbalance between osteoblast and osteoclast activity, imbalance between calcitonin and parathormone levels, reduced absorption of calcium and/or vitamin D from the digestive tract, poor diet, and lack of exercise. Whatever the cause, there is no sure way of preventing the loss, but adequate calcium and vitamin D in the diet may help reduce the effects. Exercise is also important in reducing calcium loss from bones.

Another change with age is a decrease in the rate of collagen synthesis. This means that the bones have less strength and are more brittle. Bones fracture more readily in elderly individuals and the healing process may be slow or incomplete. Tendons and ligaments become less flexible because of the changes in collagen.

The articular cartilage at the ends of bones tends to become thinner and deteriorates with age. This causes joint disorders that are commonly found in older individuals. People also appear to get shorter as they get older. This is partially due to loss of bone mass and partially due to compression of the intervertebral disks.

Age-related changes in the skeletal system cannot be prevented. An active and healthy lifestyle with appropriate exercise and an adequate diet help reduce the effect of the changes in the skeletal system.

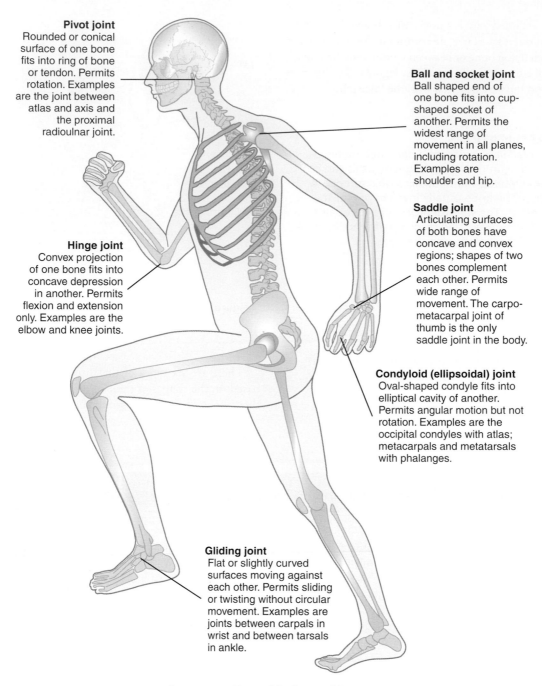

Pivot joint
Rounded or conical surface of one bone fits into ring of bone or tendon. Permits rotation. Examples are the joint between atlas and axis and the proximal radioulnar joint.

Ball and socket joint
Ball shaped end of one bone fits into cup-shaped socket of another. Permits the widest range of movement in all planes, including rotation. Examples are shoulder and hip.

Saddle joint
Articulating surfaces of both bones have concave and convex regions; shapes of two bones complement each other. Permits wide range of movement. The carpo-metacarpal joint of thumb is the only saddle joint in the body.

Hinge joint
Convex projection of one bone fits into concave depression in another. Permits flexion and extension only. Examples are the elbow and knee joints.

Condyloid (ellipsoidal) joint
Oval-shaped condyle fits into elliptical cavity of another. Permits angular motion but not rotation. Examples are the occipital condyles with atlas; metacarpals and metatarsals with phalanges.

Gliding joint
Flat or slightly curved surfaces moving against each other. Permits sliding or twisting without circular movement. Examples are joints between carpals in wrist and between tarsals in ankle.

Figure 6-32 *Types of freely movable joints.*

Representative Disorders of the
Skeletal System

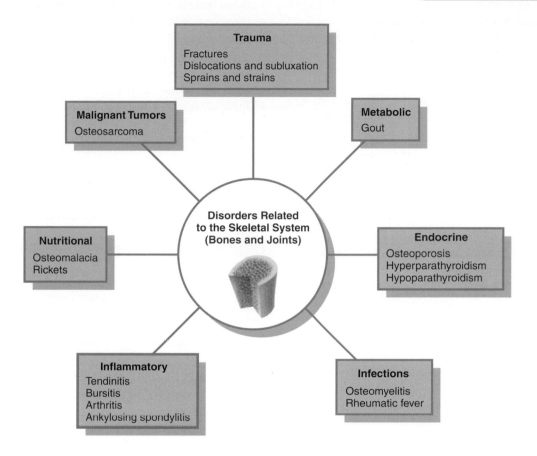

Trauma
Fractures
Dislocations and subluxation
Sprains and strains

Malignant Tumors
Osteosarcoma

Metabolic
Gout

**Disorders Related
to the Skeletal System
(Bones and Joints)**

Nutritional
Osteomalacia
Rickets

Endocrine
Osteoporosis
Hyperparathyroidism
Hypoparathyroidism

Inflammatory
Tendinitis
Bursitis
Arthritis
Ankylosing spondylitis

Infections
Osteomyelitis
Rheumatic fever

Ankylosing spondylitis (ANG-kih-loh-sing spahn-dih-LYE-tis) Inflammation of the spine that is characterized by stiffening of the spinal joints and ligaments so that movement becomes increasingly painful and difficult; also called rheumatoid spondylitis

Arthritis (ahr-THRYE-tis) Inflammation of a joint

Bursitis (burr-SYE-tis) Inflammation of a bursa, most commonly in the shoulder, often caused by excessive use of the joint and accompanied by severe pain and limitation of motion

Dislocation (dis-loh-KAY-shun) Displacement of a bone from its joint with tearing of ligaments, tendons, and articular capsule; also called luxation

Fracture (FRACK-shur) Break in the continuity of a bone

Gout (GOWT) Form of acute arthritis in which uric acid crystals develop within a joint and irritate the cartilage, causing acute inflammation, swelling, and pain; most commonly occurs in middle-aged and older men

Hyperparathyroidism (hye-purr-pair-ah-THIGH-royd-izm) Abnormally increased activity of the parathyroid gland, which causes a generalized decalcification of bone, increased blood calcium levels, and kidney stones

Hypoparathyroidism (hye-poh-pair-ah-THIGH-royd-izm) Abnormally decreased activity of the parathyroid gland, which leads to increased bone density, decreased blood calcium levels, neuromuscular excitability, and tetany

Osteomalacia (ahs-tee-oh-mah-LAY-shee-ah) Softening of bone because of inadequate amounts of calcium and phosphorus; bones bend easily and become deformed; in childhood, this is called rickets

Osteomyelitis (ahs-tee-oh-my-eh-LYE-tis) Inflammation of the bone marrow caused by bacteria

Osteoporosis (ahs-tee-oh-por-OH-sis) Decrease in bone density and mass; commonly occurs in postmenopausal women as a result of increased osteoclast activity caused by diminished estrogen levels; bones fracture easily

Osteosarcoma (ahs-tee-oh-sahr-KOH-mah) Malignant tumor derived from bone; also called osteogenic sarcoma; osteoblasts multiply without control and form large tumors in bone

Rheumatic fever (roo-MAT-ic FEEver) Childhood (usually) disease associated with the presence of hemolytic streptococci bacteria in the body and characterized by fever and joint pain

Rickets (RICK-ehts) Softening of bone that usually occurs in childhood because of inadequate amounts of absorbed calcium and phosphorus, often caused by a lack of vitamin D; bones bend easily and become deformed

Sprain (SPRAYN) Twisting of a joint with pain, swelling, and injury to ligaments, tendons, muscles, blood vessels, and nerves; most often occurs in the ankle; more serious than a strain, which is the overstretching of the muscles associated with a joint

Strain (STRAYN) Overstretching or overexertion of a muscle; less serious than a sprain

Subluxation (sub-luck-SAY-shun) Incomplete or partial dislocation of a joint

Tendinitis (ten-dih-NYE-tis) Inflammation of tendons and tendon-muscle attachments, frequently associated with calcium deposits; common cause of acute shoulder pain

CHAPTER SUMMARY

Overview of the Skeletal System

■ **Discuss five functions of the skeletal system.**

- Bones support the soft organs of the body and support the body against the pull of gravity.
- Bones protect soft body parts such as the brain, spinal cord, and heart.
- Bones store minerals, especially calcium.
- Most blood cell formation, hematopoiesis, occurs in red bone marrow.

■ **Distinguish between compact and spongy bone on the basis of structural features.**

- The microscopic unit of compact bone is the osteon or haversian system. It consists of an osteonic canal, lamellae of matrix, osteocytes in lacunae, and canaliculi. In compact bone, the osteons are packed closely together.
- Spongy bone is less dense than compact bone and consists of bone trabeculae around irregular cavities that contain red bone marrow. The trabeculae are organized to provide maximum strength to a bone.

■ **Classify bones according to size and shape.**

- Long bones are longer than they are wide; an example is the femur in the thigh.
- Short bones are roughly cube shaped; examples include the bones of the wrist.
- Flat bones have inner and outer tables of compact bone with a diploë of spongy bone in the middle; examples include the bones of the cranium.
- Irregular bones are primarily spongy bone with a thin layer of compact bone; examples include the vertebrae.

■ **Identify the general features of a long bone.**

- Long bones have a diaphysis around a medullary cavity, with an epiphysis at each end. The epiphysis is covered by articular cartilage. Except in the region of the articular cartilage, long bones are covered by periosteum and lined with endosteum. All bones have surface markings that make each one unique.

■ **Name and define three types of cells in bone.**

- Osteoblasts are bone-forming cells.
- Osteocytes are mature bone cells.
- Osteoclasts break down and reabsorb bone matrix.

■ **Discuss intramembranous and endochondral ossification.**

- Bone development is called osteogenesis.
- Intramembranous ossification involves the replacement of connective tissue membranes by bone tissue. Flat bones of the skull develop this way.
- Most bones develop by endochondral ossification. In this process, the bones first form as hyaline cartilage models, which are later replaced by bone.

■ **Describe how bones increase in length and diameter.**

- Long bones increase in length at the cartilaginous epiphyseal plate. When the epiphyseal plate completely ossifies, increase in length is no longer possible.
- Increase in diameter of long bones occurs by appositional growth. Osteoclasts break down old bone next to the medullary cavity at the same time that osteoblasts form new bone on the surface.

■ **Distinguish between the axial and appendicular skeletons, and state the number of bones in each.**

- The adult human skeleton has 206 named bones. In addition, there are varying numbers of Wormian and sesamoid bones.
- The axial skeleton, with 80 bones, forms the vertical axis of the body. It includes the skull, vertebra, ribs, and sternum.
- The appendicular skeleton, with 126 bones, includes the appendages and their attachments to the axial skeleton.

Bones of the Axial Skeleton

■ **Identify the bones of the skull and their important surface markings.**

- The skull includes the bones of the cranium, face, and the auditory ossicles.
- With the exception of the mandible, the cranial and facial bones are joined by immovable joints called sutures.
- The eight cranial bones are frontal (1), parietal (2), temporal (2), occipital (1), ethmoid (1), and sphenoid (1). The frontal, ethmoid, and sphenoid bones contain cavities called paranasal sinuses.
- The 14 bones of the face are maxillae (2), nasal (2), lacrimal (2), zygomatic (2), vomer (1), inferior nasal conchae (2), palatine (2), and mandible (1). Each maxilla contains a large maxillary paranasal sinus.
- Six auditory ossicles, three in each ear, are located in the temporal bone.
- The hyoid bone is a U-shaped bone in the neck. It does not articulate with any other bone and functions as a base for the tongue and as an attachment for muscles.

■ **Identify the general structural features of vertebrae and compare cervical, thoracic, lumbar, sacral, and coccygeal vertebrae; state the number of each type.**

- The vertebral column contains 26 vertebrae that are separated by intervertebral disks.
- Four curvatures add strength and resiliency to the column. The thoracic and sacral curvatures are concave anteriorly, and cervical and lumbar curvatures are convex anteriorly.
- All vertebrae have a body or centrum, a vertebral arch, a vertebral foramen, transverse processes, and a spinous process.
- The seven cervical vertebrae have bifid spinous processes and transverse foramina. The first two cervical vertebrae are unique; the atlas, C-1, is a ring that holds up the occipital bone; the axis, C-2, has a dens or odontoid process.
- The twelve thoracic vertebrae have facets on the bodies and transverse processes for articulation with the ribs.
- The five lumbar vertebrae have large heavy bodies to support body weight.
- The sacrum is formed from five separate bones that fuse together in the adult.
- The coccyx is the most distal part of the vertebral column. Three to five bones fuse to form the coccyx.

■ **Identify the structural features of the ribs and sternum.**

- The thoracic cage consists of the thoracic vertebrae, the ribs, and the sternum.
- The sternum has three parts: the manubrium, the body, and the xiphoid. The jugular notch in the manubrium is an important landmark. The sternal angle is when the manubrium joins the body.
- There are seven pairs of true ribs (vertebrosternal ribs), and five pairs of false ribs. The upper three pairs of false ribs are vertebrochondral, the lower two pairs are vertebral or floating ribs.

Bones of the Appendicular Skeleton

■ **Identify the bones of the pectoral girdle and their surface features.**

- The pectoral girdle is formed by two clavicles and two scapulas, one on each side; it supports the upper extremities.
- The clavicle, commonly called the collar bone, is the anterior bone. It articulates medially (proximally) with the sternum and laterally (distally) with the scapula.
- The scapula, commonly called the shoulder blade, is the posterior bone. Its features include a spine, acromion process, caracoid process, and glenoid cavity (fossa).

■ **Identify the bones of the upper extremity and their surface features.**

- The upper extremity includes the humerus, radius, ulna, carpus, metacarpus, and phalanges.
- The humerus is the bone in the arm, or brachium; in the forearm, the radius is the lateral bone and the ulna is the medial bone.
- The hand is composed of a wrist, palm, and five fingers. The wrist, or carpus, contains eight small carpal bones; the palm, or metacarpus, contains five metacarpal bones; the 14 bones of the fingers are phalanges.

■ **Identify the bones of the pelvic girdle and their surface features.**

- The pelvic girdle consists of two ossa coxae, or innominate bones, and attaches the lower extremities to the axial skeleton and provides support for the weight of the body.
- Each os coxa is formed by three bones fused together: ilium, ischium, and pubis. The two ossa coxae meet anteriorly at the symphysis pubis.
- The ilium, ischium, and pubis meet in a large cavity; the acetabulum, that provides articulation for the femur.
- The false pelvis (greater pelvis) is the area between the flared portions of the ilium bones; the true pelvis (lesser pelvis) is inferior to the false pelvis and begins at the pelvic brim.

■ **Identify the bones of the lower extremity and their surface features.**

- The bones in the lower extremity are the femur, tibia, fibula, patella, tarsal bones, metatarsals, and phalanges.
- The bone in the thigh is the femur, which articulates in the acetabulum of the os coxa. Distally, the femur articulates with the tibia.
- The medial bone in the leg is the tibia. Proximally it articulates with the femur, distally with the talus. The lateral bone in the leg is the fibula.

- Seven tarsal bones form the ankle. The talus is the tarsal bone that articulates with the tibia. The calcaneus is the heel bone. Five metatarsals form the instep of the foot, and there are 14 phalanges in the toes of each foot.
- The patella is a sesamoid bone in the anterior portion of the knee joint.

Fractures and Fracture Repair

■ **Identify these fractures by description and on diagrams: complete, incomplete, open, closed, transverse, spiral, comminuted, and displaced.**

- In a complete fracture, the break extends across the entire section of bone; in an incomplete fracture, there are pieces of the bone partially joined together.
- An open fracture has the bone protruding through the skin and is also called a compound fracture; a closed fracture is a simple fracture where the bone does not extend through the skin.
- The bone is broken at right angles to the long axis of the bone in transverse fractures; in spiral fractures, the bone is broken by twisting.
- If the bone is crushed into small pieces, it is a comminuted fracture; if the pieces of bone are not in correct alignment, it is a displaced fracture.

■ **Describe the process of fracture repair by using the terms fracture hematoma, procallus, fibrocartilaginous callus, bony callus, and remodeling.**

- Blood vessels are torn when a bone breaks and the blood from these vessels clots to form a fracture hematoma.
- New capillaries start to grow into the fracture hematoma and organize it into a procallus.
- Fibroblasts from neighboring healthy tissue migrate to the fracture area and produce collagen fibers within the procallus; chondroblasts produce fibrocartilage that transforms the procallus into a fibrocartilaginous callus.
- Osteoblasts infiltrate the fibrocartilaginous callus and produce trabeculae of spongy bone to form a bony callus.
- Remodeling is the final step. In this process, osteoblasts and osteoclasts remodel the bony callus into an original configuration of compact bone and spongy bone.

Articulations

■ **Compare the structure and function of synarthroses, amphiarthroses, and diarthroses. Give an example of each type.**

- Synarthroses are immovable joints where the bones are held together by short fibers; sutures are synarthrotic joints.
- Slightly moveable joints are amphiarthroses. In this type of joint, the bones are connected by hyaline cartilage or fibrocartilage. The symphysis pubis and intervertebral disks are examples of amphiarthrotic joints.
- Joints that are freely moveable are diarthroses. In this type of joint, the bones are held together by a fibrous joint capsule that is lined with synovial membrane. These joints are sometimes called synovial joints. There are six types of synovial joints, based on the shape of the opposing bones and the type of motion: gliding, condyloid, hinge, saddle, pivot, and ball-and-socket.

CHAPTER QUIZ

Recall

Match the definitions on the left with the appropriate term on the right.

_____ **1.** Blood cell formation

_____ **2.** Shaft of a long bone

_____ **3.** Opening in bone for blood vessels and nerves

_____ **4.** Bone cells

_____ **5.** Connective tissue is replaced by bone

_____ **6.** Increase in bone diameter

_____ **7.** Slightly movable joints

_____ **8.** Inflammation of fluid-filled sacs related to joints

_____ **9.** Narrow cleft or slit

_____ **10.** Location of red bone marrow

A. Amphiarthroses

B. Appositional growth

C. Bursitis

D. Diaphysis

E. Fissure

F. Foramen

G. Hematopoiesis or hemopoiesis

H. Intramembranous ossification

I. Osteocytes

J. Spongy bone

Thought

1. Intramembranous ossification occurs in the
   A. Femur
   B. Parietal bone
   C. Radius
   D. Carpals

2. Which of the following does *not* belong to the appendicular skeleton?
   A. Clavicle
   B. Scapula
   C. Sternum
   D. Patella

3. The bone on the lateral side of the forearm is the
   A. Humerus
   B. Ulna
   C. Fibula
   D. Radius

4. The bones that form the pectoral girdle are the
   A. Scapula and sternum
   B. Sternum and clavicle
   C. Scapula and clavicle
   D. Humerus and scapula

5. An example of a diarthrosis is the
   A. Coronal suture
   B. Elbow
   C. Symphysis pubis
   D. Joints between two vertebrae

Application

Why do you think lumbar vertebrae have large, thick bodies and thick intervertebral disks instead of small ones like the cervical vertebrae?

BUILDING YOUR MEDICAL VOCABULARY

Building a vocabulary is a cumulative process. As you progress through this book, you will use word parts, abbreviations, and clinical terms from previous chapters. Each chapter will present new word parts, abbreviations, and terms to add to your expanding vocabulary.

Word Parts and Combining Forms with Definitions and Examples

PART/COMBINING FORM	DEFINITION	EXAMPLE
acetabul/o	little cup	acetabulum: the little cup (rounded socket) in the os coxa to hold the head of the femur
ankyl/o	stiff, crooked	ankylosis: abnormal immobility of a joint
appendicul-	little attachment	appendicular: pertaining to an appendage or attachment
artic-	joint	articulate: united by a joint
arthr-	joint	arthritis: inflammation of a joint
-blast	to form, sprout	osteoblast: a cell that forms bone
burs-	pouch	bursa: a small fluid-filled sac or pouch that cushions or reduces the friction at certain points in a freely moving joint
carp-	wrist	carpal: pertaining to the wrist
-clast-	to break	osteoclast: a cell that breaks down bone
corac-	beak	coracoid: a process on a bone that looks like a beak, for example, the coracoid process of the scapula
cost-	rib	chondrocostal: pertaining to cartilage that is attached to ribs
cribr-	sieve	cribriform: having the form of a sieve, the cribriform plate of the ethmoid bone has many holes like a sieve
crist-	crest, ridge	crista galli: a crest or ridge that projects upward from the ethmoid bone

PART/COMBINING FORM	DEFINITION	EXAMPLE
ethm-	sieve	ethmoid: the sieve-like bone in the floor of the cranium
ili-	ilium	sacroiliac: pertaining to the sacrum and the ilium
kyph/o	hump	kyphosis: humpback, exaggerated thoracic spinal curvature
myel/o	spinal cord, bone marrow	myelocyte: one of the typical cells of red bone marrow; myelitis is inflammation of the spinal cord
odont-	tooth	odontoid process: projection on C2 that is shaped like a tooth
-oid	like, resembling	ethmoid: resembling a sieve
oste/o, oss-	bone	osteocyte: mature bone cell
-poie-	making	hemopoiesis: production of blood or its cells
sacr/o	sacrum	sacroiliac: pertaining to the sacrum and the ilium
sphen/o	wedge	sphenoid: resembling a wedge, sphenoid bone resembles a wedge
spondyl/o	vertebrae	spondylopathy: any disease of the bertebrae
syn-	together	synarthroses: a type of joint in which the adjoining bones are firmly held together by short fibrous elements

Clinical Abbreviations

ABBREVIATION	MEANING
ACL	anterior cruciate ligament
C1, C2, etc.	cervical vertebrae
CTS	carpal tunnel syndrome
EMG	electromyography
fx	fracture
L1, L2, etc.	lumbar vertebrae
NSAID	nonsteroidal anti-inflammatory drug
OA	osteoarthritis

ABBREVIATION	MEANING
PT	physical therapy
RA	rheumatoid arthritis
ROM	range of motion
SI	sacroiliac
SLE	systemic lupus erythematosus
Sx	symptoms
T1, T2, etc.	thoracic vertebrae
THR	total hip replacement
TKR	total knee replacement
TMJ	temporomandibular joint
Tx	traction, treatment

Clinical Terms

Arthrocentesis (ahr-throw-sen-TEE-sis) Surgical puncture to remove fluid from a joint cavity

Arthroscopy (ahr-THROHS-koh-pee) Visual examination of the inside of a joint with an endoscope and camera; orthopedists use this to remove and repair damaged tissue

Bunion (BUN-yun) Abnormal swelling of the joint between the big toe and the first metatarsal bone, resulting from a buildup of soft tissues and bone caused by chronic irritation from ill-fitting shoes

Carpal tunnel syndrome (KAHR-pull TUH-nul SIN-drohm) Condition characterized by pain and burning sensations in the fingers and hand, caused by compression of the median nerve as it passes between a wrist ligament and the bones and tendons of the wrist

Chiropractor (kye-roh-PRACK-tor) A practitioner who is not a physician and who uses physical means to manipulate the spinal column

Exostosis (eck-sohs-TOE-sis) Bony growth arising from the surface of a bone

Herniated disk (her-nee-A-ted DISK) Abnormal protusion of an intervertebral disk into the neural canal with subsequent compression of spinal nerves

Lyme disease (LYME dih-ZEEZ) A bacterial disease transmitted to humans by deer ticks; characterized by joint stiffness, headache, fever and chills, nausea, and back pain; complications include severe arthritis and cardiac problems; early stages of the disease respond well to antibiotics

Orthopedist (or-tho-PEED-ist) A physician who specializes in the treatment of bone and joint diseases

Osteoarthritis (ahs-tee-oh-ahr-THRYE-tis) A noninflammatory disease of the joints that is characterized by degeneration of the articular cartilage and changes in the synovial membrane; also called degenerative joint disease (DJD)

Rheumatoid arthritis (ROO-mah-toyd ahr-THRYE-tis) A chronic systemic disease with changes occurring in the connective tissues of the body, especially the joints; in contrast to osteoarthritis, the symptoms are usually more generalized and severe; evidence indicates it may be an autoimmune disease

Rheumatologist (room-ah-TAHL-oh-gist) A physician who specializes in the treatment of joint diseases

Spina bifida (SPY-nah BIFF-ih-dah) A developmental anomaly in which the vertebral laminae do not close around the spinal cord, leaving an opening through which the cord and meninges may or may not protrude

Talipes (TAL-ih-peez) Congenital deformity of the foot in which the patient cannot stand with the sole of the foot flat on the ground; clubfoot

VOCABULARY QUIZ

Use word parts you have learned to form words that have the following definitions.

1. Process of making bone _____

2. Presence of a little cup _____

3. Resembling a beak _____

4. Condition of having a hump _____

5. Break down bone _____

Using the definitions of word parts you have learned, define the following words.

6. Ethmoid _____

7. Odontoid _____

8. Bursa _____

9. Osteoblast _____

10. Sphenoid _____

Match each of the following definitions with the correct word.

_____ **11.** Condition of making blood

_____ **12.** Like a sieve

_____ **13.** Presence of little attachments

_____ **14.** Condition of a "together" joint

_____ **15.** Pertaining to a little joint

A. Appendicular
B. Articular
C. Cribriform
D. Hemopoiesis
E. Synarthrosis

16. What is the clinical term for inflammation of a joint? _____

17. What clinical condition is an inflammation of the bone marrow caused by bacteria? _____

18. What condition involves the accumulation of uric acid crystals within a joint? _____

19. What is the meaning of the abbreviation fx? _____

20. What is the meaning of the abbreviation TMJ _____

Muscular System

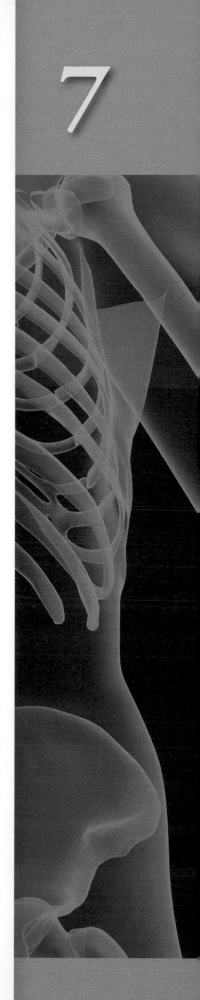

7

CHAPTER OBJECTIVES

Characteristics and Functions of the Muscular System
- List four characteristics of muscle tissue.
- List four functions of skeletal muscles.

Structure of Skeletal Muscle
- Describe the structure of a skeletal muscle, including its connective tissue coverings.
- Distinguish between direct and indirect muscle attachments.
- Define the terms *tendon, aponeurosis, origin,* and *insertion.*
- Identify the bands and lines that make up the striations on myofibers of skeletal muscle and relate these striations to actin and myosin.
- Explain why muscles have an abundant nerve and blood supply.

Contraction of Skeletal Muscle
- Explain how a muscle fiber receives a stimulus for contraction.

- Describe the sequence of events involved in the contraction of a skeletal muscle fiber.
- Compare the different types of muscle contractions.
- Describe how energy is provided for muscle contraction and how oxygen debt occurs.
- Describe and illustrate movements accomplished by the contraction of skeletal muscles.

Skeletal Muscle Groups
- Locate, identify, and describe the actions of the major muscles of the head and neck.
- Locate, identify, and describe the actions of the major muscles of the trunk.
- Locate, identify, and describe the actions of the major muscles of the upper extremity.
- Locate, identify, and describe the actions of the major muscles of the lower extremity.

KEY TERMS

Actin (AK-tin) Contractile protein in the thin filaments of skeletal muscle cells

All-or-none principle Property of skeletal muscle fiber contraction; when a muscle fiber receives a sufficient stimulus to contract, all sarcomeres shorten; with insufficient stimulus, none of the sarcomeres contract

Antagonist (an-TAG-oh-nist) Muscle that has an action opposite to the prime mover

Fasciculus (fah-SICK-yoo-lus) Small bundle or cluster of muscles or nerve fibers (cells); also called fascicle; plural fasciculi

Insertion (in-SIR-shun) End of a muscle that is attached to a relatively movable part; the end opposite the origin

Motor unit (MOH-toar YOO-nit) Single neuron and all the muscle fibers it stimulates

Myoglobin (my-oh-GLOH-bin) Iron-containing protein in the sarcoplasm of muscle cells that binds with oxygen and stores it; gives the red color to muscle

Myosin (MYE-oh-sin) Contractile protein in the thick filaments of skeletal muscle cells

Neuromuscular junction (noo-roe-MUSK-yoo-lar JUNK-shun) Area of communication between the axon terminal of a motor neuron and the sarcolemma of a muscle fiber; also called a myoneural junction

Neurotransmitter (noo-roh-TRANS-mit-ter) Chemical substance that is released at the axon terminals to stimulate a muscle fiber contraction or an impulse in another neuron

Origin (OR-ih-jin) End of a muscle that is attached to a relatively immovable part; the end opposite the insertion

Prime mover (PRYM MOO-ver) Muscle that is mainly responsible for a particular body movement; also called agonist

Sarcomere (SAR-koh-meer) Functional contractile unit in a skeletal muscle fiber

Synergist (SIN-er-gist) Muscle that assists a prime mover but is not capable of producing the movement by itself

Functional Relationships of the
Muscular System

Generates heat to maintainbody temperature.

Reproductive

Gonads produce hormones that influence muscle development and size.

Provides support for reproductive organs in the pelvis; muscle contractions contribute to orgasm in both sexes; aid in childbirth.

Integument

Covers muscles and provides barrier against harmful substances; synthesis of vitamin D for absorption of calcium necessary for bone growth, maintenance, and repair; radiates excess heat generated by muscle contraction.

Facial muscles contract and pull on the skin to provide facial expressions.

Urinary

Eliminates metabolic wastes from muscle metabolism; conserves calcium ions for muscle contraction.

Supports and protects organs of the urinary system; sphincters control voluntary urination.

Skeletal

Provides attachment sites for muscles; bones act as levers for muscle action; stores calcium necessary for muscle contraction.

Stabilizes position of bones; supplies forces for movement; tension from muscle contraction influences bone shape and size and maintains bone mass.

Digestive

Absorbs nutrients for muscle growth, maintenance, and repair; liver metabolizes lactic acid from muscle contraction.

Provides support and protection for digestive system organs; muscular sphincters control openings in the GI tract; function in chewing and swallowing; aids in defecation.

Nervous

Coordinates muscle contraction; adjusts cardiovascular and respiratory systems to maintain cardiac output and oxygen for muscle contraction.

Muscles carry out motor commands originating in the nervous system, give expressive to emotions and thoughts.

Respiratory

Supplies oxygen for muscle metabolism and contraction; removes carbon dioxide.

Muscle contractions control airflow through respiratory passages and create pressure changes necessary for ventilation.

Endocrine

Hormones influence muscle metabolism, mass, and strength; epinephrine and norepinephrine influence cardiac and smooth muscle activity.

Provides protection for some endocrine glands.

Lymphatic/Immune

Maintains interstitial fluid balance in muscle tissue; assists in defense against pathogens and repair of muscle tissue following trauma.

Skeletal muscle contraction aids in the flow of lymph; protects superficial lymph nodes.

Cardiovascular

Delivers oxygen and nutrients to muscle tissue and removes waste products and heat.

Skeletal muscle contraction aids venous return, contributes to growth of new blood vessels, promotes cardiac strength; smooth and cardiac muscle contraction contributes to vessel and heart function.

 Gives to Muscular System
⟶ Receives from Muscular System

As described in Chapter 4, there are three types of muscle tissue: skeletal, visceral, and cardiac. These are reviewed in Table 7-1. This chapter takes a closer look at skeletal muscle, which makes up about 40% of an individual's body weight. It forms more than 600 muscles that are attached to the bones of the skeleton. Skeletal muscles are under conscious control, and when they contract, they move the bones. Skeletal muscles also allow us to smile, frown, pout, show surprise, and exhibit other forms of facial expression.

Characteristics and Functions of the Muscular System

Skeletal muscle has four primary characteristics that relate to its functions:

Excitability—Excitability (eks-eye-tah-BILL-ih-tee) is the ability to receive and respond to a stimulus. To function properly, muscles have to respond to a stimulus from the nervous system.

Contractility—Contractility (kon-track-TILL-ih-tee) is the ability to shorten or contract. When a muscle responds to a stimulus, it shortens to produce movement.

Extensibility—Extensibility (eks-ten-sih-BILL-ih-tee) means that a muscle can be stretched or extended. Skeletal muscles are often arranged in opposing pairs. When one muscle contracts, the other muscle is relaxed and is stretched.

Elasticity—Elasticity (ee-lass-TISS-ih-tee) is the capacity to recoil or return to the original shape and length after contraction or extension.

Muscle contraction fulfills four important functions in the body:

- Movement
- Posture
- Joint stability
- Heat production

Nearly all movement in the body is the result of muscle contraction. Some exceptions to this are the action of cilia, the motility of the flagellum on sperm cells, and the amoeboid movement of some white blood cells. The integrated action of joints, bones, and skeletal muscles produces obvious movements such as walking and running. Skeletal muscles also produce more subtle movements that result in various facial expressions, eye movements, and respiration. Posture, such as sitting and standing, is maintained as a result of muscle contraction. The skeletal muscles are continually making fine adjustments that hold the body in stationary positions. Skeletal muscles contribute to joint stability. The tendons of many muscles extend over joints and in this way contribute to joint stability. This is particularly evident in the knee and shoulder joints, where muscle tendons are a major factor in stabilizing the joint. Heat production, to maintain body temperature, is an important byproduct of muscle metabolism. Nearly 85% of the heat produced in the body is the result of muscle contraction.

QUICK CHECK

7.1 What are the four characteristics that define muscle tissue?

7.2 List four functions of muscle contraction.

Structure of Skeletal Muscle

A whole skeletal muscle is considered an organ of the muscular system. Each organ or muscle consists of skeletal muscle tissue, connective tissue, nerve tissue, and blood or vascular tissue.

Whole Skeletal Muscle

An individual skeletal muscle may consist of hundreds, or even thousands, of muscle fibers bundled together and wrapped in a connective tissue covering. Each muscle is surrounded by a connective tissue sheath called the **epimysium** (ep-ih-MYE-see-um). Fascia, connective tissue outside the epimysium, surrounds and separates the muscles. Portions of the epimysium project inward to divide the muscle into compartments. Each compartment contains a bundle of muscle fibers. Each bundle of muscle fibers is called a **fasciculus** (fah-SIK-yoo-lus) and is surrounded by a layer of connective tissue called the **perimysium** (pair-ih-MYE-see-um). Within the fasciculus, each individual muscle cell, called a muscle fiber, is surrounded by connective tissue called the **endomysium** (end-oh-MYE-see-um). Connective tissue fibers of the endomysium are continuous with those in the perimysium, which, in turn, are continuous with the epimysium. Skeletal muscle cells (fibers), like other body cells, are soft and fragile. The connective tissue coverings furnish support and protection for the delicate cells and allow them to withstand the forces of contraction. The coverings also provide pathways

TABLE 7-1 Summary of Muscle Tissue

Feature	Skeletal	Visceral	Cardiac
Location	Attached to bones	Walls of internal organs and blood vessels	Heart
Function	Produce body movement	Contraction of viscera and blood vessels	Pump blood through heart and blood vessels
Cell shape	Cylindrical	Spindle shaped, tapered ends	Cylindrical, branching
Number of nuclei	Many	One	One
Striations	Present	Absent	Present
Type of control	Voluntary	Involuntary	Involuntary

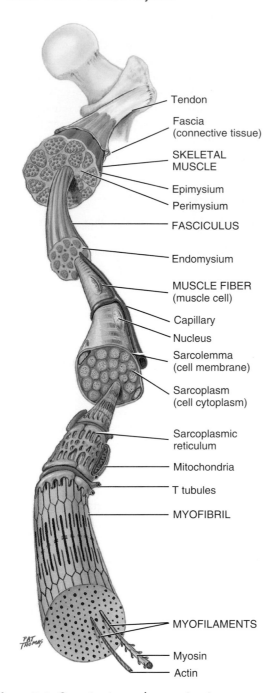

Figure 7-1 *Organization and connective tissue components of skeletal muscle.*

a muscle spans a joint and is attached to bones by tendons at both ends. One of the bones remains relatively fixed or stable while the other end moves as a result of muscle contraction. The fixed or stable end is called the **origin** of the muscle, and the more movable attachment is called the **insertion.**

Skeletal Muscle Fibers

Each skeletal muscle fiber is a single cylindrical muscle cell. The cell membrane is called the **sarcolemma** (sar-koh-LEM-mah), the cytoplasm is the **sarcoplasm** (SAR-koh-plazm), and a specialized form of smooth endoplasmic reticulum that stores calcium ions is the **sarcoplasmic reticulum** (sar-koh-PLAZ-mik reh-TIK-yoo-lum). There are multiple nuclei next to the sarcolemma at the periphery of the cell. Because the muscle cell needs energy for contraction, there are numerous mito-chondria. The sarcolemma has multiple inward extensions, or invaginations, called transverse tubules or **T tubules.**

Microscopic examination of a skeletal muscle fiber reveals alternating light **I** (isotropic) **bands** and dark **A** (anisotropic) **bands.** These are the striations visible in a light microscope as shown by the photomicrograph in Figure 7-2. Closer inspec-tion shows that the sarcoplasm is packed with filamentous **myofibrils** (mye-oh-FYE-brills) that have the same characteris-tic banding as the muscle fiber (see Figure 7-1). The myofibrils consist of still smaller protein threads called **myofilaments** (mye-oh-FILL-ah-ments). Thick filaments are formed by the protein **myosin,** whereas thin filaments are formed primarily from the protein **actin.** The overlapping arrangement of actin and myosin, illustrated in Figure 7-3, accounts for the bands and lines observed on the myofibrils and muscle fibers.

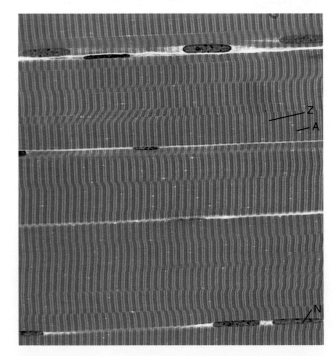

Figure 7-2 *Light micrograph of a longitudinal section of skeletal muscle. Note the cylindrical cells with multiple peripherally located nuclei (N), the darker A bands (A), and the lighter I bands (I) with the Z lines (Z). (From Gartner and Hiatt: Color Textbook of Histology. 3rd ed. Philadelphia, Elsevier/Saunders, 2007.)*

for the passage of blood vessels and nerves. The organization and connective tissue wrappings of skeletal muscle fibers are illustrated in Figure 7-1.

Skeletal Muscle Attachments

In some instances, fibers of the epimysium fuse directly with the periosteum of a bone to form a **direct** attachment. More commonly, the epimysium, perimysium, and endomysium extend beyond the fleshy part of the muscle, the **belly** or **gaster,** to form a thick ropelike **tendon** or a broad, flat, sheetlike **aponeurosis** (ah-pah-noo-ROE-sis). The tendon and aponeu-rosis form **indirect** attachments from muscles to the periosteum of bones or to the connective tissue of other muscles. Typically,

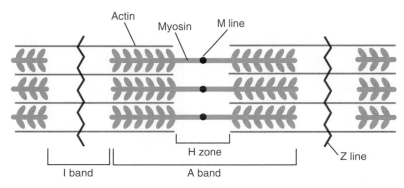

Figure 7-3 *Arrangement of myofilaments in skeletal muscle. Note the overlapping actin and myosin myofilaments. A sarcomere extends from one Z line to the next.*

The **I band** is the region where there are only thin (actin) filaments. A dark **Z line** bisects the I band. The Z line is actually a protein disk that serves as a point of attachment for the actin molecules. **A bands,** alternating with the I bands, extend the full length of the thick (myosin) filaments. Each A band is subdivided into three regions. The **zone of overlap** is at the ends of the A band where the actin overlaps the myosin. The central region of the A band where the actin does not overlap and where there are only myosin filaments is the **H zone** or **H band.** An **M line,** where thick filaments interconnect, bisects the H band. Refer to Figure 7-3 to identify these regions and look closely at Figure 7-2 to observe them on the photomicrograph.

A **sarcomere** (SAR-koh-meer), the functional unit of a myofibril, extends from one Z line to the next. The typical myofibril consists of 10,000 or more **sarcomeres** that are strung together in long chains. When a muscle cell contracts, all the sarcomeres in all the myofibrils of that cell shorten at the same time.

Nerve and Blood Supply

Skeletal muscles have an abundant supply of blood vessels and nerves. This is directly related to the primary function of skeletal muscle—contraction. Before a skeletal muscle fiber can contract, it has to receive an impulse from a nerve cell. Muscle contraction requires adenosine triphosphate (ATP), and blood vessels deliver the necessary nutrients and oxygen to produce it. Blood vessels also remove the waste products that are produced as a result of muscle contraction.

In general, an artery and at least one vein accompany each nerve that penetrates the epimysium of a skeletal muscle. Branches of the nerve and blood vessels follow the connective tissue components of the muscle so that each muscle fiber is in contact with a branch of a nerve cell and with one or more minute blood vessels called capillaries.

QUICK CHECK

7.3 When dissecting a skeletal muscle, what is the outermost connective tissue wrapping that is encountered?

7.4 What is the sarcoplasmic reticulum, and what is its function?

7.5 What band of striations has only thin (actin) myofilaments?

Contraction of Skeletal Muscle

Skeletal muscle contraction is the result of a complex series of events, based on chemical reactions, at the cellular (muscle fiber) level. This chain of reactions begins with stimulation by a nerve cell and ends when the muscle fiber is again relaxed. Contraction of a whole muscle is the result of the simultaneous contraction of many muscle fibers.

Stimulus for Contraction

Skeletal muscles are stimulated to contract by special nerve cells called **motor neurons.** As the axon of the motor neuron penetrates the muscle, the axon branches, so there is an axon terminal for each muscle fiber. A single motor neuron and all the muscle fibers it stimulates comprise what is called a **motor unit.** Some motor units include several hundred individual fibers; others contain less than 10. Because all the muscle fibers in a motor unit receive a nerve impulse at the same time, all the fibers contract at the same time.

The region in which an axon terminal meets a muscle fiber is called a **neuromuscular,** or myoneural, **junction,** which is illustrated in Figure 7-4. The axon terminal does not actually touch the sarcolemma of the muscle cell but fits into a shallow depression in the cell membrane. The fluid-filled space between the axon terminal and sarcolemma is called a **synaptic cleft** (gap). **Acetylcholine (ACh)** (ah-see-till-KOH-leen), a neurotransmitter, is contained within synaptic vesicles in the axon terminal. Receptor sites for the acetylcholine are located on the sarcolemma.

When a nerve impulse reaches the axon terminal, acetylcholine is released. The acetylcholine diffuses across the synaptic cleft and binds with the receptor sites on the sarcolemma. This reaction is the stimulus for contraction. A stimulus is a change in the cellular environment that alters the cell membrane and causes a response. In this case, the response is a muscle impulse, similar to a nerve impulse, that travels in all directions on the sarcolemma. From the sarcolemma, it travels into the T tubules, where it initiates physiologic activity within the muscle cell that results in contraction.

Meanwhile, back at the synaptic cleft, the acetylcholine that was released is rapidly inactivated by the enzyme **acetylcholinesterase** (ah-see-till-koh-lin-ES-ter-ase). This ensures that one nerve impulse will result in only one muscle impulse and only one contraction of the muscle fiber. Anything that interferes

Muscle fiber

Capillary

Axon

Axon terminal

Synaptic vesicles — contain ACh

Folded sarcolemma

Receptor sites — bind with ACh

Synaptic cleft

Nerve impulse

Figure 7-4 *Neuromuscular junction. The axon terminal fits into a depression on the sarcolemma. A nerve impulse travels down the axon to the axon terminal. The impulse causes the synaptic vesicles to release acetylcholine, which diffuses across the synaptic cleft and binds with receptors on the sarcolemma.*

with the production, release, or inactivation of acetylcholine, or its ability to bind with the receptor sites on the sarcolemma, will have an effect on muscle contraction. Muscle relaxant drugs work in this manner.

> ⊘ *QUICK* **CHECK**
> 7.6 What effect would a drug that inactivated acetylcholinesterase have on skeletal muscle contraction?
> 7.7 Where is acetylcholine stored in a myoneural junction?

Sarcomere Contraction

In a relaxed muscle fiber, myosin receptor sites on the actin thin filaments are covered or inactivated. Heads or cross-bridges on the myosin are also inactivated and are bound to ATP. Calcium is stored in the sarcoplasmic reticulum and has a low concentration in the sarcoplasm.

When an impulse travels from an axon to the sarcolemma and down the T tubule, calcium ions are released from the sarcoplasmic reticulum. This rapid influx of calcium ions into the sarcoplasm causes a change in the configuration of the troponin on the actin filaments and exposes the myosin binding sites. Simultaneously, the ATP on the myosin heads is broken down to adenosine diphosphate (ADP) so that the myosin is energized and interacts with the actin. The energized myosin heads that are now bound to actin to form cross-bridges rotate in a "power stroke" to pull the actin toward the center of the myosin. Because actin is firmly anchored to the Z line, the Z lines move closer together and the sarcomere shortens. The length of each myofilament remains the same. The actin just slides over the myosin to decrease the length of the sarcomere. When new ATP binds with

the myosin, the cross-bridges detach, and the cycle is repeated with another binding site. These events are illustrated in Figure 7-5.

Relaxed sarcomere

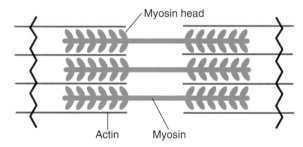

Myosin head

Actin   Myosin

Contracting sarcomere

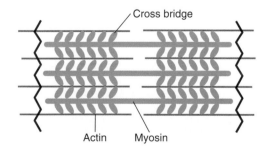

Cross bridge

Actin   Myosin

Figure 7-5 *Sliding filament theory of muscle contraction. In the relaxed sarcomere, the myosin heads are not attached to actin because binding sites are not available. In response to a stimulus, changes occur that expose the binding sites on actin so the myosin can attach to form cross bridges. The myosin heads rotate in a power stroke and the actin slides along the myosin to bring Z lines closer together, which shortens the sarcomere.*

Because the actin "slides" over the myosin, this is known as the **sliding filament theory** of muscle contraction.

Myosin continues to pull actin toward the center of the A band in a step-by-step, or ratchetlike, manner as long as there is calcium in the sarcoplasm and sufficient ATP to provide energy. The ATP is needed to energize the myosin, to cause the power stroke, and to recombine with myosin to detach the cross-bridges so another cycle can occur.

When there are no more nerve impulses at the neuromuscular junction, muscle impulses stop and calcium is actively transported from the sarcoplasm back into the sarcoplasmic reticulum. Without calcium, the actin and myosin are reconfigured into their noncontracting condition, and the muscle fiber relaxes. The events in the contraction of a skeletal muscle fiber are summarized in Table 7-2.

QUICK APPLICATIONS

The term *rigor mortis* means the "stiffness of death." Within a short time of death, the ATP in muscles breaks down so there is no ATP available to detach the cross-bridges between myosin and actin. The myofilaments remain locked in a contracted position and the body becomes rigid. A day or so later, muscle proteins begin to deteriorate and the rigor mortis disappears.

Individual muscle fibers contract according to the **all-or-none principle.** When a muscle fiber receives sufficient stimulus to contract, all the sarcomeres shorten at the same time. A greater stimulus will not elicit a greater contraction. If there is insufficient stimulus, then none of the sarcomeres contract. In other words, it is "all" or "none." The minimum stimulus necessary to cause muscle fiber contraction is a **threshold,** or **liminal, stimulus.** A lesser stimulus, one that is insufficient to cause contraction, is a **subthreshold,** or **subliminal, stimulus.**

QUICK CHECK

7.8 What inorganic ion is necessary for myosin heads to bind with receptor sites on actin?

7.9 In what two steps of muscle contraction is energy in the form of ATP required?

7.10 What term denotes the minimum stimulus necessary to cause muscle fiber contraction?

Contraction of a Whole Muscle

Whereas a single muscle fiber obeys the all-or-none principle, it is obvious that whole muscles do not. Whole muscles have a graded response; they show varying strengths of contraction. A muscle has greater contraction strength when you pick up a 25-pound weight than when you pick up a feather because more muscle fibers are contracting. Increased contraction strength is achieved by **motor unit summation** and **wave summation.** Contraction strength is decreased by fatigue, lack of nutrients, and lack of oxygen.

Within a muscle, the muscle fibers are organized into motor units. As defined previously, a **motor unit** consists of a single neuron and all the muscle fibers it stimulates. Because all the muscle fibers in a motor unit receive a threshold stimulus at the same time, a motor unit obeys the all-or-none principle, as does a muscle fiber. A stronger stimulus, however, stimulates more motor units, which increases the contraction strength of the muscle as a whole. This is called **multiple motor unit summation.**

A muscle's response to a single threshold stimulus is called a **twitch.** This is not the way a muscle in the body normally functions. This is a laboratory condition, but it provides useful information about muscle action. A myogram, a graphic recording of muscle tension in a twitch, illustrated in Figure 7-6, *A* shows that there are three distinct phases to the twitch. Initially, just after stimulation, there is no response on the myogram. This is the **lag phase.** Then the **contraction phase** begins and tension in the muscle increases to a peak. If the tension is great enough to overcome a weight load, then movement occurs. This is followed by the **relaxation phase,** when tension decreases until the resting, relaxed state is achieved.

If a second stimulus is applied during the relaxation phase, the second twitch is stronger than the first. If the muscle is stimulated at an increasingly faster rate, the relaxation time becomes shorter and shorter. Finally, all evidence of relaxation disappears and the contractions merge into a smooth, sustained contraction called **tetany** (see Figure 7-6, *B*). Tetany is a form of **multiple wave summation.** This is the usual form of muscle contraction. Neurons normally deliver a rapid succession of impulses that result in tetany, rather than a single impulse that causes a single muscle twitch.

TABLE 7-2 **Summary of Skeletal Muscle Contraction**

1. Nerve impulse travels down the axon to the axon terminals.
2. Acetylcholine is released from the axon terminals.
3. Acetylcholine diffuses across the synaptic cleft and binds with receptor sites on the sarcolemma.
4. Muscle impulse travels along the sarcolemma and into the T tubules.
5. Muscle impulse in the T tubules causes calcium ions to be released from the sarcoplasmic reticulum.
6. Calcium causes a change in the configuration of binding sites on actin and also causes ATP to break down and "energize" the myosin heads.
7. Myosin heads bind with actin to form cross-bridges.
8. Cross-bridges rotate in a "power stroke" that slides actin toward the middle of the myosin to shorten the sarcomere (muscle contraction).
9. A new ATP bind with the myosin, and myosin detaches from actin. Steps 6 through 9 repeat as long as the sarcolemma is stimulated by acetylcholine and there is sufficient ATP.
10. When the nerve impulse stops, calcium ions are actively transported (requires ATP) back into the sarcoplasmic reticulum, and the muscle relaxes with steps 9 and 10.

ATP = adenosine triphosphate

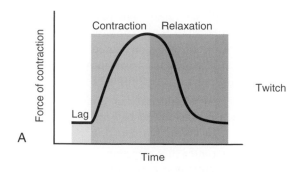

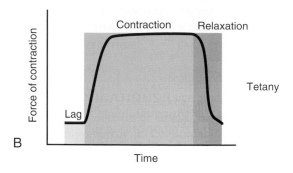

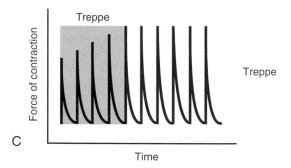

Figure 7-6 *Types of contraction in whole muscle.* ***A,*** *Twitch—a muscle's response to a single threshold stimulus.* ***B,*** *Tetany—a muscle's response to a rapid succession of stimuli resulting in a smooth, sustained contraction.* ***C,*** *Treppe—an increase in the force of contraction in response to successive threshold stimuli of the same intensity.*

QUICK APPLICATIONS

The word *tetanus* is often confusing because it means different things to different people. In reference to muscle contraction, the term denotes a wave summation that produces a steady contraction of a muscle fiber, without a relaxation phase. The word also refers to a disease, commonly called "lockjaw," that is caused by the bacterium *Clostridium tetani.* The toxin from the bacteria causes nerves to be highly excitable, which, in turn, causes uncontrollable muscle contractions, or spasms. A third use of the word is to denote a condition caused by a deficiency of calcium ions in the extracellular fluid. The lack of calcium increases nerve excitability with resulting muscle spasms, particularly of the extremities. The word *tetany* is also sometimes used to mean tetanus.

Treppe (TREP-peh), or staircase effect, is an increase in the force of muscle contraction in response to successive threshold stimuli of the same intensity. This occurs in muscle that has rested for a prolonged period of time. If the muscle is rapidly stimulated with a series of threshold stimuli of the same intensity but not rapidly enough to produce tetanus, the myogram (see Figure 7-6, *C*) will show that each contraction is slightly stronger than the previous one. The increased force of contraction is due, in part, to the increased availability of calcium ions in the sarcoplasm. Other factors that influence treppe include pH, temperature, and viscosity, which change with cellular activity. Treppe is the basis of the warm-up period for athletes.

Muscle tone (tonus) refers to the continued state of partial contraction that is present in muscles. The motor units take turns contracting and relaxing. Some motor units are stimulated to contract whereas others relax. This produces a constant tension in the muscles and keeps them ready for activity. Muscle tone is especially important in maintaining posture. It is responsible for keeping the back and legs straight, the head erect, and the abdomen from protruding; it also helps to stabilize joints. Without stimulation to contract, a muscle loses tone, becomes flaccid, and atrophies. Muscle fiber contraction to sustain muscle tone produces heat to maintain body temperature.

Not all muscle activity results in shortening the muscle to produce movement. Cross-bridge activity produces tension in the muscle. If this tension exceeds the weight of a load, then the muscle shortens and movement occurs. This is **isotonic contraction.** The tension is constant or the same, but the length of the muscle changes. If, on the other hand, the tension in the muscle increases but never exceeds the weight load, then there is no shortening and no movement. This is **isometric contraction.** If you try to lift a large boulder by yourself, your muscle tension increases but never exceeds the weight of the boulder, so there is no movement. Most body movements are the result of a combination of isotonic and isometric contractions because some muscles stabilize certain joints at the same time other muscles produce movement in different joints.

QUICK CHECK

7.11 If muscle fibers and motor units obey the all-or-none principle, what accounts for the difference in contraction strength when picking up a feather compared with picking up a book?

7.12 What are the three phases of a muscle twitch?

7.13 What term is used to describe a sustained contraction phase in response to rapid successive stimuli?

Energy Sources and Oxygen Debt

The immediate or initial source of energy for muscle contraction is **ATP** (Figure 7-7). The high-energy bond of ATP is broken to release the energy, an inorganic phosphorus (P_i), and ADP. The energy from ATP is needed for the cross-bridge power stroke, the detachment of myosin heads from the actin, and the active transport of calcium from the sarcoplasm into the sarcoplasmic reticulum. Surprisingly, muscles have limited

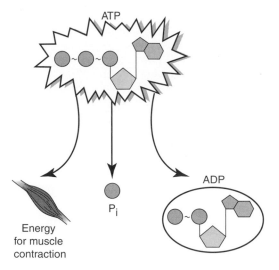

Figure 7-7 *Initial energy source for muscle contraction. ATP breaks down into inorganic phosphorus (P$_i$) and ADP. Energy for contraction is released in this reaction.*

storage facilities for ATP. In working muscles, the stored ATP is depleted in about 6 seconds, and new ATP must be regenerated if muscle contraction is to continue.

Creatine phosphate (KREE-ah-tin FOS-fate) is a unique high-energy compound that is stored in muscles (Figure 7-8). This compound provides an almost instantaneous transfer of its energy and a phosphate group to ADP molecules to regenerate ATP:

This reaction is so effective that there is very little change in ATP levels during the initial stages of muscle contraction. When ATP levels are high, this reaction is reversed to form more creatine phosphate. Muscles store enough creatine phosphate to regenerate sufficient ATP to sustain contraction for about 10 seconds.

When muscles are actively contracting for extended periods of time, **fatty acids** and **glucose** become the primary energy sources. Glucose and fatty acids are present in the blood that circulates through the muscles, and glucose is stored in muscles

as glycogen. As ATP and creatine phosphate stores are being used, more ATP is produced from the metabolism of glucose and fatty acids.

If adequate oxygen is available, fatty acids and glucose are broken down in the mitochondria through a process called **aerobic respiration** (Figure 7-9). The products are carbon dioxide, water, and large amounts of ATP:

$$\text{Fatty acids + oxygen} \rightarrow \text{carbon + water} \rightarrow \text{ATP}$$
$$\text{or glucose} \qquad\qquad \text{dioxide}$$

Limited amounts of oxygen can be stored in muscle fibers. Certain fibers, called **red fibers,** contain reddish protein pigment molecules called **myoglobin** (mye-oh-GLOH-bin). This pigment contains iron groups that attract and temporarily bind with oxygen. When the oxygen levels inside the muscle fiber diminish, the oxygen can be resupplied from myoglobin. Muscle fibers that contain very little myoglobin are called **white fibers.**

When muscles are contracting vigorously for long periods of time, myoglobin and the circulatory system are unable to deliver oxygen fast enough to maintain the aerobic pathways. Processes that do not require oxygen are necessary. Under these conditions, glucose is the primary energy source. If adequate oxygen is not available, glucose is broken down by a process called **anaerobic respiration** (Figure 7-10). The products of the anaerobic pathway are lactic acid and a small amount of ATP.

$$\text{Glucose} \rightarrow \text{lactic acid + ATP}$$

Some of the lactic acid accumulates in the muscle and causes a burning sensation. Most of it diffuses out of the muscle and into the bloodstream, which takes it to the liver. Later, when sufficient oxygen is available, the liver converts the lactic acid back to glycogen, the storage form of glucose.

The aerobic pathway produces about 20 times more ATP than the anaerobic pathway. However, the anaerobic pathway provides ATP about two and one-half times faster than the aerobic pathway. Most of the energy for vigorous activity over

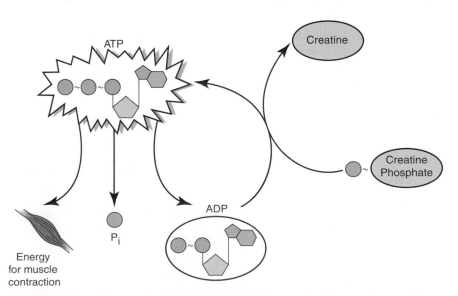

Figure 7-8 *Creatine phosphate transfers its energy and phosphate group to ADP molecules to regenerate ATP, which then breaks down to provide energy for contraction.*

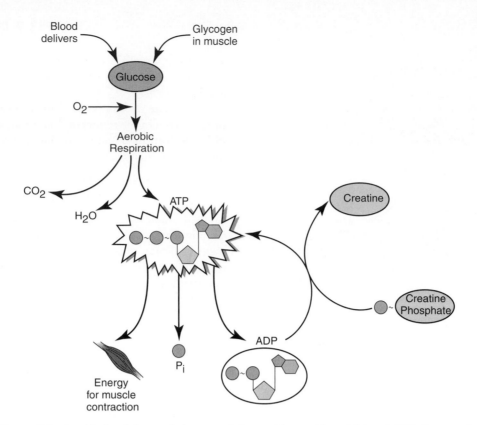

Figure 7-9 *Aerobic breakdown of glucose and fatty acids provides additional ATP for extended contraction periods.*

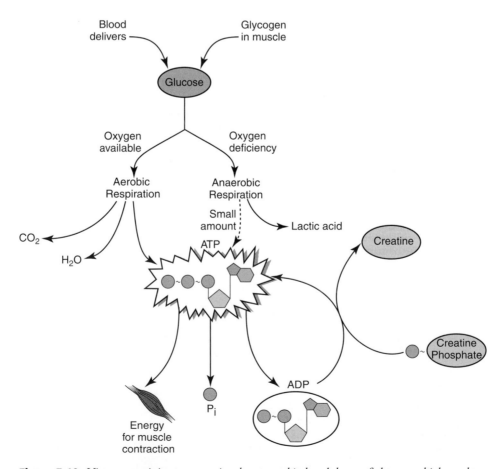

Figure 7-10 *Vigorous activity may require the anaerobic breakdown of glucose, which produces lactic acid and ATP. The accumulation of lactic acid results in oxygen debt.*

a moderate period of time comes from anaerobic respiration. Prolonged activities requiring endurance depend on aerobic mechanisms.

After periods of strenuous exercise that require anaerobic mechanisms to regenerate ATP, there is an accumulation of lactic acid in the muscle. This causes temporary muscular pain and cramping. The ATP and creatine phosphate in the muscle are depleted and need to be replenished. This creates an **oxygen debt** that must be repaid before equilibrium can be restored. Additional oxygen is needed to convert the lactic acid into glycogen, a process that occurs in the liver. Oxygen is also needed to replenish the ATP and the creatine phosphate in the muscle. Oxygen debt is defined as the additional oxygen that is required after physical activity to restore resting conditions. The debt is paid back by labored breathing that continues after the activity has stopped.

QUICK CHECK

7.14 What is the initial source of energy for muscle contraction?

7.15 What high-energy compound is stored in muscle tissue and is used to replenish ATP?

7.16 Limited amounts of oxygen for aerobic respiration is stored in muscles. What protein molecule effectively stores this oxygen?

7.17 Which is more effective in terms of ATP production— anaerobic or aerobic respiration?

Movements Produced by Skeletal Muscles

Most intact skeletal muscles are attached to bones by tendons that span joints. When the muscle contracts, one bone (the insertion) moves relative to the other bone (the origin).

Frequently muscles work in groups to perform a particular movement. If one muscle has a primary role in providing a movement, it is called a **prime mover** or **agonist.** Muscles that work with, or assist, the prime mover to cause a movement are called **synergists** (SIN-er-jists). Often muscles span more than one joint, and a synergist will stabilize one joint while the prime mover acts on the other joint. For example, the fingers can be flexed to make a fist without bending the wrist because certain muscles fix the wrist in a stabilized position. **Antagonists** are muscles that oppose, or reverse, a particular movement. The biceps brachii muscle on the anterior arm flexes the forearm at the elbow. The triceps brachii muscle on the posterior arm extends the forearm at the elbow. The two muscles are on opposite sides of the humerus and have opposite functions. They are antagonists. Figure 7-11 illustrates the actions of the biceps brachii and triceps brachii muscles.

Bones and muscles work together to perform different types of movement at the various joints. When describing muscular action or movement at joints, you need a frame of reference and descriptive terminology with definite meaning. Some commonly used terms that are used to describe particular movements are defined and illustrated in Figure 7-12.

QUICK CHECK

7.18 What term describes a muscle that works with or assists an agonist?

7.19 What term describes a muscle that opposes or reverses a given movement?

7.20 What is the muscle action that is opposite of pronation?

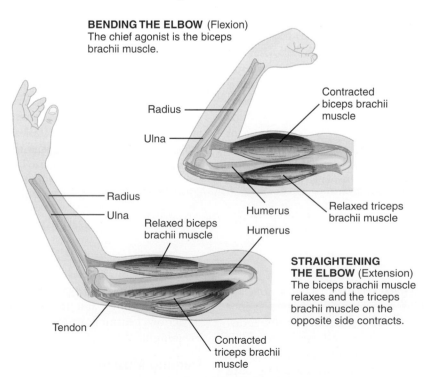

BENDING THE ELBOW (Flexion)
The chief agonist is the biceps brachii muscle.

Radius

Ulna

Contracted biceps brachii muscle

Relaxed triceps brachii muscle

Radius

Ulna

Relaxed biceps brachii muscle

Humerus

Humerus

Tendon

STRAIGHTENING THE ELBOW (Extension)
The biceps brachii muscle relaxes and the triceps brachii muscle on the opposite side contracts.

Contracted triceps brachii muscle

Figure 7-11 *An example of antagonist muscles. The biceps brachii and triceps brachii muscles are on opposite sides of the humerus and have opposite actions at the elbow.*

Flexion (FLEK-shun)
Means to bend. Flexion usually brings two
bones closer together and decreases the
angle between them. Example: bending
the elbow or the knee.

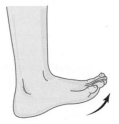

Dorsiflexion (dor-sih-FLEK-shun)
Flexion of the ankle in which the dorsum
or top of the foot is lifted upward, decreasing
the angle between the foot and leg.
Example: standing on your heels.

Extension (ek-STEN-shun)
Means to straighten. Extension is the opposite
of flexion. It increases the angle between two
bone. Example: straightening the elbow or the
knee after it has been flexed.

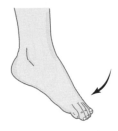

Plantar flexion (PLAN-tar FLEK-shun)
Plantar flexion is movement at the ankle that
increases the angle between the foot and leg.
Example: standing on your toes.

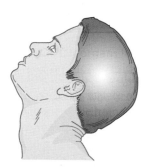

Hyperextension (hye-perk-ek-STEN-shun)
Hyperextension occurs when a part of the body
is extended beyond the anatomical position. The
joint angle becomes greater than 180°.
Example: moving the head backward.

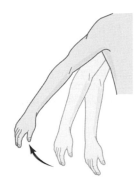

Abduction (ab-DUCK-shun)
Means to take away. Abduction moves a bone or limb away
from the midline or axis of the body. Examples: the outward
movement of the legs in "jumping jacks," moving the arms
away from the body, or spreading the fingers apart.

Figure 7-12 *Types of body movements.*

Skeletal Muscle Groups

There are more than 600 skeletal muscles in the body. A discussion of each muscle is certainly beyond the scope of this
book. Only the more significant and obvious muscles are
identified and described here. These are arranged in groups
according to location and/or function. If you identify and
learn the muscles as group associations, it will make them
easier to remember. If you can locate a muscle on your own
body, you will be able to contract the muscle and describe
its action. Learning anatomy in this manner makes it more
meaningful.

Naming Muscles

Most skeletal muscles have names that describe some feature of
the muscle. Often several criteria are combined into one name.

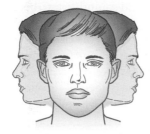

Adduction (ad-DUCK-shun)
Means to bring together. Adduction is the opposite of abduction. It moves a bone or limb toward the midline of the body. Examples: bringing the arms back to the sides of the body after they have been abducted or moving the legs back to anatomical position after abduction.

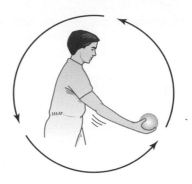

Circumduction (sir-kum-DUCK-shun)
Circumduction is the conelike, circular movement of a body segment. The proximal end of the segment remains relatively stationary while the distal end outlines a large circle. Example: the movement of the arm at the shoulder joint, with the elbow extended, so that the tips of the fingers move in a large circle.

Rotation (roh-TAY-shun)
Rotation is the movement of a bone around its own axis in a pivot joint. Example: shaking your head "no".

Inversion (in-VER-zhun)
Inversion is the movement of the sole of the foot inward or medially.

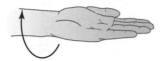

Supination (soo-pih-NAY-shun)
Supination is a specialized rotation of the forearm that turns the palm of the hand forward or anteriorly. If the elbow is flexed, supination turns the palm of the hand upward or superiorly.

Eversion (ee-VER-zhun)
Eversion is the opposite of inversion. It is the movement of the sole of the foot outward or laterally.

Pronation (proh-NAY-shun)
Pronation is the opposite of supination. It is a specialized rotation of the forearm that turns the palm of the hand backward or posteriorly. If the elbow is flexed, pronation turns the palm of the hand downward or inferiorly.

Figure 7-12—Cont'd.

Associating the muscle's characteristics with its name will help you learn and remember them. The following are some terms relating to muscle features that are used in naming muscles:

- **Size:** vastus (huge); maximus (large); longus (long); minimus (small); brevis (short)
- **Shape:** deltoid (triangular); rhomboid (like a rhombus with equal and parallel sides); latissimus (wide); teres (round); trapezius (like a trapezoid, a four-sided figure with two sides parallel)
- **Direction of fibers:** rectus (straight); transverse (across); oblique (diagonal); orbicularis (circular)
- **Location:** pectoralis (chest); gluteus (buttock or rump); brachii (arm); supra- (above); infra- (below); sub- (under or beneath); lateralis (lateral)
- **Number of origins:** biceps (two heads); triceps (three heads); quadriceps (four heads)
- **Origin and insertion:** sternocleidomastoid (origin on the sternum and clavicle, insertion on the mastoid process); brachioradialis (origin on the brachium or arm, insertion on the radius)
- **Action:** abductor (to abduct a structure); adductor (to adduct a structure); flexor (to flex a structure); extensor (to extend a structure); levator (to lift or elevate a structure); masseter (to chew)

Muscles of the Head and Neck

Muscles of Facial Expression

Humans have well-developed muscles in the face that permit a large variety of facial expressions. Because these muscles are used to show surprise, disgust, anger, fear, and other emotions, they are an important means of nonverbal communication. The following are some of the muscles used to produce facial expressions. They are identified in Figure 7-13 and Table 7-3.

The **frontalis** (frun-TAL-is) is over the frontal bone of the forehead. It is attached to the soft tissue of the eyebrow; when it contracts, it raises the eyebrows and wrinkles the forehead. The **orbicularis oris** (oar-BIK-yoo-lair-is OAR-is) is a sphincter that encircles the mouth. This muscle is used to close the mouth, to form words, and to pucker the lips as in kissing. The **orbicularis oculi** (oar-BIK-yoo-lair-is OK-yoo-lye) is another sphincter but it is around the eye (oculus). The actions of winking, blinking, and squinting use this muscle. The **buccinator** (BUCK-sin-ay-ter) is the principal muscle in the cheek area and it is used to compress the cheek when whistling, sucking, or blowing out air. It is sometimes called the trumpeter's muscle. The **zygomaticus** (zye-goh-MAT-ih-kus) extends from the zygomatic arch to the corners of the mouth. It contracts to raise the corner of the mouth when we smile.

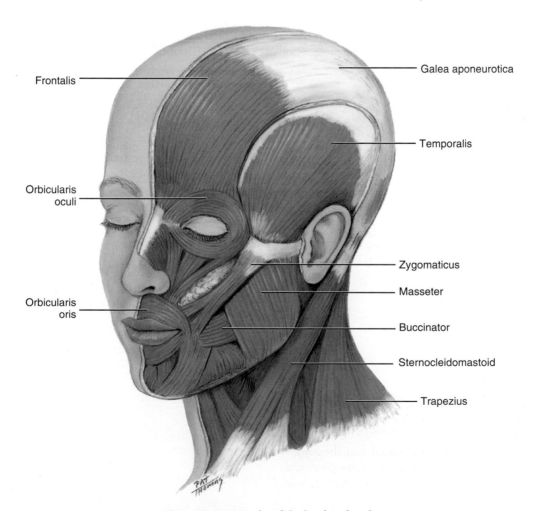

Figure 7-13 *Muscles of the head and neck.*

TABLE 7-3 Muscles of the Head and Neck

Muscle	Description	Origin	Insertion	Action
Muscles of Facial Expression				
Frontalis (frun-TAL-is)	Flat muscle that covers forehead	Galea aponeurotica	Skin of eyebrow and nose	Raises eyebrows
Orbicularis oris (oar-BIK-yoo-lair-is OAR-is) ·	Circular muscle around mouth	Maxilla and mandible	Skin around lips	Closes and purses lips
Orbicularis oculi (oar-BIK-yoo-lair-is OK-yoo-lye)	Circular muscle around eyes	Maxilla and frontal bones	Tissue of eyelid	Closes eye; winking, blinking, squinting
Buccinator (BUK-sin-ay-ter)	Horizontal cheek muscle; deep to masseter	Maxilla and mandible	Corner of mouth	Compresses cheek; trumpeter's muscle
Zygomaticus (zye-goh-MAT-ih-kus)	Extends diagonally from corner of mouth to cheekbone	Zygomatic bone	Skin and muscle at corner of mouth	Elevates corner of mouth as in smiling
Muscles of Mastication				
Temporalis (tem-por-AL-is)	Flat, fan-shaped muscle over temporal lobe	Flat portion of temporal bone	Mandible	Closes jaw
Masseter (MASS-eh-ter)	Covers lateral aspect of jaw	Zygomatic arch	Mandible	Closes jaw
Neck Muscles				
Sternocleidomastoid (stir-noh-klye-doh-MAS-toyd)	Straplike muscle that ascends obliquely over neck	Sternum and clavicle	Mastoid process	Flexes and rotates head
Trapezius (trah-PEEZ-ee-us)	Large, flat, triangular muscle on posterior neck and shoulder	Occipital bone and spine of thoracic vertebrae	Scapula	Extends head; also moves scapula

Muscles of Mastication

There are four pairs of muscles that are responsible for chewing movements or mastication. All of these muscles insert on the mandible, and they are some of the strongest muscles in the body. Two of the muscles, the **temporalis** (tem-poar-AL-is) and **masseter** (MASS-eh-ter), are superficial and are identified in Table 7-3 and Figure 7-13. The others, the lateral and medial pterygoids, are deep to the mandible and are not shown in the figure. The **temporalis** is the largest of the mastication muscles. As the name implies, it has its origin on the temporal bone. The **masseter** is located along the ramus of the mandible and is a synergist of the temporalis.

Neck Muscles

Only two of the more obvious and superficial neck muscles are considered here and identified in Figure 7-13 and Table 7-3. There are numerous muscles associated with the throat, the hyoid bone, and the vertebral column, a discussion of which is beyond the scope of this text.

 Sternocleidomastoid (stir-no-klye-doh-MAS-toyd) muscles ascend obliquely across the anterior neck from the sternum and clavicle to the mastoid process. When both of these muscles contract together, the neck is flexed and the head is bent toward the chest. When one of the muscles contracts, the head turns toward the direction opposite the side that is contracting. When the left muscle contracts, the head turns to the right. A portion of the **trapezius** (trah-PEEZ-ee-us) muscle is in the neck region and moves the head. Each trapezius muscle extends from the occipital bone at the base of the skull to the end of the thoracic vertebrae and also inserts on the scapula laterally. A portion of this muscle functions to extend the head and is antagonistic to the sternocleidomastoid.

QUICK CHECK

7.21 What muscle raises the corner of your mouth when you smile?

7.22 What bone is the insertion of the muscles of mastication?

7.23 What is the diagonal muscle along each side of the neck?

Muscles of the Trunk

The muscles of the trunk (Table 7-4) include those that move the vertebral column, the muscles that form the thoracic and abdominal walls, and those that cover the pelvic outlet.

Vertebral Column Muscles

The **erector spinae** (ee-REK-ter SPY-nee) group of muscles on each side of the vertebral column is a large muscle mass

TABLE 7-4 Muscles of the Trunk

Muscle	Description	Origin	Insertion	Action
Muscles that Move the Vertebral Column				
Erector spinae (ee-REK-ter SPY-nee)	Intrinsic back muscles on each side of vertebral column	Vertebrae	Superior vertebrae and ribs	Extends vertebral column
Deep back muscles	Short, intrinsic muscles in space between spinous and transverse processes of vertebrae	Vertebrae	Vertebrae	Move vertebral column
Muscles of the Thoracic Wall				Inspiration
External intercostals (inter-KOS-talz)	Short muscles in intercostals spaces between ribs	Ribs	Next rib below origin	Inspiration
Internal intercostals (inter-KOS-talz)	Short muscles in intercostals spaces between ribs	Ribs	Next rib above origin	Forced expiration
Diaphragm (DYE-ah-fram)	Dome-shaped muscle that forms partition between thorax and abdomen	Interior body wall	Central tendon of diaphragm	Inspiration
Abdominal Wall Muscles				
External oblique (ek-STIR-null oh-BLEEK)	Largest and most superficial of lateral abdominal wall muscles	Rib cage	Iliac crest and fascia	Compresses abdomen
Internal oblique (in-TER-null oh-BLEEK)	Underlies external oblique; fibers perpendicular to external oblique	Iliac crest and fascia	Lower ribs and fascia	Compresses abdomen
Transversus abdominus (trans-VER-sus ab-DOM-ih-nis)	Deepest muscles of abdominal wall; fibers fun horizontally	Fascia and lower ribs	Linea alba and pubis	Compresses abdomen
Rectus abdominus (REK-tus ab-DOM-ih-nis)	Long, straight muscle on each side of linea alba	Pubic bone	Ribs and sternum	Flexes vertebral column and compresses abdomen
Muscles of the Pelvic Outlet				
Pelvic diaphragm; primarily levator ani muscle	Superior (deep) muscle hammock that forms floor of pelvic cavity	Pubis and ischium	Sacrum and coccyx	Supports and maintains position of pelvic viscera
Urogenital diaphragm	Superficial to pelvic diaphragm; between two sides of pubic arch	Ischium and genitalia	Genitalia and perineum	Supports pelvic viscera and assists in function of genitalia

that extends from the sacrum to the skull. These muscles are primarily responsible for extending the vertebral column to maintain erect posture. Muscle contraction on only one side bends the vertebral column to that side. The **deep back muscles** occupy the space between the spinous and transverse processes of adjacent vertebrae. Each individual muscle is short, but as a group they extend the length of the vertebral column. They are responsible for several movements of the vertebral column.

Thoracic Wall Muscles

The muscles of the thoracic wall (Figure 7-14) are involved primarily in the process of breathing. The intercostal muscles are located in spaces between the ribs. Fibers of the **external intercostal muscles** are directed forward and downward, whereas those of the **internal intercostals** are directed backward and downward. The internal intercostal fibers are at right angles to the external intercostal fibers. External intercostal muscles contract to elevate the ribs during the inspiration phase

of breathing. The internal intercostals contract during forced expiration.

The **diaphragm** is a dome-shaped muscle that forms a partition between the thorax and the abdomen. It has three openings in it for structures that have to pass from the thorax to the abdomen. One opening is for the inferior vena cava, one is for the esophagus, and one is for the aorta. The diaphragm is responsible for the major movement in the thoracic cavity during quiet, relaxed breathing. When the diaphragm contracts, the dome is flattened. This increases the volume of the thoracic cavity and results in inspiration. When the muscle relaxes, it again resumes its dome shape and decreases the volume of the thoracic cavity, which forces air out during expiration.

 QUICK APPLICATIONS

Voluntary forceful contractions of the diaphragm increase intraabdominal pressure to assist in urination, defecation, and childbirth.

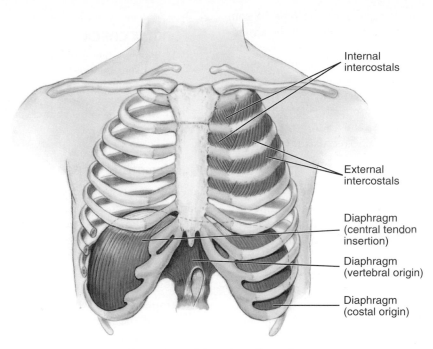

Figure 7-14 *Thoracic wall muscles.*

Abdominal Wall Muscles

The abdomen, unlike the thorax and pelvis, has no bony reinforcements or protection. The wall consists entirely of four muscle pairs, arranged in layers, and the fascia that envelops them (Figure 7-15). The aponeuroses of the muscles on opposite sides meet in the anterior midline to form the **linea alba** ("white line"), a band of connective tissue that extends from the sternum to the pubic symphysis. The outer muscle layer is the **external oblique** with fibers that run inferiorly and medially. With fibers oriented perpendicularly to the external oblique, the **internal oblique** lies just underneath it. The deepest layer of muscle is the **transversus abdominis** with its fibers running in a horizontal direction. The arrangement of the muscle layers with the fibers in each layer

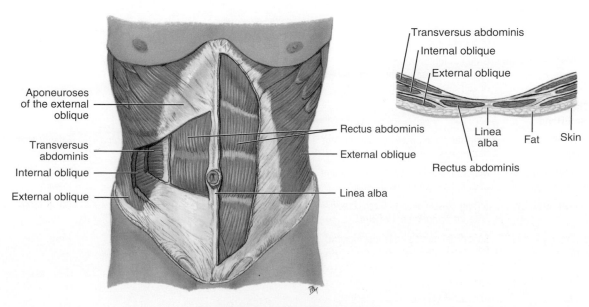

Figure 7-15 *Abdominal wall muscles.*

going in different directions is similar to the type of construction found in plywood and adds strength to the anterolateral abdominal wall. The fascia of these muscles extends anteriorly to form a broad aponeurosis along much of the anterior aspect of the abdomen. The fascia also envelops the **rectus abdominis** muscle that runs vertically from the pubic bones to the ribs and the sternum on each side of the midline. All of these muscles compress the abdominal wall and increase intraabdominal pressure. The rectus abdominis also flexes the vertebral column.

Pelvic Floor Muscles

The pelvic outlet is formed by two muscular sheets and their associated fascia. The deeper or more superior muscular sheet is the **pelvic diaphragm,** which forms the floor of the pelvic cavity. Most of the pelvic diaphragm is formed by the two **levator ani** muscles that support the pelvic viscera. They resist increased pressure in the abdominopelvic cavity and thus play a role in the control of the urinary bladder and rectum. The superficial muscle group, the **urogenital diaphragm**, fills the space within the pubic arch and is associated with the genitalia.

QUICK CHECK

7.24 What are the respiratory muscles that are located between the ribs?
7.25 What muscle of the abdominal wall extends vertically from the pubic bone to the ribs and sternum?
7.26 What is the largest muscle of the pelvic diaphragm?

Muscles of the Upper Extremity

The muscles of the upper extremity include those that attach the scapula to the thorax and generally move the scapula, those that attach the humerus to the scapula and generally move the arm, and those that are located in the arm or forearm that move the forearm, wrist, and hand (Table 7-5 and Figures 7-16 and 7-17).

Muscles That Move the Shoulder and Arm

The **trapezius** and **serratus anterior** (seh-RAY-tus anterior) are two of the muscles that attach the scapula to the axial skeleton. The trapezius is a large superficial triangular muscle

TABLE 7-5 Muscles of the Upper Extremity

Muscle	Description	Origin	Insertion	Action
Muscles that Move the Shoulder and Arm				
Trapezius (trah-PEEZ-ee-us)	Triangular muscle on posterior neck and shoulder; right and left trapezius together form trapezoid	Occipital bone and vertebrae	Scapula	Adducts, elevates, and rotates scapula; also extends head
Serratus anterior (she-RAY-tus anterior)	Deep and inferior to pectoral muscles; forms medial wall of axilla; has serrated appearance	Ribs	Medial border of scapula	Pulls scapula anteriorly and downward
Pectoralis major (pek-tor-AL-iss MAY-jer)	Large, fan-shaped muscle that covers anterior chest	Sternum, clavicle, ribs	Humerus	Adducts and flexes arm across chest
Latissimus dorsi (lah-TISS-ih-mus DOAR-sye)	Large, broad, flat muscle of lower back region; swimmer's muscle	Vertebrae	Humerus	Adducts and medially rotates arm
Deltoid (DELL-toyd)	Thick muscle that forms contour of shoulder	Clavicle and scapula	Humerus	Abducts arm
Rotator cuff	Group of four muscles that attach humerus to scapula and form cap or cuff over proximal humerus	Scapula	Humerus	Rotates arm
Muscles That Move the Forearm and Hand				
Triceps brachii (TRY-seps BRAY-kee-eye)	Only muscle in posterior compartment of arm; has three heads of origin	Humerus and scapula	Olecranon of ulna	Extends forearm
Biceps brachii (BY-seps BRAY-kee-eye)	Major muscle in anterior compartment of arm; has two heads of origin	Scapula	Radius	Flexes and supinates forearm
Brachialis (bray-kee-AL-is)	Strong muscle deep to biceps brachii in anterior compartment of arm	Humerus	Ulna	Flexes forearm
Brachioradialis (bray-kee-oh-ray-dee-AL-is)	Superficial muscle on lateral forearm	Distal humerus	Distal radius	Flexes forearm

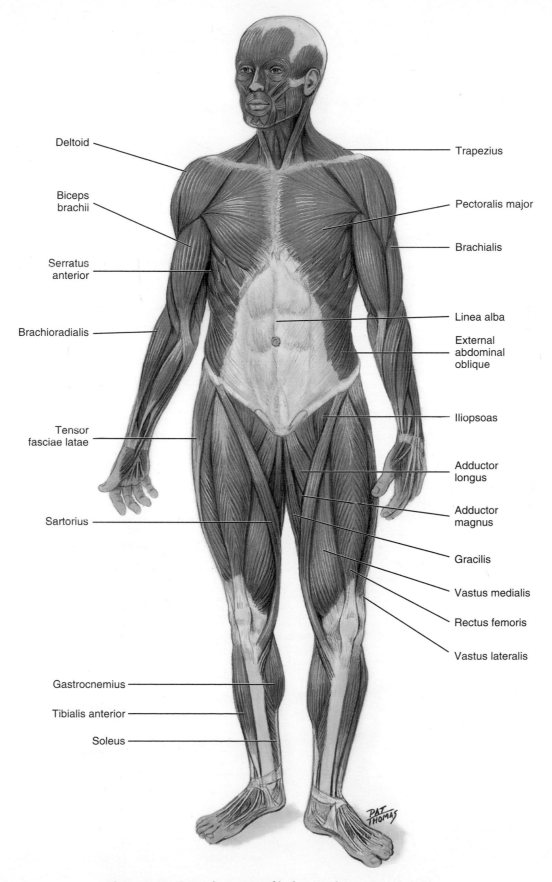

Deltoid

Biceps brachii

Serratus anterior

Brachioradialis

Tensor fasciae latae

Sartorius

Gastrocnemius

Tibialis anterior

Soleus

Trapezius

Pectoralis major

Brachialis

Linea alba

External abdominal oblique

Iliopsoas

Adductor longus

Adductor magnus

Gracilis

Vastus medialis

Rectus femoris

Vastus lateralis

Figure 7-16 *General overview of body musculature. Anterior view.*

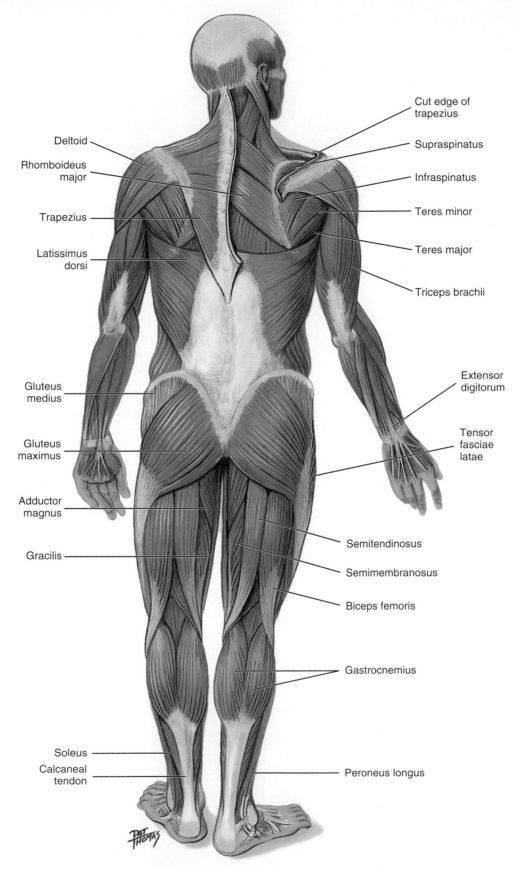

Deltoid

Rhomboideus major

Trapezius

Latissimus dorsi

Gluteus medius

Gluteus maximus

Adductor magnus

Gracilis

Soleus

Calcaneal tendon

Cut edge of trapezius

Supraspinatus

Infraspinatus

Teres minor

Teres major

Triceps brachii

Extensor digitorum

Tensor fasciae latae

Semitendinosus

Semimembranosus

Biceps femoris

Gastrocnemius

Peroneus longus

Figure 7-17 *General overview of body musculature. Posterior view.*

of the back. It is attached to the thoracic vertebrae and the scapula. When the trapezius contracts, it adducts and elevates the scapula, as in shrugging the shoulders. The serratus anterior is located on the side of the chest where it runs from the ribs to the scapula. When the serratus anterior contracts, it pulls the shoulder downward and forward, as in pushing something.

Both the **pectoralis major** (pek-tor-AL-iss MAY-jer) and the **latissimus dorsi** (lah-TISS-ih-mus DOR-sye) muscles attach the humerus to the axial skeleton. The pectoralis major is a superficial muscle on the anterior chest. It has a broad origin on the sternum, costal cartilages, and clavicle, but then the fibers converge to insert on the humerus by way of a short tendon. The primary function of the pectoralis major is to adduct and rotate the arm medially across the chest. The latissimus dorsi is a large, superficial muscle located in the lower back region. It has an extensive origin from the spines of the thoracic vertebrae, ilium, and ribs and then extends upward to insert on the humerus. The latissimus dorsi adducts and rotates the arm medially and lowers the shoulder. It is an important muscle in swimming and rowing motions.

The **deltoid** is a large, fleshy muscle that covers the shoulder and attaches the humerus to the scapula. This muscle abducts the arm to a horizontal position. It is a common site for administering injections. Another group of four muscles, the **infraspinatus, supraspinatus, subscapularis,** and **teres minor,** attaches the humerus to the scapula and moves the humerus in some way. These muscles, collectively, are called the **rotator cuff muscles** because they form a cuff or cap over the proximal humerus. A rotator cuff injury involves damage to one or more of these muscles or their tendons. The subscapularis is not shown in Figures 7-16 and 7-17 because it is on the costal surface of the scapula and not visible.

Muscles That Move the Forearm and Hand

The muscles that move the forearm are located along the humerus. The arm is divided into anterior and posterior muscle compartments. The **triceps brachii,** the primary extensor of the forearm, is the only muscle in the posterior compartment. As the name implies, it has three heads of origin. The anterior muscle compartment contains the **biceps brachii** and **brachialis** (bray-kee-AL-is). These muscles are the primary flexors of the forearm. In addition, the **brachioradialis** (bray-kee-oh-ray-dee-AL-is), a prominent muscle along the lateral side of the forearm, helps to flex the forearm. The brachioradialis extends from the lower part of the humerus to the radius and is generally considered to be a posterior forearm muscle.

The 20 or more muscles that cause most wrist, hand, and finger movements are located along the forearm. These muscles are divided into anterior and posterior compartments. Most of the anterior compartment muscles flex the wrist and fingers, whereas the posterior muscles cause extension. The muscles that flex the hand and fingers are stronger than the extensor muscles. In a normal relaxed position, the fingers are slightly flexed because the normal muscle tone is greater in the flexoid.

Muscles of the Lower Extremity

The muscles of the lower extremity include those that are located in the hip region and generally move the thigh, those that are located in the thigh and move the leg, and those that are located in the leg and move the ankle and foot (Table 7-6 and Figure 7-17).

Muscles That Move the Thigh

The muscles that move the thigh have their origins on some part of the pelvic girdle and their insertions on the femur. The largest muscle mass belongs to the posterior group, the gluteal muscles. The **gluteus maximus** forms the area of the buttocks. The **gluteus medius** is superior and deep to the gluteus maximus. The gluteus medius, rather than the gluteus maximus is usually used for intramuscular injections to avoid damage to the sciatic nerve. The **gluteus minimus** is the smallest and deepest of the gluteal muscles and is not visible in Figures 7-16 and 7-17. Although the **tensor fasciae latae** (TEN-soar FAH-she-ee LAY-tee) is a superficial muscle on the lateral side of the thigh, its function and innervation associate it with the gluteal muscles. These muscles abduct the thigh; that is, they raise the thigh sideways to a horizontal position. The gluteus maximus also extends or straightens the thigh at the hip for walking or climbing stairs.

The anterior muscle that moves the thigh is the **iliopsoas** (ill-ee-oh-SOH-as). This muscle is formed from the iliacus that originates on the iliac fossa and the psoas that originates on the lumbar vertebrae. The fibers converge into the iliopsoas and insert on the femur. The iliopsoas flexes the thigh, making it antagonistic to the gluteus maximus.

The medial muscles adduct the thigh; that is, they press the thighs together. They are often called the horse rider's muscles because when they contract they keep the rider on the horse. This group includes the **adductor longus, adductor brevis, adductor magnus,** and **gracilis** (grah-SILL-is) muscles. The adductor brevis is not illustrated in Figure 7-16 because it is deep to the adductor longus.

TABLE 7-6 **Muscles of the Lower Extremity**

Muscle	Description	Origin	Insertion	Action
Muscles that Move the Thigh				
Iliopsoas (ill-ee-oh-SOH-as)	Composite of iliacus and psoas muscles; located in groin	Iliac fossa and vertebrae	Femur	Flexes and rotates thigh
Tensor fasciae latae (TEN-soar FAH-she-ee LAY-tee)	Most lateral muscle in hip region	Anterior ilium	Tibia by way of iliotibial tract	Flexes and abducts thigh; synergist of iliopsoas
Gluteus maximus (GLOO-tee-us MAK-sih-mus)	Largest and most superficial of gluteal muscles	Ilium, sacrum, and coccyx	Femur	Extends thigh; antagonist of iliopsoas
Gluteus medius (GLOO-tee-us MIN-ih-mus)	Thick muscle deep to gluteus maximus; common site for intramuscular injections	Ilium	Femur	Abducts and rotates thigh
Gluteus minimus (GLOO-tee-us MIN-ih-mus)	Smallest and deepest of gluteal muscles	Ilium	Femur	Abducts and rotates thigh
Adductor longus (ad-DUCK-toar LONG-us)	Most anterior of adductor muscles in medial compartment of thigh	Pubis	Femur	Adducts thigh
Adductor brevis (ad-DUCK-toar BREH-vis)	Short adductor muscle deep to adductor longus	Pubis	Femur	Adducts thigh
Adductor magnus (ad-DUCK-toar MAG-nus)	Largest and deepest of adductor muscles	Pubis	Femur	Adducts thigh
Gracilis (grah-SILL-is)	Long, superficial, straplike muscle on medial aspect of thigh	Pubis	Tibia	Adducts thigh; flexes leg

 FROM THE PHARMACY

Most muscle strains and spasms respond to rest, physical therapy, antiinflammatory medications, and short-term use of skeletal muscle relaxants. Chronic muscle spasticity as a result of strokes, spinal cord injuries, multiple sclerosis, cerebral palsy, and other neurological disorders requires a long-term regimen of skeletal muscle relaxants. Central-acting drugs decrease muscle tone without interfering with voluntary movement by depressing the motor pathways in the brain stem and spinal cord. They have no direct action at the neuromuscular junction or on the contractile elements of the muscle. Unfortunately, currently there is no way to isolate the depression of motor activity from depression of other central nervous system (CNS) functions. This leads to numerous potential side effects including lethargy, drowsiness, dizziness, confusion, headache, and blurred vision. Concurrent use of alcohol and other CNS depressants should be avoided. Four skeletal muscle relaxants are described here.

Carisoprodol (Soma, Muslax, Rotalin) is used as an adjunct to rest and physical therapy for the relief of acute painful musculoskeletal conditions. Carisoprodol is metabolized in the liver and excreted by the kidney, and for this reason it should be used with caution in patients with impaired function of these organs. It should not be used in the elderly.

Chlorzoxazone (Paraflex, Parafon Forte DSC) is another skeletal muscle relaxant for the relief of acute painful musculoskeletal conditions. The clinical result of its use is a reduction of muscle spasms, decrease in pain, and increase in mobility of the affected muscles. Caution is advised for use in patients with a known history of drug allergies, and the practitioner should watch for any signs of hepatotoxicity.

Cyclobenzaprine (Flexeril) relieves skeletal muscle spasms of local origin but is ineffective for spasms caused by CNS disease. It is intended for short-term use only (2 to 3 weeks) and should not be used in patients with heart disorders or hyperthyroidism. CNS side effects such as sedation tend to be fairly common with this medication.

Baclofen (Lioresal) inhibits the transmission of reflex impulses at the level of the spinal cord. One side effect is sedation, but it is usually well tolerated. Baclofen can be used orally or may be administered via an implantable pump for continuous infusion to control spasticity. It is often used in patients with cerebral palsy and traumatic spinal cord injuries. Care should be taken to avoid abrupt discontinuation of this medication.

TABLE 7-6 Muscles of the Lower Extremity—Cont'd

Muscle	Description	Origin	Insertion	Action
Muscles That Move the Leg				
Sartorius (sar-TOAR-ee-us)	Long, straplike muscle that courses obliquely across thigh; longest muscle in body	Ilium	Tibia	Flexes thigh; flexes and rotates leg
Quadriceps femoris (KWAD-rih-seps FEM-oar-is)	Group of four muscles that form fleshy mass of anterior thigh; form a common tendon that passes over patella	Femur; except for rectus femoris, which originates on ilium	Tibial tuberosity, by way of patellar tendon	Extends leg; rectus femoris also flexes thigh
Rectus femoris (REK-tus FEM-oar-is)				
Vastus lateralis (VASS-tus lat-er-AL-is)				
Vastus medialis (VASS-tus mee-dee-AL-is)				
Vastus intermedius (VASS-tus in-ter-MEE-dee-us)				
Hamstrings Biceps femoris (BYE-seps FEM-oar-is)	Large, fleshy muscle mass in posterior leg	Ischium; one head of biceps femoris arises from femur	Biceps femoris inserts on fibula; other on tibia	Flexes leg and extends thigh; antagonist to quadriceps femoris
Semimembranosus (sem-ee-MEM-brah-noh-sus)				
Semitendinosus (sem-ee-TEN-dih-noh-sus)				
Muscles That Move the Ankle and Foot				
Tibialis anterior (tib-ee-AL-is an-TEAR-ee-or)	Superficial muscle of anterior leg	Tibia	Medial cuneiform and first metatarsal	Dorsiflexes foot
Gastrocnemius (gas-trok-NEE-mee-us)	Superficial muscle on posterior surface of leg; forms curve of calf	Femur	Calcaneus by way of Achilles tendon	Plantarflexes foot
Soleus (SOH-lee-us)	Deep to gastrocnemius on posterior leg; has a common tendon for insertion with gastrocnemius	Tibia and fibula	Calcaneus by way of Achilles tendon	Plantarflexes foot
Peroneus (pear-oh-NEE-us)	Forms lateral compartment of leg	Tibia and fibula	Tarsals and metatarsals	Plantarflexes and everts foot

Muscles That Move the Leg

Muscles that move the leg are located in the thigh region. The **quadriceps femoris** (KWAHD-rih-seps FEM-oar-is) includes four muscles that are on the anterior and lateral sides of the thigh, namely, the **vastus lateralis, vastus intermedius, vastus medialis,** and **rectus femoris.** As a group, these muscles are the primary extensors of the leg, straightening the leg at the knee, and are used in climbing, running, and rising from a chair. The vastus intermedius is deep to the rectus femoris and is not illustrated. The other muscle on the anterior surface of the thigh is the long straplike **sartorius** (sar-TOAR-ee-us) that passes obliquely over the quadriceps group. The sartorius, the longest muscle in the body, flexes and medially rotates the leg when you sit cross-legged.

The posterior thigh muscles are called the **hamstrings,** and they are used to flex the leg at the knee. All have origins on the ischium and insert on the tibia. Because these muscles extend over the hip joint as well as over the knee joint, they also extend the thigh. The strong tendons of these muscles can be felt behind the knee. These same tendons are present in hogs, and butchers used them to hang the hams for smoking and curing, so they were called "ham strings." The hamstring muscles are the **biceps femoris, semimembranosus** (sem-ee-MEM-brah-noh-sus), and **semitendinosus** (sem-ee-TEN-dih-noh-sus). A "pulled hamstring" is a tear in one or more of these muscles or their tendons.

Muscles That Move the Ankle and Foot

The muscles located in the leg that move the ankle and foot are divided into anterior, posterior, and lateral compartments. The **tibialis anterior** is the primary muscle in the anterior group. Contraction of the muscles in the anterior group, including the tibialis anterior, causes dorsiflexion of the foot. **Peroneus** (pear-oh-NEE-us) muscles occupy the lateral compartment of the leg. Contraction of these muscles everts the foot and also helps in plantarflexion. The **gastrocnemius** (gas-trok-NEE-mee-us) and **soleus** (SOH-lee-us) are the major muscles in the posterior compartment. These two muscles form the fleshy mass in the calf of the leg. They have a common tendon called the **calcaneal tendon** or **Achilles tendon.** These muscles are strong plantarflexors of the foot. They are sometimes called the toe dancer's muscles because they allow you to stand on tiptoe. Numerous other deep muscles in the leg cause flexion and extension of the toes.

QUICK **CHECK**

7.29 Where are the hamstring muscles located?

7.30 What is the function of the muscle that is located on the anterior surface of the tibia?

QUICK **APPLICATIONS**

Exercise is good for us. We all know that. But just how good is it? Benefits of regular exercise include better posture, improved muscle tone, and more efficient heart and lung function. **Stretching exercises** slowly lengthen the muscles for a greater range of motion and flexibility. **Aerobic** or **endurance exercise** strengthens the heart and improves the body's ability to use oxygen. This type of exercise includes walking, running, jogging, bicycling (either stationary or outdoor), skating, and skiing. Over time, aerobic exercise can help lower your heart rate and blood pressure. **Strengthening** or **resistance exercise** works muscles as they move against resistance. These exercises increase muscle mass by stimulating the production of protein myofibrils within each muscle cell. Increased muscle mass means you burn more calories even when you are not exercising. Strengthening exercises also help lower blood pressure and reduce cholesterol levels.

FOCUS ON AGING

One of the most "obvious" age-related changes in skeletal muscles is the loss of muscle mass. This involves a decrease in both the number of muscle fibers and the diameter of the remaining fibers. Because muscle fibers are amitotic, once they are lost they cannot be replaced by new ones. Instead, they are replaced by connective tissue, primarily adipose. The number of muscle cells lost depends on several factors, including the amount of physical activity, the nutritional state of the individual, heredity, and the condition of the motor neurons that supply the muscle tissue. There is an age-related loss of motor neurons to skeletal muscle cells, and this is considered an important cause of muscle atrophy. It is probable that exercise enhances the ability of nerves to stimulate muscle fibers and to reduce atrophy.

As muscle mass decreases, there is a corresponding reduction in muscle strength. The amount of strength loss differs, depending to a large extent on the amount of physical activity. There is evidence that the mitochondria function less effectively in nonexercised muscle cells than in exercised cells. When mitochondria are inefficient, lactic acid accumulates, which contributes to muscle weakness.

There is a tendency for the skeletal muscles of older people to be less responsive, or to respond more slowly, than those of younger people. This is because the latent, contraction, and relaxation phases of muscle action all increase in duration. The increase in response time is less in muscles that are used regularly. Continued physical activity and good nutrition are probably the best deterrents to loss of muscle mass and muscle strength and to increased muscle response time.

Representative Disorders of the
Muscular System

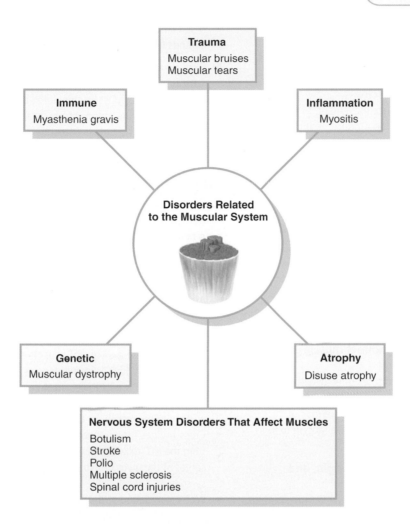

Trauma
Muscular bruises
Muscular tears

Immune
Myasthenia gravis

Inflammation
Myositis

**Disorders Related
to the Muscular System**

Genetic
Muscular dystrophy

Atrophy
Disuse atrophy

Nervous System Disorders That Affect Muscles
Botulism
Stroke
Polio
Multiple sclerosis
Spinal cord injuries

Botulism (BOTCH-yoo-lizm) Severe form of food poisoning caused by a neurotoxin produced by *Clostridium botulinum*; the toxin inhibits transmission of nerve impulses at neuromuscular junctions

Disuse atrophy (dis-YOOS AT-troh-fee) Skeletal muscle deterioration with reduction in size and muscular weakness due to bed rest, immobility, or nerve damage

Multiple sclerosis (MS) (MULL-tih pull skler-OH-sis) Disorder in which there is progressive destruction of the myelin sheaths of neurons interfering with their ability to transmit impulses to muscles; characterized by progressive loss of function

Muscular bruises (muss-kyoo-lar BROOZ-ehz) Muscle strain in which the muscle remains intact but there is mild bleeding into the surrounding tissue

Muscular dystrophy (MUSS-kyoo-lar DIS-troh-fee) Inherited, chronic, progressive wasting and weakening of muscles without involvement of the nervous system

Muscular tears (muss-kyoo-lar TAIRS) Severe muscle strain in which there is tearing of the fascia and bleeding; may require surgery and/or immobility to promote healing

Myasthenia gravis (mye-as-THEE-nee-ah GRAY-vis) Autoimmune disease, more common in females, that is characterized by weakness of skeletal muscles caused by an abnormality at the neuromuscular junction

Myositis (mye-oh-SYE-tis) Inflammation of muscle tissue

Polio (POH-lee-oh) Abbreviated term for poliomyelitis; an acute, contagious, viral disease that attacks the central nervous system, where it damages or destroys nerve cells that control muscles and causes paralysis

Spinal cord injuries (SPY-nuhl KORD IN-juhr-ees) Depending on severity, trauma to the spinal cord results in temporary or permanent disruption of cord-mediated functions, including motor impulses to skeletal muscles, with subsequent loss of motor function; extent of muscle paralysis depends on location and severity of trauma

Stroke (STROHK) Complex of symptoms caused by disrupted blood supply to the brain because of thrombosis, embolism, of hemorrhage; symptoms depend on region of brain affected, but frequently involve disturbed muscular function including paralysis; also called cerebrovascular accident (CVA)

CHAPTER SUMMARY

Characteristics and Functions of the Muscular System

■ **List four characteristics of muscle tissue.**
- Four characteristics of muscle tissue are excitability, contractility, extensibility, and elasticity.

■ **List four functions of skeletal muscles.**
- Four functions of muscles are to produce movement, to maintain posture, to provide joint stability, and to produce heat necessary to maintain body temperature.

Structure of Skeletal Muscle

■ **Describe the structure of a skeletal muscle, including its connective tissue coverings.**
- A muscle consists of many muscle fibers or cells. Each muscle is surrounded by connective tissue called epimysium.
- Inward extensions of the epimysium separate the muscle into bundles of fibers, called fasciculi, which are surrounded by perimysium.
- Each individual muscle fiber is surrounded by connective tissue called endomysium.

■ **Distinguish between direct and indirect muscle attachments.**
- In a direct attachment, the epimysium fuses directly with the periosteum of a bone.
- In indirect attachments, the connective tissue coverings of the muscles extend beyond the belly, or gaster, of the muscle to form tendons and aponeuroses.

■ **Define the terms *tendon, aponeurosis, origin,* and *insertion.***
- A tendon is a ropelike extension of the connective tissue coverings of a muscle that attaches a muscle to a bone.
- An aponeurosis is a broad flat sheet of the connective tissue coverings and usually connects two muscles.
- The origin is the more stable end of attachment, and the insertion is the more movable attachment.

■ **Identify the bands and lines that make up the striations on myofibers of skeletal muscle and relate these striations to actin and myosin.**
- A skeletal muscle fiber is a muscle cell with typical cellular organelles. The cell membrane is the sarcolemma, the cytoplasm is sarcoplasm, and the endoplasmic reticulum is called sarcoplasmic reticulum.
- The sarcoplasm contains myofibrils that are made of still smaller units called myofilaments. Thick myofilaments are composed of the protein myosin, and thin myofilaments are formed from the protein actin.
- The characteristic striations of skeletal muscle fibers (the A, I, and H bands) are due to the arrangement of myosin and actin myofilaments. The Z and M lines are points of myofilament connections.
- The A band is the entire length of the myosin filaments. The actin overlaps the myosin at the ends, making that region darker. The center of the A band, where there is only myosin, is the lighter H band. The I band is where there are only actin filaments.
- A sarcomere is the functional, or contractile, unit of a myofibril. It extends from one Z line to the next Z line.

■ **Explain why muscles have an abundant nerve and blood supply.**
- Skeletal muscles must be stimulated by a nerve before they can contract, therefore muscles have an abundant nerve supply; an abundant blood supply is necessary to deliver the nutrients and oxygen required for contraction.

Contraction of Skeletal Muscle

■ **Explain how a muscle fiber receives a stimulus for contraction.**
- A neuromuscular junction is the region where an axon terminal of a motor neuron is closely associated with a muscle fiber. Acetylcholine, a neurotransmitter released by the motor neuron, diffuses across the synaptic cleft to stimulate the sarcolemma of a muscle fiber. Acetylcholinesterase is an enzyme that inactivates acetylcholine to prevent continued contraction from a single impulse.

■ **Describe the sequence of events involved in the contraction of a skeletal muscle fiber.**
- When a muscle fiber is stimulated, an impulse travels down the sarcolemma into a T tubule, which releases calcium ions from the sarcoplasmic reticulum. Calcium alters the configuration of actin and energizes myosin by breaking down ATP.
- Cross-bridges form between the actin and myosin, and the energy from the ATP results in a power stroke that pulls actin toward the center of the myosin myofilaments. This shortens the length of the sarcomere.
- When stimulation at the neuromuscular junction stops, calcium returns to the sarcoplasmic reticulum, and actin and myosin resume their noncontracting positions.
- The minimum stimulus necessary to cause muscle fiber contraction is a threshold, or liminal, stimulus. A lesser stimulus is subthreshold, or subliminal.
- Individual muscle fibers contract according to the all-or-none principle. If a threshold stimulus is applied, the fiber will contract; a greater stimulus does not create a more forceful contraction. If the stimulus is subthreshold, the fiber does not contract.

■ **Compare the different types of muscle contractions.**
- Whole muscles show varying strengths of contraction due to motor unit and wave summation. A motor unit consists of a single motor neuron and all the muscle fibers it stimulates. Contraction strength of a whole muscle can be increased in response to increased load (stimulus) by stimulating more motor units—multiple motor unit summation.
- A muscle twitch, the response to a single stimulus, shows a lag phase, a contraction phase, and a relaxation phase. If a second stimulus of the same intensity of the first is applied before the relaxation phase is complete, a second contraction, stronger than the first, results. Repeated stimulation of the same strength that results in stronger contractions is called multiple wave summation.

- Rapid repeated stimulation that allows no relaxation results in a smooth sustained contraction that is stronger than the contraction from a single stimulus of the same intensity. This is tetany, a form of multiple wave summation. This is the usual form of muscle contraction.
- Treppe is the staircase effect that is evidenced when repeated stimuli of the same strength produce successively stronger contractions. This is due to changes in the cellular environment.
- Muscle tone refers to the continued state of partial contraction in muscles. It is important in maintaining posture and body temperature.
- Isotonic contractions produce movement but muscle tension remains constant. Isometric contractions increase muscle tension but do not produce movement. Most body movements involve both types of contractions.

Describe how energy is provided for muscle contraction and how oxygen debt occurs.

- In muscle contraction, energy is needed for the power stroke, detachment of myosin heads, and active transport of calcium.
- ATP provides the initial energy for muscle contraction. The ATP supply is replenished by creatine phosphate.
- When muscles are actively contracting for extended periods of time, glucose becomes the primary energy source to produce more ATP.
- When adequate oxygen is available, glucose is metabolized by aerobic respiration to produce ATP. If adequate oxygen is not available, the mechanism for producing ATP from glucose is anaerobic respiration. Aerobic respiration produces nearly 20 times more ATP per glucose than the anaerobic pathway, but anaerobic respiration occurs at a faster rate.
- Oxygen debt occurs when there is an accumulation of lactic acid from anaerobic respiration. Additional oxygen from continued labored breathing is needed to repay the debt and restore resting conditions.

Describe and illustrate movements accomplished by the contraction of skeletal muscle.

- A prime mover, or agonist, has the major role in producing a specific movement; a synergist assists the prime mover; an antagonist opposes a particular movement.
- Descriptive terms used to depict particular movements include *flexion, extension, hyperextension, dorsiflexion, plantarflexion, abduction, adduction, rotation, supination, pronation, circumduction, inversion,* and *eversion.*

Skeletal Muscle Groups

Locate, identify, and describe the actions of the major muscles of the head and neck.

- Muscles of facial expression include the frontalis, orbicularis oris, orbicularis oculi, buccinator, and zygomaticus.
- Chewing muscles are called muscles of mastication and include the masseter and temporalis.
- Neck muscles include the sternocleidomastoid, which flexes the neck, and the trapezius, which extends the neck.

Locate, identify, and describe the actions of the major muscles of the trunk.

- Muscles that act on the vertebral column include the erector spinae group and the deep back muscles. The erector spinae form a large muscle mass that extends from the sacrum to the skull on each side of the vertebral column. The deep back muscles are short muscles that occupy the space between the spinous and transverse processes of adjacent vertebrae.
- Muscles of the thoracic wall are involved in the process of breathing. These include the intercostal muscles and the diaphragm.
- Abdominal wall muscles include the external oblique, internal oblique, transversus abdominis, and rectus abdominis. These muscles provide strength and support to the abdominal wall.
- Pelvic floor muscles form a covering for the pelvic outlet and provide support for the pelvic viscera. The superficial muscles form the urogenital diaphragm, and the deeper muscles, the levator ani, form the pelvic diaphragm.

Locate, identify, and describe the actions of the major muscles of the upper extremity.

- The trapezius and serratus anterior muscles attach the scapula to the axial skeleton and move the scapula.
- The pectoralis major, latissimus dorsi, deltoid, and rotator cuff muscles insert on the humerus and move the arm.
- The triceps brachii, in the posterior compartment of the arm, extends the forearm. The biceps brachii and brachialis muscles, in the anterior compartment, flex the forearm.
- The brachioradialis muscle, primarily located in the forearm, flexes and supinates the forearm.
- Most of the muscles that are located on the forearm act on the wrist, hand, or fingers.

Locate, identify, and describe the actions of the major muscles of the lower extremity.

- The gluteus maximus, gluteus medius, gluteus minimus, and tensor fasciae latae muscles abduct the thigh. The iliopsoas, an anterior muscle, flexes the thigh.
- The muscles in the medial compartment include the adductor longus, adductor brevis, adductor magnus, and gracilis muscles. These muscles adduct the thigh.
- The quadriceps femoris muscle group, located in the anterior compartment of the thigh, straightens the leg at the knee.
- The sartorius muscle, a long straplike muscle on the anterior surface of the thigh, flexes and medially rotates the leg.
- The hamstring muscles, in the posterior compartment of the thigh, are antagonists to the quadriceps femoris muscle group.
- The principal muscle in the anterior compartment of the leg is the tibialis anterior, which dorsiflexes the foot.
- Contraction of the peroneus muscles, in the lateral compartment of the leg, everts the foot.
- The gastrocnemius and soleus form the bulky mass of the posterior compartment of the leg. These muscles plantarflex the foot.

CHAPTER QUIZ

Recall

Match the definitions on the left with the appropriate term on the right.

_____ **1.** Connective tissue covering around an individual muscle fiber

_____ **2.** Broad, flat sheet of tendon

_____ **3.** Muscle cell membrane

_____ **4.** More movable attachment of a muscle

_____ **5.** Functional unit of muscle; from Z line to Z line

_____ **6.** Neurotransmitter at the neuromuscular junction

_____ **7.** Minimal stimulus needed to cause contraction

_____ **8.** High-energy compound in muscle

_____ **9.** Protein pigment molecules in muscle

_____ **10.** Moving a part away from midline

A. Abduction
B. Acetylcholine
C. Aponeurosis
D. Creatine phosphate
E. Endomysium
F. Insertion
G. Myoglobin
H. Sarcolemma
I. Sarcomere
J. Threshold

Thought

1. Jonathan flexes his elbow by moving his antebrachial region upward toward his brachial region. Which one of the following bones is likely to represent the insertion of the muscle involved?

A. Humerus
B. Ulna
C. Carpal
D. Scapula

2. Arrange the given events in the sequence in which they occur in skeletal muscle contraction: (a) calcium released from sarcoplasmic reticulum; (b) acetylcholine diffuses across synaptic cleft; (c) myosin heads bind with actin; (d) impulse travels along sarcolemma and T tubules; (e) cross-bridges rotate to shorten sarcomere.

A. b, d, a, c, e
B. a, b, d, c, e
C. d, b, a, c, e
D. c, a, d, b, e

3. A ballet dancer is standing on her toes. This is an example of:

A. Abduction
B. Plantarflexion
C. Circumduction
D. Eversion

4. Which of the following muscles allows you to chew your food?

A. Zygomaticus
B. Sternocleidomastoid
C. Buccinator
D. Masseter

5. When an athlete experiences a "pulled hamstring," which of the following muscles is likely affected?

A. Vastus lateralis
B. Biceps brachii
C. Semimembranosus
D. Soleus

Application

Clarissa was "daydreaming" in class one day (certainly not an anatomy class!) when the teacher asked her a question. Embarrassed, Clarissa lowered her head, looked at the floor, and shrugged her shoulders to indicate she did not know the answer. What muscle did she use to lower her head and look down? What muscle did she use to shrug her shoulders?

BUILDING YOUR MEDICAL VOCABULARY

Building a vocabulary is a cumulative process. As you progress through this book, you will use word parts, abbreviations, and clinical terms from previous chapters. Each chapter will present new word parts, abbreviations, and terms to add to your expanding vocabulary.

Word Parts and Combining Form with Definition and Examples

PART/COMBINING FORM	DEFINITION	EXAMPLE
a-	without, lacking	atrophy: decrease in size of anormally developed organ
act-	motion	hyperactive: excessive motion
-asthenia	lack of strength	myasthenia gravis: muscles lose strength due to lack of nerve impulses to stimulate the muscle fibers
bi-	two	bilateral: occurring on two sides
delt-	triangle	deltoid: muscle that resembles a triangle
-desis	to bind	arthrodesis: bones are fused or tied together by surgery
dia-	through	diaphragm: muscle that goes through the middle, separates thoracic and abdominopelvic cavities
duct-	movement	abduction: to move away from a given point
fasci/o	fascia	fasciitis: inflammation of fascia
fibr/o	fibrous connective	fibromyalgia: chronic pain and stiffness in muscles, joints, tissue and fibrous tissue.
flex-	bend	flexion: the act of bending
iso-	same, alike	isometric: type of contraction where muscle length remains the same
kinesi/o	movement	kinesiology: scientific study of the movement of body parts
lemm-	peel, rind	sarcolemma: covering of a muscle cell, the cell membrane
masset-	chew	masseter muscle: muscle that functions in chewing

PART/COMBINING FORM	DEFINITION	EXAMPLE
metr-	measure	isometric: type of contraction where muscle length (measure) remains the same
my/o, mys-	muscle	myoatrophy: muscular atrophy
phragm-	fence, partition	diaphragm: a fence or partition through the body cavity
plant/o	sole of the foot	plantalgia: pain in the sole of the foot
-rrhexis	rupture	myorrhexis: rupture of a muscle
sarc/o	flesh, muscle	sarcoblast: cell that develops into a muscle cell
syn	together	synarthrosis: joint in which the bones are held tightly together by fibrous tissue
ton-	tone, tension	isotonic: having the same tone or tension
-troph-	nourish, develop	trophology: the science of nutrition of the body

Clinical Abbreviations

ABBREVIATION	MEANING
ACL	anterior cruciate ligament
ad lib	as desired
DTR	deep tendon reflex
EMG	electromyography
LOS	length of stay
MG	myasthenia gravis
MS	multiple sclerosis
THR	total hip replacement
TKR	total knee replacement

Clinical Terms

Cramp (KRAMP) Painful involuntary muscle spasm; often caused by myositis but can be a symptom of any irritation or ion imbalance

Electromyography (ee-lek-troh-mye-AHG-rah-fee) Process of recording the strength of muscle contraction as a result of electrical stimulation

Muscle biopsy (MUSS-uhl BYE-ahp-see) Removal of muscle tissue for microscopic examination

Myoparesis (mye-oh-pah-REE-sis) Weakness or slight paralysis of a muscle

Myopathy (mye-AHP-ah-thee) Muscle disease

Myorrhexis (mye-oh-REK-sis) Rupture of a muscle

Paralysis (pah-RALL-ih-sis) Loss or impairment of motor function due to a lesion in the neuromuscular pathway

Repetitive stress disorder (ree-PET-ah-tiv STRESS dis-OAR-der) Condition with symptoms caused by repetitive motions that involve muscles, tendons, nerves, and joints; most commonly occur as work-related or sports injuries

Shin splint (SHIN SPLINT) Strain of the long flexor muscle of the toes resulting in pain along the tibia (shinbone); usually caused by repeated stress to the lower leg

Tenomyoplasty (ten-oh-MY-oh-plas-tee) Surgical repair of a tendon and muscle; applied especially to an operation for inguinal hernia

Tenoplasty (TEN-oh-plas-tee) Surgical repair of a tendon

Tenorraphy (ten-OAR-ah-fee) Suture of a tendon

Tic (TIK) Spasmodic involuntary twitching of a muscle that is normally under voluntary control

Torticollis (tor-tih-KOHL-is) Wryneck; a contracted state of the sternocleidomastoid on one side characterized by a twisting of the nick and an unnatural position of the head.

VOCABULARY QUIZ

Use word parts you have learned to form words that have the following definitions.

1. Scientific study of movement _____

2. Partition through a space _____

3. Same measure _____

4. Chewer _____

5. Movement away from _____

Using the definitions of word parts you have learned, define the following words.

6. Deltoid _____

7. Biceps _____

8. Isotonic _____

9. Flexion _____

10. Sarcolemma _____

Match each of the following definitions with the correct word.

_____ 11. Occurring on two sides

_____ 12. Work together

_____ 13. Muscle matter

_____ 14. Without development, wasting

_____ 15. Referring to sole of the foot

A. Atrophy
B. Bilateral
C. Plantar
D. Sarcoplasm
E. Synergist

16. What is the clinical term for the rupture of a muscle? _____

17. What is the clinical term for a weakness or slight paralysis of a muscle? _____

18. What procedure records the strength of muscle contraction as a result of electrical stimulation? _____

19. What is the meaning of the abbreviation DTR? _____

20. What is the meaning of the abbreviation ACL? _____

Nervous System

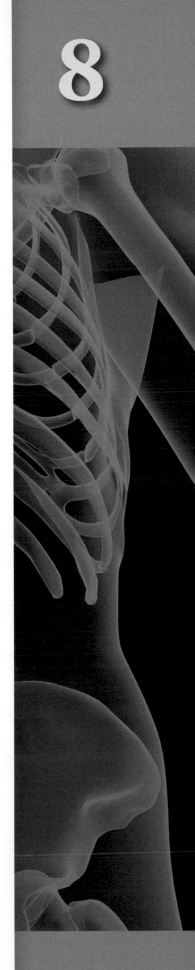

CHAPTER OBJECTIVES

Functions and Organization of the Nervous System
- Outline the organization and functions of the nervous system.
- List three categories of nervous system functions.

Nerve Tissue
- Compare the structure and functions of neurons and neuroglia.

Nerve Impulses
- Describe the characteristics of the resting membrane of a neuron.
- Describe the sequence of events that lead to an action potential when the cell membrane is stimulated.
- Explain how the impulse is conducted along the length of a neuron.
- Describe the structure of a synapse and explain how an impulse is conducted from one neuron to another across the synapse.
- List the five basic components of a reflex arc.

Central Nervous System
- Describe the three layers of meninges around the central nervous system.
- Locate and identify the major regions of the brain and describe their functions.
- Trace the flow of cerebrospinal fluid from its origin in the ventricles to its return to the blood.
- Describe the structure and functions of the spinal cord.

Peripheral Nervous System
- Describe the structure of a nerve.
- List the 12 cranial nerves and state the function of each one.
- Discuss spinal nerves and the plexuses they form.
- Compare the structural and functional differences between the somatic efferent pathways and the autonomic nervous system.
- Distinguish between the sympathetic and parasympathetic divisions of the autonomic nervous system in terms of structure, function, and neurotransmitters.

KEY TERMS

Action potential (ACK-shun po-TEN-shall) Nerve impulse; a rapid change in membrane potential that involves depolarization and repolarization

Brain stem (BRAYN STEM) Portion of the brain, between the diencephalon and spinal cord, that contains the midbrain, pons, and medulla oblongata

Cerebellum (sair-eh-BELL-um) Second largest part of the human brain, located posterior to the pons and medulla oblongata, and involved in the coordination of muscular movements

Cerebrum (se-REE-brum) The largest and uppermost part of the human brain; concerned with consciousness, learning, memory, sensations, and voluntary movements

Diencephalon (dye-en-SEF-ah-lon) Part of the brain between the cerebral hemispheres and the midbrain; includes the thalamus, hypothalamus, and epithalamus

Myelin (MY-eh-lin) White, fatty substance that surrounds many nerve fibers

Neurilemma (noo-rih-LEM-mah) Layer of Schwann cells that surrounds a nerve fiber in the peripheral nervous system and, in some cases, produces myelin; also called Schwann's sheath

Saltatory conduction (SAL-tah-toar-ee kon-DUCK-shun) Process in which a nerve impulse travels along a myelinated nerve fiber by jumping from one node of Ranvier to the next

Synapse (SIN-aps) Region of communication between two neurons

Threshold stimulus (THRESH-hold STIM-yoo-lus) Minimum level of stimulation that is required to start a nerve impulse or muscle contraction; also called liminal stimulus

Functional Relationships of the
Nervous/Sensory System

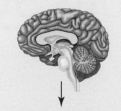

Monitors external and internal environments and mediates adjustments to maintain homeostasis.

Reproductive

Gonads produce hormones that influence CNS development and sexual behavior; menstrual hormones affect the activity of the hypothalamus.

Regulates sex drive, arousal, and orgasm; hormones involved in sperm production, menstrual cycle, pregnancy, and parturition.

Urinary

Helps maintain pH and electrolyte balance necessary for neural function; eliminates metabolic wastes harmful to nerve function.

Autonomic nervous system controls renal blood pressure and renal blood flow, which affect rate of urine formation; regulates bladder emptying.

Digestive

Absorbs nutrients for neural growth, maintenance, and repair; provides nutrients for synthesis of neurotransmitters and energy for nerve impulse conduction; liver maintains glucose levels for neural function.

Autonomic nervous system controls motility and glandular activity of the digestive tract.

Respiratory

Supplies oxygen for brain, spinal cord, and sensory organs; removes carbon dioxide; helps maintain pH.

Stimulates muscle contractions that create pressure changes necessary for ventilation; regulates rate and depth of breathing.

Integument

Provides protection for peripheral nerves; supports peripheral receptors for touch, pressure, pain, and temperature.

Influences secretions of glands in the skin, contraction of arrector pili muscles,

Skeletal

Protects brain and spinal cord; supports ear, eye, and other sensory organs; stores calcium necessary for neural function.

Innervates bones and provides sensory information about joint movement and position.

Muscular

Performs the somatic motor commands that arise in the CNS; muscle spindles provide proprioceptive sense; provides heat to maintain body temperature for neural function.

Coordinates skeletal muscle contraction; adjusts cardiovascular and respiratory systems to maintain cardiac output and oxygen for muscle contraction.

Endocrine

Hormones influence neuronal metabolism and enhance autonomic stimuli.

Regulates secretory activity of anterior pituitary and adrenal medulla; produces ADH and oxytocin.

Lymphatic/Immune

Assists in defense against pathogens and repair of neural and sensory tissue following trauma; removes excess fluid from tissues surrounding nerves.

Innervates lymphoid organs and helps regulate the immune response.

Cardiovascular

Delivers oxygen and nutrients to brain, spinal cord, and other neural and sensory tissue; removes waste products and heat; source of CSF.

Monitors and adjusts heart rate, blood pressure, and blood flow.

← Gives to Nervous/Sensory System
→ Receives from Nervous/Sensory System

The **nervous system** is the major controlling, regulatory, and communicating system in the body. It is the center of all mental activity, including thought, learning, and memory. Together with the endocrine system, the nervous system is responsible for regulating and maintaining homeostasis. Through its receptors, the nervous system keeps us in touch with our environment, both external and internal.

Like other systems in the body, the nervous system is composed of organs, principally the brain, spinal cord, nerves, and ganglia. These, in turn, consist of various tissues, including nerve, blood, and connective tissues. Together these carry out the complex activities of the nervous system.

Functions of the Nervous System

The various activities of the nervous system can be grouped together as three general, overlapping functions:

- Sensory functions
- Integrative functions
- Motor functions

Together these functions keep us in touch with our environments, maintain homeostasis, and account for thought, learning, and memory.

Millions of sensory receptors detect changes, called stimuli, that occur inside and outside the body. They monitor such things as temperature, light, and sound from the external environment. Inside the body, the internal environment, receptors detect variations in pressure, pH, carbon dioxide concentration, and the levels of various electrolytes. All of this gathered information is called **sensory input.**

Sensory input is converted into electrical signals called nerve impulses that are transmitted to the brain. In the brain, the signals are brought together to create sensations, produce thoughts, or add to memory. Decisions are made each moment based on the sensory input. This is called **integration.**

Based on the sensory input and integration, the nervous system responds by sending signals to muscles, causing them to contract, or to glands, causing them to produce secretions. Muscles and glands are called **effectors** because they cause an effect in response to directions from the nervous system. This is the **motor output** or **motor function.**

Organization of the Nervous System

There is really only one nervous system in the body, though terminology seems to indicate otherwise. Although each subdivision of the system is also called a "nervous system," all of these smaller systems belong to the single, highly integrated nervous system. Each subdivision has structural and functional characteristics that distinguish it from the others. The nervous system as a whole is divided into two subdivisions: the **central nervous system** (CNS) and the **peripheral nervous system** (PNS) (Figure 8-1).

The **brain** and **spinal cord** are the organs of the **central nervous system.** Because they are so vitally important, the brain and spinal cord, located in the dorsal body cavity, are encased in bone for protection. The brain is in the cranial vault, and the spinal cord is in the vertebral canal of the vertebral column. Although considered to be two separate organs, the brain and spinal cord are continuous at the foramen magnum.

The organs of the **peripheral nervous system** are the **nerves** and **ganglia.** Nerves are bundles of nerve fibers, much like muscles are bundles of muscle fibers. Cranial nerves (12 pairs) and spinal nerves (31 pairs) extend from the CNS to peripheral organs such as muscles and glands. Ganglia are collections, or small knots, of nerve cell bodies outside the CNS.

The PNS is further subdivided into an **afferent (sensory) division** and an **efferent (motor) division.** The afferent or sensory division transmits impulses from peripheral organs to the CNS. The efferent or motor division transmits impulses from the CNS out to the peripheral organs to cause an effect or action.

Finally, the efferent or motor division is again subdivided into the **somatic nervous system** and the **autonomic nervous system.** The somatic nervous system, also called somatomotor or somatic efferent, supplies motor impulses to the skeletal muscles. Because these nerves permit conscious control of the skeletal muscles, the somatic nervous system is sometimes called the **voluntary nervous system.** The autonomic nervous system, also called visceral efferent, supplies motor impulses to cardiac muscle, to smooth muscle, and to glandular epithelium. It is further subdivided into **sympathetic** and **parasympathetic** divisions. Because the autonomic nervous system regulates involuntary or automatic functions, it is sometimes called the **involuntary nervous system.**

QUICK CHECK

8.1 List the three types of functions attributed to the nervous system.

8.2 What are the organs of the peripheral nervous system?

Nerve Tissue

Although the nervous system is very complex, there are only two main types of cells in nerve tissue. The actual nerve cell is the **neuron.** It is the "conducting" cell that transmits impulses. It is the structural unit of the nervous system. The other type of cell is the **neuroglia,** or **glial,** cell. The word *neuroglia* means "nerve glue." These cells are nonconductive and provide a support system for the neurons. They are a special type of "connective tissue" for the nervous system.

Neurons

Neurons, or nerve cells, carry out the functions of the nervous system by conducting nerve impulses. They are highly specialized and amitotic. This means that if a neuron is destroyed, it cannot be replaced because neurons do not undergo mitosis.

Each neuron has three basic parts:

- Cell body
- One or more dendrites
- A single axon

Figure 8-2 illustrates a typical neuron. The main part of the neuron is the **cell body** or **soma.** In many ways, the cell body is similar to other types of cells. It has a nucleus with at least one nucleolus and contains many of the typical cytoplasmic organelles. It lacks centrioles, however. Because centrioles

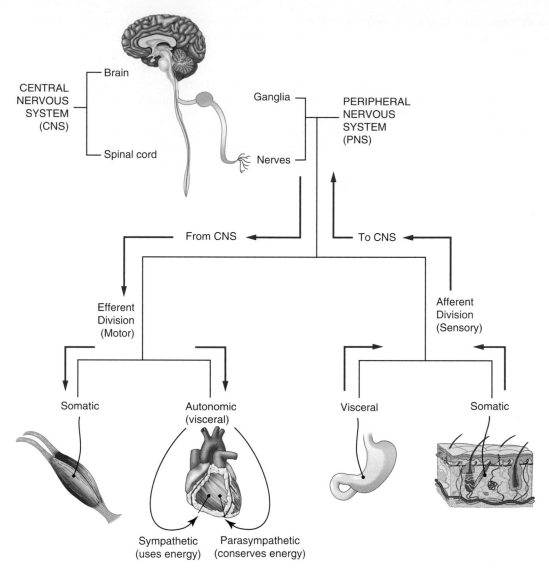

Figure 8-1 *Organization of the nervous system.*

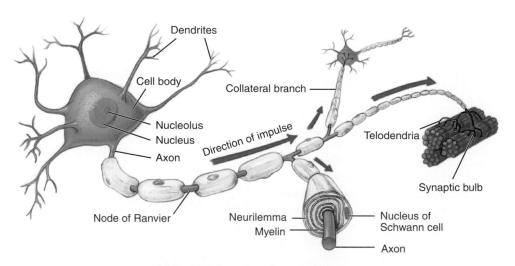

Figure 8-2 *Structure of a typical neuron.*

function in cell division, the fact that neurons do not have centrioles is consistent with the fact that neurons are not capable of mitosis. **Dendrites** and **axons** are cytoplasmic extensions, or processes, that project from the cell body. They are sometimes referred to as **fibers.** Dendrites are usually, but not always, short and branching, which increases their surface area to receive signals from other neurons. The number of dendrites on a neuron varies. They are called afferent processes because they transmit impulses to the neuron cell body. There is only one axon that projects from each cell body. It is usually elongated, and because it carries impulses away from the cell body, it is called an efferent process.

An axon may have infrequent branches called **axon collaterals.** Axons and axon collaterals terminate in many short branches or **telodendria** (tell-oh-DEN-dree-ah). The distal ends of the telodendria are slightly enlarged to form **synaptic bulbs.** Many axons are surrounded by a segmented, white, fatty substance called **myelin** (MY-eh-lin) or the **myelin sheath.** Myelinated fibers make up the white matter in the CNS, whereas cell bodies and unmyelinated fibers make up the gray matter. The unmyelinated regions between the myelin segments are called the **nodes of Ranvier** (nodes of ron-vee-AY). In the PNS, the myelin is produced by Schwann cells. The cytoplasm, nucleus, and outer cell membrane of the Schwann cell form a tight covering around the myelin and around the axon itself at the nodes of Ranvier. This covering is the **neurilemma** (noo-rih-LEM-mah), which plays an important role in the regeneration of nerve fibers. In the CNS, **oligodendrocytes** (ah-lee-go-DEN-droh-sites) produce myelin, but there is no neurilemma, which is why fibers within the CNS do not regenerate. The structure of an axon and its coverings is illustrated in Figure 8-2.

Functionally, neurons are classified as afferent, efferent, or interneurons (association neurons) according to the direction in which they transmit impulses relative to the CNS (Table 8-1). **Afferent,** or **sensory, neurons** carry impulses from peripheral sense receptors to the CNS. They usually have long dendrites and relatively short axons. **Efferent,** or **motor, neurons** transmit impulses from the CNS to effector organs such as muscles and glands. Efferent neurons usually have short dendrites and long axons. **Interneurons,** or **association neurons,** are located entirely within the CNS, where they form the connecting link between the afferent and efferent neurons. They have short dendrites and may have either a short or a long axon.

Neuroglia

Neuroglia cells do not conduct nerve impulses; instead, they support, nourish, and protect the neurons. They are far more numerous than neurons and, unlike neurons, are capable of mitosis. Table 8-2 describes the six types of neuroglia cells. Some authorities classify only the four supporting cell types found in the CNS as true neuroglia.

TABLE 8-1 Types of Neurons Classified According to Function

Type of Neuron	Structure	Function
Afferent (sensory)	Long dendrites and short axon; cell body located in ganglia in PNS; dendrites in PNS; axon extends into CNS	Transmits impulses from peripheral sense receptors to CNS
Efferent (motor)	Short dendrites and long axon; dendrites and cell body located within CNS; axons extend to PNS	Transmits impulses from CNS to effectors such as muscles and glands in periphery
Association (interneurons)	Short dendrites; axon may be short or long; located entirely within CNS	Transmits impulses from afferent neurons to efferent neurons

PNS, peripheral nervous system; CNS, central nervous system.

TABLE 8-2 Neuroglial Cell Types

Cell Type	Location	Description	Special Function
Astrocytes	CNS	Star shaped; numerous radiating processes with bulbous ends for attachment	Bind blood vessels to nerves; regulate composition of fluid around neurons
Ependymal cells	CNS (line ventricles of brain and central canal of spinal cord)	Columnar cells with cilia	Active role in formation and circulation of cerebrospinal fluid
Microglia	CNS	Small cells with long processes; modified macrophages	Protection; become mobile and phagocytic in response to inflammation
Oligodendrocytes	CNS	Small cells with few, but long, processes that wrap around axons	Form myelin sheaths around axons in CNS
Schwann cells*	PNS	Flat cells with long, flat process that wraps around axon in PNS	Form myelin sheaths around axons in PNS; active role in nerve fiber regeneration
Satellite cells*	PNS	Flat cells, similar to Schwann cells	Support nerve cell bodies within ganglia

CNS, central nervous system; PNS, peripheral nervous system.
*Some authorities do not consider these to be neuroglia because they are in the PNS.

QUICK CHECK

8.3 What types of neurons are located entirely within the CNS?

8.4 Defective oligodendrocytes interfere with the production of what substance?

8.5 What function is impaired if there is damage to afferent neurons?

QUICK APPLICATIONS

Because neurons are not capable of mitosis, primary malignant tumors of the brain are tumors of the glial cells, rather than of the neurons themselves. These tumors, called gliomas, have extensive roots, making them extremely difficult to remove.

Nerve Impulses

The functional characteristics of neurons are **excitability** and **conductivity.** Excitability is the ability to respond to a stimulus; conductivity is the ability to transmit an impulse from one point to another. All the functions associated with the nervous system, including thought, learning, and memory, are based on these two characteristics. These functional characteristics are the result of structural features of the cell membrane.

Resting Membrane

A **resting membrane** is the cell membrane of a nonconducting, or resting, neuron. The membrane is impermeable to the passive diffusion of sodium (Na^+) and potassium (K^+) ions. An active transport mechanism, the sodium-potassium pump, maintains a difference in concentration of these ions on the two sides of the membrane. Sodium ions are concentrated in the extracellular fluid, whereas the potassium ions are inside the cell. The intracellular fluid also contains proteins and other negatively charged ions. The result is a polarized membrane with more positive charges outside the cell and more negative charges inside the cell. This difference in charges on the two sides of the resting membrane is the **resting membrane potential.** Electrical measurements show the resting membrane potential to be about –70 millivolts (mV), which means that the inside of the membrane is 70 mV less positive (more negative) than the outside.

Stimulation of a Neuron

A stimulus is a physical, chemical, or electrical event that alters the neuron cell membrane and reduces its polarization for a brief time. The stimulus changes the membrane so that it becomes permeable to sodium ions, which diffuse into the cell. If the stimulus is weak, the membrane is only slightly permeable and the inward movement of sodium is offset by an outward movement of potassium. In this case, the resting membrane potential is maintained and no response is initiated.

If the stimulus is strong enough, the membrane becomes highly permeable to sodium at the point of the stimulation. Positively charged sodium ions rapidly diffuse through the membrane to the inside of the cell. This movement is driven by the concentration gradient and the electrical gradient. (Remember, the inside is negatively charged, which attracts the positive sodium ions.) As the positive ions enter the cell, the inside of the membrane becomes more positively charged, reducing the polarization found in the resting membrane. This is **depolarization.**

For just an instant, the influx of sodium ions reverses the membrane polarity with more positive charges inside the cell than outside. This is **reverse polarization**.

Very quickly, the membrane again becomes impermeable to sodium, with the sodium ions trapped inside the cell. Next the membrane becomes permeable to the intracellular potassium ions for a fraction of a millisecond, and they rapidly diffuse down the concentration gradient to the outside of the cell. Because these are positive ions, this action removes the intracellular positive charge and restores the resting membrane potential of –70 mV. This process is called **repolarization**.

The rapid sequence of events in response to a stimulus, namely, depolarization, reverse polarization, and repolarization, is called the **action potential**. Electrical measurements show the action potential to peak at approximately +30 mV (Figure 8-3). At the conclusion of the action potential, the sodium-potassium pump actively transports sodium ions out of the cell and potassium ions into the cell to completely restore resting conditions.

The minimum stimulus necessary to initiate an action potential is called a **threshold (liminal) stimulus**. A weaker stimulus, called a **subthreshold (subliminal) stimulus**, does not cause sufficient depolarization to elicit an action potential.

QUICK CHECK

8.6 In a resting cell membrane, what is the predominant extracellular ion?

8.7 What is the response of a neuron to a threshold stimulus?

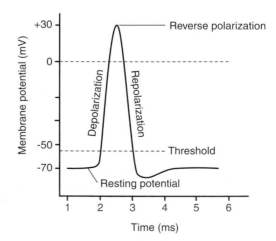

Figure 8-3 *Recording of an action potential. Note the resting potential is –70 mV and the peak action potential is +30 mV.*

QUICK APPLICATIONS

Some anesthetic agents produce their effects by inhibiting the diffusion of sodium through the cell membrane and thus blocking the initiation and conduction of nerve impulses.

Conduction Along a Neuron

Once a threshold stimulus has been applied and an action potential generated, it must be conducted along the total length of the neuron either to an effector or to another neuron.

Propagated Action Potentials

The threshold stimulus causes a localized area of reverse polarization on the membrane. In that one area, the membrane is negative on the outside and positive on the inside. The rest of the membrane is in the resting condition. When a given area reverses its polarity, the difference in potential between that area and the adjacent area creates a current flow that depolarizes the second point (Figure 8-4). When the second point reverses its polarity, current flow between the second point and the third point depolarizes the third point. This continues point by point, in domino fashion, along the entire length of the neuron, creating a **propagated action potential,** or **nerve impulse.** The following list summarizes the events that occur during nerve impulse conduction:

- Resting membrane has sodium ions on the outside and potassium ions on the inside.
- Intracellular negative ions produce a resting membrane potential of –70 mV.
- Receptors are stimulated by a physical, chemical, or electrical event.

- Stimulus alters permeability of neuron cell membrane; sodium channels open and sodium ions diffuse to the inside of the cell.
- Influx of sodium ions depolarizes the cell membrane.
- Sodium ions continue inward diffusion to reverse the membrane's polarity to +30 mV.
- Sodium channels close and potassium channels open.
- Potassium ions diffuse out of the cell.
- Outward diffusion of potassium ions restores resting membrane potential of –70 mV
- Sodium-potassium pump transports sodium ions out of the cell and potassium ions back into the cell to restore resting conditions.
- Reverse polarity at one point is the stimulus that depolarizes an adjacent point to propagate the action potential.

Saltatory Conduction

The conduction described in the previous paragraphs is representative of an unmyelinated axon. Because myelin is an insulating substance, it inhibits the flow of current from one point to another. In myelinated fibers, depolarization occurs only at the places where there is no myelin, at the nodes of Ranvier (Figure 8-5). The action potential "jumps" from node to node. This "jumping" is **saltatory** (SAL-tah-toar-ee) **conduction,** which is faster than conduction in unmyelinated fibers.

Refractory Period

The period of time during which a point on the cell membrane is "recovering" from depolarization is called the **refractory period.** While the membrane is permeable to sodium ions, it cannot respond to a second stimulus, no matter how strong the stimulus. This is the **absolute refractory period.** For a brief period after the absolute refractory period, roughly comparable to the time of altered membrane permeability to potassium, it takes a stronger than normal stimulus to reach threshold. This is the **relative refractory period.**

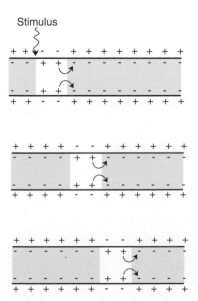

Figure 8-4 *Propagation of an action potential on an unmyelinated fiber. Depolarization occurs point by point in domino fashion along the entire length of a neuron, creating a nerve impulse.*

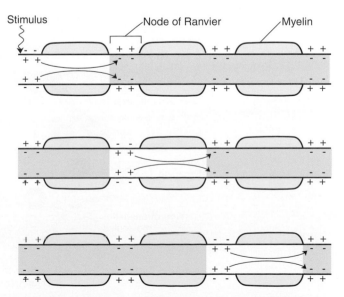

Figure 8-5 *Saltatory conduction along a myelinated fiber. The action potential "jumps" from one node of Ranvier to the next.*

All-or-None Principle

Nerve fibers, like muscle fibers, obey the **all-or-none principle.** If a threshold stimulus is applied, an action potential is generated and propagated along the entire length of the neuron at maximum strength and speed for the existing conditions. A stronger stimulus does not increase the strength of the action potential or change the rate of conduction. A weaker stimulus is subthreshold and does not evoke an action potential. If a stimulus is threshold or greater, an impulse is conducted. If the stimulus is subthreshold, there is no conduction.

QUICK CHECK

8.8 What is a distinguishing feature of a neuron on which salutatory conduction occurs?

8.9 What term is given to the period when it takes a stronger than normal stimulus to reach threshold?

Conduction Across a Synapse

A nerve impulse, or propagated action potential, travels along a nerve fiber until it reaches the end of the axon; then it must be transmitted to the next neuron. The region of communication between two neurons is called a **synapse** (SIN-aps). A synapse has three parts (Figure 8-6):

- Synaptic knob
- Synaptic cleft
- Postsynaptic membrane

The first neuron, the one preceding the synapse, is called the **presynaptic neuron;** the second neuron, the one following the synapse, is called the **postsynaptic neuron.** Synaptic knobs are tiny bulges at the end of the telodendria on the presynaptic neuron. Small sacs within the synaptic knobs, called **synaptic vesicles,** contain chemicals known as **neurotransmitters** (noo-roh-TRANS-mitters).

When a nerve impulse reaches the synaptic knob, a series of reactions releases neurotransmitters into the synaptic cleft. The neurotransmitters diffuse across the synaptic cleft and react with receptors on the postsynaptic cell membrane. This is synaptic transmission. To prevent prolonged reactions with the postsynaptic receptors, the transmitters are very quickly inactivated by enzymes. One of the best known neurotransmitters is **acetylcholine** (ah-see-till-KOH-leen), which is inactivated by the enzyme **cholinesterase** (koh-lin-ES-ter-ase). Table 8-3 lists some of the common neurotransmitters.

In **excitatory transmission,** the neurotransmitter–receptor reaction on the postsynaptic membrane depolarizes the membrane and initiates an action potential. This is excitation or stimulation. Acetylcholine is typically an excitatory neurotransmitter. Some neurotransmitters result in **inhibitory transmission.** In this case, the reaction between the neurotransmitter and the receptor opens potassium channels in the membrane so that potassium diffuses out of the cell but has no effect on the sodium channels. This action makes the inside of the membrane even more negative than it is in the resting condition; it **hyperpolarizes** the membrane, which makes it

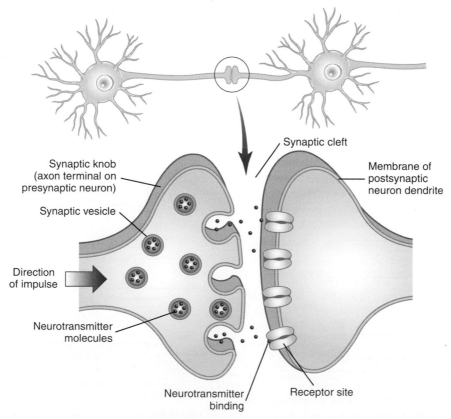

Figure 8-6 *Components of a synapse. The impulse travels from the presynaptic neuron to the postsynaptic neuron.*

TABLE 8-3 Some Common Neurotransmitters

Neurotransmitter	Location	Function	Comments
Acetylcholine	CNS and PNS	Generally excitatory but inhibitory to some visceral effectors	Found in skeletal neuromuscular junction and in many ANS synapses
Norepinephrine	CNS and PNS	May be excitatory or inhibitory depending on receptors	Found in visceral and cardiac muscle neuromuscular junctions; cocaine and amphetamines exaggerate effects
Epinephrine	CNS and PNS	May be excitatory or inhibitory depending on receptors	Found in pathways concerned with behavior and mood
Dopamine	CNS and PNS	Generally excitatory	Found in pathways that regulate emotional responses; decreased levels in Parkinson's disease
Serotonin	CNS	Generally inhibitory	Found in pathways that regulate temperature, sensory perception, mood, onset of sleep
Gamma-aminobutyric acid (GABA)	CNS	Generally inhibitory	Inhibits excessive discharge of neurons
Endorphins and enkephalins	CNS	Generally inhibitory	Inhibit release of sensory pain neurotransmitters; opiates mimic effects of these peptides

more difficult to generate an action potential. This is inhibition. Gamma-aminobutyric acid (GABA) is an inhibitory neurotransmitter in the CNS.

The billions of neurons in the CNS are organized into functional groups called **neuronal pools.** The neuronal pools receive information, process and integrate that information, and then transmit it to some other destination. Neuronal pools are arranged in pathways, or circuits, over which the nerve impulses are transmitted. The simplest pathway is the **simple series circuit** (Figure 8-7, *A*) in which a single neuron synapses with another neuron, which in turn synapses with another, and so on. Most pathways are more complex. In a **divergence circuit** (Figure 8-7, *B*), a single neuron synapses with multiple neurons within the pool. This permits the same information to diverge or to go along different pathways at the same time. This type of pathway is important in muscle contraction when many muscle fibers, or even several muscles, must contract at the same time. Another type of pathway is the **convergence circuit** (Figure 8-7, *C*). In this case, several presynaptic neurons synapse with a single postsynaptic neuron. This accounts for the fact that many different stimuli may have the same ultimate effect. For example, thinking about food, smelling food, and seeing food all have the same effect—the flow of saliva.

Reflex Arcs

The neuron is the structural unit of the nervous system; the **reflex arc** is the functional unit. The reflex arc is a type of conduction pathway. It is similar to a one-way street because it allows impulses to travel in only one direction. The simplest reflex arc consists of two neurons, but most have three or more neurons in the conduction pathway. Figure 8-8 illustrates a three-neuron reflex arc. There are five basic components in a reflex arc (Table 8-4):

- Receptor
- Sensory neuron
- Center
- Motor neuron
- Effector

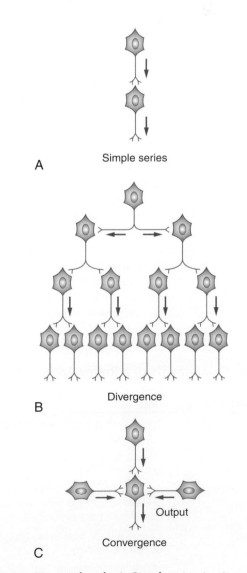

Figure 8-7 *Neuronal pools. **A**, Simple series circuit: one neuron synapses with another. **B**, Divergence circuit: a single neuron synapses with multiple neurons. **C**, Convergence circuit: several neurons synapse with a single postsynaptic neuron.*

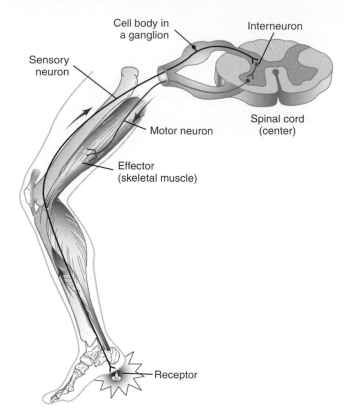

Figure 8-8 *Components of a generalized reflex arc. Note the five components of a reflex arc.*

A reflex is an automatic, involuntary response to some change, either inside or outside the body. Reflexes are important in maintaining homeostasis by making adjustments to heart rate, breathing rate, and blood pressure. Reflexes are also involved in coughing, sneezing, and reactions to painful stimuli. Everyone is familiar with the withdrawal reflex. When you step on a tack or touch a hot iron, you immediately, without conscious thought, withdraw the injured foot or hand from the source of the irritation. Clinicians frequently test an individual's reflexes to determine if the nervous system is functioning properly.

QUICK CHECK

8.10 What happens on the postsynaptic membrane in inhibitory transmission?
8.11 Many muscle fibers must contract at the same time. What type of neuronal circuit permits this?
8.12 List the five components of a reflex arc.

Central Nervous System

The CNS consists of the brain and spinal cord, which are located in the dorsal body cavity. These are vital to our well-being and are enclosed in bone for protection. The brain is surrounded by the cranium, and the spinal cord is protected by the vertebrae. The brain is continuous with the spinal cord at the foramen magnum. In addition to bone, the CNS is surrounded by connective tissue membranes, called **meninges,** and by **cerebrospinal** (seh-ree-broh-SPY-null) **fluid (CSF).**

Meninges

Three layers of meninges (men-IN-jeez) surround the brain and spinal cord (Figure 8-9). The outer layer, the **dura mater** (DOO-rah MAY-ter), is tough, white fibrous connective

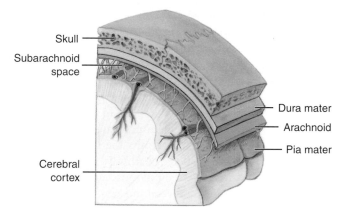

Figure 8-9 *Meninges of the central nervous system.*

TABLE 8-4 Components of a Reflex Arc

Component	Description	Function
Receptor	Site of stimulus action; receptor end of dendrite or special cell in receptor organ	Responds to some change in internal or external environment
Sensory neuron	Afferent neuron; cell body is in ganglion outside CNS; axon extends into CNS	Transmits nerve impulses from receptor to CNS
Integration center	Always within CNS; in simplest reflexes, it consists of synapse between sensory and motor neurons; more commonly one or more interneurons are involved	Process center; region in CNS where incoming sensory impulses generate appropriate outgoing motor impulses
Motor neuron	Efferent neuron; dendrites and cell body are in CNS; axon extends to periphery	Transmits nerve impulses from integration center in CNS to effector organ
Effector	Muscle or gland outside CNS	Responds to impulses from motor neuron to produce an action such as contraction or secretion

CNS, central nervous system.

tissue. It is just inside the cranial bones and lines the vertebral canal. The dura mater contains channels, called **dural sinuses,** that collect venous blood to return it to the cardiovascular system. The **superior sagittal sinus,** located in the midsagittal line over the top of the brain, is the largest of the dural sinuses.

The middle layer of meninges is the **arachnoid** (ah-RAK-noyd). The arachnoid, which resembles a cobweb in appearance, is a thin layer with numerous threadlike strands that attach it to the innermost layer. The space under the arachnoid, the **subarachnoid space,** is filled with CSF and contains blood vessels.

The **pia mater** (PEE-ah MAY-ter) is the innermost layer of meninges. This thin, delicate membrane is tightly bound to the surface of the brain and spinal cord and cannot be dissected away without damaging the surface. It closely follows all surface contours.

QUICK APPLICATIONS
Meningitis is an acute inflammation of the pia mater and the arachnoid. It is most commonly caused by bacteria. However, viral infections, fungal infections, and tumors may also cause inflammation of the meninges. Depending on the primary cause, meningitis may be mild or it may progress to a severe and life-threatening condition.

Brain

The brain is divided into the cerebrum, diencephalon, brain stem, and cerebellum.

Cerebrum

The largest and most obvious portion of the brain is the **cerebrum** (seh-REE-brum), which is divided by a deep **longitudinal fissure** (FISH-ur) into two **cerebral hemispheres.** The two hemispheres are two separate entities but are connected by an arching band of white fibers, called the **corpus callosum** (KOR-pus kah-LOH-sum), that provides a communication pathway between the two halves. An extension of dura mater, the **falx cerebri** (FALKS SAYR-eh-brye), projects into the longitudinal fissure down to the corpus callosum. The superior sagittal sinus, a dural sinus mentioned earlier, is in the superior margin of the falx cerebri. The surface of the cerebrum is marked by convolutions, or **gyri** (JYE-rye), separated by grooves, or **sulci** (SULL-see). The pia mater closely follows the convolutions and goes deep into the sulci, and then up and over the gyri.

Each cerebral hemisphere is divided into five lobes, as illustrated in Figure 8-10. Four of the lobes have the same name as the bone over them. The **frontal lobe,** under the frontal bone, is the most anterior portion of each hemisphere. The most posterior gyrus of the frontal lobe is called the precentral gyrus. The posterior boundary of the frontal lobe is the **central sulcus.** The **parietal lobe** is immediately posterior to the central sulcus, under the parietal bone. The gyrus in the parietal lobe that is adjacent to the central sulcus is called the postcentral gyrus. The **occipital lobe,** under the occipital bone, is the most posterior portion of the cerebral hemisphere. Laterally, the **temporal lobe** is inferior to the frontal and parietal lobes. The **lateral sulcus** (fissure) separates the temporal lobe from the two lobes that are superior to it. A fifth lobe, the **insula**

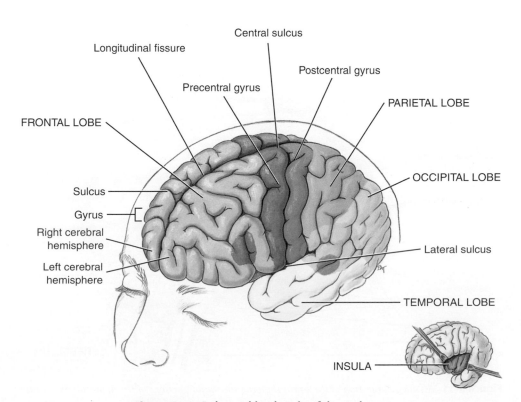

Figure 8-10 *Lobes and landmarks of the cerebrum.*

(IN-sull-ah) or **island of Reil,** lies deep within the lateral sulcus. It is covered by parts of the frontal, parietal, and temporal lobes.

The cerebral hemispheres consist of gray matter and white matter. A thin layer of **gray matter,** the **cerebral cortex,** forms the outermost portion of the cerebrum. Gray matter consists of neuron cell bodies and unmyelinated fibers. Nearly three fourths of the neuron cell bodies in the entire nervous system are found in the cerebral cortex. Additional regions of gray matter, the **basal ganglia,** are scattered throughout the white matter. The **white matter,** which makes up the bulk of the cerebrum, is just beneath the cerebral cortex. White matter is myelinated nerve fibers that form three types of communication pathways in the cerebrum. **Association fibers** transmit impulses from one gyrus to another within the same hemisphere. **Commissural fibers** transmit impulses from one cerebral hemisphere to the other. The **corpus callosum** (Figure 8-11) is a large band of commissural fibers. **Projection fibers** transmit impulses from the cerebrum to other parts of the central nervous system.

The cerebral cortex is the neural basis of what makes us "human." It is the center for sensory and motor functions. It is concerned with memory, language, reasoning, intelligence, personality, and all the other factors that we associate with human life. Even though the two cerebral hemispheres are nearly symmetrical in structure, they are not always equal in function; instead, there are areas of specialization. However, there is considerable overlap in these regions and no area really works alone; all the areas are dependent on each other for mental "consciousness"—those abilities that involve higher mental processing, such as memory, reasoning, logic, and judgment.

> **QUICK APPLICATIONS**
> In most people (about 90%), the left cerebral hemisphere dominates for language and mathematical abilities. It is the reasoning and analytical side of the brain. The right cerebral hemisphere is involved with motor skills, intuition, emotion, art, and music appreciation. It is the poetic and creative side of the brain. These people are generally right-handed. In about 10% of the people, these sides are reversed. In some cases, neither hemisphere dominates. This may result in "confusion" and learning disabilities.

It is possible to identify regions of the cerebral cortex that have specific functions. Some of these regions are described in Table 8-5. **Sensory areas** in the parietal, occipital, and temporal lobes receive information from the various sense organs and receptors throughout the body. The primary sensory area, the **somatosensory** (soh-mat-oh-SEN-soar-ee) **cortex,** is located in the **postcentral gyrus** of the parietal lobe, immediately posterior to the central sulcus. This region receives sensory input from sensory receptors in the skin and skeletal muscles. The right side of the somatosensory cortex receives input from the left side of the body and vice versa. The somatosensory cortex is highly organized with specific regions responsible for receiving sensory input from specific parts of the body. Larger areas of the cortex are devoted to the face and hands than to other parts of the body. Other areas of the cerebrum are responsible for vision, hearing, taste, and smell.

Motor areas responsible for muscle contraction are located in the frontal lobe. The primary motor area, the **somatomotor** (soh-mat-oh-MOH-ter) **cortex,** is in the **precentral gyrus,** immediately anterior to the central sulcus. Neurons in this area

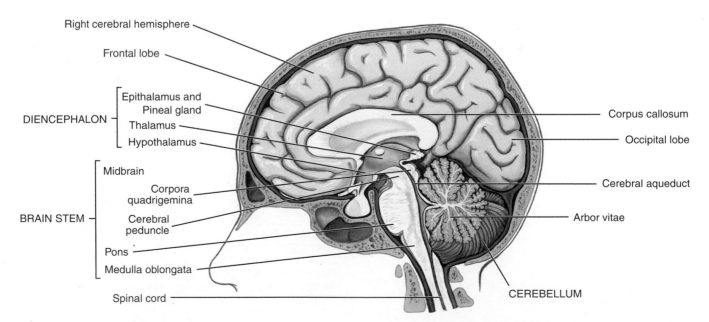

Figure 8-11 *Midsagittal section of the brain showing the major portions of the diencephalon, brain stem, and cerebellum.*

TABLE 8-5 Functional Regions of the Cerebral Cortex

Functional Region	Location	Description	Comments
Primary sensory cortex (somatosensory cortex)	Postcentral gyrus in parietal lobe	Receives sensory input from receptors in skin and skeletal muscles	Function in sensations of temperature, touch, pressure, pain
Somatosensory association area	Parietal lobe, immediately posterior to primary sensory cortex	Integrates and analyzes input received by primary sensory cortex	Analyzes and interprets current sensory information and compares it with previous experience to form a basis for recognition
Primary visual cortex	Posterior region of occipital lobe	Receives sensory input from retina of eye	Perceives current visual message
Visual association area	Anterior to primary visual cortex in occipital love	Integrates and analyzes input received by primary visual cortex	Compares present visual information with previous experience as a basis for recognition; attaches significance to what you see
Auditory cortex	Superior margin of temporal lobe, along lateral sulcus	Receives auditory impulses related to pitch, rhythm, and loudness from inner ear	Allows you to hear "sounds"
Auditory association area	Adjacent to primary auditory area in temporal love	Integrates and analyzes input received by primary auditory area	Permits recognition of sounds (e.g., speech, music, noise); involved in memory of music
Olfactory cortex	Medial aspect of temporal lobe	Receives input from olfactory (smell) receptors in nasal cavity	Permits perception of different odors
Gustatory cortex	Parietal lobe where it is overlapped by temporal lobe	Receives input from taste buds on tongue	Permits perception of different tastes
Wernicke's area	Posterior aspect of temporal lobe of one hemisphere, usually the left	Received input from auditory and visual association areas	Permits comprehension of spoken and written language
Primary motor cortex (somatomotor cortex)	Precentral gyrus in frontal lobe	Initiates efferent action potentials that control voluntary movements	Permits skeletal muscle contraction
Premotor cortex	Anterior to primary motor cortex in frontal lobe	Controls learned motor skills that involve skeletal muscles, either simultaneously or sequentially	Examples of learned motor skills are playing piano, typing, writing
Broca's area (motor speech area)	Inferior portion of frontal lobe in one hemisphere, usually the left	Programs and coordinates muscular movements necessary to articulate words	Person with injury in this area is able to understand words but is unable to speak because of inability to coordinate muscles necessary to form words
Prefrontal cortex	Anterior portion of frontal lobes	Involved with thought, reasoning, intelligence, judgment, planning, conscience	This area is well developed only in humans
Gnostic area (general interpretation area)	Region where parietal, temporal, and occipital lobes meet; found in one hemisphere (usually the left)	Integrates sensory interpretations from adjacent association areas to form thoughts; then transmits signals for appropriate responses	Stores complex memory patterns; allows person to recognize words and arrange them appropriately to express thoughts or to read and understand written ideas

allow us to consciously control our skeletal muscles. The right primary motor gyrus controls muscles on the left side of the body and vice versa. The primary motor cortex is also highly organized in a manner similar to the primary sensory cortex, with neurons in a specific region responsible for controlling movement in a specific part of the body.

Association areas of the cerebral cortex are involved in the process of recognition. They analyze and interpret sensory information, and based on previous experiences they integrate appropriate responses through the motor areas.

The **basal ganglia** are functionally related regions of gray matter that are scattered throughout the white matter of the cerebral hemispheres. These regions function as relay stations, or areas of synapse, in pathways going to and from the cortex. The major effects of the basal ganglia are to decrease muscle tone and to inhibit muscular activity. Because of these effects, they play an important role in posture and coordinating motor movements. Also, nearly all the inhibitory neurotransmitter dopamine is produced in the basal ganglia.

QUICK CHECK

8.13 What is the outermost layer of meninges?
8.14 In which central lobe is the primary somatosensory area located?
8.15 What comprises the gray matter of the cerebrum?
8.16 What is the corpus callosum?

QUICK APPLICATIONS

Parkinson's disease is a condition in which the basal ganglia do not produce enough of the inhibitory transmitter dopamine. Without dopamine, there is an excess of excitatory signals that affect certain voluntary muscles, producing rigidity and tremors.

Diencephalon

The **diencephalon** (dye-en-SEF-ah-lon) is centrally located and is nearly surrounded by the cerebral hemispheres. It includes the **thalamus, hypothalamus,** and **epithalamus.** Regions of the diencephalon are illustrated in Figure 8-11.

The **thalamus** (THAL-ah-mus), about 80% of the diencephalon, consists of two oval masses of gray matter that serve as relay stations for sensory impulses, except for the sense of smell, going to the cerebral cortex. When the impulses reach the thalamus, there is a general awareness and crude recognition of sensation. The thalamus channels the impulses to the appropriate region of the cortex for discrimination, localization, and interpretation.

The **hypothalamus** (HYE-poh-thal-ah-mus) is a small region below the thalamus. It plays a key role in maintaining homeostasis because it regulates many visceral activities. The hypothalamus also serves as a link between the nervous and endocrine systems because it regulates secretion of hormones from the pituitary gland. A slender stalk, the **infundibulum,** extends from the floor of the hypothalamus to the pituitary gland and acts as a connector between the two structures. Two visible "bumps" on the posterior portion of the hypothalamus, the **mamillary bodies,** are involved in memory and emotional responses to different odors. The following list summarizes the numerous functions of the hypothalamus.

* Regulates and integrates the autonomic nervous system.

    The hypothalamus influences the autonomic centers in the brain stem and spinal cord. In this way it regulates many visceral activities such as heart rate, blood pressure, respiratory rate, and motility of the digestive tract.

* Regulates emotional responses and behavior.

    The hypothalamus acts through the autonomic nervous system to mediate the physical responses to emotion and mind-over-body phenomena. Nuclei involved in feelings of rage, aggression, fear, pleasure, and the sex drive are localized in the hypothalamus.

* Regulates body temperature.

    Certain cells of the hypothalamus act as a body thermostat and activate appropriate responses.

* Regulates food intake.

    The feeding or hunger center in the hypothalamus is responsible for hunger sensations. After food ingestion, the satiety center inhibits the hunger center.

* Regulates water balance and thirst.

    Osmoreceptors in the hypothalamus monitor the volume and concentration of body fluids, and then initiate appropriate responses from the thirst center and antidiuretic hormone.

* Regulates sleep-wake cycles.

Centers in the hypothalamus act with other brain centers to maintain alternating periods of sleep and wakefulness.

* Regulates endocrine system activity.

    The hypothalamus produces releasing factors that control the release of hormones from the anterior pituitary gland. It also produces two hormones, antidiuretic hormone and oxytocin, that are stored in the posterior pituitary gland.

The **epithalamus** (ep-ih-THAL-ah-mus) is the most dorsal, or superior, portion of the diencephalon. The **pineal** (PIE-nee-al) **gland,** or **body,** extends from its posterior margin. This small gland is involved with the onset of puberty and rhythmic cycles in the body. It is similar to a biological clock.

QUICK APPLICATIONS

The **limbic system** consists of scattered but interconnected regions of gray matter in the cerebral hemispheres and diencephalon. The limbic system is involved in memory and in emotions such as sadness, happiness, anger, and fear. It is our emotional brain.

Brain Stem

The **brain stem** is the region between the diencephalon and the spinal cord. It consists of three regions:

* Midbrain
* Pons
* Medulla oblongata

Regions of the brain stem are illustrated in Figure 8-11 and in the midsagittal MR image of the brain in Figure 8-12.

The **midbrain** is the most superior portion of the brain stem, the region next to the diencephalon. Two **cerebral peduncles** (seh-REE-brull pee-DUNK-als) form the ventral aspect of the midbrain. These bundles of myelinated fibers contain the voluntary motor tracts that descend from the cerebral cortex. On the dorsal aspect of the midbrain, four rounded protuberances form the **corpora quadrigemina** (KOR-poar-ah kwad-rih-JEM-ih-nah). The two superior bodies, the **superior colliculi** (soo-PEER-ee-or koh-LIK-yoo-lye), function as visual reflex centers. The other two, the **inferior colliculi,** contain auditory reflex centers. A narrow channel for CSF, the **cerebral aqueduct** (seh-REE-brull AH-kweh-dukt), descends through the center of the midbrain.

The **pons** is the bulging middle portion of the brain stem. This region primarily consists of nerve fibers that form conduction tracts between the higher brain centers and the spinal cord. Four cranial nerves originate in the pons. It also contains the **pneumotaxic** (noo-moh-TACK-sik) and **apneustic** (ap-NOO-stick) **areas** that help regulate breathing movements.

The **medulla oblongata** (meh-DULL-ah ahb-long-GAH-tah), or simply **medulla,** extends inferiorly from the pons. It is continuous with the spinal cord at the foramen magnum. All the ascending (sensory) and descending (motor) nerve fibers connecting the brain and spinal cord pass through the medulla. Most of the descending fibers cross over from one side to the other. In other words, fibers descending on the left side cross over to the right and vice versa. This is called **decussation** (dee-kuh-SAY-shun). Because the fibers decussate, or cross over, the brain controls motor functions on the opposite side of the body.

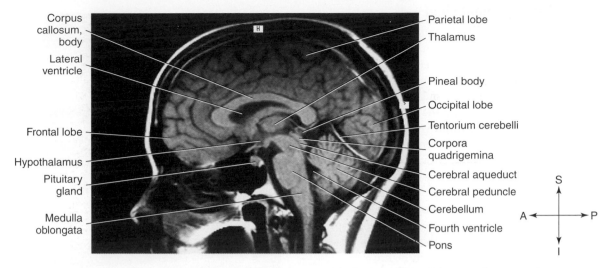

Figure 8-12 *Midsagittal magnetic resonance image of the brain. (Adapted from Applegate E: The Sectional Anatomy Learning System, Concepts, ed. 3. St. Louis, Elsevier/Saunders, 2010.)*

Five pairs of cranial nerves originate in the medulla. In addition, it contains three vital centers that control visceral activities. The **cardiac center** adjusts the heart rate and contraction strength to meet body needs. The **vasomotor center** regulates blood pressure by effecting changes in blood vessel diameter. The **respiratory center** acts with the centers in the pons to regulate the rate, rhythm, and depth of breathing. Other centers are involved in coughing, sneezing, swallowing, and vomiting.

QUICK APPLICATIONS

The **reticular formation** consists of scattered, but interconnected, neurons and fiber pathways in the midbrain and brain stem. It is a functional system that maintains alertness and filters out repetitive stimuli. Motor portions of the reticular formation help coordinate skeletal muscle activity and maintain muscle tone.

Cerebellum

The **cerebellum** (sair-eh-BELL-um), the second largest portion of the brain, is located below the occipital lobes of the cerebrum and is separated from them by the **transverse fissure.** It consists of two **cerebellar hemispheres** connected in the middle by a structure called the **vermis.** An extension of dura mater, the **falx cerebelli** (FALKS sair-eh-BELL-eye), forms a partial partition between the hemispheres. Another extension of dura mater, the **tentorium cerebelli** (ten-TOAR-ee-um sair-eh-BELL-eye), is found in the transverse fissure between the cerebellum and the occipital lobes of the cerebrum.

Like the cerebrum, the cerebellum consists of white matter surrounded by a thin layer of gray matter, the **cerebellar cortex.** Because the surface convolutions are less prominent in the cerebellum than in the cerebrum, the cerebellum has proportionately less gray matter. The branching arrangement of white matter is called **arbor vitae** (AR-bor VEE-tay).

Three paired bundles of myelinated nerve fibers, called **cerebellar peduncles,** form communication pathways between the cerebellum and other parts of the CNS. The **superior cerebellar**

peduncles connect the cerebellum to the midbrain; the **middle cerebellar peduncles** communicate with the pons; and the **inferior cerebellar peduncles** consist of pathways between the cerebellum and the medulla oblongata and spinal cord.

The cerebellum functions as a motor area of the brain that mediates subconscious contractions of skeletal muscles necessary for **coordination, posture,** and **balance.** The cerebellum coordinates skeletal muscles to produce smooth muscle movement rather than jerky, trembling motion. When the cerebellum is damaged, movements such as running, walking, and writing become uncoordinated. Posture is dependent on muscle tone, which is mediated by the cerebellum. Impulses from the inner ear concerning position and equilibrium are directed to the cerebellum, which uses that information to maintain balance.

QUICK CHECK

8.17 Name the three regions of the diencephalon.
8.18 What portion of the diencephalon regulates body temperature, water balance, and thirst?
8.19 What three vital centers are located in the medulla oblongata?
8.20 What is the primary function of the cerebellum?

QUICK APPLICATIONS

Neuroglia, particularly astrocytes, form a wall around the outside of the blood vessels in the nervous system. This astrocyte wall plus the blood vessel wall forms the blood–brain barrier. Water, oxygen, carbon dioxide, alcohol, and a few other substances are able to pass through this barrier and move between the blood and brain tissue. Other substances, such as toxins, pathogens, and certain drugs, cannot pass through this barrier. This is a protective mechanism to keep harmful substances out of the brain. It has clinical significance because drugs, such as penicillin, that may be used to treat disorders in other parts of the body have no effect on the brain because they do not cross the blood–brain barrier.

Ventricles and Cerebrospinal Fluid

A series of interconnected, fluid-filled cavities are found within the brain (Figure 8-13). These cavities are the **ventricles** of the brain, and the fluid is CSF. The **lateral ventricles** in the cerebrum are the largest of these cavities. There is one lateral ventricle in each cerebral hemisphere. A single, narrow, slitlike, midline **third ventricle** is enclosed by the diencephalon. The two lateral ventricles open into the third ventricle through the **interventricular foramina.** The **fourth ventricle** is at the level of the cerebellum and pons. A long, narrow channel, the **cerebral aqueduct,** passes through the midbrain to connect the third and fourth ventricles. The fourth ventricle is continuous with the central canal of the spinal cord. Openings in the wall of the fourth ventricle permit CSF to enter the subarachnoid space. Portions of the lateral ventricles and the third ventricle are seen in the transverse magnetic resonance image in Figure 8-14.

The CSF is a clear fluid that forms as a filtrate from the blood in specialized capillary networks, the **choroid plexus** (KOR-oyd PLEKS-us), within the ventricles. The circulation of the CSF is illustrated in Figure 8-15. Ependymal cells, a type of neuroglia cell, aid in the circulation of CSF through the ventricles and central canal. CSF then enters the subarachnoid space through foramina in the fourth ventricle. From the subarachnoid space, CSF carrying waste products filters through **arachnoid granulations** (villi) into the dural sinuses and is returned to the blood. In addition to providing support and protection for the CNS, the CSF helps to nourish the brain and maintain constant ionic conditions for the brain and spinal cord, and provides a pathway for removal of waste products.

QUICK APPLICATIONS

In hydrocephalus, an obstruction in the normal flow of CSF causes the fluid to accumulate in the ventricles. The obstruction may be a congenital defect or an acquired lesion such as a tumor. As the fluid accumulates, it causes the ventricles to enlarge and CSF pressure to increase. When this happens in an infant, before the cranial bones ossify, the cranium enlarges. In an older child or adult, the pressure damages the soft brain tissue.

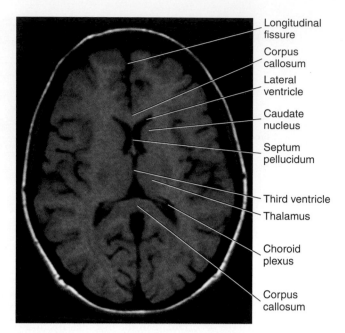

Figure 8-14 *Transverse magnetic resonance image through the third ventricle of the brain. (Adapted from Applegate E: The Sectional Anatomy Learning System, Concepts, ed. 3. St. Louis, Elsevier/ Saunders, 2010.)*

Spinal Cord

The **spinal cord,** illustrated in Figure 8-16, extends from the foramen magnum at the base of the skull to the level of the first lumbar vertebra, a distance of about 43 to 46 centimeters (17 or 18 inches). The cord is continuous with the medulla oblongata at the foramen magnum. Distally, it terminates in the **conus medullaris** (KOH-nus med-yoo-LAIR-is). Like the brain, the spinal cord is surrounded by bone, meninges, and CSF. Unlike the dura mater around the brain, the spinal dura is separated from the vertebral bones by an **epidural space** (Figure 8-17). This space is filled with loose connective tissue and adipose tissue. The meninges extend beyond the end of the spinal cord, down to the upper part of the sacrum. From there, a fibrous cord of pia mater, the **filum terminale** (FYE-lum term-ih-NAL-ee), extends down to the coccyx where it is anchored.

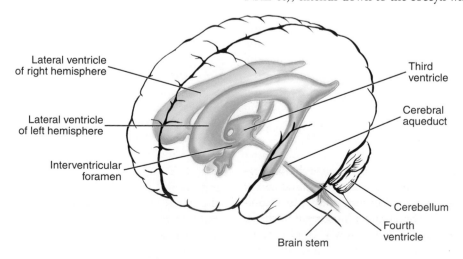

Figure 8-13 *Ventricles of the brain.*

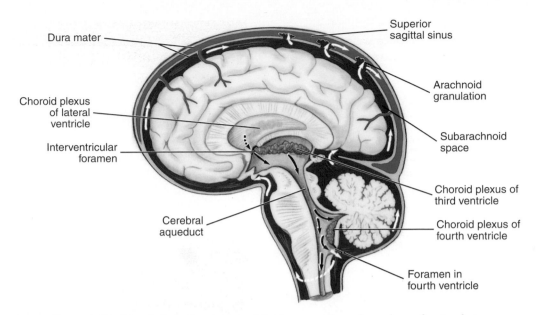

Figure 8-15 *Circulation of cerebrospinal fluid;* arrows *show the pathway for circulation.*

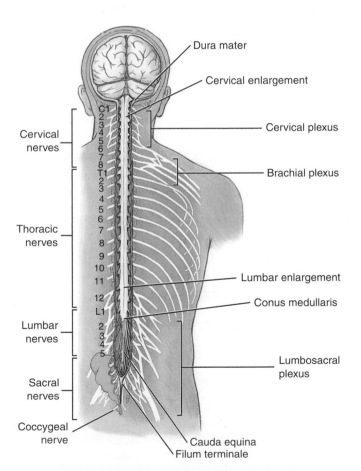

Figure 8-16 *Gross anatomy of the spinal cord.*

The spinal cord is divided into 31 segments, with each segment giving rise to a pair of spinal nerves. At the distal end of the cord, many spinal nerves extend beyond the conus medullaris to form a collection that resembles a horse's tail in shape. This is the **cauda equina** (KAW-dah ee-KWYNE-ah). There are two enlargements in the cord, one in the cervical region and one in the lumbar region. The **cervical enlargement** gives rise

to the nerves that supply the upper extremity. Nerves from the **lumbar enlargement** supply the lower extremity.

In cross section, the spinal cord appears oval (see Figure 8-17). A narrow, deep, **dorsal (posterior) median sulcus** and a shallower, but wider, **ventral (anterior) median fissure** partially divide the cord into right and left halves. Peripheral white matter surrounds a core of gray matter that resembles a butterfly or the letter H. Each side of the gray matter is divided into **dorsal, lateral,** and **ventral horns.** These contain the terminal portions of sensory neuron axons, entire interneurons, and the dendrites and cell bodies of motor neurons. The central connecting bar between the two large areas of gray matter is the **gray commissure** (KOM-ih-shur). This surrounds the **central canal,** which contains CSF. The gray matter divides the surrounding white matter into three regions on each side. These regions are the **dorsal, lateral,** and **ventral funiculi** (fuh-NIK-yoo-lee) or **columns.** The white matter contains longitudinal bundles of myelinated nerve fibers, called **nerve tracts.**

The spinal cord has two main functions. It is a conduction pathway for impulses going to and from the brain, and it serves as a reflex center.

The conduction pathways that carry sensory impulses from body parts to the brain are called **ascending tracts.** Pathways that carry motor impulses from the brain to muscles and glands are **descending tracts.** Tracts are often named according to their points of origin and termination. For example, **spinothalamic** (spy-noh-tha-LAM-ik) **tracts** are ascending (sensory) tracts that begin in the spinal cord and conduct impulses to the thalamus. They function in the sensations of touch, pressure, pain, and temperature. **Corticospinal** (kor-tih-koh-SPY-null) **tracts,** also called **pyramidal** (pih-RAM-ih-dal) **tracts,** are descending (motor) tracts that begin in the cerebral cortex and end in the spinal cord. These tracts function in the control of skeletal muscle movement. All other descending tracts are grouped together as **extrapyramidal** (eks-trah-pih-RAM-ih-dal) **tracts,** and they function in muscle movements associated with posture and balance.

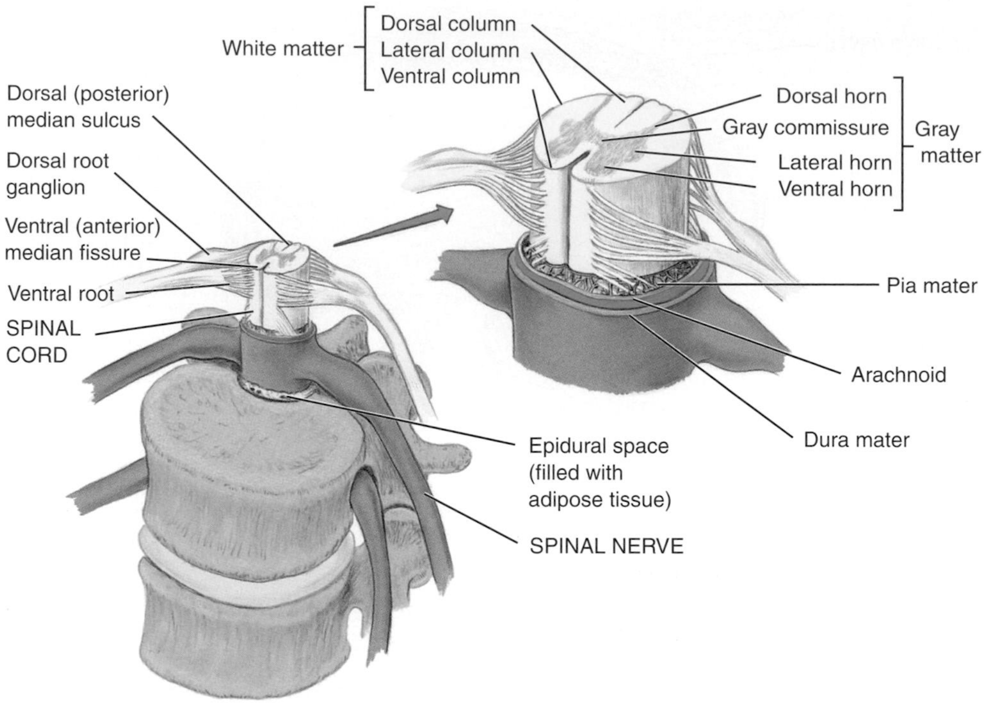

Figure 8-17 *Cross section of the spinal cord.*

In addition to serving as a conduction pathway, the spinal cord functions as a center for spinal reflexes. The reflex arc, described earlier in this chapter and illustrated in Figure 8-8, is the functional unit of the nervous system. Reflexes are responses to stimuli that do not require conscious thought and, consequently, they occur more quickly than reactions that require thought processes. For example, with the withdrawal reflex, the reflex action withdraws

the affected part before you are aware of the pain. Many reflexes are mediated in the spinal cord without going to the higher brain centers. Table 8-6 describes some clinically significant reflexes.

QUICK CHECK

8.21 What produces cerebrospinal fluid?
8.22 How many segments are in the spinal cord?
8.23 What type of nerve impulses are transmitted on the ascending tracts of the spinal cord?

QUICK APPLICATIONS

A lumbar puncture is the withdrawal of some CSF from the subarachnoid space in the lumbar region of the spinal cord. The extension of the meninges beyond the end of the cord makes it possible to do this without injury to the spinal cord. The needle is usually inserted just above or just below the fourth lumbar vertebra, and the spinal cord ends at the first lumbar vertebra. The CSF that is removed can be tested for abnormal characteristics that may indicate an injury or infection.

TABLE 8-6 Some Clinically Significant Reflexes

Reflex	Description	Indications
Patellar (knee-jerk reflex)	Stretch reflex; two-neuron path; reflex hammer strikes patellar tendon just below knee; receptors in quadriceps femoris muscle are stretched; reflex results in immediate "kick"	Reflex is blocked by damage to nerves involved and by damage to lumbar segments of spinal cord; also absent in people with chronic diabetes mellitus and neurosyphilis
Achilles tendon (ankle-jerk reflex)	Stretch reflex; two-neuron path; reflex hammer strikes Achilles tendon just above heel; gastrocnemius and soleus muscles contract to plantar flex foot	Weak or no reflex action indicates damage to nerves involved or to L5-S2 segments of spinal cord; also absent in chronic diabetes, neurosyphilis, and alcoholism
Abdominal	Stroking lateral abdominal wall produces reflex action that compresses abdominal wall and moves umbilicus toward stimulus	Absent in lesions of peripheral nerves, in lesions in thoracic segments of spinal cord, and in multiple sclerosis
Babinski's	Lateral sole of foot is stroked from heel to toe; positive sign results in dorsiflexion of big toe and spreading of other toes; negative sign results in toes curling under with slight inversion of foot	Positive Babinski's sign is normal in children younger than 18 months; negative sign is normal after 18 months of age; if motor tracts in spinal cord are damaged, positive Babinski's sign reappears

Figure labels:
White matter — Dorsal column / Lateral column / Ventral column
Dorsal (posterior) median sulcus
Dorsal root ganglion
Ventral (anterior) median fissure
Ventral root
SPINAL CORD
Dorsal horn
Gray commissure
Lateral horn
Ventral horn
Gray matter
Pia mater
Arachnoid
Dura mater
Epidural space (filled with adipose tissue)
SPINAL NERVE

FROM THE PHARMACY

There are many medications for nervous system disorders; however, the actions, underlying principles of use, and reactions are similar within each group of medications. There are anesthetics, analgesics, and psychotherapeutic drugs as well as medications to treat migraine headaches, vertigo, convulsions, and Parkinson's disease, for example. The list is extensive. Only a few examples are described here.

Alprazolam (Xanax) is a psychotherapeutic drug used to treat generalized anxiety disorders, but it is not indicated for anxiety associated with the stress of everyday life. There is a risk of dependence even after relatively short-term use. The nature of the drug itself promotes dependence, and its use should be reserved for panic or acute anxiety attacks and then used carefully. Xanax has street value and is frequently implicated when patients are admitted for polysubstance overdoses. It can also cause severe withdrawal symptoms, including seizures and death, when abruptly discontinued. Use with caution.

Fluoxetine (Prozac) is a psychotherapeutic drug used in the treatment of depression. It is presumed to inhibit the uptake of serotonin, thus increasing serotonin levels in the bloodstream. Depression correlates with low levels of serotonin, so increasing the serotonin levels has an antidepressant effect. Fluoxetine is indicated for treatment of major depressive episodes, obsessive-compulsive disorder, appetite disorders, and panic disorders. Slow elimination of the drug leads to significant accumulation in chronic use. Other newer medications of this type appear to have fewer side effects and are often better tolerated.

Lithium carbonate (Eskalith, Lithane) is a mood stabilizer that is used in the treatment of manic episodes of manic-depressive ill-ness. The major complication of lithium therapy is toxicity, and serum levels should be checked weekly until stable. The risk of toxicity is high in patients with renal disease, cardiovascular disease, dehydration, or sodium depletion. Frequent monitoring is necessary.

Carbidopa-levodopa (Sinemet) is an antiparkinsonism agent. The goal of these agents is to relieve the symptoms of the disease so the individual can carry out normal daily activities. Parkinson's disease appears to be related to decreased levels of dopamine in the brain. Dopamine cannot be administered directly because it does not cross the blood–brain barrier. Levodopa (L-dopa) is a precursor of dopamine that is able to cross the barrier and then be converted to dopamine in the brain. Unfortunately, the majority of administered L-dopa is converted to dopamine in the periphery before it has a chance to enter the brain. Carbidopa blocks the peripheral conversion and allows more L-dopa to enter the brain, where it is converted to dopamine. In this case, the combination is more effective than L-dopa alone.

Dihydroergotamine mesylate (DHE45, Migranal) is an alpha-adrenergic blocking agent that is used for the acute treatment of migraine headaches and cluster headache episodes. This preparation is derived from ergot, a fungus that grows on rye. Dihydroergotamine competes with catecholamines at receptor sites to inhibit adrenergic sympathetic stimulation. This causes constriction of the dilated cerebral blood vessels that are associated with the headache. This versatile medication can be given intravenously, intramuscularly, subcutaneously, or intranasally and has fewer rebound headache side effects than regular ergotamines.

Peripheral Nervous System

The PNS consists of the nerves that branch out from the brain and spinal cord. These nerves form the communication network between the CNS and the remainder of the body. The PNS is further subdivided into the **somatic nervous system** and the **autonomic nervous system.** The somatic nervous system consists of nerves that go to the skin and muscles and is involved in conscious activities. The autonomic nervous system consists of nerves that connect the CNS to the visceral organs such as the heart, stomach, and intestines. It mediates unconscious activities.

Structure of a Nerve

A nerve contains bundles of nerve fibers, either axons or dendrites, surrounded by connective tissue. **Sensory nerves** contain only afferent fibers—long dendrites of sensory neurons. **Motor nerves** have only efferent fibers—long axons of motor neurons. **Mixed nerves** contain both types of fibers.

Each nerve is surrounded by a connective tissue sheath called the **epineurium** (ep-ih-NOO-ree-um). Portions of the epineurium project inward to divide the nerve into compartments, each containing a bundle of nerve fibers. Each bundle of nerve fibers is called a **fasciculus** and is surrounded by a layer of connective tissue called the **perineurium** (pair-ih-NOO-ree-um). Within the fasciculus, each individual nerve fiber, with its myelin and neurilemma, is surrounded by connective tissue called the **endoneurium** (end-oh-NOO-ree-um). A nerve may also have blood vessels enclosed in its connective tissue wrappings (Figure 8-18).

Cranial Nerves

Twelve pairs of cranial nerves, illustrated in Figure 8-19, emerge from the inferior surface of the brain. All of these nerves, except the vagus nerve, pass through foramina of the skull to innervate structures in the head, neck, and facial region. The vagus nerve, cranial nerve X, has numerous branches that supply the viscera in the body. When sensory fibers are present in a cranial nerve, the cell bodies of these neurons are located in groups, called **ganglia,** outside the brain. Motor neuron cell bodies are typically located in the gray matter of the brain.

The cranial nerves are designated both by name and by Roman numerals, according to the order in which they appear

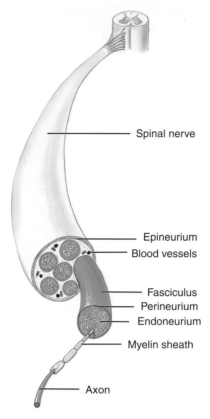

Figure 8-18 *Structure of a nerve.*

on the inferior surface of the brain. Most of the nerves have both sensory and motor components. Three of the nerves (I, II, and VIII) are associated with the special senses of smell, vision, hearing, and equilibrium and have only sensory fibers. Five other nerves (III, IV, VI, XI, and XII) are primarily motor in function but do have some sensory fibers for proprioception. The remaining four nerves (V, VII, IX, and X) consist of significant amounts of both sensory and motor fibers. Table 8-7 itemizes the cranial nerves.

Spinal Nerves

Thirty-one pairs of spinal nerves emerge laterally from the spinal cord. Each pair of nerves corresponds to a segment of the cord and they are named accordingly. This means there are **8 cervical nerves** (C1 to C8), **12 thoracic nerves** (T1 to T12), **5 lumbar nerves** (L1 to L5), **5 sacral nerves** (S1 to S5), and **1 coccygeal nerve** (Co).

Each spinal nerve is connected to the spinal cord by a **dorsal root** and a **ventral root** (see Figure 8-17). The dorsal root can be recognized by an enlargement, the **dorsal root ganglion.** The dorsal root has only sensory fibers, and the ventral root has only motor fibers. The cell bodies of the sensory neurons are in the dorsal root ganglion, but the motor neuron cell bodies are in the gray matter. The two roots join to form the spinal nerve just before the nerve leaves the vertebral column. Because all spinal nerves have both sensory and motor components, they are all mixed nerves.

Immediately after they leave the vertebral column, the spinal nerves divide into several branches that provide the nerve supply to the muscles and the skin of the body wall. Each spinal nerve innervates a particular region of the skin. Skin surface areas that are innervated by a single spinal nerve are called **dermatomes.** Figure 8-20 shows a typical dermatome map. Knowledge of the relationship between a spinal nerve and the skin segment it innervates is useful because a neurologist can identify a spinal nerve lesion from the area of the skin that is insensitive to pain such as a pinprick. In the thoracic region, the main portions of the nerves go directly to the thoracic wall where they are called **intercostal nerves.** In other regions, the main portions of the nerves form complex networks called **plexuses** (see Figure 8-16). In the plexus, the fibers are sorted and recombined so that the fibers associated with a particular body part are together even though they may originate from different regions of the cord (Table 8-8).

> ### QUICK CHECK
> 8.24 What three cranial nerves have only sensory function?
> 8.25 How many spinal nerves emerge from the spinal cord?
> 8.26 What type of nerve fibers are contained in the ventral root of a spinal nerve?

Autonomic Nervous System

General Features

The **autonomic nervous system** (ANS) is a visceral efferent system, which means it sends motor impulses to the visceral organs. It functions automatically and continuously, without conscious effort, to innervate smooth muscle, cardiac muscle, and glands. It is concerned with heart rate, breathing rate, blood pressure, body temperature, and other visceral activities that work together to maintain homeostasis.

In the somatic motor pathways, typically there is one neuron that extends from the brain or spinal cord to the effector that is innervated. Autonomic pathways have two neurons between the CNS and the visceral effector (Figure 8-21). The first neuron's cell body is in the brain or spinal cord. Its axon, the **preganglionic fiber,** leaves the CNS and synapses with a second neuron in an **autonomic ganglion.** The cell body of the second neuron is located in the autonomic ganglion. The axon of the second neuron, the **postganglionic fiber,** leaves the ganglion and goes to the effector organ.

The ANS has two parts: the **sympathetic division** and the **parasympathetic division** (Table 8-9). Many visceral organs are supplied with fibers from both divisions (**dual innervation**). In this case, one stimulates and the other inhibits. This antagonistic functional relationship serves as a balance to help maintain homeostasis.

Sympathetic Division

The sympathetic division, illustrated in Figure 8-22, is concerned primarily with preparing the body for stressful or emergency situations. Sometimes called the *fight-or-flight system*, it is an energy-expending system. It stimulates the responses that are needed to meet the emergency and inhibits the

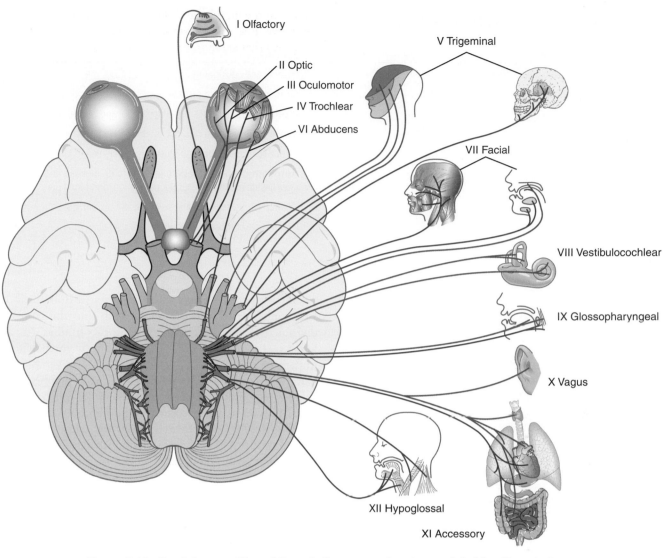

Figure 8-19 *Cranial nerves. The* **red lines** *indicate motor function, and the* **blue lines** *indicate sensory function.*

TABLE 8-7 Summary of Cranial Nerves

Number	Name	Type	Function
I	Olfactory	Sensory	Sense of smell
II	Optic	Sensory	Vision
III	Oculomotor	Primarily motor	Movement of eyes and eyelids
IV	Trochlear	Primarily motor	Movement of eyes
V	Trigeminal	Mixed	
	Ophthalmic branch		**Sensory fibers** from cornea, skin of nose, forehead, scalp
	Maxillary branch		**Sensory fibers** from cheek, nose, upper lip, and teeth
	Mandibular branch		**Motor fibers** to muscles of mastication
VI	Abducens	Primarily motor	Eye movement
VII	Facial	Mixed	**Sensory fibers** from taste receptors on anterior two thirds of tongue
			Motor fibers to muscles of facial expression, lacrimal glands, and salivary glands

Continued

TABLE 8-7 Summary of Cranial Nerves—Cont'd

Number	Name	Type	Function
VIII	Vestibulocochlear	Sensory	Hearing and equilibrium
IX	Glossopharyngeal	Mixed	**Sensory fibers** from taste receptors on posterior one third of tongue
			Motor fibers to muscles used in swallowing and to salivary glands
X	Vagus	Mixed	**Sensory fibers** from pharynx, larynx, esophagus, and visceral organs
			Somatic motor fibers to muscles of pharynx and larynx
			Autonomic motor fibers to heart, smooth muscles, and glands to alter gastric motility, heart rate, respiration, and blood pressure
XI	Accessory	Primarily motor	Contraction of trapezius and sternocleidomastoid muscle
XII	Hypoglossal	Primarily motor	Contraction of muscles in tongue

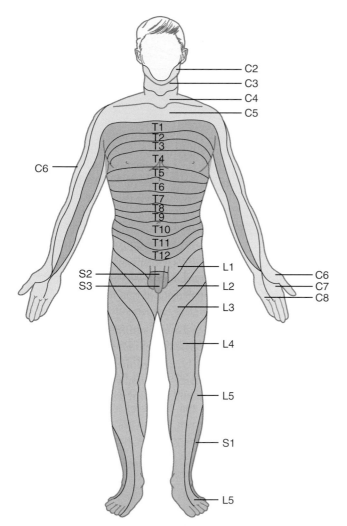

Figure 8-20 *Dermatome map showing the spinal nerve distribution to the skin surface.*

visceral activities that can be delayed momentarily. For example, during an emergency, the sympathetic system increases breathing rate, heart rate, and blood flow to skeletal muscles. At the same time, it decreases activity in the digestive tract because that is not needed to meet the emergency.

The sympathetic preganglionic fibers arise from the thoracic and lumbar regions of the spinal cord; thus the sympathetic division is sometimes called the **thoracolumbar** (thoar-ah-koh-LUM-bar) **division.** These fibers almost immediately terminate in one of the **paravertebral** (pair-ah-ver-TEE-brull) **ganglia.** A chain of these ganglia, the sympathetic chain, extends longitudinally along each side of the vertebral column. Some fibers synapse in **collateral ganglia** outside the sympathetic chain but still close to the vertebral column. Thus in the sympathetic division, the preganglionic fiber is short, but the postganglionic fiber that makes contact with the effector organ is long.

A sympathetic preganglionic fiber typically synapses with many postganglionic fibers in the ganglion. The postganglionic fibers then go to numerous organs to give a widespread, diffuse effect. At the synapses in the ganglia, the preganglionic fibers release the neurotransmitter **acetylcholine.** For this reason they are called **cholinergic** (koh-lih-NER-jik) **fibers.** Most of the postganglionic fibers release **norepinephrine** (noradrenaline) and are called **adrenergic** (add-rih-NER-jik) **fibers.** Because norepinephrine is inactivated rather slowly, these fibers provide a long-lasting effect. Table 8-10 summarizes the features of the sympathetic nervous system.

Parasympathetic Division

The parasympathetic division is most active under ordinary, relaxed conditions (see Figure 8-22). It also brings the body's systems back to a normal state after an emergency by slowing the heart rate and breathing rate, decreasing blood pressure, decreasing blood flow to skeletal muscles, and increasing digestive tract activity. Sometimes called the *rest-and-repose system,* it is an energy-conserving system.

The parasympathetic preganglionic fibers arise from the brain stem and sacral region of the spinal cord; thus, it is sometimes called the **craniosacral** (kray-nee-oh-SAY-kral) **division.** The ganglia, called **terminal ganglia,** are located near or within the visceral organs. This makes the preganglionic fiber long and the postganglionic fiber short. Typically, a parasympathetic preganglionic fiber synapses with only a few postganglionic

TABLE 8-8 Spinal Nerve Plexuses

Plexus	Location	Spinal Nerves Involved	Region Supplied	Major Nerves Leaving Plexus
Cervical	Deep in neck, under sternocleidomastoid muscle	C1-C4	Skin and muscles of neck and shoulder; diaphragm	Phrenic
Brachial	Deep to clavicle, between neck and axilla	C5-C8, T1	Skin and muscles of upper extremity	Musculocutaneous
				Ulnar
				Median
				Radial
				Axillary
Lumbosacral	Lumbar region of back	T12, L1-L5, S1-S4	Skin and muscles or lower abdominal wall, lower extremity, buttocks, external genitalia	Obturator
				Femoral
				Sciatic
				Pudendal

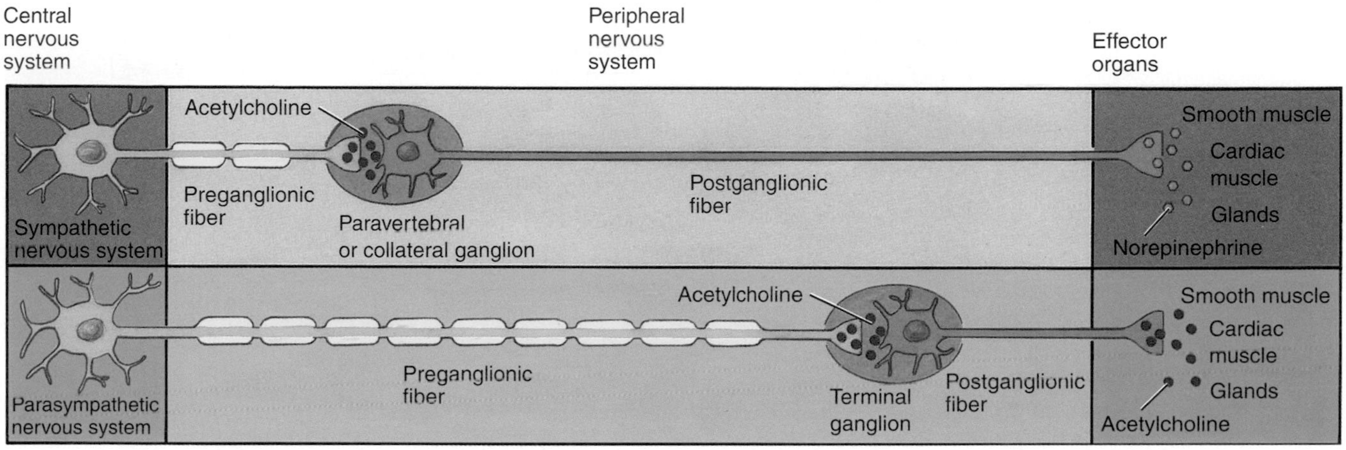

Figure 8-21 *Preganglionic and postganglionic fibers in the autonomic nervous system.*

TABLE 8-9 Comparison of Sympathetic and Parasympathetic Actions on Selected Visceral Effectors

Visceral Effectors	Sympathetic Action	Parasympathetic Action
Pupil of eye	Dilates	Constricts
Lens of eye	Lens flattens for distance vision	Lens bulges for near vision
Sweat glands	Stimulates	No innervation
Arrector pili muscles of hair	Stimulates contraction; goosebumps	No innervation
Heart	Increases heart rate	Decreases heart rate
Bronchi	Dilates	Constricts
Digestive glands	Decreases secretion of digestive enzymes	Increases secretion of digestive enzymes
Digestive tract	Decreases peristalsis	Increases peristalsis
Digestive tract sphincters	Stimulates—closes sphincters	Inhibits—opens sphincters
Blood vessels to digestive organs	Constricts	No innervation
Blood vessels to skeletal muscles	Dilates	No innervation
Blood vessels to skin	Constricts	No innervation
Adrenal medulla	Stimulates secretion of epinephrine	No innervation
Liver	Increases release of glucose	No innervation
Urinary bladder	Relaxes bladder and closes sphincter	Contracts bladder and opens sphincter

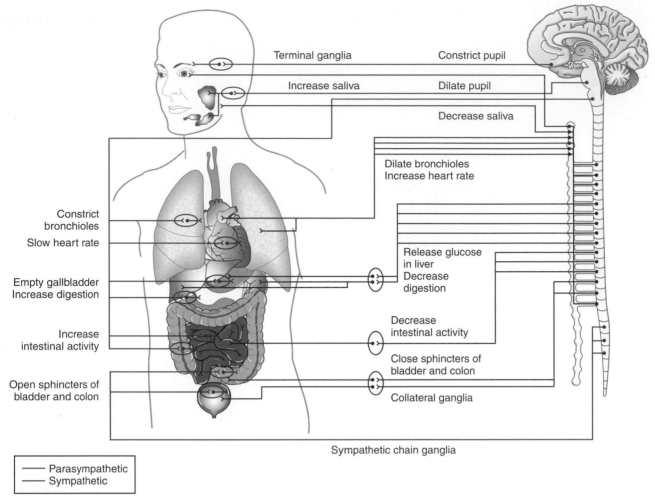

Figure 8-22 *Structure and function of the autonomic nervous system. The **red lines** indicate the sympathetic division, and the **blue lines** indicate parasympathetic innervation. Note the location of the ganglia for each division.*

TABLE 8-10 Some Comparisons Between the Sympathetic and Parasympathetic Divisions of the Autonomic Nervous System

Feature	Sympathetic Division	Parasympathetic Division
General effect	Fight or flight; stress and emergency	Rest and repose; normal activity
Extent of effect	Widespread	Localized
Duration of effect	Long lasting	Short duration
Energy	Expends energy	Conserves energy
Origin of outflow	Thoracolumbar	Craniosacral
Ganglia	Paravertebral and collateral	Terminal
Preganglionic fibers	Short	Long
Postganglionic fibers	Long	Short
Neurotransmitters	Preganglionic fibers are cholinergic; most postganglionic fibers are adrenergic	Fibers are all cholinergic
Divergence	Great divergence; a preganglionic fiber synapses with several postganglionic fibers	Little divergence; a preganglionic fiber synapses with few postganglionic fibers

fibers, so the effect is localized. Both the preganglionic and the postganglionic fibers are **cholinergic,** which means they secrete acetylcholine at the synapses. Because acetylcholine is readily inactivated, the parasympathetic effect is short term. Table 8-10 summarizes the features of the parasympathetic nervous system and compares them with the features of the sympathetic nervous system.

QUICK CHECK

8.27 In autonomic pathways, how many neurons are between the CNS and the visceral effector?
8.28 Which division of the autonomic nervous system is characterized as an energy-expending system?
8.29 Which division of the autonomic nervous system is the craniosacral division?
8.30 Which division of the autonomic nervous system uses only acetylcholine as a neurotransmitter?

 FOCUS ON AGING

Aging of the nervous system is of major importance because changes in this system affect organs in other systems and can cause disturbances of many bodily functions. For example, changes in nerves decrease stimulation of skeletal muscle, which contributes to muscle atrophy with age. Because of its widespread consequences, aging of the nervous system is one of the most distressing aspects of growing old.

Like other cells, nerve cells are lost as a person ages, even in the absence of disease processes. Because neurons are amitotic, those that are lost are not replaced. Loss of neurons is largely responsible for the decrease in brain mass that occurs with aging. Fortunately, the brain has a large reserve supply of neurons, many more than are needed to carry out its functions, so the decrease in neuron number alone is not devastating. The loss of neurons is not constant in all areas of the brain. For example, about 25% of the specialized cells in the cerebellum, which are responsible for coordinated movements, are lost during aging. This may affect balance and cause difficulty in coordinating fine movements. There are other areas of the brain in which the number of neurons remains essentially constant throughout life.

It is generally accepted that there is a decline in intelligence with aging, and this is thought to be associated with the loss of neurons. However, it is important to remember that there are wide variations in individuals regarding changes in intellect with age. Because a person is old does not mean that person is "dumb." Many elderly people retain a keen intellect until death. Along with the decline in intelligence, there is a general decline in memory. Again, this varies from person to person. In general, short-term memory seems to be affected more than long-term memory. Intellect and memory appear to be retained better in people who remain mentally and physically active.

Another change observed in older people is a decrease in the rate of impulse conduction along an axon and across a synapse. A reduction in the amount of myelin around the axon probably accounts for the diminished conduction rate along the axon. Decreases in the quantity of neurotransmitter and in the number of receptor sites cause slower conduction across the synapses. These factors contribute to the slower reflexes and the longer time required to process information that are observed in many elderly people.

Representative Disorders of the
Nervous System

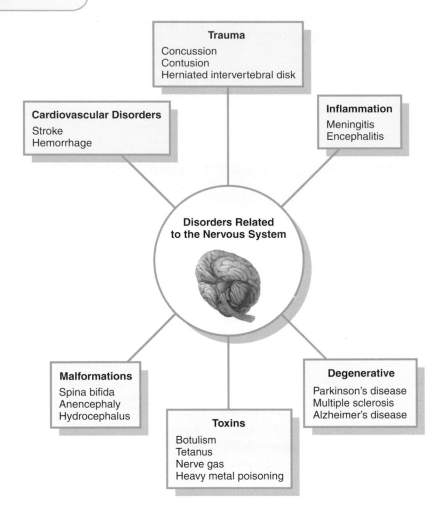

Trauma
Concussion
Contusion
Herniated intervertebral disk

Cardiovascular Disorders
Stroke
Hemorrhage

Inflammation
Meningitis
Encephalitis

**Disorders Related
to the Nervous System**

Malformations
Spina bifida
Anencephaly
Hydrocephalus

Toxins
Botulism
Tetanus
Nerve gas
Heavy metal poisoning

Degenerative
Parkinson's disease
Multiple sclerosis
Alzheimer's disease

Alzheimer's disease (ALTZ-hye-merz dih-ZEEZ)
Irreversible senile dementia characterized by increasing loss of mental abilities, including memory, recognition, reasoning, and judgment, and by mood changes, including irritability, agitation, and restlessness

Anencephaly (an-en SEF-ah-lee) Congenital disorder with cranial vault and cerebral hemispheres missing

Botulism (BOTCH-yoo-lizm) Severe form of food poisoning caused by a neurotoxin produced by *Clostridium botulinum;* toxin inhibits transmission of nerve impulses at neuromuscular junctions

Cerebral concussion (seh-REE-brull kon-KUSH-un) Loss of consciousness as the result of a blow to the head; usually clears within 24 hours; no evidence of permanent structural damage to the brain tissue

Cerebral contusion (seh-REE-brull kon-TOO-shun) Bruising of brain tissue as a result of direct trauma to the head; neurological problems persist longer than 24 hours

Encephalitis (en-sef-ah-LYE-tis) Inflammation of the brain; frequently caused by viruses

Heavy metal poisoning (HEH-vee MET-ahl POY-sun-ing) Irreversible demyelination of axons and damage to glial cells caused by chronic exposure to heavy metal ions such as arsenic, lead, or mercury

Hemorrhage (HEM-ohr-ij) Escape of blood from a ruptured blood vessel; if severe, it may lead to hypovolemia and circulatory shock; in the brain, it may be epidural, subdural, or into the brain tissue; cerebral hemorrhage is one of the three main causes of strokes

Herniated intervertebral disk (HER-nee-ayt-ed in-ter-VER-tee-bruhl DISK) Protrusion of the inner region of an intervertebral disk through the outer fibrous ring of the disc; distorts the sensory nerves and may compress nerve roots, with resulting pain from the affected area; frequently occurs in the lumbar region

Hydrocephalus (hye-droh-SEF-ah-lus) Abnormal accumulation of CSF within the ventricles of the cerebrum; in the infant, this leads to an enlargement of the cranium

Meningitis (men-in-JYE-tis) Inflammation of the meninges; may be caused by anything that activates in inflammatory process, including bacteria, viruses, fungi, and chemical toxins

Multiple sclerosis (MULL-tih-pull skler-OH-sis) Disorder in which there is progressive destruction of the myelin sheaths of central nervous system neurons, interfering with their ability to transmit impulses; characterized by progressive loss of function interspersed with periods of remission; cause is unknown and there is no satisfactory treatment

Nerve gas (NURV GAS) Toxic gas that interferes with normal nerve impulse conduction by blocking cholinesterase activity at the synapse, resulting in sustained skeletal muscle contraction; also affects smooth and cardiac muscle; has been used in warfare

Parkinson's disease (PAR-kin-sunz dih-ZEEZ) Slowly progressive disease, usually appearing after the age of 60, characterized by increased motor activity and caused by a lack of dopamine; symptoms include muscle tremors, rigidity, and difficulty in initiating movements and speech

Spina bifida (SPY-nah BIFF-ih-dah) Developmental anomaly in which the vertebral laminae do not close around the spinal cord, leaving an opening through which the cord and meninges may or may not protrude

Stroke (STROHK) Group of symptoms caused by disrupted blood supply to the brain because of thrombosis, embolism or hemorrhage; symptoms depend on the region of the brain affected, but frequently involve disturbed muscular function including paralysis; also called cerebrovascular accident (CVA)

Tetanus (TET-ah-nus) Highly fatal disease caused by the toxin from *Clostridium tetani* bacteria; the toxin attacks the central nervous system and is characterized by muscle spasms and convulsions

CHAPTER SUMMARY

Functions and Organization of the Nervous System

■ **Outline the organization and functions of the nervous system.**

- The nervous system is divided into the central nervous system and the peripheral nervous system.
- The peripheral nervous system consists of an afferent (sensory) division and an efferent (motor) division.
- The efferent (motor) division is divided into the somatic nervous system and the autonomic nervous system.
- The autonomic nervous system is further divided into the sympathetic and parasympathetic divisions.

■ **List three categories of nervous system functions.**

- The activities of the nervous system can be grouped into sensory, integrative, and motor functions.

Nerve Tissue

■ **Compare the structure and functions of neurons and neuroglia.**

- Neurons are the nerve cells that transmit impulses. They are amitotic.
- The three components of a neuron are a cell body or soma, one or more dendrites, and a single axon.
- Many neurons are surrounded by segmented myelin. The gaps in the myelin are the nodes of Ranvier. An outer covering, the neurilemma, plays a role in nerve regeneration.
- Axons terminate in telodendria, which have synaptic bulbs on the distal end.
- Functionally, neurons are classified as afferent (sensory), efferent (motor), or interneurons (association).

- Neuroglia support, protect, and nourish the neurons. They are capable of mitosis.
- Neuroglia cells include astrocytes, microglia, ependyma, oligodendrocytes, neurolemmocytes, and satellite cells.

Nerve Impulses

■ **Describe the characteristics of the resting membrane of a neuron.**

- Excitability, the ability to respond to a stimulus, and conductivity, the ability to transmit an impulse, are two functional characteristics of neurons.
- The cell membrane of a nonconducting neuron is polarized with an abundance of sodium ions outside the cell and an abundance of potassium and negatively charged proteins inside the cell. The inside of the membrane is approximately 70 mV negative to the outside.

■ **Describe the sequence of events that lead to an action potential when the cell membrane is stimulated.**

- In response to a stimulus, the cell membrane becomes permeable to sodium ions, so they rapidly enter the cell and depolarize the membrane. Continued sodium ion diffusion causes reverse polarization.
- After reverse polarization, the membrane becomes impermeable to sodium and permeable to potassium. Potassium diffuses out of the cell to repolarize the membrane.
- The action potential is a result of depolarization, reverse polarization, and repolarization of the cell membrane.
- After the action potential, active transport mechanisms move sodium out of the cell and potassium into the cell to restore resting conditions.
- A threshold stimulus is the minimum stimulus necessary to start an action potential. A weaker stimulus is subthreshold.

■ **Explain how the impulse is conducted along the length of a neuron.**

- An action potential at a given point stimulates depolarization at an adjacent point to create a propagated action potential, or nerve impulse, that continues along the entire length of the neuron.
- Saltatory conduction occurs in myelinated fibers where the action potential "jumps" from node to node.
- The refractory period is the time during which the cell membrane is recovering from depolarization; absolute refractory period is the time during which the membrane is permeable to sodium ions and cannot respond to a second stimulus no matter how strong; relative refractory period is roughly comparable to the time when the membrane is permeable to potassium. During this time, it takes a stronger than normal stimulus to initiate an action potential.
- If a threshold stimulus is applied, an action potential is generated and propagated along the entire length of the neuron at maximum strength and speed for the existing conditions.

■ **Describe the structure of a synapse and how an impulse is conducted from one neuron to another across the synapse.**

- A synapse, the region of communication between two neurons, consists of a synaptic knob, a synaptic cleft, and the postsynaptic membrane.
- At the synapse, a neurotransmitter, such as acetylcholine, diffuses across the synaptic cleft and reacts with receptors on the postsynaptic membrane.
- In excitatory transmission, the reaction between the neurotransmitter and receptor depolarizes the postsynaptic membrane and initiates an action potential.
- In inhibitory transmission, the reaction between the neurotransmitter and receptor opens the potassium channels and hyperpolarizes the membrane, which makes it more difficult to generate an action potential.
- In a single series circuit, a single neuron synapses with another neuron; in a divergence circuit, a single neuron synapses with multiple neurons; and in a convergence circuit, several presynaptic neurons synapse with a single postsynaptic neuron.

■ **List the five basic components of a reflex arc.**

- The reflex arc is a type of conduction pathway and represents the functional unit of the nervous system.
- A reflex arc utilizes a receptor, sensory neuron, center, motor neuron, and effector.

Central Nervous System

■ **Describe the three layers of meninges around the central nervous system.**

- The brain and spinal cord are covered by three layers of connective tissue membranes called meninges.
- The outer layer of the meninges is the dura mater, the middle layer is the arachnoid, and the inner layer is the pia mater.
- The subarachnoid space, between the arachnoid and pia mater, contains blood vessels and CSF.

■ **Locate and identify the major regions of the brain and describe their functions.**

- The largest portion of the brain is the cerebrum, which is divided by a longitudinal fissure into two central hemispheres. These are connected by a band of white fibers called the corpus callosum. The surface is marked by gyri separated by sulci.
- Each cerebral hemisphere is divided into frontal, parietal, occipital, and temporal lobes and an insula.
- The outer surface of the cerebrum is the cerebral cortex and is composed of gray matter. This surrounds the white matter, which consists of myelinated nerve fibers. Basal ganglia are regions of gray matter scattered throughout the white matter.
- The somatosensory area is the postcentral gyrus in the parietal lobe; the sensory area for vision is in the occipital lobe; the area for hearing is in the temporal lobe; the olfactory area is in the temporal lobe; and the sensory area for taste is in the parietal lobe.
- The somatomotor area is the precentral gyrus in the frontal lobe.
- Association areas analyze, interpret, and integrate information. They are scattered throughout the cortex.
- The diencephalon is nearly surrounded by the cerebral cortex. It includes the thalamus, hypothalamus, and epithalamus.
- The largest region of the diencephalon is the thalamus, which serves as a relay station for sensory impulses going to the cerebral cortex.
- The hypothalamus is a small region below the thalamus. It plays a key role in maintaining homeostasis.
- The epithalamus is the superior portion of the diencephalon, and the pineal gland extends from its posterior margin.
- The brain stem is the region between the diencephalon and the spinal cord.
- The midbrain, the most superior portion of the brain stem, includes the cerebral peduncles and corpora quadrigemina. The midbrain contains voluntary motor tracts, and visual and auditory reflex centers.
- The pons is the middle portion of the brain stem. It contains the pneumotaxic and apneustic areas that help regulate breathing movements.
- The medulla oblongata is inferior to the pons and is continuous with the spinal cord. It contains ascending and descending nerve fibers. It also contains the vital cardiac, vasomotor, and respiratory centers.
- The cerebellum consists of two hemispheres connected by a central region called the vermis. A thin layer of gray matter, the cortex, surrounds the white matter.
- Cerebellar peduncles connect the cerebellum to other parts of the CNS.
- The cerebellum is a motor area that coordinates skeletal muscle activity and is important in maintaining muscle tone, posture, and balance.

■ **Trace the flow of CSF from its origin in the ventricles to its return to the blood.**

- Four interconnected cavities, called ventricles, contain CSF within the brain.
- CSF is formed as a filtrate from the blood in the choroid plexus in the ventricles.
- CSF moves from the lateral ventricles, through the interventricular foramen to the third ventricle, through the cerebral aqueduct to the fourth ventricle. From the fourth ventricle, it enters the subarachnoid space and filters through arachnoid granulations into the dural sinuses and is returned to the blood.

■ **Describe the structure and functions of the spinal cord.**

- The spinal cord begins at the foramen magnum, as a continuation of the medulla oblongata, and extends to the first lumbar vertebra.

- The central core of the spinal cord is gray matter, which is divided into regions called dorsal, lateral, and ventral horns. The white matter, which surrounds the gray matter, is divided into dorsal, lateral, and ventral funiculi, or columns.
- The spinal cord is a conduction pathway and a reflex center.
- Ascending tracts in the spinal cord conduct sensory impulses to the brain. Conduction pathways that carry motor impulses from the brain to effectors are descending tracts.

Peripheral Nervous System

■ Describe the structure of a nerve.

- The cranial and spinal nerves form the peripheral nervous system. Nerves are classified as sensory, motor, or mixed, depending on the types of fibers they contain.
- Nerves are bundles of nerve fibers.
- Each individual nerve fiber is covered by endoneurium.
- A bundle of nerve fibers, surrounded by perineurium, is called a fasciculus.
- A nerve contains many fasciculi collected together and surrounded by epineurium.

■ List the 12 cranial nerves and state the function of each one.

- Twelve pairs of cranial nerves emerge from the inferior surface of the brain. Cranial nerves are designated by name and by Roman numerals.
- Names of the cranial nerves are olfactory (I), optic (II), oculomotor (III), trochlear (IV), trigeminal (V), abducens (VI), facial (VII), vestibulocochlear (VIII), glossopharyngeal (IX), vagus (X), accessory (XI), and hypoglossal (XII).

■ Discuss spinal nerves and the plexuses they form.

- All spinal nerves are mixed nerves. They are connected to the spinal cord by dorsal roots, which have only sensory fibers, and ventral roots, which have only motor fibers.
- The 31 pairs of spinal nerves are grouped according to the region of the cord from which they originate. There are 8 cervical nerves, 12 thoracic nerves, 5 lumbar nerves, 5 sacral nerves, and 1 coccygeal nerve.
- In all but the thoracic region, the main portions of the spinal nerves form complex networks called plexuses. Named nerves exit the plexuses to supply specific regions of the body.

■ Compare the structural and functional differences between the somatic efferent pathways and the ANS.

- The ANS is a visceral efferent system that innervates smooth muscle, cardiac muscle, and glands.
- An autonomic pathway consists of two neurons. A preganglionic neuron leaves the CNS and synapses with a postganglionic neuron in an autonomic ganglion.

■ Distinguish between the sympathetic and parasympathetic divisions of the ANS in terms of structure, function, and neurotransmitters.

- The sympathetic division is also called the thoracolumbar division because its preganglionic neurons originate in the thoracic and lumbar regions of the spinal cord. It is an energy expending system that prepares the body for emergency or stressful conditions.
- The parasympathetic division, also called the craniosacral division because its preganglionic neurons originate in the brain and the sacral region of the spinal cord, is an energy-conserving system that is most active when the body is in a normal relaxed condition.

CHAPTER QUIZ

Recall

Match the definitions on the left with the appropriate term on the right.

_____ **1.** Outermost covering of the central nervous system

_____ **2.** Produces cerebrospinal fluid

_____ **3.** Occurs along myelinated axons

_____ **4.** Supporting cells of the nervous system

_____ **5.** White, fatty covering of axons

_____ **6.** Controls contraction of skeletal muscles

_____ **7.** Lowest part of the brain stem

_____ **8.** Contains visual cortex

_____ **9.** Efferent process of a neuron

_____ **10.** Contains somatosensory cortex

A. Axon

B. Choroid plexus

C. Dura mater

D. Medulla oblongata

E. Myelin

F. Neuroglia

G. Occipital lobe

H. Parietal lobe

I. Saltatory conduction

J. Somatic nervous system

Thought

1. Neurons located entirely within the CNS are:
   A. Motor neurons
   B. Association neurons
   C. Efferent neurons
   D. Sensory neurons

2. Selected events in impulse conduction are given. Arrange these events in the order in which they occur: (1) sodium gates open; (2) potassium gates open; (3) stimulus alters membrane permeability; (4) active transport of sodium and potassium; (5) reverse polarization.
   A. 3, 1, 2, 4, 5
   B. 3, 1, 5, 4, 2
   C. 3, 4, 1, 5, 2
   D. 3, 1, 5, 2, 4

3. White matter:
   A. Forms the cerebral cortex
   B. Consists of neuron cell bodies
   C. Forms the outer covering of the brain
   D. Consists of myelinated fibers

4. The three regions of the brain stem are the:
   A. Thalamus, hypothalamus, pineal body
   B. Thalamus, pons, cerebellum
   C. Midbrain, pons, medulla oblongata
   D. Thalamus, midbrain, cerebral peduncles

5. Three cranial nerves that have only sensory functions are:
   A. Optic, olfactory, and trigeminal
   B. Optic, olfactory, and vestibulocochlear
   C. Optic, oculomotor, and glossopharyngeal
   D. Vagus, trigeminal, and facial

Application

A medical student is practicing technique on an anesthetized animal and accidentally severs the left phrenic nerve. What effect will this have on the animal? Which cranial nerve conveys pain impulses to the brain: (a) when you bite the tip of your tongue? (b) when a piece of dirt blows into your eye? (c) when you have a toothache from an upper molar?

BUILDING YOUR MEDICAL VOCABULARY

Building a vocabulary is a cumulative process. As you progress through this book, you will use word parts, abbreviations, and clinical terms from previous chapters. Each chapter will present new word parts, abbreviations, and terms to add to your expanding vocabulary.

Word Parts and Combining Forms with Definitions and Examples

PART/COMBINING FORM	DEFINITION	EXAMPLE
af-	toward	afferent: conducting toward a center or a specific site of reference
alges/o	sensitivity	analgesic: without pain to pain
astr/o	star	astrocyte: star-shaped cell of neuroglia
contra-	against,	contralateral: pertaining to or affecting the opposite side
corpor-	body	corpora quadrigemina: four bodies in the midbrain
dendr-	tree	dendrite: treelike processes of a neuron
ef-	away from	efferent: conducting away from a center or a specific point of reference

PART/COMBINING FORM	DEFINITION	EXAMPLE
encephal/o	within the head,	encephalalgia: pain within the brain, headache brain
esthes-	feeling	anesthesia: without feeling
-fer-	to carry	afferent: to carry toward a center or specific point of reference
gangli-	knot	ganglion: group or knot of nerve cell bodies
gli-	glue	neuroglia: nerve glue, binding tissue of the nervous system
gloss/o	tongue	glossopharyngeal: pertaining to the tongue and pharynx
idio-	individual, self	idiopathic: self-originating, occurring without a known cause
lemm-	peel, rind	neurilemma: covering of a nerve fiber
-mania	obsessive	kleptomania: obsessive preoccupation with stealing
mening-	membrane	meninges: membrane around the brain
narc-	numbness,	narcotic: drug that induces stupor, sleep
neur/o	nerve	neurilemma: covering around a nerve fiber

PART/COMBINING FORM	DEFINITION	EXAMPLE
peri-	all around	perineurium: connective tissue membrane that goes around a bundle of nerve fibers
pharyng/o	throat	glossopharyngeal: pertaining to the tongue and throat
phas-	speech	aphasia: without speech, inability to talk
pleg-	paralysis	quadriplegia: paralysis of all four extremities
plex-	interweave,	plexus: network of nerves network
schiz/o	division, split	schizonychia: splitting of nails
somn-	sleep	insomnia: inability to sleep
sulc-	furrow, ditch	sulcus: furrows or grooves of the cerebral cortex

Clinical Abbreviations

ABBREVIATION	MEANING
ADD	attention-deficit disorder
ADHD	attention-deficit/hyperactivity disorder
ALS	amyotrophic lateral sclerosis
ANS	autonomic nervous system
CNS	central nervous system
CSF	cerebrospinal fluid
CT	computed tomography
CVA	cerebrovascular accident
EEG	electroencephalogram
ICP	intracranial pressure
LP	lumbar puncture
MG	myasthenia gravis
MRI	magnetic resonance imaging
MS	multiple sclerosis
PET	positron emission tomography
PNS	peripheral nervous system
SNS	somatic nervous system
TIA	transient ischemic attack

Clinical Terms

Amyotrophic lateral sclerosis (a-my-oh-TROF-ick LAT-er-al sclair-OH-sis) A neurological disease caused by degeneration of motor neurons of the spinal cord, medulla, and cortex; marked by progressive muscular weakness and atrophy with spasticity and exaggerated reflexes; mental capabilities are not impaired; also called Lou Gehrig disease or motor neuron disease

Bell's palsy (BELL's PAUL-zee) Neuropathy of the seventh cranial nerve (facial) that causes paralysis of the muscles on one side of the face with sagging of the mouth on the affected side of the face

Cerebral palsy (seh-REE-brull PAWL-zee) Partial paralysis and lack of muscular coordination caused by damage to the cerebrum during fetal life, birth, or infancy

Cerebrovascular accident (CVA) (seh-ree-broh-VAS-kyoo-lar AK-sih-dent) Most common brain disorder; may be due to decreased blood supply to the brain or rupture of a blood vessel in the brain; commonly called a *stroke*

Computed tomography (CT) (kum-PYOO-ted toh-MAHG-rah-fee) Diagnostic procedure in which x-ray images are used to compose a computerized sectional picture of the brain

Electroencephalography (EEG) (e-lek-troh-en-sef-ah-LAHG-rah-fee) Recording the electrical activity of the brain to demonstrate seizures, brain tumors, and other diseases and injury to the brain

Magnetic resonance imaging (MRI) (mag-NET-ick REZ-oh-nans IHM-oh-jing) U Use of magnetic waves to create a sectional image of the brain; considered to be more sensitive than computed tomography in diagnosing certain brain lesions

Positron emission tomography (PET) (PAHZ-ih-tron ee-MIH-shun toh-MAHG-rah-fee) Procedure that uses a radioactive isotope, combined with a form of glucose and injected intravenously, to obtain sectional images that show how the brain uses glucose and that give information about brain function

Reye's syndrome (RS) (RAYZ SIN-drohm) Brain dysfunction that occurs primarily in children and teenagers and is characterized by edema of the brain that leads to disorientation, lethargy, and personality changes and may progress to a coma; seems to occur after chickenpox and influenza; taking aspirin is a risk factor

Shingles (SHING-gulls) Viral disease affecting peripheral nerves; characterized by blisters and pain spread over the skin in a bandlike pattern that follows the affected nerves; caused by the same herpesvirus that causes chickenpox

Tic douloureux (TICK doo-loo-ROO) Painful disorder of the fifth cranial nerve (trigeminal) that is characterized by sudden, intense, sharp pain in the face and forehead on the affected side; also known as trigeminal neuralgia

Transient ischemic attack (TIA) (TRANS-ee-ent iss-KEE-mik ah-TACK) Episode of temporary cerebral dysfunction caused by impaired blood flow to the brain; the onset is sudden, is of short duration, and leaves no long-lasting neurological impairment; common causes are blood clots and atherosclerosis

VOCABULARY QUIZ

Use word parts you have learned to form words that have the following definitions.

1. Pertaining to tongue and throat _____

2. Without feeling _____

3. Star-shaped cell _____

4. Inflammation of the brain _____

5. Without speech _____

Using the definitions of word parts you have learned, define the following words.

6. Afferent _____

7. Neuroglia _____

8. Ganglion _____

9. Sulcus _____

10. Meningitis _____

Match each of the following definitions with the correct word.

_____ **11.** Paralysis of four extremities   **A.** Corpora quadrigemina

_____ **12.** Network of nerves   **B.** Dendrites

_____ **13.** Treelike processes   **C.** Neurilemma

_____ **14.** Four bodies   **D.** Plexus

_____ **15.** Membrane of a nerve fiber   **E.** Quadriplegia

16. What is the term for the viral disease that affects the peripheral nerves and is caused by the same virus that causes chickenpox? _____

17. What is the clinical term for paralysis of facial muscles due to neuropathy of the seventh cranial nerve? _____

18. What procedure records the electrical activity of the brain? _____

19. What is the meaning of the abbreviation ICP? _____

20. What is the meaning of the abbreviation TIA? _____

The Senses

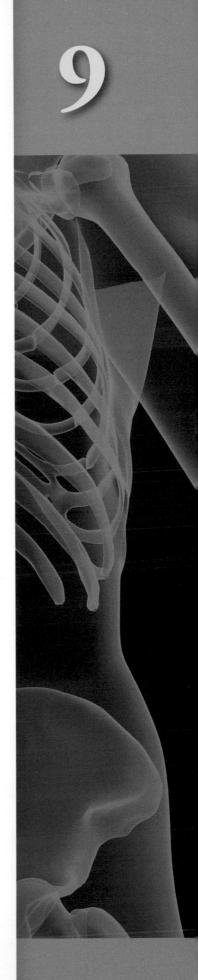

CHAPTER OBJECTIVES

Receptors and Sensations
- Distinguish between general senses and special senses.
- Classify sense receptors into five groups.
- Explain what is meant by *sensory adaptation.*

General Senses
- Describe the sense receptors for touch, pressure, proprioception, temperature, and pain.

Gustatory Sense
- Locate the four different taste sensations, and trace the impulse pathway from the stimulus to the cerebral cortex.

Olfactory Sense
- Locate the sense receptors for smell, and trace the impulse pathway to the cerebral cortex.

Visual Sense
- Describe the protective features and accessory structures of the eye.
- Describe the structure of the eye and the significance of each component.
- Explain how light focuses on the retina.
- Identify the photoreceptor cells in the retina, and describe the mechanism by which nerve impulses are triggered in response to light.
- Trace the pathway of impulses from the retina to the visual cortex.

Auditory Sense
- Describe the structure of the ear and the contribution each region makes to the sense of hearing.
- Summarize the sequence of events in the initiation of auditory impulses, and trace these impulses to the auditory cortex.

Sense of Equilibrium
- Identify and describe the structure of the components of the ear involved in static equilibrium and those involved in dynamic equilibrium.
- Summarize the events in the initiation of impulses for static equilibrium and for dynamic equilibrium and identify the cranial nerve that transmits these impulses to the cerebral cortex.

KEY TERMS

Accommodation (ah-kahm-oh-DAY-shun) Mechanism that allows the eye to focus at various distances, primarily achieved by changing the curvature of the lens

Chemoreceptor (kee-moh-ree-SEP-tor) Sensory receptor that detects the presence of chemicals; responsible for taste, smell, and monitoring the concentration of certain chemicals in body fluids

Dynamic equilibrium (dye-NAM-ik ee-kwi-LIB-ree-um) Equilibrium of motion; maintaining balance when the head or body is moving

Endolymph (EN-doh-limf) Fluid that fills the membranous labyrinth of the inner ear

Mechanoreceptor (mek-ah-noh-ree-SEP-tor) Sensory receptor that responds to a bending or deformation of the cell; examples include receptors for touch, pressure, hearing, and equilibrium

Nociceptor (noh-see-SEP-tor) Sensory receptor that responds to tissue damage; pain receptor

Perilymph (PAIR-ih-limf) Fluid inside the bony labyrinth but outside the membranous labyrinth of the inner ear

Photoreceptor (foh-toh-ree-SEP-tor) Sensory receptor that detects light; located in the retina of the eye

Proprioceptor (proh-pree-oh-SEP-tor) A type of mechanoreceptor located in muscles, tendons, and joints; provides information about body position and movements

Refraction (ree-FRAK-shun) Bending of light as it passes from one medium to another

Rhodopsin (roh-DAHP-sin) Photosensitive pigment in the rods of the retina; also called visual purple

Sensory adaptation (SEN-soh-ree add-dap-TAY-shun) Phenomenon in which some receptors respond when a stimulus is first applied but decrease their response if the stimulus is maintained; receptor sensitivity decreases with prolonged stimulation

Static equilibrium (STAT-ik ee-kwi-LIB-ree-um) Sensing and evaluating the position of the head relative to gravity

Thermoreceptor (ther-moh-ree-SEP-tor) A sensory receptor that detects changes in temperature

Sense perception depends on sensory receptors that respond to various stimuli. Some senses, such as pain, touch, pressure, and proprioception, are widely distributed in the body. These are called **general senses.** Other senses, such as taste, smell, hearing, and sight, are called **special senses** because their receptors are localized in a particular area.

Receptors and Sensations

Although there are many different kinds of sense receptors, they can be grouped into five types. The basis for these receptor types is the kind of stimulus to which they are sensitive or have a low threshold. The five types of receptors are **chemoreceptors, mechanoreceptors, nociceptors, thermoreceptors,** and **photoreceptors.** These are summarized in Table 9-1.

Perceived sensation occurs only after impulses are interpreted by the brain. Steps involved in sensory perception include the following:

- There must be a stimulus.
- A receptor must detect the stimulus and create an action potential.
- The action potential (impulse) must be conducted to the central nervous system (CNS).
- Within the CNS, the impulse must be translated into information.
- Information must be interpreted in the CNS into an awareness or perception of the stimulus.

The impulses from all the receptors are alike. The difference in perception is where they are interpreted in the brain. For example, all impulses going to one particular region are interpreted as sound, whereas those going to another region are interpreted as taste. As the brain interprets a sensation, it projects that sense back to its original source so that the "feeling" seems to come from the receptors that are stimulated. This projection allows us to locate the source of the stimulus.

Some sense receptors, when they are continually stimulated, undergo **sensory adaptation.** They have a decreased sensitivity to a continued stimulus and trigger impulses only if the strength of the stimulus is increased. An example of this is the sense of smell. A particular odor becomes unnoticed after a short time even though the odor molecules are still present in the air. The receptors have a decreased sensitivity to the continued stimulation.

TABLE 9-1 Types of Sense Receptors

Receptor	Stimulus	Example
Chemoreceptors	Changes in chemical concentration of substances	Taste and smell
Mechanoreceptors	Changes in pressure or movement in fluids	Proprioceptors in joints, receptors for hearing and equilibrium
Nociceptors	Tissue damage	Pain receptors
Thermoreceptors	Changes in temperature	Heat and cold
Photoreceptors	Light energy	Vision

General Senses

General senses, or **somatic senses,** are those that are found throughout the body. They are associated with the visceral organs as well as the skin, muscles, and joints and include the following:

- Touch
- Pressure
- Proprioception
- Temperature
- Pain

Touch and Pressure

As a group, the receptors for touch and pressure are mechanoreceptors that are sensitive to forces that deform or displace tissues. They are widely distributed in the skin. Three of the receptors involved in touch and pressure are **free nerve endings, Meissner's corpuscles** (MYZE-ners KOAR-pus-als), and **pacinian (lamellated) corpuscles** (pah-SIN-ee-an KOAR-pus-als). Receptors for touch and pressure are illustrated in Figure 9-1.

Free nerve endings are the dendritic ends of sensory neurons that are interspersed between the cells in epithelial tissue. They do not have a connective tissue covering. They are important in sensing objects, such as clothing, that are in continuous contact with the skin. Meissner's corpuscles consist of the ends of sensory nerve fibers surrounded by connective tissue and are very specific in localizing tactile sensations. They are located in the dermal papillae, just beneath the epidermis, where they are important in sensing light discriminative touch stimuli. Pacinian corpuscles are called lamellated corpuscles because several layers of connective tissue surround the nerve endings. These are common in deeper dermis and subcutaneous tissues, tendons, and ligaments. They are stimulated by heavy pressure.

Proprioception

Proprioception (proh-pree-oh-SEP-shun) is the sense of position or orientation. It allows us to sense the location and rate of movement of one body part relative to another. **Golgi tendon organs,** found at the junction of a tendon with a muscle, and **muscle spindles,** located in skeletal muscles, are important mechanoreceptors for proprioception. They are illustrated in Figure 9-1.

Temperature

Thermoreceptors are located immediately under the skin and are widely distributed throughout the body. They are most numerous on the lips and are least numerous on some of the broad surfaces of the trunk. Thermoreceptors include at least two types of free nerve endings that are sensitive to temperature changes. In general, there are up to 10 times more **cold receptors** in a given area than **heat receptors.** Extremes in temperature stimulate pain receptors. Below 10 °C, pain receptors produce a freezing sensation. As the temperature increases above 10 °C, pain impulses cease but **cold receptors** begin to be stimulated. At temperatures about 25 °C, **heat receptors** begin to be stimulated and cold receptors fade out. Finally, as temperatures approach 45 °C, heat receptors fade out and pain receptors are stimulated to produce a burning sensation. A person determines gradations in temperatures by the degree of stimulation of each type of receptor. Extreme cold and extreme heat feel almost

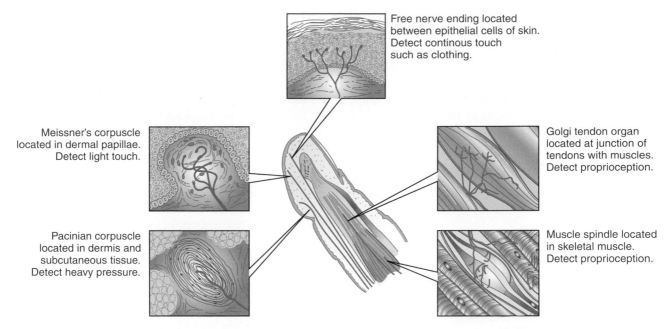

Free nerve ending located between epithelial cells of skin. Detect continous touch such as clothing.

Meissner's corpuscle located in dermal papillae. Detect light touch.

Golgi tendon organ located at junction of tendons with muscles. Detect proprioception.

Pacinian corpuscle located in dermis and subcutaneous tissue. Detect heavy pressure.

Muscle spindle located in skeletal muscle. Detect proprioception.

Figure 9-1 *General sense receptors for touch, pressure, and proprioception.*

the same—both are painful—because the pain receptors are being stimulated. Thermoreceptors are strongly stimulated by abrupt changes in temperature and then fade after a few seconds or minutes. In other words, thermoreceptors show rapid **sensory adaptation.** When a person first enters a cool swimming pool on a hot day, the abrupt change in temperature stimulates the cold receptors and there is a feeling of discomfort. After a brief time, the receptors adapt, the stimulation fades, and the water feels comfortable.

Pain

The sense of pain is initiated by **nociceptors** (noh-see-SEP-tors), which are free nerve endings that are stimulated by tissue damage. They are widely distributed throughout the skin and in the tissues of the internal organs. There are no pain receptors in the nervous tissue of the brain; however, other tissues in the head, including the meninges and blood vessels, have an abundant supply. Pain may also be referred to the head from the sinuses and the eyes. Pain receptors have a protective function because pain is usually perceived as unpleasant and is a signal to locate and to remove the source of the tissue damage. Nociceptors usually do not adapt and may continue to send signals after the stimulus is removed.

QUICK CHECK

9.1 Name three receptors for touch and pressure.

9.2 What is the term for "free nerve endings that provide a sense of pain"?

Gustatory Sense

The **gustatory sense,** or **taste,** is one of the special senses. The organs of taste, the **taste buds,** are localized in the mouth region, primarily on the surface of the tongue, where they lie along the walls of projections called **papillae.** They are illustrated in Figure 9-2. The receptors belong to the **chemoreceptor** (kee-moh-ree-SEP-tor)

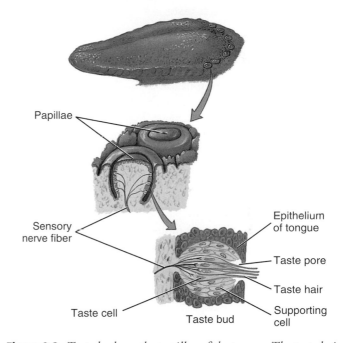

Papillae

Sensory nerve fiber

Epithelium of tongue

Taste pore

Taste hair

Taste cell

Taste bud

Supporting cell

Figure 9-2 *Taste buds on the papillae of the tongue. The taste hairs on taste (gustatory) cells are the receptors for taste.*

category because they are sensitive to chemicals in the food we eat. In order for these chemicals to be detected by a chemoreceptor, they must be dissolved in water.

Within the taste bud, specialized epithelial cells called **taste cells** or **gustatory cells** are interspersed with supporting cells and nerve fibers (see Figure 9-2). The entire taste bud opens to the surface through a **taste pore.** Tiny **taste hairs** (microvilli) project from the taste cells through the taste pore, and it is these hairs on the taste cells that function as the receptors.

Although all the taste receptors appear to be alike, there are at least four different types, each one sensitive to a particular kind of stimulus. Consequently, there are four different taste sensations: salty, sweet, sour, and bitter.

Instead of being evenly distributed over the surface of the tongue, each one of the four tastes is concentrated in a specific region. Salty and sweet are detected near the anterior region of the tongue. The receptors for sour taste are along the sides and the receptors for a bitter taste are at the back of the tongue.

When the microvilli, or taste hairs, are stimulated, an impulse is triggered on a nearby nerve fiber. Impulses from the anterior two thirds of the tongue travel along the **facial nerve** (cranial nerve VII), and those from the posterior one third travel along the **glossopharyngeal** (glos-so-fah-RIN-jee-al) **nerve** (cranial nerve IX) to the medulla oblongata. From the medulla oblongata they travel to the thalamus, then to the sensory cortex on the parietal lobe, near the lateral sulcus.

QUICK APPLICATIONS

The taste receptors with the highest degree of sensitivity are those that are stimulated by bitter substances, and a highly intense bitter taste usually causes a person to reject that substance. This is probably an important protective mechanism, because many of the deadly toxins found in poisonous plants have an intensely bitter taste.

Olfactory Sense

The receptors for **olfaction** (sense of smell) are bipolar neurons surrounded by supporting columnar epithelial cells in the **olfactory epithelium** of the nasal cavity (Figure 9-3). The **olfactory neurons** are concentrated in the superior region of the cavity. These neurons have long cilia that extend to the surface and project into the nasal cavity. The cilia are believed to be the sensitive receptors of the neuron.

Like those for taste, the olfactory receptors are **chemoreceptors.** They are stimulated by chemicals dissolved in liquids. In this case, airborne molecules responsible for odors dissolve in the fluid on the surface of the olfactory epithelium and then bind to the receptors and trigger impulses.

Axons from the olfactory neurons pass through foramina in the cribriform plate of the ethmoid bone and enter the **olfactory bulb.** Here they synapse with association neurons that conduct the impulses through the **olfactory tract** to the brain. The olfactory tracts terminate in the **olfactory cortex** in the temporal lobe.

The senses of taste and smell are closely related and complement each other. They often have a combined effect when they are interpreted in the cerebral cortex. This implies that part of what we "taste" is really smell. Also, part of what we "smell" is taste because some airborne molecules move from the nose down to the mouth and stimulate taste buds.

QUICK CHECK

9.3 What two cranial nerves transmit impulses for the sense of taste?

9.4 Impulses for what special sense are transmitted by cranial nerve I?

QUICK APPLICATIONS

Odors have the quality of being interpreted as pleasant or unpleasant. Because of this, the sense of smell is as important as taste in the selection of food. For example, a person who has became sick after eating a certain type of food is often nauseated by the smell of that same food at a later occasion.

Visual Sense

Most of us consider vision to be one of the most important senses we have. The eyes, which contain the photoreceptors, are the organs of vision. They are protected by a bony socket and are assisted in their function of vision by accessory structures that protect and move them.

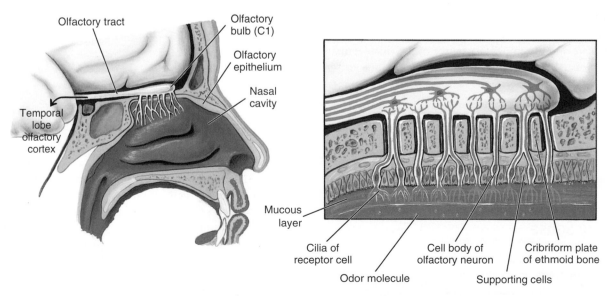

Figure 9-3 *Structure of the olfactory receptors in the nasal cavity. The impulses travel from the receptors to the olfactory bulb to the olfactory cortex in the temporal lobe of the brain.*

Protective Features and Accessory Structures of the Eye

Only a small portion of the eye is visible from the exterior. Most of it is surrounded by a protective bony **orbit,** or socket, formed by portions of seven cranial bones: the frontal, lacrimal, ethmoid, maxilla, zygomatic, sphenoid, and palatine bones. It also contains fat, various connective tissues, blood vessels, and nerves.

Eyebrows help to keep perspiration, which can be an irritant, out of the eyes. **Eyelids** function to open and close the eye and to keep foreign objects from entering the eye. The eyelids are composed of skin, connective tissue, muscle, and **conjunctiva** (kon-junk-TYE-vah). The muscles associated with the eyelids are the **orbicularis oculi,** which is a sphincter that closes the eye, and the **levator palpebrae superioris,** which elevates the eyelid to open the eye. The conjunctiva, a thin mucous membrane, lines the eyelid and then folds back to cover the anterior portion of the eyeball, except for the central portion, which is the cornea. Mucus from the conjunctiva helps keep the eye from drying out. **Eyelashes** line the margin of the eyelid and help trap foreign particles. **Sebaceous glands** associated with the eyelashes secrete an oily fluid that helps lubricate the region. Inflammation of the sebaceous glands is called a **stye.**

 QUICK APPLICATIONS

Conjunctivitis is an inflammation of the conjunctiva. This may be due to irritation, allergies, or bacterial infections. For example, acute contagious conjunctivitis, or pinkeye, is caused by a bacterial infection.

The **lacrimal** (LACK-rih-mal) **apparatus,** shown in Figure 9-4, consists of the lacrimal gland and various ducts. The **lacrimal gland** is located in the superior and lateral region of the orbit. Tears produced by the lacrimal gland flow through **lacrimal ducts** and across the surface of the eye to the medial side, where they drain into two small **lacrimal canals** (canaliculi). From the lacrimal canals, the tears flow into the **lacrimal sac,**

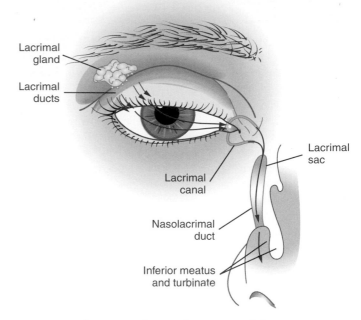

Figure 9-4 *Lacrimal apparatus of the eye.*

and then into the **nasolacrimal duct,** which opens into the nasal cavity. Tears moisten, lubricate, and cleanse the anterior surface of the eye. They also contain an enzyme (lysozyme) that helps destroy bacteria and prevent infections.

There are six muscles associated with movements of the eyeball. These muscles all originate from the bones of the orbit and insert on the tough outer layer of the eyeball. The six extrinsic eye muscles are listed in Table 9-2, along with their nerve supply and actions.

 QUICK CHECK

9.5 What is the function of the lacrimal gland?
9.6 What two muscles are associated with the eyelids?

TABLE 9-2 Muscles of the Eye

Muscle	Controlling Nerve	Function
Extrinsic (Skeletal) Muscles		
Superior rectus	Oculomotor (III)	Elevates eye or rolls it superiorly and toward midline
Inferior rectus	Oculomotor (III)	Depresses eye or rolls it inferiorly and toward midline
Medial rectus	Oculomotor (III)	Moves eye medially, toward midline
Lateral rectus	Abducens (VI)	Moves eye laterally, away from midline
Superior oblique	Trochlear (IV)	Depresses eye and turns it laterally, away from midline
Inferior oblique	Oculomotor (III)	Elevates eye and turns it laterally, away from midline
Intrinsic (Smooth) Muscles		
Ciliary	Oculomotor (III)	Causes suspensory ligament to relax; lens *becomes more convex* for close vision
Iris, circular muscles	Oculomotor (III) Parasympathetic fibers	Decreases size of pupil to allow less light to enter eye
Iris, radial muscles	Sympathetic fibers from spinal nerves	Increases size of pupil to allow more light to enter eye

FROM THE PHARMACY

Glaucoma is an eye disorder that is characterized by elevated intraocular pressure, either from excessive production of aqueous humor or from diminished drainage of the fluid. Primary medications used to treat glaucoma include beta-adrenergic blocking agents, which decrease aqueous humor production, and prostaglandins, cholinergics, and sympathomimetics, all of which increase drainage. First-line agents for glaucoma are usually beta-blockers and prostaglandins because they have fewer visual side effects than the others. **Timolol** (Timoptic) is a nonselective beta-blocker and decreases aqueous humor production. **Latanoprost** (Xalatan) is a prostaglandin that acts by absorption through the cornea, ultimately increasing the outflow of aqueous humor.

Mydriatic and cycloplegic agents are primarily used in eye examinations and the diagnosis of ophthalmic disorders. Mydriatic agents cause pupil dilation. Cycloplegic agents paralyze the ciliary muscles to prevent accommodation. **Hydroxyamphetamine and tropicamide** (Paremyd) is a combination eye drop preparation that dilates the pupil and partially paralyzes the ciliary muscle to reduce accommodation. Tropicamide is an anticholinergic agent that relaxes the ciliary muscle for accurate measurement of refractive errors, and hydroxyamphetamine is an adrenergic agent that mimics epinephrine to dilate the pupil. Together the combination produces greater pupil dilation than either one alone.

Topical antibiotics may be used to treat infections in the external auditory canal, but systemic antibiotics are indicated for inner ear infections. A wide variety of over-the-counter (OTC) preparations are available to treat impacted cerumen, bacterial infections, and other minor problems associated with the external ear. **Isopropyl alcohol in glycerin** (Swim-Ear drops) is an over-the-counter preparation used to treat swimmer's ear. OTC preparations for the ear frequently contain acidic solutions of alcohol, glycerin, or propylene glycol to help restore the normal acidic pH of the external auditory canal after swimming and bathing. The glycerin also acts as an emollient to help relieve itching and burning. Prevention is always preferable to cure. The external auditory canal's natural production of cerumen is the best barrier protection there is for the ear, and not inserting Q-tips and other objects in the ear is the best prevention for external canal problems!

Structure of the Eyeball

The eyeball, or **bulbus oculi** (BUL-bus AHK-yoo-lye), is somewhat spherical, is 2 to 3 centimeters in diameter, and has an anterior bulge. It is surrounded by orbital fat within the orbital cavity. Figure 9-5 illustrates the structure of the bulbus oculi.

The wall of the eyeball is made up of three concentric coats or tunics. The outermost layer is the **fibrous tunic.** It consists of the white, opaque **sclera** (SKLEE-rah) and the transparent **cornea.** The sclera, the white part of the eye, covers the posterior five sixths of the eyeball, and the muscles that move the eye are attached to it. The transparent cornea, which covers the anterior one sixth of the eyeball, is the "window" of the eye. It helps focus light rays entering the eye.

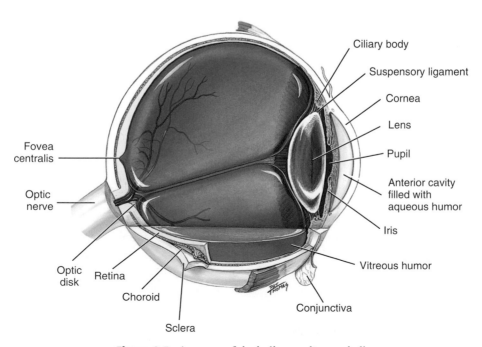

Figure 9-5 *Anatomy of the bulbus oculi or eyeball.*

QUICK APPLICATIONS

The cornea was one of the first organs transplanted. Surgical removal of deteriorating corneas and replacement with donor corneas represents a common medical procedure for several reasons. The cornea is readily accessible and relatively easy to remove. The tissue is avascular, so there is no bleeding problem or difficulty in establishing circulatory pathways. Corneas are less active than other tissues immunologically and are less likely to be rejected. Long-term success after corneal implant surgery is excellent.

The middle layer of the eyeball is the **vascular tunic.** It consists of the **choroid** (KOAR-oyd), **ciliary body,** and **iris.** The **choroid** is a highly vascular, brown pigmented layer located between the sclera and the retina in the posterior portion of the eye. It is the largest part of the middle tunic and lines most of the sclera, although it is only loosely connected to the fibrous coat and can be stripped away easily. The choroid is, however, firmly attached to the retina. The brown pigment in the choroid absorbs excess light rays that might interfere with vision. The blood vessels nourish the interior of the eye. Anteriorly, the choroid is continuous with the **ciliary body.** Numerous fingerlike **ciliary processes** within the ciliary body secrete aqueous humor, a fluid in the anterior portion of the eye. The ciliary body also contains the **ciliary muscle. Suspensory ligaments** connect the ciliary body to the transparent, biconvex **lens** of the eye. When the ciliary muscle contracts, the suspensory ligaments relax, and the lens bulges to allow focusing for close vision. The **iris** is the conspicuous, colored portion of the eye. It is a doughnut-shaped diaphragm with a central aperture, called the **pupil.** The iris contains two groups of smooth muscles: a radial group and a circular group. When the radial muscles contract, the pupil dilates; when the circular group contracts, the pupil gets smaller. These muscles of the iris continually contract and relax to change the size of the pupil, which regulates the amount of light entering the eye. Pupillary reflexes may also reflect interest or emotional state. For example, the pupils dilate when the subject is interesting or appealing. If the subject is boring or repulsive, the pupils constrict.

The innermost coat of the eyeball is the **nervous tunic,** or **retina** (RET-ih-nah), which is found only in the posterior portion of the eye. It ends at the posterior margin of the ciliary body. The retina contains several layers, which are illustrated by the diagram in Figure 9-6 and the photomicrograph in Figure 9-7. The outer layer is deeply pigmented and firmly attached to the choroid. The layer next to the pigmented layer contains the **rods and cones,** which are the receptor (photoreceptor) cells. Other layers consist of bipolar neurons and ganglion cells. The axons of the ganglion cells converge to form the **optic nerve,** which penetrates the tunics at the **optic disk** and passes through the apex of the orbital cavity to reach the brain. Because there are no receptor cells in the optic disk, it is commonly referred to as the "blind spot" of the eye. Just lateral to the optic disk, near the center of the retina, there is a yellow spot called the **macula lutea** (MACK-yoo-lah LOO-tee-ah). The region of the retina that produces the sharpest image is a depression, the **fovea centralis** (FOE-vee-ah sen-TRAL-is), in the center of the macula lutea.

QUICK APPLICATIONS

Sometimes the sensory portion of the retina breaks away from the pigmented layer, resulting in a **detached retina.** This may be due to trauma, such as a blow to the head, or to certain eye disorders. If allowed to progress, the result is distorted vision and, eventually, blindness. In many cases, the retina can be reattached by laser surgery.

The lens, suspensory ligaments, and ciliary body form a partition that divides the interior of the eyeball into two cavities. The space anterior to the lens, between the cornea and the lens,

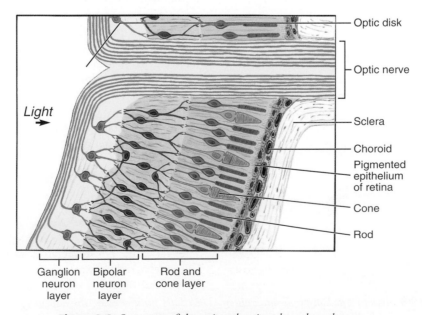

Figure 9-6 *Structure of the retina showing the rods and cones.*

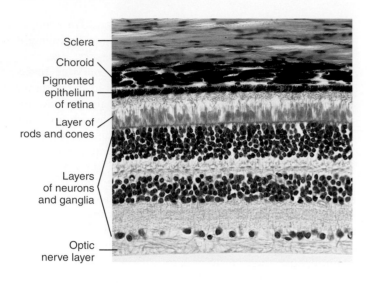

Sclera
Choroid
Pigmented epithelium of retina
Layer of rods and cones
Layers of neurons and ganglia
Optic nerve layer

Figure 9-7 *Photomicrograph of the retina. (Adapted from Gartner and Hiatt. Color Textbook of Histology, ed. 3. Philadelphia, Elsevier/ Saunders, 2007.)*

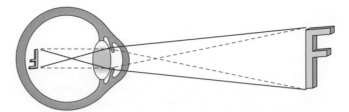

Figure 9-8 *Formation of images on the retina. The image on the retina is upside down and backward.*

is the anterior cavity and is filled with **aqueous humor** secreted by the ciliary body. Aqueous humor helps maintain the shape of the anterior part of the eye and nourishes the structures in that region. It is largely responsible for the internal pressure of the eye. The aqueous humor circulates through the anterior cavity and then is reabsorbed into blood vessels at the junction of the sclera and cornea. The posterior cavity, between the lens and the retina, is filled with a colorless, transparent, gel-like **vitreous humor.** The vitreous humor presses the retina firmly against the wall of the eye, supports the internal parts of the eye, and helps maintain its shape.

> ### QUICK CHECK
> **9.7** Why is it important that images of objects being examined for fine detail fall off the fovea centralis?
> **9.8** Why is the optic disk called the "blind spot"?
> **9.9** What is the opening in the center of the iris?

Pathway of Light and Refraction

Vision depends on light rays. When a person sees an object, light rays from the object enter the eye. Light rays have two important properties—they travel in a straight line and they can be bent. When light rays travel from one substance to another that has a different optical density, the rays bend. The bending of light rays is called **refraction.** When light rays hit a concave surface, they scatter or diverge. When the rays meet a convex surface, they get closer together or converge. There are four refractive surfaces and media in the eyes. In a normal eye, the cornea, aqueous humor, lens, and vitreous humor bend the light rays so that they focus on the retina. The image that forms on the retina is upside down and backward (Figure 9-8), but somehow the brain turns it around and interprets the image in the correct position.

When an object is at least 20 feet away, the normal relaxed eye is able to focus the image on the retina. When the object is closer than 20 feet, the eye must make adjustments to focus the image. These adjustments are called **accommodation.** The primary action in accommodation is changing the shape of the lens. For distance vision, as illustrated in Figure 9-9, the ciliary muscle is relaxed, the suspensory ligaments are taut, and the lens is flat. When the eyes accommodate for close vision, as illustrated in Figure 9-10, the ciliary muscle contracts, the suspensory ligaments become loose or relaxed, and the lens bulges or becomes more convex. The closer the object, the more the light rays have to bend to focus and the greater is the curvature of the lens.

Photoreceptors

The retina (refer to Figure 9-6 and Figure 9-7) contains two kinds of photoreceptor cells: rods and cones. These cells are next to the outermost retina layer of pigmented epithelium. **Rods,** thin cells with slender, rodlike projections, are sensitive to dim light. Even though rods are much more numerous than cones, they are absent in the fovea centralis and their number increases proportionately to the distance away from the fovea centralis. Many rods synapse with a single sensory fiber (convergence); thus vision with rods lacks fine detail. **Cones,** the receptors for color vision and visual acuity, are located primarily in the fovea centralis. They are thicker cells with short, blunt projections. Cones exhibit less convergence than rods, so in addition to color, cones provide sharp images and fine detail. Table 9-3 compares the rods and cones.

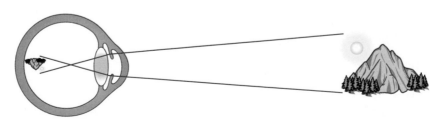

Figure 9-9 *Accommodation for distance vision. The ciliary muscles relax, the suspensory ligaments are taut, and the lens is flat.*

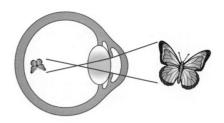

Figure 9-10 *Accommodation for close vision. The ciliary muscles contract, the suspensory ligaments relax, and the lens becomes more convex.*

Rods contain a substance, **rhodopsin** (roe-DOP-sin) (visual purple), that is very light sensitive. When even small amounts of light focus on the rods, rhodopsin breaks down into its component parts—opsin (a protein) and a derivative of vitamin A, retinal (retinene). This reaction triggers a nerve impulse. Rhodopsin is resynthesized from opsin and retinal to prepare the rods for receiving subsequent stimuli. The more rhodopsin there is in the rods, the greater is the sensitivity to light. In bright light, nearly all the rhodopsin in the rods is decomposed. After entering a dimly lit area, it takes some time for the eyes to adapt to the dim light. During this period, rhodopsin is regenerated in the rods so that they become more sensitive.

Cones function in a manner that is similar to rods. Light-sensitive pigments break down into component parts, and the reaction triggers nerve impulses. There are three different types of cones, each with a different visual pigment. All the pigments contain retinal, but the protein portion is different. One type responds best to green light, another responds best to blue light, and a third type responds best to red light. The perceived color of an object depends on the quantity and combination of cones that are stimulated. If all the pigments are stimulated, the person senses white. If none are stimulated, the person senses black.

QUICK APPLICATIONS

Color blindness occurs because there is an absence or a deficiency of one or more of the visual pigments in the cones, and the person cannot distinguish certain colors. In the most common form, red-green color blindness, the cones lack the red pigment and the person is unable to distinguish red from green. Most color blindness is inherited and occurs more frequently in males.

Visual Pathway

Visual impulses generated in the rods and cones of the retina leave the eyes in the axons that form the **optic nerves.** Just anterior to the pituitary gland, these nerves form an X-shaped structure, the **optic chiasma** (OP-tik kye-AZ-mah). Within the optic chiasma, the axons from the medial portion of each retina cross over to enter the **optic tract** on the opposite side (Figure 9-11). The right optic tract contains the fibers from the lateral portion of the right eye and the medial portion of the left

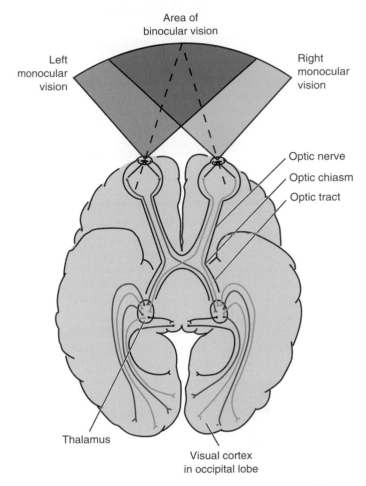

Figure 9-11 *Visual pathway. The optic nerves converge at the optic chiasma, where some axons cross to the opposite side. Impulses then travel to the thalamus and are interpreted in the visual cortex of the occipital lobe.*

TABLE 9-3 Comparison of Rods and Cones

Feature	Rods	Cones
Shape	Long, slender projections	Short, thick projections
Location	None in fovea centralis; increase in density away from fovea centralis	Concentrated in fovea centralis; decrease in density away from fovea centralis
Quantity	More numerous than cones	Less numerous than rods
Convergence	High degree of convergence	Less convergence
Pigments	Single pigment—rhodopsin	Three pigments—one each for red, green, blue
Functions	Black-and-white vision; dim light; night vision; lacks detail	Color vision; bright light; precise vision with fine detail

eye. The left optic tract contains the fibers from the lateral portion of the left eye and the medial portion of the right eye. The optic tracts lead to the **thalamus.** Just before they reach the thalamus, a few fibers leave the tracts and enter the nuclei that function in visual reflexes (superior colliculi). Most of the fibers enter the thalamus, where they synapse with neurons that carry the impulses in the **optic radiations** to the visual cortex of the occipital lobes. Because some of the fibers cross over to the other side in the optic chiasma, each occipital lobe receives an image of the entire object from each eye but from slightly different perspectives. This enables vision in three dimensions.

QUICK CHECK

9.10 **What structure in the eye accommodates for close vision by adjusting the amount of refraction?**
9.11 **What is the light-sensitive pigment in the rods?**
9.12 **Where are visual impulses interpreted?**

Auditory Sense

The **ear** is the organ of hearing (auditory or acoustic organ). It is also the organ for the sense of equilibrium, which is covered later in this chapter. The receptors for hearing, located within the ear, are **mechanoreceptors** (mek-ah-noh-ree-SEP-tors). Physical forces, in the form of sound vibrations, are responsible for initiating impulses that are interpreted as sound.

Structure of the Ear

The "ears" on the sides of the head are only a portion of the actual organ of hearing. A large part of the organ, actually the most important part, lies hidden from view in the temporal bone. Anatomically, the organ of hearing is divided into the external ear, middle ear, and inner ear. The anatomy of the ear is illustrated in Figure 9-12.

External Ear

The **external ear** consists of an auricle, or pinna, and the external auditory meatus (canal). The **auricle,** or pinna, is the fleshy part of the external ear that is visible on the side of the head and surrounds the opening into the external auditory meatus. The auricle collects sound waves and directs them toward the auditory meatus.

The **external auditory meatus** is an S-shaped tube, about 2.5 centimeters long, that extends from the auricle to the **tympanic membrane.** The skin that lines the meatus has numerous hairs and **ceruminous glands,** which secrete a waxy substance called **cerumen.** The hairs and cerumen help prevent foreign objects from reaching the eardrum. The external ear ends at the tympanic membrane.

QUICK APPLICATIONS

Surprisingly, a branch of the facial nerve passes along the inner surface of the tympanic membrane. This nerve has nothing to do with hearing but rather carries taste impulses from the tongue. The nerve may be damaged by a middle ear infection or during ear surgery.

Middle Ear

The **middle ear** is an air-filled cavity, called the **tympanic cavity,** in the temporal bone. It begins at the tympanic membrane, contains the auditory ossicles, and has an opening into the eustachian tube. The **oval window** and the **round window** in the medial wall of the middle ear connect the middle ear with the inner ear. The oval window is closed by the stapes, one of the bones in the middle ear. The round window is closed by a membrane.

The **tympanic membrane,** or eardrum, is a thin membrane that separates the external ear from the middle ear. Sound waves cause the tympanic membrane to vibrate.

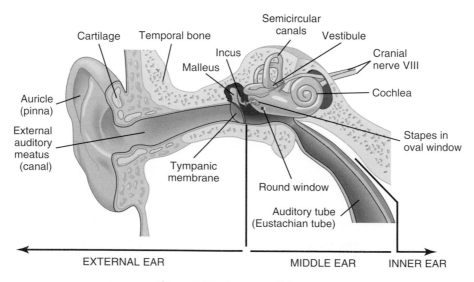

Figure 9-12 *Anatomy of the ear.*

QUICK APPLICATIONS

The eardrum is sometimes ruptured, or perforated, by shock waves from an explosion, scuba diving, trauma, or acute middle ear infections. A perforated eardrum is characterized first by acute pain, then by noise in the affected ear, and then by hearing impairment.

An **auditory tube** (eustachian tube) connects each middle ear with the throat. Its purpose is to equalize the pressure between the outside air and the middle ear cavity, a condition necessary for normal hearing. Throat infections may spread to the middle ear through the auditory tube.

QUICK APPLICATIONS

Otitis media, infection of the middle ear, occurs frequently in children. The infection usually starts in the throat and then spreads through the auditory tube into the middle ear.

The **auditory ossicles** are three tiny bones: the **malleus** (hammer), **incus** (anvil), and **stapes** (stirrup). These bones are linked together by tiny ligaments and form a bridge across the space of the tympanic cavity. The malleus is attached to the tympanic membrane, and the stapes is attached to the oval window between the middle ear and the inner ear. The incus is between the malleus and stapes. When the tympanic membrane vibrates, the ossicles transmit the vibrations across the cavity to the oval window, which transfers the motion to the fluids in the inner ear. This fluid motion excites the receptors for hearing.

Inner Ear

The **inner ear** consists of a bony labyrinth and a membranous labyrinth (Figure 9-13). The **bony** (osseous) **labyrinth** is a series of interconnecting chambers in the temporal bone. The **membranous labyrinth,** located inside the bony labyrinth, is a system of membranous tubes that are similar to the bony

labyrinth in shape but smaller. A clear fluid called **endolymph** fills the membranous labyrinth. The space between the bony and the membranous labyrinths contains a fluid called **perilymph.** The inner ear is divided into the vestibule, semicircular canals, and cochlea. The vestibule and semicircular canals function in the sense of equilibrium. The cochlea functions in the sense of hearing.

The **cochlea** (KOK-lee-ah) is the coiled portion of the bony labyrinth. The **cochlear duct** is the membranous labyrinth inside the bony labyrinth. The drawing in Figure 9-14 and the photomicrograph in Figure 9-15 show that the inside of the cochlea is divided into three regions by the **vestibular membrane** and the **basilar membrane.** The region above the vestibular membrane is the **scala vestibuli** (SKAY-lah ves-TIB-yoo-lee), which contains perilymph. It extends from the oval window to the apex of the cochlea. The region below the basilar membrane is the **scala tympani** (SKAY-lah TIM-pah-nee), which extends from the apex of the cochlea to the round window. It is continuous with the scala vestibuli at the apex and also contains perilymph. The basilar membrane contains thousands of stiff fibers that gradually increase in length from the base to the apex. The arrangement is similar to the reeds in a harmonica or the bars on a marimba. The fibers vibrate when activated by sound. The middle region of the cochlea, between the scala vestibuli and the scala tympani, is the cochlear duct, which ends as a closed sac at the apex of the cochlea. The cochlear duct contains endolymph.

The **organ of Corti** (KOAR-tee), which contains the receptors for sound, is located on the upper surface of the basilar membrane, within the cochlear duct (Figure 9-16). It consists of **supporting cells** and **hair cells.** The hair cells are specialized sensory cells that have hairlike projections (microvilli)

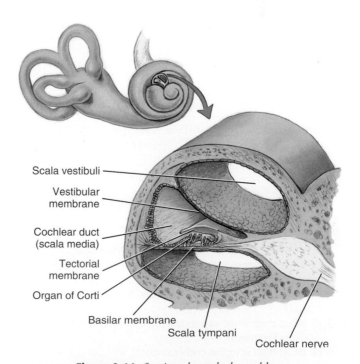

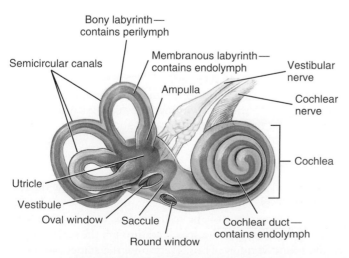

Figure 9-13 *Labyrinths of the inner ear.*

Figure 9-14 *Section through the cochlea.*

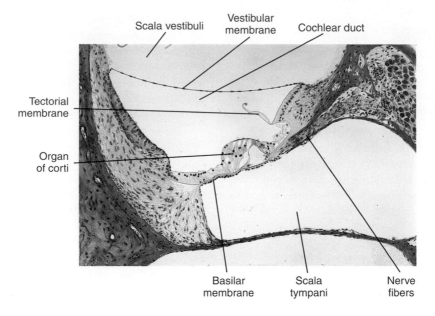

Figure 9-15 *Photomicrograph of the cochlea. The organ of Corti rests on the basilar membrane. The cochlear duct contains endolymph. The scala vestibule and scala tympani contain perilymph. (Adapted from Gartner and Hiatt. Color Textbook of Histology, ed. 3. Philadelphia, Elsevier/Saunders, 2007.)*

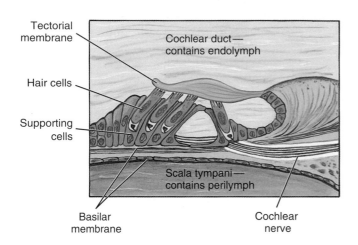

Figure 9-16 *Organ of Corti enlarged to show the hair cells and tectorial membrane.*

extending from their free surface. The tips of the projections contact a gelatinous **tectorial** (tek-TOE-ree-al) **membrane** that extends over them. Hair cells have no axons, but they are surrounded by sensory nerve fibers that form the **cochlear branch** of the **vestibulocochlear nerve** (cranial nerve VIII).

QUICK CHECK

9.13 What structure is between the external ear and the middle ear?

9.14 What structure extends from the throat to the middle ear?

9.15 What membrane separates the cochlear duct from the scala tympani?

9.16 What fluid is in the cochlear duct?

Physiology of Hearing

Sound travels through the atmosphere in waves of alternating compressions and decompressions of molecules. Low-pitched tones create low-frequency sound waves; high-pitched tones create high-frequency sound waves. Although the human ear can detect sound waves with frequencies ranging from 20 to 20,000 vibrations per second, hearing is most acute with frequencies between 2000 and 3000 vibrations per second.

Initiation of Impulses

The process of hearing begins when sound waves enter the external auditory meatus. As the waves travel through the external ear, they hit the tympanic membrane and cause it to vibrate. Because the malleus is attached to the membrane, the vibrations are transferred from the tympanic membrane to the malleus, then to the incus, and then to the stapes, which creates vibrations in the membrane of the oval window. Movement of the oval window passes the vibrations to the perilymph in the inner ear.

QUICK APPLICATIONS
Otosclerosis is an ear disorder in which spongy bone grows around the oval window and fuses with the stapes. This immobilizes the stapes and results in a form of conduction deafness.

Figure 9-17 illustrates an uncoiled cochlea to show the relationships of its components. The vibrations in the perilymph, caused by movement of the oval window, travel through the scala vestibuli, around the apex, and through the scala tympani to the round window. Thus, every time the stapes pushes the oval window inward toward the inner ear, movement in the

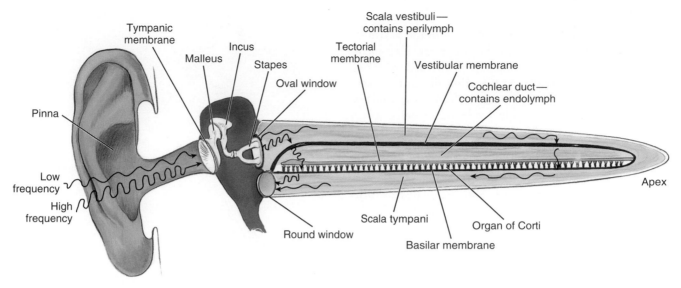

Figure 9-17 *Uncoiled cochlea showing the pathway of pressure waves.*

perilymph pushes the round window outward toward the middle ear to relieve the pressure in the fluid. As vibrations travel through the perilymph, they create corresponding oscillations in the vestibular membrane, which causes movement of the endolymph within the scala media (cochlear duct), and finally the vibrations are transferred to the basilar membrane.

When the basilar membrane moves up and down, the organ of Corti moves with it and the hairs on the hair cells rub against the tectorial membrane. As the hairs contact the membrane, they bend, and this mechanical deformation initiates the nerve impulses that result in hearing. The following list summarizes the sequence of events in the initiation of auditory impulses.

- The tympanic membrane vibrates in response to sound waves.
- The malleus, incus, and stapes transfer vibrations to the oval window membrane.
- Movement of the oval window membrane starts oscillations in the perilymph in the cochlea.
- Oscillations in the perilymph cause vibrations in the vestibular and basilar membranes.
- When the basilar membrane moves, the hairs on the hair cells in the organ of Corti rub against the tectorial membrane and bend.
- Bending of the hairs on the hair cells stimulates the formation of impulses.
- Impulses are transmitted to the auditory cortex of the temporal lobe on the cochlear branch of cranial nerve VIII, the vestibulocochlear nerve.

Pitch and Loudness

Hair cells along the length of the organ of Corti have varying sensitivities to different frequencies. Also, different regions of the basilar membrane vibrate in response to different frequencies. The portion of the membrane near the base of the cochlea vibrates in response to high frequencies. Near the apex of the cochlea, the basilar membrane responds to low frequencies. Pitch

is detected by the portion of the basilar membrane that vibrates in response to the sound and the sensitivity of the hair cells.

Loudness is determined by the intensity of the sound waves. The basilar membrane vibrates more (i.e., has a greater magnitude of oscillation) in response to loud sounds. This means that more hair cells are stimulated and more impulses travel to the auditory cortex.

Auditory Pathway

Auditory impulses leave the cochlea on the **cochlear nerve,** a branch of cranial nerve VIII, the **vestibulocochlear nerve.** The fibers synapse in the **medulla,** where some cross over to the opposite side, and then continue to the inferior colliculi of the **midbrain** for auditory reflexes. Next, the impulses are transmitted to the **thalamus,** where they synapse with neurons that transmit the impulses to the auditory cortex of the **temporal lobe.**

QUICK CHECK

9.17 What initiates oscillations of the perilymph in the cochlea?

9.18 What triggers an impulse on the hair cell of the organ of Corti?

9.19 How is the pitch of a sound detected?

Sense of Equilibrium

The sense of equilibrium is a combination of two different senses: the sense of **static equilibrium** and the sense of **dynamic equilibrium.** Static equilibrium is involved in evaluating the position of the head relative to gravity. It occurs when the head is motionless or moving in a straight line. Dynamic equilibrium occurs when the head is moving in a rotational or angular direction.

Static Equilibrium

The organs of static equilibrium are located in the **vestibule** portion of the bony labyrinth of the inner ear. The membranous labyrinth inside the vestibule is divided into two saclike structures, the **utricle** (YOO-trih-kull) and the **saccule** (SACK-yool) (see Figure 9-13). Each of these contains a small structure called a **macula** (MACK-yoo-lah), which is the organ of static equilibrium. The macula consists of sensory hair cells, similar to those in the organ of Corti, and supporting cells (Figure 9-18). The projections, or hairs, of the hair cells are embedded in a gelatinous mass that covers the macula. Grains of calcium carbonate, called **otoliths** (OH-toe-liths), are embedded on the surface of the gelatinous mass.

When the head is in an upright position, the hairs are straight. When the head tilts or bends forward, the otoliths and the gelatinous mass move in response to gravity. As the gelatinous mass moves, it bends some of the hairs on the receptor cells. This action initiates an impulse that travels to the CNS by way of the vestibular branch of the vestibulocochlear nerve. The CNS interprets the information and sends motor impulses out to appropriate muscles to maintain balance.

Dynamic Equilibrium

The sense organs for dynamic equilibrium, the equilibrium of rotational or angular movements, are located in the membranous labyrinth of the **semicircular canals.** There are three semicircular canals, positioned at right angles to each other, in three different planes (see Figure 9-13). Each membranous canal is surrounded by perilymph and contains endolymph. At the base of each canal, near where it attaches to the utricle, there is a swelling called the **ampulla.** The sensory organs of the semicircular canals are located within the ampullae. Each of these organs, called a **crista ampullaris**

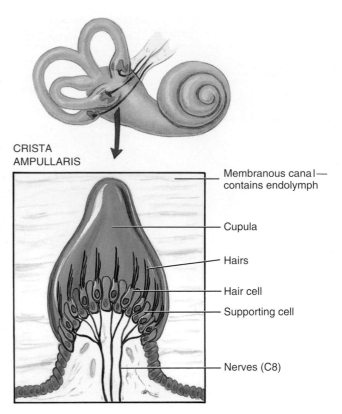

CRISTA AMPULLARIS

— Membranous canal— contains endolymph

— Cupula

— Hairs

— Hair cell

— Supporting cell

— Nerves (C8)

Figure 9-19 *Structure of the crista ampullaris in the semicircular canals. The crista ampullaris is the organ for dynamic equilibrium.*

(KRIS-tah amp-yoo-LAIR-is), contains sensory hair cells and supporting cells (Figure 9-19). The crista ampullaris is covered by a dome-shaped gelatinous mass called the **cupula** (KEW-pew-lah). The hairs of the hair cells are embedded in the cupula.

When the head turns rapidly, the semicircular canals move with the head but the endolymph tends to remain stationary. The fluid pushes against the cupula, and it tilts to one side. As the cupula tilts, it bends some of the hairs on the hair cells, which triggers a sensory impulse. Because the three canals are in different planes, their cristae are stimulated differently by the same motion. This creates a mosaic of impulses that are transmitted to the CNS on the **vestibular branch** of the **vestibulocochlear nerve.** The CNS interprets the information and initiates appropriate responses to maintain balance. The cerebellum is particularly important in mediating the sense of balance and equilibrium.

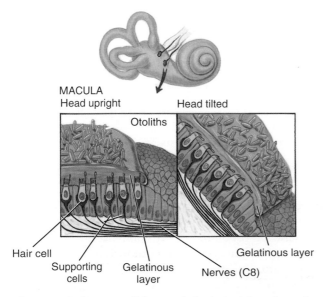

MACULA
Head upright

Head tilted

Otoliths

Hair cell

Supporting cells

Gelatinous layer

Nerves (C8)

Gelatinous layer

Figure 9-18 *Structure of the macula in the utricle and saccule. The macula is the organ of static equilibrium.*

> **QUICK CHECK**
>
> 9.20 In what portion of the bony labyrinth are the organs of static equilibrium located?
> 9.21 What are the sense organs for dynamic equilibrium and where are they located?
> 9.22 What branch of cranial nerve VIII transmits impulses for equilibrium?

FOCUS ON AGING

There is a general decline in all of the special senses as the body ages. The most significant changes in the eye occur in the lens. It tends to become thicker and less elastic, which makes it less able to change shape to accommodate for near vision. This condition, called **presbyopia** or farsightedness of aging, probably is the most common age-related dysfunction of the eye. The lens also tends to become cloudy or opaque, forming cataracts. About 90% of people older than age 70 have some degree of cataract formation; however, it is not always significant enough to affect vision. The cornea tends to become more translucent and less spherical, which contributes to an increase in astigmatism in older people. Older people require more light to see well because atrophy of the muscles in the iris reduces the ability of the pupil to dilate and decreases the amount of light that reaches the retina. The chemical processes that rebuild the visual pigment, rhodopsin, are slower in older people, so dark adaptation takes longer and is not as complete as in young people.

Most age-related changes in the external ear and middle ear have little effect on hearing. A buildup of cerumen, or earwax, in the external ear may contribute to hearing loss in the low-frequency range. The joints between the auditory ossicles in the middle ear may become less movable, which interferes with the transmission of sound waves to the inner ear, but generally it is not clinically significant. Most of the gradual loss of hearing that usually begins by the age of 40 is due to degeneration of the receptor cells in the spiral organ of Corti in the inner ear. Another factor is the decrease in the number of nerve fibers in the vestibulocochlear nerve. The reduction in fibers in the cochlear branch contributes to hearing loss. A decrease in vestibular fibers affects balance and equilibrium.

Taste and smell, both chemical senses, show a decline with age; however, the mechanism is unclear. Diminished perception may be due to degeneration of the receptor cells, to changes in the way the impulses are processed in the brain, or to other factors. It is likely that decreases in sensory perception are due to a combination of several factors.

Representative Disorders of
the Special Senses

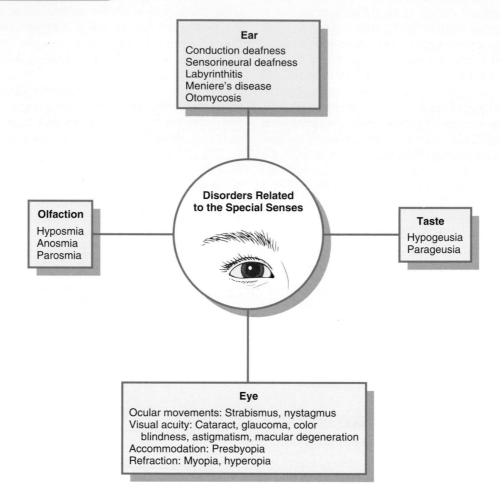

Ear
Conduction deafness
Sensorineural deafness
Labyrinthitis
Meniere's disease
Otomycosis

Olfaction
Hyposmia
Anosmia
Parosmia

**Disorders Related
to the Special Senses**

Taste
Hypogeusia
Parageusia

Eye
Ocular movements: Strabismus, nystagmus
Visual acuity: Cataract, glaucoma, color
   blindness, astigmatism, macular degeneration
Accommodation: Presbyopia
Refraction: Myopia, hyperopia

Anosmia (an-AHZ-mee-ah) Lack of sense of smell

Astigmatism (ah-STIG-mah-tizm) Defective curvature of the cornea or lens of the eye resulting in a distorted image on the retina

Cataract (KAT-ah-rakt) Lens of the eye or its capsule becomes opaque, resulting in blurry or dim vision

Color blindness (KULL-uhr BLIND-ness) Inability to distinguish between certain colors; complete inability to see colors is rare; usually the term denotes a deficiency in color vision; most common form is red-green color deficiency

Conduction deafness (kon-DUK-shun DEFF-nis) Hearing loss associated with impaired transmission of sound waves from the external ear to the fluids in the cochlea

Glaucoma (glaw-KOH-mah) Disease of the eye characterized by increased intraocular pressure from an accumulation of aqueous humor due either to an increased production or to decreased drainage; the increased pressure causes pathological changes in the optic disk and typical defects in the field of vision; if untreated, glaucoma leads to blindness

Hyperopia (hye-per-OH-pee-ah) Defect in vision in which light rays focus beyond the retina; farsightedness

Hypogeusia (hye-poh-GOO-zee-ah) Diminished sense of taste

Hyposmia (hye-PAHZ-mee-ah) Diminished sense of smell

Labyrinthitis (lab-ih-rin-THIGH-tis) Inflammation of the labyrinth of the inner ear resulting in dizziness and loss of balance; otitis interna

Macular degeneration (MACK-yoo-lahr dee-jen-er-A-shun) Gradually progressive condition that results in the loss of central vision because of the breakdown of cells in the macula lutea; frequently affects older people

Meniere's disease (men-ih-ARZ dih-ZEEZ) A chronic disease of the inner ear characterized by recurring attacks of dizziness, tinnitus, and fluctuating hearing loss; attacks vary in duration and frequency

Myopia (mye-OH-pee-ah) Defect in vision in which light rays focus in front of the retina; nearsightedness

Nystagmus (nihs-TAG-muss) Involuntary, rapid, rhythmic movements of the eyeball

Otomycosis (oh-toh-my-KOH-sis) Fungal infection of the external ear; also called swimmer's ear

Parageusia (par-ah-GOO-zee-ah) An abnormal or perverted sense of taste

Parosmia (par-AHZ-mee-ah) Abnormal or perverted sense of smell

Presbyopia (prez-bih-OH-pee-ah) Impairment of vision caused by aging

Sensorineural deafness (sen-soh-ree NEW-rahl DEFF-nis) Hearing loss as a result of damage along any part of the auditory pathway

from the receptor cells in the cochlea to the auditory cortex of the cerebrum; noise-related damage to the receptor cells in the cochlear is this type of deafness

Strabismus (strah-BIHZ-muss) Deviation of one eye from the other when looking at an object; caused by a weakened or hypertonic muscle in one of the eyes; squint and cross-eyes are examples

CHAPTER SUMMARY

Receptors and Sensations

■ **Distinguish between general senses and special senses.**
- Receptors for the general senses are widely distributed in the body.
- Receptors for the special senses are localized.

■ **Classify sense receptors into five groups.**
- Receptors are classified as chemoreceptors, mechanoreceptors, nociceptors, thermoreceptors, or photoreceptors.

■ **Explain what is meant by sensory adaptation.**
- Sensory adaption occurs when a continued stimulus decreases the sensitivity of the receptors.

General Senses

■ **Describe the sense receptors for touch, pressure, proprioception, temperature, and pain.**
- The receptors for touch and pressure are mechanoreceptors, which include free nerve endings, Meissner's corpuscles, and Pacinian corpuscles.
- Proprioception is the sense of position or orientation. The receptors are mechanoreceptors including Golgi tendon organs and muscle spindles.
- Temperature changes are detected by thermoreceptors, which are free nerve endings. Some are sensitive to heat, and others are sensitive to cold; temperature extremes also stimulate pain receptors.
- Receptors for pain are called nociceptors, which are free nerve endings that are stimulated by tissue damage.

Gustatory Sense

■ **Locate the four different taste sensations, and follow the impulse pathway from stimulus to the cerebral cortex.**
- Receptors for taste are chemoreceptors located in the taste buds.
- Taste buds, the organs for taste, are located on the walls of the papillae that are on the surface of the tongue.
- Taste buds contain the taste cells and supporting cells. Taste hairs on the taste cells function as the receptors.
- Sweet, salty, sour, and bitter are the four taste sensations. Sweet is located at the tip of the tongue, salty is on the anterior sides of the tongue, sour is on the posterior sides of the tongue, and bitter is at the back of the tongue.
- Impulses for taste are transmitted along the facial nerve or the glossopharyngeal nerve to the sensory cortex of the parietal lobe.

Olfactory Sense

■ **Locate the sense receptors for smell, and trace the impulse pathway to the cerebral cortex.**
- The sense of smell is called *olfaction.*
- The receptors for the sense of smell are chemoreceptors and are located in the olfactory epithelium of the nasal cavity.
- The olfactory neurons enter the olfactory bulb. Impulses are transmitted along the olfactory tracts to the olfactory cortex in the temporal lobe.
- The senses of taste and smell are closely related and complement each other.

Visual Sense

■ **Describe the protective features and accessory structures of the eye.**
- Protective features of the eye include the bony orbit; the eyebrows, eyelids, and eyelashes, which help to protect the eye from foreign particles and irritants; and the lacrimal apparatus, which produces tears that moisten and cleanse the eye. Tears contain an enzyme that helps destroy bacteria.

■ **Describe the structure of the eye and the significance of each component.**
- The sclera and cornea are parts of the outermost layer, of fibrous tunic, of the eye. They give shape to the eye, and the cornea refracts light rays.
- The middle, or vascular, tunic includes the choroid, ciliary body, and iris. The choroid absorbs excess light rays; the ciliary body changes the shape of the lens; and the iris regulates the size of the pupil.
- The retina is the innermost layer, or nervous tunic, of the eye. It contains the receptor cells.
- The lens, suspensory ligaments, and ciliary body form a partition that divides the interior of the eye into two cavities.
- The anterior cavity is filled with aqueous humor and the posterior cavity is filled with vitreous humor. Both the aqueous and vitreous humors refract light rays.

■ **Explain how light focuses on the retina.**
- Refraction is the bending of light rays as they travel between substances of differing optical densities. The refractive media in the eye are the cornea, aqueous humor, lens, and vitreous humor.

- In the normal relaxed eye, the four refractive media sufficiently bend the light rays from objects at least 20 feet away to focus on the retina.
- When the eyes accommodate for close vision, the ciliary muscle contracts, the suspensory ligaments become less taut, and the lens becomes more convex so light rays have sufficient refraction to focus on the retina.

■ **Identify the photoreceptor cells in the retina, and describe the mechanism by which nerve impulses are triggered in response to light.**

- Rods are the photoreceptors for black-and-white vision and for vision in dim light. Cones are the receptors for color vision and visual acuity. The rods and cones are located in the retina.
- Rods contain rhodopsin, which breaks down into opsin and retinal when it is exposed to light. This reaction triggers a nerve impulse.
- Cones, concentrated in the fovea centralis, function in a manner similar to rods. Color recognition is possible because one type of cone is sensitive to green light, one is sensitive to blue light, and the third responds to red light.

■ **Trace the pathway of impulses from the retina to the visual cortex.**

- Visual impulses triggered by the rods and cones travel on the optic nerve to the optic chiasma. From the optic chiasma, the impulses travel on the optic tracts to the thalamus. From there, the impulses travel on optic radiations to the visual cortex in the occipital lobe.

Auditory Sense

■ **Describe the structure of the ear and the contribution each region makes to the sense of hearing.**

- The outer, or external, ear, which includes the auricle and external auditory meatus, ends at the tympanic membrane. The auricle collects the sound waves and directs them toward the external auditory meatus, which serves as a passageway to the tympanic membrane.
- The middle ear contains three tiny bones called auditory ossicles: the malleus, incus, and stapes. The auditory ossicles transmit sound vibrations from the tympanic membrane to the oral window.
- The inner ear consists of a bony labyrinth that surrounds a membranous labyrinth. It includes the vestibule, semicircular canals, and cochlea. The cochlea is the part that functions in hearing. The membranous labyrinth of the cochlea is the cochlear duct, which contains endolymph. The organ of Corti, located on the basilar membrane in the cochlear duct, contains the receptors for sound.

■ **Summarize the sequence of events in the initiation of auditory impulses, and trace these impulses to the auditory cortex.**

- Sound waves cause vibration of the tympanic membrane. Auditory ossicles transmit the vibrations through the middle ear to the oval window.
- Movement of the oval window passes the vibrations to the perilymph in the scala vestibuli and scala tympani in the inner ear. This creates corresponding oscillations in the vestibular and basilar membranes of the cochlear duct.
- As the basilar membrane moves up and down, the hairs on the hair cells of the organ of Corti rub against the tectorial membrane. Mechanical deformation of the hairs triggers the nerve impulses.
- The interpretation of pitch is mediated by the portion of the basilar membrane that vibrates, and loudness is interpreted by the number of hair cells that are stimulated.
- Cranial nerve VIII transmits auditory impulses to the medulla oblongata. From there, the impulses travel to the thalamus, and then to the auditory cortex of the temporal lobe.

Sense of Equilibrium

■ **Identify and describe the structure of the components of the ear involved in static equilibrium and those involved in dynamic equilibrium.**

- Static equilibrium occurs when the head is motionless. It is involved in evaluating the position of the head relative to gravity. The organ of static equilibrium is the macula, located within the utricle and saccule, which are portions of the membranous labyrinth inside the vestibule.
- Dynamic equilibrium is the equilibrium of motion and occurs when the head is moving. The receptors for dynamic equilibrium are located in the crista ampullaris within the ampullae at the base of the semicircular canals.

■ **Summarize the events in the initiation of impulses for static equilibrium and for dynamic equilibrium, and identify the cranial nerve that transmits these impulses to the cerebral cortex.**

- For both static and dynamic equilibrium, as the head moves, hairs on the receptor cells bend and trigger an impulse, which is transmitted to the central nervous system on cranial nerve VIII.

CHAPTER QUIZ

Recall

Match the definitions on the left with the appropriate term on the right.

_____ 1. Sense of position or orientation
_____ 2. Free nerve endings for sense of pain
_____ 3. Gland that produces tears
_____ 4. Contains intrinsic eye muscles
_____ 5. Bending of light rays
_____ 6. Light-sensitive pigment
_____ 7. Ossicle adjacent to oval window
_____ 8. Contains auditory receptors
_____ 9. Sense organ for dynamic equilibrium
_____ 10. Fluid within membranous labyrinth

A. Cochlear duct
B. Crista ampullaris
C. Endolymph
D. Iris
E. Lacrimal
F. Nociceptor
G. Proprioception
H. Refraction
I. Rhodopsin
J. Stapes

Thought

1. Which of the following is *not* a general sense?
   A. Taste
   B. Touch
   C. Pressure
   D. Pain

2. Damage to which cranial nerve will interfere with the sense of taste from the tip of the tongue?
   A. Gustatory
   B. Facial
   C. Glossopharyngeal
   D. Hypoglossal

3. In which lobe of the brain is the sense of smell interpreted?
   A. Frontal
   B. Parietal
   C. Occipital
   D. Temporal

4. Dr. Ike Ular, an ophthalmologist, suspects that his patient might have glaucoma. One of the tests he performed was to measure the pressure in the anterior cavity of the eye. The fluid in this cavity is:
   A. Endolymph
   B. Vitreous humor
   C. Aqueous humor
   D. Perilymph

5. Which of the following is the first to vibrate in response to sound waves?
   A. Tectorial membrane
   B. Tympanic membrane
   C. Vestibular membrane
   D. Basilar membrane

Application

Ima Student had an appointment with an otolaryngologist because she had difficulty hearing. A week previously she had a sinus infection and sore throat, but apparently that condition had resolved. After performing numerous tests, the physician explained that she had a middle ear infection with an accumulation of fluid and that it was the result of the sinus infection and sore throat.

1. How would fluid in the middle ear reduce the ability to perceive sound?

2. Explain how a sore throat and sinus infection can result in a middle ear infection.

BUILDING YOUR MEDICAL VOCABULARY

Building a vocabulary is a cumulative process. As you progress through this book, you will use word parts, abbreviations, and clinical terms from previous chapters. Each chapter will present new word parts, abbreviations, and terms to add to your expanding vocabulary.

Word Parts and Combining Form with Definition and Examples

PART/COMBINING FORM	DEFINITION	EXAMPLE
acous-	sound, hearing	acoustic: pertaining to sound or hearing
audi-	to hear	audiologist: specialist in the evaluation and rehabilitation of those with impaired hearing
blephar-	eyelid	blepharectomy: surgical excision of an eyelid
coch-	snail	cochlea: spiral or snail-shaped portion of the inner ear
-cusis	hearing	presbycusis: impairment of hearing as a result of aging
dacry/o	tear,	dacryoadenalgia; pain in a lacrimal gland lacrimal apparatus
fove-	pit	fovea centralis: small pit in the retina
gust-	taste	gustation: act of tasting or sense of taste
irid/o	iris	iridemia: hemorrhage from the iris of the eye
kerat/o	cornea	keratometer: instrument for measuring the curves of the cornea
lacr-	tears	lacrimation: secretion and discharge of tears
lith-	stone	otolith: little stones of calcium carbonate in the mucula of the inner ear
lute-	yellow	macula lutea: yellowish depression on the retina
macul-	spot, depression	macula lutea: yellowish depression on the retina
meat-	passage	external auditory meatus: passage in the external ear that leads to the tympanic membrane
myring/o	tympanic membrane,	myringoplasty: surgical repair of the tympanic membrane eardrum
ocul/o	eye	oculomotor: pertaining to eye movements

PART/COMBINING FORM	DEFINITION	EXAMPLE
olfact/o	smell	olfactometer: instrument for testing the sense of smell
opt/o	eye	optometer: device for measuring refraction of the eye
ophthalm/o	eye	ophthalmomycosis: fungal disease of the eye
-opia	visual condition	hyperopia: visual condition in which light fays focus behind the retina; farsightedness
ot/o	ear	otalgia: earache
phac/o	lens of the eye	phacocele: hernia of the eye lens
presby-	old age	presbyopia: reduced accommodation in the eye that occurs normally with aging
scler/o	hard	sclera: the tough, white outer coat of the eyeball
tympan/o	drum	tympanic membrane: eardrum
vitre/o	glass	vitreous humor: glassy, jelly-like substance in the posterior cavity of the eye

Clinical Abbreviations

ABBREVIATION	MEANING
AMD	age-related macular degeneration
AOM	acute otitis media
aq	aqueous, water
db	decibel
EENT	eye, ear, nose, and throat
ENT	ear, nose, and throat
EOM	extraocular movement, extraocular muscle
HEENT	head, eyes, ears, nose, and throat
IOL	intraocular lens
IOP	intraocular pressure
LASIK	laser in situ keratomileusis
PERRLA	pupils equal, round, reactive to light and accommodation
POAG	primary open-angle glaucoma
REM	rapid eye movement

Clinical Terms

Audiometry (aw-dee-AHM-eh-tree) Hearing test

Blepharitis (bleff-ahr-EYE-tis) Inflammation of the edges of the eyelid

Diplopia (dip-LOH-pee-ah) Double vision

Emmetropia (emm-eh-TROH-pee-ah) Normal or perfect vision

Mydriasis (mih-DRY-ah-sis) Dilation of the pupil

Nyctalopia (nick-tah-LOH-pee-ah) A Condition in which the individual has difficulty seeing at night; night blindness

Ophthalmologist (off-thal-MAHL-oh-jist) Physician who specializes in the diagnosis and treatment of eye defects, injuries, and diseases; qualified to perform eye surgery

Optician (ahp-TISH-an) Specialist in the translation and filling of ophthalmic prescriptions; not qualified to examine eyes or to prescribe eyeglasses

Optometrist (ahp-TOM-eh-trist) Professional person trained to examine the eyes and prescribe eyeglasses to correct irregularities in vision; not a physician and not qualified to diagnose or treat diseases of the eye or to perform eye surgery

Otalgia (oh-TAL-jee-ah) Earache

Otosclerosis (oh-toe-sklee-ROH-sis) Progressive formation of bony tissue around the oval window, immobilizing the stapes; results in conduction deafness

Otoscopy (oh-TAHS-koh-pee) Visual examination of the external auditory canal and the tympanic membrane using an otoscope

Presbycusis (prez-bih-KUS-is) Impairment of hearing resulting from aging

Snellen chart (SNELL-en chart) Chart, printed with lines of black letters that are graduated in size from smallest on the bottom to largest on the top, used for testing visual acuity

Tinnitus (tin-EYE-tus) Ringing or buzzing sound in the ears

Tonometry (toh-NAHM-eh-tree) Measurement of the tension or pressure within the eye, which is useful in detecting glaucoma

Tympanitis (tim-pan-EYE-tis) Inflammation of the tympanic membrane

Vertigo (VER-tih-goh) Feeling of dizziness, loss of balance, and lightheadedness caused by a disturbance of the semicircular canals, utricle, saccule, or vestibular nerve

VOCABULARY QUIZ

Use word parts you have learned to form words that have the following definitions.

1. Study of hearing _____

2. Inflammation of the eyelid _____

3. Glassy fluid _____

4. Pertaining to taste _____

5. Like a snail _____

Using the definitions of word parts you have learned, define the following words.

6. Olfactory _____

7. Macula lutea _____

8. Intraocular _____

9. Tympanectomy _____

10. Lacrimal _____

Match each of the following definitions with the correct word.

_____ 11. Vision of old age

_____ 12. Inflammation of the cornea

_____ 13. Surgical repair of the ear

_____ 14. Surgical excision of a portion of the iris

_____ 15. Central pit

A. Fovea centralis
B. Iridectomy
C. Keratitis
D. Otoplasty
E. Presbyopia

16. What is the clinical term for "swimmer's ear," a fungal infection of the external ear? _____

17. What is the clinical term for "impairment of vision caused by aging"? _____

18. What procedure measures the intraocular pressure? _____

19. What is the meaning of the abbreviation REM? _____

20. What is the meaning of the abbreviation ENT? _____

Endocrine System

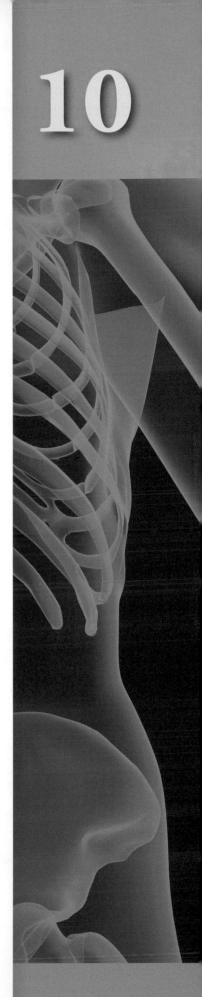

CHAPTER OBJECTIVES

Introduction to the Endocrine System
- Compare the actions of the nervous system and the endocrine system.

Characteristics of Hormones
- Compare the major chemical classes of hormones.
- Discuss the general mechanisms of hormone action.

Endocrine Glands and Their Hormones
- Identify the major endocrine glands and discuss their hormones and functions: pituitary, thyroid, parathyroid, adrenal, pancreas, gonads, thymus, and pineal.

- Name and describe the function of at least one hormone from the (a) gastric mucosa, (b) small intestine, (c) heart, and (d) placenta.

Prostaglandins
- Differentiate between hormones and prostaglandins.

KEY TERMS

Adenohypophysis (add-eh-noe-hye-PAH-fih-sis) Anterior portion of the pituitary gland

Endocrine gland (EN-doh-krin GLAND) Gland that secretes its product directly into the blood; opposite of exocrine gland

Exocrine gland (EKS-oh-krin GLAND) Gland that secretes its product to a surface or cavity through ducts; opposite of endocrine gland

Hormone (HOAR-mohn) Substance secreted by an endocrine gland

Negative feedback (NEG-ah-tiv FEED-bak) Mechanism of response in which a stimulus initiates reactions that reduce the stimulus

Neurohypophysis (noo-roh-hye-PAH-fih-sis) Posterior portion of the pituitary gland

Prostaglandins (prahss-tih-GLAN-dins) Group of substances, derived from fatty acids, that are produced in small amounts and have an immediate, short-term, localized effect; sometimes called local hormones

Target tissue (TAR-get TISH-yoo) Tissue (cells) that responds to a particular hormone because it has receptor sites for that hormone

Functional Relationships of the
Endocrine System

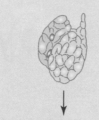

Influences development and
growth; adjusts rates of metabolism.

Reproductive

Gonads produce hormones that
feedback to influence pituitary function.

Hormones have a major role in differentiation
and development of reproductive organs,
sexual development, sex drive, gamete
production, menstrual cycle,
pregnancy, parturition, and lactation.

Integument

Provides barrier against entry of pathogens;
mechanical protection of superficial glands.

Sex hormones regulate activity of sebaceous
glands, distribution of subcutaneous adipose
and growth of hair.

Urinary

Helps maintain pH and electrolyte balance
necessary for endocrine function; eliminates
inactivated hormones and other metabolic
wastes; releases renin and erythropoietin.

Aldosterone, ADH, and atrial natriuretic
hormone regulate urine formation in the kidneys.

Skeletal

Protects glands in head, thorax, and pelvis.

Hormones from the pituitary, thyroid, and
gonads stimulate bone growth; calcitonin
and parathyroid hormone regulate calcium
uptake and release from bone.

Digestive

Absorbs nutrients for endocrine
metabolism and hormone synthesis.

Hormones influence motility and
glandular activity of the digestive tract,
gallbladder secretion, and secretion of enzymes
from the pancreas; insulin and glucagon
adjust glucose metabolism in the liver.

Muscular

Provides protection for some endocrine glands.

Hormones influence muscle metabolism,
mass, and strength; epinephrine and
norepinephrine adjust cardiac and s
mooth muscle activity.

Respiratory

Supplies oxygen, removes
carbon dioxide, and helps maintain pH for
metabolism in endocrine glands; converts
angiotensin I into active angiotensin II.

Epinephrine promotes bronchodilation;
thyroxine and epinephrine
stimulate cell respiration.

Nervous

Regulates secretory activity of anterior
pituitary and adrenal medulla; produces
ADH and oxytocin.

Hormones influence neuronal metabolism
and enhance autonomic stimuli.

Lymphatic/Immune

Assists in defense against infections in endocrine glands.

Hormones from thymus influence development of
lymphocytes; glucocorticoids suppress immune respone.

Cardiovascular

Delivers oxygen and nutrients to endocrine glands; removes
carbon dioxide and heat; transports hormones from glands to
target tissue; heart secretes atrial natriuretic hormone.

Hormones adjust heart rate, contraction strength, blood volume,
blood pressure, and red blood cell production.

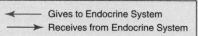

◀────── Gives to Endocrine System
─────▶ Receives from Endocrine System

The organs that make up the endocrine system are the glands that secrete hormones into the blood. Unlike the organs in other systems, endocrine glands are scattered throughout the body. In addition, they are small and unimpressive; however, as you study this chapter, you will discover that they are extremely important. The study of endocrine glands and hormones is called endocrinology.

Introduction to the Endocrine System

Comparison of the Endocrine and Nervous Systems

The endocrine system, along with the nervous system, functions in the regulation of body activities. The nervous system acts through electrical impulses and neurotransmitters to cause muscle contraction and glandular secretion. The effect is of short duration, measured in seconds, and localized. The endocrine system acts through chemical messengers called hormones that influence growth, development, and metabolic activities. The action of the endocrine system is measured in minutes, hours, or weeks and is more generalized than the action of the nervous system.

Comparison of Exocrine and Endocrine Glands

There are two major categories of glands in the body—**exocrine** and **endocrine.** Exocrine glands have ducts that carry their secretory product to a surface or into a cavity. These have a variety of functions and include the sweat, sebaceous, and mammary glands and the glands that secrete digestive enzymes. The endocrine glands do not have ducts to carry their product to a surface. They are called ductless glands. The word "endocrine" is derived from the Greek terms *endo,* meaning "within," and *krine,* meaning "to separate or secrete." The secretory products of endocrine glands are called **hormones** and are secreted directly into the blood and then carried throughout the body where they influence only those cells that have receptor sites for that hormone. Other cells are not affected. Endocrine glands have an extensive network of blood vessels, and organs with the richest blood supply include some of the endocrine glands such as the thyroid and adrenal glands.

Characteristics of Hormones

Each hormone produced in the body is unique. Each one is different in its chemical composition, structure, and action. Despite the differences, there are similarities in these molecules.

Chemical Nature of Hormones

Chemically, hormones may be classified as either **proteins** or **steroids.** All of the hormones in the human body, except the sex hormones and those from the adrenal cortex, are proteins or protein derivatives. This means that their fundamental building blocks are **amino acids.** Protein hormones are difficult to administer orally because they are quickly inactivated by the acid and pepsin in the stomach. These hormones must be administered by injection. Sex hormones and those from the adrenal cortex are steroids, which are lipid derivatives. These lipid-soluble hormones may be taken orally.

Mechanism of Hormone Action

Hormones are carried by the blood throughout the entire body, yet they affect only certain cells. The specific cells that respond to a given hormone have **receptor sites** for that hormone. This is sort of a lock-and-key mechanism. If the key fits the lock, then the door will open. If a hormone fits the receptor site, then there will be an effect (Figure 10-1). If a hormone and a receptor site do not match, then there is no reaction. All the cells that have receptor sites for a given hormone make up the **target tissue** for that hormone. In some cases, the target tissue

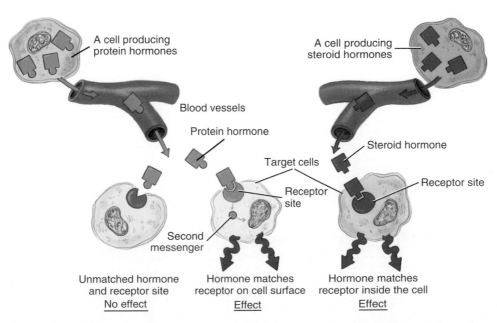

A cell producing protein hormones

A cell producing steroid hormones

Blood vessels

Protein hormone

Steroid hormone

Target cells

Receptor site

Receptor site

Second messenger

Unmatched hormone and receptor site
No effect

Hormone matches receptor on cell surface
Effect

Hormone matches receptor inside the cell
Effect

Figure 10-1 *Hormone–receptor action. There must be a match between hormone and receptor. Receptors for protein hormones are on the cell surface. Receptors for steroid hormones are inside the cell.*

is localized in a single gland or organ. In other cases, the target tissue is diffuse and scattered throughout the body so that many areas are affected. Hormones bring about their characteristic effects on target cells by modifying cellular activity.

Receptor sites may be located on the **surface of the cell membrane** or in the **interior of the cell.** Protein hormones, in general, are unable to diffuse through the cell membrane so they react with receptor sites on the surface of the cell. The hormone-receptor reaction on the cell membrane activates an enzyme within the membrane, called adenyl cyclase, which diffuses into the cytoplasm. Within the cell, adenyl cyclase catalyzes the removal of phosphates from adenosine triphosphate (ATP) to produce cyclic adenosine monophosphate (cyclic AMP). Cyclic AMP activates enzymes within the cytoplasm that alter the cellular activity. The protein hormone, which reacts at the cell membrane, is called the **first messenger.** Cyclic AMP, which brings about the action attributed to the hormone, is called the **second messenger.** This type of action is relatively rapid because the precursors are already present and they just need to be activated in some way.

Steroids, which are lipid soluble, diffuse through the cell membrane and react with receptors inside the cell. The hormone–receptor complex that is formed enters the nucleus, where it has a direct effect on specific genes within the DNA. These genes act as templates for the synthesis of messenger RNA (mRNA), which diffuses into the cytoplasm. The mRNA in the cytoplasm directs the synthesis of proteins at the ribosomes. The proteins that are formed represent the cell's response to the hormone. This method of hormone action is relatively slow because mRNA and proteins actually have to be synthesized rather than just activated.

Control of Hormone Action

Hormones are very potent substances, which means that very small amounts of a hormone may have profound effects on metabolic processes. Because of their potency, hormone secretion must be regulated within very narrow limits to maintain homeostasis in the body.

Many hormones are controlled by some form of a negative feedback mechanism. In this type of system, a gland is sensitive to the concentration of a substance that it regulates. A negative feedback system causes a reversal of increases and decreases in body conditions to maintain a state of stability or homeostasis.

Some endocrine glands secrete hormones in response to **other hormones.** The hormones that cause secretion of other hormones are called **tropic** hormones. A hormone from gland "A" causes gland "B" to secrete its hormone. For example, thyroid-stimulating hormone from the anterior pituitary gland causes the thyroid gland to secrete the hormone thyroxin.

A third method of regulating hormone secretion is by direct **nervous stimulation.** A nerve stimulus causes a gland to secrete

its hormone. A physiologic example of this mechanism is the sympathetic nerve stimulation of the adrenal medulla, which responds by secreting epinephrine (adrenaline).

Endocrine Glands and Their Hormones

The endocrine system is made up of the endocrine glands that secrete hormones. Figure 10-2 illustrates that the eight major endocrine glands are scattered throughout the body; however, they are still considered to be one system because they have similar functions, similar mechanisms of influence, and many important interrelationships. Some glands also have nonendocrine regions that have functions other than hormone secretion. The pancreas is one of these glands. It has a major exocrine portion that secretes digestive enzymes and an endocrine portion that secretes hormones. The ovaries and testes secrete hormones and also produce the ova and sperm. Some organs, such as the stomach, intestines, and heart, produce hormones, but their primary function is not hormone secretion. These organs are discussed in more detail in the chapters dealing with their predominant function. Table 10-1 summarizes the major endocrine glands and their hormones.

Pituitary Gland

The **pituitary gland** or **hypophysis** is a small gland about 1 centimeter in diameter, or about the size of a pea. It is nearly surrounded by bone as it rests in the sella turcica, a depression

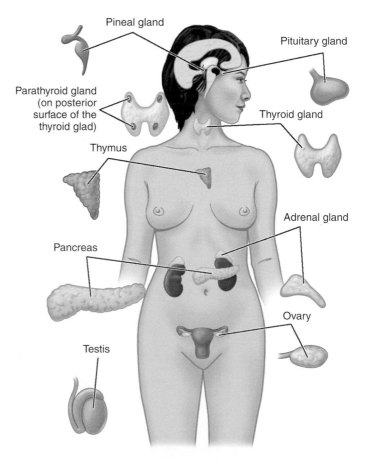

Figure 10-2 *Major endocrine glands.*

> ### QUICK CHECK
> **10.1** What is the difference between an exocrine gland and an endocrine gland?
> **10.2** What are the two chemical classes of hormones?
> **10.3** Where are the receptors for steroid hormones located?

TABLE 10-1 The Principal Endocrine Glands and Their Hormones

Gland	Hormone	Target Tissue	Principal Actions
Hypothalamus	Releasing and inhibiting hormones	Anterior lobe of pituitary gland	Stimulates or inhibits secretion of specific hormones
Anterior lobe of pituitary	Growth hormone (GH)	Most tissues in body	Stimulates growth by promoting protein synthesis
	Thyroid-stimulating hormone (TSH)	Thyroid gland	Increases secretion of thyroid hormone; increases size of thyroid gland
	Adrenocorticotropic hormone (ACTH)	Adrenal cortex	Increases secretion of adrenocortical hormones, especially glucocorticoids such as cortisol
	Follicle-stimulating hormone (FSH)	Ovarian follicles in female; seminiferous tubules of testis in male	Follicle maturation and estrogen secretion in female; spermatogenesis in male
	Luteinizing hormone (LH); also called interstitial cell–stimulating hormone (ICSH) in males	Ovary in females; testis in males	Ovulation; progesterone production in female; testosterone production in male
	Prolactin	Mammary gland	Stimulates milk production
Posterior lobe of pituitary	Antidiuretic hormone (ADH)	Kidney	Increases water reabsorption (decreases water lost in urine)
	Oxytocin	Uterus; mammary gland	Increases uterine contractions; stimulates ejection of milk from mammary gland
Thyroid gland	Thyroxine and triiodothyronine	Most body cells	Increases metabolic rate; essential for normal growth and development
	Calcitonin	Primarily bone	Decreases blood calcium by inhibiting bone breakdown and release of calcium; antagonistic to parathyroid hormone
Parathyroid gland	Parathyroid hormone (PTH) or parathormone	Bone, kidney, digestive tract	Increases blood calcium by stimulating bone breakdown and release of calcium; increases calcium absorption in the digestive tract; decreases calcium lost in urine
Adrenal cortex	Mineralocorticoids (cortisol)	Kidney	Increases sodium reabsorption and potassium secretion in kidney tubules; secondarily increases water retention
	Androgens and estrogens	Most body tissues	Increases blood glucose levels; inhibits inflammation and immune response
Adrenal medulla	Epinephrine and norepinephrine	Heart, blood vessels, liver, adipose	Helps cope with stress; increases heart rate and blood pressure; increases blood flow to skeletal muscle; increases blood glucose level
Pancreas (islets of Langerhans)	Glucagon	Liver	Increases breakdown of glycogen to increase blood glucose levels
	Insulin	General, but especially liver, skeletal muscle, adipose	Decreases blood glucose levels by facilitating uptake and use of glucose by cells; stimulates glucose storage as glycogen and production of adipose
Testes	Testosterone	Most body cells	Maturation and maintenance of male reproductive organs and secondary sex characteristics
Ovaries	Estrogens	Most body cells	Maturation and maintenance of male reproductive organs and secondary sex characteristics
	Progesterone	Uterus and breast	Prepares uterus for pregnancy; stimulates development of mammary gland; menstrual cycle
Pineal gland	Melatonin	Hypothalamus	Inhibits gonadotropin-releasing hormone, which consequently inhibits reproductive functions; regulates daily rhythms such as sleep and wakefulness
Thymus	Thymosin	Tissues involved in immune response	Immune system development and function

in the sphenoid bone. The gland is connected to the hypothalamus of the brain by a slender stalk called the infundibulum. There are two distinct regions in the gland. The anterior portion consists of epithelial cells derived from the embryonic oral cavity and is called the **adenohypophysis** (add-eh-noe-hye-PAH-fih-sis). The posterior region is an extension of the brain and consists of neurons and neuroglia. It is called the **neurohypophysis** (noo-roh-hye-PAH-fih-sis).

Just as there are two distinct regions or lobes with different embryonic derivations, there are separate regulating mechanisms that influence the secretory activity of the two parts. Figure 10-3 illustrates the control of the anterior lobe by the hypothalamus of the brain. Releasing and inhibitory hormones from the hypothalamus are secreted in response to neural stimulation. They enter capillaries in the hypothalamus and are transported by veins along the infundibulum to another capillary network in the anterior lobe of the pituitary. The two sets of capillaries and the veins between them are called the hypothalamic-pituitary portal system. Within the adenohypophysis, the releasing hormones affect the glandular epithelium to cause secretion of anterior pituitary hormones. Inhibitory hormones have the opposite effect.

Some of the neurosecretory cells in the hypothalamus have long axons that descend through the infundibulum and terminate in the posterior lobe. A stimulus to the nerve cell in the hypothalamus triggers the axon end in the posterior lobe to secrete a hormone that enters blood vessels and is transported to the target tissue. These hormones actually are made in the neuron cell bodies within the hypothalamus; then they travel down the axon to the neurohypophysis, where they are stored and subsequently released. Figure 10-4 illustrates the relationship of the hypothalamus to the posterior lobe of the pituitary gland.

Hormones of the Anterior Lobe (Adenohypophysis)

Growth Hormone

Growth hormone (GH) is a protein that stimulates the growth of bones, muscles, and other organs by promoting protein synthesis (Figure 10-5). This hormone affects the appearance of an individual because it influences height. If there is a hyposecretion (too little) of the hormone in a child, that person may

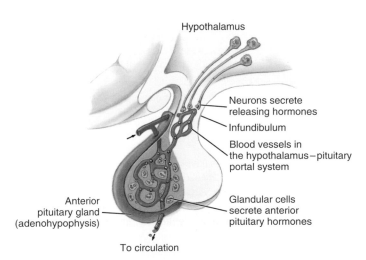

Figure 10-3 *Relationship of the hypothalamus to the adenohypophysis (anterior pituitary gland).*

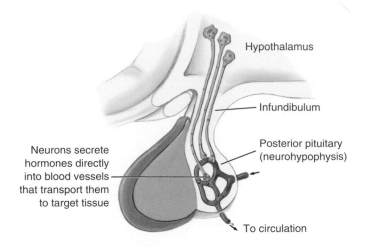

Figure 10-4 *Relationship of the hypothalamus to the neurohypophysis (posterior pituitary gland).*

become a pituitary dwarf of normal proportions but small stature. Hypersecretion of the hormone in a child results in exaggerated bone growth, and the individual becomes exceptionally tall or a giant. After ossification is complete and an increase in bone length is no longer possible, excess GH may cause an enlargement in the diameter of the bones. The result is a condition called acromegaly (ack-roh-MEG-ah-lee), in which the bones of the hands and face become abnormally large.

Thyroid-Stimulating Hormone

Thyroid-stimulating hormone (TSH), or thyrotropin (thye-roh-TROH-pin), causes the glandular cells of the thyroid to secrete thyroid hormone (see Figure 10-5). When there is a hypersecretion of TSH, the thyroid gland enlarges and secretes too much thyroid hormone. Hyposecretion of TSH results in atrophy of the thyroid gland and too little hormone.

Adrenocorticotropic Hormone

Adrenocorticotropic (ah-dree-noh-kor-tih-koh-TROH-pik) hormone (ACTH) reacts with receptor sites in the cortex of the adrenal gland to stimulate the secretion of cortical hormones, particularly cortisol (see Figure 10-5). ACTH also affects the melanocytes in the skin and increases pigmentation. Hypersecretion and hyposecretion of ACTH are reflected in the activity of the adrenal cortex.

Gonadotropic Hormones

Gonadotropic (go-nad-oh-TROH-pik) hormones react with receptor sites in the gonads—ovaries and testes—to regulate the development, growth, and function of these organs. Follicle-stimulating hormone (FSH) stimulates the development of eggs or ova in the ovaries and of sperm in the testes (see Figure 10-5). In addition, it stimulates estrogen production in the female. Luteinizing hormone (LH) causes ovulation and the production and secretion of the female sex hormones—progesterone and estrogen. In the male, LH is sometimes called interstitial cell–stimulating hormone (ICSH) because it stimulates the interstitial cells of the testes to produce and secrete the male sex hormone testosterone (see Figure 10-5). Without the gonadotropins FSH

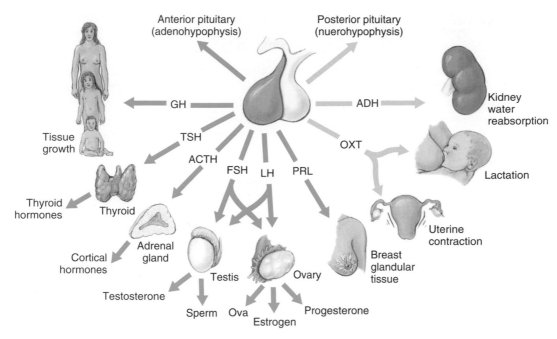

Figure 10-5 *Effects of hormones from the pituitary gland.*

and LH, the ovaries and testes decrease in size, ova and sperm are not produced, and sex hormones are not secreted.

Prolactin

Prolactin (PRL) or lactogenic hormone promotes the development of glandular tissue in the female breast during pregnancy and stimulates milk production after the birth of the infant (see Figure 10-5). This hormone does not cause the milk to be ejected from the breast. A hormone from the posterior pituitary and other neural influences are responsible for the ejection of the milk. Hyposecretion of prolactin generally presents no problem except in women who choose to breast-feed their babies. Hypersecretion is more common and is usually the result of pituitary tumors. This causes inappropriate lactation and lack of menstruation in females. In males, it results in impotence.

Hormones of the Posterior Lobe (Neurohypophysis)

Antidiuretic Hormone

Antidiuretic (ant-eye-dye-yoo-RET-ik) hormone (ADH) promotes the reabsorption of water by the kidney tubules, with the result that less water is lost as urine (see Figure 10-5). This mechanism conserves water for the body. Insufficient amounts of ADH cause excessive water loss in the urine with large amounts of a very dilute urine being produced. This condition is called diabetes insipidus. ADH, especially in large amounts, also causes blood vessels to constrict, which increases blood pressure. For this reason, ADH is sometimes called vasopressin. Ingestion of alcoholic beverages inhibits ADH secretion and results in increased urine output.

QUICK APPLICATIONS

Certain drugs, called diuretics, counteract the effects of ADH and result in fluid loss. These drugs are sometimes prescribed for patients with high blood pressure or those with edema caused by congestive heart failure because the drugs have the effect of removing fluid from the body by increasing urine production.

FROM THE PHARMACY

Most of the medications for the endocrine system are natural or synthetic preparations to replace, increase, or decrease hormones that occur naturally in the human body. It follows that the medication has the same effect as the hormone involved. A few examples are presented to illustrate this concept.

Oxytocin (Pitocin, Syntocinon) is a polypeptide hormone from the posterior pituitary gland. The pharmaceutical preparation is identical in action to the hormone produced naturally in the body. It acts on the smooth muscles of the uterus, causes vasoconstriction, and stimulates the flow of milk from the postpartum mammary gland. Oxytocin is used to stimulate or induce labor, to assist in the expulsion of the placenta, to control postpartum bleeding, and to stimulate the ejection of milk.

Vasopressin (Pitressin) is a polypeptide hormone from the posterior pituitary gland. It is identical to antidiuretic hormone. Peptide hormones are not absorbed from the GI tract because they

Continued

FROM THE PHARMACY—CONT'D

are inactivated by proteolytic enzymes; therefore they cannot be taken orally and must be administered intravenously or by subcutaneous injection, or possibly through nasal sprays. Vasopressin is used to treat diabetes insipidus and to relieve postsurgical abdominal distention. It is primarily used as a hospital-based medication that is administered intravenously or by subcutaneous injection. **Desmopressin** (DDAVP) is a synthetic hormone analogue that is used more frequently than vasopressin. It is administered intravenously, by subcutaneous injection, or as a nasal spray to treat diabetes insipidus and nighttime bed-wetting in children.

Levothyroxine sodium (Levoxyl, Levothroid, Synthroid) is a synthetic preparation of the T₄ thyroid hormone that is used to treat decreased activity of the thyroid gland (hypothyroidism). This medication needs to be started slowly to gradually reverse the hypothyroid condition, and it should be monitored and adjusted to meet the body's need. Overdosing has the same toxic systemic effects as hyperthyroidism. Abuse potential exists, particularly in women who desire the increased metabolic rate the medication provides that promotes weight loss.

Prednisone (Deltasone, Orasone) is a synthetic preparation of glucocorticoid, an adrenocortical steroid. Glucocorticoids may be used to replace missing hormones in adrenal insufficiency. They are more commonly used for suppression of inflammatory, allergic, or immunologic responses. Glucocorticoid use should be monitored carefully to detect possible side effects including hyperglycemia (most significant), hypokalemia, hypertension, and tachycardia. Prolonged use can cause adrenal suppression, osteoporosis, and weight gain. Steroids should not be abruptly discontinued if they have been used daily for more than 2 weeks.

Oxytocin

Oxytocin (ahk-see-TOH-sin) causes contraction of the smooth muscle in the wall of the uterus. It also stimulates the ejection of milk from the lactating breast (see Figure 10-5). A commercial preparation of this hormone, called Pitocin, is sometimes used to induce labor. Oxytocin, or similar synthetic drugs, may be used to hasten the delivery of the placenta, to control bleeding after delivery, or to stimulate milk ejection.

QUICK CHECK

10.4 How do the mechanisms that regulate the anterior and posterior regions of the pituitary differ?
10.5 What is the target tissue for ACTH?
10.6 Jerry has a severe case of the flu with excessive vomiting and diarrhea and is in a condition of dehydration. How does this affect the amount of ADH secreted by the posterior pituitary gland.

Thyroid Gland

The thyroid (THYE-royd) gland is a very vascular organ that is located in the neck (see Figure 10-2). It consists of two lobes, one on each side of the trachea, just below the larynx or voice box. The two lobes are connected by a narrow band of tissue called the isthmus. The ultrasound image in Figure 10-6 shows the lobes of the thyroid gland adjacent to the trachea. Internally, the gland consists of follicles filled with a colloid and parafollicular cells interspersed between the follicles. The follicles, composed of simple cuboidal epithelium, secrete the hormones that contain iodine, such as thyroxine (thye-RAHK-sin) and triiodothyronine (trye-eye-oh-doh-THYE-roh-neen). Parafollicular cells secrete calcitonin (kal-sih-TOH-nin).

Thyroxine and Triiodothyronine

About 95% of the active thyroid hormone is thyroxine, and most of the remaining 5% is triiodothyronine. Both of these require iodine for their synthesis. The iodine is actively transported into the thyroid gland, where its concentration may

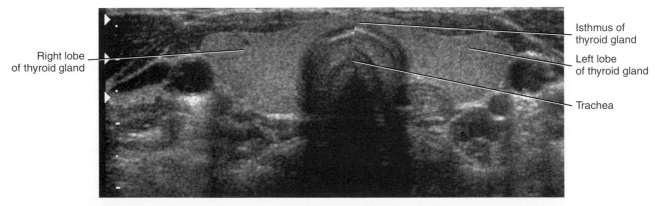

Figure 10-6 *Bilateral ultrasound image of the thyroid gland. (Adapted from Applegate E: The Sectional Anatomy Learning System, Concepts, ed. 3. St. Louis, Elsevier/Saunders, 2010.)*

become as much as 25 times that of the blood, and then it is incorporated into the hormone molecules. Thyroid hormone secretion is regulated by releasing hormones from the hypothalamus and by the circulating thyroid hormones that exert an inhibiting influence on the anterior pituitary and hypothalamus. This regulatory mechanism is illustrated in Figure 10-7.

If there is an iodine deficiency, the thyroid cannot make sufficient hormone. This stimulates the anterior pituitary to secrete TSH, which causes the thyroid gland to increase in size in a vain attempt to produce more hormone. However, it cannot produce more hormone because it does not have the necessary raw materials, namely, iodine. This type of thyroid enlargement is called simple goiter or iodine deficiency goiter.

The thyroid hormones containing thyroxine and triiodothyronine help to regulate the metabolism of carbohydrates, proteins, and lipids in the body. They do not have a single target organ; instead, they affect most of the cells in the body. They increase the rate at which cells release energy from carbohydrates, they enhance protein synthesis, they are necessary for normal growth and development, and they stimulate the nervous system.

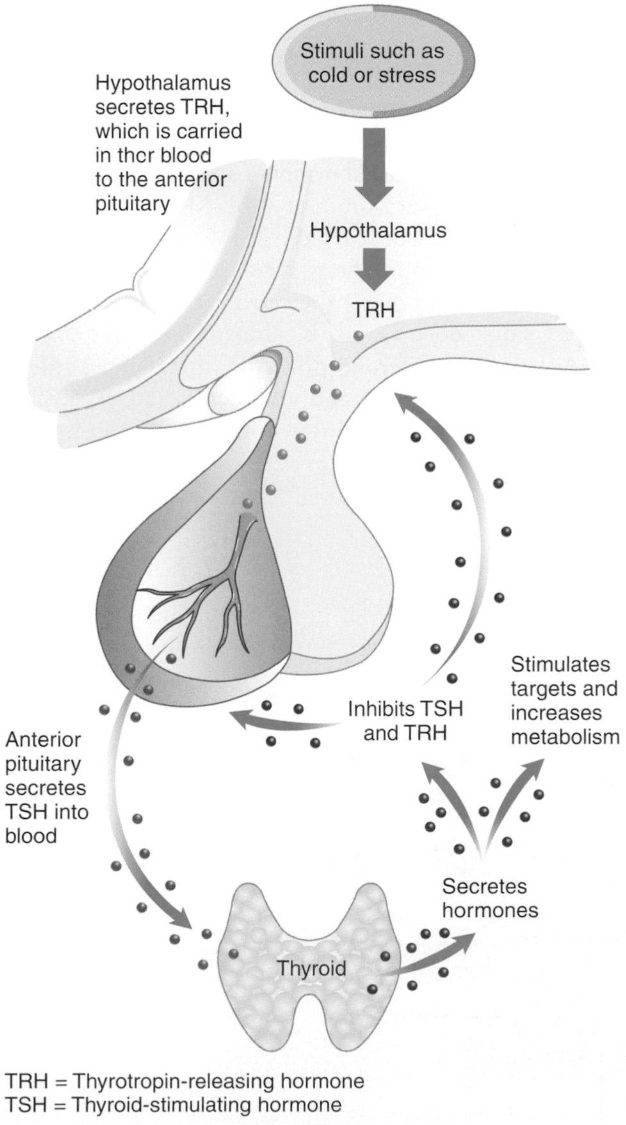

In the infant, a deficiency or lack of thyroid hormone, hypothyroidism, results in a condition called cretinism (KREE-tin-izm). A cretin is a mentally retarded dwarf with abnormal skeletal features. If the condition is detected early enough, hormone therapy may stimulate growth, prevent the abnormal skeletal features, and increase the metabolic rate. However, therapy must be started within 2 months or less after birth to prevent severe mental retardation. Hypothyroidism in the adult results in a condition called myxedema (miks-eh-DEE-mah), which is characterized by lethargy, weight gain, loss of hair, decreased body temperature, low metabolic rate, and slow heart rate. Hormone therapy, in appropriate doses, usually alleviates these symptoms.

Hyperthyroidism results from an enlarged thyroid gland that produces too much hormone. This condition is characterized by a high metabolic rate, hyperactivity, insomnia, nervousness, irritability, and chronic fatigue. An individual with hyperthyroidism often has protruding eyes or exophthalmos (eks-off-THAL-mus) because there is swelling in the tissues behind the eyes. Removal, or destruction by radioactive iodine, of a portion of the thyroid may effectively reduce the symptoms of hyperthyroidism.

QUICK APPLICATIONS

When thyroxine and triiodothyronine, with their incorporated iodine, are released into the blood, more than 99% combines with plasma proteins. This iodine is called **protein-bound iodine (PBI)**. The amount of PBI can be measured by a laboratory procedure and is widely used as a test of thyroid function.

Hypothalamus secretes TRH, which is carried in the blood to the anterior pituitary

Stimuli such as cold or stress

Hypothalamus

TRH

Anterior pituitary secretes TSH into blood

Inhibits TSH and TRH

Stimulates targets and increases metabolism

Secretes hormones

Thyroid

TRH = Thyrotropin-releasing hormone
TSH = Thyroid-stimulating hormone

Figure 10-7 *Interaction of hypothalamus, anterior pituitary gland, and thyroid gland.*

Calcitonin

Calcitonin is secreted by the parafollicular cells of the thyroid gland. This hormone opposes the action of the parathyroid glands by reducing the calcium level in the blood. If blood calcium levels become too high, calcitonin is secreted until calcium ion levels decrease to normal. Calcitonin reduces blood calcium levels by reducing the rate at which calcium is released from bone, by increasing the rate of calcium excretion by the kidneys, and by reducing calcium absorption in the intestines. A deficiency of calcitonin does not seem to increase blood calcium levels.

Parathyroid Glands

Four small masses of epithelial tissue are embedded in the connective tissue capsule on the posterior surface of the thyroid glands. These are the parathyroid glands, and they secrete **parathyroid hormone** (PTH, or parathormone). PTH is the most important regulator of blood calcium levels. The hormone is secreted in response to low blood calcium levels, and its effect is to increase those levels. It does this by increasing osteoclast activity in bones so that calcium is released from the bones into the blood, by increasing calcium reabsorption from the kidney tubules into the blood, which decreases the amount lost in the urine, and by increasing the absorption of dietary calcium in the intestines. Vitamin D is also necessary for dietary calcium to be absorbed in the intestines. PTH is antagonistic, or has the opposite effect, to calcitonin from the thyroid gland (Figure 10-8).

Hypoparathyroidism, or insufficient secretion of PTH, leads to increased nerve excitability. The low blood calcium levels trigger spontaneous and continuous nerve impulses, which then stimulate muscle contraction. Tumors of the parathyroid gland may cause excessive secretion of PTH or hyperparathyroidism. This leads to an increased osteoclast activity that removes calcium from the bones and increases the level in the blood. The excess calcium in the blood may precipitate in abnormal locations or cause kidney stones.

QUICK CHECK

10.7 What is the basic function of the hormone from the follicular cells of the thyroid gland?

10.8 What hormone is produced by the parafollicular cells of the thyroid gland?

10.9 What is the basic function of the hormone from the parathyroid glands?

QUICK APPLICATIONS

Hypersecretion of parathyroid hormone sometimes causes a bone disease called **osteitis fibrosa cystica**. Bone mass decreases as a result of osteoclast activity, decalcification occurs, cystlike cavities appear in the bone, and spontaneous fractures result.

Adrenal (Suprarenal) Glands

The adrenal, or suprarenal (soo-prah-REE-null), glands are paired with one gland located near the upper portion of each kidney. The glands are embedded in the fat that surrounds the kidneys. Each gland is divided into an outer region—the adrenal cortex—and an inner region—the adrenal medulla (Figure 10-9). The cortex and medulla of the adrenal gland, like the anterior and posterior lobes of the pituitary, develop from different embryonic tissues and secrete different hormones. The adrenal cortex is essential to life, but the medulla may be removed with no life-threatening effects because its functions are similar to those of the sympathetic nervous system.

The hypothalamus of the brain influences both portions of the adrenal gland but by different mechanisms. The medulla receives direct stimulation from nerve impulses that originate in the hypothalamus and then travel through the brain stem, spinal cord, and sympathetic nerves (Figure 10-10). The hypothalamus affects the cortex by secreting ACTH-releasing hormone, which stimulates the anterior pituitary to secrete ACTH, which then stimulates the adrenal cortex (Figure 10-10).

Hormones of the Adrenal Cortex

The adrenal cortex consists of three different regions, with each region producing a different group or type of hormones. Chemically, all the cortical hormones are steroids.

Mineralocorticoids (min-er-al-oh-KOR-tih-koyds) are secreted by the outermost region of the adrenal cortex. As a group, these hormones help regulate blood volume and the concentration of mineral electrolytes in the blood. The principal mineralocorticoid is **aldosterone** (al-DAHS-ter-ohn). Although aldosterone primarily affects the kidneys, it also acts on the intestines, salivary glands, and sweat glands. In general, its effect is to conserve sodium ions and water in

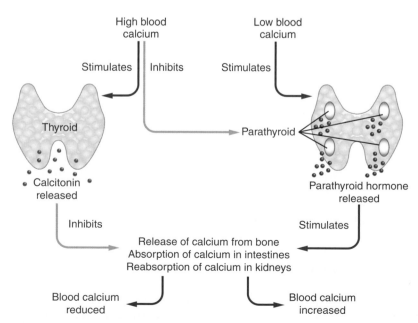

Figure 10-8 *Effects of calcitonin and parathyroid hormone on blood calcium levels.*

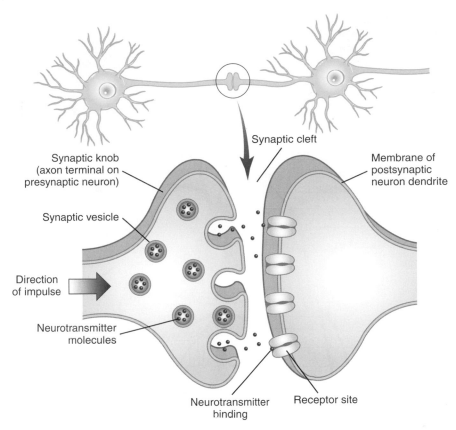

Synaptic cleft

Synaptic knob
(axon terminal on
presynaptic neuron)

Membrane of
postsynaptic
neuron dendrite

Synaptic vesicle

Direction
of impulse

Neurotransmitter
molecules

Neurotransmitter
binding

Receptor site

Figure 10-9 *Adrenal gland.*

the body and to eliminate potassium ions. Sodium and potassium levels are important in maintaining blood pressure, nerve impulse conduction, and muscle contraction. When the kidney tubules reabsorb sodium ions and excrete potassium in response to aldosterone, they also conserve water and reduce urine output, which increases blood volume. Aldosterone is secreted in direct response to sodium and potassium ions. The rate of aldosterone secretion increases when there is an increase in blood potassium level or a decrease in blood sodium level. ACTH from the anterior pituitary has a minor effect on aldosterone secretion (Figure 10-11). Tumors of the adrenal cortex may lead to a hypersecretion of mineralocorticoids, which may lead to potassium depletion. When this occurs, neurons and muscle fibers become less responsive to stimuli, which may lead to muscle weakness, cramps, or paralysis.

Glucocorticoids (gloo-koh-KOR-tih-koyds) are secreted by the middle region of the adrenal cortex. As a group, these hormones help regulate nutrient levels in the blood. The principal glucocorticoid is **cortisol,** also called hydrocortisone. The overall effect of the glucocorticoids is to increase blood glucose levels. Cortisol does this by increasing the cellular metabolism of proteins and fats as energy sources, thus conserving glucose. It also stimulates the liver cells to produce glucose from amino acids and fats. These actions help to maintain appropriate blood glucose levels between meals. In times of prolonged stress, cortisol is secreted in greater than normal amounts to help increase glucose levels to provide energy to respond to the stress. Glucocorticoid secretion is controlled by ACTH from the anterior pituitary

gland under the influence of a releasing hormone from the hypothalamus (see Figure 10-11). Cortisol also helps to counteract the inflammatory response. For this reason, it is used clinically to reduce the inflammation in certain allergic reactions, bursitis and arthritis, infections, and some types of cancer.

Gonadocorticoids (go-nad-oh-KOR-tih-koyds), or sex hormones, are the third group of steroids secreted by the adrenal cortex. These are secreted by the innermost region. Male hormones, **androgens,** and female hormones, **estrogens,** are secreted in minimal amounts in both sexes by the adrenal cortex, but their effect is usually masked by the hormones from the testes and ovaries. In females, the masculinization effect of androgen secretion may become evident after menopause, when estrogen levels from the ovaries decrease.

Hyposecretion of hormones from the adrenal cortex leads to a condition known as Addison's disease. The lack of mineralocorticoids causes low blood sodium levels, high blood potassium levels, and dehydration, while the lack of glucocorticoids causes low blood glucose levels. In addition, there is increased pigmentation in the skin. If untreated, hyposecretion of the adrenal cortex leads to death in a few days.

Hypersecretion of hormones from the adrenal cortex, whether from a tumor or excessive ACTH, causes Cushing's syndrome. This is characterized by elevated blood glucose levels, retention of sodium ions and water with subsequent puffiness or edema, loss of potassium ions, and, in females, masculinization. These effects are directly related to the actions of the three groups of hormones secreted by the cortex.

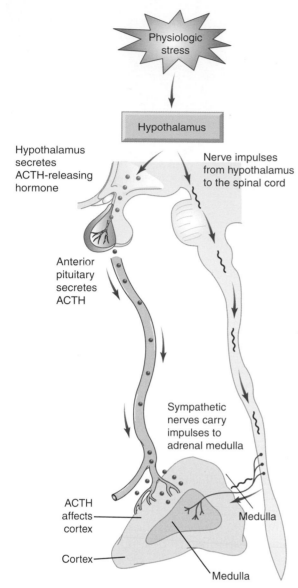

Figure 10-10 *Hypothalamic control of the adrenal gland. The cortex is controlled by releasing hormones from the hypothalamus and ACTH from the anterior pituitary gland. The medulla is stimulated by sympathetic nerves.*

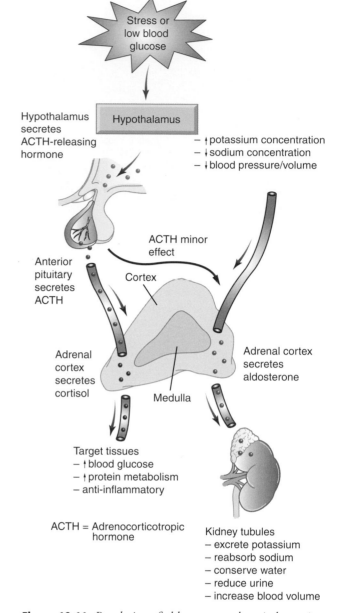

Figure 10-11 *Regulation of aldosterone and cortisol secretion.*

QUICK APPLICATIONS

Tumors that result in hypersecretion of gonadocorticoids may have dramatic effects in prepubertal boys and girls. There is a rapid onset of puberty and sex drive in males. Females develop the masculine distribution of body hair, including a beard, and the clitoris enlarges to become more like a penis.

Hormones of the Adrenal Medulla

The adrenal medulla develops from neural tissue and secretes two hormones: **epinephrine** (adrenaline) and **norepinephrine** (noradrenaline). About 80% of the medullary secretion is epinephrine. These two hormones are secreted in response to stimulation by sympathetic nerves, particularly during

stressful situations (Figure 10-12). Epinephrine, a cardiac stimulator, and norepinephrine, a vasoconstrictor, together cause increases in heart rate, in the force of cardiac muscle contraction, and in blood pressure. They divert blood supply to the skeletal muscles and decrease the activity of the digestive tract, dilate the bronchioles and increase the breathing rate, and increase the rate of metabolism to provide energy. They prepare the body for strenuous activity and are sometimes called the fight-or-flight hormones. Their effect on the body is similar to the sympathetic nervous system, but the effect lasts up to 10 times longer because the hormones are removed from the tissues slowly. The effects of epinephrine are summarized in Figure 10-12. A lack of hormones from the adrenal medulla usually produces no significant effects. Hypersecretion, usually from a tumor, causes prolonged or continual sympathetic responses.

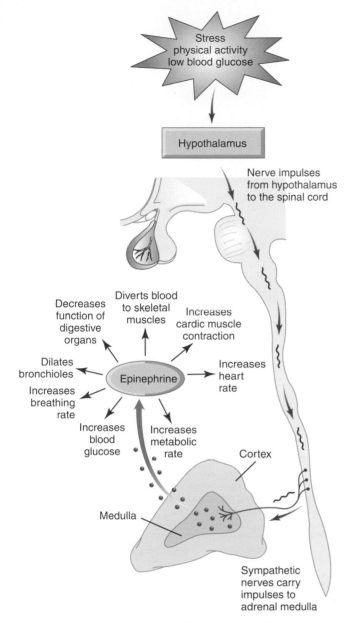

Figure 10-12 *Epinephrine—its effects and control of its secretion.*

endocrine portion that secretes hormones into the blood. The endocrine portion consists of over a million small groups of cells, called pancreatic islets or islets of Langerhans, that are interspersed throughout the exocrine tissue. The pancreatic islets contain alpha cells that secrete the hormone **glucagon** and beta cells that secrete the hormone **insulin.** Both of these hormones have a role in regulating blood glucose levels. It is important to maintain blood glucose levels within a normal range because this is the primary source of energy for the nervous system. If blood glucose levels fall too low, the nervous system does not function properly. If blood glucose levels become too high, the kidneys produce large quantities of urine, and dehydration may result.

Glucagon

Alpha cells in the pancreatic islets secrete the hormone glucagon in response to a low concentration of glucose in the blood. Glucagon's principal action is to raise blood glucose levels (Figure 10-13). It does this by mobilizing glucose and fatty acids from their storage forms. It also stimulates the liver to break down glycogen into glucose and to manufacture glucose from noncarbohydrate sources such as proteins and fats. These mechanisms prevent hypoglycemia from occurring between meals or when glucose is being used rapidly.

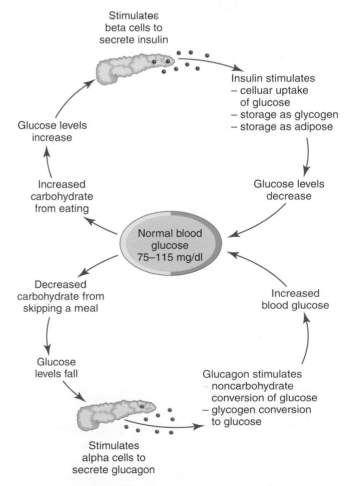

Figure 10-13 *Effects of insulin and glucagon.*

QUICK CHECK

10.10 What effect does aldosterone have on urine output? Explain.

10.11 What is the primary glucocorticoid from the adrenal cortex?

10.12 What two hormones are secreted by the adrenal medulla?

Pancreas—Islets of Langerhans

The pancreas is a long, soft organ that lies transversely along the posterior abdominal wall, posterior to the stomach, and extends from the region of the duodenum to the spleen. This gland has an exocrine portion that secretes digestive enzymes that are carried through a duct to the duodenum and an

Insulin

Beta cells in the pancreatic islets secrete the hormone insulin in response to a high concentration of glucose in the blood. The action of insulin is opposite or antagonistic to glucagon. It promotes cellular uptake and use of glucose for energy. It stimulates the liver and muscle to remove glucose from the blood and to store it as glycogen. When the liver has stored all the glycogen possible, glucose is converted to fat or adipose. Insulin inhibits the manufacture of glucose from noncarbohydrate sources. As a result of these actions, insulin decreases the blood glucose concentration (see Figure 10-13). Hypoactivity of insulin may be caused by insufficient insulin secretion, insufficient receptor sites on target cell membranes, or defective receptor sites that do not recognize insulin. These dysfunctions lead to diabetes mellitus, which is characterized by abnormally high blood glucose levels. Hyperinsulinism is usually caused by an overdose of insulin, rarely by islet cell tumors. The result is low blood glucose levels or hypoglycemia.

Gonads (Testes and Ovaries)

The gonads, the primary reproductive organs, are the testes in the male and the ovaries in the female. These organs are responsible for producing the sperm and ova, but they also secrete hormones and are considered to be endocrine glands. A brief description of their endocrine functions is given here. Reproductive functions and a more thorough discussion of the hormones appear in Chapter 19.

Testes

Male sex hormones, as a group, are called **androgens** (AN-droh-jenz). The principal androgen is **testosterone** (tess-TAHS-ter-ohn), which is secreted by the testes. A small amount is also produced by the adrenal cortex. Production of testosterone begins during fetal development, continues for a short time after birth, nearly ceases during childhood, and then resumes at puberty. This steroid hormone is responsible for the following:

- Growth and development of the male reproductive structures
- Increased skeletal and muscular growth
- Enlargement of the larynx accompanied by voice changes
- Growth and distribution of body hair
- Increased male sexual drive

Testosterone secretion is regulated by a negative feedback system that involves releasing hormones from the hypothalamus and gonadotropins from the anterior pituitary (Figure 10-14).

Ovaries

Two groups of female sex hormones are produced in the ovaries: the **estrogens** (ESS-troh-jenz) and **progesterone** (proh-JESS-ter-ohn). These steroid hormones contribute to the development and function of the female reproductive organs and sex characteristics. At the onset of puberty, estrogens promote:

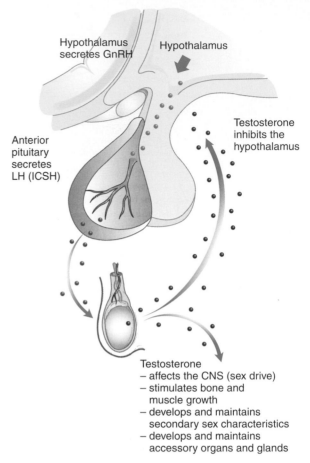

GnRH = Gonadotropin-releasing hormone
LH = Luteinizing hormone
ICSH = Interstitial cell-stimulating hormone
CNS = Central nervous system

Figure 10-14 *Relationship of the hypothalamus, anterior pituitary gland, and testes in testosterone production.*

- Development of the breasts
- Distribution of fat evidenced in the hips, legs, and breasts
- Maturation of reproductive organs such as the uterus and vagina

Progesterone causes the uterine lining to thicken in preparation for pregnancy. Together, progesterone and estrogens are responsible for the changes that occur in the uterus during the female menstrual cycle. Like testosterone in the male, estrogen and progesterone activity is controlled by a negative feedback mechanism involving releasing hormones from the hypothalamus and the gonadotropins FSH and LH from the anterior pituitary.

QUICK CHECK

10.13 What hormone is secreted in response to elevated blood glucose levels?
10.14 What hormone is produced by the alpha cells in the pancreas?
10.15 What effect will injections of testosterone have on gonadotropins from the anterior pituitary gland?

Pineal Gland

The **pineal** (PIE-nee-al) gland, also called pineal body or **epiphysis cerebri,** is a small cone-shaped structure that extends posteriorly from the third ventricle of the brain. The gland is often visible on radiographs because it becomes infiltrated with calcium deposits after puberty. It was once believed that this indicated an atrophy of the gland; however, this does not seem to be the case. Recent evidence indicates that the gland is active throughout the life of an individual. The presence of calcium suggests high metabolic activity.

The pineal gland consists of portions of neurons, neuroglial cells, and specialized secretory cells called **pinealocytes** (PIE-nee-al-oh-cytes). The pinealocytes synthesize the hormone **melatonin** (mell-ah-TOH-nihn) and secrete it directly into the cerebrospinal fluid, which takes it into the blood. Melatonin secretion is rhythmic in nature, with high levels secreted at night and low levels secreted during the day.

The function of the pineal gland and melatonin in humans has been the subject of controversy and speculation for centuries. Even the ancient Greeks wrote about it. Evidence accumulated during the 1980s indicates that melatonin has a regulatory role in sexual and reproductive development. Melatonin acts on the hypothalamus to inhibit gonadotropin-releasing hormone (GnRH). Without GnRH, the anterior pituitary gland does not secrete gonadotropins and the lack of these hormones inhibits gonad development.

Another function of melatonin involves the organization and regulation of circadian rhythms, or daily changes in physiologic processes that follow a regular pattern. An example of this is the sleepiness/wakefulness cycle. Increased plasma melatonin levels, which occur at night, are associated with sleepiness. The hormone also seems to play a role in hunger/satiety cycles, mood changes, and jet lag. The high nighttime level of melatonin seems to be a mechanism to daily "reset" the biological clock.

QUICK APPLICATIONS
Melatonin production appears to be related to the amount of light that enters through the eye. People who work at night and sleep during the day have a reversed cycle of melatonin production. The high melatonin levels occur during the day while they are asleep and the low levels are at night when they are working and light is entering the eye.

Other Endocrine Glands

In addition to the major endocrine glands, other organs have some hormonal activity as part of their function. These include the thymus, stomach, small intestines, heart, and placenta.

The **thymus** gland is located near the midline in the anterior portion of the thoracic cavity. It is posterior to the sternum and slightly superior to the heart. Through the production of the hormone **thymosin** (THYE-mohsin), the thymus gland assists in the development of certain white blood cells, called lymphocytes, that help protect the body against foreign organisms. In this way, the thymus gland plays an important role in the body's immune mechanism. If an infant is born without a thymus gland, the immune system does not develop properly and the body is highly susceptible to infections. The thymus gland is most active just before birth and during early childhood. It is relatively large in young children, and then gradually diminishes in size as an individual reaches adulthood.

The lining of the stomach, the **gastric mucosa,** produces a hormone, called **gastrin,** in response to the presence of food in the stomach. This hormone stimulates the production of hydrochloric acid and the enzyme pepsin, which are used in the digestion of food. In this case, the stomach produces the hormone and is also the target organ.

The mucosa of the **small intestine** secretes the hormones **secretin** and **cholecystokinin.** Secretin stimulates the pancreas to produce a bicarbonate-rich fluid that neutralizes the stomach acid. Cholecystokinin stimulates contraction of the gallbladder, which releases bile. It also stimulates the pancreas to secrete digestive enzymes.

Surprisingly, the heart acts as an endocrine organ in addition to its major role of pumping blood. Special cells in the wall of the upper chambers of the heart, called atria, produce a hormone called **atrial natriuretic hormone,** or **atriopeptin.** The primary effect of this hormone is the loss of sodium and water in the urine. The result of this action is a decrease in both blood volume and blood pressure.

The **placenta** develops in the pregnant female as a source of nourishment and gas exchange for the developing fetus. It also serves as a temporary endocrine gland. One of the hormones it secretes is **human chorionic gonadotropin** (hCG), which signals the mother's ovaries to secrete hormones to maintain the uterine lining so that it does not degenerate and slough off in menstruation. hCG reaches high levels early in pregnancy, and then decreases. The placenta also produces **estrogen** and **progesterone** during pregnancy.

Prostaglandins

Prostaglandins (prahs-tih-GLAN-dins) are potent chemical regulators that are produced in minute amounts and are found widely distributed in cells throughout the body. They are similar to hormones but different enough that they are not classified as hormones. These hormonelike substances are derivatives of arachidonic acid, one of the essential fatty acids, and a dietary deficiency of this fatty acid results in an inability to synthesize prostaglandins. While hormones are produced by specialized cells grouped together in structures called endocrine glands, prostaglandins are produced by cells widely distributed throughout the entire body. In contrast to a hormone, which is transported in the blood and may have an effect distant from its point of origin, a prostaglandin has a localized effect on or near the cell in which it is made. For this reason it is sometimes called a local hormone. In addition to being localized, the effect is immediate and short term. These compounds cannot be stored in the body but must be synthesized "on demand" and are readily inactivated.

Numerous and varied effects are attributed to prostaglandins. These are often confusing because the same substance may have opposite effects on different tissues. Some modulate hormone action; others affect smooth muscle contraction; still others are involved in blood clotting mechanisms—some promote clotting and others inhibit the process. Prostaglandins foster many aspects of the inflammatory process, including the development of fever and pain. Nonsteroidal antiinflammatory drugs, such as aspirin, ibuprofen, and acetaminophen, block the synthesis of prostaglandins, which reduces the fever and pain. Prostaglandins also appear to inhibit the gastric secretion of hydrochloric acid. Drugs that inhibit prostaglandins will increase gastric acid secretion and make a person susceptible to peptic ulcers. Some of the symptoms of premenstrual syndrome (severe menstrual cramps and pain) and premature labor are attributed to elevated levels of prostaglandins.

QUICK CHECK

10.16 What gland secretes melatonin?
10.17 What gland has an important role in the development of the immune system?
10.18 What organ secretes atrial natriuretic hormone?

 FOCUS ON AGING

With age, most endocrine glands show some degree of glandular atrophy, with increased amounts of fibrous tissue and fat deposits. However, the glands remain responsive to stimulation and secrete adequate amounts of hormones. Exceptions to this generalization are the gonads, which are discussed in Chapter 19. There is some evidence of a decline in the rate of hormone secretion, but this may be due to changes in the target tissues that decrease the cellular need for the hormone. There is also evidence of a reduction in target-tissue receptor sites or in their sensitivity. Whatever the reason for the decline in hormone secretion, it is accompanied by a decreased rate of metabolic destruction so that the blood levels of circulating hormones remain relatively constant throughout senescence. There is no evidence that age-related structural changes in endocrine glands have functional significance or contribute to the overall aging process.

Representative Disorders of the
Endocrine System

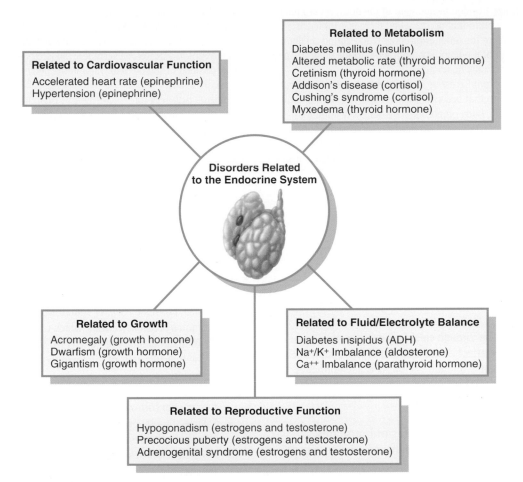

Related to Cardiovascular Function

Accelerated heart rate (epinephrine)
Hypertension (epinephrine)

Related to Metabolism

Diabetes mellitus (insulin)
Altered metabolic rate (thyroid hormone)
Cretinism (thyroid hormone)
Addison's disease (cortisol)
Cushing's syndrome (cortisol)
Myxedema (thyroid hormone)

**Disorders Related
to the Endocrine System**

Related to Growth

Acromegaly (growth hormone)
Dwarfism (growth hormone)
Gigantism (growth hormone)

Related to Fluid/Electrolyte Balance

Diabetes insipidus (ADH)
Na+/K+ Imbalance (aldosterone)
Ca++ Imbalance (parathyroid hormone)

Related to Reproductive Function

Hypogonadism (estrogens and testosterone)
Precocious puberty (estrogens and testosterone)
Adrenogenital syndrome (estrogens and testosterone)

Accelerated heart rate (ack-SELL-uhr-ayt-ehd HART RAYT) Unusually rapid heart rate due to increased amounts of epinephrine from the adrenal medulla; may also have other causes

Acromegaly (ack-roh-MEG-ah-lee) Enlargement of the extremities caused by excessive growth hormone in the adult

Addison's disease (ADD-ih-sunz dih-ZEEZ) Syndrome resulting from insufficient secretion of glucocorticoids and mineralcorticoids from the adrenal cortex; potentially life-threatening hypoglycemia and fluid and electrolyte disturbances may develop

Adrenogenital syndrome (ah-DREE-noh-jen-ih-tal SIN-drohm) Group of symptoms associated with alterations in secondary sex characteristics as a result of increased sex hormones from the adrenal cortex; increased androgens in the female lead to masculinization and increased amounts of estrogens in the male lead to breast enlargement (gynecomastia)

Altered metabolic rate (ALL-turhd meh-tah-BAHL-ik RAYT) Metabolic changes due to alterations in thyroid hormone activity; hypersecretion of thyroid hormones is characterized by a high metabolic rate, hyperactivity, insomnia, nervousness, irritability, and chronic fatigue;

hyposecretion is characterized by lethargy, weight gain, loss of hair, and low metabolic rate; in the infant it may lead to cretinism

Ca++ Imbalance Calcium imbalance due to alterations in parathyroid hormone secretion; parathyroid hormone is the most important regulator of blood calcium and its effect is to increase blood calcium

Cretinism (KREE-tin-izm) Dwarfism caused by a deficiency of thyroid hormone in childhood and usually accompanied by mental retardation

Cushing's syndrome (KOOSH-ingz SIN-drohm) Group of symptoms caused by prolonged exposure to high levels of cortisol; characterized by excessive deposition of fat in the subscapular area and in the face ("moon face"), high blood pressure, generalized weakness, and loss of muscle mass because of excessive protein catabolism; may be caused by overproduction of cortisol or by taking glucocorticoid hormone medications to treat inflammatory diseases such as asthma and rheumatoid arthritis

Diabetes insipidus (di-ah-BEE-teez in-SIP-ih-dus) Metabolic disorder caused by a deficient quantity of antidiuretic hormone, resulting in a large quantity of dilute urine (polyuria) and great thirst (polydipsia)

Diabetes mellitus (di-ah-BEE-teez mell-EYE-tus) Disorder caused by a deficiency of insulin from the beta cells of the pancreatic islets; characterized by a disturbance in the metabolism of blood glucose and manifested by polyuria, polyphagia, and polydipsia

Dwarfism (DWAR-fizm) Underdevelopment of the body, usually refers to the small body stature that occurs when there is insufficient growth hormone from the anterior pituitary gland during childhood; in this case the size of the limbs is proportional to the body size and mental development may be normal

Gigantism (jye-GAN-tizm) Excessive size and stature; usually refers to the abnormal growth that occurs if there is hypersecretion of growth hormone in the child; body proportions are generally normal

Hypertension (hye-purr-TEN-shun) Abnormally increased blood pressure due to increased amounts of epinephrine from the adrenal medulla; elevated blood pressure may also be due to other causes

Hypogonadism (hye-poh-GO-nad-izm) Sterility and lack of secondary sex characteristics caused by lack of estrogens in the female and androgens in the male

Myxedema (miks-eh-DEE-mah) Condition of swelling attributable to an accumulation of mucus in the skin; results from deficiency of thyroid hormone in the adult

Na⁺/K⁺ Imbalance Sodium and potassium imbalances due to alterations in aldosterone secretion; aldosterone causes a conservation of sodium and water in the body with a decrease of potassium ions

Precocious puberty (pree-KOH-she-us PEW-burr-tee) Unusually early sexual maturation caused by overproduction of estrogens in the female and androgens in the male.

CHAPTER SUMMARY

Introduction to the Endocrine System

■ **Compare the actions of the nervous system and the endocrine system.**

- The nervous system acts through electrical impulses and neurotransmitters; the effect is localized and of short duration.
- The endocrine system acts through chemicals called hormones; the effect is generalized and of long-term duration.

Characteristics of Hormones

■ **Compare the major chemical classes of hormones.**

- Hormones are classified chemically as either proteins or steroids.
- Most hormones in the body are proteins or protein derivates. The receptors for these hormones are located on the cell membrane.
- The sex hormones and those from the adrenal cortex are steroids. The receptors for these hormones are located within the cytoplasm of the cell.

■ **Discuss the general mechanisms of hormone action.**

- Hormones react with receptor sites on selected cells. The cells that have receptor sites for a specific hormone make up the target tissue for that hormone.
- Protein hormones react with receptors on the surface of the cell; steroids react with receptors inside the cell.
- Many hormones are regulated by a negative feedback mechanism; some hormones are secreted in response to other hormones; and a third method for regulating hormone secretion is by direct nerve stimulation.

Endocrine Glands and Their Hormones

■ **Identify the major endocrine glands and discuss their hormones and functions; pituitary, thyroid, parathyroid, adrenal, pancreas, gonads, thymus, and pineal.**

- The pituitary gland is divided into an anterior lobe, or adenohypophysis, which is regulated by releasing hormones from the hypothalamus, and a posterior lobe, or neurohypophysis, which is regulated by nerve stimulation.

- Hormones of the anterior lobe (adenohypophysis) include the following:
  - Growth hormone (GH), or somatotropic hormone (STH), promotes protein synthesis, which results in growth.
  - Thyroid-stimulating hormone (TSH) stimulates the activity of the follicular cells of the thyroid gland. These cells secrete thyroxine.
  - Adrenocorticotropic hormone (ACTH) stimulates activity of the adrenal cortex, particularly the secretion of cortisol.
  - Follicle-stimulating hormone (FSH) is a gonadotropin that stimulates the development of ova in the ovaries and sperm in the testes. It also stimulates the production of estrogen in the female.
  - Luteinizing hormone (LH), another gonadotropin, causes ovulation and secretion of progesterone and estrogen in females. In males it stimulates the production of testosterone.
  - Prolactin promotes the development of glandular tissue in the breast and stimulates the production of milk.
- Hormones of the posterior lobe (neurohypophysis) include the following:
  - Antidiuretic hormone (ADH) promotes reabsorption of water in the kidney tubules.
  - Oxytocin causes uterine muscle contraction and ejection of milk from the lactating breast.
- Thyroid gland hormones from the follicular cells include thyroxine, triiodothyronine, and calcitonin.
  - Both thyroxine and triiodothyronine require iodine for synthesis. About 95% of active thyroid hormone is thyroxine. Thyroid hormone secretion is regulated by a negative feedback mechanism that involves the amount of circulating hormone, the hypothalamus, and TSH from the adenohypophysis. Thyroid hormones affect the metabolism of carbohydrates, proteins, and lipids.
  - Calcitonin is produced by parafollicular cells in the thyroid gland. It reduces calcium levels in the blood.

- Parathyroid glands are embedded on the posterior surface of the thyroid gland. Parathyroid hormone, antagonistic to calcitonin, increases blood calcium levels.
- The adrenal (suprarenal) gland is separated into a cortex and a medulla.
  - All hormones from the adrenal cortex are steroids and are regulated by a negative feedback mechanism involving the hypothalamus and ACTH from the adenohypophysis. Hormones from the adrenal cortex are classified as mineralocorticoids, glucocorticoids, or sex steroids. The principal mineralocorticoid is aldosterone, which promotes sodium ion reabsorption in the kidney tubules. The principal glucocorticoid is cortisol, which increases blood glucose levels and helps to counteract the inflammatory response. Sex steroids from the adrenal cortex have minimal effect compared with the hormones from the ovaries and testes.
  - The two hormones produced by the adrenal medulla are epinephrine and norepinephrine. About 80% of the product is epinephrine. Epinephrine and norepinephrine prepare the body for strenuous activity and stress. The effect of epinephrine and norepinephrine is similar to that of the sympathetic nervous system but lasts up to 10 times longer.
- The endocrine portion of the pancreas consists of the pancreatic islets, or islets of Langerhans. Alpha cells in the islets produce glucagon, and the beta cells produce insulin. The principal action of glucagon is to raise blood glucose levels. Insulin, antagonistic to glucagon, decreases blood glucose levels.
- Testes produce the male sex hormones, which are collectively called androgens. The principal androgen is testosterone, which is responsible for the development and maintenance of male secondary sex characteristics.
- The ovaries produce estrogen and progesterone. Estrogen is responsible for the development and maintenance of female secondary sex characteristics. Progesterone maintains the uterine lining for pregnancy.

- The thymus, located near the midline in the anterior portion of the thoracic cavity, produces the hormone thymosin. Thymosin plays an important role in the development of the immune system of the body.
- The pineal gland extends posteriorly from the third ventricle of the brain. Secretory cells, called pinealocytes, synthesize and secrete the hormone melatonin. This hormone inhibits gonadotropin-releasing hormone from the hypothalamus, which inhibits reproductive junctions. Melatonin also regulates circadian rhythms.

■ **Name and describe the function of at least one hormone from the (a) gastric mucosa, (b) small intestine, (c) heart, and (d) placenta.**
- The gastric mucosa, the lining of the stomach, produces gastrin, a hormone that stimulates the production of hydrochloric acid and the enzyme pepsin.
- Cells in the duodenum of the small intestine secrete cholecystokinin and secretin. Cholecystokinin stimulates contraction of the gallbladder and stimulates the pancreas to produce an enzyme-rich fluid. Secretin stimulates the pancreas to produce a fluid that is rich in bicarbonate ions.
- Special cells in the wall of the heart produce atrial natriuretic hormone. The primary effect of this hormone is the loss of sodium and water in the urine, which decreases blood pressure.
- The placenta is a temporary endocrine gland that secretes human chorionic gonadotropin (hCG), which maintains the uterine lining during the first part of pregnancy.

Prostaglandins

■ **Differentiate between hormones and prostaglandins.**
- Prostaglandins are hormone-like molecules that are derived from arachidonic acid.
- They are produced by cells widely distributed throughout the body; their effect is localized near their origin; and their effect is immediate and short term.
- Prostaglandins are not stored; they are synthesized when needed.

CHAPTER QUIZ

Recall

Match the definitions on the left with the appropriate term on the right.

_____ 1. Product of an endocrine gland
_____ 2. Cells that have receptors for a given hormone
_____ 3. Lipid-soluble hormones
_____ 4. Produced by the adenohypophysis
_____ 5. Produced by the thyroid gland
_____ 6. Effect is to conserve sodium and water
_____ 7. Secreted by the adrenal medulla
_____ 8. Lowers blood glucose levels
_____ 9. Has short-term, localized, immediate effect
_____ 10. Source of melatonin

A. Aldosterone
B. Calcitonin
C. Epinephrine
D. Growth hormone
E. Hormone
F. Insulin
G. Pineal gland
H. Prostaglandin
I. Steroid
J. Target tissue

Thought

1. Steroid hormones:
   A. React with receptors on the surface of the cell membrane
   B. Trigger the release of a "second messenger," usually adenosine triphosphate
   C. Diffuse through the cell membrane to react with receptors
   D. Include the tropic hormones from the adenohypophysis

2. Which of the following hormones is *not* produced by the anterior lobe of the pituitary gland?
   A. Thyroid-stimulating hormone
   B. Antidiuretic hormone
   C. Prolactin
   D. Luteinizing hormone

3. Which of the following hormones is released in response to a releasing hormone from the hypothalamus?
   A. Oxytocin
   B. Thyroxine
   C. Aldosterone
   D. Follicle-stimulating hormone

4. The hormone that is antagonistic (has the opposite effect) to calcitonin is:
   A. Triiodothyronine
   B. Produced by the parathyroid gland
   C. Thymosin
   D. Produced by the pancreas

5. Susan has eaten a meal that is high in carbohydrates and finished the meal with a large banana split. As a result, her blood glucose level tends to increase. Which of the following will help maintain glucose homeostasis in this instance?
   A. Glucagon is released to lower the blood glucose level.
   B. Cortisol mobilizes glucose to remove it from the blood.
   C. Glucagon is released to stimulate the liver to break down glycogen.
   D. Insulin is released to promote cellular uptake of glucose.

Application

Candy, a high school senior, is pregnant. Also, she really does not pay much attention to what she eats—mostly fast foods, potato chips, and candy bars. She never drinks milk; instead, she usually has soft drinks, particularly colas. Under these circumstances, what is likely to happen to the amount of parathyroid hormone released from the gland? Explain.

Joe Cool is studying endocrinology and wants to compare two hormones—insulin and estrogen. Based on your knowledge of the chemical nature of hormones, their receptors, and mechanisms of action, predict which of the two hormones will cause a quicker metabolic response from their target cells. Explain.

BUILDING YOUR MEDICAL VOCABULARY

Building a vocabulary is a cumulative process. As you progress through this book, you will use word parts, abbreviations, and clinical terms from previous chapters. Each chapter will present new word parts, abbreviations, and terms to add to your expanding vocabulary.

Word Parts and Combining Form with Definition and Examples

PART/ COMBINING FORM	DEFINITION	EXAMPLE
acr/o	extremities	acromegaly: abnormal enlargement of the extremities of the skeleton
aden-	gland	adenoma: tumor of a gland
-agon	assemble, gather together	glucagon: raises circulating blood glucose level by stimulating its release from glycogen and gathering it in the bloodstream
-amine	nitrogen compound	histamine: nitrogen compound found in all body tissues

PART/ COMBINING FORM	DEFINITION	EXAMPLE
andr/o	male, maleness	androgens: steroid hormones that promote male characteristics
calc-	calcium	calcitonin: thyroid hormone that reduces blood calcium levels
cortic-	outer region, cortex	glucocorticoid: hormones from the adrenal cortex that raise blood sugar levers
crin-	to secrete	endocrine: glands that secrete their product into the bloodstream
dips-	thirst	polydipsia: excessive thirst
-gen-	to produce	androgens: hormones that produce male characteristics
-gest-	to carry, pregnancy	progesterone: hormone that helps sustain pregnancy
glyc/o	sugar, glucose	glycosuria: presence of glucose in the unine

PART/ COMBINING FORM	DEFINITION	EXAMPLE
gonad/o	gonad, primary sex organ	gonadotropin: any hormone that has a stimulating effect on the primary sex organs
lact-	milk	prolactin: hormone that stimulates the production of milk
-megaly	enlargement	acromegaly: abnormal enlargement of the extremities of the skeleton
oxy-	swift, rapid	oxytocin: hormone that stimulates a rapid birth
para-	beside	parathyroid: beside the thyroid; glands embedded in the thyroid
-physis	to grow	hypophysis: gland that grows down from the brain and is attached to it by a stalk; pituitary gland
pin-	pine cone	pineal body: body that resembles a pine cone
-ren-	kidney	adrenal: a gland that is on the top of the kidney
test-	eggshells, eggs	testes: egg-shaped male gonads
-toc-	birth	oxytocin: hormone that stimulates rapid birth
trop-	to change, influence	gonadotropin: hormone that influences the gonads
-uria	urine condition	polyuria: excessive urination

Clinical Abbreviations

ABBREVIATION	MEANING
ACTH	adrenocorticotropic hormone
ADH	antidiuretic hormone
BMR	basal metabolic rate
FBS	fasting blood sugar
FSH	follicle-stimulating hormone
GH	growth hormone (somatotropin)
GTT	glucose tolerance test
hCG	human chorionic gonadotropin
ICSH	interstitial cell–stimulating hormone

ABBREVIATION	MEANING
IDDM	insulin-dependent diabetes mellitus; type 1 diabetes or juvenile-onset diabetes
LH	luteinizing hormone (luteotropin)
MSH	melanocyte-stimulating hormone (melanotropin)
NIDDM	non–insulin-dependent diabetes mellitus; type 2 diabetes or maturity-onset diabetes
OXT	oxytocin
PBI	protein-bound iodine
PRL	prolactin (lactogenic hormone)
PTH	parathyroid hormone
T_3	triiodothyronine (thyroid hormone)
T_4	tetraiodothyronine (thyroid hormone)
TFT	thyroid function test
TSH	thyroid-stimulating hormone

Clinical Terms

Adenoma (add-eh-NOH-mah) Tumor of a gland

Anabolic steroids (an-ah-BAHL-ic STEER-oyds) Group of synthetic derivatives of testosterone that are used clinically to promote growth and repair of body tissues; used illegally by athletes to increase strength and muscle mass

Endocrinopathy (en-doh-krin-AHP-ah-thee) Any disease caused by a disorder of the endocrine system

Endocrinology (en-doh-krin-AHL-oh-jee) Study of the endocrine glands

Exophthalmic (eks-off-THAL-mick) Pertaining to an abnormal protrusion or bulging of the eye

Fasting blood sugar (FASS-ting BLOOD SHOOG-ahr) Measures circulating blood glucose level after fasting at least 4 hours

Glucose tolerance test (GTT) (GLOO-kohs TAHL-er-ans test) Blood sugar test performed at specified intervals after the patient has been given a certain amount of glucose; blood samples are drawn, and the blood glucose level of each sample is determined

Pheochromocytoma (fee-oh-kroh-moh-sye-TOE-mah) Benign tumor of the adrenal medulla that results in increased secretion of epinephrine and norepinephrine

Polydipsia (pah-lee-DIP-see-ah) Excessive thirst

Polyphagia (pah-lee-FAY-gee-ah) Excessive ingestion of food

Polyuria (pah-lee-YOO-ree-ah) Excessive urination

Progeria (pro-JEER-ih-ah) Condition of premature old age occurring in childhood, which may be due to hormone dysfunction

VOCABULARY QUIZ

Use word parts you have learned to form words that have the following definitions.

1. Substance for quick birth _____

2. Like a pine cone _____

3. Toward the kidney _____

4. Steroid before pregnancy _____

5. Passing through urine _____

Using the definitions of word parts you have learned, define the following words.

6. Adenoma _____

7. Endocrinology _____

8. Testosterone _____

9. Parathyroid _____

10. Adrenocorticotropin _____

Match each of the following definitions with the correct word.

_____ **11.** Responds to low glucose levels in the blood

_____ **12.** Surgical excision of the adrenal gland

_____ **13.** Excessive thirst

_____ **14.** Hormones that produce male characteristics

_____ **15.** Hormone necessary for milk production

A. Adrenalectomy
B. Androgens
C. Glucagon
D. Polyphagia
E. Prolactin

16. What is the clinical term for a "tumor of a gland"? _____

17. What is the clinical term for "excessive hunger or ingestion of food"? _____

18. What is the clinical term for "any disorder of the endocrine system"? _____

19. What is the meaning of the abbreviation IDDM? _____

20. What is the meaning of the abbreviation OXT? _____

Cardiovascular System: The Heart

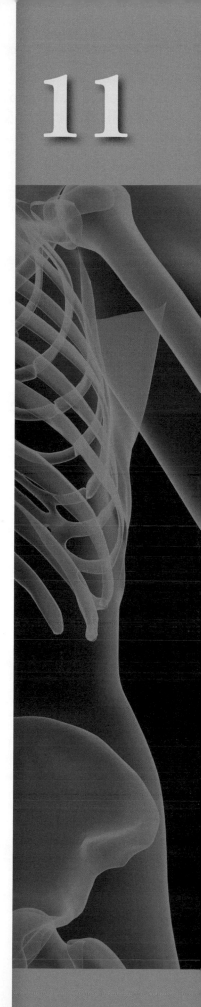

CHAPTER OBJECTIVES

Overview of the Heart
- Describe the size and location of the heart.
- Describe the coverings around the heart.

Structure of the Heart
- Identify the layers of the heart wall and state the type of tissue in each layer.
- Label a diagram of the heart, identifying the chambers, valves, and associated vessels.
- Trace the pathway of blood flow through the heart, including chambers, valves, and pulmonary circulation.
- Identify the major vessels that supply blood to the myocardium and return the deoxygenated blood to the right atrium.

Physiology of the Heart
- Describe the components and function of the conduction system of the heart.
- Identify the waves or deflections of an electrocardiogram.
- Summarize the events of a complete cardiac cycle.
- Correlate the heart sounds heard with a stethoscope with the events of the cardiac cycle.
- Explain what is meant by *stroke volume* and *cardiac output,* and describe the factors that affect these values.

KEY TERMS

Atrioventricular valve (ay-tree-oh-ven-TRIK-yoo-lar VALVE) Valve between an atrium and a ventricle in the heart

Cardiac cycle (KAR-dee-ak SYE-kul) Complete heartbeat consisting of contraction and relaxation of both atria and both ventricles

Cardiac output (KAR-dee-ak OUT-put) Volume pumped from one ventricle in 1 minute; usually measured from the left ventricle

Conduction myofibers (kon-DUCK-shun my-o-FYE-bers) Cardiac muscle cells specialized for conducting action potentials to the myocardium; part of the conduction system of the heart; also called Purkinje fibers

Conduction system (kon-DUCK-shun SIS-tem) Specialized cardiac muscle cells that coordinate contraction of the heart chambers

Diastole (dye-AS-toh-lee) Relaxation phase of the cardiac cycle; opposite of systole

Semilunar valve (seh-mee-LOO-nar VALVE) Valve between a ventricle of the heart and the vessel that carries blood away from the ventricle; also pertains to the valves in veins

Starling's law of the heart (STAR-lings LAW OF THE HART) Principle that the more cardiac muscle fibers are stretched, the greater will be the contraction strength of the heart

Stroke volume (STROAK VAHL-yoom) Volume of blood ejected from one ventricle during one contraction; normally about 70 milliliters

Systole (SIS-toh-lee) Contraction phase of the cardiac cycle; opposite of diastole

Functional Relationships of the
Cardiovascular System

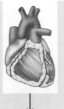

Delivers oxygen and nutrients; removes carbon dioxide and other metabolic wastes; dissipates heat.

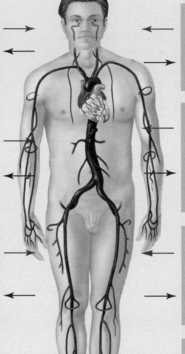

Reproductive

Gonads produce hormones that help maintain healthy blood vessels; testosterone stimulates erythropoiesis.

Transports reproductive hormones; provides nutrients for developing fetus and removes wastes; vasodilation responsible for erection in penis and clitoris.

Urinary

Helps maintain blood pH and electrolyte composition; adjusts blood volume and pressure; initiates renin-angiotensin-aldosterone mechanism.

Delivers wastes to be excreted in the urine; adjusts blood flow to maintain kidney function; transports hormones that regulate reabsorption in the kidneys.

Digestive

Absorbs nutrients for blood cell formation; absorption of nutrients affects plasma composition; liver metabolism affects blood glucose content.

Delivers hormones that affect the motility and glandular activity of the digestive tract; transports absorbed nutrients to liver.

Respiratory

Helps maintain blood pH; provides oxygen and removes wastes for cardiac tissue; breathing movements assist in venous return.

Transports oxygen and carbon dioxide between lungs and tissues.

Integument

Provides barrier against entry of pathogens; provides mechanical protection of superficial blood vessels; radiates heat for thermoregulation.

Transports clotting factors and phagocytic cells to sites of skin wounds; hemoglobin in blood contributes to skin color.

Skeletal

Protects heart and thoracic vessels; produces blood cells in bone marrow; supplies calcium for cardiac muscle contraction and blood clotting.

Delivers oxygen and nutrients and removes metabolic wastes; delivers erythropoietin to bone marrow; delivers hormones that regulate skeletal growth to osetoclasts and osteoblasts.

Muscular

Skeletal muscle contraction assists venous return, promotes growth of new vessels; smooth and cardiac muscle contraction contributes to vessel and heart function.

Removes heat and other waste products generated by muscle contraction; delivers oxygen for metabolism to sustain energy for muscle contraction.

Nervous

Adjusts heart rate to maintain adequate cardiac output; regulates blood pressure; controls blood flow patterns in systemic circulation.

Endothelial cells of capillaries help maintain blood-brain barrier and generate CSF.

Lymphatic/Immune

Defends against pathogens and toxins in the blood; fights infections in the heart; returns interstitial fluid to the blood.

Transports the agents of the immune response.

Endocrine

Hormones adjust heart rate, contraction strength, blood volume, blood pressure, and red blood cell production.

Delivers oxygen and nutrients to endocrine glands; removes carbon dioxide and heat; transports hormones from glands to target tissue; heart secretes natriuretic hormone.

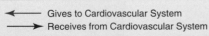
⟵——— Gives to Cardiovascular System
⟶——— Receives from Cardiovascular System

The heart is a muscular pump that provides the force necessary to circulate the blood to all the tissues in the body. Its function is vital because, to survive, the tissues need a continuous supply of oxygen and nutrients, and metabolic waste products have to be removed from them. Deprived of these necessities, cells soon undergo irreversible changes that lead to death. Although blood is the transport medium, the heart is the organ that keeps the blood moving through the vessels. The normal adult heart pumps about 5 liters of blood every minute throughout life. If it loses its pumping effectiveness for even a few minutes, the individual's life is jeopardized.

Overview of the Heart

Form, Size, and Location of the Heart

Knowledge of the heart's position in the thoracic cavity is important in hearing heart sounds, in conducting electrocardiograms (ECGs), and in performing cardiopulmonary resuscitation (CPR). The heart is located in the middle mediastinal region of the thoracic cavity between the two lungs as illustrated in Figure 11-1. It is posterior to the sternum and anterior to the vertebral column, and it rests on the diaphragm. About two thirds of the heart mass is to the left of the body's midline and one third is to the right. The **apex,** or pointed end of the heart, is directed inferiorly, anteriorly, and to the left. It extends downward to the level of the fifth intercostal space. The opposite end, the **base,** is larger and less pointed than the apex and has several large vessels attached to it. The base is directed superiorly, posteriorly, and to the right. Its most superior portion is at the level of the second rib. The size of the heart varies with the size of the individual. On average, it is about 9 centimeters (cm) wide and 12 cm long, about the size of a closed fist. Figure 11-2 shows a reconstructed three-dimensional

computed tomography (CT) image of the heart and associated vessels with the surrounding structures removed.

Coverings of the Heart

The heart and the proximal portions of the vessels attached to its base are enclosed by a loose-fitting, double-layered sac called the **pericardium** (pair-ih-KAR-dee-um), or pericardial sac. The outer layer of the pericardium is formed of tough, white fibrous connective tissue and is called the **fibrous pericardium.** It is attached to the diaphragm, the posterior portion of the sternum, the vertebrae, and the large vessels at the base of the heart. The fibrous pericardium is lined with a layer of serous membrane called the **parietal pericardium.** Where the pericardium is attached to the vessels at the base of the heart, the parietal pericardium reflects onto the surface of the heart to form the **visceral pericardium,** or **epicardium.** The small potential space between the parietal and visceral layers of the pericardium is the **pericardial cavity.** It contains a thin layer of serous fluid that reduces friction between the membranes as they rub against each other during heart contractions.

QUICK CHECK

11.1 The apex of the heart is located on which side of the sternum?
11.2 What is another term for the "visceral pericardium"?

QUICK APPLICATIONS
Pericarditis is an inflammation of the pericardium. This may interfere with production of the serous fluid that lubricates the surfaces of the parietal and visceral layers. Painful adhesions may form that interfere with contraction of the heart.

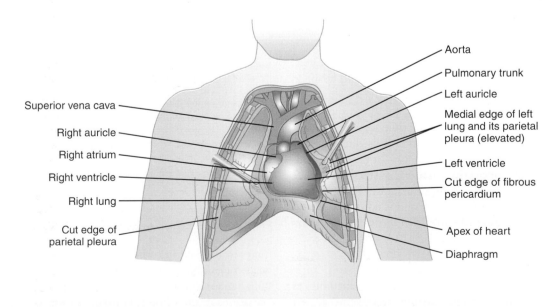

Figure 11-1 *Frontal view of the mediastinum, showing the position of the heart.*

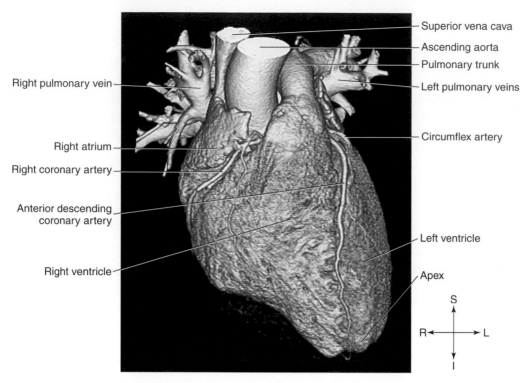

Figure 11-2 *Reconstructed CT image of the heart and associated vessels. (Adapted from Applegate E: The Sectional Anatomy Learning System, Concepts, ed. 3. St. Louis, Elsevier/Saunders, 2010.)*

Structure of the Heart

Layers of the Heart Wall

The heart wall is formed by three layers of tissue: an outer epicardium, a middle myocardium, and an inner endocardium. The **epicardium** (eh-pih-KAR-dee-um), which is the same as the visceral pericardium, is a serous membrane that consists of connective tissue covered by simple squamous epithelium. It is a thin protective layer that is firmly anchored to the underlying muscle. Blood vessels that nourish the heart wall are located in the epicardium.

The thick middle layer is the **myocardium** (my-oh-KAR-dee-um). It forms the bulk of the heart wall and is composed of cardiac muscle tissue. Refer to Chapter 4 for a review of the different types of muscle tissue. Cardiac muscle cells are elongated and branched with one or two nuclei per cell. The arrangement of myofilaments makes the fibers appear striated. The cells are connected by **intercalated** (in-TER-kuh-lay-ted) disks and wrapped around the heart in a spiral fashion to make up the myocardium. Contractions of the myocardium provide the force that ejects blood from the heart and moves it through the vessels.

The smooth inner lining of the heart wall is the **endocardium** (en-doh-KAR-dee-um), a layer of simple squamous epithelium overlying connective tissue. Its smooth surface permits blood to move easily through the heart. The endocardium also forms the valves of the heart and is continuous with the lining of the blood vessels. Figure 11-3 illustrates the layers of the heart wall.

Chambers of the Heart

The internal cavity of the heart is divided into four chambers as illustrated in Figure 11-4; the right atrium, right ventricle, left atrium, and left ventricle. The two atria are thin-walled chambers

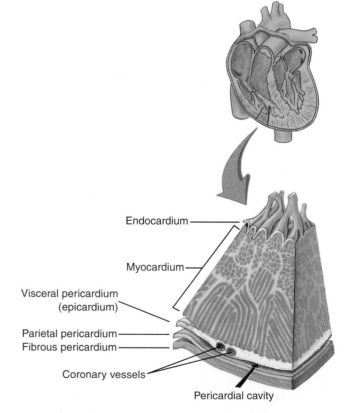

Figure 11-3 *Layers of the heart wall.*

that receive blood from the veins. The two ventricles are thick-walled chambers that forcefully pump blood out of the heart. Differences in thickness of the heart chamber walls are due to variations in the amount of myocardium present, which reflects the amount of force each chamber is required to generate.

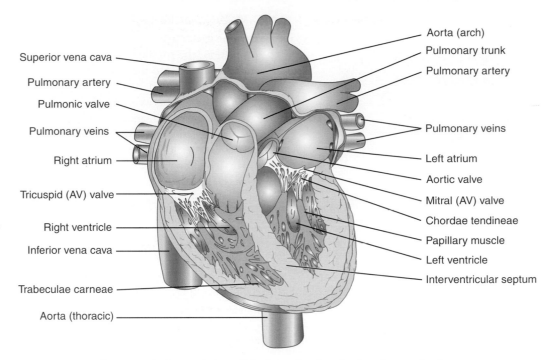

Superior vena cava

Pulmonary artery

Pulmonic valve

Pulmonary veins

Right atrium

Tricuspid (AV) valve

Right ventricle

Inferior vena cava

Trabeculae carneae

Aorta (thoracic)

Aorta (arch)

Pulmonary trunk

Pulmonary artery

Pulmonary veins

Left atrium

Aortic valve

Mitral (AV) valve

Chordae tendineae

Papillary muscle

Left ventricle

Interventricular septum

Figure 11-4 *Internal view of the heart showing the chambers and valves.*

The **right atrium** (AY-tree-um) receives deoxygenated blood from the superior vena cava, the inferior vena cava, and the coronary sinus. The superior vena cava returns blood to the heart from the head, neck, and upper extremities. The inferior vena cava returns blood to the heart from the thorax, abdomen, pelvis, and lower extremities. The coronary sinus is a small venous structure on the posterior surface of the heart that returns blood to the right atrium from the myocardium of the heart wall. The **left atrium** receives oxygenated blood from the lungs through four pulmonary veins—two on the right and two on the left. Because the atria are "receiving" chambers rather than "pumping" chambers, their myocardium is relatively thin, which is reflected in their thin walls. Both atria have small extensions, called **auricles,** that protrude anteriorly. The right and left atria are separated by a partition called the **interatrial septum.** There is a thin region, called the **fossa ovalis,** in the interatrial septum. This represents an opening, the foramen ovale, that is present between the atria in the fetal heart.

The **right ventricle** (VEN-trih-kull) receives blood from the right atrium and pumps it out to the lungs, where it picks up a new supply of oxygen. The **left ventricle,** which forms the apex of the heart, receives blood from the left atrium and pumps it out to the tissues of the whole body, where the oxygen transported by the blood is used in metabolic activities. The ventricles are "pumping" chambers, and this is reflected by a thick myocardium. Because the left ventricle pumps blood to the whole body and the right ventricle only sends blood to the lungs, the left ventricle has to generate a lot more pumping force than the right ventricle. This is reflected in the fact that the left ventricular wall has a thicker myocardium than the right ventricular wall. Both ventricles hold about the same volume of blood. In both ventricles, the myocardium is marked by ridges called **trabeculae carneae** (trah-BEK-yoo-lee KAR-nee-ee). Two or three fingerlike masses of myocardium, called

papillary (PAP-ih-lair-ee) **muscles,** project from the wall of the ventricle into the chamber (see Figure 11-4). The thick, muscular partition between the right and left ventricles is the **interventricular septum.**

QUICK CHECK

11.3 Which chamber of the heart has the thickest myocardium?

11.4 What two heart chambers contain oxygen-rich blood?

Valves of the Heart

Pumps need a set of valves to keep the fluid flowing in one direction, and the heart is no exception. The heart has two types of valves that keep the blood flowing in the correct direction. The valves between the atria and ventricles are called **atrioventricular (AV)** (ay-tree-oh-ven-TRIK-yoo-lar) valves, whereas those at the bases of the large vessels leaving the ventricles are called **semilunar (SL)** (seh-mee-LOO-nar) valves. Figure 11-5 illlustrates the valves of the heart.

Atrioventricular Valves

The AV valves permit the flow of blood from the atria into the corresponding ventricle but prevent the backflow of blood from the ventricles into the atria. Each valve consists of a fibrous connective tissue ring, which reinforces the junction between the atrium and ventricle, and double folds of endocardium that form the **cusps** of the valve (Figure 11-5). The valve cusps are attached to the papillary muscles in the ventricles by connective tissue strings called **chordae tendineae** (KOR-dee ten-DIN-ee) (see Figure 11-4). As blood returns to the atria while the ventricles are relaxed, it pushes open the valve cusps and the blood flows into the ventricles (see Figure 11-6). When the ventricles contract, the

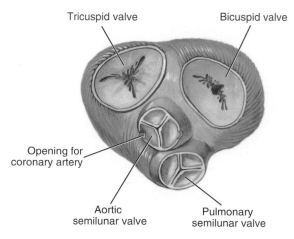

Tricuspid valve

Bicuspid valve

Opening for
coronary artery

Aortic
semilunar valve

Pulmonary
semilunar valve

Figure 11-5 *Valves of the heart as viewed from above.*

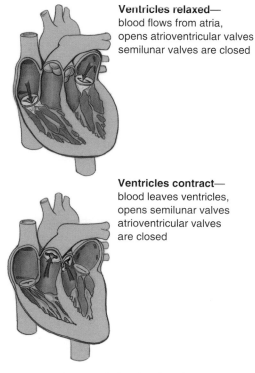

Ventricles relaxed—
blood flows from atria,
opens atrioventricular valves
semilunar valves are closed

Ventricles contract—
blood leaves ventricles,
opens semilunar valves
atrioventricular valves
are closed

Figure 11-6 *Open and closed valves during ventricular relaxation and contraction.*

force of the blood against the cusps causes them to close and prevents the backward flow of blood into the atria. With myocardial contraction, the papillary muscles exert tension on the chordae tendineae attached to the cusps and prevent the valve from opening back into the atria.

The AV valve between the right atrium and right ventricle, located at the fourth intercostal space, has three cusps and is called the **tricuspid valve.** The valve between the left atrium and left ventricle has only two cusps, is at the level of the fourth costal cartilage, and is called the **bicuspid,** or **mitral, valve**. Figure 11-7 illustrates the surface projection of the heart valves.

Semilunar Valves

The SL valves are located at the bases of the large vessels that carry blood from the ventricles. Each valve consists of three cuplike cusps of fibrous connective tissue and endothelium (see Figure 11-5). Contraction of the ventricular myocardium increases the pressure of the blood so that it pushes open the valves and the blood leaves the heart. As the ventricles relax and pressure decreases, the blood starts to flow back down the large vessels toward the ventricles. When the blood flows toward the ventricles, it enters the "cups" of the valve cusps and causes them to meet in the center of the vessel. This closes the opening and prevents the flow of blood back into the ventricles (see Figure 11-6).

The valve at the exit of the right ventricle is in the base of the pulmonary trunk and is called the **pulmonary semilunar valve.** It is at the level of the third costal cartilage. The valve at the exit of the left ventricle is in the base of the ascending aorta. It is called the **aortic semilunar valve** and is at the level of the third intercostal space (see Figure 11-7).

QUICK APPLICATIONS

Sometimes disease processes damage the heart valves so that they are unable to function properly. Incompetent valves permit a "backflow" of blood, and the heart has to pump the same blood over and over to get it into the vessels. In valvular stenosis, the valves are stiff and have narrow openings. The heart has to work harder to pump blood out through the small opening.

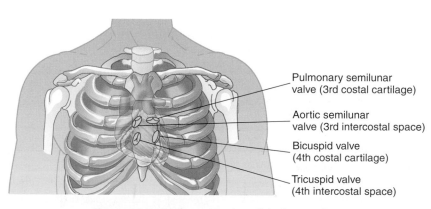

Pulmonary semilunar
valve (3rd costal cartilage)

Aortic semilunar
valve (3rd intercostal space)

Bicuspid valve
(4th costal cartilage)

Tricuspid valve
(4th intercostal space)

Figure 11-7 *Surface projection of the heart valves.*

Pathway of Blood Through the Heart

Although it is convenient to describe the flow of blood through the right side of the heart and then through the left side, it is important to realize that both atria contract at the same time and that both ventricles contract at the same time. The heart functions as two pumps, one on the right and one on the left, that work simultaneously. The "right pump" pumps the blood to the lungs (pulmonary circulation) at the same time that the "left pump" pumps blood to the rest of the body (systemic circulation). The sequence in which the chambers contract is described in more detail in Cardiac Cycle.

Figure 11-8 depicts the flow of blood through the heart. Venous blood from the systemic circulation, which is relatively low in oxygen and high in carbon dioxide content, enters the **right atrium** through the superior vena cava and inferior vena cava. This blood flows through the **tricuspid valve** into the **right ventricle.** From the right ventricle, it passes through the **pulmonary semilunar valve** into the **pulmonary trunk** and then into the **pulmonary arteries,** which carry the blood to the **lungs.** In the capillaries of the lungs, the blood releases carbon dioxide and picks up a new supply of oxygen; then **pulmonary veins** carry the blood to the **left atrium.** From the left atrium, it flows through the **bicuspid valve** into the **left ventricle,** and then through the **aortic semilunar valve**

into the **ascending aorta.** Oxygen-rich blood flowing into the aorta is distributed to all parts of the body through the systemic circulation.

Blood Supply to the Myocardium

The myocardium of the heart wall is working muscle that needs a continuous supply of oxygen and nutrients to function with efficiency. Unlike skeletal muscle, cardiac muscle cannot build up an oxygen debt to be repaid at a later time. It needs a continuous oxygen supply or it dies. For this reason, cardiac muscle has an extensive network of blood vessels to bring oxygen to the contracting cells and to remove waste products.

> ### QUICK APPLICATIONS
>
> If a branch of a coronary artery becomes blocked, blood supply to that region of the heart is cut off and the muscle cells in that area die of lack of oxygen. This is a **myocardial infarction (MI),** also called a coronary or a heart attack. The extent of the damage and chances of recovery depend on the location of the blockage and the length of time that elapses before medical intervention occurs.

Two main coronary arteries branch from the ascending aorta just distal to the aortic SL valve. Figure 11-9 is a CT image that shows the aortic SL valve and coronary arteries. The **right coronary artery** extends to the right from the ascending aorta and continues in the right AV sulcus to the posterior surface of the heart. Its branches supply blood to most of the myocardium in the right ventricle. The **left coronary artery,** which extends to the left from the ascending aorta, continues for about 2 cm and then divides into two major branches. The **anterior interventricular** (descending) **artery** descends in the anterior interventricular sulcus. The **circumflex artery** continues in the left AV sulcus to the posterior surface. The left coronary artery and its branches supply blood to most of the myocardium in the left ventricle. The

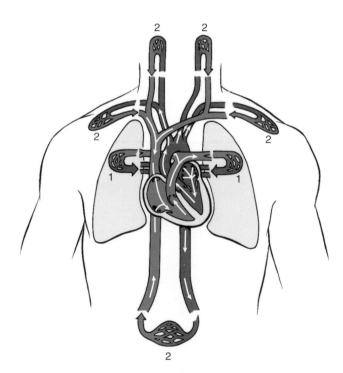

Blue = deoxygenated blood   Red = oxygenated blood

1 = capillary beds of lungs
  where gas exchange occurs

2 = capillary beds of body tissues
  where gas exchange occurs

Figure 11-8 *Pathway of the blood through the heart: SVC and IVC → Right atrium → Tricuspid valve → Right ventricle → Pulmonary SL valve → Pulmonary trunk → Pulmonary arteries → Capillaries of lungs → Pulmonary veins → Left atrium → Bicuspid valve → Left ventricle → Aortic semilunar valve → Ascending aorta → Systemic circulation.*

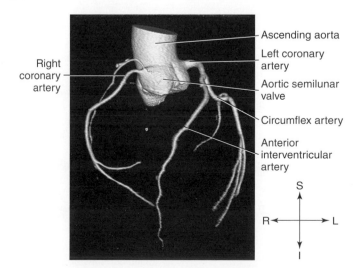

Right coronary artery

Ascending aorta

Left coronary artery

Aortic semilunar valve

Circumflex artery

Anterior interventricular artery

Figure 11-9 *CT image of the coronary arteries. (Adapted from Applegate E: The Sectional Anatomy Learning System, Concepts, ed. 3. St. Louis, Elsevier/Saunders, 2010.)*

arteries have numerous branches and they anastomose freely to provide alternative pathways for blood flow to the myocardium. Blood flow through the coronary arteries is greatest when the myocardium is relaxed. When the ventricles contract, they compress the arteries, which reduces the flow.

After blood passes through the capillaries in the myocardium, it enters a system of **cardiac** (coronary) **veins.** The cardiac veins lie next to the coronary arteries, and the vessels are usually surrounded by deposits of fat. Most of the cardiac veins drain into the coronary sinus, which opens into the right atrium. The coronary arteries and cardiac veins are illustrated in Figure 11-10.

QUICK CHECK

11.5 Certain bacterial diseases may damage heart valves. Damage to the left semilunar valve interferes with blood flow into what chamber or vessel?

11.6 What two structures work together to prevent atrioventricular valves from opening back into the atria?

11.7 What are the two main branches of the left coronary artery?

QUICK APPLICATIONS

When heart muscle is damaged, the dying cells release enzymes into the bloodstream. These enzymes can be measured and are useful in confirming an MI. The enzymes assayed are creatine kinase (CK) and lactate dehydrogenase (LDH).

Physiology of the Heart

The work of the heart is to pump blood to the lungs through the pulmonary circulation and to the rest of the body through the systemic circulation. This is accomplished by systematic contraction and relaxation of the cardiac muscle in the myocardium. The intercalated disks permit impulses to travel rapidly between adjacent cells so that they function together as a single electrical unit instead of as individual cells. Effective contractions of the heart depend on this characteristic, and they are coordinated by the conduction system of the heart.

Conduction System

An effective cycle for productive pumping of blood requires that the heart be synchronized accurately. Both atria need to contract simultaneously, followed by contraction of both ventricles. Contraction of the chambers is coordinated by specialized cardiac muscle cells that make up the **conduction system** of the heart. These cells contain only a few myofibrils, and instead of contracting they act somewhat like neural tissue by initiating and distributing impulses throughout the myocardium to coordinate the events of the cardiac cycle.

Components of the Conduction System
Sinoatrial Node

The conduction system includes several components (Figure 11-11). The first part of the conduction system is the **sinoatrial (SA) node** (sye-noh-AY-tree-al node), which is located in the posterior wall of the right atrium, near the entrance of the superior vena cava. Without any neural stimulation, the SA

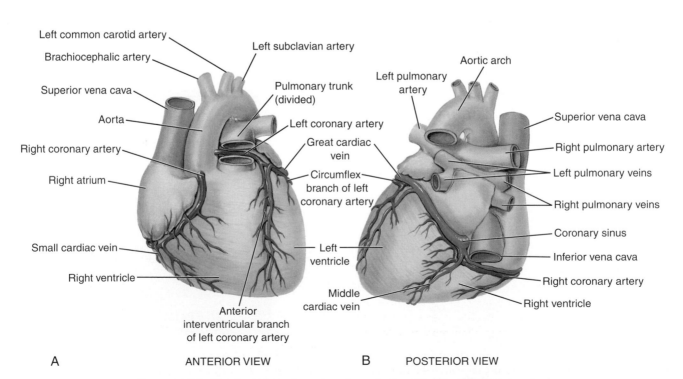

Figure 11-10 *Blood supply to the myocardium. **A,** Anterior view. **B,** Posterior view.*

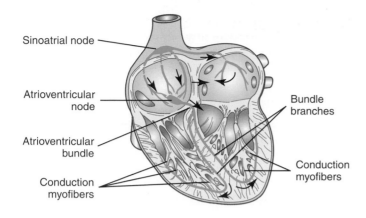

Figure 11-11 *Conduction system of the heart. Impulses travel from the sinoatrial (SA) node (pacemaker) → Atrioventricular (AV) node → AV bundle → Right and left bundle branches → Conduction myofibers → Myocardium.*

node rhythmically initiates impulses (action potentials) 70 to 80 times per minute. Because it establishes the basic rhythm of the heartbeat, it is called the **pacemaker** of the heart. The impulses from the SA node rapidly travel throughout the atrial myocardium and cause the two atria to contract simultaneously. At the same time, the impulses reach the second part of the conduction system.

Atrioventricular Node

The **AV node,** the second part of the conduction system, is located in the floor of the right atrium, near the interatrial septum. The cells in the AV node conduct impulses more slowly than do other parts of the conduction system so there is a brief time delay as the impulses travel through the node. This allows time for the atria to finish their contraction phase before the ventricles begin contracting.

Atrioventricular Bundle, Bundle Branches, and Conduction Myofibers

From the AV node, the impulses rapidly travel through the **atrioventricular bundle** (bundle of His) to the **right** and **left bundle branches.** The bundle branches extend along the right and left sides of the interventricular septum to the apex. These branch profusely to form **conduction myofibers** (Purkinje fibers) that transmit the impulses to the myocardium. The AV bundle, bundle branches, and conduction myofibers rapidly transmit impulses throughout all the ventricular myocardium so that both ventricles contract at the same time. As the ventricles contract, blood is forced out through the SL valves into the pulmonary trunk and the ascending aorta. After the ventricles complete their contraction phase, they relax and the SA node initiates another impulse to start another cardiac cycle.

All parts of the conduction system and cardiac muscle cells are capable of initiating impulses to start a cardiac cycle. The SA node is called the pacemaker because impulses spontaneously originate in the SA node faster than in any other part of the heart. If the SA node is unable to function, another area such as the AV node becomes the pacemaker. The resulting heart rate is slower than normal.

QUICK APPLICATIONS

Variations in normal contraction patterns are called *arrhythmias.* One type of arrhythmia occurs when a conduction myofiber or a heart muscle cell independently depolarizes to threshold and triggers a **premature heart contraction.** The cell responsible for the premature contraction is called an **ectopic focus.** Sometimes ectopic foci form feedback loops within the conduction system, causing myocardial contractions to occur at a rapid rate. If not treated properly, this often leads to ventricular tachycardia, fibrillation, and death.

FROM THE PHARMACY

Three representative preparations for the treatment of heart disorders are presented here. One is a cardiac glycoside used to treat congestive heart failure, another relieves the pain of angina pectoris, and the third is an antidysrhythmic agent.

Digoxin (Lanoxin, Cardoxin) is a preparation of digitalis, which is a cardiac glycoside derived from the foxglove plant. For hundreds of years, the foxglove plant was used by commoners (it was called a housewife's recipe) as an herbal treatment for "dropsy," or fluid accumulation. In the eighteenth century, digitalis was accepted by the medical community for the treatment of edema, and it is still used today to treat congestive heart failure. It affects heart function by increasing the strength and regularity of contractions to increase cardiac output. It also slows the impulse conduction speed through the heart and decreases the heart rate to allow more time for ventricular filling.

Nitroglycerin (Nitrostat, Nitrogard) is a vasodilator used to relieve the pain of angina pectoris. By dilating the coronary arteries, it provides more efficient blood flow through the coronary arteries. By dilating the peripheral resistance arteries, it reduces the demand on the heart "pump" and thus lessens the myocardial oxygen requirement. Nitroglycerin may also be used to treat congestive heart failure and hypertension.

Quinidine (Quinaglute, Cardioquin, Quinora) is a group I antidysrhythmic drug. Group I drugs bind to sodium channels, thus interfering with the production of action potentials. The result is a decrease in conduction velocity. On the ECG there is a broadening of the QRS complex that indicates slower conduction through the conduction myofibers of the ventricles. Abnormal or ectopic pacemaker sites appear to be more sensitive to quinidine than the normal pacemaker, the SA node. By reducing or eliminating impulses from ectopic sites, the SA node is able to reestablish normal cardiac rhythms.

Electrocardiogram

Impulses conducted through the heart during a cardiac cycle produce electric currents that can be detected and measured on the surface of the body. A recording of this electrical activity is called an **electrocardiogram (ECG** or **EKG).**

A normal ECG consists of waves or deflections that correlate with the depolarization and repolarization events of the cardiac cycle (Figure 11-12). The **P wave** is a small upward deflection produced by depolarization of the atrial myocardium as the impulses travel through the myocardium just before contraction. The **QRS complex** is a large upward deflection produced by depolarization of the ventricular myocardium immediately preceding contraction of the ventricles. The greater magnitude of the QRS complex is due to the greater muscle mass in the ventricles. The third deflection, the **T wave,** is due to the repolarization

of the ventricles that occurs just before the ventricles relax. A deflection that corresponds to atrial repolarization is not evident because it occurs at the same time as the QRS complex.

> ### QUICK CHECK
> 11.8 What will be the effect on heart rate if the sinoatrial node is not functioning?
> 11.9 What is the purpose of the slower impulse conduction or delay in the atrioventricular node?
> 11.10 What portion of the ECG roughly corresponds to atrial depolarization?

Cardiac Cycle

The cardiac cycle refers to the alternating contraction and relaxation of the myocardium in the walls of the heart chambers, coordinated by the conduction system, during one heartbeat (Figure 11-13). The two atria contract at the same time; then they relax while the two ventricles simultaneously contract. The contraction phase of the chambers is called **systole** (SIS-toh-lee); the relaxation phase is called **diastole** (dye-AS-toh-lee). When the terms systole and diastole are used alone, they refer to action of the ventricles.

With a heart rate of 75 beats per minute, one cardiac cycle lasts 0.8 second. The cycle begins with **atrial systole,** when both atria contract (see Figure 11-13). During this time, the AV valves are open, the ventricles are in diastole, and blood is forced into the ventricles. Atrial systole lasts for 0.1 second; then the atria relax (**atrial diastole**) for the remainder of the cycle, 0.7 second.

When the atria finish their contraction phase, the ventricles begin contracting. **Ventricular systole** lasts for 0.3 second. Pressure in the ventricles increases as they contract. This closes

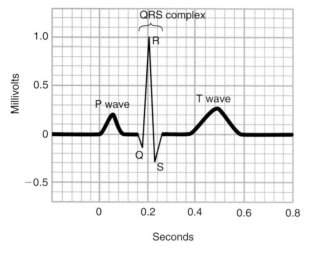

Figure 11-12 *Electrocardiogram showing the P wave, QRS complex, and T wave.*

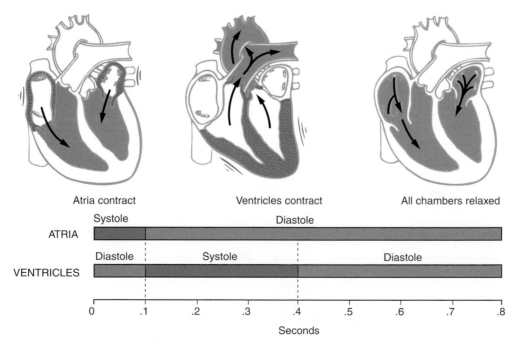

Figure 11-13 *Cardiac cycle. Atrial systole is followed by ventricular systole. All chambers are simultaneously in diastole for one half of the cycle.*

the AV valves and opens the SL valves, and blood is forced into the pulmonary trunk and ascending aorta that carry blood away from the heart. During this time, the atria are in diastole and are filling with blood returned through the venae cavae. After ventricular systole, when the ventricles relax, the SL valves close, the AV valves open, and blood flows from the atria into the ventricles. All chambers are in simultaneous diastole for 0.4 second, and about 70% of ventricular filling occurs during this period. The remaining blood enters the ventricles during atrial systole.

Heart Sounds

The sounds associated with the heartbeat are due to vibrations in the tissues and blood caused by closure of the valves. A **stethoscope** is used to listen to these sounds, usually described as *lubb-dupp*. The **first heart sound,** the *lubb,* is caused by closure of the AV valves near the beginning of ventricular systole. The **second heart sound,** the *dupp,* is caused by closure of the SL valves near the beginning of ventricular diastole. It has a higher pitch than the first heart sound. The time between the first and second heart sounds represents the period of ventricular systole. The time between the second heart sound and the first heart sound of the next beat represents the period of ventricular diastole. Because diastole lasts longer than systole, there is a pause between the *dupp* of the first beat and the *lubb* of the second beat so that the sequence is *lubb-dupp*, pause, *lubb-dupp*, pause, *lubb-dupp*, pause, and so on. Abnormal heart sounds, called **murmurs,** are caused by faulty valves.

Cardiac Output

Cardiac output is the volume of blood pumped by a ventricle in 1 minute. Because the primary function of the heart is to pump blood, this is a measure of its effectiveness. Cardiac output is calculated by multiplying the volume (in milliliters [ml]) pumped out in one cardiac cycle (stroke volume) times the number of cycles or heartbeats in 1 minute (min) (heart rate):

Cardiac output = Stroke volume × Heart rate
(ml/min)   (ml/cycle)   (cycles/min)

In a normal resting adult, the stroke volume averages about 70 milliliters and the heart rate averages about 72 beats per minute. Cardiac output is calculated as follows:

Cardiac output = 70 ml/beat × 72 beats/min = 5040 ml/min

This is approximately equal to the total volume of blood in the body. This means that when the body is at rest, the heart pumps the body's total blood volume out to the systemic circulation every minute. When the needs of the body's cells change, the cardiac output must change also. For example, with strenuous activity the skeletal muscles need more oxygen. In response, the heart rate increases to provide additional cardiac output so that more oxygen is delivered to the muscle cells. Anything that affects either the stroke volume or the heart rate changes the cardiac output. Various control mechanisms operate on the stroke volume and the heart rate to adjust the cardiac output as the needs of the body change.

Stroke Volume

Stroke volume is the amount of blood pumped from a ventricle each time the ventricle contracts. Stroke volume depends on the amount of blood in the ventricle when it contracts (end-diastolic volume) and the strength of the contraction (Figure 11-14). The end-diastolic volume, the amount of blood in the ventricle at the end of diastole (or beginning of systole), is directly related to venous return. The more blood returned by the veins, the greater is the volume in the ventricle to be pumped out again. In this way, increased venous return increases end-diastolic volume, which increases stroke volume.

The amount of blood in the ventricle also affects contraction strength. There is a direct relationship between venous return, end-diastolic volume, and contraction strength. This relationship is known as **Starling's law of the heart.** As blood fills the ventricles, the cardiac muscle fibers stretch to accommodate the increasing volume. In response to stretch, the fibers contract with a greater force, which increases the amount of blood ejected from the ventricle (stroke volume). Conversely, if venous return decreases, end-diastolic volume decreases, there is less stretch in the muscle fibers, and contraction strength decreases.

The autonomic nervous system also affects stroke volume by altering the contraction strength. Sympathetic stimulation increases the contraction strength of the ventricular myocardium. When sympathetic stimulation is removed, the contraction strength decreases.

Heart Rate

The SA node, acting alone, produces a constant rhythmic heart rate. Regulating factors act on the SA node to increase or decrease the heart rate to adjust cardiac output to meet the

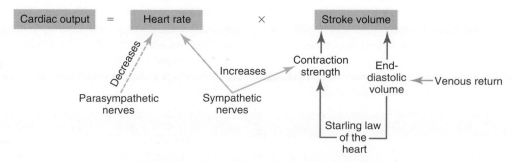

Figure 11-14 *Factors that affect cardiac output.*

changing needs of the body. Most changes in the heart rate are mediated through the **cardiac center** in the medulla oblongata of the brain. This center has both sympathetic and parasympathetic components to regulate the action of the heart. Factors such as blood pressure levels and the relative need for oxygen determine which component is active. Generally, the sympathetic impulses increase the heart rate and cardiac output, whereas parasympathetic impulses decrease the heart rate.

Baroreceptors are stretch receptors in the wall of the aorta and in the walls of the internal carotid arteries, which deliver blood to the brain. As blood pressure increases, the vessels are stretched, which increases the frequency of impulses going from the receptors to the medulla oblongata. This prompts the cardiac center to increase parasympathetic stimulation and to decrease sympathetic stimulation, so heart rate, cardiac output, and blood pressure decrease. As blood pressure decreases, the frequency of impulses also decreases. In response, the cardiac center increases sympathetic impulses and decreases parasympathetic impulses to increase the pressure.

Other factors also influence heart rate. Emotions such as excitement, anxiety, fear, or anger cause increased sympathetic stimulation, which results in increased heart rate. Depression influences the parasympathetic system and decreases the heart rate. Epinephrine, secreted from the adrenal medulla in response to stress, increases the heart rate. Increased carbon dioxide concentration and decreased pH of the blood, detected by chemoreceptors, increase the heart rate to deliver more blood to the tissues. Correct concentrations of potassium, calcium, and sodium ions are important to maintaining a regular heartbeat. Body temperature affects the metabolic rate of all cells, including cardiac muscle. Elevated body temperature increases the heart rate, whereas decreased temperature reduces the rate. Body temperature sometimes is deliberately decreased below normal (hypothermia) to reduce heart action during surgery.

QUICK CHECK

11.11 In a normal cardiac cycle, what is the length of time for ventricular systole?

11.12 What causes the heart sounds that are heard in a stethoscope?

11.13 What is the definition of *stroke volume*?

11.14 What two main factors directly determine stroke volume?

FOCUS ON AGING

There are numerous "age-related" changes in the heart. How many of these are due to an actual aging process and how many are due to other factors are questions worth considering. Is it possible to lessen the effects of aging by adjusting lifestyle? Cardiac changes that were once thought to be the result of aging are now believed to be the consequence of a sedentary lifestyle that many consider their "reward" after retirement. Other cardiac changes are due to a lifetime of habits that, while seemingly enjoyable at the time, take their toll later in life. It is difficult to isolate the aging process of the heart because it is so closely related to diet, exercise, and disease processes, but even when these factors are excluded, a clinical pattern of cardiac aging emerges.

In the absence of cardiovascular disease, the heart, particularly the left ventricle, tends to become slightly smaller in elderly people. This is partially due to a decrease in the number and size of cardiac muscle cells and partially due to the reduced demands placed on it by decreasing physical activity. However, because of cardiovascular disease, the heart is often enlarged.

There is a general thickening of the endocardium and valves of the heart as part of the aging process. The valves tend to become more rigid and incompetent. Thus, heart murmurs are detected more frequently in the elderly. Structural changes also occur throughout the conduction system as conducting myofibers are replaced with fibrous tissue. Usually this does not alter the resting pulse rate, but there is a greater than normal increase in heart rate in response to activity. Arrhythmias also are more frequent.

There appears to be no significant change in resting heart rate or stroke volume; thus, cardiac output remains about the same in the elderly. However, there is a decline in cardiac reserve. There is decreasing ability for the heart to respond to stress, either sudden or prolonged. Sympathetic controls of the heart are less effective, and the aging heart rate becomes more variable. Physically active people have less change in this respect than the sedentary elderly.

Numerous disease processes that occur more frequently in the aging individual have an effect on the heart. Most notable of these is arteriosclerosis, or hardening of the arteries. This puts additional stress on the heart and aggravates the normal age-related changes. Prevention is better and less expensive than cure. Adopt a "healthy heart" lifestyle.

CHAPTER SUMMARY

Overview of the Heart

■ **Describe the size and location of the heart.**
- The heart, about the size of a closed fist, is located in the middle mediastinum between the second and sixth ribs. The apex points downward and to the left so that two thirds of the mass is on the left side.

■ **Describe the coverings around the heart.**
- The heart is enclosed in a double-layered pericardial sac. The outer layer is fibrous connective tissue. The inner layer is parietal serous membrane. The visceral layer of the serous membrane forms the surface of the heart and is called the epicardium. The space between the parietal and visceral layers of the serous membrane is the pericardial cavity.

Structure of the Heart

■ **Identify the layers of the heart wall, and state the type of tissue in each layer.**
- The outermost layer of the heart wall is the visceral layer of the serous pericardium and is called the epicardium.
- The middle layer of the heart wall is the cardiac muscle tissue. It is the thickest layer and is called the myocardium.
- The innermost layer is simple squamous epithelium and is called the endocardium.

■ **Label a diagram of the heart, identifying the chambers, valves, and associated vessels.**
- The right atrium is a thin-walled chamber that receives deoxygenated blood from the superior vena cava, inferior vena cava, and coronary sinus.
- The right ventricle receives blood from the right atrium and pumps it out to the lungs to receive oxygen. It has a thick myocardium.
- The left atrium is a thin-walled chamber that receives oxygenated blood from the lungs through the pulmonary veins.
- The left ventricle has the thickest myocardium. It receives the oxygenated blood from the left atrium and pumps it out to systemic circulation.
- There are two types of valves associated with the heart: atrioventricular (AV) valves and semilunar (SL) valves. AV valves are located between the atria and ventricles and prevent blood from flowing back into the atria when the ventricles contract. SL valves are located at the exits from the ventricles and prevent blood from flowing back into the ventricles when the ventricles relax.
- The AV valve of the right side is the tricuspid valve. On the left, it is the bicuspid, or mitral, valve. The pulmonary SL valve is located at the exit of the right ventricle. The aortic SL valve is at the exit of the left ventricle.

■ **Trace the pathway of blood flow through the heart, including chambers, valves, and pulmonary circulation.**
- Deoxygenated blood enters the right atrium through the superior vena cava, inferior vena cava, and coronary sinus. From the right atrium, the blood goes through the tricuspid valve into the right ventricle and then is pumped through the pulmonary SL valve into the pulmonary trunk. From there, pulmonary arteries take the blood to the capillaries of the lungs, where the blood gives off CO_2 and picks up O_2.
- The oxygenated blood enters the pulmonary views and flows into the left atrium.
- From the left atrium, the blood goes through the bicuspid valve into the left ventricle and then is pumped through the aortic SL valve into the ascending aorta to enter systemic circulation.

■ **Identify the major vessels that supply blood to the myocardium and return the deoxygenated blood to the right atrium.**
- The right and left coronary arteries, branches of the ascending aorta, supply blood to the myocardium in the wall of the heart.
- Blood from the capillaries in the myocardium enters the cardiac veins, which drain into the coronary sinus. From there it enters the right atrium.

Physiology of the Heart

■ **Describe the components and function of the conduction system of the heart.**
- The conduction system of the heart consists of specialized cardiac muscle cells that act in a manner similar to neural tissue. The conduction system coordinates the contraction and relaxation of the heart chambers.
- The sinoatrial (SA) node has the fastest rate of depolarization; therefore, it is the pacemaker in the conduction system. Other components in the conduction system are the AV node, AV bundle, bundle branches, and conduction myofibers.

■ **Identify the waves or deflections of an electrocardiogram.**
- An electrocardiogram is a recording of the electrical activity of the heart. The P wave is produced by depolarization of the atrial myocardium. The QRS wave is produced by depolarization of the ventricular myocardium and repolarization of the atria. The T wave is due to repolarization of the ventricles.

■ **Summarize the events of a complete cardiac cycle.**
- Systole is the contraction phase of the cardiac cycle and diastole is the relaxation phase. At a normal heart rate, one complete cardiac cycle lasts for 0.8 second. Atrial systole lasts for 0.1 second, followed by ventricular systole for 0.3 second. All chambers are in diastole at the same time for 0.4 second. Most ventricular filling occurs while all chambers are relaxed.

■ **Correlate the heart sounds heard with a stethoscope with the events of the cardiac cycle.**
- Heart sounds are due to vibrations in the blood caused by the valves closing. The first heart sound is caused by closure of the AV valves. The second heart sound is caused by closure of the SL valves.

■ **Explain what is meant by *stroke volume* and *cardiac output*, and describe the factors that affect these values.**
- Cardiac output equals stroke volume times heart rate. Anything that affects either component affects the output.
- Stroke volume is the amount of blood ejected from the ventricles during one cardiac cycle. It is influenced by end-diastolic volume and contraction strength. End-diastolic volume depends on venous return, and contraction strength depends on end-diastolic volume and stimulation by the autonomic nervous system.
- Heart rate directly influences cardiac output. The cardiac center in the medulla oblongata has both sympathetic and parasympathetic components that adjust the heart rate to meet the changing needs of the body. Peripheral baroreceptors and chemoreceptors send impulses to the cardiac center, where appropriate responses adjust heart rate. Emotions and body temperature also affect heart rate. These effects are usually coordinated through the cardiac center.

CHAPTER QUIZ

Recall

Match the definitions on the left with the appropriate term on the right.

_____ **1.** Contraction phase of cardiac cycle

_____ **2.** Phase of cardiac cycle when most ventricular filling occurs

_____ **3.** Cardiac muscle

_____ **4.** Valve between right atrium and right ventricle

_____ **5.** Valve at the exit of the left ventricle

_____ **6.** Heart chamber with thickest walls

_____ **7.** Heart chamber with oxygen-poor blood

_____ **8.** Volume of blood pumped by left ventricle in 1 minute

_____ **9.** Location of cardiac center in the brain

_____ **10.** Pacemaker of the heart

A. Aortic semilunar valve
B. Cardiac output
C. Diastole
D. Left ventricle
E. Medulla oblongata
F. Myocardium
G. Right ventricle
H. Sinoatrial node
I. Systole
J. Tricuspid valve

Thought

1. The atrioventricular valve on the same side of the heart as the pulmonary semilunar valve is: the
A. Tricuspid valve
B. Bicuspid valve
C. Mitral valve
D. Aortic valve

2. The chamber that receives oxygen-rich blood through the pulmonary veins is the:
A. Right atrium
B. Right ventricle
C. Left atrium
D. Left ventricle

3. In a cardiac cycle that lasts 0.8 second, both ventricles are in simultaneous systole for:
A. 0.1 second
B. 0.2 second
C. 0.3 second
D. 0.4 second

4. If an individual has a heart rate of 70 beats per minute and a stroke volume of 70 milliliters, what is the cardiac output?
A. 70 mL/min
B. 140 mL/min
C. 1400 mL/min
D. 4900 mL/min

5. Increased venous return:
A. Increases both end-diastolic volume and contraction strength
B. Increases both cardiac rate and stroke volume
C. Increases end-diastolic volume but decreases contraction strength
D. Increases both cardiac rate and contraction strength

Application

1. Which of the following occur when the semilunar valves are open?
A. Coronary arteries fill.
B. Blood enters the aorta.
C. AV valves are closed.
D. Blood enters the ventricles.
E. Blood enters the pulmonary trunk.
F. Atria contract.
G. Ventricles are in systole.

2. A heart rate of 45 beats per minute and an absence of P waves on the ECG suggest damage to which component of the conduction system?

BUILDING YOUR MEDICAL VOCABULARY

Building a vocabulary is a cumulative process. As you progress through this book, you will use word parts, abbreviations, and clinical terms from previous chapters. Each chapter will present new word parts, abbreviations, and terms to add to your expanding vocabulary.

Word Parts and Combining Form with Definition and Examples

PART/COMBINING FORM	DEFINITION	EXAMPLE
aort/o	lift up	aorta: vessel that lifts blood up and out of the heart
atri/o	entrance room	atrium: entrance chamber of the heart, where blood is returned to the heart
brady-	slow	bradycardia: slow heart rate
cardi-	heart	cardiac: pertaining to the heart
coron-	crown	coronary: encircling in the manner of a crown; coronary arteries encircle the base of the heart
cusp-	point	tricuspid: heart valve with three points
diastol-	expand, separate	diastole: period when ventricles expand or dilate as blood flows into them
ech/o	sound	echocardiography: use of high-frequency sound waves to produce images that show the structure and movement of the heart
lun-	moon shaped	semilunar valve: heart valve that is half-moon shaped
meg-	large	cardiomegaly: enlargement of the heart
sept-	partition	atrial septum: wall or partition between the right and left atria of the heart
son/o	sound	ultrasonography: diagnostic technique that projects and receives high-frequency sound waves as they echo off parts of the body
sphygm/o	pulse	sphygmomanometer: instrument for recording the energy of a pulse wave

PART/COMBINING FORM	DEFINITION	EXAMPLE
-sten-	narrowing	mitral stenosis: narrowing of the left atrioventricular orifice
steth/o	chest	stethoscope: instrument used to hear the sounds in the chest and internal organs
systol-	contraction	systole: period of contraction
tachy-	fast, rapid	tachycardia: rapid heart rate
valvul/o	valve	valvuloplasty: surgical repair of a valve, especially a valve of the heart

Clinical Abbreviations

ABBREVIATION	MEANING
ASD	atrial septal defect
ASHD	arteriosclerotic heart disease
AV	atrioventricular
CABG	coronary artery bypass graft
CAD	coronary artery disease
CHF	congestive heart failure
CPR	cardiopulmonary resuscitation
CXR	chest x-ray (radiograph)
ECG	electrocardiogram
ECHO	echocardiography
HCVD	hypertensive cardiovascular disease
IHD	ischemic heart disease
LA	left atrium
LCA	left coronary artery
LV	left ventricle
MI	myocardial infarction; mitral insufficiency
MVP	mitral valve prolapse
PDA	patent ductus arteriosus
PVC	premature ventricular contraction
PVT	paroxysmal ventricular tachycardia
RA	right atrium

ABBREVIATION	MEANING
RCA	right coronary artery
RV	right ventricle
SA	sinoatrial
SSCP	substernal chest pain
TIA	transient ischemic attack
VSD	ventricular septal defect
VT	ventricular tachycardia

Clinical Terms

Artificial pacemaker (ahr-tih-FISH-al PAYSE-may-ker) Electronic device that stimulates the initiation of an impulse within the heart

Auscultation (ahs-kool-TAY-shun) Physical assessment procedure using a stethoscope to listen to sounds within the chest, abdomen, and other parts of the body

Bradycardia (bray-dee-KAR-dee-ah) Abnormally slow heart rate, usually less than 60 beats per minute

Cardiac arrest (KAR-dee-ack ah-REST) Cessation of an effective heartbeat; heart may be completely stopped or quivering ineffectively in fibrillation

Cardiac catheterization (KAR-dee-ack kath-eh-ter-ih-ZAY-shun) Process of inserting a thin, flexible tube, called a catheter, into a vein or an artery and guiding it into the heart for the purpose of detecting pressures and patterns of blood flow

Cardiomegaly (kar-dee-oh-MEG-ah-lee) Enlargement of the heart

Cardiomyopathy (kar-dee-oh-my-AHP-ah-thee) Any primary disease of the heart muscle

Congestive heart failure (kahn-JES-tiv HART FAIL-yer) Condition in which the heart's pumping ability is impaired and results in fluid accumulation in vessels and tissue spaces; various stages of difficult breathing occur as fluid accumulates in pulmonary vessels and lung tissue

Coronary artery bypass grafting (CABG) Surgical procedure in which a blood vessel from another part of the body is used to bypass the blocked region of a coronary artery

Cor pulmonale (kor pul-moh-NAY-lee) Hypertrophy of the right ventricle caused by hypertension in the pulmonary circulation

Defibrillation (dee-fib-rih-LAY-shun) Procedure in which an electric shock is applied to the heart with a defibrillator to stop an abnormal heart rhythm

Echocardiography (eck-oh-kar-dee-AHG-rah-fee) Noninvasive clinical procedure using pulses of high-frequency sound waves (ultrasound) that are transmitted into the chest, and echoes returning from the valves, chambers, and surfaces of the heart are plotted and recorded; provides information about valvular or structural defects and coronary artery disease

Fibrillation (fib-rih-LAY-shun) Rapid, random, ineffectual, and irregular contractions of the heart at 350 or more beats per minute

Mitral valve prolapse (MY-tral valve PRO-laps) Improper closure of the mitral valve when the heart is pumping blood; also called floppy valve syndrome

Pericardiocentesis (pair-ih-kar-dee-oh-sen-TEE-ses) Surgical puncture into the pericardial cavity with aspiration of fluid

Tachycardia (tack-ee-KAR-dee-ah) Abnormally rapid heart rate, usually greater than 100 beats per minute

Valvular heart disease (VAL-vyoo-lar HART dih-ZEEZ) Any disorder of the heart valves including insufficiency, stenosis, and prolapse

VOCABULARY QUIZ

Use word parts you have learned to form words that have the following definitions.

1. Heart entrance chamber

2. Rapid heartbeat

3. Has three points

4. Vessel that lifts up from heart

5. Half-moon shaped

Using the definitions of word parts you have learned, define the following words.

6. Cardiomegaly

7. Mitral stenosis

8. Valvulitis

9. Bradycardia

10. Systole

Match each of the following definitions with the correct word.

_____ **11.** Membrane around the heart

_____ **12.** Disease condition of heart muscle

_____ **13.** Surgical repair of a valve

_____ **14.** Inflammation of heart lining

_____ **15.** Recording of electrical activity of heart

A. Cardiomyopathy
B. Electrocardiogram
C. Endocarditis
D. Pericardium
E. Valvuloplasty

16. What is the clinical term for "any primary disease of the heart muscle"? _____

17. What is the clinical term for "acute chest pain caused by decreased blood supply to the heart"? _____

18. What procedure applies electric shock to the heart to stop an abnormal heart rhythm? _____

19. What is the meaning of the abbreviation CABG? _____

20. What is the meaning of the abbreviation SA? _____

Cardiovascular System: Blood Vessels

CHAPTER OBJECTIVES

Classification and Structure of Blood Vessels
- Describe the structure and function of arteries, capillaries, and veins.

Physiology of Circulation
- Describe how oxygen, carbon dioxide, and glucose move across capillary walls.
- Describe the mechanisms and pressures that move fluids across capillary walls.
- Discuss the factors that affect blood flow through arteries, capillaries, and veins.

- Discuss four primary factors that affect blood pressure and how blood pressure is regulated.

Circulatory Pathways
- Trace blood through the pulmonary circuit from the right atrium to the left atrium.
- Identify the major systemic arteries and veins.
- Describe the blood supply to the brain.
- Describe five features of fetal circulation that make it different from adult circulation.

KEY TERMS

Central venous pressure (SEN-tral VAYN-us PRESH-ur) Blood pressure in the right atrium

Diastolic pressure (dye-ah-STAHL-ik PRESH-ur) Blood pressure in the arteries during relaxation of the ventricles

Korotkoff sounds (koh-ROT-kof SOUNDZ) Sounds heard in the stethoscope while taking blood pressure

Metarteriole (met-ahr-TEER-ee-ohl) Microscopic vessel that connects an arteriole directly to a venule without an intervening capillary network; an arteriovenous shunt

Peripheral resistance (per-IF-er-al ree-SIS-tans) Opposition to blood flow caused by friction of the blood vessel walls

Pulse (PULS) Expansion and recoil of arteries caused by contraction and relaxation of the heart

Systolic pressure (sis-TAHL-ik PRESH-ur) Blood pressure in the arteries during contraction of the ventricles

Vasoconstriction (vaz-oh-kon-STRIK-shun) Narrowing of blood vessels; decrease in the size of the lumen of blood vessels

Vasa vasorum (VAS-ah vah-SOR-um) Small blood vessels that supply nutrients to the tissues in the walls of the large blood vessels

Vasodilation (vaz-oh-dye-LAY-shun) Enlarging of blood vessels; increase in the size of the lumen of blood vessels

Blood vessels are the channels or conduits through which blood is distributed to body tissues. The vessels make up two closed systems of tubes that begin and end at the heart (Figure 12-1). One system, the **pulmonary vessels,** transports blood from the right ventricle to the lungs and back to the left atrium. The other system, the **systemic vessels,** carries blood from the left ventricle to the tissues in all parts of the body and then returns the blood to the right atrium. Based on their structure and function, blood vessels are classified as arteries, capillaries, or veins.

Classification and Structure of Blood Vessels

Arteries

Arteries carry blood away from the heart. Pulmonary arteries transport blood that has a low oxygen content from the right ventricle to the lungs. Systemic arteries transport oxygenated blood from the left ventricle to the body tissues. Blood is pumped from the ventricles into large elastic arteries that branch repeatedly into smaller and smaller arteries until the branching results in microscopic arteries called **arterioles**

(ar-TEER-ee-ohlz). The arterioles play a key role in regulating blood flow into the tissue capillaries. About 10% of the total blood volume is in the systemic arterial system at any given time.

The wall of an artery consists of three layers (Figure 12-2). The innermost layer, the **tunica intima** (also called **tunica interna**), is simple squamous epithelium surrounded by a connective tissue basement membrane with elastic fibers. The middle layer, the **tunica media,** is primarily smooth muscle and is usually the thickest layer. It not only provides support for the vessel but also changes vessel diameter to regulate blood flow and blood pressure. The outermost layer, which attaches the vessel to the surrounding tissues, is the **tunica externa** or **tunica adventitia.** This layer is connective tissue with varying amounts of elastic and collagenous fibers. The connective

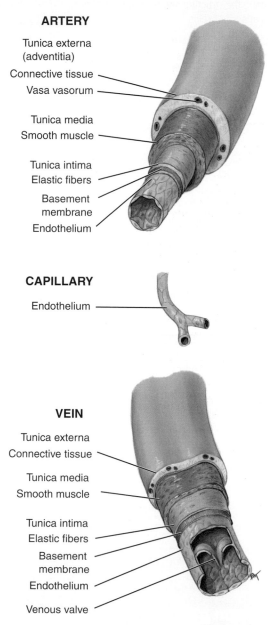

Figure 12-2 *Structure of blood vessels. Arteries have thick walls with significant smooth muscle and connective tissue. Capillary walls have only a single layer of thin endothelium. Veins have thinner walls than arteries and have valves.*

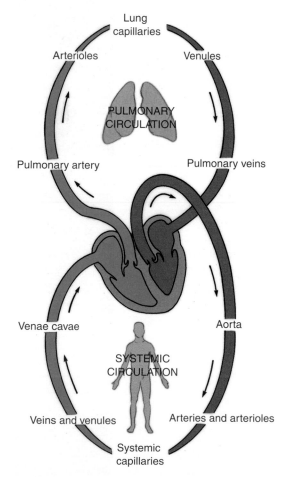

Figure 12-1 *Scheme of circulation. In pulmonary circulation, arteries take blood from the right ventricle to the lungs and veins return the blood to the left atrium. In systemic circulation, arteries take blood from the left ventricle to the body tissues and veins return the blood to the right atrium.*

tissue in this layer is quite dense where it is adjacent to the tunica media, but it changes to loose connective tissue near the periphery of the vessel. The tunica externa of the larger arteries contains small blood vessels, called **vasa vasorum** (VAS-ah vah-SOR-um), that provide the blood supply for the tissues of the vessel wall.

QUICK APPLICATIONS

An **aneurysm** is a bulge, or bubble, that develops at a weakened region in the wall of an artery. This is especially dangerous if it is in the aorta or arteries of the brain. If diagnosed soon enough, the aneurysm sometimes may be removed and the vessel surgically repaired. Because the wall is weakened, an aneurysm is subject to rupture. Little can be done when this happens because the massive bleeding usually leads to death before medical care can be obtained.

Capillaries

Capillaries, the smallest and most numerous of the blood vessels, form the connection between the vessels that carry blood away from the heart (arteries) and the vessels that return blood to the heart (veins). They are the continuation of the smallest arterioles. Arterioles are the smallest vessels with three distinguishable layers in their wall. When the arterioles branch into capillaries, the middle and outer layers of the wall disappear so that the capillary wall is only a thin endothelium (simple squamous epithelium) with a basement membrane. This thin wall permits the exchange of materials between the blood in the capillary and the adjacent tissue cells. This exchange is the primary function of capillaries.

The diameter of a capillary is so small that erythrocytes must pass through them in single file. This slows the blood flow to allow ample time for the transport of substances across the capillary endothelium.

Capillary distribution varies with the metabolic activity of body tissues. Tissues such as skeletal muscle, liver, and kidney have extensive capillary networks because they are metabolically active and require an abundant supply of oxygen and nutrients. Other tissues, such as connective tissue, have a less abundant supply of capillaries. The epidermis of the skin and the lens and cornea of the eye completely lack a capillary network. About 5% of the total blood volume is in the systemic capillaries at any given time. Another 10% is in the lungs.

QUICK APPLICATIONS

It is estimated that if all the capillaries in the body were placed end to end, they would encircle the earth at the equator two and one-half times!

Blood flow from the arterioles into the capillaries is regulated by smooth muscle cells in the arterioles where they branch to form the capillaries. These **precapillary sphincters** constrict to reduce blood flow into the capillary bed and relax to increase the blood flow. When the precapillary sphincters contract, blood passes directly from small arterioles into small venules through **metarterioles** (met-ahr-TEER-ee-ohlz), or **arteriovenous anastomoses** (ar-teer-ee-oh-VAY-nus ah-NAS-toh-moh-ses) (Figure 12-3). This arrangement allows blood to be diverted from one capillary bed to another for distribution to the regions that need it most at any given time.

Veins

Veins carry blood toward the heart. After blood passes through the capillaries, it enters the smallest veins, called **venules.** From the venules, it flows into progressively larger and larger

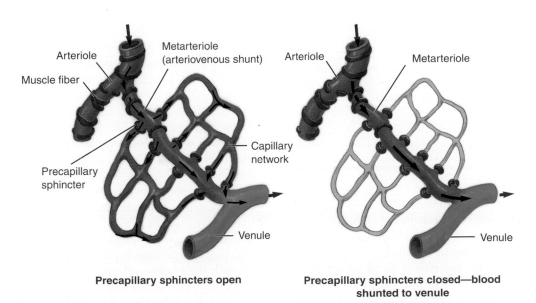

Precapillary sphincters open   Precapillary sphincters closed—blood shunted to venule

Figure 12-3 *Organization of a capillary network. When the precapillary sphincters are open, blood flows into the capillaries. When the precapillary sphincters are closed, blood flows directly from the arteriole into the venule through a metarteriole.*

veins until it reaches the heart. In the pulmonary circuit, the pulmonary veins transport blood from the lungs to the left atrium of the heart. This blood has a high oxygen content because it has just been oxygenated in the lungs. Systemic veins transport blood from the body tissues to the right atrium of the heart. This blood has a reduced oxygen content because the oxygen has been used for metabolic activities in the tissue cells.

The walls of veins have the same three layers as the arteries (see Figure 12-2). Although all the layers are present, there is less smooth muscle and connective tissue. This makes the walls of veins thinner than those of arteries, which is related to the fact that blood in the veins has less pressure than blood in the arteries. Because the walls of the veins are thinner and less rigid than arteries, veins can hold more blood. Almost 70% of the total blood volume is in the veins at any given time. Medium and large veins have **venous valves,** similar to the semilunar valves associated with the heart, that help keep the blood flowing toward the heart. Venous valves are especially important in the arms and legs, where they prevent the backflow of blood in response to the pull of gravity.

QUICK **CHECK**

12.1 At any given time, which type of vessels contain the greatest volume of blood?

12.2 What is the purpose of valves in the veins?

12.3 What is the structure of a capillary wall?

12.4 What happens to blood flow in a capillary bed when the precapillary sphincters contract?

QUICK **APPLICATIONS**

Varicose veins are veins that are twisted and dilated with accumulated blood. These frequently occur in the legs. Conditions that hinder venous return, such as pregnancy, obesity, and standing for long periods of time, allow blood to accumulate in the veins of the extremities. This stretches the veins so the valve flaps no longer overlap and they permit the backflow of blood. Superficial veins are more susceptible because they receive less support from surrounding tissue.

Physiology of Circulation

Role of the Capillaries

In addition to forming the connection between the arteries and veins, capillaries have a vital role in the exchange of gases, nutrients, and metabolic waste products between the blood and the tissue cells. Tissue cells are surrounded by a small amount of extracellular fluid, called **interstitial fluid** (in-ter-STISH-al FLOO-id). This fluid is formed from the fluids and solutes that leave the capillaries (Figure 12-4). Substances that transfer between the blood and tissue cells must pass through the interstitial fluid, and therefore the interstitial fluid plays an intermediary role in the exchange process.

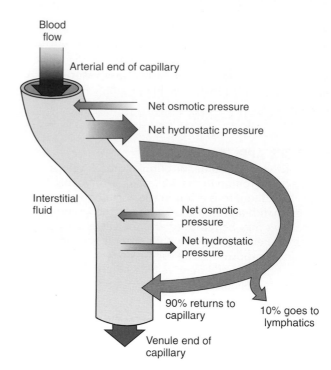

Figure 12-4 *Capillary microcirculation. Net osmotic pressure moves fluid into the capillary. Net filtration pressure moves fluid out of the capillary.*

Substances pass through the capillary wall by diffusion, filtration, and osmosis. Oxygen, carbon dioxide, and glucose move across the capillary wall by diffusion from higher concentrations to lower concentrations. Blood entering the capillary has higher oxygen and glucose concentrations than the interstitial fluid and tissue cells; thus, the oxygen and glucose diffuse through the capillary wall and enter the interstitial fluid and then the tissue cells. Metabolically active tissue cells produce carbon dioxide as a waste product, so there is a higher concentration inside the cells. The carbon dioxide diffuses down the concentration gradient from the tissue cells to the interstitial fluid and then into the capillary.

Filtration involves hydrostatic pressure to force (push) water molecules and certain dissolved substances through the capillary wall. In capillaries, the hydrostatic pressure is blood pressure generated by ventricular contractions. Because blood pressure is higher at the arteriolar end of the capillary than it is at the venule end, more filtration occurs at the arteriolar end.

Protein molecules are generally too large to pass through the capillary wall, so they remain in the plasma and create an osmotic pressure in the blood. There are relatively few protein molecules in the interstitial fluid, so its osmotic pressure is negligible. Osmotic pressure attracts (or pulls) water into the capillary. At the arteriolar end of the capillary, hydrostatic pressure is greater than the osmotic pressure, so fluid leaves the capillary. Hydrostatic (blood) pressure decreases as the blood moves through the capillary. At the venule end, the attraction produced by the osmotic pressure overcomes the hydrostatic pressure and draws fluid into the capillary (see Figure 12-4).

The net result of the capillary microcirculation created by hydrostatic and osmotic pressures is that substances leave the blood at the arteriole end of the capillary and return at the venule end. Normally, more fluid leaves the capillaries than is returned to them. About 90% of the fluid is returned at the venule end; the remaining 10% is collected by lymphatic vessels and is returned to the general circulation in venous blood. Because of the capillary microcirculation, the interstitial fluid is continually changing and nutrients, gases, and waste products are moved between the tissue cells and blood.

QUICK **CHECK**

12.5 What are the two types of pressures that are involved in the capillary microcirculation?

12.6 What are the primary molecules that create the osmotic pressure in the blood?

12.7 Why is there not a significant osmotic pressure in the interstitial fluid?

QUICK **APPLICATIONS**

Edema is an abnormal accumulation of interstitial fluid, or swelling. This may be caused by a disruption of normal capillary microcirculation. Factors that may lead to edema include an increase in capillary blood pressure, a decrease in the quantity of plasma proteins, and an increase in the permeability of the capillary wall so that proteins leak out.

Blood Flow

Relationship of Blood Flow to Pressure

Blood flow refers to the movement of blood through the vessels from the arteries to the capillaries and then into the veins. Pressure is a measure of the force that the blood exerts against vessel walls. Pressure moves the blood through the vessels. Similar to all fluids, blood flows from a high pressure area to a region with lower pressure. Because the contraction of the ventricles provides this force, or pressure, it is greatest during ventricular systole when the blood is pumped from the left ventricle into the aorta. Figure 12-5 illustrates the progressive decrease in pressure through the arteries, capillaries, and veins, so that it is lowest as the venae cavae enter the right atrium. The pressure in the right atrium is often called the **central venous pressure.** Blood flows in the same direction as the pressure gradient.

Velocity of Blood Flow

The rate, or velocity, of blood flow varies inversely with the total cross-sectional area of the blood vessels. As the total cross-sectional area of the vessels increases, the velocity of flow decreases (Figure 12-6). Therefore, velocity is greatest in the aorta and progressively decreases as the blood flows through increasing numbers of smaller and smaller vessels. Blood flow is slowest in the capillaries, which have the largest total cross-sectional area. This allows time for the exchange of gases and nutrients. Velocity increases again as the blood enters decreasing numbers of progressively larger and larger veins during the return to the heart.

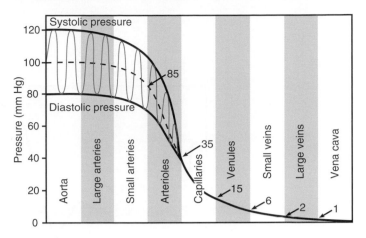

Figure 12-5 *Blood pressure in various types of systemic vessels. Pressure is greatest in the aorta during ventricular systole and progressively decreases until it is lowest when the venae cavae enter the right atrium.*

Relationship of Blood Flow to Resistance

Resistance is a force that opposes the flow of a fluid. For example, it is more difficult to get a milkshake through a straw than water through a straw because the milkshake offers more resistance. It opposes the flow. In blood vessels, most of the resistance is due to vessel diameter. As vessel diameter decreases, the resistance increases and blood flow decreases. The autonomic nervous system plays a role in regulating blood flow by changing the vascular resistance through vasodilation and vasoconstriction. During exercise, autonomic control causes vasoconstriction in the viscera and skin (increases resistance) and vasodilation in the skeletal muscles (decreases resistance). As a result, blood flow to the viscera and skin decreases and blood flow to the skeletal muscles increases.

QUICK **APPLICATIONS**

When you are in the sun for an extended period, the cutaneous blood vessels dilate to bring more blood to the skin's surface, which helps keep the body cool. This action decreases the amount of blood in other parts of the body and may diminish the blood supply to the brain. If you are sunbathing and stand up abruptly, you may feel dizzy. This is because the blood momentarily remains in the dilated cutaneous vessels instead of returning to the heart. This causes a decrease in blood pressure. The dizziness is a signal that the brain is not receiving enough oxygen.

Contraction and relaxation of the precapillary sphincters control the flow of blood into the capillaries by changing the vessel diameter to modify the resistance. When precapillary sphincters contract, the diameter of the vessel decreases and local resistance increases. Blood flow to that area decreases. When the sphincters relax, vessel diameter increases, resistance decreases, and blood flows through the capillaries. An increase in carbon dioxide, a decrease in pH, or a decrease in oxygen causes the

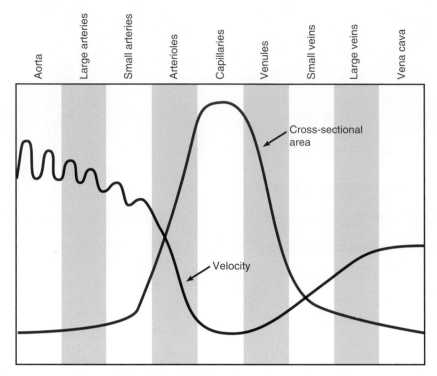

Figure 12-6 *Relation of blood flow to total vessel cross-sectional area. As cross-sectional area increases, velocity of blood flow decreases. Capillaries have the greatest total cross-sectional area so the velocity of flow is slowest. This allows time for capillary microcirculation.*

precapillary sphincters to relax, which allows increased blood flow to deliver more oxygen to the tissues and to carry away the waste products.

Venous Blood Flow

Very little pressure remains by the time blood leaves the capillaries and enters the venules. Blood flow through the veins is not the direct result of ventricular contraction. Instead, venous return depends on **skeletal muscle action, respiratory movements,** and **constriction of the veins.**

QUICK APPLICATIONS

The difference in blood pressure between arteries and veins is obvious when the vessels are cut. Blood flows smoothly and freely from a vein, but it spurts forcefully from an artery.

When skeletal muscles contract, they thicken, which squeezes the veins adjacent to them. The squeezing, or milking, action creates a localized high pressure area and the blood flows from that high-pressure region to an area with lower pressure. Figure 12-7 illustrates this action. Valves in the veins offer very little resistance to blood flowing toward the heart, but they close to prevent blood from flowing in the opposite direction. Muscular contraction during exercise enhances venous return. A lack of muscular movement allows blood to accumulate, or pool, in the extremities rather than to return to the heart for circulation.

Respiratory movements create pressure gradients that enhance the movement of venous blood. When the diaphragm contracts during inspiration, it exerts pressure in the abdomen and decreases the pressure in the thoracic cavity. Blood in the abdominal veins moves from the higher pressure in the abdomen to the lower pressure in the thorax. Valves prevent the blood from flowing in the opposite direction into the legs. During exercise, when the breathing rate increases, respiratory movements increase the rate at which blood is returned to the heart, which increases the cardiac output necessary to meet the needs of the muscular activity.

Sympathetic reflexes cause contraction of the smooth muscle in the walls of the veins. This venous constriction along with valves that prevent backflow moves blood toward the heart. As mentioned previously, about 70% of the total blood volume is in the veins at any given time, which makes them an important blood reservoir. If there is blood loss for some reason, and blood pressure decreases, sympathetic reflexes stimulate venoconstriction. This moves blood out of the venous reservoir and helps restore blood pressure to normal.

QUICK CHECK

12.8 In what vessels is blood flow the slowest?
12.9 What three factors create the pressure necessary for blood flow in the veins?
12.10 What factor is responsible for most of the resistance in blood vessels?

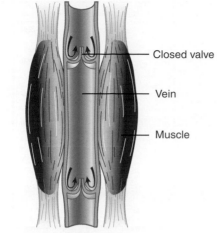

Muscles relaxed—valves closed,
pressure reduced and little blood flow.

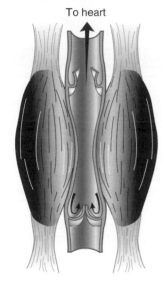

Muscles contracted—upper valve opens
and blood moves upward. The closed
lower valve prevents backflow.

Figure 12-7 *Effects of skeletal muscle contraction on venous blood flow. Muscle contraction creates a localized high pressure area that moves blood into an adjacent low pressure region. Valves prevent backflow.*

Pulse and Blood Pressure

Meaning of Pulse

Pulse is the alternating expansion and recoil of an artery in response to the surge of blood ejected from the left ventricle during contraction. This pulse can be felt in places where an artery is near the surface of the body and passes over something firm such as a bone. Figure 12-8 illustrates the location of nine commonly used **pulse points,** where the pulse may be detected by placing the fingers over a superficial artery. Taking a person's pulse provides information about the rate, strength, and rhythmicity of the heartbeat.

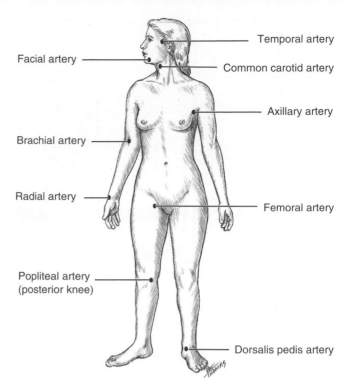

Figure 12-8 *Locations of commonly used pulse points.*

Arterial Blood Pressure and Its Determination

In common use, the term **blood pressure** refers to arterial blood pressure, the pressure in the aorta and its branches. The pressure in the arteries is greatest during ventricular contraction (systole) when blood is forcefully ejected from the left ventricle into the aorta. This is called **systolic pressure.** Arterial pressure is lowest when the ventricles are in the relaxation phase (diastole) of the cardiac cycle just before the next contraction. This is called **diastolic pressure.** The difference between the systolic pressure and diastolic pressure is called **pulse pressure.** The standard units of measurement for blood pressure are millimeters of mercury (mm Hg). Each unit of pressure will lift a column of mercury 1 millimeter. A blood pressure of 110 mm Hg will lift a column of mercury 110 millimeters.

In many clinical settings, a **sphygmomanometer** (sfig-moh-mah-NAHM-eh-ter) is used to measure blood pressure in the brachial artery (Figure 12-9). An inflatable cuff, connected to the sphygmomanometer and an air supply, is wrapped around the patient's arm, just above the elbow. The cuff is inflated with sufficient air pressure to collapse the brachial artery. A sensor is placed over the brachial artery. While the examiner is listening through the stethoscope and watching the column of mercury on the sphygmomanometer, air is slowly released through the pressure valve on the cuff. As the pressure gradually decreases, sounds representing the first flow of blood through the collapsed artery are heard in the stethoscope. The sounds heard through the stethoscope as a result of blood flow are called **Korotkoff** (koh-ROT-kof) **sounds.**

The pressure at which the first Korotkoff sound is heard represents the systolic pressure, which normally averages about 120 mm Hg. As the pressure in the cuff continues to decline,

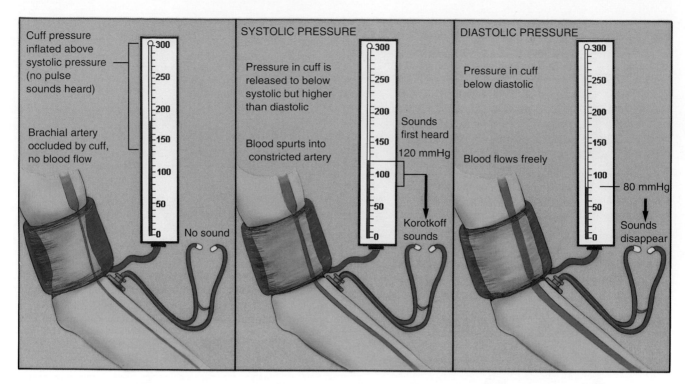

Figure 12-9 *Measurement of blood pressure using a sphygmomanometer and stethoscope.*

the Korotkoff sounds change in tone and loudness until they completely stop. This represents the point at which the artery is completely open and blood flows freely, and this is recorded as the diastolic pressure, which normally averages about 80 mm Hg. The blood pressure is recorded as systolic pressure over diastolic pressure, in this case 120/80 mm Hg. Pulse pressure in this instance is 40 mm Hg.

Factors That Affect Blood Pressure

Four major factors interact to affect blood pressure. These are **cardiac output, blood volume, peripheral resistance,** and **viscosity.** Each one has a direct relationship to blood pressure.

Cardiac output is the amount of blood pumped by the heart in 1 minute (see Chapter 11). It is determined by multiplying the heart rate by the stroke volume. Anything that increases either the heart rate or the stroke volume will increase cardiac output and also increase blood pressure. When either heart rate or stroke volume decreases, cardiac output decreases, which decreases blood pressure.

The volume of blood in the body directly affects blood pressure. Although blood volume varies with age, body size, and gender, the normal average is about 5 liters for adults. Blood volume may be reduced by severe hemorrhage, vomiting, diarrhea, or reduced fluid intake. Any changes in blood volume are accompanied by corresponding changes in blood pressure. When blood volume decreases, blood pressure also decreases. Circulatory shock occurs when blood cannot circulate normally because the vessels are not adequately filled. If this is the result of a severe hemorrhage, it is called hypovolemic shock. Heart rate increases in an attempt to maintain cardiac output resulting in a rapid pulse rate. When the blood volume is restored by a transfusion or fluid intake, blood pressure returns to normal.

If the body retains too much fluid, blood volume and blood pressure increase.

Peripheral resistance is the opposition to blood flow caused by friction of the vessel walls. Increased peripheral resistance causes an increase in blood pressure. Vasoconstriction decreases the vessel diameter and increases resistance. This increases blood pressure. In arteriosclerosis, blood vessels lose their elasticity and their resistance increases. This increases blood pressure.

Viscosity is a physical property of blood that refers to the ease with which the molecules and cells slide across each other. Viscosity opposes the flow of a fluid. Syrup does not flow as easily as water, so the syrup is said to be more viscous. Normally the viscosity of blood remains fairly constant, but it changes when either the number of blood cells or the concentration of plasma proteins changes. If the number of erythrocytes increases (polycythemia), the blood becomes more viscous and blood pressure increases.

Regulation of Arterial Blood Pressure

Arterial blood pressure is maintained within normal ranges by changes in cardiac output and peripheral resistance. Pressure receptors (baroreceptors) are located in the walls of the large arteries in the thorax and neck. They are abundant in the carotid sinus, located at the bifurcation of the common carotid arteries in the neck, and in the aortic arch. These receptors respond when the walls are stretched by sudden increases in pressure. The action potentials from the baroreceptors are transmitted to the cardiac and vasomotor centers in the medulla oblongata. The centers in the medulla respond by sending out signals that decrease heart rate (decrease cardiac output) and cause vasodilation (decrease peripheral resistance). These actions return blood pressure toward normal.

Decreases in blood pressure reduce the frequency of action potentials from the receptors, which increases heart rate and causes vasoconstriction. Baroreceptors are important for moment-by-moment short-term pressure regulation.

QUICK APPLICATIONS

Sometimes there is a feeling of dizziness when rising rapidly from lying down to a standing position. This is because the baroreceptors have not had time to respond to the decrease in blood pressure caused by the downward pull of gravity on the blood. The dizziness is a signal that the brain is not receiving enough blood.

Chemoreceptors near the carotid sinus and aortic arch respond to changes in carbon dioxide concentration, hydrogen ion concentration (pH), and oxygen concentration. When carbon dioxide or hydrogen ion concentrations increase, or oxygen concentration decreases, these receptors send impulses to the medulla oblongata. The medulla oblongata responds by sending out impulses that increase heart rate and peripheral resistance to increase blood pressure. This increases blood flow to the lungs and to the tissues. Chemoreceptors have a significant role in blood pressure regulation only in emergency situations.

Certain hormones also have an effect on blood pressure. Epinephrine and norepinephrine from the adrenal medulla have an effect similar to the sympathetic nervous system— they increase cardiac output and blood pressure. Antidiuretic hormone from the posterior pituitary gland reduces fluid loss through urine and increases body fluid volume. This increases blood volume, which increases blood pressure.

When blood pressure decreases, the kidneys secrete **renin,** an enzyme, into the blood. Renin acts on certain blood proteins to produce **angiotensin.** Active angiotensin is a powerful vasoconstrictor that increases blood pressure toward normal readings. Angiotensin also promotes the secretion of **aldosterone** from the adrenal cortex. Aldosterone acts on the kidneys to conserve sodium ions and water so that blood volume increases, which increases blood pressure. The **renin-angiotensin-aldosterone** mechanism has a significant role in the long-term regulation of blood pressure.

QUICK CHECK

12.11 Why does exercise increase cardiac output?

12.12 External pressure on the common carotid artery, just below the carotid sinus, decreases the blood pressure in the carotid sinus. How does this affect heart rate?

12.13 What are the four major factors that interact to affect blood pressure?

12.14 What substance is secreted by the kidneys that has an influence on blood pressure?

FROM THE PHARMACY

Hypertension is sometimes called the silent killer. Although it seems relatively harmless, prolonged hypertension damages the heart, brain, and kidneys, leading to premature death. It is an incurable but controllable disease. The first step in controlling hypertension is to adopt a healthy lifestyle that includes exercise, weight control, smoking cessation, alcohol restriction, and a decrease in sodium intake. If changes in lifestyle do not reduce blood pressure sufficiently, there is a wide variety of drugs available to treat hypertension. These drugs fall generally into five categories: diuretics; adrenergic antagonists that stimulate or inhibit the nervous system to reduce cardiac output and/or peripheral resistance; renin-angiotensin system inhibitors; calcium channel blockers; and vasodilators. One example of each category is presented.

Furosemide (Lasix) is an example of a diuretic. The action of all diuretics is to promote fluid loss from the body. This reduces blood volume, which tends to reduce blood pressure. Furosemide acts by blocking the active transport of chloride, sodium, and potassium ions in the ascending loop of Henle. There is also a direct dilating effect on arterioles that helps reduce blood pressure.

Bisoprolol fumarate (Zebeta) is an example of an adrenergic antagonist, specifically a selective β1-adrenergic blocker. It acts by blocking sympathetic stimulation of cardiac muscle to decrease heart rate and contraction strength, which decreases blood pressure. One of the side effects of this drug is "beta blocker blues" manifested by depression and fatigue. Sometimes bisoprolol fumarate is combined with the diuretic hydrochlorothiazide for use as an antihypertensive. The trade name of this combined preparation is Ziac.

Captopril (Capoten) is an example of a renin-angiotensin system inhibitor, specifically an angiotensin-converting enzyme (ACE) inhibitor. This drug inhibits the conversion of angiotensin I (inactive) to angiotensin II (active vasoconstrictor). Inhibiting the vasoconstriction of peripheral vessels decreases peripheral resistance and thus decreases blood pressure. It may be used alone or in combination with thiazide diuretics to treat hypertension.

Diltiazem hydrochloride (Cardizem, Tiazac) is an example of a calcium channel blocker. This group of drugs blocks the inward flow of calcium through the calcium channels in the membranes of cardiac and smooth muscle. In the heart, calcium channel blockers reduce the contraction strength and slow the heart rate. In the peripheral blood vessels, they reduce the contraction strength of the smooth muscle in the vessel walls. This action decreases peripheral resistance, which decreases blood pressure.

Diazoxide (Hyperstat) is an example of a vasodilator. The antihypertensive effect is the result of direct action on the smooth muscle in the walls of peripheral blood vessels. This action inhibits the constriction of the peripheral vessels (causes vasodilation) with a resulting decrease in peripheral resistance and lower blood pressure.

Circulatory Pathways

The blood vessels of the body are functionally divided into two distinct circuits: the pulmonary circuit and the systemic circuit. The pump for the pulmonary circuit, which circulates blood through the lungs, is the right ventricle. The left ventricle is the pump for the systemic circuit, which provides the blood supply for the tissue cells of the body.

Pulmonary Circuit

The pulmonary circuit takes blood from the right side of the heart to the lungs, and then returns it to the left side of the heart (Figure 12-10). Oxygen-poor blood, which has increased levels of carbon dioxide, is returned to the **right atrium** from the tissue cells of the body. It passes through the **tricuspid valve** into the **right ventricle.** During ventricular systole, the blood is ejected through the **pulmonary semilunar valve** into the **pulmonary trunk,** which divides into the right and left **pulmonary arteries.** Each pulmonary artery enters a lung and repeatedly divides into smaller and smaller vessels until they become capillaries. The **capillaries of the lungs** form networks that surround the air sacs, or alveoli, of the lungs. Here carbon dioxide diffuses from the capillary blood into the alveoli of the lungs, and oxygen diffuses from the alveoli into the blood. The newly oxygenated blood enters pulmonary venules, which form progressively larger veins, until two **pulmonary veins** emerge from each lung and carry the blood to the **left atrium.** In the pulmonary circuit, the arteries carry deoxygenated blood away from the heart, and the veins carry oxygenated blood to the heart.

Systemic Circuit

The systemic circulation provides the functional blood supply to all body tissues. It carries oxygen and nutrients to the cells and picks up carbon dioxide and waste products. Systemic circulation carries oxygenated blood from the left ventricle, through the arteries, to the capillaries in the tissues of the body. From the tissue capillaries, the oxygen-poor blood returns through a system of veins to the right atrium of the heart. The major systemic arteries are illustrated in Figure 12-11 and are schematically represented in Figure 12-12.

Major Systemic Arteries

All systemic arteries are branches, either directly or indirectly, from the aorta. The aorta ascends from the left ventricle, curves posteriorly and to the left, and then descends through the thorax and abdomen. This geography divides the aorta into three portions: **ascending aorta, aortic arch,** and **descending aorta.** The descending aorta is further subdivided into the **thoracic aorta** and **abdominal aorta.** Table 12-1 summarizes the divisions of the aorta, the major vessels that branch from each region, and the organs supplied by each branch. The major branches of the aorta are illustrated in Figure 12-13.

Ascending Aorta

The ascending aorta, which begins at the aortic semilunar valve, is the portion that ascends from the left ventricle. The **right** and **left coronary arteries** are the only tributaries that branch from this portion. The coronary arteries, which supply oxygenated blood to the myocardium, branch from the ascending aorta just distal to the semilunar valve. When the semilunar valve is open, while blood is forcefully ejected from the heart during ventricular systole, the valve cusps cover the openings into the coronary arteries. When the valve is closed during ventricular diastole, the openings into the coronary arteries are clear and blood flows into them.

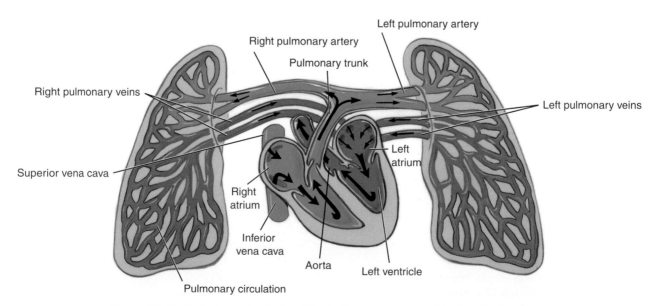

Figure 12-10 *Pulmonary circulation. Blue indicates oxygen-poor blood and red indicates oxygen-rich blood. The arrows indicate the direction of blood flow. Pulmonary circulation takes oxygen-poor blood from the right side of the heart to the lungs and returns oxygen-rich blood to the left side of the heart.*

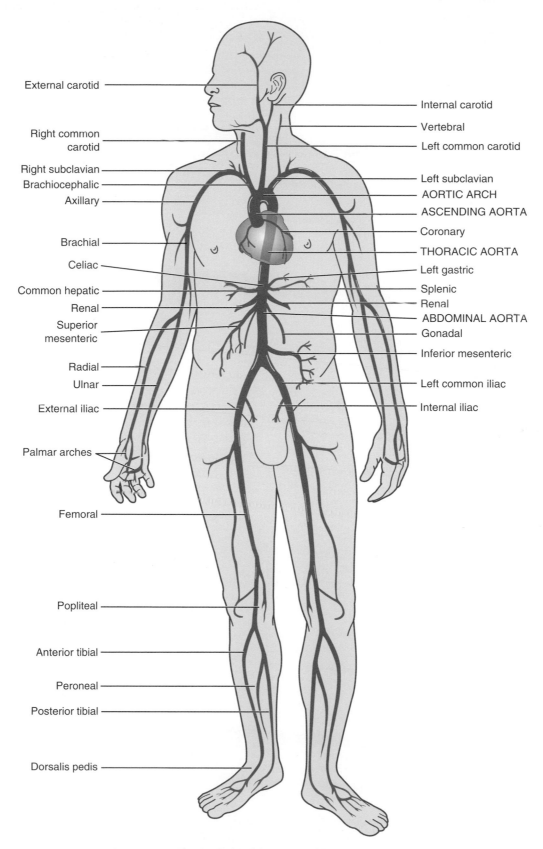

Figure 12-11 *Major systemic arteries.*

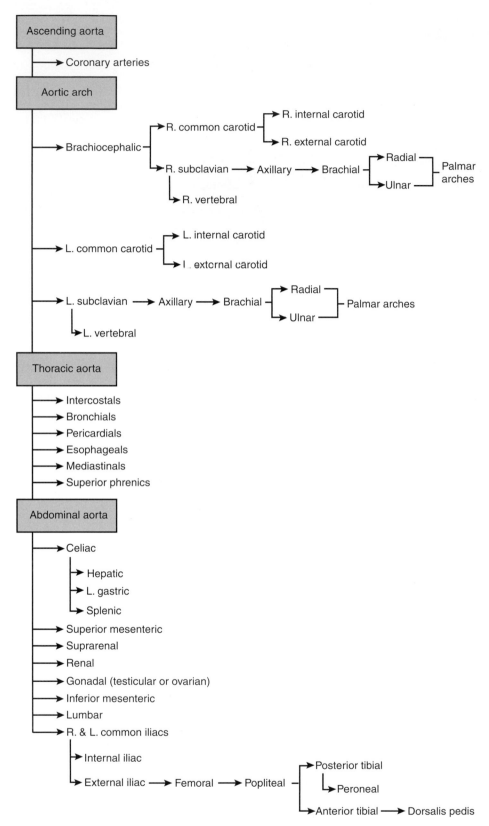

Figure 12-12 *Schematic diagram of the major systemic arteries. The diagram shows the major arteries that branch from each region of the aorta and the subsequent branching of the arteries.* **Arrows** *indicate the direction of blood flow.*

TABLE 12-1 Summary of Major Arteries and Regions They Supply

Artery	Region Supplied	Comments
From the Ascending Aorta		
Right and left coronary	Myocardium of heart wall	Branches of ascending aorta
From the Aortic Arch		
Brachiocephalic	Head and arm on right side	First branch of aortic arch; divides into right common carotid and right subclavian
Common carotid	Head and neck	Right is a branch of brachiocephalic; left comes directly from aortic arch; divides into external and internal carotids
External carotid	Face, scalp, pharynx, larynx, superficial neck	
Internal carotid	Brain	Branches from subclavian; ascends neck in transverse foramina; enters cranium through foramen magnum
Axillary	Axilla or armpit	Continuation of subclavian
Brachial	Arm	Continuation of ancillary
Radial	Lateral side of forearm and hand	Formed from brachial at elbow
Ulnar	Medial side of forearm and hand	Formed from brachial at elbow
Palmar arches	Hand and fingers	Formed by anastomosis of radial and ulnar arteries; branches extend into fingers
From the Thoracic Aorta		
Intercostals	Intercostals and other muscles of thoracic wall	Numerous, paired vessels
Bronchials	Bronchi and other passageways of respiratory tract	Numerous, paired vessels
Pericardials	Pericardium	
Esophageals	Esophagus	
Mediastinals	Structures in mediastinum	
Superior phrenics	Diaphragm	
From the Abdominal Aorta		
Celiac	Liver, stomach, pancreas, spleen	Short trunk that branches into hepatic, left gastric, and splenic
Hepatic	Liver	Branch of celiac
Left gastric	Stomach	Branch of celiac
Splenic	Spleen, pancreas, stomach	Branch of celiac
Superior mesenteric	Small intestine and part of large intestine	Numerous branches located between layers of mesentery
Suprarenal	Suprarenal (adrenal) glands	Paired, one right and one left
Renal	Kidneys	Paired, one right and one left
Gonadal	Gonads (ovaries/testes)	Paired, one right and one left
Inferior mesenteric	Distal portion of large intestine	Single vessel with several branches
Lunbar	Spinal cord and lumbar region of back	Four or five pairs branch from aorta
Common iliac	Pelvis and lower extremities	Two vessels formed by bifurcation of aorta
Internal iliac	Muscles of pelvic wall and urinary and reproductive organs in pelvis	Branch of common iliac
External iliac	Lower extremities	Branch of common iliac; continues as femoral in thigh
Femoral	Muscles of thigh	Continuation of external iliac; becomes popliteal in knee region
Popliteal	Knee and leg	Continuation of femoral
Anterior tibial	Anterior muscles of leg	Formed from popliteal
Posterior tibial	Posterior muscles of leg	Formed from popliteal
Peroneal	Lateral muscles of leg	Branches from posterior tibial
Dorsalis pedis	Ankle and dorsal part of foot	Continuation of anterior tibial

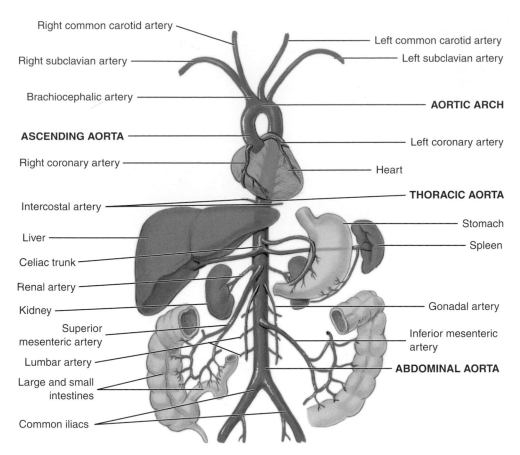

Figure 12-13 *Major branches of the aorta.*

Aortic Arch

The aortic arch, which is a continuation of the ascending aorta, curves posteriorly and to the left. There are three main branches from this region of the aorta. The first and most anterior branch is the **brachiocephalic** (bray-kee-oh-seh-FAL-ik) **artery,** which supplies blood to the right side of the head and neck, right shoulder, and right upper extremity. This vessel divides into the **right subclavian** (sub-KLAY-vee-an) **artery** and the **right common carotid** (kah-ROT-id) **artery.** The middle branch from the aortic arch is the **left common carotid artery,** which supplies the left side of the head and neck. The third and most posterior branch is the **left subclavian artery,** which takes blood to the left shoulder and left upper extremity. Figure 12-14 shows the branches of the aortic arch in a reconstructed CT image and in an angiogram.

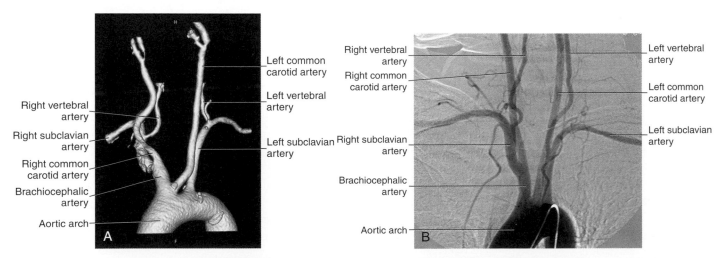

Figure 12-14 *Branches of the aortic arch. **A,** Reconstructed CT image. **B,** Angiogram. The brachiocephalic artery is the first or anterior branch, the left common carotid artery is the middle branch, and the left subclavian artery is the posterior branch. (Adapted from Applegate E: The Sectional Anatomy Learning System, Concepts, ed. 3. St. Louis, Elsevier/Saunders, 2010.)*

Descending Aorta

The descending aorta is a continuation of the aortic arch. It descends through the thoracic and abdominal regions immediately anterior to the vertebral column and slightly to the left of it. The descending aorta is subdivided into the **thoracic aorta** above the diaphragm and the **abdominal aorta** below the diaphragm.

Small, paired **intercostal arteries** branch from the thoracic aorta. These vessels are found in the spaces between the ribs, where they supply the intercostal muscles. Other small arteries supply the viscera in the thorax. After the descending aorta goes through the diaphragm, it is called the abdominal aorta. Major branches of the abdominal aorta are the **celiac** (SEE-lee-ak) **artery, superior mesenteric** (MES-en-tair-ik) **artery, renal** (REE-nal) **arteries, gonadal** (go-NAD-al) **arteries, inferior mesenteric artery,** and **lumbar arteries.** Approximately at the L4 vertebral level, the abdominal aorta bifurcates to form the **right and left common iliac** (ILL-ee-ak) **arteries.**

Arteries of the Head and Neck

The arterial blood supply for the head and neck comes from branches of the aortic arch. Note from Figure 12-13 that one of the branches of the short brachiocephalic artery is the right common carotid artery. There is no brachiocephalic artery on the left side. The left common carotid artery branches directly from the aortic arch.

The common carotid arteries ascend along each side of the neck to the angle of the mandible where each one divides into an **external carotid artery** and an **internal carotid artery** (see Figure 12-11). At the base of the internal carotid artery there is a slight enlargement called the **carotid sinus,** which contains baroreceptors and chemoreceptors to monitor pressure and chemical conditions of the blood. The external carotid arteries have numerous branches that carry blood to the skin and muscles of the neck, face, and scalp. The internal carotid arteries enter the cranial cavity through the carotid canals in the temporal bones and provide most of the blood supply to the brain. At the base of the brain, each internal carotid artery divides into an anterior and middle cerebral artery.

Another source of blood for the brain is through the **vertebral arteries,** which are branches of the subclavian arteries (see Figures 12-11 and 12-14). The vertebral arteries ascend the neck through the foramina in the transverse processes of the cervical vertebrae and enter the cranial cavity through the foramen magnum. The vertebral arteries are a secondary supply. Blood from the vertebral arteries is not sufficient to sustain life if the carotid supply is blocked. Inside the cranial cavity, the right and left vertebral arteries join to form a single **basilar** (BASE-ih-lar) **artery,** which passes over the pons and then divides into two posterior cerebral arteries (Figure 12-15).

Small communicating arteries connect the anterior, middle, and posterior cerebral arteries in a way that forms a system of vessels called the **circle of Willis,** or circulus arteriosus cerebri (Figure 12-16). This circular arrangement of vessels provides alternate pathways for blood flow when a segment is blocked.

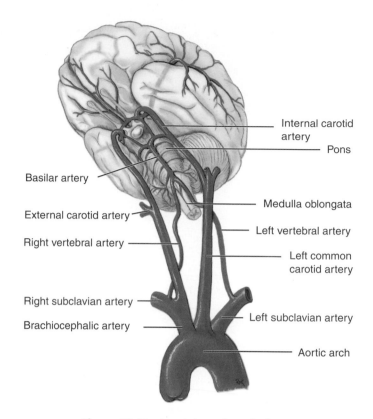

Figure 12-15 *Arterial supply to the brain.*

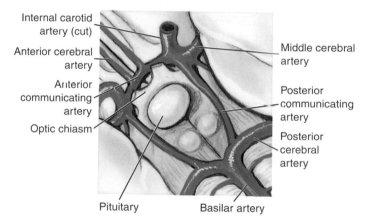

Figure 12-16 *Circle of Willis at the base of the brain. The circle is formed by the anastomosis of branches of the internal carotid and vertebral arteries.*

Arteries of the Upper Extremity

The blood supply to the shoulder and upper extremity is provided by the **subclavian arteries** (see Figure 12-11). The right subclavian artery is a branch of the brachiocephalic artery, but the left subclavian artery branches directly from the aortic arch. After passing under the clavicle, the subclavian artery continues in the axilla as the **axillary** (AK-sih-lair-ee) **artery,** which then continues in the arm as the **brachial** (BRAY-kee-al) **artery.** At the elbow, the brachial divides to form the **ulnar artery** on the medial side of the forearm and the **radial artery** on the lateral side. Branches from the radial and ulnar arteries join to form a network of arteries, the **palmar arches,** that supply the hand. The radial artery is frequently used for measurement of the pulse rate. Figure 12-17 is an arteriogram that shows these vessels.

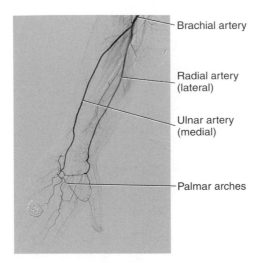

- Brachial artery
- Radial artery (lateral)
- Ulnar artery (medial)
- Palmar arches

Figure 12-17 *Arteriogram of the vessels of the forearm. (Adapted from Applegate E: The Sectional Anatomy Learning System, Concepts, ed. 3. St. Louis, Elsevier/Saunders, 2010.)*

Arteries of the Abdominal Viscera

The first major branch from the abdominal aorta, shown in Figures 12-11 and 12-13, is the **celiac** (SEE-lee-ak) **trunk** (artery). This short, unpaired vessel divides to form the **common hepatic artery,** which goes to the liver, the **left gastric artery,** which goes to the stomach, and the **splenic artery,** which supplies the pancreas and spleen.

The **superior mesenteric artery** is the second branch from the abdominal aorta. It supplies the small intestine and the proximal two thirds of the large intestine. Next there are three paired branches: the **suprarenal, renal,** and **gonadal arteries.** The suprarenal arteries go to the suprarenal, or adrenal, glands. The renal arteries supply the kidneys, and the gonadal arteries supply the testes in the male and ovaries in the female. The final visceral branch from the abdominal aorta is the unpaired **inferior mesenteric artery,** which supplies the distal one third of the large intestine. Several **lumbar arteries** arise from the abdominal aorta and provide the blood supply for the muscles and spinal cord in the lumbar region (see Figures 12-11 and 12-13).

Arteries of the Lower Extremity

At its termination, the aorta divides into the **right** and **left common iliac arteries** (see Figure 12-11). Each common iliac artery branches to form a larger **external iliac artery** and a smaller internal iliac artery. The **internal iliac artery** enters the pelvis and has numerous branches that supply the urinary bladder, uterus, vagina, muscles of the pelvic floor and wall, and external genitalia. The external iliac arteries continue into the thigh as the **femoral** (FEM-or-al) **arteries,** which continue into the posterior knee region as the **popliteal** (pop-lih-TEE-al) **arteries.** The popliteal arteries branch to form the **anterior tibial arteries** and the **posterior tibial arteries.** The anterior tibial artery supplies the anterior portion of the leg, and then continues as the **dorsalis pedis artery** that supplies the ankle and foot. A strong pulse in the dorsalis pedis artery usually indicates good circulation because this point is farthest from the heart. The posterior tibial artery supplies the posterior

QUICK CHECK

12.15 Damage to the celiac trunk interferes with blood flow to what organs?
12.16 Where is the origin of the vertebral arteries?
12.17 What major pair of arteries supplies most of the blood to the brain?
12.18 What are the three major unpaired visceral arteries that originate from the abdominal aorta?

portion of the leg. A branch of the posterior tibial, the **peroneal** (pair-oh-NEE-al) **artery,** goes to the lateral portion of the leg.

Major Systemic Veins

After blood delivers oxygen to the tissues and picks up carbon dioxide, it returns to the heart through a system of veins. The capillaries, where the gaseous exchange occurs, merge into venules and these converge to form larger and larger veins until the blood reaches either the **superior vena cava** or the **inferior vena cava,** both of which drain into the right atrium. The extremities have superficial veins that empty into deep veins. Venous tributaries that provide alternate pathways for blood flow form complex interconnecting networks between the superficial and deep veins. This makes it difficult to follow venous routes; however, the major deep veins usually follow the corresponding arteries and have the same name. The major systemic veins are illustrated in Figure 12-18 and are schematically represented in Figure 12-19. They are summarized in Table 12-2.

Veins of the Head and Neck

The **external jugular** (JUG-yoo-lar) **veins** are superficial vessels that drain blood from the skin and muscles of the face, scalp, and neck. The external jugular veins descend through the neck and empty into the **subclavian veins.** A small tributary may also drain into the internal jugular vein on each side.

The **internal jugular veins** receive blood from the veins and venous sinuses of the brain and from the deep regions of the face. The internal jugular veins are large vessels that descend through the neck and join the subclavian veins to form the **brachiocephalic veins.** The right and left brachiocephalic veins merge to form the **superior vena cava.**

Vertebral veins also drain blood from the posterior regions of the brain. These vessels descend through the neck within the foramina in the transverse processes of the cervical vertebrae. The vertebral veins empty into the subclavian veins.

Veins of the Shoulder and Arms

The deep veins of the shoulder and arm follow a pattern similar to the arteries. The **radial vein,** on the lateral side of the forearm, and the **ulnar vein,** on the medial side, join in the region of the elbow to form the **brachial vein.** This vessel follows the path of the brachial artery and continues as the **axillary vein** and then as the **subclavian vein.**

The superficial veins of the upper extremity form complex networks just underneath the skin and drain into the deep veins mentioned in the previous paragraph. The major superficial veins are the **basilic** (bah-SILL-ik) and the **cephalic** (seh-FAL-ik). The basilic vein ascends the forearm

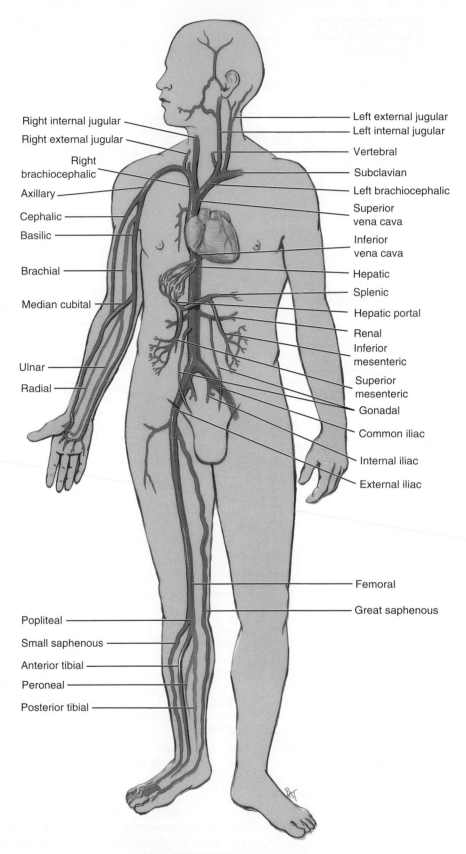

Right internal jugular
Right external jugular
Right brachiocephalic
Axillary
Cephalic
Basilic
Brachial
Median cubital
Ulnar
Radial
Popliteal
Small saphenous
Anterior tibial
Peroneal
Posterior tibial

Left external jugular
Left internal jugular
Vertebral
Subclavian
Left brachiocephalic
Superior vena cava
Inferior vena cava
Hepatic
Splenic
Hepatic portal
Renal
Inferior mesenteric
Superior mesenteric
Gonadal
Common iliac
Internal iliac
External iliac
Femoral
Great saphenous

Figure 12-18 *Major systemic veins.*

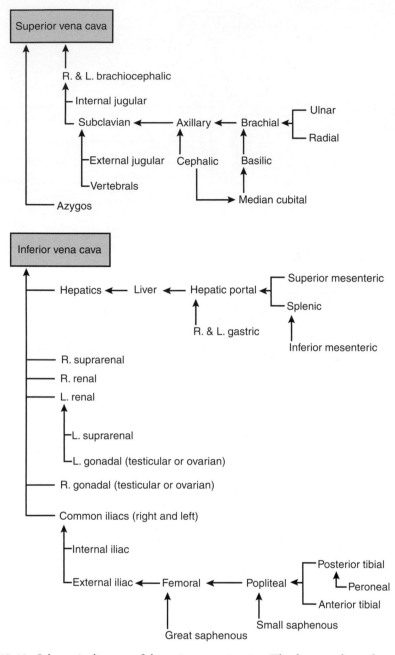

Figure 12-19 *Schematic diagram of the major systemic veins. The diagram shows the pathway of blood return from the extremities and internal organs to the superior vena cava and inferior vena cava.* **Arrows** *indicate the direction of blood flow.*

and arm on the medial side. In the upper part of the arm, it then penetrates the tissues to join the brachial vein and form the axillary vein. The cephalic ascends the forearm and arm on the lateral side. In the shoulder region, the cephalic vein penetrates deep into the tissues and joins the axillary vein to form the subclavian vein. A **median cubital** (KYOO-bih-tal) **vein** is usually prominent on the anterior surface of the arm at the bend of the elbow. This vessel ascends from the cephalic vein to the basilic vein and is frequently used as a site for drawing blood.

Veins of the Thoracic and Abdominal Walls

The **azygos** (AZ-ih-gus) **vein** drains blood from most of the muscle tissue of the thoracic and abdominal walls. This vessel begins in the dorsal abdominal wall, ascends along the right side of the vertebral column, and empties into the superior vena cava.

Veins of the Abdominal and Pelvic Organs

Blood from the abdominal organs, pelvic cavity, and lower extremities is returned to the right atrium by the **inferior vena cava** (see Figure 12-18). The **hepatic** (liver), **right suprarenal** (adrenal glands), **renal** (kidneys), and **right gonadal** (ovaries or testes) **veins** empty directly into the inferior vena cava. On the left side, the suprarenal and gonadal veins empty into the left renal vein, which carries the blood to the inferior vena cava.

Blood from the other abdominal organs enters the **hepatic portal system** and is carried to the liver before going to the inferior vena cava. The hepatic portal system begins with

TABLE 12-2 Summary of Major Veins and Regions They Drain

Vein	Region Drained	Comments
Blood Returned to Heart by Superior Vena Cava		
Brachiocephalic	Head, neck, upper extremity	Right and left brachiocephalics join to form superior vena cava
Internal jugular	Brain and portions of face and neck	Drains regions supplied by internal carotid artery; joins with subclavian to form brachiocephalic
Subclavian	Shoulder and upper extremity	Joins with internal jugular to form brachiocephalic; receives blood from axillary
Axillary	Axilla (armpit)	Receives blood from brachial and cephalic; drains into subclavian
Brachial	Deep vein of arm	Receives blood from ulnar, radial, and basilic; drains into axillary
Radial	Deep vein on lateral side of forearm	Drains into brachial
Ulnar	Deep vein on medial side of forearm	Drains into brachial
Vertebral	Brain	Descends neck in transverse foramina; drains into subclavian
External jugular	Face and scalp	Drains region supplied by external carotid artery; drains into subclavian and also has branches into internal jugular
Cephalic	Superficial vein on lateral side of forearm and arm	Drains into axillary
Basilic	Superficial vein on medial side of forearm and arm	Drains into brachial
Azygos	Thoracic and abdominal walls	Drains into superior vena cava; receives blood from intercostals, internal thoracic, and lumbar veins
Blood Returned to Heart by Inferior Vena Cava		
Hepatic	Liver	Receives blood from venous sinusoids in liver
Hepatic portal	Digestive tract	Receives blood from gastric, splenic, and superior mesenteric veins and takes it to liver
Gastric	Stomach	Drains into hepatic portal vein
Superior mesenteric	Small intestine and proximal portion of large intestine	Joins with splenic vein to form hepatic portal vein
Splenic	Spleen, pancreas, and portion of stomach	Joins with superior mesenteric vein to form hepatic portal vein
Inferior mesenteric	Distal portion of large intestine	Drains into splenic vein
Suprarenal	Suprarenal (adrenal) glands	On right, vein empties into inferior vena cava; on left, it drains into renal vein
Renal	Kidneys	Drains into inferior vena cava; on left, renal vein receives blood from left suprarenal and left gonadal veins
Gonadal	Gonads (ovaries/testes)	On right, they drain into inferior vena cava; on left, they empty into renal vein
Common iliac	Pelvis and lower extremities	Right and left join to form inferior vena cava
Internal iliac	Muscles of pelvic wall and urinary and reproductive organs in pelvis	Joins with external iliac to form common iliac vein
External iliac	Lower extremities	Joins with internal iliac to form common iliac vein; continuation of femoral vein from thigh
Femoral	Deep region of thigh	Receives blood from popliteal and great saphenous veins
Popliteal	Knee and leg	Receives blood from anterior and posterior tibial veins and small saphenous vein; continues in thigh as femoral vein
Anterior tibial	Anterior muscles of leg	Deep vein of leg that joins with posterior tibial to form popliteal vein
Posterior vein	Posterior muscles of leg	Deep vein of leg that joins with anterior tibial to form popliteal vein; receives blood from peroneal vein
Peroneal	Lateral muscles of leg	Drains into posterior tibial vein
Small saphenous	Superficial tissues of posterior and lateral leg	Superficial vein that drains into popliteal vein
Great saphenous	Superficial tissues of anterior and medial leg and thigh	Longest vein in body; superficial vein that empties into femoral vein

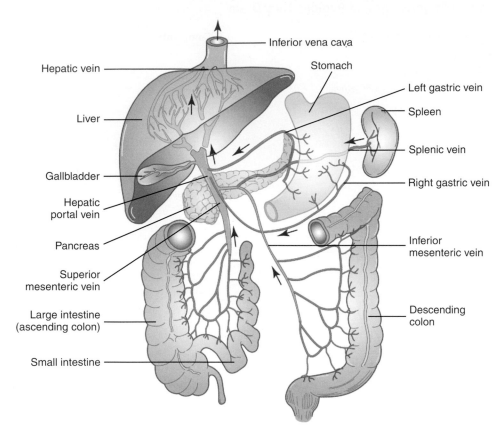

Inferior vena cava

Hepatic vein

Stomach

Left gastric vein

Liver

Spleen

Splenic vein

Gallbladder

Right gastric vein

Hepatic portal vein

Pancreas

Inferior mesenteric vein

Superior mesenteric vein

Large intestine (ascending colon)

Descending colon

Small intestine

Figure 12-20 *Hepatic portal circulation. Blood from organs associated with digestion returns to the inferior vena cava via the hepatic portal vein and liver. The splenic vein and superior mesenteric vein merge to form the hepatic portal vein. Arrows indicate the direction of blood flow.*

capillaries in the viscera and ends with capillaries in the liver (Figure 12-20). The **splenic vein** and the **superior mesenteric vein** join to form the **hepatic portal vein,** which enters the liver. The superior mesenteric vein drains the small intestine and proximal portion of the large intestine, and the splenic vein drains the spleen and pancreas. The **inferior mesenteric vein** drains blood from the distal portion of the large intestine, and then empties into the splenic vein. The **right** and **left gastric veins** are small vessels from the stomach that empty into the portal vein. Blood in the hepatic portal vein is rich in nutrients from the intestines, but it may also contain harmful substances that are toxic to the tissues. The liver removes the nutrients from the blood and either stores them or modifies them so that they can be used by the cells of the body. The liver also removes the toxic substances from the blood and either neutralizes them or alters them so that they are less harmful. Blood from the liver is collected into hepatic veins, which empty into the inferior vena cava.

The **internal iliac veins,** illustrated in Figure 12-18, drain blood from the pelvic viscera and pelvic wall. These vessels join the **external iliac veins** from the lower extremity to form the **common iliac veins.** The two common iliac veins merge to form the inferior vena cava, which empties into the right atrium.

Veins of the Lower Extremity

The arrangement of veins in the lower extremity is similar to that in the upper extremity (see Figure 12-18). There is a deep set of veins that follows the pathway of the arteries. A superficial set of veins forms a complex network just underneath the skin, and then penetrates the tissues to drain into the deep veins.

In the leg, the deep veins are the **anterior tibial vein,** which drains the dorsal foot and anterior muscles in the leg, and the **posterior tibial vein,** which drains the plantar region of the foot and posterior muscles in the leg. The **peroneal vein,** which drains the lateral muscles in the leg, empties into the posterior tibial vein. In the knee region, the anterior and posterior tibial veins join to form the single **popliteal vein,** which continues through the thigh as the **femoral vein.** The **external iliac vein** is a continuation of the femoral vein.

The two major superficial veins of the lower extremity are the small and great **saphenous** (sah-FEE-nus) **veins.** The **small saphenous vein** ascends under the skin along the posterior leg. Near the knee, it penetrates the tissues to empty into the popliteal vein. The **great saphenous vein** is the longest vein in the body. It begins near the medial malleolus of the tibia and ascends through the leg and thigh until it empties into the femoral vein. A portion of the great saphenous vein is often removed and used for grafts during coronary artery bypass surgery. The saphenous veins have numerous interconnecting tributaries and also have many branches that connect with the deep veins. These provide alternate pathways by which blood can be returned to the inferior vena cava from the lower extremities.

QUICK CHECK

12.19 Blood from the superior mesenteric vein drains into what vessel?
12.20 An embolism from a clot in the great saphenous vein circulates until it reaches a vessel smaller that the clot. What vessel is most likely to become blocked by the embolus?
12.21 What vessels merge to form the inferior vena cava?
12.22 What vessel takes blood from the spleen and small intestines to the liver?

Fetal Circulation

Most circulatory pathways in the fetus are similar to those in the adult, but there are some notable differences because the lungs, the gastrointestinal tract, and the kidneys are not functioning before birth. The fetus obtains its oxygen and nutrients from the mother and also depends on maternal circulation to remove the carbon dioxide and waste products. Figure 12-21 illustrates the features of fetal circulation.

The exchange of gases, nutrients, and waste products occurs through the **placenta,** which is attached to the uterine wall of the mother and connected to the umbilicus (navel) of the fetus by the **umbilical cord.** The umbilical cord contains two **umbilical arteries** and one **umbilical vein.** Umbilical arteries, branches of the internal iliac arteries, carry blood that is loaded with carbon dioxide and waste products from the fetus to the placenta. In the placenta, the carbon dioxide and waste products diffuse from the fetal blood into the maternal blood. At the same time, oxygen and nutrients diffuse from the maternal blood in the placenta into the umbilical vein, which carries the oxygen-rich and nutrient-rich blood to the fetal circulation. Diffusion normally takes place across capillary walls, and there is no mixing of fetal and maternal blood in the placenta.

The umbilical vein carries blood to the fetal liver where it divides into two branches. One small branch supplies blood for nourishment of the liver cells. Most of the blood enters the other branch, the **ductus venosus** (DUK-tus veh-NO-sus), which bypasses the immature liver and goes directly to the inferior vena cava and then to the right atrium.

Because the fetal lungs are collapsed and nonfunctional, two structures permit most of the blood to circumvent the pulmonary circuit. An opening in the interatrial septum, the **foramen ovale,** allows some of the blood to go directly from the right atrium into the left atrium and into the systemic

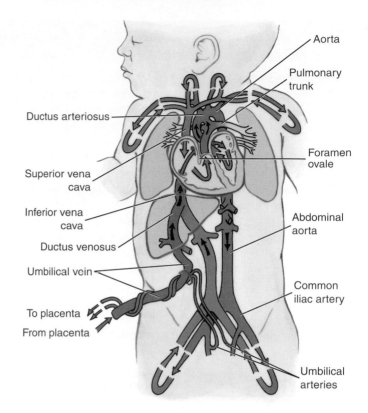

Figure 12-21 *Fetal circulation. The ductus venosus bypasses the liver and shunts blood directly from the umbilical vein to the inferior vena cava. The foramen ovale is an opening in the interatrial septum to bypass the lungs. The ductus arteriosus is a shunt between the pulmonary trunk and the aorta.* ARROWS *indicate the direction of blood flow.*

circulation. The rest of the blood enters the right ventricle and is pumped into the pulmonary trunk. A short vessel, the **ductus arteriosus** (DUK-tus ar-teer-ee-OH-sus), connects the pulmonary trunk to the aorta. Most of the blood that enters the pulmonary trunk enters the systemic circulation by this route. A small amount of blood goes to the lungs to maintain their viability.

At or shortly after birth, when pulmonary, renal, and digestive functions are established, the special structures in the fetal circulatory pathway are no longer necessary. Changes occur that make the structures nonfunctional, and the circulatory pattern becomes like that of an adult. Table 12-3 summarizes the special features of fetal circulation and the changes that occur.

QUICK CHECK

12.23 In fetal circulation, where does the exchange of oxygen and carbon dioxide normally occur?
12.24 In fetal circulation, what vessel provides a shunt between the umbilical vein and the inferior vena cava?

FOCUS ON AGING

Problems with blood vessels are relatively rare in children and young adults. However, signs of vascular aging begin to appear as early as age 40. One of the most common and significant changes is a decrease in the amount of elastic fibers and an increase in the amount of collagen. Furthermore, the collagen is altered and is less flexible than normal. The result is a decreased elasticity of the vessel walls called **arteriosclerosis,** or hardening of the arteries. The overall effect is an increased resistance to each surge of blood from ventricular contraction, which produces an increased systolic pressure. The vessels are less able to maintain diastolic pressure, so pulse pressure increases. Increased peripheral resistance from the loss of vessel elasticity puts an extra load on the heart, which then has to work harder and it enlarges to compensate.

The internal diameter of blood vessels tends to decrease with age because lipids gradually accumulate in the wall. This condition in which fatty, cholesterol-filled deposits, called atheromas, develop in the vessel walls is called **atherosclerosis.** Atheromas are especially common in areas of high blood pressure or turbulence, such as the coronary arteries, aorta, iliac arteries, carotid arteries, and cerebral arteries. The decreased lumen caused by the atheromas contributes to increased peripheral resistance. The decreased blood flow that results may not be significant until a time when maximum flow is needed. Atherosclerosis is often the primary agent in cerebrovascular accident, myocardial infarction, thrombus formation, and embolisms.

The walls of veins may become thicker with age because of an increase in connective tissue and calcium deposits. The valves also tend to become stiff and incompetent. Varicose veins develop. Because of the low blood pressure in veins, these changes probably are not significant for cardiovascular function. They may be of concern because of the possibility of phlebitis and thrombus formation.

While some degree of arteriosclerosis probably is inevitable in the aging process, a significant amount of vascular disease can be prevented by a proper diet, regular walking or other aerobic exercise, and the elimination of cigarette smoking. In other words, lifestyle probably has more effect on the cardiovascular system than aging.

TABLE 12-3 Summary of Structures in Fetal Circulation

Structure	Location	Function	Fate After Birth
Umbilical arteries	Two vessels in umbilical cord	Transports blood from fetus to placenta to pick up oxygen and nutrients and to get rid of carbon dioxide	Degenerates to become lateral umbilical ligaments
Umbilical vein	Single vessel in umbilical cord	Transports oxygen- and nutrient-rich blood from placenta to fetus	Becomes round ligament (ligamentum teres) of liver
Ductus venosus	Continuation of umbilical vein to inferior vena cava	Carries blood directly from umbilical vein to inferior vena cava; bypasses liver	Becomes ligamentum venosum of liver
Foramen ovale	In septum between right and left atria	Allows blood to go directly from right atrium into left atrium to bypass pulmonary circulation	Closes after birth to become fossa ovalis
Ductus arteriosus	Between pulmonary trunk and aorta	Permits blood in pulmonary trunk to go directly into descending aorta and to bypass pulmonary circulation	Becomes a fibrous cord: ligamentum arteriosum

CHAPTER SUMMARY

Classification and Structure of Blood Vessels

■ **Describe the structure and function of arteries, capillaries, and veins.**

- Arteries carry blood away from the heart. The wall of an artery consists of the tunica intima (simple squamous epithelium), tunica media (smooth muscle), and tunica externa (connective tissue).
- Capillaries form the connection between arteries and veins. Their walls are simple squamous epithelium. Capillaries function in the exchange of materials between the blood and the tissue cells. Small arterioles and precapillary sphincters regulate blood flow into the capillaries.
- Veins carry blood toward the heart. The walls of veins have the same three layers as the arteries, but the layers are thinner. Veins have valves to prevent the backflow of blood.

Physiology of Circulation

■ **Discuss how oxygen, carbon dioxide, and glucose move across capillary walls.**

- Oxygen, carbon dioxide, and glucose move across the capillary wall by simple diffusion from a region of high concentration to a region of low concentration.

■ **Describe the mechanisms and pressures that move fluids across capillary walls.**

- A combination of hydrostatic pressure and osmotic pressure determines fluid movement across the capillary wall. In general, fluid moves out of the capillary at the arteriole end and returns at the venous end.

■ **Discuss the factors that affect blood flow through arteries, capillaries, and veins.**

- Blood flows in the same direction as the pressure gradient.
- Pressure is lowest as the venae cavae enter the right atrium. Pressure in the right atrium is called the central venous pressure.
- Velocity of blood flow varies inversely with the total cross-sectional area of the blood vessels. As the area increases, the velocity decreases. Since capillaries are so numerous, they have the greatest cross-sectional area and the slowest blood flow.
- Resistance is a force that opposes blood flow. As resistance increases, blood flow decreases. The autonomic nervous system regulates blood flow by changing the resistance of the vessels through vasoconstriction and vasodilation.
- Very little pressure from ventricular contraction remains by the time the blood reaches the veins. Venous blood flow depends on skeletal muscle action, respiratory movements, and contraction of smooth muscle in venous walls.

■ **Discuss four primary factors that affect blood pressure and how blood pressure is regulated.**

- Pulse refers to the rhythmic expansion of an artery that is caused by ejection of blood from the ventricle. It can be felt where an artery is close to the surface and rests on something firm. These locations are called pulse points.
- Systolic pressure is the pressure in the arteries during ventricular contraction (systole). Normal systolic pressure is about 120 mm Hg. Diastolic pressure is the pressure in the arteries during ventricular relaxation (diastole). Normal diastolic pressure is about 80 mm Hg. Pulse pressure is the difference between systolic pressure and diastolic pressure. This is usually about 40 mm Hg.
- A sphygmomanometer is often used to measure blood pressure in the brachial artery.
- Cardiac output, blood volume, peripheral resistance, and viscosity of the blood affect blood pressure.
- Baroreceptors in the aortic arch and carotid sinus detect stretching in the vessel walls when blood pressure increases. Signals are relayed back to the heart and blood vessels that react to reduce the pressure. Baroreceptors are important in short-term blood pressure regulation.
- Chemoreceptors detect carbon dioxide, hydrogen ion, and oxygen concentrations. This is significant only in situations of severe stress.
- Antidiuretic hormone plays a role in regulating blood volume, which has an effect on blood pressure.
- In response to decreases in blood pressure, the kidneys secrete renin, which stimulates the production of angiotensin. Angiotensin causes vasoconstriction and promotes the release of aldosterone. Both actions result in increased blood pressure.

Circulatory Pathways

■ **Trace blood through the pulmonary circuit from the right atrium to the left atrium.**

- Pulmonary circulation transports oxygen-poor blood from the right ventricle to the lungs, where the blood picks up oxygen then it returns the oxygen-rich blood to the left atrium.

■ **Identify the major systemic arteries and veins.**

- Systemic arteries carry oxygenated blood from the left ventricle to the capillaries in the tissues of the body. Figures 12-11 and 12-12 identify the major systemic arteries.
- Systemic veins carry oxygen-poor, carbon dioxide–laden blood from the tissues to the right atrium of the heart. Figures 12-18 and 12-19 identify the major systemic veins.

■ **Describe the blood supply to the brain.**

- The principal blood supply to the brain is through the internal carotid arteries and the vertebral arteries. Branches of these vessels from the circle of Willis at the base of the brain.

■ **Describe five features of fetal circulation.**

- There are some differences in the circulatory pathways of the fetus because the lungs, gastrointestinal tract, and kidneys are not functioning.
- The two umbilical arteries carry fetal blood to the placenta; the umbilical vein carries blood from the placenta to the fetus; the placenta functions in the exchange of gases and nutrients between the maternal and fetal blood; the ductus venosus allows blood to bypass the immature fetal liver; the foramen ovale and ductus arteriosus permit blood to bypass the fetal lungs.

CHAPTER QUIZ

Recall

Match the definitions on the left with the appropriate term on the right.

_____ 1. Carry blood away from the heart	**A.** Arteries
_____ 2. Smallest blood vessels	**B.** Capillaries
_____ 3. Primarily smooth muscle	**C.** Celiac trunk
_____ 4. Fluid between cells	**D.** Foramen ovale
_____ 5. Heard through the stethoscope	**E.** Interstitial fluid
_____ 6. Fetal opening in the interatrial septum	**F.** Korotkoff sounds
_____ 7. Frequently used for drawing blood	**G.** Median cubital vein
_____ 8. First major branch from the abdominal aorta	**H.** Sphygmomanometer
_____ 9. Returns blood from the head, neck, and arms to the heart	**I.** Superior vena cava
_____ 10. Used to measure blood pressure	**J.** Tunica media

Thought

1. The layer of the arteriole wall that provides contractility and elasticity for vasoconstriction and vasodilation is the:
   A. Tunica externa
   B. Tunica adventitia
   C. Tunica media
   D. Tunica interna

2 At any given time, most of the blood in the body is found in the:
   A. Capillaries
   B. Veins
   C. Heart
   D. Arteries

3 When a precapillary sphincter relaxes:
   A. Blood flows into a capillary bed
   B. Blood flows through the arteriovenous anastomoses
   C. Blood flows into metarterioles
   D. Blood backs up in the arterioles

4 Which of the following factors opposes the flow of blood?
   A. Decreased blood viscosity
   B. Increased diameter of blood vessels
   C. Decreased total cross-sectional area of blood vessels
   D. Increased resistance in blood vessels

5 A portion of a blood clot in the femoral vein breaks loose. In what vessel is this embolus likely to lodge?
   A. Inferior vena cava
   B. Capillary in the lungs
   C. Capillary in the liver
   D. Capillary in the brain

Application

Trace a drop of blood from the inferior vena cava to the muscles on the lateral side of the right forearm. List all vessels, heart chambers, and valves.

Briefly describe two ways in which angiotensin increases blood pressure.

BUILDING YOUR MEDICAL VOCABULARY

Building a vocabulary is a cumulative process. As you progress through this book, you will use word parts, abbreviations, and clinical terms from previous chapters. Each chapter will present new word parts, abbreviations, and terms to add to your expanding vocabulary.

Word Parts and Combining Form with Definition and Examples

PART/COMBINING FORM	DEFINITION	EXAMPLE
aneur-	widening	aneurysm: local dilation (widening) of an arterial wall
angi/o	vessel	angiorrhaphy: suture of a blood vessel
arteri/o	artery	arteriosclerosis: hardening of an artery
ather/o	yellow fatty plaque	atherogenesis: formation of fatty plaques in the wall of an artery
brachi/o	arm	brachiocephalic: pertaining to the arm and the head
carotid	put to sleep	carotid artery: blockage puts the patient to sleep, becomes unconscious
cephal/o	head	brachiocephalic: pertaining to the arm and the head
edem-	to swell	edema: accumulation of excess fluid, causing swelling
embol-	stopper, wedge	embolism: sudden blockage of an artery by foreign matter that has been brought to the site by blood flow
isch-	deficiency	ischemia: deficiency of blood supply to a region
phleb-	vein	phlebitis: inflammation of a vein
scler/o	hard	arteriosclerosis: hardening of an artery
sten/o	narrowing	stenosis: narrowing of a body passage or opening

PART/COMBINING FORM	DEFINITION	EXAMPLE
thromb/o	clot	thromboangiitis: inflammation of a blood vessel with the formation of a clot
vas/o	vessel	vasa vasorum: blood vessel of a blood vessel, small vessels that supply nutrients to the wall of a large vessel
ven/o	vein	venipuncture: surgical puncture of a vein

Clinical Abbreviations

ABBREVIATION	MEANING
ABP	arterial blood pressure
ACE	angiotensin-converting enzyme
CV	cardiovascular
CVP	central venous pressure
CXR	chest x-ray film
DSA	digital subtraction angiography
DVT	deep vein thrombosis
H&H	hemoglobin and hematocrit
HBP	high blood pressure
HTN	hypertension
IMA	inferior mesenteric artery
IMV	inferior mesenteric vein
IVC	inferior vena cava
IVCP	inferior vena cava pressure
PTA	percutaneous transluminal angioplasty
PTCA	percutaneous transluminal coronary angioplasty
PVD	peripheral vascular disease
SMA	superior mesenteric artery
SMV	superior mesenteric vein
SVC	superior vena cava
TEE	transesophageal echocardiography

Clinical Terms

Angiography (an-jih-AHG-rah-fee) Procedure in which a radiopaque substance is injected into the bloodstream and then radiographs are taken; used to determine the condition of blood vessels

Angioplasty (AN-jih-oh-plas-tee) Surgical repair of a blood vessel or vessels

Antiarrhythmic (an-tye-ah-RITH-mik) Any agent administered to control irregularities of the heartbeat

Antihypertensive (an-tye-hye-per-TEN-siv) Any agent administered to reduce high blood pressure

Arterectomy (ahr-teh-RECK-toh-mee) Surgical removal of an artery

Arteriosclerosis (ahr-tee-rih-oh-skleh-ROH-sis) Condition of hardening of an artery; an artery becomes less elastic and does not expand under pressure

Atherectomy (ath-er-ECK-toh-mee) Surgical removal of plaque from the interior lining of an artery

Atheroma (ath-er-OH-mah) Abnormal mass of fatty or lipid material (plaque) with a fibrous covering within an arterial wall

Hemangioma (hee-man-jee-OH-mah) Benign tumor of a blood vessel

Hemorrhoids (HEM-oh-royds) Varicose veins in the anal canal resulting from a persistent increase in venous pressure

Hypoperfusion (hye-poh-per-FEW-zhun) Deficiency of blood flow through an organ or body part

Ischemia (iss-KEE-mee-ah) Deficiency in blood supply caused by either the constriction or the obstruction of a blood vessel

Phlebitis (fleh-BYE-tis) Inflammation of veins, which may be caused by pooling and stagnation of blood; often leads to the formation of blood clots within the vessel

Phlebotomy (fleh-BAH-toh-mee) Incision into a vein

Raynaud's phenomenon (ray-NOHZ fee-NAHM-eh-non) Intermittent bilateral attacks of ischemia of fingers and toes characterized by severe pallor and often accompanied by paresthesia and pain; typically brought on by cold or emotional stimuli

VOCABULARY QUIZ

Use word parts you have learned to form words that have the following definitions.

1. Pertaining to the arm and head _____

2. Condition of arterial hardening _____

3. Inflammation of a vein _____

4. Yellow fatty tumor _____

5. Condition of stoppage _____

Using the definitions of word parts you have learned, define the following words.

6. Angiology _____

7. Arteritis _____

8. Phlebotomy _____

9. Edema _____

10. Angiostenosis _____

Match each of the following definitions with the correct word.

_____ 11. Crushing a blood vessel to stop hemorrhage

_____ 12. Surgical repair of a blood vessel

_____ 13. Disease of a blood vessel

_____ 14. Surgical excision of an artery

_____ 15. Tumor of a blood vessel

A. Angioma
B. Angiopathy
C. Angioplasty
D. Arterectomy
E. Vasotripsy

16. What is the clinical term for "inflammation of veins"? _____

17. What is the clinical term for "deficiency of blood flow through an organ"? _____

18. What procedure surgically removes plaque from the interior lining of an artery? _____

19. What is the meaning of the abbreviation H&H? _____

20. What is the meaning of the abbreviation SMA? _____

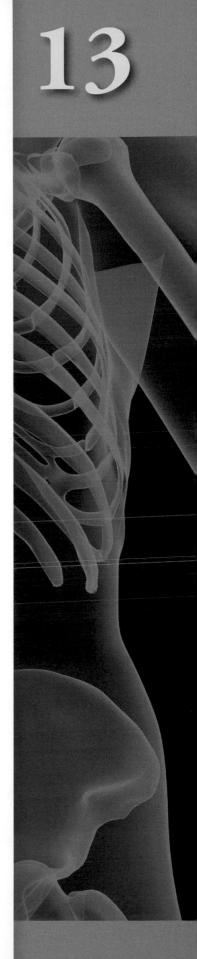

Cardiovascular System: Blood

13

CHAPTER OBJECTIVES

Functions and Characteristics of the Blood
- Describe the physical characteristics of blood.
- Describe the functions of blood

Composition of the Blood
- Describe the composition of blood plasma.
- Identify the formed elements of the blood and state at least one function for each formed element.
- Discuss the life cycle of erythrocytes.

- Differentiate between five types of leukocytes on the basis of their structure.

Hemostasis
- Describe the mechanisms that reduce blood loss after trauma.

Blood Typing and Transfusions
- Characterize the different blood types and explain why some are incompatible for transfusions.

KEY TERMS

Agglutination (ah-GLOO-tih-nay-shun) Clumping of red blood cells or microorganisms; typically an antigen-antibody reaction

Agglutinin (ah-GLOO-tih-nin) Specific substance in plasma that is capable of causing a clumping of red blood cells

Agglutinogen (ah-gloo-TIN-oh-jen) Genetically determined antigen on the cell membrane of a red blood cell that determines blood types

Coagulation (koh-ag-yoo-LAY-shun) Process of blood clotting

Diapedesis (dye-ah-peh-DEE-sis) Process by which white blood cells squeeze between the cells in a vessel wall to enter the tissue spaces outside the blood vessel

Erythrocyte (ee-RITH-roh-syte) Red blood cell

Erythropoiesis (ee-rith-roh-poy-EE-sis) The process of red blood cell formation

Erythropoietin (ee-rith-roh-POY-eh-tin) Hormone released by the kidneys that stimulates red blood cell production

Hemocytoblast (hee-moh-SYTE-oh-blast) Stem cell in the bone marrow from which the blood cells arise

Hemopoiesis (hee-moh-poy-EE-sis) Blood cell production, which occurs in the red bone marrow; also called hematopoiesis

Leukocyte (LOO-koh-syte) White blood cell

Phagocytic (fag-oh-SIT-ik) Capable of phagocytosis in which solid particles are engulfed and taken in by a cell

Thrombocyte (THROM-boh-syte) One of the formed elements of the blood; functions in blood clotting; also called platelet

The body consists of metabolically active cells that need a continuous supply of nutrients and oxygen. Metabolic waste products need to be removed from the cells to maintain a stable cellular environment. Blood is the primary transport medium that is responsible for meeting these cellular demands. A central pump, the heart, provides the force to move the blood through a system of vessels that extend throughout the body. The blood, heart, and blood vessels make up the cardiovascular system. This chapter focuses on the blood and how it supports cellular activities.

Functions and Characteristics of the Blood

Blood is one of the connective tissues. As a connective tissue, it consists of cells and cell fragments (**formed elements**) suspended in an intercellular matrix (**plasma**). Blood is the only liquid tissue in the body. The total blood volume in an average adult is 4 to 5 liters in the female and 5 to 6 liters in the male. It accounts for approximately 8% of the total body weight. Blood is slightly heavier and four to five times more viscous than water. It is slightly alkaline with a normal pH range between 7.35 and 7.45.

The activities of the blood may be categorized as **transportation, regulation**, and **protection**. These functional categories overlap and interact as the blood carries out its role in providing suitable conditions for cellular functions. The following activities of blood are transport functions:

- It carries oxygen and nutrients to the cells.
- It transports carbon dioxide and nitrogenous wastes from the tissues to the lungs and kidneys where these wastes can be removed from the body.
- It carries hormones from the endocrine glands to the target tissues.

The following activities of blood are in the regulation category:

- It helps regulate body temperature by removing heat from active areas, such as skeletal muscles, and transporting it to other regions or to the skin where it can be dissipated.
- It plays a significant role in fluid and electrolyte balance because the salts and plasma proteins contribute to the osmotic pressure.
- It functions in pH regulation through the action of buffers in the blood.

Functions of the blood that are in the protection category include the following:

- Its clotting mechanisms prevent fluid loss through hemorrhage when blood vessels are damaged.
- Certain cells in the blood, the phagocytic white blood cells (WBCs), help to protect the body against microorganisms that cause disease by engulfing and destroying the agent.
- Antibodies in the plasma help protect against disease by their reactions with offending agents.

Composition of the Blood

When a sample of blood is spun in a centrifuge, the cells and cell fragments are separated from the liquid intercellular matrix (Figure 13-1). Because the **formed elements** are heavier than the liquid matrix, they are packed in the bottom of the tube by the centrifugal force. The straw-colored liquid on the top is the **plasma**. Figure 13-1 illustrates that the plasma accounts for about 55% of the blood volume and red blood cells (RBCs) make up the remaining 45% of the volume. The percentage attributed to the RBCs is called the **hematocrit** (hee-MAT-oh-krit), or **packed cell volume (PCV)**. The WBCs and platelets form a thin white layer, called the "buffy coat," between the plasma and RBCs.

Plasma

Plasma, the liquid portion of the blood, is about 90% water. The remaining portion consists of more than 100 different organic and inorganic solutes dissolved in the water.

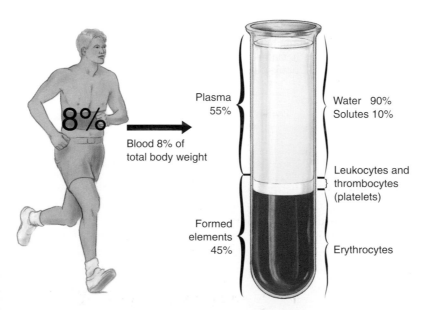

8%

Blood 8% of total body weight

Plasma 55%

Formed elements 45%

Water 90%
Solutes 10%

Leukocytes and thrombocytes (platelets)

Erythrocytes

Figure 13-1 *Composition of the blood.*

TABLE 13-1 Major Solutes in Plasma

Plasma Proteins	Nitrogenous Molecules	Other
Albumins—60%	Amino acids	Nutrients
Globulins—36%	Urea	Hormones
Fibrinogen—4%	Uric acids	Oxygen
		Carbon dioxide
		Electrolytes

Table 13-1 itemizes some of the major solutes in plasma. Because plasma is a transport medium, its solutes are continuously changing as substances are added or removed by the cells. With a healthy diet, the plasma is normally in a state of dynamic balance that is maintained by various homeostatic mechanisms.

Plasma Proteins

Plasma proteins are the most abundant of the solutes in the plasma. These proteins normally remain in the blood and interstitial fluid and are not used for energy. The three major classes of plasma proteins are albumins, globulins, and fibrinogen. Many of the plasma proteins are synthesized in the liver, and each one has a different function.

Albumins (al-BYOO-mins) account for about 60% of the plasma proteins. Albumin molecules are produced in the liver and are the smallest of the plasma protein molecules. Because they are so abundant, they contribute to the osmotic pressure of the blood and play an important role in maintaining fluid balance between the blood and interstitial fluid. If the osmotic pressure of the blood decreases, fluid moves from the blood into the interstitial spaces, which results in edema. This also decreases blood volume and, in severe cases, may reduce blood pressure. When blood osmotic pressure increases, fluid moves from the interstitial spaces into the blood and increases blood volume. This increases blood pressure and decreases the amount of water available to the cells.

Globulins (GLOB-yoo-lins) account for about 36% of the plasma proteins. There are three types of globulins: alpha, beta, and gamma. Alpha and beta globulins are produced in the liver and function in transporting lipids and fat-soluble vitamins in the blood. Gamma globulins are the antibodies that function in immunity. These are produced in lymphoid tissue.

The remaining 4% of the plasma proteins is **fibrinogen** (fye-BRIN-oh-jen), which is the largest of the plasma protein molecules. It is produced in the liver and functions in blood clotting. During the clotting process, a series of reactions converts the soluble fibrinogen into insoluble fibrin, which forms the foundation of a blood clot. When blood clots in a test tube, the liquid that remains is called **serum**. It is similar to plasma but has no fibrinogen because the fibrinogen is converted to fibrin.

Nonprotein Molecules That Contain Nitrogen

Along with the proteins, some other plasma solutes contain nitrogen. These include amino acids, urea, and uric acid. The amino acids are the products of protein digestion. They are absorbed into the blood and are transported to the cells that need them. Urea and uric acid are waste products of protein and nucleic acid catabolism. These molecules are transported to the kidneys for excretion.

Nutrients and Gases

The simple nutrients that are the end products of digestion are transported in the blood so that they form a fraction of the plasma solutes. These nutrients include the amino acids from protein digestion, glucose and other simple sugars from carbohydrate digestion, and fatty acids from lipid digestion.

Oxygen and carbon dioxide are the main respiratory gases that are found as solutes in the plasma. About 3% of the oxygen and 7% to 10% of the carbon dioxide are transported as dissolved gases. Nitrogen is another gas that dissolves in the plasma, but it has little, if any, function in the human body.

Electrolytes

Most of the electrolytes that are solutes in the plasma are inorganic ions, and they contribute to the osmotic pressure of the plasma. In addition, some are important in maintaining membrane potentials and others are significant in regulating the pH of body fluids. Common electrolytes found in the plasma include sodium (Na^+), potassium (K^+), calcium (Ca^{2+}), chloride (Cl^-), bicarbonate (HCO_3-), and phosphate (PO_4^{3-}).

QUICK CHECK

13.1 What are the three categories of blood functions?

13.2 Under average normal conditions, what percentage of blood volume is plasma?

13.3 List the three types of plasma proteins.

13.4 Liver disease may lead to a decrease in plasma proteins. How would this affect the functions of the blood?

Formed Elements

The formed elements are cells and cell fragments suspended in the plasma. The three classes of formed elements are the **erythrocytes** (ee-RITH-roh-sytes), or RBCs, the **leukocytes** (LOO-koh-sytes), or WBCs, and the **thrombocytes** (THROM-boh-sytes), or platelets. Table 13-2 summarizes the formed elements in the blood and they are shown in the photomicrograph in Figure 13-2.

The production of these formed elements, or blood cells, is called **hematopoiesis** (hemopoiesis). Before birth, hematopoiesis (hee-ma-to-poy-EE-sis) occurs primarily in the liver and spleen, but some cells develop in the thymus, lymph nodes, and red bone marrow. After birth, most production is limited to specific regions of red bone marrow, but some white blood cells are produced in lymphoid tissue. All types of

TABLE 13-2 Formed Elements in the Blood

Erythrocytes		Biconcave disks; no nucleus 7 to 8 μm in diameter; 4.5 to 6.0 million/mm³ Function to transport oxygen and carbon dioxide
Leukocytes		Nucleated cells; 5000 to 9000/mm³ Function as part of body's defense against disease
Neutrophils	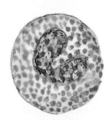	Nucleus with 2 to 5 lobes; indistinct granules in cytoplasm; 12 to 15 μm in diameter 60% to 70% of total WBCs Function in phagocytosis
Eosinophils		Bilobed nucleus; red-staining granules in cytoplasm; 10 to 12 μm in diameter 2% to 4% of total WBCs Function to counteract histamine in allergic reactions; destroy parasitic worms
Basophils		U-shaped or bilobed nucleus; granules in cytoplasm stain blue; 10 to 12 μm in diameter Less than 1% of total WBCs Function to release histamine and the anticoagulant heparin; called mast cells in the tissues
Lymphocytes	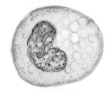	Agranulocyte; small cell with large round nucleus; 6 to 8 μm in diameter 20% to 25% of total WBCs Function in immunity; product antibodies
Monocytes		Agranulocyte; large cells with bean-shaped nucleus; may be 20 μm in diameter 3% to 8% of total WBCs Function in phagocytosis; engulf relatively large particles; called macrophages in tissues
Thrombocytes		Cell fragments of megakaryocytes; 2 to 5 μm in diameter 250,000 to 500,000/mm³ Function in hemostasis by forming platelet plug and releasing factors necessary for blood clotting

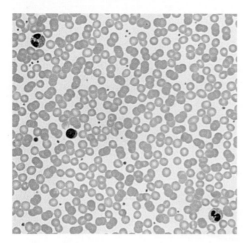

Figure 13-2 *Light micrograph of circulating blood (original magnification, 270×). Note the abundance of erythrocytes, two neutrophils, one lymphocyte, and the platelets that appear as small dots interspersed among the erythrocytes. (Adapted from Gartner and Hiatt: Color Textbook of Histology, ed. 3. Philadelphia, Elsevier/ Saunders, 2007.)*

formed elements develop from a single cell type called a **hemocytoblast** (hee-moh-SYTE-oh-blast). Seven different cell lines, each controlled by a specific growth factor, develop from the hemocytoblast.

Erythrocytes

Characteristics and Functions

Erythrocytes (ee-RITH-roh-sytes), or RBCs, are the most numerous of the formed elements. Although the number varies, the normal range for a healthy adult male is 4.5 to 6 million RBCs per cubic millimeter (mm^3) of blood. The normal RBC range for females is slightly less. Erythrocytes are tiny **biconcave disks** about 7.5 micrometers (μm) in diameter. They are thin in the middle and thicker around the periphery. The shape of the RBC provides a combination of flexibility for moving through tiny capillaries along with a maximum surface area for the diffusion of gases. Mature RBCs are **anucleate**. During development the nucleus and most other organelles are lost from the cell, presumably to give more room for hemoglobin. Because the mature cells are anucleate, they cannot undergo mitosis, which means that replacement cells have to develop from the stem cells. Erythrocytes normally move from the bone marrow into the blood while they are still immature, but after they lose the nucleus. These immature erythrocytes that are circulating in the blood are called **reticulocytes**.

QUICK APPLICATIONS

The **reticulocyte count** in circulating blood gives information about the rate of hematopoiesis. Normally 0.5% to 1.5% of the RBCs in normal blood are reticulocytes. A number below 0.5% indicates a slowdown in production. Values above 1.5% indicate a greater than normal rate of RBC formation.

The primary function of erythrocytes is to transport oxygen and, to a lesser extent, carbon dioxide. This function is directly related to the **hemoglobin** (hee-moh-GLOH-bin) within the RBC. About one third of each erythrocyte consists of hemoglobin. This molecule has two parts: **heme** and **globin**. The heme portion is formed from a pigment that contains iron. The globin portion is a protein. In the lungs, the heme portion combines with oxygen to form **oxyhemoglobin** (ok-see-hee-moh-GLOH-bin), which is bright red. About 97% of the oxygen used by tissue cells is transported as oxyhemoglobin. In the tissues, the oxygen is released to diffuse into the tissue cells. This produces a reduced form, called **deoxyhemoglobin** (dee-ok-see-hee-moh-GLOH-bin), which is darker.

QUICK APPLICATIONS

About 20% of a cigarette smoker's hemoglobin is nonfunctional for transporting oxygen because it is bound to carbon monoxide from the cigarette smoke.

Production of Erythrocytes

Erythrocyte production is regulated by a negative feedback mechanism that uses the hormone **erythropoietin** (ee-rith-roh-POY-ee-tin) to stimulate erythrocyte production (Figure 13-3). The liver produces erythropoietin in an inactive form and secretes it into the blood. The kidneys produce a **renal erythropoietic factor** (REF), which activates the erythropoietin. When blood oxygen concentration is low, the kidneys release REF into the blood, which activates the erythropoietin, which then stimulates the red bone marrow to produce RBCs. The additional RBCs combine with oxygen to increase the blood oxygen concentration. As blood oxygen concentration increases, levels of REF and active erythropoietin decrease and RBC production decreases.

Iron, vitamin B_{12}, and folic acid are essential to normal RBC production. The iron is necessary for the synthesis of normal hemoglobin. Iron deficiency anemia results when there is a lack of iron in the diet. This results in a reduced amount of hemoglobin, which decreases the blood's oxygen-carrying capacity. All cells in the body require vitamins B_{12} and folic acid for normal formation. This is especially significant in erythrocytes because of the large numbers produced every day. Certain cells in the stomach produce **intrinsic factor**, a factor necessary for the absorption of vitamin B_{12} in the intestines. Without intrinsic factor, vitamin B_{12}, even though present in the diet, is not absorbed and the cells are defective. This condition is called **pernicious anemia**.

Destruction of Erythrocytes

Normal erythrocytes live for approximately 120 days. During this time, they travel thousands of miles as they circulate throughout the body. Normally, the erythrocytes have a flexible cell membrane that allows them to bend and squeeze through the capillaries. As they age, however, their membrane loses its elasticity and becomes fragile. When they are

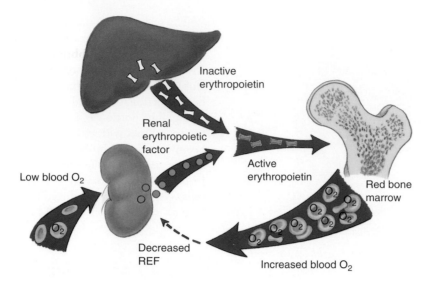

Figure 13-3 *Regulation of erythrocyte production. The liver secretes inactive erythropoietin into the blood. In response to low blood O_2, the kidneys release REF into the blood. This activates the erythropoietin, which stimulates the bone marrow to produce RBCs.*

defective or worn out, macrophages, which are phagocytic cells in the spleen and liver, remove them from circulation, and they are replaced by an equal number of new cells. Under typical conditions, over 2 million erythrocytes are destroyed and replaced every second!

When RBCs are destroyed, the hemoglobin is separated into its heme and globin components (Figure 13-4). The protein portion of the hemoglobin in the erythrocyte is broken down into its constituent amino acids, which are added to the supply of amino acids that are available in the body. The heme portion of the molecule is broken down into an iron compound

and bilirubin, a yellow bile pigment. The liver then recycles the iron and sends it to the bone marrow for new hemoglobin. Bilirubin becomes part of the bile, which is secreted by the liver, and is carried in the bile duct to the small intestine.

QUICK APPLICATIONS
Hemolytic anemia is a reduction in erythrocytes caused by excessive destruction. The excessive destruction leads to jaundice because of the accumulation of bilirubin in the blood.

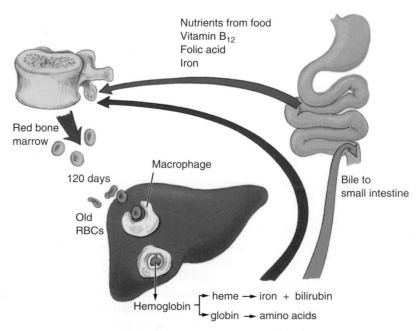

Figure 13-4 *Life cycle of red blood cells and breakdown of hemoglobin. Bone marrow produces RBCs. Old RBCs are removed from circulation by macrophages in the liver. The hemoglobin is broken down and the iron is reused to make new RBCs.*

QUICK CHECK

13.5 What are the anucleate, immature red blood cells that are circulating in the blood?

13.6 What stimulates the release of renal erythropoietic factor?

13.7 What is the approximate life span of an erythrocyte?

13.8 How does an increase in oxygen delivery to the kidneys affect the active erythropoietin in the blood?

Leukocytes

Characteristics and Functions

Leukocytes (LOO-koh-sytes), or WBCs, are generally larger than erythrocytes, but they are fewer in number. An average leukocyte count ranges between 5000 and 9000 per cubic millimeter. All leukocytes are derived from hemocytoblast stem cells, but they do not lose their nuclei or accumulate hemoglobin during development. The lack of hemoglobin makes them appear whitish.

Even though they are considered to be blood cells, leukocytes do most of their work in the tissues. They use the blood as a transport medium. Some are phagocytic, others produce antibodies, some secrete histamine and heparin, and others neutralize histamine. Leukocytes are able to move through the capillary walls into the tissue spaces, by a process called **diapedesis** (dye-ah-peh-DEE-sis). In the tissue spaces, they provide a defense against organisms that cause disease and they affect inflammatory responses.

Types of Leukocytes

There are two main groups of leukocytes in the blood. The cells that develop granules in the cytoplasm are called **granulocytes**, and those that do not have granules are called **agranulocytes**. **Neutrophils, eosinophils**, and **basophils** are granulocytes. **Monocytes** and **lymphocytes** are agranulocytes.

Neutrophils (NOO-troh-fills) are the most common type of leukocyte. These granulocytes make up 60% to 70% of the total number of WBCs. They are characterized by a multilobed nucleus (usually three to five lobes) and inconspicuous granules in the cytoplasm that stain pink with a neutral stain. Neutrophils are the first leukocytes to respond to tissue damage, where they engulf bacteria by phagocytosis. The number of neutrophils increases significantly in acute infections.

Eosinophils (ee-oh-SIN-oh-fills), which make up 2% to 4% of the WBCs, are characterized by a nucleus with two lobes and large granules in the cytoplasm that stain red with acid stains. These granulocytes neutralize histamine, and their number increases during allergic reactions. They also destroy parasitic worms.

Basophils (BAY-soh-fills), the least numerous of the leukocytes, usually account for less than 1% of the total WBCs. These granulocytes are about the same size as an eosinophil and have a nucleus that has two lobes or is U-shaped. The cytoplasm has large granules that stain dark blue with basic stains. Basophils that leave the blood and enter the tissues are called **mast cells**. In the tissues, mast cells secrete histamine and heparin. Histamine dilates blood vessels to increase blood flow to damaged tissues. It also dilates blood vessels in allergic reactions. Heparin is an anticoagulant; it inhibits blood clot formation.

Lymphocytes (LIM-foh-sytes) account for 20% to 25% of the WBCs in the blood. These agranulocytes have a large spherical nucleus that is surrounded by a small amount of cytoplasm. They are especially abundant in lymphoid tissue and have an important role in the body's defense against disease. One group, the T lymphocytes, directly attacks invading microorganisms such as bacteria and viruses. Another group, the B lymphocytes, responds by producing antibodies that react with microorganisms or with bacterial toxins.

Monocytes (MON-oh-sytes), the largest of the WBCs, make up 3% to 8% of the leukocytes in the blood. These agranulocytes have a U-shaped or bean-shaped nucleus surrounded by abundant cytoplasm. When monocytes leave the blood and enter the tissues, they are called **macrophages** (MACK-roh-fahj-es). In damaged tissues, the macrophages engulf bacteria and cellular debris to finish the cleanup process started by the neutrophils.

Thrombocytes

Thrombocytes (THROM-boh-sytes), or **platelets**, are not complete cells but are small fragments of very large cells called **megakaryocytes**. Megakaryocytes (meg-ah-KAIR-ee-oh-sytes) develop from hemocytoblasts in the red bone marrow. Platelets are one third to one half the size of an erythrocyte, and an average platelet count ranges from 250,000 to 500,000 platelets/mm³ of blood. Thrombocytes become sticky and clump together to form platelet plugs that close breaks and tears in blood vessels. They also initiate the formation of blood clots.

QUICK CHECK

13.9 What is the most common leukocyte?

13.10 Name the two agranulocytes.

13.11 WBCs circulate in the blood, but their effective work is in the tissue spaces. By what process do the cells get from the blood to the tissue spaces?

Hemostasis

Blood vessels that are torn or cut permit blood to escape into the surrounding tissues or to the outside of the body. This has damaging effects on the tissues and, in cases of excessive blood loss, may result in death. Whenever blood vessels are injured, several reactions occur that attempt to minimize blood loss and tissue damage. The stoppage of bleeding is called **hemostasis** (hee-moh-STAY-sis). It includes three separate but interrelated processes: vascular constriction, platelet plug formation, and coagulation.

Vascular Constriction

The first response to blood vessel injury is contraction of the smooth muscle in the vessel walls. This creates a **vascular constriction**, or **spasm**, that restricts the flow of blood through the opening in the vessel. The initial vascular spasm lasts for only a few minutes but allows enough time for the other aspects of hemostasis to begin. As platelets accumulate at the site of the injury, they secrete **serotonin**, a chemical that stimulates smooth muscle contraction and that prolongs the vascular spasm.

Platelet Plug Formation

Normally platelets do not stick to each other or to the endothelium that lines blood vessel walls. When the lining of the blood vessel breaks, the underlying connective tissue is exposed. Collagen in the connective tissue attracts the platelets and they accumulate in the damaged region, where they adhere to the connective tissue and to each other. This creates a mass of platelets, or **platelet plug**, that obstructs the tear in the vessel. Normal daily activities create numerous tears in minute blood vessels, and these are closed by platelet plugs so that there is no blood loss or damage to surrounding tissues.

Coagulation

The third and most effective mechanism in hemostasis is the formation of a blood clot, or **coagulation** (koh-ag-yoo-LAY-shun). The blood contains factors, called **procoagulants**, that promote clotting. It also contains **anticoagulants** that inhibit clotting. Normally the anticoagulants predominate and override the procoagulants so that the blood remains fluid and does not clot. When vessels are damaged, the procoagulants increase their activity, which results in the formation of a clot.

The formation of a blood clot involves a complex series of chemical reactions and includes numerous clotting factors that are present in the plasma. Even though it is a complex process, it can be summarized in three main steps, as illustrated in Figure 13-5.

1. Platelets and damaged tissues release chemicals that initiate a series of reactions that result in the formation of prothrombin activator.
2. In the presence of calcium ions and prothrombin activator, **prothrombin** (pro-THROM-bin) in the plasma is converted from an inactive form to active thrombin.
3. Thrombin, in the presence of calcium ions, acts as an enzyme to convert inactive and soluble fibrinogen (fye-BRIN-oh-jen) into active and insoluble fibrin. The fibrin threads form a mesh that adheres to the damaged tissue and traps blood cells and platelets to form the clot.

Platelets and all the necessary clotting factors must be available for successful clot formation. The liver produces most of the clotting factors, and many of them require vitamin K for their synthesis. Numerous reactions in the clotting process also require calcium ions. A low platelet count

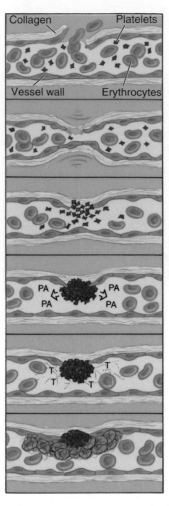

Figure 13-5 *Hemostasis. The first response to vessel injury is a vascular spasm. This is followed by platelet plug formation. The third and most effective mechanism of hemostasis is the formation of a blood clot.*

(thrombocytopenia), deficiency of vitamin K or calcium, and liver dysfunction can impair the clotting process.

After a clot has formed, the fibrin strands contract. This process, called clot retraction, causes the clot to condense or shrink. Clot retraction pulls the edges of the damaged tissue closer together, reduces the flow of blood to the area, reduces the probability of infection, and enhances healing. Fibroblasts migrate into the clot and form fibrous connective tissue that repairs the damaged area. As healing occurs, the clot is dissolved by a process called **fibrinolysis** (fye-brin-AHL-ih-sis).

QUICK APPLICATIONS

The "K" in vitamin K is for *Koagulation*, the German word for "clotting." In other words, vitamin K is the "Koagulation" vitamin. Although this vitamin is necessary in the diet, it is also produced by bacteria in the large intestine and absorbed into the blood.

FROM THE PHARMACY

Many of the pharmaceutical preparations for the blood are directed at managing the blood clotting process. There are anticoagulants to inhibit blood clotting and antihemophiliac agents to promote clotting. The anticoagulants may act indirectly by preventing the synthesis of clotting factors in the liver or they may act directly in the blood by preventing the activation of clotting factors. Antihemophiliac drugs are directed at replacing the specific clotting factors that are absent. Two examples of anticoagulants and the Rho(D) immune globulin are presented here.

Heparin is an anticoagulant that acts directly in the blood to prevent activation of clotting factors. It is used for anticoagulant therapy in the prophylaxis and treatment of all types of thromboses and emboli. It is also used to prevent blood clotting during cardiac and vascular surgery, during blood transfusion, in dialysis procedures, as treatment to disseminated intravascular coagulation, and in blood samples for laboratory purposes. Newer low-molecular-weight heparin preparations are marketed by names such as **enoxaparin** (Lovenox) and **dalteparin**. One of the advantages of these is that they can be administered as twice-daily subcutaneous injections

rather than continuous IV infusions. This means that therapy for thrombosis can be done on an outpatient basis, which is becoming the new standard of care.

Warfarin sodium (Coumadin) is an oral anticoagulant drug that acts in the liver to prevent the synthesis of clotting factors. This drug is used for long-term prophylaxis and treatment of deep venous thrombosis, pulmonary thromboembolism, and thromboembolism associated with chronic atrial fibrillation or myocardial infarction. It is usually used over a long period of time in maintenance dosages. The dosage needs to be closely monitored with daily, then weekly, then monthly laboratory testing to prevent potential uncontrolled bleeding.

Rho(D) immune globulin (Gamulin Rh, RhoGAM, HypRho-D) is an immunoglobulin solution that contains anti-Rh antibodies. It effectively prevents the Rh− mother from making antibodies against her Rh+ fetus. It is used to prevent sensitization to the Rh factor and thus to prevent hemolytic disease of the newborn in future pregnancies.

QUICK CHECK

13.12 What are the processes of hemostasis?

13.13 A sample of bone marrow has fewer-than-normal megakaryocytes. How will this affect blood clotting?

13.14 A person with liver disease bruises easily. Why?

Blood Typing and Transfusions

When an individual loses a large quantity of blood, the volume must be restored to prevent shock and death. Sometimes it is sufficient to replace the volume with plasma or a special preparation of solutes. Other times, erythrocytes must also be replaced to restore the oxygen-carrying capacity of the blood. A **transfusion** is the transfer of blood, plasma, or other solution into the blood of another individual.

Early transfusion attempts produced varied results. Some transfusions were successful and the recipient benefitted from the procedure. Other transfusions resulted in reactions that were detrimental to the recipient, frequently causing death. In these cases, the RBCs clumped together and obstructed blood vessels and damaged the kidneys. The reactions also caused hemolysis (hee-MAHL-ih-sis), or rupture of the RBCs. These unsuccessful attempts led to the discovery of blood types and procedures for accurately typing blood for safe transfusions.

Agglutinogens and Agglutinins

The clumping of RBCs in unsuccessful transfusions is caused by interactions between antigens and antibodies. Antigens are molecules, usually proteins, that elicit a response from

antibodies. Antibodies are protein molecules usually found in gamma globulin that are produced by certain lymphocytes in response to a foreign antigen. Antibodies are very specific, which means that a particular antibody will combine with only one certain type of antigen and no others. There are thousands of different antigens and antibodies in the body, but blood types are based on those specifically related to the RBCs.

Specific blood type antigens, called **agglutinogens** (ah-gloo-TIN-oh-jens), are found in the cell membrane of erythrocytes. Antibodies, called **agglutinins** (ah-GLOO-tih-nins), are in the plasma and are formed after birth. When agglutinins in the plasma combine with agglutinogens on the surface of the RBC, the result is **agglutination** (ah-GLOO-tih-nay-shun), a clumping of the RBCs. The agglutinogens on the red blood cells are organized into blood groups. Although many blood groups are recognized, the ABO and Rh groups are the most important.

ABO Blood Groups

The ABO blood groups are based on the presence or absence of certain agglutinogens (antigens) on the surface of the RBC membrane. These agglutinogens, A and B, are inherited; consequently, blood types are also inherited. Type A blood has type A agglutinogens; type B blood has type B agglutinogens; type AB blood has both type A and type B agglutinogens; and type O blood has neither type A nor type B agglutinogens (Figure 13-6). Certain agglutinins develop in the plasma shortly after birth. Specifically, a person with type A agglutinogens (type A blood) develops anti-B agglutinins; a person with type B agglutinogens (type B blood) develops anti-A agglutinins; a person with both type A and type B agglutinogens (type AB blood) develops neither anti-A nor

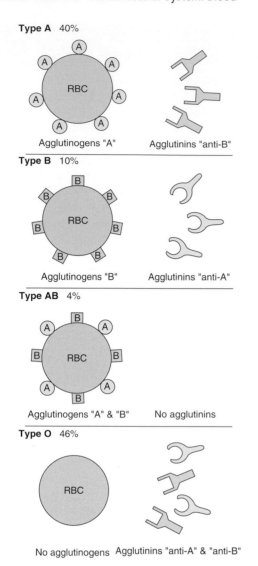

Type A 40%

Agglutinogens "A"   Agglutinins "anti-B"

Type B 10%

Agglutinogens "B"   Agglutinins "anti-A"

Type AB 4%

Agglutinogens "A" & "B"   No agglutinins

Type O 46%

No agglutinogens   Agglutinins "anti-A" & "anti-B"

Figure 13-6 *Agglutinogens (antigens) and agglutinins (antibodies) involved in the ABO blood groups.*

anti-B agglutinins; and a person with neither type A nor type B agglutinogens (type O blood) develops both anti-A and anti-B agglutinins (see Figure 13-6).

A **donor** is a person who gives blood, and a **recipient** is the person who receives blood. Because agglutinins of one type will react with the same type agglutinogen—anti-A agglutinins react with type A agglutinogens, and anti-B agglutinins react with type B agglutinogens—these combinations must be avoided. The major concern in blood transfusions is that the agglutinins in the plasma of the recipient's blood not react with, or agglutinate, the cells of the donor's blood. A person (recipient) with type A blood should not receive type B blood (donor) because the anti-B agglutinins in the type A recipient will agglutinate the type B agglutinogens in the donor's blood (Figure 13-7). Similar conditions exist when the recipient has type B blood and the donor has type A blood.

Because type AB blood has neither anti-A nor anti-B agglutinins to react with donor agglutinogens, it appears that a person with this type can receive blood of any type. For this reason, type AB blood is called the **universal recipient**. Type O blood has neither type A nor type B agglutinogens on the RBC, so it is called the **universal donor**. The terms universal recipient and universal donor are misleading because the agglutinins of the donor may react with agglutinogens of the recipient. Usually these reactions are not serious because the donor's agglutinins are diluted in the recipient's blood. In emergency, life-or-death situations, type O blood may be given to a person with another type or a person with type AB blood may receive other types because the alternative is death. It is always best to use donor blood of the same type as the recipient blood, including the same Rh factor, which is discussed in the next section. Table 13-3 indicates the preferred and permissible donor types for each recipient type.

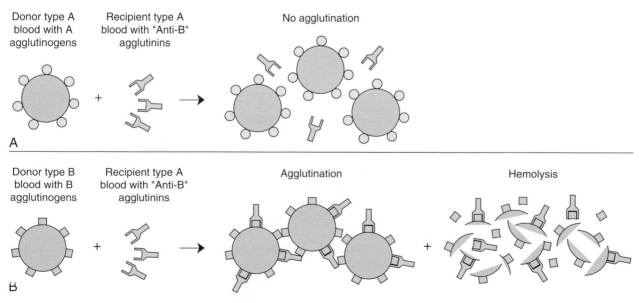

Donor type A blood with A agglutinogens   Recipient type A blood with "Anti-B" agglutinins   No agglutination

A

Donor type B blood with B agglutinogens   Recipient type A blood with "Anti-B" agglutinins   Agglutination   Hemolysis

B

Figure 13-7 *Agglutination reactions, **A**, Type A donor and type A recipient results in no agglutination. **B**, Type B donor and type A recipient results in agglutination and hemolysis.*

TABLE 13-3 Preferred and Permissible Blood Types for Transfusions

Recipient Blood Type	Preferred Donor Blood Type	Additional Types Acceptable
A+	A+	A−, O+, O−
A−	A−	O−
B+	B+	B−, O+, O−
B−	B−	O−
AB+	AB+	AB−, A+, A−, B+, B−, O+, O−
AB−	AB−	A−, B−, O−
O+	O+	O−
O−	O−	None

Rh Blood Groups

Even after the ABO blood groups were well established and accurate blood typing procedures were developed, there were still unexplained cases of transfusion reactions. This led to more research, which led to the discovery of the Rh factor. It is called Rh because it was first studied in the rhesus monkey.

People are Rh positive (Rh+) if they have Rh agglutinogens on the surface of their red blood cells. About 85% of the population are Rh+. The other 15% do not have the Rh agglutinogens and they are Rh negative (Rh−). The presence or absence of Rh agglutinogens is an inherited trait. Normally, neither Rh+ nor Rh− individuals have anti-Rh agglutinins. If an Rh− person is exposed to Rh+ blood, either through a blood transfusion or by transfer of blood between a mother and fetus, the Rh− individual develops anti-Rh agglutinins. If that individual is exposed to Rh+ blood a second time, a transfusion reaction results. When transfusions are given, it is necessary to match both the Rh type and the ABO type (see Table 13-3).

Hemolytic disease of the newborn (formerly called erythroblastosis fetalis) is a special problem associated with the Rh blood groups. This occurs in some pregnancies when the mother is Rh− and the fetus is Rh+ (Figure 13-8). If some of the fetal Rh+ blood mixes with the maternal Rh− blood, usually through ruptured blood vessels in the placenta during birth, the mother is sensitized and develops anti-Rh agglutinins over a period of time. In a subsequent pregnancy with an Rh+ fetus, the anti-Rh agglutinins from the mother pass through the placenta and react with the Rh agglutinogens on the fetal RBCs. The reaction causes agglutination and hemolysis of fetal RBCs, which reduces the oxygen-carrying capacity of the blood. Clumps of RBCs may obstruct blood vessels and cause damage, especially in the kidney. Excessive destruction of the fetal blood causes increased bilirubin concentration, which produces jaundice. Oxygen deficiency and high bilirubin concentrations may cause brain damage.

If hemolytic disease of the newborn develops, the Rh+ blood of the fetus or newborn is slowly removed and replaced with Rh− blood. This provides RBCs that will not be affected by the anti-Rh agglutinins, increases the oxygen-carrying capacity of the infant's blood, and reduces the bilirubin level. Gradually, the transfused Rh− blood is destroyed by normal physiologic processes. Remember, the life span of an RBC is 120 days. Hematopoiesis in the infant's red bone marrow replaces it with Rh+ cells, but by this time the maternal agglutinins have disappeared.

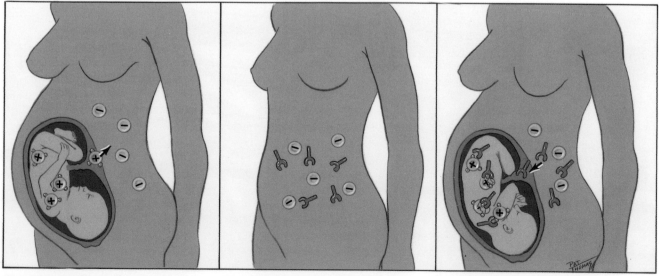

First pregnancy Rh− mother exposed to Rh+ agglutinogens.

After exposure, Rh− mother produces anti-Rh agglutinins.

Second pregnancy with Rh+ fetus. Anti-Rh agglutinins cause agglutination of fetal red blood cells.

Figure 13-8 *Development of hemolytic disease of the newborn.*

Hemolytic disease of the newborn often may be prevented by treating the Rh– mother with a special preparation of anti-Rh gamma globulin (RhoGAM) during pregnancy or immediately after each delivery of an Rh+ baby. This inactivates any Rh+ agglutinogens that may enter the mother's blood, preventing the development of anti-Rh agglutinins. The agglutinins in the RhoGAM soon disappear and present no problem for future pregnancies.

QUICK CHECK

13.15 What term is used for the blood type antigens that are present on the surface of RBCs?

13.16 What ABO blood type is called the universal donor?

13.17 A person with type A+ blood wants to donate blood to a friend who is AB+. Theoretically, is this transfusion possible? Why?

13.18 What happens the first time an Rh– person is given Rh+ blood?

FOCUS ON AGING

The blood appears to be rather resistant to the aging process and under normal conditions blood values remain normal. The volume and composition remain stable. Blood cells retain their normal size, shape, and structure. The amount of red bone marrow decreases with age so the capability for blood cell formation decreases, but the hematopoietic mechanisms are still adequate for normal replacement so that blood cell counts and hemoglobin levels stay within normal ranges. Unusual circumstances, such as hemorrhage, may put a strain on the hematopoietic mechanism; therefore it will take longer to restore normal blood values after a hemorrhagic event.

There is an increase in blood-related disorders with age, but these are secondary to other etiologic factors and are not primary disorders in the blood. For example, there is an increased incidence of anemia in the elderly, but this is often due to nutritional deficiencies rather than to a problem within the blood itself. There appears to be an increase in the incidence of leukemia, but this is caused by a breakdown in the immune system. Abnormal thrombus and embolus formation in the elderly is usually due to atherosclerosis in the blood vessels and is not a problem of the clotting factors within the blood. These examples illustrate that increases in the incidence of blood-related disorders in the elderly are usually secondary to disease processes elsewhere in the body.

Representative Disorders of the
Cardiovascular System

Anemia (ah-NEE-mee-ah) Deficiency in red blood cells or hemoglobin; most common form is iron deficiency anemia caused by a lack of iron to make hemoglobin; other types are aplastic anemia, hemolytic anemia, pernicious anemia, sickle cell anemia, and thalassemia

Aneurysm (AN-yoo-rizm) Saclike protrusion formed by a localized dilation in the wall of a weakened blood vessel, usually an artery

Angina pectoris (an-JYE-nah PECK-tohr-is) Acute chest pain caused by decreased blood supply to the heart muscle

Arrhythmia (ah-RITH-mee-ah) Variation from the normal rhythm of the heart beat; often described as palpitations

Atherosclerosis (ath-er-oh-skleh-ROH-sis) Form of arteriosclerosis characterized by the buildup of fatty plaques in the wall of the vessel

Bacterial endocarditis (back-TEER-ee-al en-doh-kar-DYE-tis) Inflammation of the endocardium, particularly the heart valves and chordate tendineae, caused by any number of bacteria; characterized by the presence of vegetations on the surface of the endocardium

Coronary artery disease (Kor-oh-nair-ee ART-er-ee dih-ZEEZ) Degenerative changes in the coronary circulation, usually caused by the formation of fatty deposits in the walls of the coronary vessels; also called CAD

Heart block (HART BLOCK) Impairment of conduction of impulses from the sinoatrial node to heart muscle

Hemophilia (hee-moh-FILL-ih-ah) Excessive bleeding caused by a congenital lack of one or more of the factors necessary for blood clotting; treatment of hemophilia is directed at replacing the missing clotting factors to control and prevent bleeding

Hypovolemia (hye-poh-voh-LEE-mee-ah) Abnormally decreased amount of circulating plasma in the body

Leukemia (loo-KEE-mee-ah) Chronic or acute progressive cancer of blood-forming tissues

Myocardial infarction (mye-oh-KAR-dee-ahl in-FARK-shun) Destruction of a region of heart muscle as a result of oxygen deprivation because of a blockage in blood vessels to that area; sometimes called a heart attack or myocardial infarction

Patent foramen ovale (PAY-tent foh-RAY-men oh-VAL-ee) Congenital heart defect occurring when the opening between the right and left atria fails to close after birth, permitting mixing of blood between the right and left sides of the heart

Polycythemia (pahl-ee-sye-THEE-mee-ah) Any type of increase in the number of red blood cells

Rheumatic heart disease (roo-MAT-ik HART dih-ZEEZ) Consistent clinical effect of rheumatic fever in which, over a period of time, the heart valves become thickened and often calcified, seriously affecting cardiac function

Septal defects (SEP-tull DEE-fekts) Openings in the atrial, ventricular, and/or atrioventricular septa of the heart, causing disturbances in blood flow patterns through the heart and pulmonary circulation

Tetralogy of Fallot (teh-TRALL-oh-jee of fah-LOH) Complex group of congenital heart defects that combines four structural anomalies: the pulmonary trunk is abnormally narrow, obstructing blood flow to the lungs; the interventricular septum is incomplete; the aortic opening overrides the interventricular septum and receives blood from both ventricles; the right ventricle is enlarged

Thrombus (THRAHM-bus) Blood clot; may occlude a blood vessel and halt blood flow to the region; may break loose and become an embolus

Varicose veins (VAIR-ih-kohs VANES) Abnormally swollen, distended, and knotted veins, usually in the subcutaneous tissues of the leg

Vasculitis (vas-kew-LYE-tis) Inflammation of a vessel; also called angiitis

CHAPTER SUMMARY

Functions and Characteristics of Blood

■ **Describe the physical characteristics of blood.**

• Blood is a liquid connective tissue; measures about 5 liters; accounts for 8% body weight; is slightly heavier than water; is 4 to 5 times more viscous than water; and has a pH of 7.35 to 7.45.

■ **Describe the functions of blood.**

• Blood transports gases, nutrients, and waste products; helps regulate body temperature, fluid and electrolyte balance, and pH; helps prevent fluid loss and disease.

Composition of Blood

■ **Describe the composition of blood plasma.**

• Blood is 55% plasma and 45% formed elements.
• Plasma is 90% water; the remaining 10% includes various solutes such as plasma proteins, gases, electrolytes, nutrients, and other molecules.

• Plasma proteins include the albumins, globulins, and fibrinogen. Albumins account for 60% of the plasma proteins. They maintain the osmotic pressure of the blood. Globulins account for 36% of the plasma proteins. They function in lipid transport and in immune reactions. Fibrinogen accounts for 4% of the plasma proteins. It functions in the formation of blood clots.

• Amino acids, urea, and uric acid are nonprotein molecules that contain nitrogen and may be present in blood plasma.

• Simple nutrients that are the end products of digestion are transported in the plasma.

• Oxygen and carbon dioxide are gases that are transported in the plasma.

• Sodium, potassium, calcium, chloride, bicarbonate, and phosphate ions are common electrolytes in the blood plasma. Electrolytes contribute to the osmotic pressure of the blood, maintain membrane potentials, and regulate the pH of body fluids.

■ **Identify the formed elements of the blood and state at least one function for each formed element.**

- Three categories of formed elements are erythrocytes, leukocytes, and thrombocytes.
- Erythrocytes (red blood cells [RBCs]) are anucleate, biconcave disks, about 7.5 μm in diameter; there are 4.5 to 6 million/mm³ of blood; they contain hemoglobin. The primary function of erythrocytes is to transport oxygen.
- Leukocytes (white blood cells [WBCs]) have a nucleus and do not have hemoglobin; they average between 5000/mm³ and 9000/mm³ of blood; they move through capillary walls by diapedesis. WBCs provide a defense against disease and mediate inflammatory reactions.
- Thrombocytes, or platelets, are fragments of megakaryocytes; they average 250,000 to 500,000/mm³ of blood; they function in blood clotting.

■ **Discuss the life cycle of erythrocytes.**

- Erythrocyte production is regulated by erythropoietin, which is activated by renal erythropoietin factor. Iron, vitamin B$_{12}$, and folic acid are essential for RBC production.
- The life span of RBCs is about 120 days; then they are destroyed by the spleen and liver. The iron and protein portions are reused; the pigment portion is converted to bilirubin and is secreted in bile.

■ **Differentiate between five types of leukocytes on the basis of their structure.**

- Neutrophils are granulocytes with light-colored granules; they are the most numerous leukocyte and are phagocytic.
- Eosinophils are granulocytes with red granules; they help counteract the effects of histamine.
- Basophils are granulocytes with blue granules; they secrete histamine and heparin. In the tissues they are called mast cells.
- Lymphocytes are agranulocytes that have a special role in immune processes; some attack bacteria directly, others produce antibodies.
- Monocytes are large phagocytic agranulocytes. In the tissues they are called macrophages.

Hemostasis

■ **Describe the mechanisms that reduce blood loss after trauma.**

- Hemostasis, the stoppage of bleeding, includes vascular constriction, platelet plug formation, and coagulation.
- The initial reaction in hemostasis is vascular constriction, which reduces the flow of blood through a torn or severed vessel.
- Collagen from damaged tissues attracts platelets, which form a platelet plug to fill the gap in a broken vessel to reduce blood loss.
- Coagulation, or blood clot formation, starts with the formation of prothrombin activator, continues with the conversion of prothrombin to thrombin, and ends with the conversion of soluble fibrinogen to insoluble fibrin. Calcium and vitamin K are necessary for successful clot formation. After a clot forms, it condenses, or retracts, to pull the edges of the wound together. As healing takes place, the clot dissolves by fibrinolysis.

Blood Typing and Transfusions

■ **Characterize the different blood types and explain why some are incompatible for transfusions.**

- Blood type antigens on the surface of RBCs are called agglutinogens. Antibodies that react with agglutinogens are in the plasma and are called agglutinins.
- The ABO blood types are based on the agglutinogens present on the surface of the RBCs. Type A blood has type A agglutinogens and anti-B agglutinins; type B blood has type B agglutinogens and anti-A agglutinins; type AB blood has both type A and type B agglutinogens but neither agglutinin; type O blood has neither agglutinogen but has both anti-A and anti-B agglutinins.
- In transfusion reactions involving mismatched blood, the recipient's agglutinins react with the donor's agglutinogens. Type AB blood is called the universal recipient and type O is the universal donor.
- People who have Rh+ blood have Rh agglutinogens; Rh− individuals do not have Rh agglutinogens. Normally, neither type has anti-Rh agglutinins. Exposure to Rh+ blood causes an Rh− individual to develop anti-Rh agglutinins and subsequent exposures may result in a transfusion reaction.
- Hemolytic disease of the newborn is a risk when the mother is Rh− and the developing fetus is Rh+. If the mother has previously developed anti-Rh agglutinins, they may cross the placenta and enter the fetal blood, causing agglutination and hemolysis. If hemolytic disease of the newborn develops, the fetal blood is temporarily replaced with Rh− blood.

CHAPTER QUIZ

Recall

Match the definitions on the left with the appropriate term on the right.

_____ **1.** Most abundant plasma protein
_____ **2.** Plasma protein that functions in blood clotting
_____ **3.** Precursor, or stem cell, from which blood cells develop
_____ **4.** Red blood cell
_____ **5.** Process by which WBCs move through the capillary wall
_____ **6.** Found on the cell membrane of RBCs
_____ **7.** Hormone that stimulates RBC production
_____ **8.** Pigmented protein that binds with oxygen
_____ **9.** 45% of blood volume
_____ **10.** Granular leukocyte

A. Agglutinogen
B. Albumin
C. Basophil
D. Diapedesis
E. Erythrocyte
F. Erythropoietin
G. Fibrinogen
H. Formed elements
I. Hemocytoblast
J. Hemoglobin

Thought

1. The plasma protein most responsible for maintaining blood osmotic pressure is:
   A. Gamma globulin
   B. Prothrombin
   C. Fibrinogen
   D. Albumin

2. Immature erythrocytes are called:
   A. Hemocytes
   B. Reticulocytes
   C. Thrombocytes
   D. Leukocytes

3. When oxygen delivery to the kidneys decreases:
   A. The liver increases production of albumins
   B. The kidneys decrease production of prothrombin
   C. The liver decreases production of erythropoietin
   D. The kidneys increase production of renal erythropoietic factor

4. Theoretically, a person with type A blood should be able to receive:
   A. Type AB blood and type A blood
   B. Type B blood and type AB blood
   C. Type O blood and type B blood
   D. Type A blood and type O blood

5. Blood clotting is a multistep process. Which of the following represents the correct sequence of the steps in blood clotting?
   A. Formation of prothrombin activator, formation of thrombin, formation of fibrin
   B. Formation of thrombin, formation of fibrin, formation of prothrombin activator
   C. Formation of fibrin, formation of prothrombin activator, formation of thrombin
   D. Formation of prothrombin activator, formation of fibrin, formation of thrombin

Application

Why do individuals with advanced kidney disease often have a low hematocrit?

Why do individuals with advanced liver disease often bruise easily, bleed freely, and have a slow clotting time?

BUILDING YOUR MEDICAL VOCABULARY

Building a vocabulary is a cumulative process. As you progress through this book, you will use word parts, abbreviations, and clinical terms from previous chapters. Each chapter will present new word parts, abbreviations, and terms to add to your expanding vocabulary.

Word Parts and Combining Form with Definition and Examples

PART/COMBINING FORM	DEFINITION	EXAMPLE
agglutin-	clumping, sticking together	agglutinogen: antigen that causes cells to stick together
anti-	against	anticoagulant: substance that inhibits blood clotting
coagul-	clotting	coagulation: process of forming a blood clot
-emia	blood condition	leukemia: blood condition in which there are malignant leukocytes in the blood
erythr-	red	erythrocyte: red blood cell
fibr-	fiber	fibrinogen: substance that forms a fibrous mesh in blood clotting
-globin	protein	hemoglobin: protein in red blood cells
hem/o	blood	hemolysis; destruction of red blood cells
kary/o	nucleus	megakaryocyte: large cell with a large lobular nucleus in the bone marrow
leuk/o	white	leukocyte: white blood cell
-lysis	destruction	hemolysis: destruction of red blood cells
mon/o	single, one	monocyte: white blood cell with a single, rather than a mulitlobed, nucleus
-penia	deficiency, lack of	leukopenia: deficiency of leukocytes

PART/COMBINING FORM	DEFINITION	EXAMPLE
-pheresis	removal	plasmapheresis: removal of plasma from withdrawn blood
-phil	love, affinity for	eosinophil: leukocyte with granules that have an affinity for the red dye eosin
plast-	growth, development	aplastic anemia: blood condition resulting from a failure of blood cell production in the bone marrow
-poiesis	formation of	hemopoiesis: formation of blood cells
poikil/o	irregular	poikilocytosis: irregularity in the shape of red blood cells
-rrhage	burst forth, flow	hemorrhage: excessive bleeding or flow of blood
-stasis	control	hemostasis: control or stoppage of bleeding
thromb/o	clot	thrombocyte: formed element of the blood that functions in blood clotting

Clinical Abbreviations

ABBREVIATION	MEANING
ABMT	autologous bone marrow transplantation
AML	acute myelogenous leukemia
BMT	bone marrow transplantation
CBC	complete blood count
CML	chronic myelocytic leukemia
DIC	disseminated (diffuse) intravascular coagulation
EBV	Epstein-Barr virus
ESR	erythrocyte sedimentation rate
Hb	hemoglobin
HCL	hairy cell leukemia
Hct	hematocrit
HLA	human leukocyte antigen

ABBREVIATION	MEANING
MCH	mean corpuscular hemoglobin
PCV	packed cell volume
PMN	polymorphonuclear
PT	prothrombin time
RBC	red blood cell
sed rate	erythrocyte sedimentation rate
segs	segmented mature white blood cells
WBC	white blood cell

Clinical Terms

Coagulopathy (koh-ag-yoo-LAHP-ah-thee) Any disorder of blood coagulation

Ecchymosis (eck-ih-MOH-sis) Blue or purplish patch in the skin caused by intradermal hemorrhage; larger than a petechia; a bruise; plural, ecchymoses

Embolus (EMM-boh-lus) Moving clot or other plug; an object, often a blood clot, that moves in the blood until it obstructs a small vessel and blocks circulation

Erythrocytosis (ee-rith-roh-sye-TOH-sis) Increase in the number of red blood cells as a result of factors other than a disorder of the hematopoietic mechanism; secondary polycythemia

Hematocrit (hee-MAT-oh-krit) Laboratory procedure that measures the percentage of red blood cells in whole blood; also used for the value determined by the procedure

Leukocytosis (loo-koh-sye-TOH-sis) Increase in the number of white blood cells in the blood, which may result from hemorrhage, fever, infection, inflammation, or other factors

Leukopenia (loo-koh-PEE-nee-ah) Decrease in the number of white blood cells in the blood

Multiple myeloma (MULL-tih-pull my-eh-LOH-mah) Malignant tumor of the bone marrow

Petechia (pee-TEE-kee-ah) Pinpoint, purplish red spot in the skin caused by intradermal hemorrhage; plural, petechiae

Plasmapheresis (plaz-mah-feh-REE-sis) Removal of plasma from withdrawn blood with the retransfusion of the formed elements into the donor

Purpura (PER-pyoo-rah) Group of disorders characterized by multiple pinpoint hemorrhages and accumulation of blood under the skin

Reticulocyte (reh-TICK-yoo-loh-syte) Immature red blood cell with a network of granules in its cytoplasm

Thrombocytopenia (thrahm-boh-syte-oh-PEE-nee-ah) Lower-than-normal number of thrombocytes, or platelets, in the blood

Thrombus (THRAHM-bus) Blood clot

VOCABULARY QUIZ

Use word parts you have learned to form words that have the following definitions.

1. Destruction of blood _____

2. Deficiency of clotting cells _____

3. Formation of blood _____

4. Affinity for basic dye _____

5. Destruction of fibrin _____

Using the definitions of word parts you have learned, define the following words.

6. Agglutination _____

7. Anticoagulant _____

8. Fibrinogen _____

9. Leukocyte _____

10. Thrombocyte _____

Match each of the following definitions with the correct word.

_____ 11. Substance before thrombin

_____ 12. Globe-shaped protein in blood

_____ 13. Condition of too many cells in blood

_____ 14. Flowing of blood

_____ 15. Cell with large nucleus

A. Hemoglobin
B. Hemorrhage
C. Megakaryocyte
D. Polycythemia
E. Prothrombin

16. What is the clinical term for "any disorder of blood clotting"? _____

17. What is the clinical term for "a decrease in the number of white blood cells in the blood"? _____

18. What laboratory procedure measures the percentage of red blood cells in whole blood? _____

19. What is the meaning of the abbreviation PCV? _____

20. What is the meaning of the abbreviation CBC? _____

Lymphatic System and Body Defense

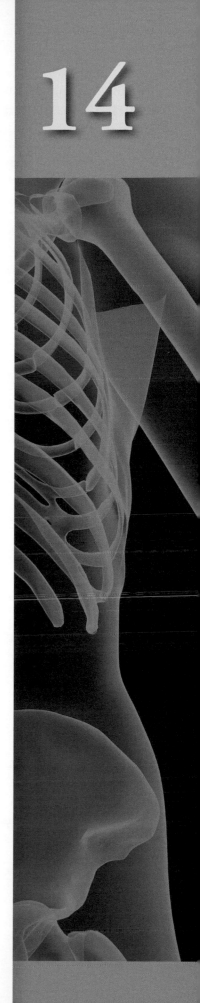

CHAPTER OBJECTIVES

Functions of the Lymphatic System
- State three functions of the lymphatic system.

Components of the Lymphatic System
- List the components of the lymphatic system
- Describe the origin of lymph.
- Describe the lymphatic vessels and the mechanisms that move the lymph through the vessels.
- Describe the structure of lymph nodes, tonsils, spleen, and thymus and explain their importance to the body.

Resistance to Disease
- List four nonspecific mechanisms that provide resistance to disease and explain how each functions.
- State the two characteristics of specific defense mechanisms and identify the two principal cells involved in specific resistance.

- Distinguish between self and nonself as they relate to disease resistance.
- Describe the development of lymphocytes.
- Briefly describe the mechanism of cell-mediated immunity and list four subgroups of T cells.
- Briefly describe the mechanism of antibody-mediated immunity and list two subgroups of B cells.
- Distinguish between the primary response and the secondary response to a pathogen.
- List five classes of immunoglobulins and state the role each has in immunity.
- Give examples of active natural immunity, active artificial immunity, passive natural immunity, and passive artificial immunity.

KEY TERMS

Active immunity (AK-tiv ih-MYOO-nih-tee) Immunity that is produced as the result of an encounter with an antigen, with subsequent production of memory cells

Antibody (AN-tih-bahd-ee) Substance produced by the body that inactivates or destroys another substance that is introduced into the body; immunoglobulin

Antibody-mediated immunity (An-tih-bahd-ee MEE-dee-ate-ed ih-MYOO-nih-tee) Immunity that is the result of B-cell action and the production of antibodies; also called humoral immunity

Antigen (AN-tih-jen) Substance that triggers an immune response when it is introduced into the body

Artificial immunity (art-ih-FISH-al ih-MYOO-nih-tee) Immunity that requires some deliberate action, such as a vaccination, to achieve exposure to the potentially harmful antigen

Cell-mediated immunity (SELL MEE-dee-ate-ed ih-MYOO-nih-tee) Immunity that is the result of T-cell action; also called cellular immunity

Immunity (ih-MYOO-nih-tee) Specific defense mechanisms that provide resistance to invading organisms

Immunoglobulin (ih-myoo-noh-GLAHB-yoo-lin) Substance produced by the body that inactivates or destroys another substance that is introduced into the body; antibody

Natural immunity (NAT-yoor-al ih-MYOO-nih-tee) Immunity acquired through normal processes of daily living

Passive immunity (PASS-iv ih-MYOO-nih-tee) Immunity that results when an individual receives the immune agents from some source other than his or her own body

Primary response (PRY-mair-ee ree-SPONS) Initial reaction of the immune system to a specific antigen

Resistance (ree-SIS-tans) Body's ability to counteract the effects of pathogens and other harmful agents

Secondary response (SEK-on-dair-ee ree-SPONS) Rapid and intense reaction to antigens on second and subsequent exposures attributable to memory cells

Susceptibility (sus-sep-tih-BILL-ih-tee) Lack of resistance to disease

Functional Relationships of the
Lymphatic/Immune System

Provides specific defense mechanisms against pathogens; surveillance mechanisms detect and destroy cancer cells; lymph vessels remove excess interstitial fluid.

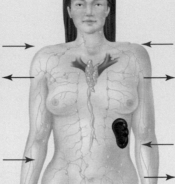

Reproductive

Acidic vaginal secretions provide barrier against reproductive tract infections; blood-testis barrier prevents immune destruction of sperm cells.

Provides specific defense against pathogens that enter the body through the reproductive tract; surveillance against cancer cells.

Urinary

Provides lymphoid tissue and immune cells with oxygen and removes wastes; tonsils located in pharynx; breathing movements assist lymph flow.

Tonsils combat pathogens that enter through nasal passages; IgA protects respiratory mucosa.

Digestive

Absorbs nutrients needed by lymphoid tissue and immune cells; stomach acid provides nonspecific defense against ingested pathogens.

Lacteals absorb fats and fat soluble vitamins from the intestine.

Respiratory

Helps maintain water, electrolyte, and pH balance of body fluids for effective immune cell function; acid pH of urine acts as barrier against urinary tract pathogens.

Immune cells provide specific defense against urinary tract pathogens.

Cardiovascular

Provides oxygen and nutrients to lymphoid organs and removes wastes; transports agents involved in immune response.

Returns interstitial fluid to the blood to maintain blood volume; protects against infections; spleen disposes of old RBCs.

Integument

Provides mechanical barrier against entry of pathogens.

Supplies antibodies to skin surface for specific defense.

Skeletal

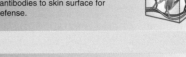

Leukocytes involved in immune response arise from stem cells in bone marrow; protects thymus and spleen.

Maintains interstitial fluid balance in bone tissue; assists in defense against pathogens and repair of bone tissue following trauma.

Muscular

Skeletal muscle contraction moves lymph within vessels; protects superficial lymph nodes.

Maintains interstitial fluid balance in muscle tissue; assists in defense against pathogens and repair of muscle tissue following trauma.

Nervous

Innervates lymphoid organs and helps regulate immune response.

Assists in defense against pathogens and repair of neural and sensory tissue following trauma; removes excess interstitial fluid from tissues surrounding nerves.

Endocrine

Hormones from thymus influence lymphocyte development; glucocorticoids suppress immune response.

Protects against infection in endocrine glands.

← Gives to Lymphatic/Immune System
→ Receives from Lymphatic/Immune System

The lymphatic system sometimes is considered to be a part of the circulatory system because it transports a fluid through vessels and empties it into venous blood. Because it consists of organs that work together to perform certain functions, it may be treated as a separate system. The lymphatic system has a major role in the body's defense against disease, so that topic is included in this chapter.

Functions of the Lymphatic System

The lymphatic system has three primary functions. First, it returns excess interstitial fluid to the blood. Chapter 12 described capillary microcirculation in which fluid leaves the capillary at the arteriole end and returns at the venous end. Of the fluid that leaves the capillary, about 90% is returned. The 10% that does not return becomes part of the interstitial fluid that surrounds the tissue cells. Small protein molecules may "leak" through the capillary wall and increase the osmotic pressure of the interstitial fluid. This further inhibits the return of fluid into the capillary, and fluid tends to accumulate in the tissue spaces. If this continues, blood volume and blood pressure decrease significantly and the volume of tissue fluid increases, which results in edema. Lymph capillaries pick up the excess interstitial fluid and proteins and return them to the venous blood. Figure 14-1 illustrates the relationship between the lymphatic system and the cardiovascular system.

The second function of the lymphatic system is the absorption of fats and fat-soluble vitamins from the digestive system and the subsequent transport of these substances to the venous circulation. The mucosa that lines the small intestine is covered with fingerlike projections called **villi**. There are blood capillaries and special lymph capillaries, called **lacteals** (lak-TEELS), in the center of each villus. Most nutrients are absorbed by the blood capillaries, but the fats and fat-soluble vitamins are absorbed by the lacteals. The lymph in the lacteals has a milky appearance because of its high fat content and is called **chyle** (KYL).

The third function of the lymphatic system is defense against invading microorganisms and disease. Lymph nodes and other lymphatic organs filter the lymph to remove microorganisms and other foreign particles. Lymphatic organs contain lymphocytes that destroy invading organisms.

Components of the Lymphatic System

The lymphatic system consists of a fluid (lymph), vessels that transport the lymph, and organs that contain lymphoid tissue.

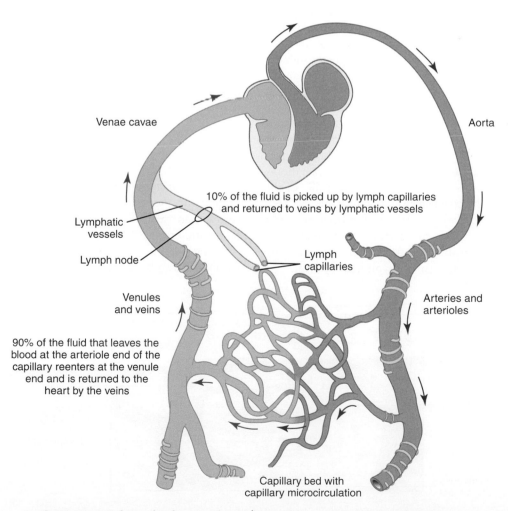

Figure 14-1 *Relationship between the cardiovascular system and the lymphatic system.*

Lymph

Lymph is a fluid similar in composition to blood plasma. It is derived from blood plasma as fluids pass through capillary walls at the arterial end. About 90% of the fluid reenters the capillary at the venule end. The remaining 10% is added to the interstitial fluid. As the interstitial fluid begins to accumulate, it is picked up and removed by tiny lymphatic vessels which eventually return it to the blood. As soon as the interstitial fluid enters the lymph capillaries, it is called **lymph**. Returning the fluid to the blood prevents edema and helps to maintain normal blood volume, plasma protein concentration, and blood pressure.

Lymphatic Vessels

Lymphatic vessels, unlike blood vessels, only carry fluid away from the tissues. The smallest lymphatic vessels are the **lymph capillaries**, which begin in the tissue spaces as blind-ended sacs (Figure 14-2). Lymph capillaries are found in all regions of the body except the bone marrow, central nervous system, and tissues that lack blood vessels, such as the epidermis. The wall of the lymph capillary is composed of endothelium in which the simple squamous cells overlap to form a simple one-way valve. This arrangement permits fluid to enter the capillary but prevents lymph from leaving the vessel.

The microscopic lymph capillaries merge to form **lymphatic vessels**. These vessels are similar to veins in structure, but they have thinner walls and more valves than veins. Small lymphatic vessels join to form larger tributaries, called **lymphatic trunks**, which drain large regions. Lymphatic trunks merge until the lymph enters the two **lymphatic ducts**. The **right lymphatic duct** receives lymph from the vessels in the upper right quadrant of the body (Figure 14-3). This includes the right side of the head and neck, the right upper extremity, and the right side of the thorax. The right lymphatic duct empties into the right subclavian vein. The **thoracic duct** collects the lymph from the remaining regions of the body (Figure 14-3). The thoracic duct begins in the upper abdomen and ascends through the thorax to empty into the left subclavian vein. The beginning of the thoracic duct is called the **cisterna chyli** (sis-TER-nah KY-lee). The cisterna chyli collects lymph from two lumbar trunks that drain the lower limbs and from the intestinal trunk that drains the digestive organs. **Lymph nodes** that filter the lymph are located along the various routes of the lymphatic system.

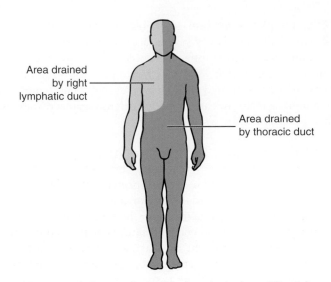

Figure 14-3 *Regions drained by lymphatic ducts. The right lymphatic duct drains the upper right quadrant of the body. The thoracic duct drains the remainder.*

Like veins, the lymphatic tributaries have thin walls and have valves to prevent backflow of blood. There is no pump in the lymphatic system like the heart in the cardiovascular system. The pressure gradients to move lymph through the vessels have to come from external sources. These pressure gradients come from skeletal muscle contraction, from respiratory movements, and from contraction of the smooth muscle within the wall of the vessel. Because there is no rhythmic heart to pump it along, lymph transport is sporadic and much slower than the transport of blood in the veins. Anything that interferes with the flow of lymph, such as an obstruction or surgical ligation, may cause tissue fluid to accumulate, resulting in edema.

Lymphatic Organs

Lymphatic organs are characterized by clusters of **lymphocytes** and other cells, such as macrophages, enmeshed in a framework of short, branching connective tissue fibers. The lymphocytes originate in the red bone marrow with other types of blood cells and are carried in the blood from the bone marrow to the lymphatic organs. When the body is exposed to microorganisms and other foreign substances, the lymphocytes proliferate within the lymphatic organs and then travel in the blood to the

Figure 14-2 *Lymph capillaries in the tissue spaces.*

site of the invasion. This is part of the immune response that attempts to destroy the invading agent. The lymph nodes, tonsils, spleen, and thymus are examples of lymphatic organs.

Lymph Nodes

Lymph nodes are small, bean-shaped structures that are usually less than 2.5 centimeters (cm) in length. They are widely distributed throughout the body along the lymphatic pathways, where they filter the lymph before it is returned to the blood. Lymph nodes are not present in the central nervous system. There are three superficial regions on each side of the body where lymph nodes tend to cluster. These areas are the inguinal nodes in the groin, the axillary nodes in the armpit, and the cervical nodes in the neck (Figure 14-4).

✛ *QUICK* APPLICATIONS

The development of breast cancer often necessitates the removal of part or all of the breast tissue, a surgical procedure called a **mastectomy**. There is an extensive network of lymphatic vessels associated with the breast, and the cancer cells from the breast can spread to surrounding lymph nodes through these vessels. For this reason, the axillary nodes may be removed with the breast tissue. Sometimes this procedure interferes with lymph drainage from the arm and the fluid accumulates, resulting in swelling, or **lymphedema**.

The typical lymph node is surrounded by a connective tissue **capsule** and divided into compartments called **lymph nodules** (Figure 14-5). The lymph nodules are dense masses of lymphocytes and macrophages and are separated by spaces called **lymph sinuses**. Several **afferent lymphatic vessels**, which carry lymph into the node, enter the node on the convex side. The lymph moves through the lymph sinuses and enters an **efferent lymphatic vessel**, which carries the lymph away from the node. Because there are more afferent vessels than efferent vessels, the passage of lymph through the sinuses is slowed down, which allows time for the cleansing process. The efferent vessel leaves the node at an indented region called the **hilum**.

Lymph nodes are the only structures that filter the lymph, and this is their primary function. As the lymph moves through the sinuses, infectious agents, damaged cells, cancerous cells, and cellular debris become trapped in the fibrous mesh so that the lymph is cleansed before it enters the blood. The lymphocytes react against the bacteria, viruses, and cancerous cells to destroy them. Macrophages, also present in the node, engulf the destroyed pathogens, the damaged cells, and the cellular debris.

✛ *QUICK* APPLICATIONS

The lymphatic system is one route by which cancer cells can spread from a primary tumor site to other areas of the body. As the cells travel with the lymph, they pass through the lymph nodes, where the lymph is filtered. At first, this traps the cancer cells within the lymph node, and the cells are destroyed. Eventually, the number of cancer cells may overwhelm the filtration ability of the lymph nodes, and some of the cells pass through the nodes to establish secondary tumors.

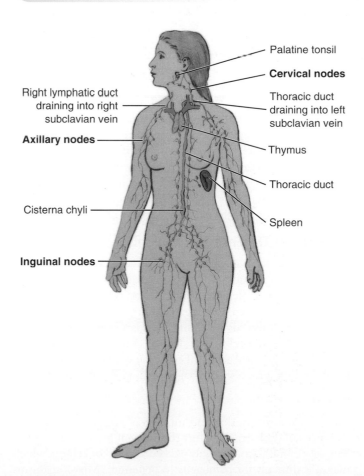

Figure 14-4 *Location of the clusters of superficial lymph nodes.*

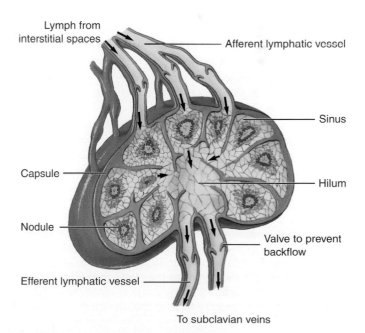

Figure 14-5 *Structure of a lymph node. Note that there are several afferent vessels entering the node, but only one or two efferent vessels leaving the node. This slows the passage of lymph to allow time for the cleansing process.*

Tonsils

Tonsils are clusters of lymphatic tissue just under the mucous membranes that line the nose, mouth, and throat (pharynx). There are three groups of tonsils. The **pharyngeal tonsils** are located near the opening of the nasal cavity into the pharynx. When these tonsils become enlarged, they may interfere with breathing and are called **adenoids**. The **palatine tonsils** are the ones that are commonly called "the tonsils." These are located near the opening of the oral cavity into the pharynx. **Lingual tonsils** are located on the posterior surface of the tongue, which also places them near the opening of the oral cavity into the pharynx. Lymphocytes and macrophages in the tonsils provide protection against harmful substances and pathogens that may enter the body through the nose or mouth.

Spleen

The **spleen** is located in the upper left abdominal cavity, just beneath the diaphragm, and posterior to the stomach. It is similar to a lymph node in shape and structure but it is much larger. The spleen is the largest lymphatic organ in the body. Surrounded by a connective tissue capsule, which extends inward to divide the organ into lobules, the spleen consists of two types of tissue called **white pulp** and **red pulp**. The white pulp is lymphatic tissue consisting mainly of lymphocytes around arteries. The red pulp consists of venous sinuses filled with blood and cords of lymphatic cells, such as lymphocytes and macrophages. Blood enters the spleen through the splenic artery, moves through the sinuses where it is filtered, and then leaves through the splenic vein.

The spleen filters blood in much the same way that the lymph nodes filter lymph. Lymphocytes in the spleen react to pathogens in the blood and attempt to destroy them. Macrophages then engulf the resulting debris, the damaged cells, and the other large particles. The spleen, along with the liver, removes old and damaged erythrocytes from the circulating blood. Like other lymphatic tissue, it produces lymphocytes, especially in response to invading pathogens. The sinuses in the spleen are a reservoir for blood. In emergencies such as hemorrhage, smooth muscle in the vessel walls and in the capsule of the spleen contracts. This squeezes the blood out of the spleen into the general circulation. If the spleen must be removed (splenectomy), its functions will be performed by other lymphatic tissue and the liver.

QUICK APPLICATIONS

The spleen is a rather soft and fragile organ, and although it is somewhat protected by the ribs, it is often ruptured in abdominal injuries. Because the spleen is a reservoir for blood, this results in severe internal hemorrhage and shock, which may lead to death if it is not stopped. A **splenectomy**, surgical removal of the spleen, may be necessary to stop the bleeding.

Thymus

The **thymus** is a soft organ with two lobes that is located anterior to the ascending aorta and posterior to the sternum. It is relatively large in infants and children, but after puberty it begins to decrease in size so that in older adults it is quite small.

The primary function of the thymus is the processing and maturation of special lymphocytes called **T lymphocytes** or **T cells**. While in the thymus, the lymphocytes do not respond to pathogens and foreign agents. After the lymphocytes have matured, they enter the blood and go to other lymphatic organs where they help provide defense against disease. The thymus also produces a hormone, **thymosin**, that stimulates the maturation of lymphocytes in other lymphatic organs.

QUICK CHECK

14.1 Because there is no "heart" to pump fluid through the lymphatic vessels, what provides the pressure gradients for lymph flow?

14.2 Why do lymph nodes enlarge during infections?

14.3 What is the "crisis" or danger when the spleen ruptures from an abdominal injury?

14.4 Which tonsils are removed in a tonsillectomy?

Resistance to Disease

The human body is continually exposed to disease-producing organisms, called **pathogens**, and other harmful substances. If these enter the body, they may disrupt normal homeostasis and cause disease. The body's ability to counteract the effects of pathogens and other harmful agents is called **resistance** and is dependent on a variety of defense mechanisms. **Susceptibility** is a lack of resistance. Some defense mechanisms, called **nonspecific mechanisms**, act against all harmful agents and provide **nonspecific resistance**. Other defense mechanisms only act against certain agents and are called **specific mechanisms**. These provide **specific resistance**, or **immunity** (Figure 14-6). To maintain a state of health, all the body's defense mechanisms must act together to provide protection against invading pathogens, foreign cells that are transplanted into the body, and the body's own cells that have become cancerous.

Nonspecific Defense Mechanisms

Nonspecific defense mechanisms are directed against all pathogens and foreign substances regardless of their nature. They present the initial defense against invading agents. The first line of defense is the barrier against entry into the body. If the foreign agent succeeds in passing the barrier and entering the body, then the second line of defense comes into action. This includes the chemical action of complement proteins and interferon, and the processes of phagocytosis and inflammation.

Barriers

Intact, or unbroken, skin and mucous membranes form effective mechanical barriers against the entry of foreign substances. The cells of the skin are closely packed and full of keratin, which makes it difficult for pathogens to penetrate. Microorganisms that normally grow on the surface of the skin offer a barrier

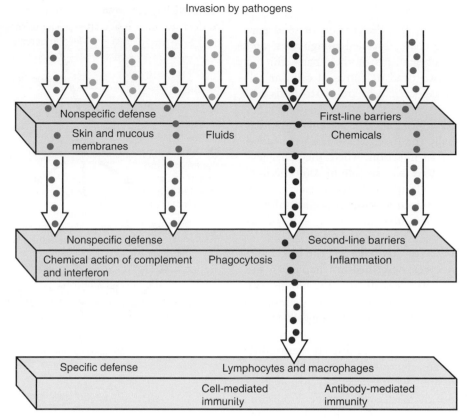

Figure 14-6 *Overview of defense mechanisms. Pathogens that are stopped from entering the body by the first-line barriers of nonspecific defense are represented by* green dots. *Others get through the first-line barriers, but are stopped by nonspecific second-line barriers. These pathogens are represented by* blue dots. *Those that penetrate nonspecific defense mechanisms are subject to specific defense mechanisms. These pathogens are represented by the* red dots.

against pathogens because they inhibit the growth of other bacteria by competing for space and nutrients. Mucous membranes that line the respiratory and digestive tracts are not as tough as skin, but the mucus produced by these membranes traps foreign particles before they gain entry. Some mucous membranes, especially those in the respiratory tract, have cilia that propel the mucus with the entrapped particles upward to be expelled or swallowed.

Fluids, such as tears flowing across the eyes, saliva that is swallowed, and urine passing through the urethra, are examples of mechanical barriers that flush pathogens out of the body before they have a chance to damage the tissues.

Lysozymes in the tears, saliva, and nasal secretions destroy bacteria. Sebaceous secretions and the salts in perspiration also have an antimicrobial action. Hydrochloric acid in the stomach inhibits the growth of bacteria that are swallowed. These are all examples of chemical barriers that deter microbial invasion.

Chemical Action

Various body chemicals, including complement proteins, stimulate phagocytosis and inflammation. Others, including interferon, are produced as a direct response to microbial invasion. These are part of the second line of defense that continues the battle against disease if microorganisms succeed in getting through the barriers of the first defense.

Complement is a group of proteins normally found in the plasma in an inactive form. Certain complement proteins become activated when they come in contact with a foreign substance. A series of reactions, similar to the cascade reactions in the clotting process, follows. The reactions proceed in an orderly sequential manner, with step 1 necessary for step 2, then step 2 necessary for step 3, and so on. As each complement protein is activated, it activates the next complement protein, until the final protein is activated. The final activated complement enhances phagocytosis and inflammation. It also causes bacterial cells to rupture.

Interferon (in-ter-FEER-on) has particular significance because it offers protection against viruses. When a cell becomes infected with a virus, the cell usually stops its normal functions. The virus uses the cell's metabolic machinery for one goal—viral replication. When the cell is full of viruses, it ruptures and releases a myriad of viruses to infect new cells. In this way, the viral infection is established. When a virus infects a cell, that cell produces interferon, which diffuses into neighboring uninfected cells. Interferon stimulates the uninfected cells to produce a protein that blocks viral replication. In this way, the uninfected cells are protected from the virus. Interferon does not protect the cell in which it was produced or the cells in which the virus is already established, but it does protect the neighboring uninfected cells.

Phagocytosis

Phagocytosis is the ingestion and destruction of solid particles by certain cells. The cells are called **phagocytes**, and the particles may be microorganisms or their parts, foreign particles, an individual's own dead or damaged cells, or cell fragments. The primary phagocytic cells are **neutrophils** and **macrophages**.

Neutrophils, described in Chapter 13, are small granular leukocytes. They are usually the first cells to leave the blood and migrate to the site of an infection, where they phagocytize the invading bacteria. This is a "suicide mission" because the neutrophils die after engulfing only a few bacteria. Pus is primarily an accumulation of dead neutrophils, cellular debris, and bacteria. The number of neutrophils greatly increases in acute infections.

Macrophages are monocytes that have left the blood and entered the tissues. Monocytes, described in Chapter 13, are large agranular leukocytes. When they leave the blood, they become macrophages by increasing in size and developing additional lysosomes. Macrophages usually appear at the scene of an infection after the neutrophils and are responsible for clearing away cellular debris and dead neutrophils during the latter stages of an infection. Macrophages are also present in uninfected tissues, where they may phagocytize the invading agents before there is tissue damage. For example, they are present in the lymph nodes, where they cleanse the lymph as it filters through the node. They perform a similar cleansing action on the blood as it passes through the liver and spleen.

Inflammation

Inflammation, briefly discussed in Chapter 4, is a nonspecific defense mechanism that occurs in response to tissue damage from microorganisms or trauma. **Localized inflammation** is contained in a specific region. It is evidenced by **redness** (rubor), **warmth** (calor), **swelling** (tumor), and **pain** (dolor). A combination of these effects frequently causes loss of function, at least temporarily, and the irritation sometimes makes inflammation more harmful than beneficial. In spite of this, it usually is a worthwhile process because it is aimed at localizing the damage and destroying its source. Inflammation also sets the stage for tissue repair. The unpleasant signs and symptoms also have a protective function because they warn that tissue damage has occurred so that the source of the damage may be removed. Figure 14-7 briefly describes and illustrates the inflammatory process.

Systemic inflammation is not contained in a localized region but is widespread throughout the body. The warmth, redness, swelling, pain, and loss of function associated with localized inflammation may be present at specific sites, but the systemic nature of the inflammation is evidenced by three additional responses.

- Bone marrow is stimulated to produce more white blood cells, especially neutrophils and monocytes, so there is a condition of leukocytosis.
- Chemical mediators include **pyrogens** (PYE-roh-jenz) that influence the hypothalamus and cause an increase in body

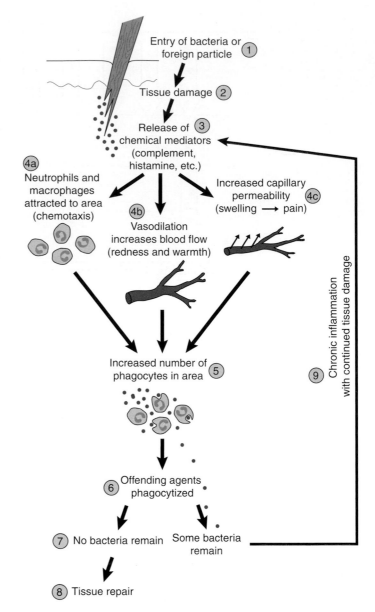

Figure 14-7 *Steps in inflammation. 1. Bacteria or foreign particles enter the body. 2. Tissues are damaged. 3. Damaged tissues release chemical mediators. 4. Chemical mediators have three effects: (a) attract neutrophils and macrophages, (b) increase blood flow through vasodilation, (c) increase capillary permeability. 5. Effect of chemical mediators is to bring additional phagocytes to damaged area. 6. Phagocytes are successful and destroy bacteria. 7. Area is cleansed and no bacteria remain. 8. Tissues are repaired. 9. If phagocytes are not successful, steps 3 through 6 continue and result in chronic inflammation.*

temperature, a **fever**. The fever speeds up the metabolic reactions in the body, including those directed at destroying the invading pathogens.

- Vasodilation and increased capillary permeability may become so generalized that there is a drastic and dangerous decrease in blood pressure. Systemic inflammation is a medical crisis and needs immediate attention.

Figure 14-8 summarizes the components of nonspecific defense mechanisms.

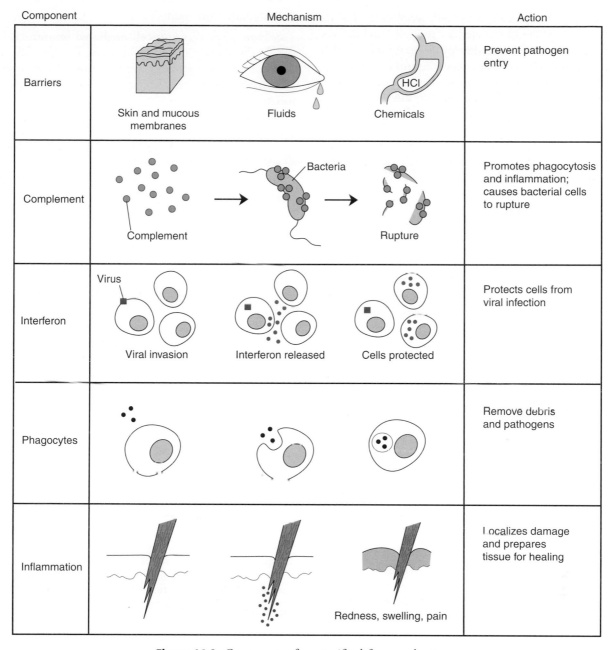

Component	Mechanism			Action
Barriers	Skin and mucous membranes	Fluids	HCl — Chemicals	Prevent pathogen entry
Complement	Complement	Bacteria	Rupture	Promotes phagocytosis and inflammation; causes bacterial cells to rupture
Interferon	Virus — Viral invasion	Interferon released	Cells protected	Protects cells from viral infection
Phagocytes				Remove debris and pathogens
Inflammation			Redness, swelling, pain	Localizes damage and prepares tissue for healing

Figure 14-8 *Components of nonspecific defense mechanisms.*

QUICK CHECK

14.5 If pathogens succeed in passing through the first-line barriers, what other nonspecific defense mechanisms are available to prevent disease?

14.6 What type of infection may result in elevated levels of interferon?

14.7 What are the four signs of inflammation?

14.8 What are the three effects of the chemical mediators in inflammation?

Specific Defense Mechanisms

In contrast to the nonspecific defense mechanisms that react against all types of foreign agents, the specific defense mechanisms are programmed to be selective. This characteristic is called **specificity**. Another characteristic of specific defense mechanisms is **memory**. Once the system has been exposed to a particular invading agent, components of the specific defense mechanisms "remember" that agent and launch a quicker attack if it enters the body again. Specific defense mechanisms provide the third line of defense against microbial invasion. This third line of defense is **specific resistance**, or **immunity**. The primary cells involved are **lymphocytes** and **macrophages**. Nonspecific mechanisms and immune responses take place at the same time, and resistance to disease depends on the interaction of all the mechanisms.

Recognition of Self Versus Nonself

For the immune system to function properly, lymphocytes have to distinguish between self and nonself. During their development and maturation process, the lymphocytes learn to recognize the proteins and other large molecules that belong to the body. They interpret these as "self." Molecules that are not recognized as self are interpreted as "nonself," and defense mechanisms are set in motion to destroy them. A molecule that is interpreted as nonself and that triggers an immune response is called a foreign **antigen**. Antigens are usually some form of protein or large polysaccharide molecules on the surface of cell membranes. Normally, antigens that cause problems are foreign molecules that enter the body, but sometimes the body fails to recognize its own molecules and triggers an immune reaction against self. This damages normal body tissues and is the basis of **autoimmune diseases** such as rheumatoid arthritis.

Development of Lymphocytes

Like all other blood cells, lymphocytes develop from stem cells in the bone marrow (Figure 14-9). During fetal development, the bone marrow releases immature and undifferentiated (unspecialized) lymphocytes into the blood. Some of these go to the thymus gland where they acquire the ability to distinguish between self and nonself molecules. These lymphocytes differentiate to become **T lymphocytes**, or **T cells**, in the thymus gland. For several months after birth, the thymus gland continues to process the T lymphocytes for specific activities in immune reactions. Differentiated T cells leave the thymus, enter the blood, and are distributed to lymphoid tissue, especially the lymph nodes. About 70% of the circulating lymphocytes are T cells.

Lymphocytes that do not go to the thymus travel in the blood to some other area, probably the fetal liver, where they differentiate into **B lymphocytes**, or **B cells**. After they acquire the ability to distinguish between self and nonself and are prepared for their special roles in immune responses, B cells enter the blood and are also distributed to lymphoid tissues. B cells account for about 30% of the circulating lymphocytes.

In the differentiation and maturation process, thousands of different types of T cells and B cells are produced. Each type has receptor sites that fit with specific antigens. Thus they provide specific resistance. There has to be an exact match between the receptor site on the lymphocyte and the antigen of the invader before an immune reaction occurs.

Cell-Mediated Immunity

T cells are responsible for **cell-mediated immunity**, in which the T cells directly attack the invading antigen. Cell-mediated immunity is most effective against virus-infected cells, cancer cells, foreign tissue cells (transplant rejection), fungi, and protozoan parasites.

When the antigen is introduced into the body, it is phagocytized by a macrophage, which then presents the antigen to the T-cell population (Figure 14-10). T cells that have receptor sites for that specific antigen recognize it and become activated. Both the macrophage and the activated T cells secrete chemicals that stimulate division of the activated T cells. This results in large numbers of cells that are all alike, a clone of activated T cells. There are four subgroups within the **clone** of activated T cells, and each group has a specific function. **Killer T cells** directly destroy the cells with the offending antigen. **Helper T cells** secrete substances that stimulate B cells and promote the immune response. **Suppressor T cells** have the opposite effect; they inhibit B cells and the immune response. The helper and suppressor T cells are regulatory cells that control the immune response. Normally, in a correctly operating immune system, there are twice as many helper cells as suppressor cells. The fourth group of cells is the population of **memory T cells**. These cells "remember" the specific antigen and stimulate a faster and more intense response if the same antigen is introduced another time.

> ## QUICK APPLICATIONS
>
> One of the substances secreted by activated T cells is **interleukin-2**, which stimulates production of both T cells and B cells. Researchers are using interleukin-2 made by genetic engineering techniques to stimulate the immune system. The converse of this is that decreasing the activity of interleukin-2 can suppress the immune system. **Cyclosporine**, a drug that inhibits the production of interleukin-2, is used to prevent the rejection of transplanted organs.

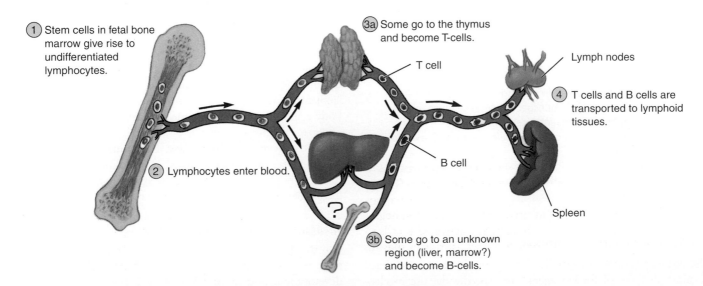

① Stem cells in fetal bone marrow give rise to undifferentiated lymphocytes.

② Lymphocytes enter blood.

③a Some go to the thymus and become T-cells.

T cell

Lymph nodes

④ T cells and B cells are transported to lymphoid tissues.

B cell

Spleen

③b Some go to an unknown region (liver, marrow?) and become B-cells.

Figure 14-9 *Development of lymphocytes.*

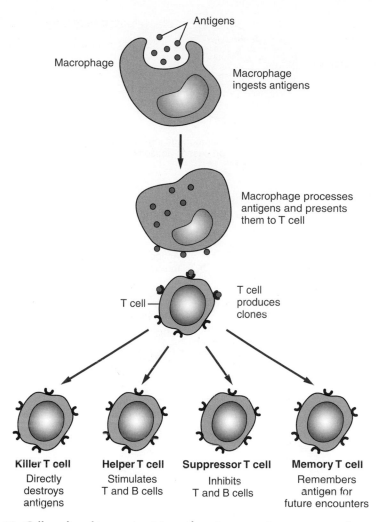

Figure 14-10 *Cell-mediated immunity. Macrophage ingests antigens, processes them, and presents them to the T cell. The T cell produces four clones.*

Antibody-Mediated Immunity (Humoral Immunity)

B cells are responsible for **antibody-mediated immunity**. Like T cells, each type of B cell can respond to only one specific type of antigen. There must be a match between the receptor on the B cell and the antigen. Unlike T cells, B cells do not directly assault the antigen. Instead, they are responsible for the production of antibodies that react with the antigen or substances produced by the antigen. Because the antibodies are found in body fluids, it is sometimes called **humoral immunity**. Antibody-mediated immunity is most effective against bacteria, viruses that are outside body cells, and toxins. It is also involved in allergic reactions.

When an antigen enters the body, a macrophage engulfs and processes it, and then presents it to B cells and helper T cells (Figure 14-11). The B cells and helper T cells that have receptors for that specific antigen are activated. The activated helper T cells secrete substances that stimulate the activated B cells to rapidly divide and to form a clone of cells consisting of **plasma cells** and **memory B cells**.

Plasma cells rapidly produce large quantities of protein molecules, called **antibodies**, that are transported in the blood and lymph to the site of the infection where they inactivate the invading antigens. This initial action is the **primary response**. When the antigens are destroyed, macrophages clean up the debris, and suppressor T cells decrease the immune response. Memory B cells remain dormant in lymphatic tissue until the same antigen again enters the system. The memory cells recognize the antigen and launch a rapid and intense response against it. This is called a **secondary response** (Figure 14-12). The purpose of vaccinations is to provide an initial exposure so memory cells are available for a rapid and intense reaction against subsequent exposure to the antigen.

All antibodies have a similar structure but one portion of the molecule differs so that each antibody is capable of reacting with only a specific antigen. They belong to the class of proteins called **globulins**, and because they are involved in immune reactions, they are called **immunoglobulins** (ih-myoo-noh-GLOB yoo-lins), abbreviated **Ig**. There are several classes of antibodies or immunoglobulins, designated as IgA, IgG, IgM, IgE, and IgD. Immunoglobulins of the IgG class are called gamma globulins. Each class has a specific role in immunity, which is summarized in Table 14-1.

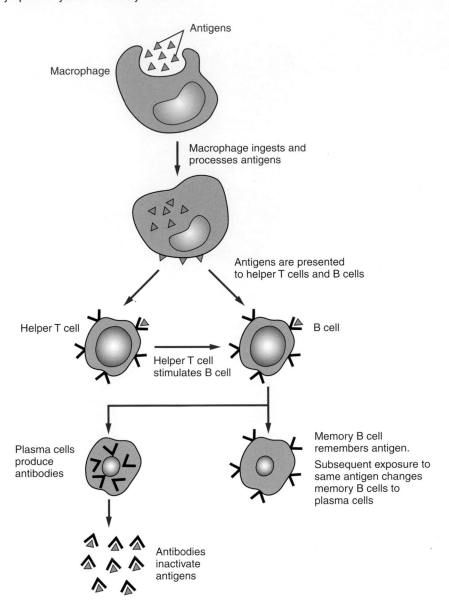

Figure 14-11 *Antibody-mediated immunity. Macrophage ingests and processes antigen, then presents it to helper T cells and B cells. Helper T cells stimulate B cells to divide and produce two clones consisting of memory B cells and plasma cells. Plasma cells produce antibodies that inactivate the antigen.*

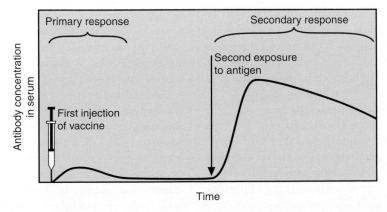

Figure 14-12 *Comparison of primary and secondary responses. The secondary response is more rapid and intense than the primary response.*

TABLE 14-1 **Classes of Antibodies**

Class	Percentage of Total	Location	Function
IgG	75% to 85%	Blood plasma	Major antibody in primary and secondary immune responses; inactivates antigen; neutralizes toxins; crosses placenta to provide immunity for newborn; responsible for Rh reactions
IgA	5% to 15%	Saliva, mucus, tears, breast milk	Protects mucous membranes on body surfaces; provides immunity for newborn
IgM	5% to 10%	Attached to B cells; released into plasma during immune response	Causes antigens to clump together; responsible for transfusion reactions in ABO blood typing system
IgD	0.2%	Attached to B cells	Receptor sites for antigens on B cells; binding with antigen results in B-cell activation
IgE	0.5%	Produced by plasma cells in mucous membranes and tonsils	Binds to mast cells and basophils, causing release of histamine; responsible for allergic reactions

> ### QUICK CHECK
>
> 14.9 What is the function of macrophages in immunity?
> 14.10 What are the four clones of T cells produced by activated T cells?
> 14.11 How do helper T cells influence humoral immunity?
> 14.12 What is the function of plasma cells?
> 14.13 What class of immunoglobulins is responsible for most primary and secondary immune responses?

Acquired Immunity

There are four ways to acquire specific resistance, or immunity. The terms **active** and **passive** refer to whose immune system reacts to the antigen. Active immunity occurs when the individual's own body produces memory T cells and B cells in response to a harmful antigen. Active immunity takes several days to develop and lasts for a long time because memory cells are produced. Passive immunity results when the immune agents develop in another person (or animal) and are transferred to an individual who was not previously immune. Passive immunity provides immediate protection but is effective for only a short time because no memory cells are produced in the individual.

The terms **natural** and **artificial** refer to how the immunity is obtained. Natural immunity occurs when the immunity is acquired through normal, everyday living, without any deliberate action. Artificial immunity is the result of some type of deliberate action being taken to acquire the immunity, such as receiving a vaccination. Combining the terms gives the four types of acquired immunity (Figure 14-13): active natural immunity, active artificial immunity, passive natural immunity, and passive artificial immunity.

Active Natural Immunity

Active natural immunity results when a person is exposed to a harmful antigen, contracts the disease, and recovers. Exposure to the pathogen stimulates production of memory cells. In subsequent exposures, the memory cells recognize the pathogen and launch a rapid assault before the disease develops.

An example of this is the child who gets chickenpox, recovers, and never contracts it again although he or she is exposed many times.

Active Artificial Immunity

Active artificial immunity develops when a specially prepared antigen is deliberately introduced into an individual's system. This is called **vaccination**. The prepared antigen, called a **vaccine** (vak-SEEN), usually consists of weakened (attenuated), inactivated, or dead pathogens or their toxins. The antigens stimulate the immune system but are altered so they do not produce the symptoms of the disease. Nearly everyone is familiar with vaccines; examples include vaccines for mumps, diphtheria, whooping cough, and tetanus. After a period of time the number of antibodies against a given antigen may decline. A **booster** is an additional dose of vaccine given to increase the number of antibodies.

Passive Natural Immunity

Passive natural immunity results when antibodies are transferred from one person to another through natural means. This occurs only in the prenatal and postnatal relationship between mother and child. Some antibodies (IgG) can cross the placenta and enter fetal blood. This provides protection for the child for a short time after birth, but eventually the antibodies deteriorate and the infant must rely on its own immune system. IgA antibodies are transferred from mother to infant through the mother's milk. This accounts for less than 1% of an infant's immunity, but offers some intestinal protection not available through the placenta.

Passive Artificial Immunity

Passive artificial immunity results when antibodies that developed in another person (or animal) are injected into an individual. **Antiserum** is the general term used for the preparation that contains the antibodies. Antisera may contain antibodies that act against microorganisms (for example, hepatitis and rabies), bacterial toxins (for example, tetanus and botulism), or venoms (for example, poisonous snakes and spiders). Passive artificial immunity provides immediate but short-term protection.

Natural   Artificial

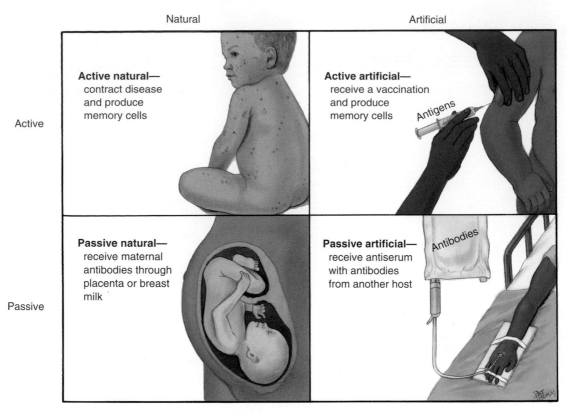

Figure 14-13 *Acquired immunity.*

QUICK CHECK

14.14   Gamma globulins may cross the placenta to provide some protection for the newborn. What type immunity is this?

14.15   What type of immunity is produced by vaccines for whooping cough and tetanus?

QUICK APPLICATIONS

Hypersensitivity reactions, commonly known as *allergies*, are conditions in which the body reacts with an exaggerated immune response and produces tissue damage and disordered function rather than immunity. These reactions vary in intensity from the rhinitis familiar to hay fever sufferers to systemic anaphylaxis, which may be life threatening.

FROM THE PHARMACY

Pharmaceutical preparations affecting the immune system include those that stimulate immunity, those that suppress the immune response (immunosuppressants), and those that modulate the response (immunomodulators). Nearly everyone is familiar with the vaccines and toxoids that are available to provide artificially acquired active immunity against diphtheria, tetanus, whooping cough, mumps, measles, hepatitis B, influenza, and other diseases. Vaccines contain dead or attenuated whole microbes that are not pathogenic but can induce formation of antibodies. Toxoids contain antigenic microbial byproducts that have been detoxified but still induce antibody production. Sera and antitoxins contain antibodies from another individual or animal and are used to provide artificially acquired passive immunity after exposure to the antigen. This includes the antivenins used to treat bites from poisonous snakes and spiders, the antitoxins for the treatment of diphtheria and tetanus, and the immune globulins for the treatment of hepatitis and the management of problems with Rh factor in pregnancy (see Chapter 13).

Immunosuppressants decrease or prevent an immune response and are used to reduce the risk of rejection of kidney, liver, and heart transplants. They also may be used to treat certain autoimmune diseases. The various agents interfere with the immune response at different places in the pathway. The accompanying flow chart illustrates the pathway for the response after a transplant.

Immunomodulating agents modify a biologic response to unwanted antigens. Some of these stop virus replication and prevent viral penetration into healthy cells. Others stop the division of cancer cells. Another type stimulates the bone marrow to inhibit the decrease in neutrophil counts with chemotherapy. Recombinant DNA technology, which was developed in the 1980s, allows quantities of these agents to be produced for clinical trials. The ongoing research into immunomodulating agents holds promise for the future treatment of viral diseases, AIDS, and cancer.

FROM THE PHARMACY—CONT'D

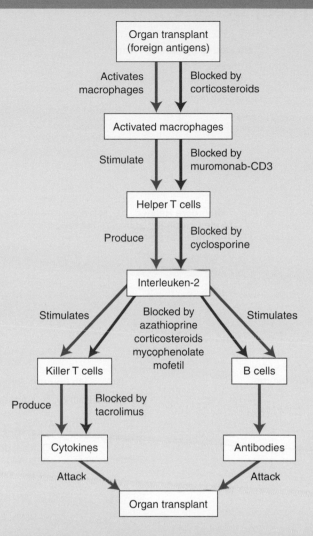

FOCUS ON AGING

Most lymphoid tissues, such as the spleen, thymus, tonsils, and lymph nodes, undergo structural changes with age. They reach their maximum development at the time of puberty, and then slowly regress after that period. There is a reduction in the amount of bone marrow, but enough stem cells remain to produce adequate blood cells for replacement of old cells. There does not appear to be a significant decrease in the number of lymphocytes in the elderly.

The thymus progressively degenerates after puberty so that most of the adult gland is connective tissue. Structural changes in the gland are accompanied by a decreased production of the hormone thymosin, which affects the differentiation and functional activity of T lymphocytes. Consequently, there is an increase in the number of immature T cells and a decrease in the number and/or activity of mature T cells.

The number of B lymphocytes does not appear to change significantly with age; however, there is a decrease in antigen-antibody reactions. This indicates that the B cells are less responsive to the antigens and do not form plasma-cell clones to produce antibodies. However, this may be because the helper T cells are less active and do not stimulate the B cells as they normally do.

In general, in older people there is a decrease in immune sensitivity and an increase in autoimmune reactions in which the immune system fails to recognize the body's own cells. The elderly are more susceptible to infectious diseases and autoimmune disorders than are younger people. The declining immune system also accounts, in part, for the increased occurrence of cancer in the elderly.

Representative Disorders of the
Lymphatic System and Body Defense

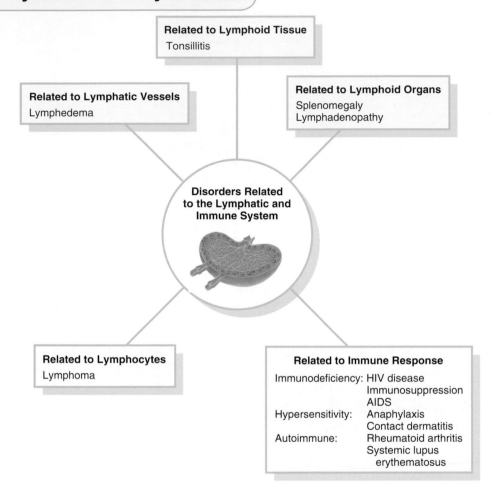

Related to Lymphoid Tissue
Tonsillitis

Related to Lymphatic Vessels
Lymphedema

Related to Lymphoid Organs
Splenomegaly
Lymphadenopathy

**Disorders Related
to the Lymphatic and
Immune System**

Related to Lymphocytes
Lymphoma

Related to Immune Response
Immunodeficiency: HIV disease
                 Immunosuppression
                 AIDS
Hypersensitivity: Anaphylaxis
                 Contact dermatitis
Autoimmune: Rheumatoid arthritis
                 Systemic lupus
                   erythematosus

Acquired immunodeficiency syndrome (AIDS) (ah-KWEYERD ih-myoo-noh-dee-FISH-en-see SIN-drohm) Fatal late stage of HIV infection characterized by profound immunosuppression; characterized by opportunistic infections and malignancies, such as Kaposi's sarcoma, that rarely affect other people

Anaphylaxis (an-ah-fill-AKS-is) Exaggerated or unusual hypersensitivity to a foreign protein or other substances, characterized by a systemic vasodilation with a dramatic decrease in blood pressure that can be life threatening

Contact dermatitis (KAHN-takt der-mah-TYE-tis) Rash caused by direct contact between the skin and a substance to which the individual is sensitive; characterized by itching, swelling, blistering, oozing, and scaling; most common form is poison ivy

Human immunodeficiency virus (HIV) disease (HYOO-man im-yoo-noh-dee-FISH-ehn-see VYE-rus) Impairment of the immune system caused by a virus that destroys helper T cells; in time, the number of circulating antibodies declines and cellular immunity is reduced, leaving the body defenseless against numerous microbial invaders

Immunosuppression (im-yoo-noh-soo-PRESH-un) Inhibition of the formation of antibodies to antigens that may be present; suppression of the immune system

Lymphadenopathy (lim-fed-eh-NAH-pah-thee) Disease of the lymph nodes

Lymphedema (lim-fah-DEE-mah) Swelling of tissues because of fluid accumulation resulting from obstruction of lymph vessels or disorders of the lymph nodes

Lymphoma (lim-FOH-mah) Malignant tumor of lymph nodes and lymph tissue

Rheumatoid arthritis (ROO-mah-toyd ahr-THRYE-tis) Chronic systemic disease with changes occurring in the connective tissues of the body, especially the joints; in contrast to osteoarthritis, the symptoms are usually more generalized and severe; evidence indicates it may be an autoimmune disease

Splenomegaly (spleh-noh-MEG-ah-lee) Enlargement of the spleen

Systemic lupus erythematosus (sih-STEM-ik LOO-pus air-ith-eh-mah-TOE-sis) Chronic autoimmune connective tissue disease characterized by injury to the skin, joints, kidneys, nervous system, and mucous membranes, but can affect any organ of the body

Tonsillitis (tahn-sih-LYE-tis) Inflammation and enlargement of the tonsils, especially the palatine tonsils

CHAPTER SUMMARY

Functions of the Lymphatic System

■ **State three functions of the lymphatic system.**
- The lymphatic system returns excess interstitial fluid to the blood.
- The lymphatic system absorbs fats and fat-soluble vitamins from the digestive system.
- The lymphatic system provides defense against invading microorganisms and disease.

Components of the Lymphatic System

■ **List the components of the lymphatic system.**
- The lymphatic system consists of the lymph, lymphatic vessels, and lymphatic organs.

■ **Describe the origin of lymph.**
- As soon as interstitial fluid enters lymphatic vessels, it is called lymph. Lymphatic vessels return lymph to the blood plasma.

■ **Describe the lymphatic vessels and the mechanisms that move lymph through the vessels.**
- Lymphatic vessels are similar to veins in structure. They have thin walls and have valves to prevent backflow of lymph. These vessels carry fluid away from the tissue spaces and return it to the venous system. The right lymphatic duct drains lymph from the upper right quadrant of the body. The thoracic duct, which begins in the abdomen as the cisterna chyli, drains the other three-fourths of the body.
- Pressure gradients that move fluid through the lymphatic vessels come from skeletal muscle action, respiratory movements, and contraction of smooth muscle in vessel walls.

■ **Describe the structure of lymph nodes, tonsils, spleen, and thymus and explain their importance to the body.**
- Lymph nodes consist of dense masses of lymphocytes that are separated by spaces called lymph sinuses. Lymph enters a node through afferent vessels, filters through the sinuses, and leaves through an efferent vessel. Three areas in which lymph nodes tend to cluster are the inguinal nodes in the groin, axillary nodes in the armpit, and cervical nodes in the neck. There are no lymph nodes associated with the central nervous system. Lymph nodes filter and cleanse the lymph before it returns to the blood.
- Tonsils are clusters of lymphatic tissue in the region of the nose, mouth, and throat and they provide protection against pathogens that may enter through these regions. The pharyngeal tonsils, also called adenoids, are near the opening of the nasal cavity into the pharynx; palatine tonsils are near the opening of the oral cavity into the pharynx; and the lingual tonsils are at the base of the tongue, also near the opening to the oral cavity into the pharynx.
- The spleen is located in the upper right quadrant of the abdomen, posterior to the stomach. It is much like a lymph node, but larger. It contains masses of lymphocytes and macrophages that are supported by a fibrous framework. The spleen filters blood in much the same way the lymph nodes filter lymph. The spleen also is a reservoir for blood.
- The thymus is located anterior to the ascending aorta and posterior to the sternum. The principal function of the thymus is the processing and maturation of T cells. It also produces thymosin that stimulates the maturation of lymphocytes in other organs.

Resistance to Disease

■ **List four nonspecific mechanisms that provide resistance to disease and explain how each functions.**
- Disease-producing organisms are called pathogens. The ability to counteract pathogens is resistance, and a lack of resistance is susceptibility.
- Nonspecific defense mechanisms include barriers, chemical action, phagocytosis, and inflammation.
- Barriers are factors that deter microbial invasion. They may be of a mechanical nature (unbroken skin), fluid (tears), or chemical (lysozymes).
- If microorganisms succeed in passing through the barriers, internal defenses, such as chemical action, respond. Complement is a chemical defense that promotes phagocytosis and inflammation. Interferon has particular significance because it offers protection against viruses. It is produced by virus-infected cells to provide protection for the neighboring cells.
- Phagocytosis is the ingestion and destruction of solid particles by certain cells, particularly neutrophils and macrophages. Neutrophils are small cells that are the first to migrate to an infected area. Macrophages are monocytes that leave the blood and enter the tissue spaces. They phagocytize cellular debris. They also perform a cleansing action of the lymph and blood.
- Inflammation is characterized by redness, warmth, swelling and pain. It includes a series of events, described in Figure 14-7, that occur in response to tissue damage. The overall purpose of inflammation is to destroy bacteria, cleanse the area of debris, and promote healing. Systemic inflammation is characterized by leukocytosis, fever, and a dangerous decrease in blood pressure.

■ **State the two characteristics of specific defense mechanisms and identify the two principal cells involved in specific resistance.**
- Specificity and memory are two features of specific defense mechanisms. The principal cells involved are lymphocytes and macrophages.

■ **Distinguish between self and nonself as they relate to disease resistance.**
- Proteins and other large molecules that are recognized as belonging to the body of an individual are interpreted as "self." Others are interpreted as "nonself." Antigens are molecules that trigger an immune response. Usually they are foreign proteins that enter the body and are interpreted as nonself.

■ **Describe the development of lymphocytes.**
- During fetal development, the bone marrow releases immature lymphocytes into the blood. Some of the immature lymphocytes go to the thymus gland where they differentiate to become T lymphocytes (T cells). About 70% of the circulating lymphocytes are T cells. Lymphocytes that differentiate in some region other than the thymus are B lymphocytes (B cells). These account for about 30% of the circulating lymphocytes.

■ **Briefly describe the mechanism of cell-mediated immunity and list four subgroups of T cells.**
- Cell-mediated immunity is the result of T cell action.

- When an antigen enters the body, it is phagocytized by a macrophage, which presents the antigen to an appropriate T cell. The T cell responds by producing clones of T cells.
- Killer T cells directly destroy the cells with the offending antigen; helper T cells secrete substances that promote the immune response; suppressor T cells inhibit the immune response and help regulate it; memory T cells stimulate a faster and more intense response if the same antigen enters the body again.

■ **Briefly describe the mechanism of antibody-mediated immunity and list two subgroups of B cells.**

- B cells are responsible for antibody-mediated immunity.
- When an antigen enters the body, a macrophage phagocytizes it and presents to the appropriate B cell. The B cell responds by forming clones of plasma cells and memory B cells. Plasma cells produce large quantities of antibodies that inactivate the invading antigens.

■ **Distinguish between the primary response and the secondary response to a pathogen.**

- The initial action against a pathogen is a primary response.
- If the same pathogen enters the body a second time, memory B cells launch a rapid and intense response. This is called a secondary response, which is faster and more intense than the initial primary response.

■ **List five classes of immunoglobulins and state the role each has in immunity.**

- Antibodies belong to a class of proteins called globulins. Since they are involved in the immune response, they are called immunoglobulins.
- IgA, IgG, IgM, IgE, and IgD are five classes of immunoglobulins. Each class has a specific role in immunity as indicated in Table 14-1.

■ **Give examples of active natural immunity, active artificial immunity, passive natural immunity, and passive artificial immunity.**

- Active immunity occurs when the body produces memory cells; passive immunity results when the immune agents are transferred into an individual. Natural immunity is acquired through normal activities; artificial immunity requires some deliberate action.
- Active natural immunity results when a person is exposed to a harmful antigen, contracts the disease, and recovers.
- Active artificial immunity develops when a prepared antigen is deliberately introduced into the body (vaccination) and stimulates the immune system.
- Passive natural immunity results when antibodies are transferred from mother to child through the placenta or milk.
- Passive artificial immunity results when antibodies are injected into an individual.

CHAPTER QUIZ

Recall

Match the definitions on the left with the appropriate term on the right.

_____ **1.** Lymph capillaries
_____ **2.** Collects lymph from three fourths of the body
_____ **3.** Enlarged pharyngeal tonsils
_____ **4.** Lack of resistance
_____ **5.** Provides localized protection against viruses
_____ **6.** Cause an increase in body temperature
_____ **7.** Principal cells involved in cell-mediated immunity
_____ **8.** Protein that triggers an immune response
_____ **9.** Produce antibodies
_____ **10.** Rapid, intense reaction against an antigen

A. Adenoids
B. Antigen
C. Interferon
D. Lacteals
E. Plasma cells
F. Pyrogens
G. Secondary response
H. Susceptibility
I. T cells
J. Thoracic duct

Thought

1. Lymphatic vessels carry lymph away from the tissues and eventually return it to the blood in the:
A. Inferior vena cava
B. Superior vena cava
C. Subclavian arteries
D. Subclavian veins

2. Lymph is filtered by the:
A. Spleen
B. Lymph nodes
C. Liver
D. Tonsils

3. Which of the following describes, or is a characteristic of, nonspecific defense mechanisms?
A. Phagocytosis
B. Memory
C. B cells
D. Immunoglobulins

4. Which of the following statements about T lymphocytes is *false*?
A. They are responsible for humoral immunity.
B. Some are regulatory cells that control the immune response.
C. Some directly destroy antigens.
D. Some stimulate B cells.

5. Active immunity is produced when an:
A. Individual receives an injection of gamma globulin
B. Infant receives antibodies through the placenta or breast milk
C. Individual receives an injection of a vaccine
D. Individual is injected with an antiserum

Application

In treating a woman for breast cancer, the surgeon removed some of the axillary lymph nodes on the right side. After surgery, the patient experiences edema in the right upper extremity. Explain why this side effect occurred.

Booster shots may be given as part of a vaccination regimen. A booster shot is a second dose of the same vaccine given sometime after the original vaccination. Why are booster shots given if they are just a repeat of the same vaccine?

BUILDING YOUR MEDICAL VOCABULARY

Building a vocabulary is a cumulative process. As you progress through this book, you will use word parts, abbreviations, and clinical terms from previous chapters. Each chapter will present new word parts, abbreviations, and terms to add to your expanding vocabulary.

Word Parts and Combining Form with Definition and Examples

PART/COMBINING FORM	DEFINITION	EXAMPLE
aden-	gland	lymph<u>adenitis</u>: inflammation of one or more lymph nodes (lymph glands)
-ectomy	surgical removal	tonsill<u>ectomy</u>: surgical removal of the tonsils
immun/o	protection	<u>immuno</u>suppression: inhibition of the immune response because of drugs or disease
lymph/o	lymph	<u>lymph</u>edema: interstitial fluid collects within the spaces between cells as a result of obstruction of lymphatic vessels and nodes
-lytic	to reduce, destroy	thrombo<u>lytic</u>: pertaining to an agent that dissolves or destroys a clot
onc/o	tumor	<u>onc</u>ogenic: giving rise to or causing the formation of a tumors
-pexy	fixation	spleno<u>pexy</u>: surgical fixation of the spleen
splen/o	spleen	<u>spleno</u>megaly: abnormal enlargement of the spleen
-tic	pertaining to	lympha<u>tic</u>: pertaining to lymph
tox-	poison	<u>tox</u>ic: pertaining to a poison
thym/o	thymus	<u>thym</u>ectomy: surgical removal of the thymus gland

Clinical Abbreviations

ABBREVIATION	MEANING
AIDS	acquired immunodeficiency syndrome
CMV	cytomegalovirus
DCIS	ductal carcinoma in situ
ELISA	enzyme-linked immunosorbent assay
HIV	human immunodeficiency virus
HL	Hodgkin's lymphoma
HSV	herpes simplex virus
HZ	herpes zoster
IDC	invasive ductal carcinoma
Ig	immunoglobulin
ILC	invasive lobular carcinoma
KS	Kaposi's sarcoma
NHL	non-Hodgkin's lymphoma
SLE	systemic lupus erythematosus

Clinical Terms

Allergen (AL-er-jen) Substance capable of causing a specific hypersensitivity in the body; a type of antigen

Autoimmune disease (aw-toh-ih-MYOON dih-ZEEZ) Condition in which the body's immune system becomes defective and produces antibodies against itself

Immunoelectrophoresis (ih-myoo-noh-ee-lek-troh-for-EE-sis) Test that separates immunoglobulins IgG, IgM, IgE, IgA, and IgD

Immunologist (ih-myoo-NAHL-oh-jist) Specialist in the study, diagnosis, and treatment of immune system disorders

Immunotherapy (ih-myoo-noh-THAIR-ah-pee) Passive immunization of an individual by preformed antibodies

Interferon (in-ter-FEER-on) Substance that is produced by the body in response to the presence of a virus, which, in turn, offers some protection against that virus by inhibiting its multiplication

Interleukins (in-ter-LOO-kins) Proteins that stimulate the growth of T-cell lymphocytes and activate immune responses

Kaposi's sarcoma (CAP-oh-sees sar-KOH-mah) Proliferation of malignant neoplastic lesions characterized by bluish-red nodules in the skin, usually beginning in the lower extremities and then spreading to more proximal sites; frequently occurs in AIDS patients

Lymphadenitis (lim-fad-en-EYE-tis) Inflammation of the lymph glands (nodes)

Lymphangiogram (lim-FAN-jee-oh-gram) Procedure in which a dye is injected into lymph vessels in the foot and radiographs are taken to show the path of lymph flow as it moves into the chest region

Metastasis (meh-TASS-tah-sis) Spread of a malignant tumor to a secondary site

Monoclonal antibody (mah-noh-KLOH-nal AN-tih-bahd-ee) Antibody produced in a laboratory to attack antigens, useful in immunotherapy and cancer treatment

Mononucleosis (mah-noh-noo-klee-OH-sis) Acute infectious disease, caused by the Epstein-Barr virus, with enlarged lymph nodes, increased numbers of agranulocytes in the bloodstream, fatigue, sore throat, and enlarged, tender lymph nodes

Oncologist (ahn-KAHL-oh-jist) A Specialist in the diagnosis and treatment of malignant disorders

Pyrogen (PIE-roh-jen) Agent that causes fever

VOCABULARY QUIZ

Use word parts you have learned to form words that have the following definitions.

1. Surgical excision of a tonsil _____

2. Fixation of a movable spleen _____

3. Stopping flow of lymph _____

4. Study of body protection _____

5. Inflammation of the thymus _____

Using the definitions of word parts you have learned, define the following words.

6. Thymoma _____

7. Splenomegaly _____

8. Lymphangiology _____

9. Lymphadenitis _____

10. Lymphocytopenia _____

Match each of the following definitions with the correct word.

_____ 11. Poison substance

_____ 12. Surgical removal of the thymus

_____ 13. Incision into a lymph gland

_____ 14. Condition of spleen congested with blood

_____ 15. Formation of lymph

A. Lymphadenotomy
B. Lymphopoiesis
C. Splenemia
D. Thymectomy
E. Toxin

16. What is the term for "a specialist in diagnosing and treating malignant disorders"? _____

17. What is the clinical term for "enlargement of the spleen"? _____

18. What radiologic procedure shows the path of lymph flow? _____

19. What is the meaning of the abbreviation CMV? _____

20. What is the meaning of the abbreviation AIDS? _____

Respiratory System

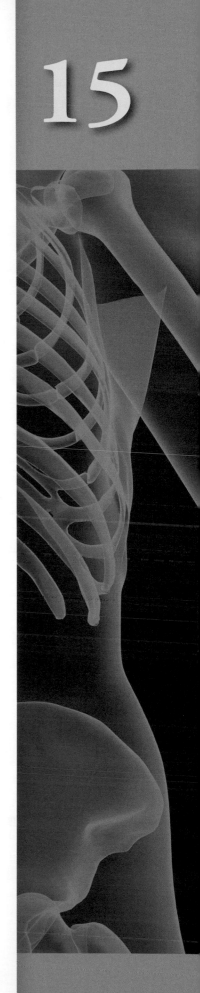

CHAPTER OBJECTIVES

Functions and Overview of Respiration
- Define five activities or functions of the respiratory process.

Ventilation
- Describe the structures and features of the upper respiratory tract.
- Describe the structures and features of the lower respiratory tract.
- Describe the structure of the lungs, including shape, lobes, tissue, and membranes.
- Name and define three pressures involved in pulmonary ventilation, and relate these pressures to the sequence of events that result in inspiration and expiration.
- Define four respiratory volumes and four respiratory capacities, state their average normal values, and describe factors that influence them.

Basic Gas Laws and Respiration
- Discuss factors that govern the diffusion of gases into and out of the blood.
- Distinguish between external respiration and internal respiration.

Transport of Gases
- Describe two methods of oxygen transport in the blood.
- Describe three ways in which carbon dioxide is transported in the blood.

Regulation of Respiration
- Name two regions in the brain that make up the respiratory center and two nerves that carry impulses from the center.
- Describe the role of chemoreceptors, stretch receptors, higher brain centers, and temperature in regulating breathing.

KEY TERMS

Alveolus (al-VEE-oh-lus) Microscopic dilations of terminal bronchioles in the lungs where diffusion of gases occurs; air sacs in the lungs

Bronchial tree (BRONG-kee-al TREE) Bronchi and all their branches that function as passageways between the trachea and the alveoli

Bronchopulmonary segment (brong-koh-PUL-moh-nair-ee SEG-ment) Portion of a lung surrounding a tertiary, or segmental, bronchus; lobule of the lung

Carbaminohemoglobin (kar-bah-meen-oh-HEE-moh-gloh-bin) Compound that is formed when carbon dioxide combines with the protein portion of hemoglobin; accounts for approximately 23% of the carbon dioxide transport in the blood

External respiration (eks-TER-nal res-per-RAY-shun) Exchange of gases between the lungs and the blood

Hering-Breuer reflex (HER-ing BREW-er REE-fleks) Stretch reflex in the lungs that prevents overinflation of the lungs

Internal respiration (in-TER-nal res-per-RAY-shun) Exchange of gases between the blood and tissue cells

Lower respiratory tract (LOW-er RES-per-ah-tor-ee TRACT) Portion of the respiratory tract that is below the larynx, which includes the trachea, bronchial tree, and lungs

Oxyhemoglobin (ahk-see-HEE-moh-gloh-bin) Compound that is formed when oxygen combines with the heme portion of hemoglobin; form in which most of the oxygen is transported in the blood

Pleura (PLOO-rah) Serous membrane that surrounds the lungs; consists of a parietal layer and a visceral layer

Respiratory membrane (RES-per-ah-tor-ee MEM-brayn) Any surface in the lungs where diffusion occurs; consists of the layers that the gases must pass through to get into or out of the alveoli

Surfactant (sir-FAK-tant) Substance, produced by certain cells in lung tissue, that reduces surface tension between fluid molecules that line the respiratory membrane and helps keep the alveolus from collapsing

Upper respiratory tract (UP-per RES-per-ah-tor-ee TRACT) Portion of the respiratory tract that includes the nose, pharynx, and larynx

Ventilation (ven-tih-LAY-shun) Movement of air into and out of the lungs; breathing

Functional Relationships of the
Respiratory System

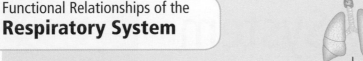

Provides oxygen and removes carbon dioxide.

Reproductive

Sexual arousal stimulates changes in rate and depth of breathing.

Supplies oxygen and removes carbon dioxide to maintain metabolism in tissues of reproductive system; helps maintain pH for gonadal hormone function.

Urinary

Helps maintain water, electrolyte, and pH balance of body fluids for effective respiratory function; eliminates waste products generated by respiratory organs.

Assists in the regulation of pH by removing carbon dioxide.

Digestive

Absorbs nutrients that are necessary for maintenance of cells in the lungs and other tissues of the respiratory tract.

Provides oxygen for metabolism of cells in digestive system; removes carbondioxide; helps maintain pH of body fluids for effective enzyme function.

Lymphatic/Immune

Tonsils combat pathogens that enter through respiratory passageways; IgA protects respiratory mucosa.

Provides lymphoid tissue and immune cells with oxygen and removesdioxide; pharynx contains the tonsils; breathing movements assist in flow of lymph.

Cardiovascular

Transports oxygen and carbon dioxide between lungs and tissues.

Breathing movements assist in venous return; helps maintain blood pH;supplies oxygen and removes carbon dioxide for cardiac tissue.

Integument

Helps maintain body temperature for metabolism; hairs of nasal cavity filter particles that may damage the upper respiratory tract.

Furnishes oxygen and removes carbon dioxide by gaseous exchange

Skeletal

Encases the lungs for protection; provides passageways for air through the nasal cavity.

Supplies oxygen for bone tissue metabolism and removes carbon dioxide.

Muscular

Muscle contractions control airflow through respiratory passages and create pressure changes necessary for ventilation.

Supplies oxygen for muscle metabolism and removes carbon dioxide.

Nervous

Innervates muscles involved in breathing; controls rate and depth of breathing.

Supplies oxygen for brain, spinal cord, and sensory tissue; removes carbonhelps maintain pH for neural function.

Endocrine

Thyroxine and epinephrine promote cell respiration; epinephrine stimulates bronchodilation.

Supplies oxygen, removes carbon dioxide, and helps maintain pH for metabolismendocrine glands; converts angiotensin I into angiotensin II.

← Gives to Respiratory System
→ Receives from Respiratory System

When the **respiratory system** is mentioned, people generally think of breathing, but this is only one of the activities of the respiratory system. The body cells need a continuous supply of oxygen for the metabolic processes that are necessary to maintain life. The respiratory system works with the cardiovascular system to provide this oxygen and to remove the waste products of metabolism. It also helps to regulate the pH of the blood.

Functions and Overview of Respiration

Respiration is the sequence of events that results in the exchange of oxygen and carbon dioxide between the atmosphere and the body cells. Every 3 to 5 seconds nerve impulses stimulate the breathing process, or **ventilation**, which moves air through a series of passages into and out of the lungs. After this, there is an exchange of gases between the lungs and the blood. This is called **external respiration**. The blood **transports** the gases to and from the tissue cells. The exchange of gases between the blood and tissue cells is **internal respiration**. Finally, the cells use the oxygen for their specific activities. This is cellular metabolism, or **cellular respiration**, which is discussed in Chapter 17. Together these activities constitute respiration.

Ventilation

Ventilation, or breathing, is the movement of air through the conducting passages between the atmosphere and the lungs. The air moves through the passages because of pressure gradients that are produced by contraction of the diaphragm and thoracic muscles.

Conducting Passages

The conducting passages are divided into the **upper respiratory tract** and the **lower respiratory tract** (Figure 15-1). The upper respiratory tract includes the nose, pharynx, and larynx. The lower respiratory tract consists of the trachea, bronchial tree, and lungs. These passageways open to the outside and are lined with mucous membrane. In some regions, the membrane has hairs that help filter the air. Other regions may have cilia to propel mucus.

Nose and Nasal Cavities

The framework of the **nose** consists of bone and cartilage. Two small nasal bones and extensions of the maxillae form the bridge of the nose, which is the bony portion. The remainder of the framework is cartilage. This is the flexible portion. Connective tissue and skin cover the framework.

The interior chamber of the nose is the **nasal cavity**. It is divided into two parts by the **nasal septum**, a vertical partition formed by the vomer and the perpendicular plate of the ethmoid bone. Air enters the nasal cavity from the outside through two openings—the **nostrils**, or **external nares** (NAY-reez). The openings from the nasal cavity into the pharynx are the **internal nares**. The **palate** forms the floor of the nasal cavity and separates the nasal cavity from the oral cavity. The anterior portion of the palate is called the **hard palate** because

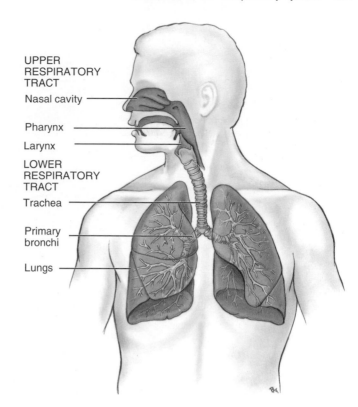

UPPER RESPIRATORY TRACT
Nasal cavity
Pharynx
Larynx
LOWER RESPIRATORY TRACT
Trachea
Primary bronchi
Lungs

Figure 15-1 *Conducting passages of the respiratory system. The upper respiratory tract includes the nose, pharynx, and larynx. The lower respiratory tract consists of the trachea, bronchial tree, and lungs.*

it is supported by bone. The posterior portion has no bony support, so it is called the **soft palate**. The soft palate terminates in a projection called the **uvula** (YOO-vyoo-lah), which helps direct food into the oropharynx. Three **nasal conchae** (KONG-kee), bony ridges that project medially into the nasal cavity from each lateral wall, increase the surface area of the cavity to warm and moisten the air and also to help direct air flow. Dust and other nongaseous particles in the air tend to become trapped in the mucous membrane around the conchae. Figure 15-2 illustrates the features of the nasal cavity.

Paranasal sinuses are air-filled cavities in the frontal, maxillae, ethmoid, and sphenoid bones. These sinuses, which have the same names as the bones in which they are located, surround the nasal cavity and open into it. They function to reduce the weight of the skull, to produce mucus, and to influence voice quality by acting as resonating chambers. The sinuses are lined with mucous membrane that produces mucus, which drains into the nasal cavity. During infections and allergies, the membranes in the passages that drain the sinuses become inflamed and swollen. The swelling may block the passages and cause the mucus to accumulate in the sinuses. As the mucus accumulates, pressure within the sinuses increases, resulting in a sinus headache.

As air passes through the nasal cavity, it is filtered, warmed, and moistened. The mucous membrane that lines most of the nasal cavity is ciliated pseudostratified columnar epithelium, which filters the air. Goblet cells in the mucous membrane produce mucus that traps microorganisms, dust, and other foreign particles. Cilia propel the mucus with the trapped

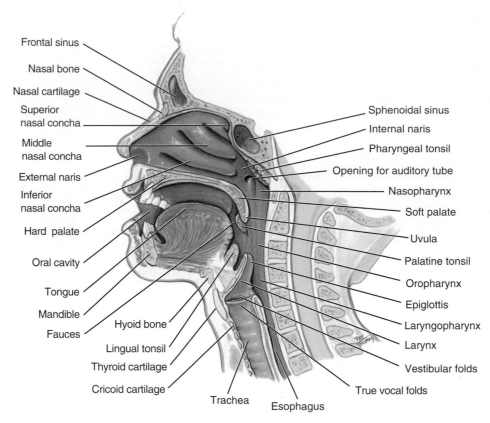

Figure 15-2 *Features of the upper respiratory tract. The upper respiratory tract includes the nose, pharynx, and larynx.*

particles toward the pharynx where it is swallowed. Acid in the gastric juice destroys most of the microorganisms that are swallowed. Extensive capillary networks under the mucous membrane warm and moisten the air before it reaches the rest of the respiratory tract.

Pharynx

The **pharynx** (FAIR-inks), commonly called the throat, is a passageway, about 13 cm long, that extends from the base of the skull to the level of the sixth cervical vertebra. It serves both the respiratory and the digestive systems by receiving air from the nasal cavity and air, food, and water from the oral cavity. Inferiorly, it opens into the larynx and esophagus. The pharynx is divided into three regions according to location (see Figure 15-2).

The **nasopharynx** (nay-zoh-FAIR-inks) is the portion of the pharynx that is posterior to the nasal cavity and extends inferiorly to the uvula. Air enters this region from the nasal cavity through the internal nares. The mucous membrane in the nasopharynx is similar to the lining of the nasal cavity. The **auditory** (eustachian) **tubes** from the two middle ear cavities open into the nasopharynx. The auditory tubes help to equalize the air pressure on both sides of the tympanic membrane (eardrum). Collections of lymphoid tissue, called **pharyngeal tonsils**, or **adenoids**, are located in the posterior wall of the nasopharynx.

The **oropharynx** (ohr-oh-FAIR-inks) is the portion of the pharynx that is posterior to the oral cavity. It extends from the uvula down to the level of the hyoid bone and receives air, food, and water from the oral cavity. During swallowing,

the soft palate and uvula move upward to prevent the material from going into the nasopharynx. The opening between the oral cavity and oropharynx is called the **fauces** (FAW-seez), and it is bordered by masses of lymphoid tissue called **tonsils**. The **palatine tonsils** are in the lateral walls of the oropharynx, adjacent to the fauces, and the **lingual tonsils** are located on the surface of the posterior portion of the tongue, also in the region of the fauces. Because they are lymphoid tissue, all of the tonsils in the pharynx function in immune responses and help prevent infections.

The most inferior portion of the pharynx is the **laryngopharynx** (lah-ring-goh-FAIR-inks) that extends from the hyoid bone down to the lower margin of the larynx. Both the oropharynx and laryngopharynx are lined with a mucous membrane of stratified squamous epithelium.

Larynx

The **larynx** (LAIR-inks), commonly called the voice box, is the passageway for air between the pharynx above and the trachea below. It is about 5 cm long and extends from the fourth to the sixth vertebral levels. It is formed by nine pieces of cartilage that are connected to each other by muscles and ligaments. Six pieces of the cartilage are grouped into three pairs: the **arytenoid, corniculate**, and **cuneiform cartilage**. The other three pieces of cartilage are single, unpaired segments: the **thyroid, cricoid**, and **epiglottis cartilage**. All of the cartilage is hyaline cartilage, except the epiglottis, which is elastic cartilage. The larynx is also supported by ligaments that attach to the hyoid bone.

The three largest cartilaginous portions of the larynx are the **thyroid cartilage**, the **cricoid** (KRY-koyd) **cartilage**, and the **epiglottis** (eh-pih-GLOT-is) (see Figure 15-2). The thyroid cartilage, consisting of two shield-shaped plates, is the most superior of the cartilage. It forms an anterior projection in the neck called the Adam's apple. This is more pronounced in males than in females. The cricoid cartilage is the most inferior of the laryngeal cartilage. It forms the base of the larynx and is attached to the trachea. The epiglottis is a long, leaf-shaped structure. Its inferior margin is attached to the thyroid cartilage, but the upper portion is a movable flap that projects superiorly. During swallowing, the epiglottis covers the opening into the larynx to prevent food and water from entering.

Inside the larynx, two pairs of ligaments, covered by mucous membrane, extend from the **arytenoid cartilage** to the posterior surface of the thyroid cartilage. The upper pair are the **vestibular folds**, or **false vocal cords**. They work with the epiglottis to prevent particles from entering the lower respiratory tract. The lower pair are the **true vocal cords**, which function in sound production. Muscles control the length and tension of the true vocal cords. They are relaxed during normal breathing; however, when they are under tension, exhaled air moving by them causes them to vibrate and produce sound. The length of the vocal cords determines the pitch of the sound, and the force of the moving air regulates the loudness. The opening between the true vocal cords is the **glottis**, which leads to the trachea.

Trachea

The **trachea**, commonly called the windpipe, is a tube that extends from the cricoid cartilage of the larynx, at the level of the sixth cervical vertebra, into the mediastinum where it divides into the right and left bronchi at the level of the fifth thoracic vertebra (Figure 15-3). The trachea is the beginning of the lower respiratory track and is about 12 to 15cm long. The anterior and lateral walls of the trachea are supported by 15 to 20 C-shaped pieces of hyaline cartilage that hold the trachea open despite the pressure changes that occur during breathing (Figure 15-4). The posterior open part of the C-shaped cartilage is closed by smooth muscle and connective tissue and is next to the esophagus. During swallowing, the esophagus bulges into the soft part of the trachea.

The mucous membrane that lines the trachea is ciliated pseudostratified columnar epithelium similar to that in the nasal cavity and nasopharynx. Goblet cells produce mucus that traps airborne particles and microorganisms, and the cilia propel the mucus upward, where it is either swallowed or expelled. Continued irritation from cigarette smoke and other air pollutants damages the cilia, and the mucus with the trapped particles is not removed. Microorganisms thrive in the accumulated mucus, which results in respiratory infections. Irritation and inflammation of the mucous membrane stimulate the cough reflex.

QUICK CHECK

15.1 What are the five activities that comprise respiration?
15.2 Name the three regions of the pharynx.
15.3 What are the three regions of the upper respiratory tract?

QUICK APPLICATIONS

Foreign objects that become lodged in the larynx or trachea are usually expelled by coughing. If a person cannot speak or make a sound because of the obstruction, it means that the airway is completely blocked. This is a life-threatening situation. The Heimlich maneuver is a procedure in which the air in the person's own lungs is used to forcefully expel the object.

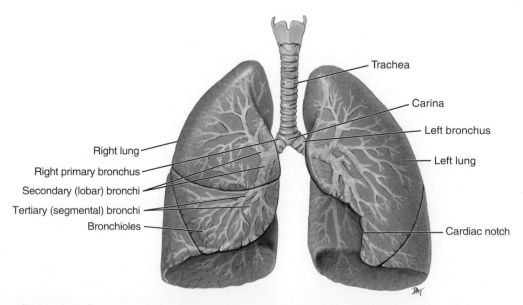

Figure 15-3 *Features of the lower respiratory tract. The lower respiratory tract includes the trachea, bronchial tree, and lungs.*

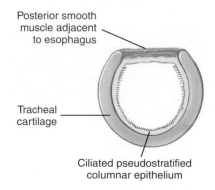

Posterior smooth
muscle adjacent
to esophagus

Tracheal
cartilage

Ciliated pseudostratified
columnar epithelium

Figure 15-4 *C-shaped cartilage ring of the trachea.*

Bronchi and Bronchial Tree

In the mediastinum, at the level of the fifth thoracic vertebra, the trachea divides into the **right** and **left primary bronchi.** In the region of the tracheal bifurcation, the hyaline cartilage forms a ridge called the **carina** (kah-RYE-nah). The right primary bronchus is shorter, more vertical, and wider in diameter than the left bronchus. Because the right bronchus is wider and more vertical than the left, foreign particles tend to enter it more frequently.

After the bronchi enter the lungs, they branch several times into smaller and smaller passages to form the **bronchial tree** (see Figure 15-3). The primary bronchi divide to form **secondary (lobar) bronchi**; then the secondary bronchi branch into **tertiary (segmental) bronchi**. There are 3 secondary and 10 tertiary bronchi on the right side but only 2 secondary and 8 tertiary bronchi on the left side. The branching continues, finally giving rise to the **bronchioles.** The terminal bronchioles branch into smaller respiratory bronchioles, which lead into microscopic **alveolar ducts.** Alveolar ducts terminate in clusters of tiny air sacs called **alveoli.** The alveoli are surrounded by extensive capillary beds from pulmonary circulation (Figure 15-5). It is here that external respiration occurs.

The cartilage and mucous membrane of the primary bronchi are similar to those in the trachea. As the branching continues through the bronchial tree, the amount of hyaline cartilage

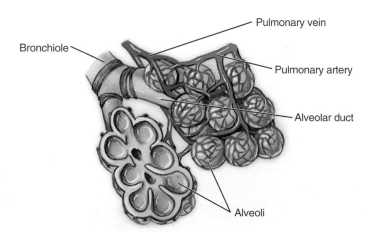

Bronchiole

Pulmonary vein

Pulmonary artery

Alveolar duct

Alveoli

Figure 15-5 *Terminal branching of the bronchial tree. Respiratory bronchioles branch into alveolar ducts that terminate in alveoli.*

in the walls decreases until it is absent in the smallest bronchioles. As the amount of cartilage decreases, the amount of smooth muscle increases. The mucous membrane also undergoes a transition from ciliated pseudostratified columnar epithelium to simple cuboidal epithelium to simple squamous epithelium. Because there is abundant smooth muscle and no cartilage in the walls of the bronchioles, they can constrict to a very small size when the smooth muscle contracts, which occurs during an asthma attack. This restricts the flow of air and makes breathing difficult.

The alveolar ducts and alveoli consist primarily of simple squamous epithelium, which permits rapid diffusion of oxygen and carbon dioxide. Exchange of gases between the air in the lungs and the blood in the capillaries occurs across the walls of the alveolar ducts and alveoli.

QUICK APPLICATIONS

There are several different forms of **asthma**, but they all have sensitive conducting passages. In many cases the agent that triggers the attack is an allergen in the air. The most obvious and dangerous symptom involves the constriction of the smooth muscle around the bronchial tree. The airways become narrow and breathing is difficult. Treatment includes the use of bronchodilators to dilate the respiratory passages to permit airflow.

Lungs

The two **lungs**, which contain all the components of the bronchial tree beyond the primary bronchi, occupy most of the space in the thoracic cavity. The lungs are soft and spongy because they are mostly air spaces surrounded by the alveolar cells and elastic connective tissue. They are separated from each other by the mediastinum, which contains the heart. Each lung is roughly cone shaped, rests on the diaphragm, and extends upward just above the midpoint of the clavicle. The only point of attachment for each lung is at the **hilum**, or **root**, on the medial side. This is where the bronchi, blood vessels, lymphatics, and nerves enter the lungs.

The **right lung** is shorter, is broader, and has a greater volume than the left lung. It is divided into three lobes (superior, middle, and inferior) by two fissures. Each lobe is supplied by one of the secondary (lobar) bronchi. The lobes are further subdivided into **bronchopulmonary** (brong-koh-PUL-moh-nair-ee) **segments (lobules)** by connective tissue septa that are not visible on the surface. Because each segment has its own bronchus and blood supply, which do not cross the septa, a segment can be surgically removed with relatively little damage to the rest of the lung.

The **left lung** is longer and narrower than the right lung. It has an indentation, called the **cardiac notch**, on its medial surface for the apex of the heart. The left lung is divided into two lobes by a single fissure. Figure 15-6 is a reconstructed three-dimensional computed tomography (CT) image of the lungs.

Each lung is enclosed by a double-layered **serous membrane**, called the **pleura** (Figure 15-7). The **visceral pleura** is firmly attached to the surface of the lung. At the hilum, the

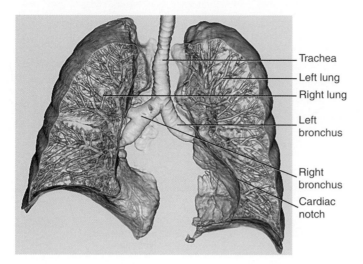

Figure 15-6 *Reconstructed three-dimensional computed tomography image of the trachea, bronchial tree, and lungs. (Adapted from Applegate E: The Sectional Anatomy Learning System, Concepts, ed. 3. St. Louis, Elsevier/Saunders, 2010.)*

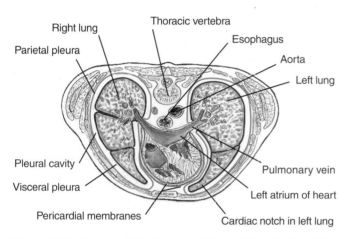

Figure 15-7 *Features of the lungs and pleura. Note the three lobes in the right lung and two lobes in the left lung.* **Red** *indicates visceral pleura and* **blue** *indicates parietal pleura.*

visceral pleura is continuous with the **parietal pleura** that lines the wall of the thorax. The small space between the visceral and parietal pleurae is the **pleural cavity**. It contains a thin film of serous fluid that is produced by the pleura. The fluid acts as a lubricant to reduce friction as the two layers slide against each other, and it helps to hold the two layers together as the lungs inflate and deflate.

QUICK APPLICATIONS

Pleuritis, or pleurisy, is an inflammation of the pleura and is often painful because the sensory nerves in the parietal pleura are irritated. As the condition progresses, the permeability of the membrane changes, which results in an accumulation of fluid in the pleural cavity, making breathing difficult.

QUICK CHECK

15.4 How many secondary (lobar) bronchi are associated with the right lung?

15.5 What is the name for the serous membrane that encloses the lungs?

15.6 What is external respiration and where does it occur?

Mechanics of Ventilation

Pulmonary ventilation is commonly referred to as breathing. It is the process of air flowing into the lungs during inspiration (inhalation) and out of the lungs during expiration (exhalation). Air flows because of pressure differences between the atmosphere and the gases inside the lungs.

One of the fundamental properties of gases is Boyle's law, which states that at constant temperature, when the volume of a gas increases the pressure decreases; conversely, when the volume decreases the pressure increases. This is stated in equation form as $P_1V_1 = P_2V_2$, where P represents pressure and V represents volume. A gas expands to fill a given container, and when it expands (volume increases) the pressure of the gas decreases. In ventilation, the containers are the atmosphere, the lungs, and the pleural cavity. Ventilation depends on changes in pressures and volumes within the containers.

Pressures in Pulmonary Ventilation

Air, like other gases, flows from a region with higher pressure to a region with lower pressure. Muscular breathing movements and recoil of elastic tissues create the changes in pressure that result in ventilation. Pulmonary ventilation involves three different pressures (Figure 15-8): atmospheric pressure, intrapulmonary (intraalveolar) pressure, and intrapleural pressure.

Atmospheric pressure is the pressure of the air outside the body. At sea level, this pressure is normally 760 mm Hg. **Intrapulmonary** (in-trah-PUL-mon-air-ee) **pressure**, also called **intraalveolar** (in-trah-al-VEE-oh-lar) **pressure**, is the pressure inside the alveoli of the lungs. When the lungs are at rest, between breaths, this pressure equals atmospheric pressure. The intrapulmonary pressure varies as the thoracic cavity changes size with each breath, and it is responsible for air moving into and out of the lungs. When intrapulmonary pressure is less than atmospheric pressure, air flows into the lungs. When it is greater than atmospheric pressure, air flows out of the lungs.

Intrapleural (in-trah-PLOO-ral) **pressure** is the pressure within the pleural cavity, between the visceral and parietal pleurae (see Figure 15-8, *A*). This pressure also changes with each breath, but under normal conditions it is slightly less than both the atmospheric pressure and the intrapulmonary pressure. It represents a partial vacuum or negative pressure and is an important factor in keeping the lungs inflated. Because the pressure inside the lungs is greater than the intrapleural pressure, the lungs always expand to fill the space and press against the thoracic wall. If the intrapleural pressure becomes greater than the intrapulmonary pressure, the lungs collapse and are nonfunctional.

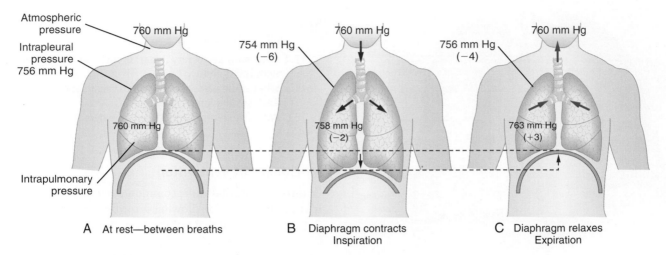

Atmospheric pressure 760 mm Hg

Intrapleural pressure 756 mm Hg

760 mm Hg

Intrapulmonary pressure

A   At rest—between breaths

760 mm Hg

754 mm Hg (−6)

758 mm Hg (−2)

B   Diaphragm contracts
Inspiration

760 mm Hg

756 mm Hg (−4)

763 mm Hg (+3)

C   Diaphragm relaxes
Expiration

Figure 15-8 *Pressures in pulmonary ventilation. (**A**) illustrates the lungs at rest. In inspiration (**B**), the intrapulmonary pressure is less than atmospheric pressure and air flows into the lungs. In expiration (**C**), intrapulmonary pressure is greater than atmospheric pressure and air flows out of the lungs. Intrapleural pressure is always less than either intrapulmonary or atmospheric pressure.*

QUICK APPLICATIONS

The accumulation of air in the pleural cavity is called **pneumothorax**. This condition can occur in pulmonary disease, such as emphysema, carcinoma, tuberculosis, or lung abscesses, when rupture of a lesion allows air to escape from the alveoli into the pleural cavity. It also may follow trauma in which the chest wall is perforated and atmospheric air enters the cavity. Air in the pleural cavity increases the intrapleural pressure and causes the lungs to collapse.

Inspiration

Inspiration, also called inhalation, is the process of taking air into the lungs. It is the active phase of ventilation because it is the result of muscle contraction. In normal, quiet breathing, the primary muscle involved in inspiration is the **diaphragm**, a dome-shaped muscle that separates the thoracic cavity from the abdominal cavity. When the diaphragm contracts, it drops, or becomes flatter, and increases the size (volume) of the thoracic cavity. When the diaphragm relaxes, the volume of the cavity decreases.

The parietal and visceral layers of the pleura tend to adhere to each other because there is an attraction between the water molecules in the serous fluid between the two layers. Because the two layers of pleura stick together, increasing the volume of the thoracic cavity causes the lungs to expand, or to increase in volume. The fact that the intrapleural pressure is less than the pressure within the lungs also contributes to lung expansion because the lungs enlarge (inflate) into the lower pressure region.

Between breathing cycles, when the lungs are at rest, the intrapulmonary pressure within the lungs is equal to the atmospheric pressure outside the body. Contraction of the diaphragm causes the thoracic volume to increase, and the intrapulmonary

pressure decreases below atmospheric pressure (Boyle's law). Air flows from the region of higher atmospheric pressure outside the body into the region of lower intrapulmonary pressure within the lungs (see Figure 15-8, *B*). Air continues to flow into the alveoli until intraalveolar pressure equals atmospheric pressure.

During labored breathing, the **external intercostal muscles** and other muscles of respiration work with the diaphragm to create a greater increase in the volume of the thoracic cavity. This results in a greater lung expansion and allows more air to flow into the lungs.

Expiration

Expiration, or **exhalation**, is the process of letting air out of the lungs during the breathing cycle. In normal, quiet breathing, it is a passive process involving the relaxation of respiratory muscles and the elastic recoil of tissues. Forceful expiration requires the active contraction of the **internal intercostal muscles**.

When the diaphragm and other muscles used in inspiration relax, the volume of the thoracic cavity decreases to its normal resting size. Following Boyle's law, this decrease in lung volume causes an increase in the intrapulmonary pressure (see Figure 15-8, *C*). Air now flows from the region of higher intrapulmonary pressure within the lungs to the region of lower atmospheric pressure outside the body until the two pressures are equal.

As air leaves the lungs during expiration, the alveoli become smaller. The interior surfaces of the alveoli are coated with a thin layer of fluid. The fluid molecules are attracted to each other (surface tension), which tends to cause the surfaces to adhere to each other. This makes it harder to inflate the lungs during inspiration and creates a tendency for the lungs to collapse. Normally this is prevented by a substance called **surfactant** (sir-FAK-tant), a lipoprotein substance that is produced by certain cells within the lung tissue and that reduces the attraction between the fluid molecules. Without surfactant, the alveoli collapse and become nonfunctional.

QUICK APPLICATIONS

Surfactant is not produced until the late stages of fetal life. Newborns that are born prematurely may not have enough surfactant, and the forces of surface tension collapse the alveoli. The newborn must reinflate the alveoli with each breath, which requires tremendous energy. The lack of surfactant accounts for many of the signs and symptoms of infant respiratory distress syndrome (IRDS). The condition is treated by using positive-pressure respirators that maintain pressure within the alveoli to keep them inflated.

Respiratory Volumes and Capacities

Under normal conditions, the average adult takes 12 to 15 breaths per minute. A breath is one complete respiratory cycle that consists of one inspiration and one expiration. The amount of air that is exchanged during one cycle varies with age, sex, size, and physical condition.

An instrument called a **spirometer** (spy-ROM-eh-ter) is used to measure the volume of air that moves into and out of the lungs, and the process of taking the measurements is called **spirometry**. Figure 15-9 illustrates a graphic record, called a **spirogram**, produced by a spirometer. Respiratory (pulmonary) volumes are an important aspect of pulmonary function testing because they can provide information about the physical condition of the lungs. The four respiratory volumes measured by spirometry are the **tidal volume, inspiratory reserve volume, expiratory reserve volume**, and **residual volume**. These are described, with their normal values, in Table 15-1 and are illustrated by the spirogram in Figure 15-9.

Respiratory capacity (pulmonary capacity) is the sum of two or more volumes. Four respiratory capacities that are measured are the **vital capacity, inspiratory capacity, functional residual capacity**, and **total lung capacity** (see Table 15-1 and Figure 15-9). In normal, healthy lungs, the vital capacity equals about 80% of the total lung capacity.

Factors such as age, sex, body build, and physical conditioning have an influence on lung volumes and capacities. Lungs usually reach their maximum capacity in early adulthood and then decline with age. Females generally have 20% to 25% less lung volume than males. Tall people tend to have greater lung capacity than short individuals. Slender people have greater capacity than do obese people. Physical conditioning can increase lung capacity as much as 40%. Muscular diseases and factors that reduce the elasticity of the lungs reduce the capacity.

 QUICK CHECK

15.7 Why does the lung collapse when air enters the pleural cavity?

15.8 What is the significance of surfactant in the lungs?

15.9 During a recent episode of pneumonia, Marie had fluid accumulate in her alveoli. Explain how this affected her vital capacity.

15.10 When the diaphragm contracts, what happens to intrapulmonary volume and pressure?

Basic Gas Laws and Respiration

The diffusion of gases from the alveoli to the blood (external respiration) and from the blood to the tissues (internal respiration) depends on two fundamental properties of gases. These are known as **Dalton's law of partial pressures** and **Henry's law**.

Properties of Gases

Dalton's Law of Partial Pressures

Dalton's law of partial pressures, or simply Dalton's law, states that the total pressure exerted by a mixture of gases is equal to the sum of the pressures exerted by each gas independently. If P represents pressure, then

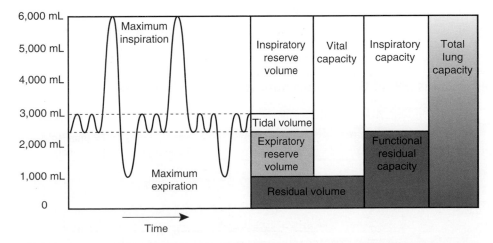

Figure 15-9 *Respiratory volumes and capacities. A respiratory capacity includes two or more respiratory volumes. Tidal volume is the amount of air that is inhaled and exhaled in one normal breathing cycle.*

TABLE 15-1 Respiratory Volumes and Capacities

Term	Abbreviation	Normal Value	Description
Lung Volumes			**The four separate components of total lung capacity**
Tidal volume	TV	500 ml	Amount of air that is inhaled and exhaled in a normal, quiet breathing cycle
Inspiratory reserve volume	IRV	3100 ml	Maximum amount of air that can be forcefully inhaled after a tidal inspiration
Expiratory reserve volume	ERV	1200 ml	Maximum amount of air that can be forcefully exhaled after a tidal espiration
Residual volume	RV	1200 ml	Amount of air that remains in lungs after a maximum expiration
Lung Capacities			**Measurements that are the sum of two or more lung volumes**
Vital capacity	VC	4800 ml	Maximum amount of air that can be exhaled after a maximum inspiration; equals TV + IRV + ERV
Inspiratory capacity	IC	3600 ml	Maximum amount of air that can be inhaled; equals TV + IRV
Functional residual capacity	FRC	2400 ml	Amount of air remaining in lungs after a tidal experience; equals RV + ERV
Total lung capacity	TLC	6000 ml	Amount of air in lungs after a maximum inspiration; equals RV + TV + IRV + ERV

$$P_{gas1} + P_{gas2} + P_{gas3} + P_{gas4} = P_{Total}$$

Further, the pressure exerted by each individual gas, its **partial pressure**, is proportional to its percentage in the total mixture. For example, if a gas mixture contains 75% nitrogen and 25% oxygen, and the total pressure is 160 mm Hg, then the partial pressure caused by nitrogen is 120 mm Hg (75% of 160) and the partial pressure resulting from oxygen is 40 mm Hg (25% of 160). Air is a mixture of gases, namely, nitrogen, oxygen, carbon dioxide, and water vapor. At sea level the total pressure is 760 mm Hg. Because the air is about 21% oxygen, the partial pressure of oxygen in the air is 159.6 mm Hg. Table 15-2 compares the composition of atmospheric air and alveolar air.

Henry's Law

According to Henry's law, when a mixture of gases is in contact with a liquid, each gas dissolves in the liquid in proportion to its own solubility and partial pressure. The greater the solubility, the more gas that will dissolve in the liquid. Of the atmospheric gases, carbon dioxide is the most soluble, oxygen is intermediate in solubility, and nitrogen is the least soluble.

More gas dissolves in a liquid if the partial pressure of the gas is greater. Nearly everyone is familiar with what happens to a can of soda if it is left open. It goes "flat." When the soda was made, carbon dioxide, under high pressure, dissolved in the liquid. When the can is opened, the pressure is reduced and the carbon dioxide "undissolves" and escapes.

TABLE 15-2 Partial Pressures (PP) of Gases in the Atmosphere and Alveolar Air*

	ATMOSPHERE			ALVEOLAR AIR		
Gas	Percentage	PP (mm Hg) (sea level)	PP (mm Hg) (6000 ft)	Percentage	PP (mm Hg) (sea level)	PP (mm Hg) (6000 ft)
Nitrogen	78.60%	567.0	479.0	74.9%	569	456
Oxygen	20.90%	159.0	127.0	13.7%	104	83
Carbon dioxide	0.04%	0.3	0.2	5.2%	40	32
Water vapor	0.46%	3.7	2.8	6.2%	47	38
Total	100.0%	760.0	609.0	100.0%	760	609

*These values are approximate and vary with the weather.

QUICK APPLICATIONS

Henry's law is familiar to deep-sea divers. As the diver descends, the total pressure around the diver increases; consequently, the total pressure of nitrogen increases. Normally, very little nitrogen dissolves in the blood, but the increased partial pressure makes more of it dissolve. If the diver ascends too rapidly, nitrogen gas comes out of solution and forms bubbles in body fluids. At first, the bubbles escape into the joints, which causes severe pain but is not particularly damaging. This condition is called the **bends** because the person bends over in pain. A serious situation arises when the nitrogen bubbles travel in the bloodstream where they may cause infarctions and cerebral damage.

External Respiration

External respiration is the exchange of oxygen and carbon dioxide between the air in the lungs and the blood in the surrounding capillaries. Oxygen diffuses from the alveoli of the lungs into the blood, and carbon dioxide diffuses from the blood into the air in the alveoli. The surfaces in the lungs where diffusion occurs constitute the **respiratory membrane**. Some diffusion takes place in the respiratory bronchioles, so these passages do contribute a small amount of surface area to the respiratory membrane.

The respiratory membrane consists of the layers that the gases must travel through to get into or out of the alveoli (Figure 15-10). These layers are:

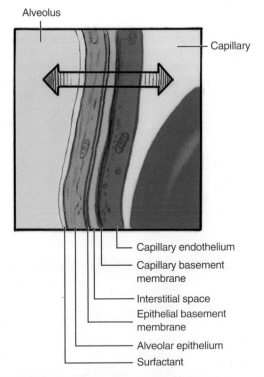

Figure 15-10 *Components of the respiratory membrane. Note the six extremely thin layers (each one cell thick or less) that constitute the respiratory membrane.*

- Thin layer of fluid that lines the alveolus
- Simple squamous epithelium in the alveolar wall
- Basement membrane of the epithelium
- Small interstitial space
- Basement membrane of capillary epithelium
- Simple squamous epithelium (endothelium) of the capillary wall

The rate of gaseous exchange across the respiratory membrane depends on the surface area of the membrane, the thickness of the membrane, the solubility of the gas, and the difference in partial pressure of the gas on the two sides of the membrane. Most of the approximately 70 square meters of surface area included in the respiratory membrane comes from the more than 300 million alveoli in healthy adult lungs. Diseases such as emphysema destroy alveolar walls and reduce the surface area of the respiratory membrane. This adversely affects the diffusion of oxygen and carbon dioxide. Normally, the membrane is very thin, but in patients with pulmonary edema fluids accumulate in the alveoli and the gases must diffuse through a fluid lining that is thicker than a normal lining. The diffusion rate decreases because the respiratory membrane is thicker. Increasing the breathing rate or increasing the volume of air exchanged with each breath increases the amount of oxygen in the alveoli and decreases the amount of carbon dioxide. This increases the differences in partial pressures on the two sides of the membrane and increases the rate of diffusion. Conversely, anything that reduces either the breathing rate or the volume also reduces the diffusion rate.

Internal Respiration

Internal respiration is the exchange of gases between the tissue cells and the blood in the tissue capillaries. After oxygen diffuses into the blood and carbon dioxide diffuses out of the blood in external respiration, the blood returns to the left side of the heart, which pumps it to the tissue capillaries. This blood has a higher concentration of oxygen and a lower concentration of carbon dioxide than the body tissue cells. The tissue cells use oxygen for metabolism and produce carbon dioxide in the process. This creates a lower oxygen content and a higher carbon dioxide content in the cells than in the capillaries. The concentration gradients that exist drive the oxygen from the capillaries into the tissue cells and the carbon dioxide from the tissue cells into the capillaries. Between the tissue cells and the capillaries, both gases pass through the interstitial fluid. External respiration and internal respiration are illustrated in Figure 15-11.

QUICK CHECK

15.11 Emphysema, characterized by shortness of breath and inability to tolerate physical exertion, destroys alveolar walls. How does this affect external respiration?

15.12 If the atmospheric pressure is 650 mm Hg and the air is 21% oxygen and 79% nitrogen, what is the partial pressure of oxygen?

15.13 What are the six layers of the respiratory membrane?

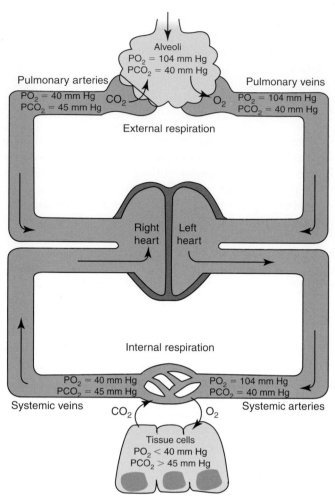

Figure 15-11 *External and internal respiration. In external respiration, oxygen enters the blood and carbon dioxide enters the alveolus. In internal respiration, oxygen leaves the blood and enters tissue cells, and carbon dioxide enters the blood.*

Transport of Gases

The blood transports the respiratory gases, oxygen and carbon dioxide, between the lungs and tissue cells. Erythrocytes (red blood cells) have the major role in transporting oxygen. Plasma has the major role in transporting carbon dioxide.

Oxygen Transport

After oxygen diffuses across the respiratory membrane from the alveolus into the capillary, it first dissolves in the plasma. About 3% of the oxygen remains in the plasma as a dissolved gas and is transported this way. The remainder quickly diffuses from the plasma into the red blood cells where it combines with the heme portion of the hemoglobin molecules to form a compound called **oxyhemoglobin** (ahk-see-HEE-moh-gloh-bin). Because the oxygen is bound to hemoglobin, this increases the amount of oxygen in a given amount of blood without increasing the partial pressure of oxygen. About 97% of the oxygen is transported as oxyhemoglobin. The reaction between oxygen and hemoglobin (Hb), which occurs in the lungs where oxygen content (partial pressure) is high, is called

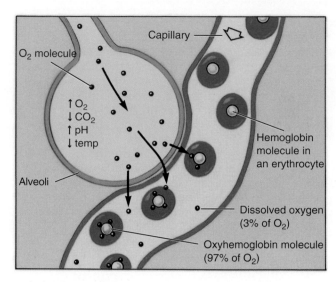

Figure 15-12 *Loading of oxygen in the lungs. A small amount of oxygen dissolves in the plasma. The remaining 97% combines with hemoglobin to form oxyhemoglobin.*

loading (Figure 15-12). The conditions in the lungs that promote loading of oxygen are high oxygen content, low carbon dioxide content, elevated pH, and decreased temperature.

The bonds between oxygen and hemoglobin are relatively unstable and are reversible. When the blood reaches the tissue capillaries, where the oxygen content (partial pressure) is low, the bonds break and oxygen is released to the tissues. This is called **unloading** (Figure 15-13). Not all the oxygen is released to the tissue cells in unloading. Only about 25% of the oxygen in the blood is delivered to the tissue cells. The other 75% remains attached to hemoglobin. This means that oxygen-poor, or deoxygenated, blood is carrying 75% of its maximum oxygen load on the return trip from the tissue cells to the lungs.

Several factors influence the unloading of oxygen in the tissues. More oxygen is released from oxyhemoglobin when oxygen levels (partial pressure) are low, carbon dioxide levels

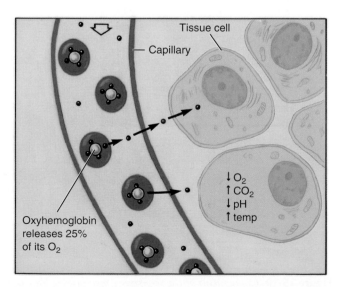

Figure 15-13 *Unloading of oxygen in the tissues. Oxyhemoglobin releases 25% of its oxygen, which then diffuses into the tissue cells.*

(partial pressure) are increased, temperature is increased, and pH is more acidic. Cells use oxygen for metabolism. They produce carbon dioxide and heat as byproducts of metabolism. The carbon dioxide reacts with water to form carbonic acid, which makes the cellular environment more acidic. Cells that are metabolically active, such as skeletal muscle, create an environment that favors the unloading of oxygen from oxyhemoglobin.

QUICK APPLICATIONS

Carbon monoxide poisoning is one of the most common types of gas poisoning. Most cases are accidental due to poor ventilation in areas where carbon fuels are burned. It is present in the exhaust of gasoline and diesel engines, in the smoke from gas and wood fires, and in the manufactured gas such as propane and natural gas that are used in homes. Carbon monoxide has no odor so its presence is not readily detected. When the gas is inhaled, it combines with hemoglobin to form carboxyhemoglobin and takes the place of oxygen in the RBCs. Unlike the bonds between oxygen and hemoglobin, which are relatively weak so that the oxygen is readily released in the tissues, the bonds in carboxyhemoglobin are very strong and the carbon monoxide is not easily released. When the hemoglobin is combined with carbon monoxide, it cannot transport oxygen and tissues are deprived of their normal oxygen supply. Central nervous system damage and asphyxiation may result. In the case of carbon monoxide poisoning, prevention is the best cure. Adequate ventilation and proper maintenance when using appliances and devices that burn coal and other carbon derivative are common sense preventive actions. Carbon monoxide detectors are available and these provide an early warning system if they are used appropriately.

Carbon Dioxide Transport

Carbon dioxide, which is a byproduct of cellular metabolism, diffuses from the tissue cells into the blood in the capillaries. The blood transports the carbon dioxide to the lungs by three mechanisms (Figure 15-14): dissolved in the plasma, combined with hemoglobin, and as part of bicarbonate ions. When carbon dioxide diffuses from the tissue cells into the blood, about 7% of it dissolves in the plasma. In the lungs, where carbon dioxide levels are low, the carbon dioxide leaves the plasma and diffuses into the alveoli. It is removed from the body in the exhaled air.

Approximately 23% of the carbon dioxide passes through the plasma, diffuses into the red blood cells, and combines with hemoglobin. Carbon dioxide combines with the protein portion of the hemoglobin molecule to form **carbaminohemoglobin** (kar-bah-meen-oh-HEE-moh-gloh-bin). Because oxygen and carbon dioxide react with different parts of the molecule, hemoglobin can carry both at the same time. In the lungs, this reaction reverses and carbon dioxide detaches from the hemoglobin. The carbon dioxide diffuses out of the red blood cell into the plasma and then diffuses from the plasma into the alveoli and is exhaled (Figure 15-15).

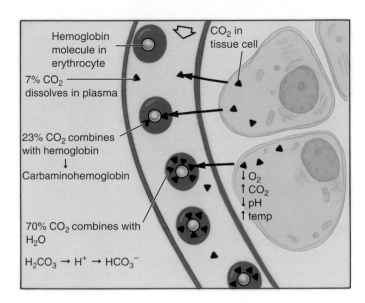

Figure 15-14 *Carbon dioxide transport from the tissues to the lungs. About 7% of the carbon dioxide dissolves in the plasma, 23% combines with hemoglobin to form carbaminohemoglobin, and 70% combines with water and forms bicarbonate ions.*

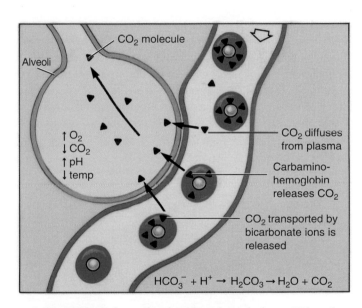

Figure 15-15 *Release of carbon dioxide in the lungs. When the blood transporting carbon dioxide gets to the capillaries in the lungs, carbon dioxide is released from the blood components, diffuses into the alveoli, and is exhaled.*

Most of the carbon dioxide, approximately 70%, is transported in the form of **bicarbonate ions**. The carbon dioxide diffuses into the red blood cell where it combines with water to form **carbonic acid**. An enzyme inside the red blood cell, **carbonic anhydrase**, speeds up this reaction so that it happens quite rapidly. The carbonic acid dissociates into hydrogen ions and bicarbonate ions. The carbon dioxide is contained within the bicarbonate ions:

Most of the hydrogen ions combine with hemoglobin, so they do not cause a dramatic and potentially dangerous drop in the pH of the blood. The bicarbonate ions diffuse out of the red blood cell into the plasma where they are transported to the lungs.

In the lungs, where the carbon dioxide content is relatively low, the above reactions reverse. The bicarbonate ions reenter the red blood cell and combine with hydrogen ions to form carbonic acid, which dissociates into water and carbon dioxide. The carbon dioxide diffuses into the alveoli and is exhaled (see Figure 15-15). An increase in carbon dioxide levels in the blood causes an increase in the number of hydrogen ions, which reduces the pH. Conversely, a decrease in carbon dioxide levels in the blood causes a decrease in the number of hydrogen ions, which makes the blood more alkaline and increases the pH.

> **QUICK CHECK**
>
> **15.14** In what two ways is oxygen transported in the blood?
> **15.15** In what three ways is carbon dioxide transported in the blood?
> **15.16** Patients with severe chronic lung disease often have elevated carbon dioxide levels in the blood. How does this affect the body's pH?

Regulation of Respiration

The normal breathing rate in adults averages between 12 and 20 breaths per minute. The rate is higher, up to 40 breaths per minute, in children. The basic rate is established by the respiratory center in the brain stem, but environmental conditions, both external and internal, and emotions induce variations in the rate.

Respiratory Center

Groups of neurons in the pons and medulla oblongata, regions of the brain stem, collectively make up the **respiratory center** (Figure 15-16). This center, which contains both **inspiratory** and **expiratory areas**, controls the rate and depth of breathing. The inspiratory area sends impulses along the **phrenic nerve** to the diaphragm and, for deeper breathing, along the **intercostal nerves** to the external intercostal muscles. When the respiratory center sends out impulses, the muscles contract and inspiration results. The inspiratory neurons fatigue quickly and quit sending impulses to the muscles. When the impulses cease, the muscles relax and expiration occurs. When more forceful expirations are necessary, the expiratory area sends impulses to the internal intercostal muscles. If the respiratory center in the brain stem is damaged, the impulses cease and breathing stops.

Factors That Influence Breathing

Even though the respiratory center establishes the basic rhythm of breathing, it is influenced by factors that cause variations in the rate and depth of breathing (Figure 15-17). Some conditions are detected by receptors that relay the information to the respiratory center. Other factors act on the respiratory center directly.

Chemoreceptors

Chemoreceptors in the medulla oblongata respiratory center are sensitive to changes in carbon dioxide and hydrogen ion concentrations in the blood and cerebrospinal fluid. They are not sensitive to changes in oxygen levels. If carbon dioxide and hydrogen ion concentrations increase, the receptors stimulate the respiratory center to increase the rate and depth of breathing. This

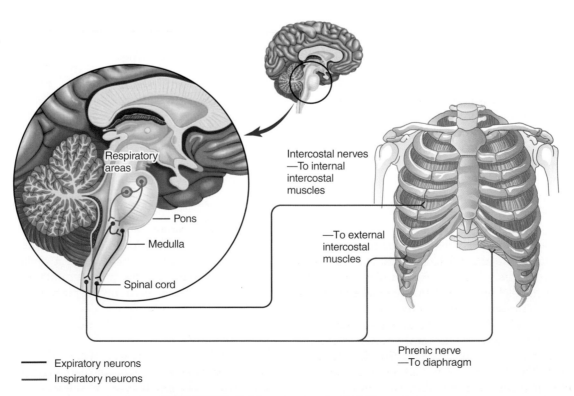

Figure 15-16 *Respiratory center in the brain stem.*

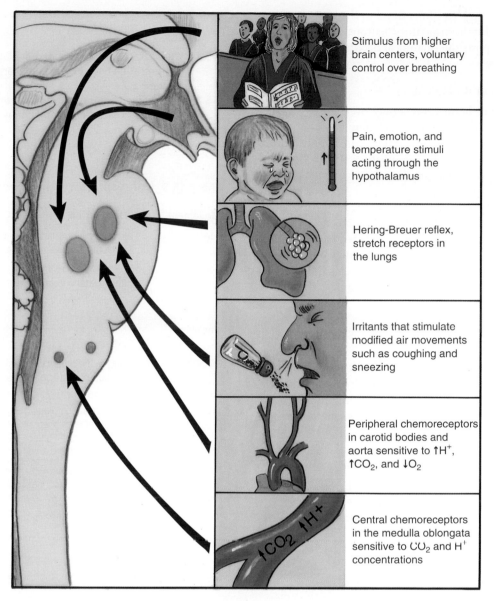

Stimulus from higher brain centers, voluntary control over breathing

Pain, emotion, and temperature stimuli acting through the hypothalamus

Hering-Breuer reflex, stretch receptors in the lungs

Irritants that stimulate modified air movements such as coughing and sneezing

Peripheral chemoreceptors in carotid bodies and aorta sensitive to $\uparrow H^+$, $\uparrow CO_2$, and $\downarrow O_2$

Central chemoreceptors in the medulla oblongata sensitive to CO_2 and H^+ concentrations

Figure 15-17 *Factors that influence breathing.*

decreases the concentrations back to normal levels. In contrast, low carbon dioxide and hydrogen ion levels decrease the rate and depth of breathing. Breathing may even stop for brief periods of time until concentrations increase to normal levels.

Receptors that are sensitive to changes in oxygen levels are located in the aortic and carotid bodies. These receptors are also sensitive to hydrogen ion and carbon dioxide levels. The receptors send sensory impulses to the respiratory center, which responds by altering the rate and depth of breathing. A decrease in oxygen level is usually not a strong stimulus for breathing. The primary effect seems to be to make the receptors in the respiratory center more sensitive to changes in carbon dioxide levels. Blood oxygen levels become an important stimulus under conditions, such as those created by emphysema, that result in chronic high carbon dioxide concentrations. Oxygen deficiency may also become a stimulus for breathing when oxygen levels decrease, but carbon dioxide levels are also low or unchanged. Examples of this are sudden

exposure to high altitudes and cases of shock when blood pressure is alarmingly low.

QUICK APPLICATIONS
Patients with severe chronic lung disease often have elevated carbon dioxide levels in the blood and the respiratory drive comes from the receptors for low arterial oxygen concentration. If these patients are given too much oxygen, they may literally stop breathing because the stimulus to breathe (low oxygen) has been removed.

Stretch Receptors and the Hering-Breuer Reflex

Stretch receptors in the lungs initiate the Hering-Breuer reflex that prevents overinflation of the lungs. As the alveoli expand during inspiration, stretch receptors in the lungs are stimulated.

Impulses from the stretch receptors travel to the medulla oblongata, where they inhibit the inspiratory neurons and cause expiration. This reflex supports the rhythm of breathing by inhibiting extended inspiration.

Stimulus From Higher Brain Centers

Impulses from higher brain centers may override the respiratory center temporarily. These impulses may be either voluntary or involuntary; however, the voluntary controls are limited. If you try to voluntarily hold your breath, you can do so for only a limited time. When carbon dioxide levels reach a certain critical point, the impulses from the higher brain centers are ignored and the respiratory center resumes regular breathing.

Involuntary impulses from higher brain centers may stimulate rapid breathing in response to emotions such as anxiety or excitement. Chronic pain also may result in involuntary stimulation from the higher brain centers. In contrast, sudden pain or sudden cold may cause a gasp or a momentary cessation of breathing.

Temperature

An increase in body temperature (for example, the elevated body temperature that occurs during a fever or while performing strenuous exercise) increases the breathing rate. The increased body temperature is associated with increased metabolism, which uses more oxygen and generates more carbon dioxide. When body temperature decreases, metabolic rate diminishes and breathing rate also decreases.

Nonrespiratory Air Movements

In addition to the normal air movements that occur during breathing and that result in pulmonary ventilation, there are a number of modifications called **nonrespiratory air movements**. Some of these are reflexes that clear air passages, others are voluntary, and some express emotions (Table 15-3).

QUICK CHECK

15.17 What effect does an elevated hydrogen ion concentration have on the medullary respiratory centers?

15.18 What nerve carries impulses from the medullary respiratory center to the diaphragm?

TABLE 15-3 **Nonrespiratory Air Movements**

Movement	Description
Sneezing	Spasmodic contraction of expiratory muscles that forces air through nose and mouth
Coughing	Long inspiration followed by closure of glottis; then a strong expiration forces glottis open and sends a blast of air through upper respiratory tract
Sighing	Long inspiration followed by a shorter but forceful expiration
Hiccuping	Spasmodic contraction of diaphragm followed by sudden closure of glottis to produce a sharp sound
Crying	An inspiration followed by many short expirations; glottis remains open and vocal cords vibrate; usually accompanied by tears and characteristic facial expressions
Laughing	Same basic movements as crying but facial expressions differ; may be indistinguishable from crying
Yawning	A deep inspiration through a widely opened mouth

 FROM THE PHARMACY

There are many preparations available to treat the disorders and discomforts associated with the respiratory system. Everyone is familiar with over-the-counter medications for colds, coughs, and sinusitis, but there are times when prescriptions are needed for severe acute and chronic conditions. Three groups of medications are presented here.

Antitussives suppress the cough reflex center in the medulla. Side effects include dry respiratory secretions, drowsiness, and constipation. Common nonnarcotic drugs are benzonatate (Tessalon Perles) and diphenhydramine hydrochloride (Benadryl).

Mucolytics break down mucus and promote coughing to remove mucus from the trachea, bronchi, and lungs. Common drugs are acetylcysteine (Mucomyst) and guaifenesin (Glycotuss).

Bronchodilators expand the lumina of the air passages for better air flow and are primarily used to treat chronic respiratory diseases such as asthma. Figure 15-14 illustrates the pathogenesis of asthma. Antiasthmatic medications are aimed at keeping the airway open and fall roughly into four groups.

1. Sympathetic agonists stimulate the sympathetic response to decrease mucus secretions and relax bronchial smooth muscle.
2. Xanthine derivatives, such as theophylline, inhibit the release of histamine and slow-reacting substance of anaphylaxis (SRS-A) from the mast cells.
3. Cromolyn sodium inhibits release of histamine from mast cells. This is not a bronchodilator, but is a prophylactic used to prevent the inflammatory response.
4. Corticosteroids prevent the release of enzymes that promote the inflammatory response, reduce mucus secretions, and reduce histamine release.

FOCUS ON AGING

Various harmful substances, including cigarette smoke, air pollution, and pathogens, continually bombard the respiratory system and take their toll. There is no way to avoid all of these irritants except to stop breathing! Some, like cigarette smoke, can be decreased, but others are inescapable. Because of the continual contact between the respiratory system and the environment, it is difficult to distinguish between the changes in the tissues of the breathing apparatus, including the lungs, that are due to aging and those that are the result of disease or other factors outside the body. Modifications in the lining of the respiratory tract probably are caused by environmental, rather than solely aging, factors. Long-term exposure to irritants results in deterioration of the cilia, which hinders their cleansing action and movement of mucus. As a consequence, the occurrence of emphysema and chronic bronchitis increases with age. Diminishing effectiveness of the immune system makes the elderly more susceptible to pneumonia and other microbial diseases. However, excluding external influences, there are changes that take place as a result of "normal" aging.

One of the most common signs of respiratory aging, in general, is when a person is unable to maintain the same level of physical activity that was experienced in younger years. This is a gradual decline and may not be noticeable until phrases such as, "I used to be able to..." become part of the conversation. The cardiovascular and muscular systems have an effect on endurance, and the skeletal system has an effect on thoracic volume, but the major change is a decreased ability of the respiratory system to acquire and deliver oxygen to the arterial blood.

The functional impairment in oxygen delivery is the result of structural changes that take place in the respiratory tissues. One type of structural change is a loss of elasticity. The cartilage in the walls of the trachea and bronchi undergoes a progressive calcification. Smooth muscle fibers in the bronchioles are replaced by fibrous tissue so that they are less able to stretch and contract. Modifications in lung tissue cause the alveoli to lose some of their elastic recoil. The cumulative effect of these changes is a gradual decrease in tidal volume and vital capacity and an increase in the volume of residual air in the lungs. Another type of structural change is deterioration of the walls between adjacent alveoli. This increases the size of each individual alveolus but reduces the total surface area of the respiratory membrane for diffusion of gases. A lower percentage of the oxygen in alveolar air is able to diffuse into the lung capillaries. These two types of structural changes result in the functional change of decreased ability to acquire and deliver oxygen to the arterial blood, which reduces the capacity for physical activity.

Representative Disorders of the
Respiratory System

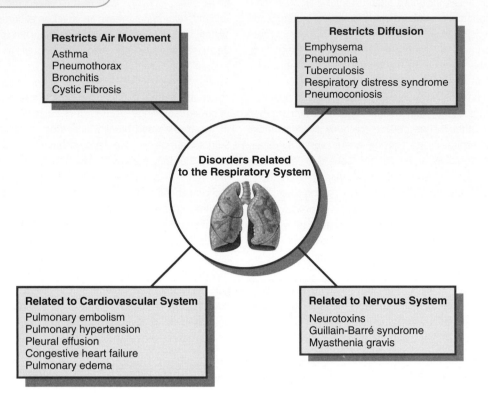

Restricts Air Movement
Asthma
Pneumothorax
Bronchitis
Cystic Fibrosis

Restricts Diffusion
Emphysema
Pneumonia
Tuberculosis
Respiratory distress syndrome
Pneumoconiosis

**Disorders Related
to the Respiratory System**

Related to Cardiovascular System
Pulmonary embolism
Pulmonary hypertension
Pleural effusion
Congestive heart failure
Pulmonary edema

Related to Nervous System
Neurotoxins
Guillain-Barré syndrome
Myasthenia gravis

Asthma (AZ-mah) Condition characterized by recurrent attacks of difficult breathing with wheezing because of spasmodic constriction of the bronchi; often caused by allergy to certain antigens

Bronchitis (brong-KYE-tis) Inflammation of one or more bronchi; may be acute or chronic; characterized by restricted air movements

Congestive heart failure (kahn-JES-tive HART FAIL-yer) Condition in which the heart's pumping ability is impaired and results in fluid accumulation in vessels and tissue spaces; various stages of difficult breathing occur as fluid accumulates in pulmonary vessels and lung tissue

Cystic fibrosis (SIS-tick fye-BROH-sis) Hereditary disorder associated with the accumulation of excessively thick and adhesive mucus, which obstructs bronchioles and restricts air movements

Emphysema (em-fih-SEE-mah) Lung disorder in which the terminal bronchioles become plugged with mucus; eventually there is a loss of elasticity in lung tissue, which makes breathing, especially expiration, difficult

Guillain-Barré syndrome (gee-YAN bah-RAY SIN-drohm) Relatively rare disorder that affects the peripheral nervous system, particularly the spinal nerves; characterized by muscular weakness or flaccid paralysis, usually beginning in the lower extremities and progressing upward; respiratory manifestations are the result of respiratory muscle involvement

Myasthenia gravis (mye-as-THEE-nee-ah GRAY-vis) Autoimmune disease, more common in females, that is characterized by weakness of skeletal muscles caused by an abnormality at the neuromuscular junction; ventilator deficiency may result when respiratory muscles are involved

Neurotoxins (new-row-TAHK-sins) Substances that are poisonous or destructive to nerve tissue; many affect acetylcholine activity at synapses and neuromuscular junctions, resulting in paralysis; respiratory involvement occurs when the diaphragm is affected

Pleural effusion (PLOO-rahl eh-FEW-shun) Accumulation of fluid in the space between the visceral and parietal layers of the pleura; may be due to trauma, inflammatory processes, or cardiac dysfunction

Pneumoconiosis (new-moh-koh-nee-OH-sis) General term for lung pathology that occurs after long-term inhalation of pollutants, characterized by chronic inflammation, infection, and bronchitis

Pneumonia (new-MOW-nee-ah) Inflammation of the lung usually caused by bacteria or viruses; inflammation and edema from the immune response cause the alveoli and terminal bronchioles to fill with fluid, which restricts ventilation and perfusion

Pneumothorax (new-moh-THOH-raks) Accumulation of air in the pleural space, resulting in collapse of the lung on the affected side

Pulmonary edema (PULL-moh-nair-ee eh-DEE-mah) Swelling and fluid in the air sacs and bronchioles; often caused by the inability of the heart to pump blood; the blood then backs up in the pulmonary blood vessels and fluid leaks into the alveoli and bronchioles

Pulmonary embolism (PULL-moh-nair-ee EM-boh-lizm) Obstruction of a pulmonary artery or one of its branches by an embolus, usually a blood clot, which interrupts blood supply and interferes with gas exchange

Pulmonary hypertension (PULL-moh-nair-ee hye-per TEN-shun) Excessive pressure in the pulmonary arteries, resulting in respiratory and cardiac dysfunction

Respiratory distress syndrome (RES-per-ah-tor-ee dis-TRES SIN-drohm) Condition resulting from abnormalities in the lung's surfactant that causes the alveoli to collapse; often occurs in infants when surfactant production fails to reach normal levels

Tuberculosis (too-ber-kyoo-LOH-sis) Infectious, inflammatory disease, usually in the lungs, that results in the formation of a fluid that restricts diffusion; original lesion may become dormant and then reactivate at a later time; may spread to other parts of the body and may become chronic

CHAPTER SUMMARY

Functions and Overview of Respiration

■ **Define five activities of the respiratory process.**

- The entire process of respiration includes ventilation, external respiration, transport of gases, internal respiration, and cellular respiration.

Ventilation

■ **Describe the structure and features of the upper respiratory tract.**

- The upper respiratory tract includes the nose, pharynx, and larynx.
- The nasal cavity opens to the outside through the external nares and into the pharynx through the internal nares. It is separated from the oral cavity by the palate. The frontal, maxillary, ethmoidal, and sphenoidal sinuses are air-filled cavities that open into the nasal cavity. Air is warmed, moistened, and filtered as it passes through the nasal cavity.
- The region of the pharynx is divided into the nasopharynx, oropharynx, and laryngopharynx. Pharyngeal tonsils are located in the wall of the nasopharynx, but the palatine and lingual tonsils are located in the oropharynx. The auditory tubes open into the nasopharynx. The opening from the oral cavity into the oropharynx is the fauces.
- The larynx is formed by nine cartilages that are connected to each other by muscles and ligaments. The three largest cartilages are the thyroid, cricoids, and epiglottis. There are two pairs of folds in the larynx. The upper pair are the vestibular folds. The lower pair are the true vocal cords. The opening between the vocal cords is the glottis.

■ **Describe the structures and features of the lower respiratory tract.**

- The lower respiratory tract consists of the trachea, bronchial tree, and lungs.
- The framework of the trachea is supported by a series of C-shaped pieces of hyaline cartilage. The mucous membrane that lines the trachea has goblet cells and cilia. The goblet cells secrete mucus that traps inhaled particles, and the cilia provide a cleansing action to remove the mucus with the particles.
- The trachea divides into right and left primary bronchi, which then divide into secondary (lobar) bronchi, and these into tertiary (segmental) bronchi. The branching pattern continues into smaller and smaller passageways until they terminate in tiny air sacs called alveoli.

■ **Describe the structure of the lungs, including shape, lobes, tissue, and membranes.**

- The right lung is shorter, broader, and has a greater volume than the left lung.
- The right lung is divided into three lobes; the left lung has two lobes.
- The left lung has an indentation, called the cardiac notch, for the apex of the heart.
- The lungs consist of the bronchial tree, except for the primary bronchi, which are outside the lungs. The alveoli of the lungs consist of simple squamous epithelium, which permits rapid diffusion of oxygen and carbon dioxide.
- The parietal pleura lines the wall of the thorax; the visceral pleura is firmly attached to the surface of the lung. The pleural cavity is the space between the two layers of pleura.

■ **Name and define three pressures involved in pulmonary ventilation and relate these pressures to the sequence of events that result in inspiration and expiration.**

- Air flows because of pressure differences between the atmosphere and the gases inside the lungs. Atmospheric pressure is the pressure of the air outside the body. Intraalveolar (interpulmonary) pressure is the pressure inside the alveoli of the lungs. Intrapleural pressure is the pressure within the pleural cavity, the space between the visceral and parietal pleura.
- During inspiration, the diaphragm contracts and the thoracic cavity increases in volume. An increase in thoracic volume decreases the pressure below atmospheric pressure so the air flows into the lungs.
- During expiration, the relaxation of the diaphragm and elastic recoil of tissues decrease the thoracic volume. The decrease in thoracic volume increases the intraalveolar pressure so that air flows out of the lungs. Surfactant reduces the surface tension inside the alveoli so they do not adhere to each other and collapse.

■ **Define four respiratory volumes and four respiratory capacities. State their average normal values for an adult male, and describe factors that influence them.**

- The four respiratory volumes measured by spirometry are tidal volume (500 ml), inspiratory reserve volume (3100 ml), expiratory reserve volume (1200 ml), and residual volume (1200 ml).
- A respiratory capacity is the sum of two or more volumes. Four respiratory capacities are the vital capacity (4800 ml), inspiratory capacity (3600 ml), functional residual capacity (2400 ml), and total lung capacity (6000 ml).

- Age, sex, body build, and physical conditioning have an influence on lung volumes and capacities.

Basic Gas Laws and Respiration

■ **Discuss factors that govern the diffusion of gases into and out of the blood.**

- Dalton's law of partial pressures states that the total pressure exerted by a mixture of gases is equal to the sum of the pressures exerted by each gas independently, and the partial pressure exerted by each gas is proportional to its percentage in the total mixture.
- Henry's law states that when a mixture of gases is in contact with a liquid, each gas dissolves in the liquid in proportion to its own solubility and partial pressure.

■ **Distinguish between external respiration and internal respiration.**

- External respiration is the exchange of gases between the lungs and the pulmonary capillaries. The surfaces in the lungs where diffusion occurs are called the respiratory membranes.
- The rate at which external respiration occurs varies with the surface area and thickness of the respiratory membrane, the solubility of the gas, and the difference in partial pressure of the gas on the two sides of the membrane.
- Internal respiration is the exchange of gases between the tissue cells and the blood in the tissue capillaries. Oxygen diffuses from the blood into the tissue cells and carbon dioxide diffuses from the tissue cells into the blood.

Transport of Gases

■ **Describe the methods of oxygen transport in the blood.**

- Approximately 3% of the oxygen is transported as a dissolved gas in the plasma. The remaining 97% is carried by hemoglobin molecules as oxyhemoglobin. Loading occurs in the lungs when oxygen combines with hemoglobin. It takes place when oxygen levels are high and carbon dioxide levels are low. Unloading occurs in the tissues when hemoglobin releases oxygen. It occurs when oxygen levels are low, carbon dioxide levels are high, temperature is increased, and pH is decreased.

■ **Describe three ways in which carbon dioxide is transported in the blood.**

- Approximately 7% of the carbon dioxide is transported as a gas dissolved in the plasma. Another 23% of the carbon dioxide combines with the protein portion of hemoglobin and is transported as carbaminohemoglobin. The remaining 70% of the carbon dioxide is transported as bicarbonate ions in the plasma.
- In the lungs, where carbon dioxide levels are relatively low, reactions occur that release the carbon dioxide from its transport forms. The carbon dioxide diffuses into the alveoli and is exhaled.

Regulation of Respiration

■ **Name two regions in the brain that make up the respiratory center and two nerves that carry impulses from the center.**

- The respiratory center includes groups of inspiratory and expiratory neurons in the medulla oblongata and pons.
- The inspiratory area sends impulses along the phrenic nerve to the diaphragm and along the intercostal nerves to the external intercostal muscles. When inspiratory impulses cease, the muscles relax, and expiration occurs. When more forceful expiration is necessary, the expiratory center sends impulses along the intercostal nerves to the internal intercostal muscles.

■ **Describe the role of chemoreceptors, stretch receptors, higher brain centers, and temperature in regulating breathing.**

- Central chemoreceptors in the medulla oblongata are sensitive to increases in carbon dioxide and hydrogen ion levels. Peripheral chemoreceptors in the aortic and carotid bodies detect decreases in oxygen levels, but this is not a strong stimulus for breathing.
- Stretch receptors in the lungs initiate the Hering-Breuer reflex that prevents overinflation of the lungs.
- Voluntary or involuntary impulses from the higher brain centers may override the respiratory center temporarily, but after a limited time the respiratory center resumes control.
- An increase in temperature increases the breathing rate.

CHAPTER QUIZ

Recall

Match the definitions on the left with the appropriate term on the right.

_____ 1. Tiny air sacs in the lungs
_____ 2. Exchange of gases between air and blood
_____ 3. Reduces surface tension inside alveoli
_____ 4. Supported by C-shaped hyaline cartilage
_____ 5. Serous membrane around the lungs
_____ 6. Primary muscle of inspiration
_____ 7. Air exchanged in normal, quiet breathing
_____ 8. Form in which most oxygen is transported
_____ 9. Form in which most carbon dioxide is transported
_____ 10. Initiate the Hering-Breuer reflex

A. Alveoli
B. Bicarbonate ions
C. Diaphragm
D. External respiration
E. Oxyhemoglobin
F. Pleura
G. Stretch receptors
H. Surfactant
I. Tidal volume
J. Trachea

Thought

1. The most superior portion of the pharynx:
   A. Is called the laryngopharynx
   B. Contains the palatine tonsils
   C. Has openings for the auditory (eustachian) tubes
   D. Opens into the oral cavity through the fauces

2. Which of the following does *not* pertain to the left lung?
   A. It is surrounded by a serous membrane.
   B. It has three lobes.
   C. It has a cardiac notch for the apex of the heart.
   D. It is longer and narrower than the right lung.

3. When intraalveolar pressure exceeds atmospheric pressure:
   A. The lung collapses
   B. Air is forced into the lungs
   C. Intrapleural pressure also becomes greater than atmospheric pressure
   D. Expiration occurs

4. Vital capacity is the:
   A. Sum of tidal volume, inspiratory reserve volume, and expiratory reserve volume
   B. Maximum amount of air that can be inhaled
   C. Sum of reserve volume, tidal volume, inspiratory reserve volume, and expiratory reserve volume
   D. Amount of air that remains in the lungs after a maximum expiration

5. Which of the following statements is *true* about the partial pressure of carbon dioxide?
   A. It is higher in the pulmonary veins than in the systemic arteries.
   B. It is higher in systemic arteries than in systemic veins.
   C. It is higher in systemic veins than in pulmonary arteries.
   D. It is higher in pulmonary arteries than in pulmonary veins.

Application

Why does a person who hyperventilates for several seconds experience a period of apnea before normal breathing resumes? Explain.

As a result of a street fight, Jason was rushed to the emergency department with a deep stab wound in the left thorax.

Examination revealed pneumothorax and a collapsed lung on the left side. Explain why the lung collapsed and why only the left lung was affected.

BUILDING YOUR MEDICAL VOCABULARY

Building a vocabulary is a cumulative process. As you progress through this book, you will use word parts, abbreviations, and clinical terms from previous chapters. Each chapter will present new word parts, abbreviations, and terms to add to your expanding vocabulary.

Word Parts and Combining Form with Definition and Examples

PART/COMBINING FORM	DEFINITION	EXAMPLE
a-	lack of	apnea: lack of breathing
alveol-	tiny cavity	alveolus: tiny air sacs in the lungs
anthrac-	coal	anthracosis: lung disease caused by inhalation of coal dust
atel-	imperfect or incomplete	atelectasis; collapsed or airless state of a portion of all of the lung; incomplete dilation of a lung
bronchi-	bronchi	bronchiectasis: chronic dilation of the lung bronchi associated with secondary infection
-capnia	carbon dioxide	hypercapnia: excess carbon dioxide in the blood
-coni-	dust	pneumoconiosis: group of lung diseases resulting from inhalation of industrial dust particles such as coal, silica, asbestos, and iron ore
cric-	ring	cricoid: laryngeal cartilage that resembles a ring
dys-	difficult	dyspnea: labored or difficult breathing
-ectasis	dilation	bronchiectasis: chronic dilation of the bronchi
eu-	good	eupnea: normal or good breathing

PART/COMBINING FORM	DEFINITION	EXAMPLE
laryng/o	larynx	laryngotomy: incision into the larynx
-ole	little, small	bronchiole: little bronchus; small branches of the bronchi
phon-	voice	phonation: voice production, utterance of vocal sounds
phren-	diaphragm	phrenic: pertaining to the diaphragm; the phrenic nerve innervates the diaphragm
-pnea	breathing	dyspnea: difficult or labored breathing
pneum/o	lung, air	pneumomycosis: any fungal disease of the lung
-ptysis	spitting	hemoptysis: coughing and spitting of blood as a result of bleeding in any part of the respiratory tract
pulmon-	lung	pulmonary: pertaining to the lungs
-rrhea	flow or discharge	rhinorrhea: free discharge of a thin nasal mucus
rhin/o	nose	rhinoplasty: plastic surgery of the nose
spir/o	breath	spirometer: instrument for measuring breathing ability
thyr-	shield	thyroid: resembling a shield; the thyroid cartilage of the larynx resembles a shield
-tion	act of, process of	respiration: act of breathing
ventilat-	to fan or blow	ventilator: apparatus that assists breathing, also called a respirator

Clinical Abbreviations

ABBREVIATION	MEANING
ARDS	acute respiratory distress syndrome
CF	cystic fibrosis
COPD	chronic obstructive pulmonary disease
CPAP	continuous positive airway pressure
DPT	diphtheria, pertussis, tetanus
IPPB	intermittent positive-pressure breathing
OSA	obstructive sleep apnea
PEEP	positive end-expiration pressure
PFT	pulmonary function test
RDS	respiratory distress syndrome
SIDS	sudden infant death syndrome
SOB	shortness of breath
TB	tuberculosis
URI	upper respiratory infection
VC	vital capacity

Clinical Terms

Aspiration (as-pih-RAY-shun) Process of removing substances by means of suction

Atelectasis (at-eh-LECK-tah-sis) Collapse of the alveoli; the lung is airless

Bronchogenic carcinoma (brong-koh-JEN-ik kar-sin-OH-mah) Cancerous tumors arising from a bronchus; lung cancer; smoking is the primary etiologic agent; spreads readily to the liver, brain, and bones

Bronchoscopy (brong-KAHS-koh-pee) Fiberoptic endoscope inserted into the bronchi for diagnosis, biopsy, or collection of specimens

Chronic obstructive pulmonary disease (COPD) (KRAHN-ik ob-STRUCK-tiv PULL-mon-air-ee dih-ZEEZ) Chronic condition of obstructed airflow through the bronchial tubes and lungs, usually accompanied by dyspnea; includes emphysema and chronic bronchitis

Coryza (koh-RYE-zah) Common cold, characterized by sneezing, nasal discharge, coughing, and malaise; caused by a rhinovirus

Croup (KROOP) Acute respiratory syndrome in infants and children, characterized by obstruction of the larynx, barking cough, and strained, high-pitched, noisy breathing

Hemoptysis (hee-MAHP-tih-sis) Spitting of blood as a result of bleeding from any part of the respiratory tract

Laryngitis (lair-in-JYE-tis) Inflammation of the mucous membranes of the larynx

Pertussis (per-TUSS-is) Whooping cough; a highly contagious bacterial infection of the pharynx, larynx, and trachea; characterized by explosive coughing spasms ending in a "whooping" sound

Pharyngitis (fair-in-JYE-tis) Inflammation of the mucous membranes of the pharynx; sore throat

Phrenitis (freh-NYE-tis) Inflammation of the diaphragm

Pneumonectomy (new-moh-NECK-toh-mee) Surgical removal of all or part of a lung, such as a lobe; removal of a lobe is also called lobectomy

Rhinitis (rye-NYE-tis) Inflammation of the nasal mucosa

Rhinoplasty (RYE-noh-plas-tee) Plastic surgery on the nose; medical term for a "nose job"

Smoker's respiratory syndrome (SMOH-kers reh-SPY-rah-tor-ee SIN-drohm) Group of respiratory symptoms seen in smokers; includes coughing, wheezing, vocal hoarseness, pharyngitis, dyspnea, and susceptibility to respiratory infections

Thoracocentesis (thor-ah-koh-sen-TEE-sis) Surgical procedure through the chest wall into the pleural cavity to remove fluid

Tracheotomy (tray-kee-AHT-oh-mee) Surgical incision into the trachea

VOCABULARY QUIZ

Use word parts you have learned to form words that have the following definitions.

1. Lack of breathing _____

2. Resembles a ring _____

3. Surgical repair of the nose _____

4. Dilation of bronchi _____

5. Rapid breathing _____

Using the definitions of word parts you have learned, define the following words.

6. Hemoptysis _____

7. Thyroid _____

8. Rhinitis _____

9. Dyspnea _____

10. Pneumoconiosis _____

Match each of the following definitions with the correct word.

_____ 11. Discharge from the nose

_____ 12. Presence of tiny cavities

_____ 13. Process of voice production

_____ 14. Condition caused by inhaling coal dust

_____ 15. Surgical removal of a lung

A. Alveoli
B. Anthracosis
C. Phonation
D. Pulmonectomy
E. Rhinorrhea

16. What is the general term for "lung pathology that occurs after long-term inhalation of pollutants"? _____

17. What is the clinical term for "collapse of the alveoli"? _____

18. What is the surgical procedure in which a lung is removed? _____

19. What is the meaning of the abbreviation COPD? _____

20. What is the meaning of the abbreviation URI? _____

Digestive System

16

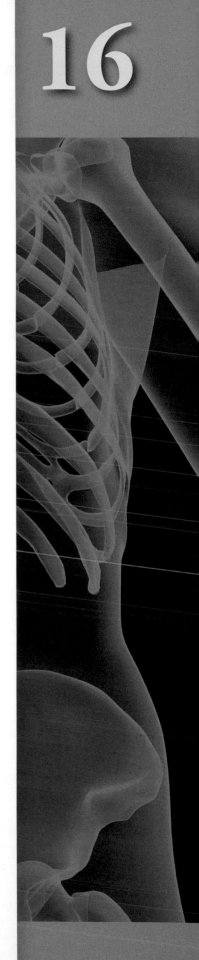

CHAPTER OBJECTIVES

Overview and Functions of the Digestive System
- Define *digestion*.
- List the components of the digestive tract.
- List the accessory organs of the digestive system.
- List six functions of the digestive system.

General Structure of the Digestive Tract
- Describe the general histology of the four layers, or tunics, in the wall of the digestive tract.

Components of the Digestive Tract
- Describe the features and functions of the oral cavity and teeth.
- Name and describe the location of the three major types of salivary glands and describe the functions of the saliva they produce.
- Describe the features and functions of the pharynx and esophagus.

- Describe the structure and histologic features of the stomach and its role in digestion.
- Describe the structure and histologic features of the small intestine and its role in digestion and absorption.
- Describe the structure, histologic features, and functions of the large intestine.

Accessory Organs of Digestion
- Describe the structure and functions of the liver, gallbladder, and pancreas.

Chemical Digestion
- Summarize carbohydrate, protein, and lipid digestion by stating the intermediate and final products and the enzymes that facilitate the digestive process.

Absorption
- Compare the absorption of simple sugars and amino acids with that of lipid-related molecules.

KEY TERMS

Absorption (ab-SOARP-shun) Passage of digestive end products from the gastrointestinal tract into the blood or lymph

Chylomicrons (kye-loh-MY-krons) Small fat droplets that are covered with a protein coat and found in the epithelial cells of the mucosa of the small intestine

Chyme (KYME) Semifluid mixture of food and gastric juice that leaves the stomach through the pyloric sphincter

Digestion (dye-JEST-shun) Process of converting food into chemical substances that can be utilized by the body

Mesentery (MEZ-en-tair-ee) Extensions of peritoneum that are associated with the intestine

Peristalsis (pair-ih-STALL-sis) Rhythmic contractions of the intestine that move food along the digestive tract

Plicae circulares (PLY-kee sir-kyoo-LAIR-eez) Circular folds in the mucosa and submucosa of the small intestine

Rugae (ROO-jee) Longitudinal folds in the mucosa of the stomach

Functional Relationships of the
Digestive System

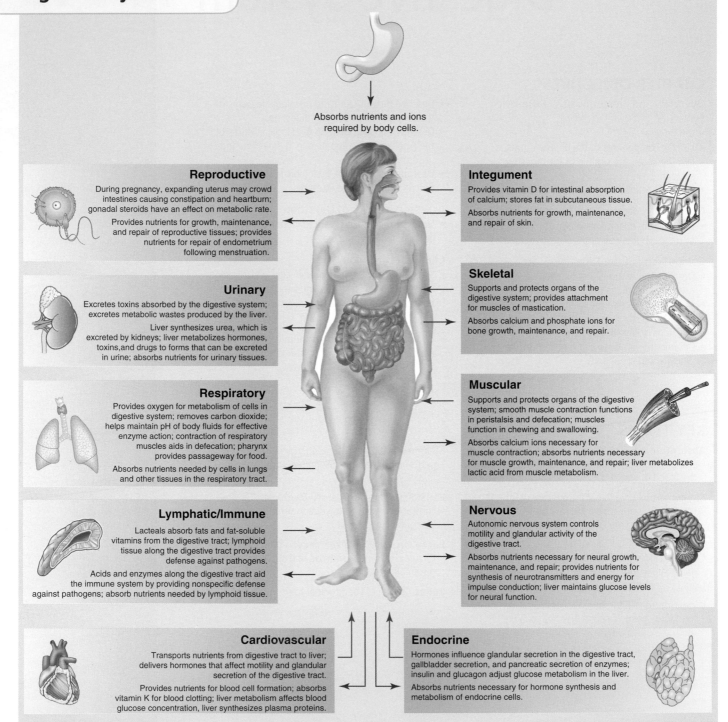

Absorbs nutrients and ions
required by body cells.

Reproductive

During pregnancy, expanding uterus may crowd
intestines causing constipation and heartburn;
gonadal steroids have an effect on metabolic rate.

Provides nutrients for growth, maintenance,
and repair of reproductive tissues; provides
nutrients for repair of endometrium
following menstruation.

Urinary

Excretes toxins absorbed by the digestive system;
excretes metabolic wastes produced by the liver.

Liver synthesizes urea, which is
excreted by kidneys; liver metabolizes hormones,
toxins, and drugs to forms that can be excreted
in urine; absorbs nutrients for urinary tissues.

Respiratory

Provides oxygen for metabolism of cells in
digestive system; removes carbon dioxide;
helps maintain pH of body fluids for effective
enzyme action; contraction of respiratory
muscles aids in defecation; pharynx
provides passageway for food.

Absorbs nutrients needed by cells in lungs
and other tissues in the respiratory tract.

Lymphatic/Immune

Lacteals absorb fats and fat-soluble
vitamins from the digestive tract; lymphoid
tissue along the digestive tract provides
defense against pathogens.

Acids and enzymes along the digestive tract aid
the immune system by providing nonspecific defense
against pathogens; absorb nutrients needed by lymphoid tissue.

Cardiovascular

Transports nutrients from digestive tract to liver;
delivers hormones that affect motility and glandular
secretion of the digestive tract.

Provides nutrients for blood cell formation; absorbs
vitamin K for blood clotting; liver metabolism affects blood
glucose concentration, liver synthesizes plasma proteins.

Integument

Provides vitamin D for intestinal absorption
of calcium; stores fat in subcutaneous tissue.

Absorbs nutrients for growth, maintenance,
and repair of skin.

Skeletal

Supports and protects organs of the
digestive system; provides attachment
for muscles of mastication.

Absorbs calcium and phosphate ions for
bone growth, maintenance, and repair.

Muscular

Supports and protects organs of the digestive
system; smooth muscle contraction functions
in peristalsis and defecation; muscles
function in chewing and swallowing.

Absorbs calcium ions necessary for
muscle contraction; absorbs nutrients necessary
for muscle growth, maintenance, and repair; liver metabolizes
lactic acid from muscle metabolism.

Nervous

Autonomic nervous system controls
motility and glandular activity of the
digestive tract.

Absorbs nutrients necessary for neural growth,
maintenance, and repair; provides nutrients for
synthesis of neurotransmitters and energy for
impulse conduction; liver maintains glucose levels
for neural function.

Endocrine

Hormones influence glandular secretion in the digestive tract,
gallbladder secretion, and pancreatic secretion of enzymes;
insulin and glucagon adjust glucose metabolism in the liver.

Absorbs nutrients necessary for hormone synthesis and
metabolism of endocrine cells.

← Gives to Digestive System
→ Receives from Digestive System

Overview and Functions of the Digestive System

The digestive system includes the **digestive tract** and its **accessory organs**, which process food into molecules that can be absorbed and used by the cells of the body. Food is broken down, bit by bit, until the molecules are small enough to be absorbed and the waste products are eliminated. The digestive tract, also called the **alimentary canal** or **gastrointestinal (GI) tract**, consists of a long continuous tube that extends from the mouth to the anus. It includes the mouth, pharynx, esophagus, stomach, small intestine, and large intestine (Figure 16-1). The tongue and teeth are accessory structures located in the mouth. The salivary glands, liver, gallbladder, and pancreas are not part of the digestive tract but are major accessory organs that secrete fluids into the digestive tract and have a role in digestion.

Food undergoes three types of processes in the body: digestion, absorption, and metabolism. Digestion and absorption occur in the digestive tract. After the nutrients are absorbed, they are available to all cells in the body and are used by the cells in metabolism. The process of metabolism is discussed in Chapter 17.

The digestive system prepares nutrients for use by the body's cells through six activities, or functions.

- **Ingestion**—The first activity of the digestive system is to take in food. This process, called ingestion, has to take place before anything else can happen.
- **Mechanical digestion**—The large pieces of food that are ingested have to be broken into smaller particles that can be acted upon by various enzymes. This is called mechanical digestion, which begins in the mouth with chewing, or **mastication** (mas-tih-KAY-shun), and continues with churning and mixing actions in the stomach.
- **Chemical digestion**—The complex molecules of carbohydrates, proteins, and fats are transformed by chemical digestion into smaller molecules that can be absorbed and used by the cells. Chemical digestion, through a chemical process called **hydrolysis**, uses water to break apart the complex molecules. **Digestive enzymes** speed up the hydrolysis process, which is otherwise very slow.
- **Movements**—After ingestion and mastication, the food particles move from the mouth into the pharynx, and then into the esophagus. This movement is called **deglutition** (dee-gloo-TISH-un), or swallowing. **Mixing movements** occur in the stomach as a result of smooth muscle contraction. These repetitive contractions usually occur in small segments of the digestive tract and mix the food particles with enzymes and other fluids. The movements that propel the food particles through the digestive tract are called **peristalsis**. These are rhythmic waves of contractions that move the food particles through the various regions where mechanical and chemical digestion occur.
- **Absorption**—The simple molecules that are produced from chemical digestion pass through the lining in the small intestine into the blood or lymph capillaries. This is called absorption.
- **Elimination**—The food molecules that cannot be digested need to be eliminated from the body. The removal of indigestible wastes through the anus, in the form of feces, is **defecation** (def-eh-KAY-shun).

General Structure of the Digestive Tract

The long continuous tube that is the digestive tract is about 9 meters in length. It opens to the outside through the mouth at the proximal end and through the anus at the distal end. Although there are variations in each region, the basic structure of the wall is the same throughout the entire length of the tube. The wall of the digestive tract has four layers or **tunics** (Figure 16-2): mucosa, submucosa, muscular layer (muscularis), and serous layer or serosa.

The **mucosa**, or mucous membrane layer, is the innermost tunic of the wall. It lines the lumen of the digestive tract. The mucosa consists of epithelium, an underlying loose connective tissue layer called the lamina propria, and a thin layer of smooth muscle called the muscularis mucosa. In certain regions, the mucosa develops folds that increase the surface area. Specialized cells in the mucosa secrete mucus, digestive enzymes, and hormones. Ducts from other glands pass through the mucosa to the lumen. In the mouth and anus,

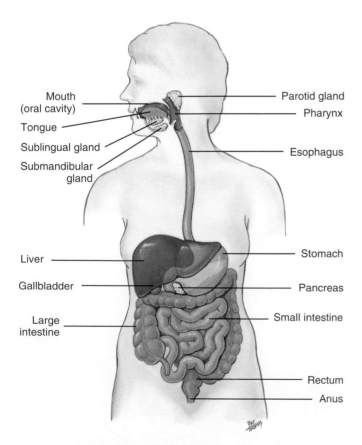

Mouth (oral cavity)
Tongue
Sublingual gland
Submandibular gland
Parotid gland
Pharynx
Esophagus
Liver
Gallbladder
Large intestine
Stomach
Pancreas
Small intestine
Rectum
Anus

Figure 16-1 *Organs of the digestive system.*

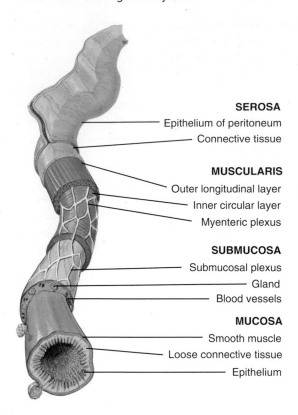

SEROSA
— Epithelium of peritoneum
— Connective tissue

MUSCULARIS
— Outer longitudinal layer
— Inner circular layer
— Myenteric plexus

SUBMUCOSA
— Submucosal plexus
— Gland
— Blood vessels

MUCOSA
— Smooth muscle
— Loose connective tissue
— Epithelium

Figure 16-2 *Basic histology of the digestive tract. Progressing from inner to outer, the tissue layers of the digestive tract are the mucosa, submucosa, muscularis, and serosa (adventitia if above the diaphragm).*

where thickness for protection against abrasion is needed, the epithelium is stratified squamous. The stomach and intestines have a thin, simple columnar epithelial layer to enhance secretion and absorption.

The **submucosa** is a thick layer of loose connective tissue that surrounds the mucosa. This layer also contains blood and lymphatic vessels, nerves, and some glands. Abundant blood vessels supply necessary nourishment to the surrounding tissues. Blood and lymph carry away absorbed nutrients that are the end products of digestion. The nerves in the submucosa form a network called the **submucosal plexus** (Meissner's plexus) that provides autonomic nerve impulses to the muscle layers of the digestive tract.

The **muscular layer** (labeled **muscularis** in Figure 16-2), which is superficial to the submucosa, consists of two layers of smooth muscle. The **inner circular layer** has fibers arranged in a circular manner around the circumference of the tube. When these muscles contract, the diameter of the tube is decreased. In the **outer longitudinal layer** the fibers extend lengthwise along the long axis of the tube. When these fibers contract, their length decreases and the tube shortens. There is a network of autonomic nerve fibers, called the **myenteric** (mye-en-TAIR-ik) **plexus** (Auerbach's plexus), between the circular and longitudinal muscle layers. The myenteric plexus and the submucosal plexus are important for controlling the movements and secretions of the digestive tract. In general, parasympathetic impulses stimulate movement and secretion and sympathetic impulses inhibit these activities.

The fourth and outermost layer in the wall of the digestive tract is called the **adventitia** if it is above the diaphragm and the **serosa** if it is below the diaphragm. The adventitia is composed of connective tissue. The serosa, which is below the diaphragm, has a layer of epithelium covering the connective tissue. It is actually the **visceral peritoneum** and secretes serous fluid for lubrication so that the abdominal organs move smoothly against each other without friction.

Components of the Digestive Tract

Mouth

The mouth, or **oral cavity**, is the first part of the digestive tract. It is adapted to ingest food, break it into small particles by mastication, and mix it with saliva. The lips, cheeks, and palate form the boundaries of the mouth. The oral cavity contains the teeth and tongue and receives the secretions from the salivary glands (Figure 16-3).

Lips and Cheeks

The lips and cheeks help hold food in the mouth and keep it in place for chewing. They are also used in the formation of words for speech.

The lips are folds of skeletal muscle covered with a thin transparent epithelium. Their reddish color is due to the many blood vessels underlying the epithelium. The lips contain numerous sensory receptors that are useful for judging the temperature and texture of foods.

The cheeks form the lateral boundaries of the oral cavity. The main component of the cheeks is the **buccinator muscle** and other muscles of facial expression. On the exterior, the muscles are covered by skin and subcutaneous tissue. Inside the oral cavity, the muscles are lined with a moist mucous membrane of stratified squamous epithelium. The multiple layers of epithelium provide protection against abrasion from food particles.

Palate

The **palate** is the roof of the oral cavity. It separates the oral cavity from the nasal cavity. The anterior portion, the **hard palate**, is supported by bone. The posterior portion,

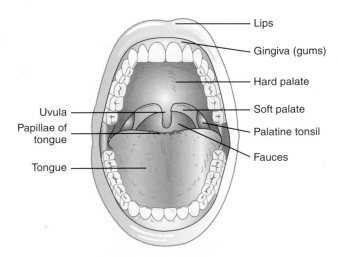

— Lips
— Gingiva (gums)
— Hard palate
— Soft palate
— Palatine tonsil
— Fauces

Uvula —
Papillae of tongue —

Tongue —

Figure 16-3 *Features of the oral cavity.*

the **soft palate**, is skeletal muscle and connective tissue. Posteriorly, the soft palate ends in a projection called the **uvula**. During swallowing, the soft palate and uvula move upward to direct food away from the nasal cavity and into the oropharynx.

QUICK APPLICATIONS

Cleft palate is a condition in which the bones in the hard palate do not fuse completely during prenatal development. This leaves an opening between the nasal and oral cavities. An infant with this problem has difficulty creating enough suction for proper feeding. Cleft palate usually can be corrected surgically.

Tongue

The largest and most movable organ in the oral cavity is the **tongue**. Most of the tongue consists of skeletal muscle. The major attachment for the tongue is the posterior region, or **root**, which is anchored to the hyoid bone. The anterior portion is relatively free but is connected to the floor of the mouth, in the midline, by a membranous fold of tissue called the **frenulum** (FREN-yoo-lum). An abnormally short lingual frenulum restricts tongue movements. The dorsal surface of the tongue is covered by tiny projections called **papillae**. In addition to providing friction for manipulating food in the mouth, the papillae contain the taste buds (see Chapter 9). Masses of lymphoid tissue, the **lingual tonsils**, are embedded in the posterior dorsal surface. These provide defense against bacteria that enter the mouth.

The muscles in the tongue allow it to manipulate the food in the mouth for mastication, move the food around to mix it with saliva, shape it into a ball-like mass called a bolus, and direct the bolus toward the pharynx for swallowing. It is a major sensory organ for taste and is one of the major organs used in speech. When tongue movements are restricted, the ability to properly enunciate words is difficult. Surgically loosening the frenulum may correct this problem.

Teeth

Two different sets of teeth develop in the mouth. The first set begins to appear at approximately 6 months of age and continues to develop until about 2 ½ years of age. This set, called the **primary** or **deciduous teeth**, contains 5 teeth in each quadrant for a total of 20 teeth. Figure 16-4, *A* illustrates the types of primary teeth. Starting at 6 years of age, the primary teeth begin to fall out and they are replaced by the **secondary** or **permanent teeth**. This set contains 8 teeth in each quadrant (16 in each jaw) for a total of 32 teeth. These teeth are illustrated in Figure 16-4, *B*.

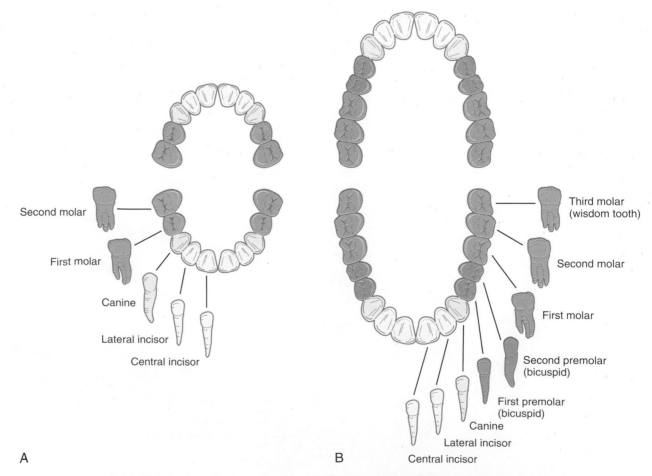

A

B

Figure 16-4 *Deciduous (**A**) and permanent (**B**) teeth. The deciduous dentition on the left has 20 teeth. The permanent dentition on the right contains 32 teeth.*

Different teeth are shaped to handle food in different ways. The **incisors** are chisel shaped and have sharp edges for biting food; **cuspids (canines)** are conical and have points for grasping and tearing food; **bicuspids (premolars)** and **molars** have flat surfaces with rounded projections for crushing and grinding. Note the location of each type of tooth in Figure 16-4. Table 16-1 summarizes the ages at which the different types of teeth erupt and when the deciduous teeth are shed. They are listed in the sequence in which they usually erupt. The times listed represent an average range and there may be exceptions.

TABLE 16-1 Ages at Which Teeth Erupt and Are Shed

Tooth Type	Age at Eruption	Age at Shedding
Deciduous Teeth		
Central incisors	6 to 8 months	5 to 7 years
Lateral incisors	8 to 10 months	6 to 8 years
First molars	12 to 16 months	9 to 11 years
Canines	16 to 20 months	8 to 11 years
Second molars	20 to 30 months	9 to 11 years
Permanent Teeth		
First molars	6 to 7 years	
Central incisors	6 to 8 years	
Lateral incisors	7 to 9 years	
Canines	9 to 10 years	
First premolars	9 to 11 years	
Second premolars	10 to 12 years	
Second molars	11 to 13 years	
Third molars	15 to 25 years	

Although the different types of teeth have different shapes, each one has three parts: a crown, a neck, and a root. The **crown** is the visible portion of the tooth, covered by **enamel**, and the **root** is the portion that is embedded in the alveolar processes (sockets) of the mandible and maxilla. The **neck**, a small region in which the crown and root meet, is adjacent to the **gingiva**, or **gum**. Gingivitis is an inflammation of the gums, which become sore and may bleed. Peridontal disease results when the condition is neglected and bacteria invade the bone around the teeth.

The central core of a tooth is the **pulp cavity**. It contains the **pulp**, which consists of connective tissue, blood vessels, and nerves. In the root, the pulp cavity is called the **root canal**. Nerves and blood vessels enter the root through an **apical foramen**. The pulp cavity is surrounded by **dentin**, which forms the bulk of the tooth. Dentin is a living cellular substance similar to bone. In the root, the dentin is surrounded by a thin layer of calcified connective tissue called **cementum**, which attaches the root to the **periodontal ligaments**. The ligaments have fibers that firmly anchor the root in the alveolar process. **Enamel**, the hardest substance in the body, surrounds the dentin in the crown of the tooth. Figure 16-5 shows a longitudinal section of a tooth and illustrates the major features.

✂ QUICK APPLICATIONS

Caries, or dental cavities, are caused by the demineralization of the teeth resulting from the action of bacteria that live in the mouth. The bacteria metabolize sugars in the mouth, producing acids that dissolve the calcium salts of the tooth. If the bacteria reach the pulp cavity, it is necessary to perform a **root canal** procedure. In this procedure, the pulp cavity with its nerve is destroyed, and the cavity is completely filled with a solid filling material.

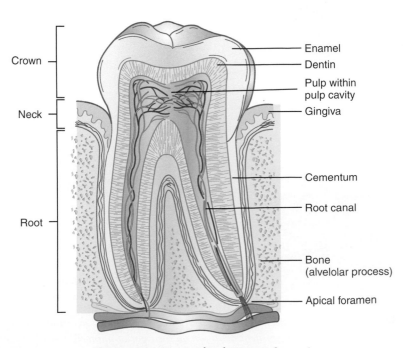

Figure 16-5 *Longitudinal section of a tooth.*

Salivary Glands

Three pairs of major **salivary glands** and numerous smaller ones secrete saliva into the oral cavity, where it is mixed with food during mastication (Figure 16-6). The **parotid glands** are the largest of the salivary glands. One gland is located on each side, between the skin and masseter muscle, just anterior and inferior to the ear. The duct from each parotid gland (parotid duct) opens into the oral cavity next to the second upper molar. **Submandibular glands** are located on the floor of the mouth along the medial surface of the mandible. The submandibular ducts open into the oral cavity by the lingual frenulum. Small **sublingual glands** are in the floor of the mouth, anterior to the submandibular glands, and under the tongue. They have numerous small ducts that open into the cavity along the frenulum.

Saliva contains water, mucus, and the enzyme **amylase**. Functions of saliva include the following:

- It has a cleansing action on the teeth.
- It moistens and lubricates food during mastication and swallowing.
- It dissolves certain molecules so that foods can be tasted.
- It begins the chemical digestion of starches through the action of amylase, which breaks down polysaccharides into disaccharides.

> ### *QUICK* CHECK
>
> **16.1** What two layers of the digestive tract contain a nerve plexus?
> **16.2** How many teeth are in a full deciduous set?
> **16.3** What part of a tooth is covered by enamel?
> **16.4** Which salivary glands are the largest?

Pharynx

The **pharynx** is a fibromuscular passageway that connects the nasal and oral cavities to the larynx and esophagus (see Figure 15-2). It serves both the respiratory and the digestive systems as a channel for air and food. The upper region, the **nasopharynx**, is posterior to the nasal cavity. It contains the pharyngeal tonsils, or adenoids, functions as a passageway for air, and has no function in the digestive system. The middle region posterior to the oral cavity is the **oropharynx**. This is the region food enters when it is swallowed. The opening from the oral cavity into the oropharynx is called the **fauces**. Masses of lymphoid tissue, the **palatine tonsils**, are near the fauces. The lower region, posterior to the larynx, is the **laryngopharynx**, or hypopharynx. The laryngopharynx opens into both the esophagus and the larynx.

Food is forced into the pharynx by the tongue. When food reaches the opening, sensory receptors around the fauces respond and initiate an involuntary swallowing reflex. This reflex action has several parts. The uvula is elevated to prevent food from entering the nasopharynx. The epiglottis drops downward to prevent food from entering the larynx and to direct the food into the esophagus. Peristaltic movements propel the food from the pharynx into the esophagus.

Esophagus

The **esophagus** is a collapsible muscular tube, about 25 cm long, that serves as a passageway for food between the pharynx and stomach. As it descends, it is posterior to the trachea and anterior to the vertebral column. It passes through an opening in the diaphragm, called the esophageal hiatus, and then empties into the stomach. The mucosa has glands that secrete mucus to keep the lining moist and well lubricated to ease the passage of food. The **lower esophageal sphincter**, sometimes called the cardiac sphincter, controls the movement of food between the esophagus and the stomach.

Stomach

The **stomach**, which receives food from the esophagus, is located in the upper left quadrant of the abdomen. Its capacity varies, but in the adult it averages about 1.5 liters, although in some individuals it may hold up to 4 liters.

Structure

The stomach is divided into the cardiac, fundic, body, and pyloric regions (Figure 16-7). The **cardiac region** is a small region around the opening from the esophagus. The **fundus**,

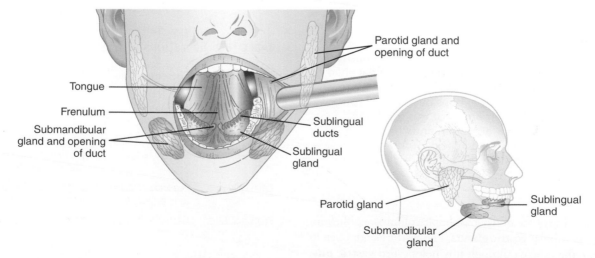

Figure 16-6 *Locations of the salivary glands.*

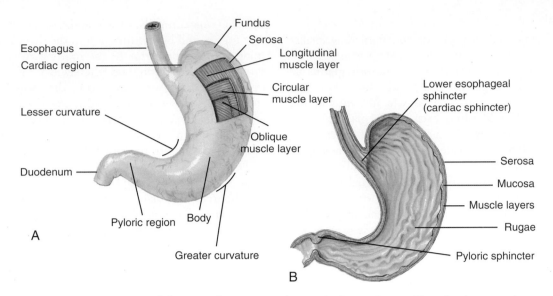

Figure 16-7 *Features of the stomach.* **A**, *External view.* **B**, *Internal view. Note the three muscle layers and the rugae.*

the most superior region, balloons above the cardiac region to form a temporary storage area. The **body** is the main portion of the stomach, which curves to the right and creates two curvatures. The **lesser curvature** is concave and is directed superiorly and to the right. On the opposite side, the convex **greater curvature** is directed inferiorly and to the left. As the body approaches the exit from the stomach, it narrows into the **pyloric region**. A circular band of smooth muscle forms the **pyloric sphincter**, which acts as a valve between the stomach and small intestine.

The muscular layer in the wall of the stomach has three layers, instead of two as found in other parts of the digestive tract. The additional third layer is innermost, located just under the submucosa, and is formed of oblique muscle fibers. The next muscle layer is the circular layer, and the outermost layer consists of longitudinal muscle fibers. The oblique layer adds another dimension to the mixing action of the stomach. When the stomach is empty, the mucosa and submucosa show longitudinal folds, called **rugae** (ROO-jee). These folds allow the stomach to expand, and as it fills the rugae become less apparent.

QUICK APPLICATIONS

A **hiatal hernia** occurs when a portion of the stomach protrudes into the thoracic cavity through a weakened area of the diaphragm. Frequently, it develops when a small region of the fundus balloons backward through the esophageal hiatus. Symptoms of this condition include pain in the upper abdomen and "heartburn" caused by the reflux of stomach acid into the esophagus, especially when the person is lying down.

Gastric Secretions

The mucosal lining of the stomach is simple columnar epithelium with numerous tubular **gastric glands**. The gastric glands open to the surface of the mucosa through tiny holes called **gastric pits**.

Four different types of cells make up the gastric glands: mucous cells, parietal cells, chief cells, and endocrine cells. The secretions of the exocrine gastric glands—composed of the mucous, parietal, and chief cells—make up the gastric juice. Approximately 2 to 3 liters of gastric juice are produced every day. The products of the endocrine cells are secreted directly into the bloodstream and are not a part of the gastric juice.

Mucous cells produce two types of **mucus** in the stomach. One type is thick and alkaline and forms a protective coating for the stomach lining. The other type is thin and watery. It mixes with the food and creates a fluid medium for chemical reactions. **Parietal cells** secrete **hydrochloric acid** and **intrinsic factor**. The hydrochloric acid kills bacteria and provides an acidic environment for the action of enzymes in the stomach. Intrinsic factor aids in the absorption of vitamin B_{12}. **Chief cells** secrete **pepsinogen**, an inactive form of the enzyme pepsin. Hydrochloric acid converts the inactive pepsinogen into the active enzyme pepsin, which begins the chemical digestion of proteins. The **endocrine cells** secrete the hormone **gastrin**, which functions in the regulation of gastric activity. Table 16-2 summarizes the various cells and secretions of the gastric glands.

TABLE 16-2 Secretions of Gastric Glands

Cell Type	Secretion	Function
Mucous cells	Mucus (thick, alkaline) Mucus (thin, watery)	Protects stomach lining Medium for chemical reactions
Parietal cells	Hydrochloric acid Intrinsic factor	Kills bacteria; activates pepsinogen Absorption of B_{12}
Chief cells	Pepsinogen (active form is pepsin)	Begins digestion of proteins into polypeptides
Endocrine cells	Gastrin (a hormone)	Stimulates gastric gland secretion

The churning action of the muscles in the stomach wall breaks the food particles of the bolus that was swallowed into smaller sizes and mixes them with the gastric juice. This produces a semifluid mixture called **chyme** (KYME), which leaves the stomach through the pyloric sphincter and enters the small intestine.

QUICK APPLICATIONS

Vomiting is the forceful ejection of the stomach contents through the mouth. It can be initiated by extreme stretching of the stomach or by the presence of irritants such as bacterial toxins, alcohol, spicy foods, and certain drugs. The vomiting action is a coordinated reflex controlled by the vomiting center of the medulla oblongata.

Regulation of Gastric Secretions

The regulation of gastric secretions is accomplished through neural and hormonal mechanisms. Gastric juice is produced all the time, but the amount varies subject to the regulatory factors. Regulation of gastric secretions may be divided into cephalic, gastric, and intestinal phases.

The **cephalic phase**, illustrated in Figure 16-8, *A*, begins when an individual thinks pleasant thoughts about food or sees, smells, or tastes food. This phase anticipates food and prepares the stomach to receive it. The sensory input stimulates centers in the medulla oblongata, which sends impulses along the parasympathetic neurons in the vagus nerve (cranial nerve X) to the stomach. These impulses cause an increase in the secretion of gastric juice. The impulses also increase the secretion of the hormone gastrin, which enters the blood and circulates back to the stomach to increase the activity of the gastric glands.

The **gastric phase**, which accounts for more than two thirds of the gastric juice secretion, begins when food reaches the stomach (Figure 16-8, *B*). The presence of food in the stomach and the distention of the stomach wall stimulate local reflexes that result in gastrin secretion. Gastrin, in turn, stimulates the secretion of gastric juice, which contains hydrochloric acid and pepsinogen. The hydrochloric acid acidifies the stomach contents and activates the pepsinogen into pepsin, which breaks down proteins into peptides. Stomach distention also sends signals to the brain, which responds by transmitting impulses back to the gastric glands along the vagus nerve.

The passage of chyme through the pyloric sphincter into the first part (duodenum) of the small intestine triggers the **intestinal phase** of regulation (Figure 16-8, *C*). Distention and the presence of acid chyme in the duodenum stimulate the secretion of intestinal hormones, which in turn inhibit gastric secretions. These factors also initiate responses in the medulla oblongata that inhibit gastric secretions. These hormonal and neural inhibitory responses help prevent excess acid chyme from entering the small intestine. When the chyme is neutralized and moves away from the duodenum, the inhibitory responses stop and gastric secretion is again stimulated. The intestinal phase regulates the entry of chyme into the small intestine.

Stomach Emptying

Peristalsis in the stomach pushes chyme toward the pyloric region. As the chyme accumulates, the pyloric sphincter relaxes and a small amount of chyme is pumped into the small intestine. The rate at which the stomach empties depends on the nature of the contents and the receptivity of the small intestine. The stomach is usually empty within 4 hours after a meal. Liquids tend to pass through the stomach quickly. Solids stay in the stomach until they are well mixed with gastric juice. Carbohydrates move through rather quickly, proteins take a little longer, and fatty foods may stay in the stomach as long as 4 to 6 hours. The presence of chyme in the first part of the small intestine decreases the receptivity and slows the emptying process.

If the stomach empties too slowly, the rate at which nutrients are digested and absorbed is diminished. There is also danger that the highly acid chyme will damage the stomach lining. On the other hand, if the stomach empties too quickly, before particles are thoroughly mixed with the gastric juice, the efficiency of digestion is reduced and the acid chyme may damage the intestinal mucosa. The neural and hormonal mechanisms that control gastric secretions also regulate stomach emptying.

QUICK CHECK

16.5 The trachea is supported by C-shaped cartilages with the open part of the "C" directed posteriorly. Why is this important for swallowing?

16.6 What keeps food from going back into the esophagus from the stomach?

16.7 What muscle relaxes to permit chyme to pass from the stomach into the small intestine?

16.8 What effect does the vagus nerve have on gastric secretions?

Small Intestine

The small intestine is about 2.5 centimeters in diameter and 6 meters long. It extends from the pyloric sphincter to the ileocecal valve, where it empties into the large intestine. The small intestine finishes the process of digestion, absorbs the nutrients, and passes the residue on to the large intestine. The liver, gallbladder, and pancreas are accessory organs of the digestive system that are closely associated with the small intestine. These are described later in this chapter.

Structure and Features of the Small Intestine

The small intestine follows the general structure of the digestive tract in that the wall has four layers: mucosa with simple columnar epithelium and goblet cells; submucosa; smooth muscle with inner circular and outer longitudinal layers; and serosa. The mucosa and submucosa have circular folds, called **plicae circulares** (PLY-kee sir-kyoo-LAIR-eez), which increase the surface area for absorption (Figure 16-9). Fingerlike extensions of the mucosa, called **villi**, project from the circular folds, and this further increases the surface area. Each villus surrounds a blood capillary network and a lymph capillary, or **lacteal**. These function in the absorption of nutrients. **Intestinal**

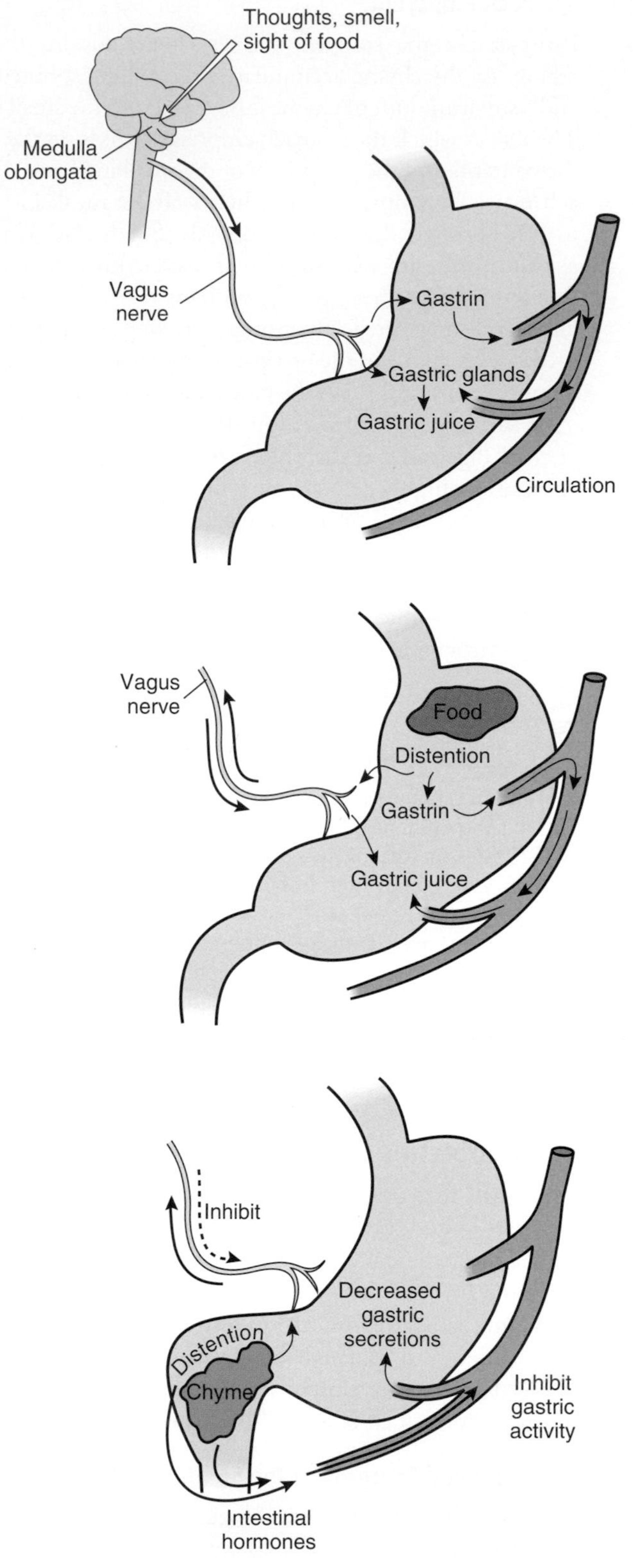

Thoughts, smell, sight of food

Medulla oblongata

Vagus nerve

Gastrin

Gastric glands

Gastric juice

Circulation

A Cephalic phase

- Prepares stomach to receive food
- Begins with thought, sight, smell of food
- Parasympathetic impulses from medulla oblongata travel along vagus nerve to stomach
- Impulses stimulate secretion of gastric juice and gastrin
- Gastrin enters blood and circulates back to stomach to stimulate gastric glands

Vagus nerve

Food

Distention

Gastrin

Gastric juice

B Gastric phase

- Begins when food reaches stomach
- Food and stomach distention stimulate secretion of gastrin
- Gastrin enters blood and circulates back to stomach to stimulate secretion of gastric juice
- Stomach distention also sends signals to brain and impulses retarn along vagus nerve to increase gastric juice

Inhibit

Decreased gastric secretions

Distention

Chyme

Inhibit gastric activity

Intestinal hormones

C Intestinal phase

- Regulates entry of chyme into duodenum
- Begins when chyme passes through pyloric valve
- Presence of chyme in duodenum stimulates intestinal hormones
- Intestinal hormones enter blood and circulate to inhibit gastric secretions
- Impulses from the medulla oblongata also inhibit gastric secretions

Figure 16-8 *Regulation of gastric secretions.* **A***, Cephalic phase.* **B***, Gastric phase.* **C***, Intestinal phase.*

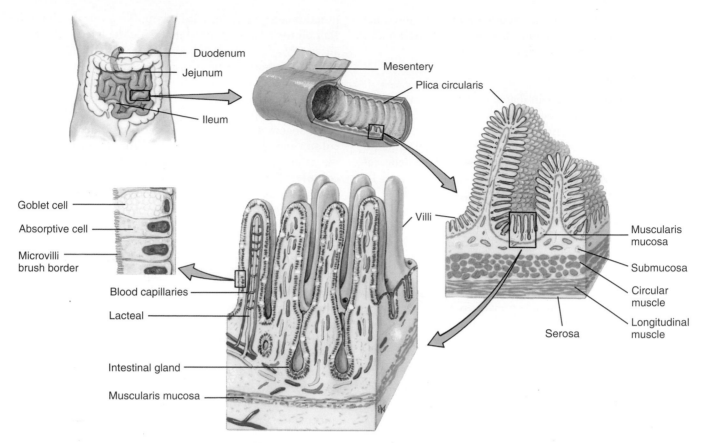

Figure 16-9 *Wall of the small intestine.*

glands extend downward between adjacent villi. The surface epithelium on the villi has tiny hairlike cytoplasmic extensions, called **microvilli**, that form a **brush border**, which again increase surface area.

Although the structure is similar throughout, the length of the small intestine is divided into three regions: duodenum, jejunum, and ileum. The **duodenum** is the first part and is about 25 centimeters long. It begins at the pyloric sphincter and continues in a C-shaped curve to the jejunum. The duodenum is behind the parietal peritoneum and is the most fixed portion of the small intestine. It receives the chyme from the stomach and secretions from the liver and pancreas. A distinguishing feature of the duodenum is the presence of mucous glands in the submucosa. These are called **duodenal** (doo-oh-DEE-nal) **glands** (Brunner's glands).

The second portion of the small intestine is the **jejunum**, which is about 2.5 meters long. This is continuous with the third portion, the **ileum**, which is about 3.5 meters long. There is no distinct separation between the jejunum and ileum. They are similar in structure, are mobile, and are suspended from the dorsal abdominal wall by a fold of peritoneum, called **mesentery**. There is a gradual decrease in the number and length of the villi and an increase in the number of goblet cells in the mucosa from the beginning of the jejunum to the terminal portion of the ileum.

Secretions of the Small Intestine

Intestinal glands secrete large amounts of watery fluid that is neutral or slightly alkaline in pH. It keeps the chyme in a liquid form and provides both an appropriate environment for the many chemical reactions of digestion and a fluid medium for the absorption of nutrients. The fluid is readily reabsorbed by the capillaries in the microvilli.

Goblet cells in the mucosa throughout the small intestine and duodenal glands in the submucosa of the duodenum secrete mucus. The alkaline mucus protects the intestinal wall from the acid chyme and digestive enzymes.

Digestive enzymes, which have a significant role in the final stages of chemical digestion, are located in the microvilli of the mucosal epithelial cells. These enzymes, called brush border enzymes, include the following: **peptidase**, which acts on segments of proteins called peptides; **maltase, sucrase,** and **lactase**, which act on disaccharides (double sugars); and an **intestinal lipase**, which acts on neutral fats. **Enterokinase** (enter-oh-KYE-nayz), although not actually a digestive enzyme, is produced by the mucosal epithelial cells. This enzyme activates a protein-splitting enzyme from the pancreas.

In addition to mucus and digestive enzymes, intestinal cells secrete at least two hormones—secretin and cholecystokinin. **Secretin** (see-KREE-tin) stimulates the pancreas to secrete a fluid that has a high bicarbonate ion concentration. This fluid helps to neutralize chyme so that the intestinal enzymes can function. **Cholecystokinin** (koh-lee-sis-toh-KYE-nin) stimulates the release of bile from the gallbladder and the secretion of digestive enzymes from the pancreas. It also inhibits gastric motility and secretions.

The most important factor for regulating secretions in the small intestine is the presence of chyme. This is largely a local reflex action in response to chemical and mechanical irritation from the chyme and in response to distention of the intestinal wall. This is a direct reflex action; thus the greater the amount of chyme, the greater the secretion.

Large Intestine

The **large intestine** is larger in diameter than the small intestine but is only about 1.5 meters long. It begins at the **ileocecal** (ill-ee-oh-SEE-kul) **junction**, where the ileum enters the large intestine, and ends at the anus. The ileocecal junction has a circular band of smooth muscle fibers called the **ileocecal sphincter**, and a valve called the **ileocecal valve**. Figure 16-10 illustrates the features of the large intestine.

Structure and Features of the Large Intestine

The wall of the large intestine has the same types of tissue that are found in other parts of the digestive tract, but there are some distinguishing characteristics. The mucosa has large numbers of goblet cells but does not have any villi. The longitudinal muscle layer, although present, is incomplete. The longitudinal muscle is limited to three distinct bands, called **teniae coli** (TEE-nee-aye KOH-lye), that run the entire length of the colon. Contraction of the teniae coli exerts pressure on the wall and creates a series of pouches, called **haustra** (HAWS-trah), along the colon. **Epiploic** (ep-ih-PLOH-ik) **appendages**, pieces of fat-filled connective tissue, are attached to the outer surface of the colon.

Regions of the Large Intestine

The large intestine consists of the cecum, colon, rectum, and anal canal (see Figure 16-10).

The **cecum** is the proximal portion of the large intestine. It is a blind pouch that extends inferiorly from the ileocecal junction. The **vermiform appendix** is attached to the cecum. In humans, the appendix has no function in digestion but does contain some lymphatic tissue.

The **colon** is the longest portion of the large intestine and is divided into ascending, transverse, descending, and sigmoid portions. The **ascending colon** begins at the ileocecal

junction and travels upward, along the posterior abdominal wall on the right side, until it reaches the liver. Here it turns anteriorly and to the left, becomes the **transverse colon**, and continues across the anterior abdomen toward the spleen on the left side. At this point, the colon turns sharply downward and travels inferiorly along the posterior abdominal wall as the **descending colon**. At the pelvic brim, the descending colon makes a variable S-shaped curve, called the **sigmoid colon**, and then becomes the rectum. The curve between the ascending and transverse portions is the **hepatic (right colonic) flexure**. The curve between the transverse and descending portions is the **splenic (left colonic) flexure**.

The **rectum** continues from the sigmoid colon to the anal canal and has a thick muscular layer. It follows the curvature of the sacrum and is firmly attached to it by connective tissue. The rectum ends about 5 centimeters below the tip of the coccyx, at the beginning of the anal canal.

The last 2 to 3 centimeters of the digestive tract is the **anal canal**, which continues from the rectum and opens to the outside at the **anus**. The mucosa of the rectum is folded to form longitudinal **anal columns**. The smooth muscle layer is thick and forms the **internal anal sphincter** at the superior end of the anal canal. This sphincter is under involuntary control. There is an **external anal sphincter** at the inferior end of the anal canal. This sphincter is composed of skeletal muscle and is under voluntary control.

Functions of the Large Intestine

Unlike the small intestine, the large intestine produces no digestive enzymes. Chemical digestion is completed in the small intestine before the chyme reaches the large intestine. There are no villi for the absorption of nutrients. This process is also accomplished in the small intestine. The primary functions of the large intestine are the absorption of fluid and electrolytes and the elimination of waste products.

The chyme that enters the large intestine contains materials that were not digested or absorbed in the small intestine—water, electrolytes, and bacteria. Some of the water and electrolytes

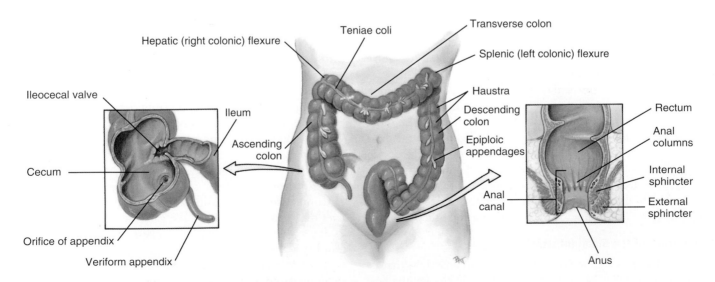

Figure 16-10 *Features of the large intestine.*

are absorbed in the cecum and ascending colon. Although the quantity is relatively small, this absorptive function of the large intestine is important in maintaining fluid balance in the body. The residue that remains from the chyme becomes the feces.

The large intestine has the same types of mixing and peristaltic movements as occur in other parts of the digestive tract, but they are more sluggish and occur less frequently. They are more likely to occur after a meal as a result of reflexes initiated in the small intestine. As the rectum fills with feces, the defecation reflex is triggered and the waste products are eliminated.

The only secretory product in the large intestine is mucus from the numerous goblet cells. The mucus protects the intestinal wall against abrasion and irritation from the chyme. It also helps hold the particles of fecal matter together. Because mucus is slightly alkaline, it helps control the pH of the material in the large intestine.

QUICK CHECK

16.9 What are the three features of the lining of the small intestine that increase the surface area for absorption?

16.10 Name two hormones that are secreted by cells of the small intestine.

16.11 What is another name for the right colonic flexure of the colon?

16.12 What is the term used for the three bands of longitudinal muscle fibers in the wall of the colon?

Accessory Organs of Digestion

The salivary glands, liver, gallbladder, and pancreas are not part of the digestive tract, but they have a role in digestive activities and are considered accessory organs. Because the salivary glands are so closely associated with the mouth and their primary function is performed in the mouth, they are considered part of the oral cavity. The liver and pancreas have functions in

addition to digestion, and the gallbladder is closely related to the liver; thus these three organs are described as separate accessory organs in this section.

Liver

The liver is a large reddish brown organ that is located primarily in the right hypochondriac and epigastric regions of the abdomen, just beneath the diaphragm. It is the largest gland in the body, weighs about 1.5 kilograms, and has numerous functions.

Structure and Features of the Liver

On the surface, the liver is divided into two major lobes and two minor lobes. On the anterior surface, the **falciform** (FALL-sih-form) **ligament**, a double fold of peritoneum that attaches the liver to the abdominal wall, separates the **right lobe** from the **left lobe** (Figure 16-11, *A*). Two additional small lobes are evident on the visceral surface (Figure 16-11, *B*). The **caudate lobe** is between the **ligamentum venosum** and the **inferior vena cava**. The **quadrate lobe** is between the **ligamentum teres** and the **gallbladder**. The **porta** is also on the visceral surface. The porta is where the **hepatic artery** and **hepatic portal vein** enter the liver and where the **hepatic ducts** exit.

The substance of the liver is divided into functional units called **liver lobules** (Figure 16-12). A liver lobule consists of **hepatocytes** (liver cells) that radiate outward from the **central vein** like spokes of a wheel. The central veins of adjacent lobules unite to form larger vessels, until they form the **hepatic veins**, which drain into the inferior vena cava. Tiny channels, called **bile canaliculi**, are interwoven with the liver cells and carry the bile that is produced by the hepatocytes toward the periphery of the lobule. Bile canaliculi merge to form larger **right and left hepatic ducts**. These two ducts combine to form the common hepatic duct, which transports bile out of the liver. The plates of hepatocytes are separated from each other by venous channels, called **sinusoids**, which carry blood from the periphery of the lobule toward the central vein. The sinusoids are lined with special phagocytic cells, called Kupffer cells, that remove

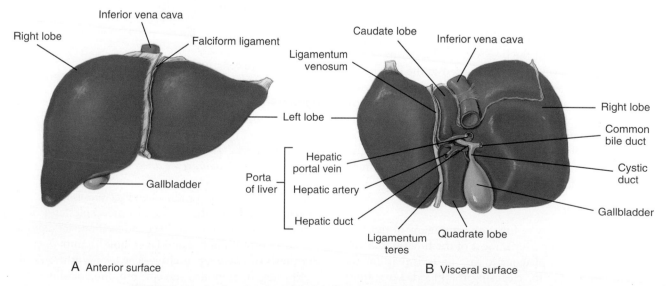

Figure 16-11 *Surface features of the liver.* ***A,*** *Anterior surface.* ***B,*** *Visceral surface.*

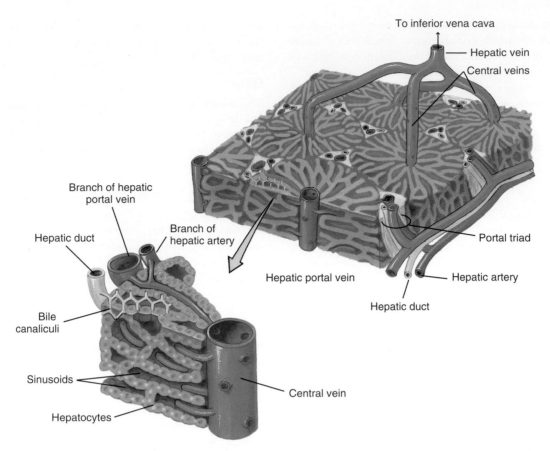

Figure 16-12 *Structure of liver lobules. Note the portal triads at the periphery of the lobules. A portal triad consists of branches of the portal vein, hepatic artery, and hepatic duct.*

foreign particles from the blood as it flows through the sinusoids. **Portal triads**, which consist of a branch of the hepatic portal vein, a branch of the hepatic artery, and a branch of a hepatic duct, are located around the periphery of the lobule.

Blood Supply to the Liver

The liver receives blood from two sources (Figure 16-13). Freshly oxygenated blood is brought to the liver by the **common hepatic artery**, a branch of the celiac trunk from the abdominal aorta. The common hepatic artery branches into smaller and smaller vessels until it forms the small hepatic arteries in the portal triads at the periphery of the liver lobules. Blood that is rich in nutrients from the digestive tract is carried to the liver by the **hepatic portal vein** (see Figure 12-20). The portal vein divides until it forms the small portal vein branches in the portal triads at the periphery of the liver lobules. Venous blood from the hepatic portal vein and arterial blood from the hepatic arteries mix together as the blood flows through the sinusoids toward the **central vein**. The central veins of the liver lobules merge to form larger **hepatic veins** that drain into the inferior vena cava.

Functions of the Liver

The liver has a wide variety of functions and many of these are vital to life. Hepatocytes perform most of the functions attributed to the liver, but the phagocytic Kupffer cells that line the sinusoids are responsible for cleansing the blood. Liver functions include the following:

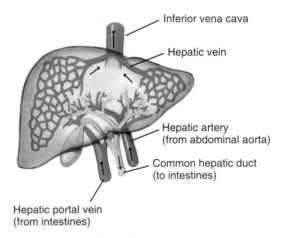

Figure 16-13 *Blood supply for the liver. Blood is brought to the liver by the hepatic artery and hepatic portal vein. After flowing through the sinusoids, the blood enters hepatic veins, which take it to the inferior vena cava. The common hepatic duct transports bile. Arrows indicate the direction of flow.*

Secretion—The liver produces and secretes bile.

Synthesis of bile salts—Bile salts are cholesterol derivatives that are produced in the liver and facilitate fat digestion and the absorption of fats and fat-soluble vitamins.

Synthesis of plasma protein—The liver synthesizes albumin, fibrinogen, globulins (except immunoglobulins), and clotting factors.

Storage—The liver stores glucose in the form of glycogen and also stores iron and vitamins A, B$_{12}$, D, E, and K.

Detoxification—The liver alters the chemical composition of toxic compounds, such as ammonia, to make them less harmful. It also changes the configuration of certain drugs, such as penicillin, and excretes them in the bile to remove them from the body.

Excretion—Hormones, drugs, cholesterol, and bile pigments from the breakdown of hemoglobin are excreted in the bile.

Carbohydrate metabolism—The liver has a major role in maintaining blood glucose levels. It removes excess glucose from the blood and converts it to glycogen for storage; it breaks down glycogen into glucose when more is needed; it converts noncarbohydrate molecules into glucose.

Lipid metabolism—The liver functions in the breakdown of fatty acids, in the synthesis of cholesterol and phospholipids, and in the conversion of excess carbohydrates and proteins into fats.

Protein metabolism—The liver converts certain amino acids into different amino acids as needed for protein synthesis. It also converts ammonia, produced in the breakdown of proteins, into urea, which is less toxic and can be excreted in the bile.

Filtering—The phagocytic **Kupffer cells** that line the sinusoids remove bacteria, damaged red blood cells, and other particles from the blood.

Bile

About 1 liter of bile, a yellowish-green fluid, is produced by liver cells each day. The bile enters the tiny bile canaliculi, which merge to form hepatic ducts that are a part of the portal triads. The largest hepatic duct, the common hepatic duct, leaves the liver at the porta (see Figure 16-13). Bile is slightly alkaline, with a pH of 7.6 to 8.6, so it helps neutralize the acid chyme. The main components of bile are water, bile salts, bile pigments, and cholesterol. The bile salts are useful secretory products of the liver, but the bile pigments and cholesterol are waste products excreted in the bile and eliminated from the body.

Although they are not enzymes, bile salts have a function in the digestion of fats. Bile salts act as **emulsifying agents** that break large fat globules into tiny fat droplets. This increases the surface area of the fat for more efficient enzyme action in fat digestion. Bile salts also facilitate the absorption of fat-soluble vitamins and the end products of fat digestion.

Bile pigments are produced in the breakdown of hemoglobin from damaged red blood cells. The hemoglobin is broken down into heme, which contains iron, and globin, a protein. The liver recycles the iron from the heme and the globin. The remainder of the heme portion is converted into bile pigments, which are normally excreted in the bile. Bile pigments are responsible for the color of the urine and feces. The principal bile pigment is **bilirubin** (bill-ih-ROO-bin).

Cholesterol is a product of lipid metabolism. Bile salts act on cholesterol to make it soluble; then it is excreted in the bile.

Gallbladder

The **gallbladder** is a pear-shaped sac that is attached to the visceral surface of the liver by the **cystic duct** (see Figure 16-11). The cystic duct joins the common hepatic duct from the liver to form the **common bile duct**, which empties into the duodenum. Simple columnar epithelium lines the gallbladder, and this is surrounded by smooth muscle covered with visceral peritoneum. When the muscle layer contracts, bile is ejected from the gallbladder into the cystic duct.

The principal functions of the gallbladder are to store and concentrate bile. Bile is continuously produced by the liver and then travels through the hepatic duct and common bile duct to the duodenum. There is a sphincter (sphincter of Oddi) where the common bile duct enters the duodenum. If the small intestine is empty, the sphincter is closed and the bile backs up through the cystic duct into the gallbladder for concentration and storage until it is needed. When chyme with fatty contents enters the duodenum, the hormone **cholecystokinin** (koh-lee-sis-toh-KYE-nin) stimulates the gallbladder to contract and the sphincter to open. This permits bile to flow from the gallbladder, through the cystic duct and common bile duct, and then into the duodenum.

Gallstones are formed in the gallbladder when cholesterol precipitates from the bile and hardens into stones because there is a lack of bile salts. Problems develop when the stones leave the gallbladder and lodge in the bile duct. This obstructs the flow of bile into the small intestine and interferes with fat absorption. Surgery may be required to remove the gallstones. Figure 16-14 shows a sonogram of a liver with multiple gallstones.

Pancreas

The **pancreas** is an elongated and flattened organ that is located along the posterior abdominal wall behind the parietal peritoneum (Figure 16-15). One end of the pancreas, the **head**, is on the right side within the curve of the duodenum; the other end, the **tail**, is on the left side next to the spleen.

The pancreas has both endocrine and exocrine functions. The endocrine portion consists of the scattered **islets of Langerhans**, which secrete the hormones insulin and glucagon into the blood. These hormones and their functions are discussed in Chapter 10. The exocrine portion is the major part of the gland. It consists of **pancreatic acinar** (AS-ih-nar) **cells** that secrete digestive enzymes into tiny ducts interwoven between the cells (Figure 16-16). The tiny ducts merge to form the main **pancreatic duct**, which extends the full length

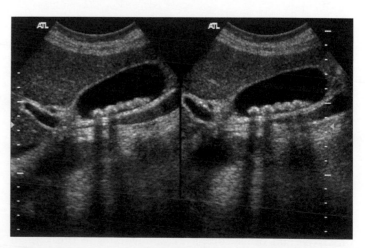

Figure 16-14 *Sonogram of a gallbladder with multiple gallstones. (Adapted from Applegate E: The Sectional Anatomy Learning System, Concepts, ed. 3. St. Louis, Elsevier/Saunders, 2010.)*

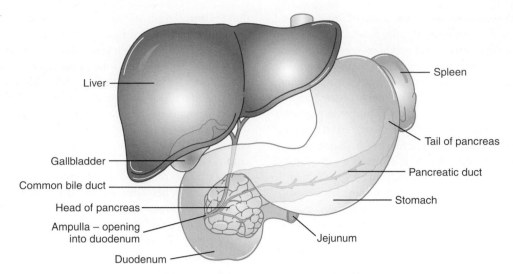

Figure 16-15 *Location and features of the pancreas. The head of the pancreas is enclosed by the duodenum and the tail is adjacent to the spleen.*

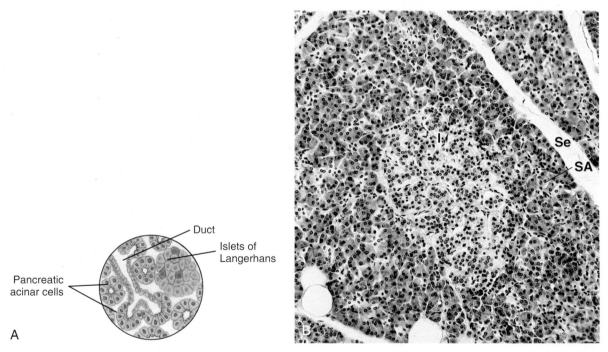

Figure 16-16 *Exocrine and endocrine cells in the pancreas. **A**, Drawing of acinar cells and islets of Langerhans in the pancreas. **B**, Photomicrograph of human pancreas showing numerous acinar cells and a large islet of Langerhans. (Adapted from Gartner and Hiatt: Color Textbook of Histology, ed. 3. Philadelphia, Elsevier/Saunders, 2007.)*

of the pancreas and empties into the duodenum. The pancreatic duct usually joins the common bile duct to form a single point of entry into the duodenum. Both ducts are controlled by the hepatopancreatic sphincter (sphincter of Oddi) (see Figure 16-15).

Pancreatic juice has a high concentration of bicarbonate ions and contains digestive enzymes that act on carbohydrates, proteins, and lipids. **Pancreatic amylase** acts on starch and other complex carbohydrates to break them into simpler sugars called disaccharides. Protein-splitting enzymes from the pancreas include **trypsin**, which breaks the proteins into shorter chains of amino acids, called peptides. Like other enzymes that act on proteins, trypsin is secreted in an inactive form, **trypsinogen**, which is activated by enterokinase when it reaches the duodenum. The pancreas also secretes peptidase enzymes that break peptides into amino acids. **Pancreatic lipase** breaks fats into fatty acids and monoglycerides.

Pancreatic secretion of digestive juice is regulated by the nervous system and by hormones. When parasympathetic impulses from the nervous system stimulate secretion of gastric juice, some impulses go to the pancreas and stimulate the secretion of pancreatic juice. When acid chyme enters the duodenum, the intestinal mucosa produces the hormone secretin, which travels in the blood to the pancreas. Secretin stimulates the pancreas to produce a fluid that has a high concentration of bicarbonate ions to neutralize the acids in the duodenum. Proteins and fats in the chyme stimulate the intestinal mucosa to secrete the hormone **cholecystokinin**, which also travels in the blood to the pancreas. This hormone stimulates the pancreas to produce a pancreatic juice that is rich in digestive enzymes. These digestive enzymes travel through the pancreatic duct to the duodenum, where they perform their actions.

QUICK CHECK

16.13 What are the components of a portal triad?
16.14 What two vessels carry blood to the liver?
16.15 What is the principal bile pigment?
16.16 What hormone causes contraction of the gallbladder?
16.17 How does secretin affect the pancreas?

Chemical Digestion

Chemical digestion breaks down large complex molecules into smaller molecules that can be absorbed by the cells of the intestinal mucosa. The reactions in chemical digestion are **hydrolysis** reactions, which use water to split molecules. These reactions proceed at a slow rate. The purpose of the various digestive enzymes is to speed up the hydrolysis reactions of chemical digestion. The enzymes do not alter the reactions; they just make them occur more rapidly. Table 16-3 reviews the hormones and digestive enzymes that have been discussed in previous sections of this chapter.

Carbohydrate Digestion

Starches and other complex carbohydrates are first broken down into disaccharides, or double sugars, by the action of salivary amylase and pancreatic amylase. The disaccharides **sucrose, maltose**, and **lactose** are the result of this stage of digestion. Sucrase, maltase, and lactase, enzymes from the small intestine, act on the disaccharides to convert them to monosaccharides, or simple sugars, that can be absorbed. The digestion of maltose yields two molecules of glucose; sucrose produces one molecule of glucose and one of fructose; lactose yields a molecule each of glucose and galactose. The end products of complete carbohydrate digestion are the monosaccharides **glucose, fructose**, and **galactose**. Glucose is also called dextrose and fructose is sometimes called levulose. Carbohydrate digestion is summarized in Figure 16-17.

Protein Digestion

The first digestive enzyme to act on proteins is pepsin in the stomach. Pepsin is secreted by the gastric glands in an inactive form, pepsinogen. This is activated by the action of hydrochloric acid. When chyme reaches the duodenum, trypsin from the

FROM THE PHARMACY

Some of the medications that are used to treat gastrointestinal (GI) disorders and diseases are introduced here. Many of the medications are available over-the-counter and the user should exercise caution not to overuse them. Overuse may lead to additional significant problems, such as failure to absorb other prescribed medications and nutrients. There are three major groups of GI medications: (1) those that help restore or maintain the mucosa of the GI tract; (2) those that affect the motility of the tract; and (3) laxatives that primarily affect motility in the colon.

Drugs that restore and maintain mucosa are used to treat gastric hyperacidity and peptic ulcer disease. This group includes antacids and histamine antagonists. **Antacids** neutralize the acid and raise the pH in the stomach. They are used to treat indigestion, to treat esophageal irritation and inflammation from reflux of stomach contents into the esophagus, and in the prevention and treatment of peptic ulcers. Some common examples include aluminum hydroxide (Amphojel), magnesium hydroxide (Milk of Magnesia), calcium carbonate (Tums), sodium bicarbonate (baking soda), and bismuth subsalicylate (Pepto-Bismol). **Histamine antagonists** reduce gastric acid secretions by inhibiting the release of histamine at the receptor sites in the stomach, and they also form a complex that adheres to the ulcer site. Examples include cimetidine (Tagamet) and famotidine (Pepcid).

Drugs that affect the motility of the GI tract include anticholinergics and antidiarrheals. **Anticholinergics** may be used to treat the abdominal discomfort associated with chronic GI diseases such as ulcers and colitis. They decrease gastric motility, reduce gastric acid secretions, decrease intestinal motility, and reduce pancreatic secretions. Examples include atropine sulfate and anisotropine (Valprin). **Antidiarrheals** reduce peristalsis in the intestines and reduce the amount of fluid in the intestinal contents. *Lactobacillus* (Bacid, Lactinex) is a nonprescription product that reestablishes intestinal flora and is used specifically for diarrhea caused by antibiotics. Loperamide (Imodium) is an example of an over-the-counter preparation that may be used for nonspecific diarrhea.

Laxatives change the consistency of the feces and increase the rate at which the feces move through the colon. Many people use them for self-treatment of constipation and they have a high rate of overuse, especially in the elderly. There are five categories of laxatives based on their mechanism of action: (1) bulk-forming (Metamucil); (2) fecal softeners (Colace); (3) hyperosmolar or saline (Milk of Magnesia and Fleet Enema); (4) lubricant (Purge, which is castor oil); and (5) stimulant or irritant (Feen-a-Mint and ExLax).

TABLE 16-3 Enzymes and Hormones of the Digestive System

Secretion	Source	Action
Enzymes		
Amylase	Salivary glands Pancreas	Digestion of complex carbohydrates into disaccharides
Pepsin	Stomach	Digestion of proteins into polypeptides
Sucrase Maltase Lactase	Small intestine	Digestion of disaccharides into glucose, fructose, and galactose
Peptidase	Small intestine Pancreas	Digestion of peptides into amino acids
Lipase	Small intestine Pancreas	Digestion of fats into monoglycerides and fatty acids
Enterokinase	Small intestine	Activates trypsinogen
Hormones		
Gastrin	Stomach	Stimulates activity of gastric glands
Cholecystokinin	Small intestine	Stimulates gallbladder to contract and release bile; stimulates pancreas to secrete digestive enzymes

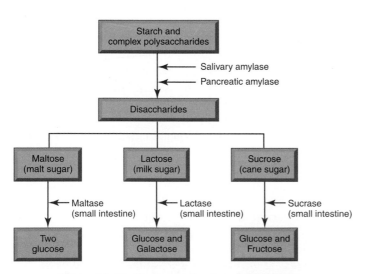

Figure 16-17 *Summary of carbohydrate digestion.*

pancreas acts on the proteins. Trypsin is secreted in the inactive form, trypsinogen, which is activated by enterokinase in the small intestine. Pepsin and trypsin break down proteins into shorter chains of amino acids called **peptides**. Peptidase enzymes from the small intestine and pancreas break the peptide bonds to produce **amino acids**. The amino acids are the absorbable end products of protein digestion. Protein digestion is summarized in Figure 16-18.

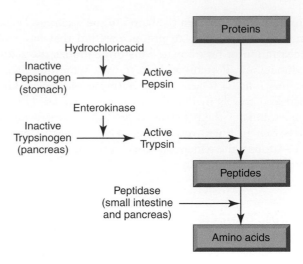

Figure 16-18 *Summary of protein digestion.*

Lipid Digestion

The small intestine is the only place in which lipid (fat) digestion occurs because the necessary enzymes are produced by the pancreas and enter the small intestine through the pancreatic duct. Triglycerides, glycerol molecules with three long-chain fatty acids attached, are the most abundant dietary fats. Fat molecules tend to attract each other to form large globules, which reduces the surface area for enzyme action. After the fats enter the duodenum, they are **emulsified** by bile. Emulsification does not break any chemical bonds, but it reduces the attraction between molecules so that they disperse. Pancreatic lipases act on the surfaces of the emulsified fat droplets. Lipase action breaks two fatty acid chains from the triglyceride molecules, yielding **monoglycerides** and **free fatty acids**. Lipid (fat) digestion is summarized in Figure 16-19.

Absorption

Approximately 10 liters of food, beverage, and secretions enter the digestive tract every day. Usually less than 1 liter enters the large intestine. The other 9 liters or more are absorbed in the small intestine. Absorption takes place along the entire length of the small intestine, but most of it occurs in the jejunum. By the time the chyme reaches the distal part of the ileum and

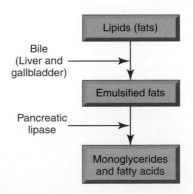

Figure 16-19 *Summary of lipid digestion.*

the large intestine, all that remains is some water, undigestible materials, and bacteria.

Water is absorbed throughout the length of the digestive tract by osmosis. Most water-soluble vitamins are easily absorbed by diffusion. Electrolytes are absorbed by diffusion and active transport. Most absorption of simple sugars and amino acids is by active transport across the cell membranes of the microvilli (Figure 16-20). This requires cellular energy in the form of adenosine triphosphate (ATP). After transport across the cell membranes, simple sugars (monosaccharides) and amino acids passively move into the blood capillaries in the villi to be transported to the liver in the hepatic portal vein.

Monoglycerides and free fatty acids, the end products of fat digestion, are coated with bile salts to form tiny droplets, called **micelles** (mye-SELZ), in the lumen of the small intestine. The micelles come into close contact with the microvilli, and because their contents (the monoglycerides and fatty acids) are lipid soluble, they diffuse across the cell membrane. The bile salts remain in the lumen to form additional micelles. Once inside the epithelial cells of the mucosa, a few short-chain fatty acids move directly into the blood capillaries of the villi. The monoglycerides and most of the fatty acids recombine to form triglycerides. These move to the Golgi apparatus where they combine with proteins to form **chylomicrons** (kye-loh-MYE-krons). The chylomicrons pass from the cells into the lacteals (lymph capillaries) in the villi. The mixture of lymph and digested fats is called **chyle**. Lymph carries the chylomicrons through the lymphatic vessels into the thoracic duct, which empties into the left subclavian vein. At this point, the products of fat digestion enter the blood. Fat-soluble vitamins (A, D, E, and K) are absorbed with fats in the micelles. Figure 16-20 and Table 16-4 summarizes absorption in the small intestine.

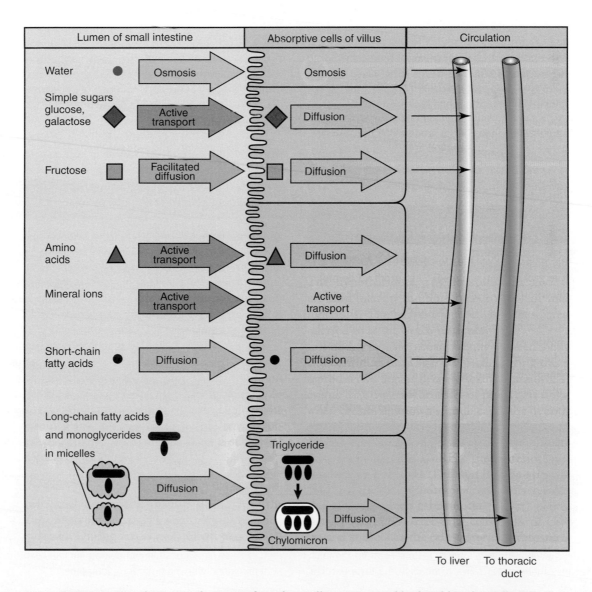

Figure 16-20 *Absorption of nutrients from the small intestine into blood and lymph capillaries.*

TABLE 16-4 Absorption of Nutrients in the Small Intestine

Nutrient	Absorptive Mechanism	Transport Route
Water	Osmosis	Blood capillaries in villi
Glucose and galactose	Active transport into epithelial cells, then diffusion into capillaries	Blood capillaries in villi to hepatic portal circulation
Fructose	Facilitated diffusion into epithelial cells, then simple diffusion into capillaries	Blood capillaries in villi to hepatic portal circulation
Amino acids	Active transport into epithelial cells, then simple diffusion into capillaries	Blood capillaries in villi to hepatic portal circulation
Short-chain fatty acids	Simple diffusion into epithelial cells, then into capillaries	Blood capillaries in villi to hepatic portal circulation
Long-chain fatty acids, monoglycerides, and fat-soluble vitamins	Combine with bile salts to form micelles, then simple diffusion into epithelial cells; within cells they form chylomicrons, which diffuse into lymph capillaries	Lymph capillaries (lacteals) of a villus
Electrolytes	Active transport and diffusion into epithelial cells, then into blood capillaries	Blood capillaries in villi to hepatic portal circulation
Water-soluble vitamins	Most are absorbed by diffusion into epithelial cells, then into blood capillaries; vitamin B_{12} requires intrinsic factor	Blood capillaries in villi to hepatic portal circulation

QUICK CHECK

16.18 What type of reaction is responsible for chemical digestion?

16.19 How does enterokinase affect chemical digestion?

16.20 How does cholecystokinin affect the pancreas?

16.21 How will obstruction of the common bile duct affect chemical digestion?

16.22 What membrane transport mechanism is responsible for the movement of glucose from the lumen of the small intestine into the cells of the villi?

QUICK APPLICATIONS

Cirrhosis is a chronic liver disease that may develop as a result of chronic alcoholism or severe hepatitis. The hepatic cells of the liver are destroyed and replaced with fibrous connective tissue so that the liver no longer functions properly. One consequence of cirrhosis is the buildup of bilirubin in the blood because it is not properly incorporated into the bile and excreted. The word *cirrhosis* means "orange-colored condition," which refers to the discoloration of the liver in this disease.

FOCUS ON AGING

Throughout life, the digestive system normally functions day after day with relatively few problems. There may be an occasional episode of gastrointestinal tract inflammation, called gastroenteritis, caused by eating something that "doesn't agree," by irritation from excessively spicy foods, or by eating food that is contaminated by bacteria or toxins. Appendicitis tends to be fairly common in teenagers, but the prevalence decreases with age because the opening into the appendix tends to become smaller and possibly eventually closes. Ulcers and gallbladder problems are associated with middle age, often considered to be the high-stress time of life. Most of the difficulties in the digestive system before "old age" are due to external problems rather than to structural changes within the system itself.

Structural changes in the digestive system occur as part of the normal aging process. These changes affect the overall operation of the system and may influence the nutritional state of the aging individual. In the mouth, teeth may become loose, as a result of periodontal disease, and have to be extracted. Because of dental problems, chewing may be uncomfortable. Salivary glands decrease their production of saliva, which reduces the salivary cleansing action and leads to a dry mouth (*xerostomia*). Thus, food is not adequately moistened for chewing and swallowing. Taste sensations diminish, partially because there is less saliva to dissolve the taste particles

and partially because there are fewer taste receptors. Loneliness and the problems in the oral cavity associated with aging may make eating a chore rather than a pleasure.

The mucosa in the stomach and intestines undergoes some atrophy with advancing age. In the stomach this may lead to a deficiency in hydrochloric acid and gastric juice for digestion. Pernicious anemia may develop because there is a lack of intrinsic factor from the gastric mucosa. In the small intestine, mucosal atrophy may lead to fewer enzymes and shorter villi; however, this does not appear to impair digestion and absorption in normal healthy people. The wall of the large intestine becomes thinner and weakens. This makes older people more susceptible to diverticulosis, in which the wall bulges outward to form balloon-like pockets. Constipation is a common complaint in the elderly; statistically, however, there seems to be no basis for it. This is more likely due to lifestyle and habits rather than to structural changes in the digestive system.

Although structural and functional changes take place in the digestive system as part of the aging process, digestion and absorption are not altered noticeably in healthy older persons. A balanced diet, exercise, and a positive outlook on life will keep the digestive system in good working order for a long time.

Representative Disorders of the
Digestive System

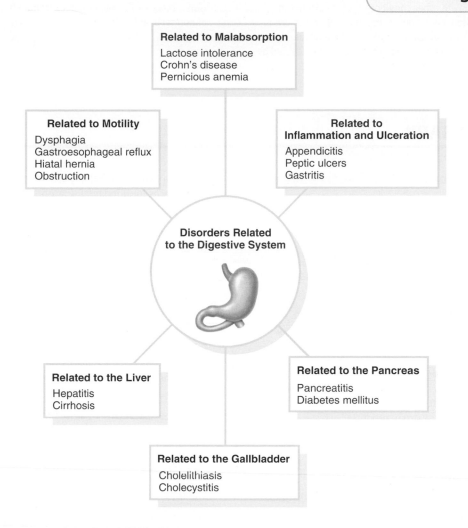

Related to Malabsorption
Lactose intolerance
Crohn's disease
Pernicious anemia

Related to Motility
Dysphagia
Gastroesophageal reflux
Hiatal hernia
Obstruction

Related to Inflammation and Ulceration
Appendicitis
Peptic ulcers
Gastritis

Disorders Related to the Digestive System

Related to the Liver
Hepatitis
Cirrhosis

Related to the Pancreas
Pancreatitis
Diabetes mellitus

Related to the Gallbladder
Cholelithiasis
Cholecystitis

Appendicitis (ah-pen-dih-SYS-tis) Inflammation of the vermiform appendix; usually requires surgical removal of the organ

Cholecystitis (kohl-ee-sis-TYE-tis) Inflammation of the gallbladder; sometimes caused by obstruction of the cystic duct by gallstones

Cholelithiasis (kohl-ee-lith-EYE-ah-sis) Formation or presence of gallstones

Cirrhosis (sih-ROH-sis) Chronic liver disease marked by degeneration of liver cells with eventual resistance to blood flow through the organ

Crohn's disease (KROHNZ dih-ZEEZ) Chronic, relapsing inflammation of the intestinal tract, usually the terminal ileum; genetic disorder that produces abdominal cramps and diarrhea with malabsorption of nutrients

Diabetes mellitus (di-ah-BEE-teez mell-EYE-tus) Disorder caused by a deficiency of insulin from the beta cells of the pancreatic islets; characterized by a disturbance in the utilization of blood glucose and manifested by polyuria, polyphagia, and polydipsia

Dysphagia (dis-FAY-jee-ah) Difficulty in swallowing because of inflammation, paralysis, or obstruction

Gastritis (gas-TRY-tis) Inflammation of the stomach lining; may be acute or chronic

Gastroesophageal reflux (gas-troh-ee-sahf-ah-JEE-al REE-flucks) Reflux of chyme from the stomach into the esophagus; irritation from the acid in the chyme may cause an inflammation of the esophagus

Hepatitis (hep-ah-TYE-tis) Inflammation of the liver; may be caused by alcoholism, parasites, and viruses

Hiatal hernia (hi-A-tahl HER-nee-ah) Protrusion of a structure, usually a portion of the stomach, through the opening in the diaphragm for the esophagus

Lactose intolerance (LACK-tose in-TAHL-er-ans) Inability to digest the sugar lactose because of a deficiency of the intestinal enzyme lactase; symptoms include abdominal pain and diarrhea after drinking milk

Obstruction (ob-STRUCK-shun) Any blockage of the digestive tract; pyloric obstruction is a narrowing of the opening between the stomach and duodenum; intestinal obstruction interferes with the flow of chyme through the tract; causes include hernia, telescoping of one part of the intestine into another, twisting of the intestine, diverticulosis, tumors, and loss of peristaltic motor activity

Pancreatitis (pan-kree-ah-TYE-tis) Inflammation of the pancreas resulting from autodigestion of pancreatic tissue by its own enzymes; associated with other conditions such as alcoholism, biliary tract obstruction, peptic ulcers, trauma, and certain drugs

Peptic ulcers (PEP-tick UL-ser) Deterioration of the mucous membrane lining of esophagus, stomach, or duodenum caused by the action of the acid in gastric juice

Pernicious anemia (per-NISH-us ah-NEE-mee-ah) Type of anemia caused by a lack of intrinsic factor from the gastric mucosa, which is necessary for the absorption of vitamin B_{12}; results in a deficiency of normal blood cells

CHAPTER SUMMARY

Overview and Functions of the Digestive System

■ **Define *digestion*.**

- Digestion is the process of converting food into chemical substances that can be used by the body.

■ **List the components of the digestive tract.**

- The digestive tract includes the mouth, pharynx, esophagus, stomach, small intestine, and large intestine.

■ **List the accessory organs of the digestive system.**

- The accessory organs of the digestive system are the salivary glands, liver, gallbladder, and pancreas.

■ **List six functions of the digestive system.**

- Functions of the digestive system include ingestion, mechanical digestion, chemical digestion, mixing and propelling movements, absorption, and elimination of waste products.

General Structure of the Digestive Tract

■ **Describe the general histology of the four layers, or tunics, in the wall of the digestive tract.**

- The basic structure of the wall of the digestive tube is the same throughout the entire length, although there are variations in each region.

- The wall of the digestive tract consists of a mucosa, submucosa, muscular layer, and outer adventitia (above the diaphragm) or serosa (below the diaphragm).

- Nerve plexuses are located in the submucosa and in the muscular layer.

Components of the Digestive Tract

■ **Describe the features and functions of the oral cavity and teeth.**

- The lips and cheeks are muscles covered with epithelium and lined with mucous membrane. The palate is the roof of the mouth. The anterior portion is supported by bone; the posterior portion is muscle and connective tissue.

- The tongue is composed of skeletal muscle. The dorsal surface is covered with papillae, some of which contain taste buds. The tongue manipulates food in the mouth, contains sensory receptors for taste, and is used in speech.

- The primary teeth are the deciduous teeth that fall out and are replaced by the secondary or permanent teeth. There are 20 teeth in the complete primary set and 32 teeth in the complete secondary set. The incisors have sharp edges for biting; cuspids

have points for grasping and tearing; bicuspids and molars have flat surfaces for grinding. Each tooth has a crown, a neck, and a root. Enamel covers the crown.

■ **Name and describe the location of the three major types of salivary glands and describe the functions of the saliva they produce.**

- The parotid, submandibular, and sublingual glands secrete saliva, which contains the enzyme amylase.

- The parotid glands are anterior and inferior to the ear; the submandibular glands are along the medial surface of the mandible; and the sublingual glands are under the tongue.

- Saliva contains water, mucus, and amylase.

- Saliva has a cleansing action, moistens food, dissolves substances for taste, and begins digestion of carbohydrates.

■ **Describe the features and functions of the pharynx and esophagus.**

- The pharynx is a passageway that transports food to the esophagus. It is divided into the nasopharynx, oropharynx and laryngopharynx.

- The esophagus is posterior to the trachea and anterior to the vertebral column. The lower esophageal sphincter, also called the cardiac sphincter, controls the passage of food into the stomach.

■ **Describe the structure and histologic features of the stomach and its role in digestion.**

- The stomach is divided into a fundus, cardiac region, body, and pyloric region and has a greater curvature and a lesser curvature.

- The mucosal lining has folds called rugae, and there are three layers of smooth muscle in the wall.

- Mucous cells secrete mucus; parietal cells secrete hydrochloric acid and intrinsic factor; chief cells secrete pepsinogen; and endocrine cells secrete gastrin.

- The semifluid mixture of food and gastric juice that leaves the stomach is called chyme.

- The regulation of gastric secretions is divided into cephalic, gastric, and intestinal phases. Thoughts and smells of food start the cephalic phase; the presence of food in the stomach initiates the gastric phase; and the presence of acid chyme in the small intestine starts the intestinal phase.

- Relaxation of the pyloric sphincter allows chyme to pass from the stomach into the small intestine. The rate at which stomach emptying occurs depends on the nature of the chyme and the receptivity of the small intestine.

■ **Describe the structure and histologic features of the small intestine and its role in digestion and absorption.**

- The absorptive surface area of the small intestine is increased by plicae circulares, villi, and microvilli. Each villus contains a blood capillary network and a lymph capillary called a lacteal.
- The small intestine is divided into the duodenum, jejunum, and ileum. The duodenum has mucous glands in the submucosa; the jejunum has numerous, long villi; and the ileum has a large number of goblet cells.
- Cells in the small intestine produce peptidase, which acts on proteins; maltase, sucrase, and lactase, which act on disaccharides; and lipase, which acts on neutral fats.
- The small intestine produces two hormones, secretin and cholecystokinin. Secretin stimulates the pancreas and cholecystokinin stimulates the gallbladder and digestive enzymes from the pancreas.

■ **Describe the structure, histologic features, and functions of the large intestine.**

- The mucosa of the large intestine does not have villi but has a large number of goblet cells. The longitudinal muscle layer is limited to three bands called teniae coli. Haustra and epiploic appendages are also characteristic.
- The large intestine consists of the cecum, colon, rectum, and the anal canal. The colon is divided into the ascending colon on the right side, transverse colon across the anterior abdomen, descending colon on the left, and sigmoid colon across the pelvic brim.
- The functions of the large intestine include the absorption of water and electrolytes, and the elimination of feces.

Accessory Organs of Digestion

■ **Describe the structure and functions of the liver, gallbladder, and pancreas.**

- Externally, the liver is divided into right, left, caudate, and quadrate lobes by the falciform ligament, inferior vena cava, gallbladder, ligamentum venosum, and ligamentum teres. The porta of the liver is where the hepatic artery and hepatic portal vein enter the liver and the hepatic ducts exit.
- The functional units of the liver are lobules with sinusoids that carry blood from the periphery to the central vein of the lobule. Blood is brought to the liver by the hepatic portal vein and the hepatic artery. The blood from both vessels flows through the sinusoids into the central vein. Central veins merge to form the hepatic veins, which drain into the interior vena cava.
- The liver has numerous functions that include secretion, synthesis of bile salts, synthesis of plasma proteins, storage, detoxification, excretion, carbohydrate metabolism, lipid metabolism, protein metabolism, and filtering blood.
- The main components of the bile produced by the liver are water, bile salts, bile pigments, and cholesterol. Bile salts act

as emulsifying agents in the digestion and absorption of fats. The principal bile pigment is bilirubin, which is formed from the breakdown of hemoglobin. Cholesterol and bile pigments are excreted from the body in the bile.

- The gallbladder is attached to the visceral surface of the liver by the cystic duct, which joins the hepatic duct to form the common bile duct. The common bile duct empties into the duodenum. The gallbladder stores and concentrates the bile.
- The pancreas is retroperitoneal along the posterior body wall and extends from the duodenum to the spleen. Most of the pancreas is exocrine and composed of acinar cells, which produce digestive enzymes. The islets of Langerhans are endocrine and produce insulin and glucagon.
- Pancreatic enzymes include amylase, which acts on starch; trypsin, which acts on proteins; peptidase, which acts on peptides; and lipase, which acts on lipids.
- The hormone secretin stimulates the pancreas to secrete a bicarbonate-rich fluid; cholecystokinin stimulates the production of pancreatic enzymes.

Chemical Digestion

■ **Summarize carbohydrate, protein, and lipid digestion by stating the intermediate and final products and the enzymes that facilitate the digestive process.**

- Carbohydrates are first broken down into disaccharides by amylase. Disaccharides are then broken down into monosaccharides by sucrase, maltase, and lactase. The end products of carbohydrate digestion are the monosaccharides glucose, fructose, and galactose.
- Pepsin and trypsin break proteins into shorter chains called peptides. Peptidase breaks peptides into amino acids. The end products of protein digestion are amino acids.
- Fats are first emulsified by bile. Lipase acts on emulsified fats and breaks them down into monoglycerides and free fatty acids. Monoglycerides and free fatty acids are the end products of lipid digestion.

Absorption

■ **Compare the absorption of simple sugars and amino acids with that of lipid-related molecules.**

- Most nutrient absorption takes place in the jejunum.
- Water is absorbed by osmosis in all regions.
- Simple sugars and amino acids are absorbed into the blood capillaries in the villi of the small intestine and then transported to the liver in the hepatic portal vein.
- Fatty acids, monoglycerides, and fat-soluble vitamins enter the lacteals in the villi of the small intestine and circulate in the lymph until the lymph enters the left subclavian vein.

CHAPTER QUIZ

Recall

Match the definitions on the left with the appropriate term on the right.

_____ **1.** Fold of peritoneum that attaches small intestines to the dorsal body wall

_____ **2.** Semifluid mixture that leaves the stomach

_____ **3.** Movements that propel food particles through the digestive tract

_____ **4.** Longitudinal folds in the stomach

_____ **5.** Swallowing

_____ **6.** Circular folds in the small intestine

_____ **7.** Middle region of the small intestine

_____ **8.** Fold of peritoneum that attaches the liver to the anterior abdominal wall

_____ **9.** Phagocytic cells in the liver

_____ **10.** Chemical reactions of digestion

A. Chyme
B. Deglutition
C. Falciform ligament
D. Hydrolysis
E. Jejunum
F. Kupffer cells
G. Mesentery
H. Peristalsis
J. Rugae

Thought

1. The simple columnar epithelial cells in the stomach are a part of the:
  A. Serosa
  B. Adventitia
  C. Submucosa
  D. Mucosa

2. The inactive enzyme pepsinogen is secreted by:
  A. Chief cells in the stomach
  B. Crypts in the small intestine
  C. Acinar cells in the pancreas
  D. Parietal cells in the stomach

3. If the pH in the duodenum decreases to 4.0, secretion of which of the following will be increased?
  A. Enterokinase
  B. Bile
  C. Secretin
  D. Pepsinogen

4. An obstruction of the common bile duct:
  A. Blocks the flow of cholecystokinin into the duodenum
  B. Is likely to interfere with fat digestion
  C. Blocks the action of enterokinase
  D. Is likely to interfere with carbohydrate digestion

5. Which one of the following does *not* increase the surface area for absorption in the small intestine?
  A. Plicae circulares
  B. Villi
  C. Teniae coli
  D. Microvilli (brush border)

Application

Surgical removal of the stomach will most likely interfere with the absorption of which vitamin? Explain.

John has chronic stomach ulcers. His gastroenterologist suggests a treatment that involves cutting branches of the vagus nerve that go to the stomach. Explain the rationale of this treatment.

BUILDING YOUR MEDICAL VOCABULARY

Building a vocabulary is a cumulative process. As you progress through this book, you will use word parts, abbreviations, and clinical terms from previous chapters. Each chapter will present new word parts, abbreviations, and terms to add to your expanding vocabulary.

Word Parts and Combining Form with Definition and Examples

PART/COMBINING FORM	DEFINITION	EXAMPLE
-algia	pain	dentalgia: tooth pain, toothache
amyl-	starch	amylase: enzyme that breaks down starch
-ary	pertaining to	biliary: pertaining to bile
-ase	enzyme	lipase: enzyme that breaks down lipids
bili-	bile, gall	bilirubin: bile pigment
cec-	cecum	ileocecal valve: valve at the junction of the ileum and cecum
cheil/o	lip	cheiloplasty: plastic surgery of the lip
-chezia	defecation, feeces	hematochezia: blood in the feces
chole-	gall, bile	cholecystitis: inflammation of the gallbladder
col-	colon, large intestine	colectomy: excision of the colon, or a portion of it
cyst/o	bladder	cholecystorraphy: suture or repair of the gallbladder
dent/o	tooth	dentoalveolar: pertaining to a tooth and its alveolus
-emesis	vomit	hematemesis: vomiting of blood
enter/o	intestine	enteritis: inflammation of the intestine
gastr/o	stomach	gastralgia: pain in the stomach
gingiv/o	gums	gingivectomy: surgical excision of diseased tissue of the gums
gloss/o	tongue	glossopathy: any disease of the tongue

PART/COMBINING FORM	DEFINITION	EXAMPLE
hepat/o	liver	hepatitis: inflammation of the liver
lingu/o	tongue	linguopapillitis: inflammation of the papillae on the tongue
-orexia	appetite	anorexia: lack of appetite
prandi-	meal	postpandial: after a meal
proct/o	rectum, anus	proctology: branch of medicine concerned with disorders of the rectum and anus
sial/o	saliva	sialadenitis: inflammation of a salivary gland
-stalsis	contraction	peristalsis: wavelike contractions of the digestive tube
verm-	worm	vermiform: worm-shaped

Clinical Abbreviations

ABBREVIATION	MEANING
BaE	barium enema
BM	bowel movement
CLD	chronic liver disease
EGD	esophagogastroduodenoscopy
FOBT	fecal occult blood test
GB	gallbladder
GERD	gastric esophageal reflux disease
GI	gastrointestinal
IBD	inflammatory bowel disease
LES	lower esophageal sphincter
LFT	liver function test
NG tube	nasogastric tube
NPO	nothing by mouth
PUD	peptic ulcer disease
TPN	total parenteral nutrition
UGI	upper gastrointestinal

Clinical Terms

Anorexia (ann-oh-REK-see-ah) Lack or loss of appetite for food

Aphagia (ah-FAY-jee-ah) Inability to swallow

Ascites (ah-SYE-teez) Accumulation of serous fluid in the peritoneal cavity

Borborygmus (bor-boh-RIG-mus) Rumbling noise caused by propulsion of gas through the intestines

Bulimia (boo-LIM-ee-ah) Emotional disorder characterized by binge eating and often terminating in self-induced vomiting

Colostomy (koh-LAHS-toh-mee) Surgical procedure in which an opening from the colon is created through the abdominal wall; the opening serves as a substitute anus

Diarrhea (dye-ah-REE-ah) Frequent passage of unformed watery feces

Diverticula (dye-ver-TICK-yoo-lah) Pouchlike herniations through the muscular wall of a tubular organ such as the colon

Edentulous (ee-DEN-too-lus) Without teeth; term used after natural teeth are lost

Emesis (EM-eh-sis) Vomiting

Eructation (ee-ruck-TAY-shun) Belching or burping; the expulsion of gas through the mouth

Flatus (FLAY-tus) Gas in the stomach or intestines, which may result from gas released during the breakdown of foods, from swallowing air, or from drinking carbonated beverages

Gavage (gah-VAHJ) Procedure in which liquid or semiliquid food is fed through a tube

Hematemesis (hee-mat-EM-eh-sis) Blood in the vomit

Intussusception (in-tuh-sus-SEP-shun) Telescoping of one part of the intestine into the opening of an immediately adjacent part

Laparoscopy (lap-ah-RAHS-koh-pee) Procedure in which the inside of the abdomen is viewed with a lighted instrument, or a surgical procedure performed through the instrument

Pyrosis (pye-ROH-sis) Regurgitation of stomach acid into the esophagus; heartburn

Volvulus (VAHL-vyoo-lus) Twisting of the bowel on itself that causes an obstruction

VOCABULARY QUIZ

Use word parts you have learned to form words that have the following definitions.

1. Enzyme that breaks down starch _____

2. Lack of appetite _____

3. Inflammation of the intestine _____

4. Pain in the stomach _____

5. Pertaining to below the tongue _____

Using the definitions of word parts you have learned, define the following words.

6. Biliary _____

7. Cholecystitis _____

8. Dentalgia _____

9. Hepatitis _____

10. Sialadenitis _____

Match each of the following definitions with the correct word.

_____ 11. After a meal

_____ 12. Surgical excision of the gallbladder

_____ 13. Vomiting blood

_____ 14. Pain in the rectum and anus

_____ 15. Inflammation of the gums

A. Cholecystectomy
B. Gingivitis
C. Hematemesis
D. Postprandial
E. Proctalgia

16. What term denotes "difficulty in swallowing"? _____

17. What is the name of the procedure in which liquid or semiliquid food is fed through a tube? _____

18. What term denotes "the twisting of the bowel on itself that causes an obstruction"? _____

19. What is the meaning of the abbreviation GERD? _____

20. What is the meaning of the abbreviation IBD? _____

Metabolism and Nutrition

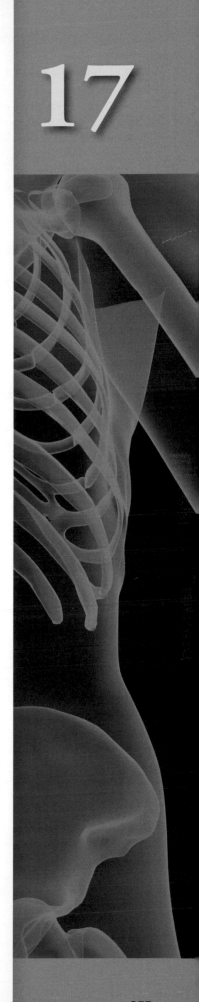

17

CHAPTER OBJECTIVES

Metabolism of Absorbed Nutrients
- Define the terms *metabolism* and *nutrition*.
- Distinguish between anabolism and catabolism.
- Describe the basic steps in glycolysis, the citric acid cycle, and electron transport.
- Define *glycogen, glycogenesis, glycogenolysis,* and *gluconeogenesis.*
- Describe the pathway by which proteins are used in the body.
- Describe the pathway by which fatty acids are broken down to produce adenosine triphosphate.
- Name two key molecules in the metabolism and interconversion of carbohydrates, proteins, and fats.
- Explain how energy from food is measured.
- State four factors that influence basal metabolism.

Body Temperature
- Distinguish between core temperature and shell temperature.

- Explain three ways in which core temperature is maintained when the environmental temperature is cold.
- State four ways in which heat is lost from the body.
- Describe the general mechanism by which the body maintains core temperature and identify the region of the brain that integrates this mechanism.

Basic Elements of Nutrition
- List the functions of carbohydrates, proteins, and fats in the body.
- State the caloric value of carbohydrates, proteins, and fats.
- Explain the importance of fiber in the diet.
- Distinguish between essential and nonessential amino acids and between complete and incomplete proteins.
- Discuss the functions of vitamins, minerals, and water in the body.

KEY TERMS

Acetyl-CoA (as-SEE-tul CO-A) Molecule formed from pyruvic acid in the mitochondria when oxygen is present

Aerobic (air-ROE-bik) Requiring molecular oxygen

Anabolism (ah-NAB-oh-lizm) Building up, or synthesis, reactions that require energy and make complex molecules out of two or more smaller ones; opposite of catabolism

Anaerobic (an-air-OH-bik) Not requiring molecular oxygen

Basal metabolic rate (BAY-sal met-ah-BAHL-ik RAYT) Amount of energy that is necessary to maintain life and keep the body functioning at a minimal level

Beta oxidation (BAY-tah ox-ih-DAY-shun) Catabolic reaction in which a two-carbon segment is removed from a fatty acid and is converted to acetyl-CoA

Catabolism (kah-TAB-oh-lizm) Reactions that break down complex molecules into two or more smaller ones with the release of energy; opposite of anabolism

Citric acid cycle (SIT-rik AS-id SYE-kul) Aerobic series of reactions that follows glycolysis in glucose metabolism to release energy and carbon dioxide; also called Krebs cycle

Complete protein (kum-PLEET PRO-teen) Protein that contains all of the essential amino acids

Core temperature (KOR TEM-per-ah-chur) Temperature deep in the body; temperature of the internal organs

Deamination (dee-am-ih-NAY-shun) Catabolic reaction in which an amino group is removed from an amino acid to form ammonia and a keto acid; occurs in the liver as part of protein catabolism

Gluconeogenesis (gloo-koh-nee-oh-JEN-eh-sis) Process of forming glucose from noncarbohydrate nutrient sources such as proteins and lipids

Glycogenesis (gly-koh-JEN-eh-sis) Series of reactions that convert glucose or other monosaccharides into glycogen for storage

Glycogenolysis (gly-koh-jen-AHL-ih-sis) Series of reactions that convert glycogen into glucose

Glycolysis (gly-KAHL-ih-sis) Anaerobic series of reactions that produces two molecules of pyruvic acid from one molecule of glucose; first series in the catabolism of glucose

Lipogenesis (lip-oh-JEN-eh-sis) Series of reactions in which lipids are formed from other nutrients

Thermogenesis (ther-moh-JEN-eh-sis) Production of heat in response to food intake

Metabolism (meh-TAB-oh-lizm) is the aggregate of all the chemical reactions that take place in the body. Some of these reactions result in digestion and absorption, which are described in Chapter 16. After nutrients are absorbed by the cells, chemical reactions within the cells synthesize new materials for cellular use or secretion and produce energy for body activities. These reactions constitute **cellular metabolism**, which is the subject of this chapter. Each metabolic reaction, including those in cellular metabolism, requires a specific enzyme. The enzymes speed up the reactions so the rate at which they occur is consistent with life. Without the enzymes, the reaction rates are too slow to sustain life.

Nutrition (noo-TRIH-shun) is the acquisition, assimilation, and use of the nutrients that are contained in food. The energy for body activities ultimately comes from the food that is taken into the body. Because metabolism is the use of nutrients, the basic concepts of nutrition are included in this chapter.

Metabolism of Absorbed Nutrients

Chemical reactions either require energy or release energy. Metabolic processes are divided into two categories on this basis. **Anabolism** (ah-NAB-oh-lizm) includes the building up, or synthesis, reactions that require energy. **Catabolism** (kah-TAB-oh-lizm) involves breaking down large molecules into smaller ones and releases energy.

Anabolism

Anabolism is the constructive portion of metabolism. These reactions use chemical energy, usually in the form of **adenosine triphosphate (ATP)**, to make large molecules from smaller ones. Glucose molecules are linked together in long chains to make the larger glycogen molecules. Amino acids are connected by peptide bonds to form complex protein molecules. Glycerol and fatty acids combine to form triglycerides. In each of these examples, the two smaller molecules react to produce a water molecule and a larger molecule. Anabolic reactions such as these represent **dehydration synthesis** because a water molecule is removed and a larger molecule is synthesized. Chemical energy is required to form the new bonds in these reactions.

Catabolism

Catabolism is the part of metabolism in which large molecules are broken down into smaller ones. The **hydrolysis** reactions of digestion are examples of catabolism. Water is used to split the large molecule into two smaller parts. Another example of catabolism is the cellular use of nutrients. After the products of digestion are absorbed into the cells of the body, they are "burned," or oxidized, in the processes of **cellular respiration**. The chemical bonds break, releasing chemical energy that is used to drive the reactions of anabolism.

QUICK CHECK
17.1 Is the hydrolysis of maltose into two molecules of glucose an example of catabolism or of anabolism?
17.2 What type of reaction is the opposite of hydrolysis?
17.3 What is the primary chemical energy molecule that is used in cellular respiration?

Energy from Foods

Within the body's cells, the reactions of cellular respiration break chemical bonds and release energy. Some of this energy is captured and stored as chemical energy in molecules of ATP. The remainder is given off as heat, which is used to maintain body temperature. ATP is the energy exchange molecule in the cell (Figure 17-1). When energy is needed for active transport, muscle contraction, or synthesis reactions, a high-energy bond in ATP breaks and releases energy, a phosphate, and **adenosine diphosphate (ADP)**. When energy is released from catabolic reactions, it combines with the ADP and a phosphate to form new ATP. Cellular respiration uses the absorbed end products of digestion and, through a series of reactions, generates ATP.

Carbohydrates

The absorbed end products of carbohydrate digestion are monosaccharides, or simple sugars. The most important of these six-carbon molecules in cellular respiration is glucose.

The first series of reactions in the catabolism of glucose is **glycolysis** (gly-KAHL-ih-sis). These reactions take place in the cytoplasm of the cell and they are **anaerobic**, which means that they do not require oxygen. During glycolysis, a six-carbon molecule of glucose is split into two, three-carbon molecules of **pyruvic acid** (Figure 17-2). These reactions require two ATP molecules to get them started, but they yield four ATP molecules, for a net gain of two ATP molecules.

The fate of the pyruvic acid that is produced in glycolysis depends on whether oxygen is absent or present (Figure 17-3). In the absence of oxygen, the pyruvic acid is converted to **lactic acid**, which is the end product of anaerobic respiration. Most of the lactic acid diffuses out of the cells that produce it, enters the blood, and is transported to the liver. When sufficient oxygen becomes available, the liver converts lactic acid back to glucose. The oxygen necessary for this conversion is called **oxygen debt**.

If oxygen is present when pyruvic acid is produced, the pyruvic acid enters the aerobic phase of cellular respiration, which takes place in the mitochondria. Pyruvic acid enters the mitochondria and goes through a series of reactions that removes one of the carbons to form carbon dioxide and produces a two-carbon molecule of **acetyl-CoA** (Figure 17-4). Within the mitochondria, acetyl-CoA enters a cyclic series of reactions called the **citric acid cycle** (Krebs cycle). To start these reactions, the two-carbon acetyl-CoA combines with a four-carbon molecule in the cycle to form the six-carbon molecule of citric acid. During the cycle, two carbons are removed to form

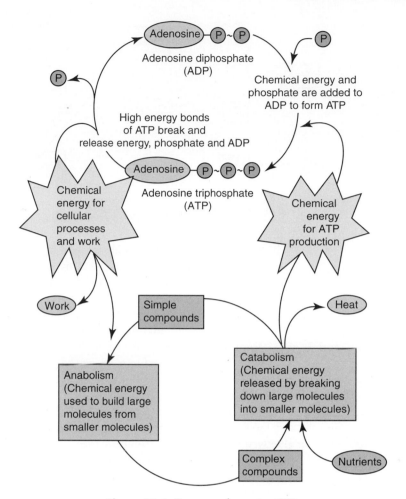

Figure 17-1 *Energy exchange in ATP.*

carbon dioxide, energy is released, hydrogen ions are removed, and a four-carbon molecule is produced. This four-carbon molecule combines with another acetyl-CoA and the cycle is repeated. The carbon dioxide is exhaled in the act of breathing. Carrier molecules (nicotinamide adenine dinucleotide and flavin adenine dinucleotide) transport the hydrogen ions to the **electron-transport chain**, a series of molecules along the inner mitochondrial membrane. Here, in reactions that require oxygen, the hydrogen ions combine with oxygen to form water, and in the process ATP is produced. The final result of the aerobic reactions in the mitochondria is the production of carbon dioxide, water, and energy. Some of the energy forms ATP; the remainder is released as heat. The complete breakdown of a

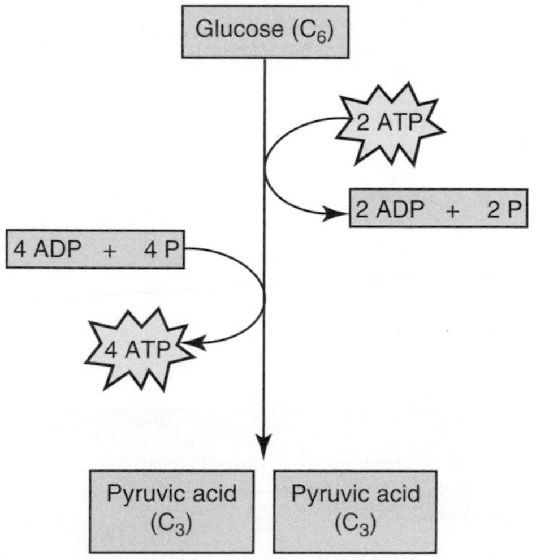

Figure 17-2 *Glycolysis. In glycolysis, a glucose molecule is split into two pyruvic acid molecules. These reactions are anaerobic and occur in the cytoplasm. There is a net gain of two ATP molecules.*

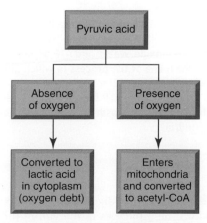

Figure 17-3 *Fate of pyruvic acid depending on presence or absence of molecular oxygen.*

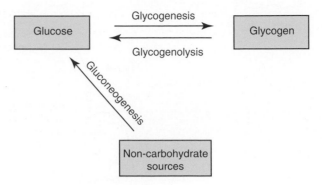

Figure 17-4 *Anaerobic breakdown of pyruvic acid in the mitochondria.*

six-carbon glucose molecule yields 36 to 38 molecules of ATP: 2 from the anaerobic glycolysis phase, and 34 to 36 from the aerobic phase within the mitochondria.

After meals, when glucose is absorbed from the digestive tract, blood glucose levels increase. The liver removes the excess glucose from the blood and stores it as **glycogen**. Some glycogen is stored in skeletal muscle and later used for skeletal muscle activity. The production of glycogen from glucose is called **glycogenesis** (gly-koh-JEN-eh-sis) (Figure 17-5). When glycogen storage is at full capacity, glucose may be converted to fat by a process called **lipogenesis** (lip-oh-JEN-eh-sis). When blood glucose levels start to decline before the next meal, glycogen breaks down to form glucose. **Glycogenolysis** (gly-koh-jen-AHL-ih-sis) is the conversion of glycogen into glucose. Glycogenesis and glycogenolysis help keep blood glucose levels relatively constant. In addition, some noncarbohydrate nutrient sources may be converted to glucose through **gluconeogenesis** (gloo-koh-nee-oh-JEN-eh-sis).

> ### QUICK CHECK
> **17.4** What are the end products of glycolysis?
> **17.5** Where in the cell does the catabolism of pyruvic acid into carbon dioxide, water, and ATP occur?
> **17.6** What is the name of the process in which glucose is converted to glycogen?
> **17.7** Which produces more ATP for each glucose—anaerobic glycolysis in the cytoplasm or aerobic reactions in the mitochondria?

Proteins

The end products of protein digestion are amino acids, which are absorbed into the capillaries in the villi of the small intestine. From the villi, the amino acids enter the hepatic portal circulation to the liver and then enter the general circulation pathways in the body so they are available to all cells.

Figure 17-5 *Blood glucose homeostasis.*

Amino acids are used by the cells for a wide variety of functions. Most of these involve the synthesis of proteins that are needed by the body's cells. Examples of these anabolic reactions are proteins to build new tissues or to replace damaged tissues and the synthesis of hemoglobin, hormones, enzymes, and plasma proteins. The creation of proteins from amino acids involves dehydration synthesis reactions in which water is removed to form peptide bonds between the amino acids.

If there is an excess of amino acids or insufficient carbohydrates and fats, then the amino acids may be used as an energy source. This is an example of catabolism. The principal reaction that prepares an amino acid for use as an energy source is **deamination** (dee-am-ih-NAY-shun), which occurs in the liver. Deamination removes the amino group ($-NH_2$) from the amino acid to form a keto acid and ammonia. Ammonia is toxic to cells, so the liver converts it to urea, which is less toxic and is excreted in the urine.

Depending on the amino acid involved, the keto acid is used in one of several pathways (Figure 17-6). Some are converted to pyruvic acid or to acetyl-CoA, which then enters the

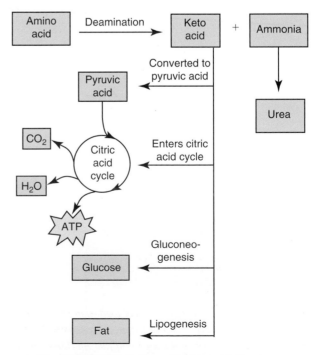

Figure 17-6 *Catabolism of amino acids for energy.*

citic acid cycle. Others enter directly into the citric acid cycle. These pathways ultimately produce carbon dioxide, water, and energy. If energy is not needed immediately, excess amino acids may be converted into glucose by gluconeogenesis or into fat molecules by lipogenesis.

Lipids

Approximately 40% of the calories in the normal American diet are derived from fats. Furthermore, an average of 30% to 50% of the ingested carbohydrates are converted to triglycerides and stored. All cells in the body, except brain cells, can use fatty acids from triglycerides instead of glucose as an energy source. Lipids are an important factor in the body's metabolism.

Before triglycerides can be used as an energy source, they are changed into glycerol and fatty acids by hydrolysis. The glycerol enters the glycolysis pathway and continues through the citric acid cycle to yield carbon dioxide, water, and energy. Fatty acids undergo a series of reactions, called **beta oxidation** (BAY-tah oxih-DAY-shun). Beta oxidation removes two-carbon segments from the end of a fatty acid chain and converts them into acetyl-CoA. The process keeps repeating, two carbon atoms at a time, until the entire fatty acid chain is converted into acetyl-CoA molecules. The acetyl-CoA molecules enter the citric acid cycle to yield carbon dioxide, water, and energy. Figure 17-7 illustrates the catabolic pathways for lipids.

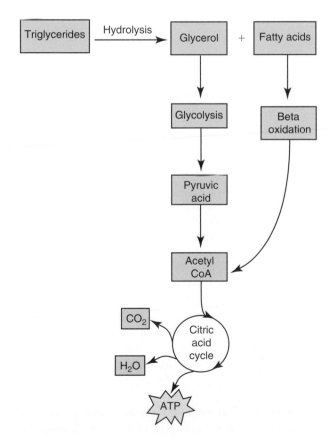

Figure 17-7 *Catabolism of lipids. Beta oxidation removes a two-carbon segment from the end of a fatty acid and converts the segment to acetyl Co-A. This process continues, two carbons at a time, until the entire fatty acid chain is converted into acetyl Co-A molecules.*

QUICK APPLICATION

Ketone bodies form when there is an excess of acetyl-CoA. In fasting and other conditions in which fats instead of carbohydrates are used as an energy source, ketone bodies form from the acetyl-CoA that is produced by beta oxidation. A buildup of ketone bodies in the blood is called ketosis. Because most of the ketone bodies are acids, the pH of the blood decreases and acidosis develops. This can lead to death if medical intervention is delayed and the pH of the blood becomes too low.

Interconversion of Carbohydrates, Proteins, and Lipids

The preceding paragraphs indicate that carbohydrates, proteins, and fats can be converted to products that enter the citric acid cycle to ultimately yield energy. Pyruvic acid and acetyl-CoA are key molecules in these conversion pathways. Because the reactions are reversible, these molecules are intermediates in the pathways for the interconversion of carbohydrates, fats, and proteins. The formation of triglycerides from carbohydrates and proteins is called lipogenesis. Figure 17-8 summarizes the metabolic pathways for the nutrients that produce energy.

QUICK APPLICATION

Chronic alcoholism frequently leads to cirrhosis of the liver. Liver enzymes convert the alcohol into acetyl-CoA and in the process form reduced nicotinamide adenine dinucleotide (NADH) molecules, which enter the electron-transport chain to produce ATP. High levels of NADH in the cell from the metabolism of alcohol inhibit glycolysis and the Krebs cycle; consequently, sugars and amino acids are converted into fat instead of being used for energy. The fat is deposited in the liver cells, which die as a result of the fat accumulation. Scar tissue replaces the cells that die, resulting in cirrhosis of the liver. This condition can lead to death because the scar tissue is unable to perform the functions of viable liver cells.

QUICK CHECK

17.8 What is produced from deamination reactions in the catabolism of proteins?

17.9 What molecule is produced from the beta oxidation of fatty acids?

17.10 What two molecules are the keys, or central to, the interconversion of carbohydrates, proteins, and lipids?

Uses for Energy

The energy derived from food is measured in units called **kilocalories**. Each kilocalorie, sometimes called a "large" Calorie, is equal to 1000 calories (**Calorie**, with a capital C, is used in this chapter to represent a kilocalorie). One Calorie is the amount of energy required to raise the temperature of 1 kilogram of

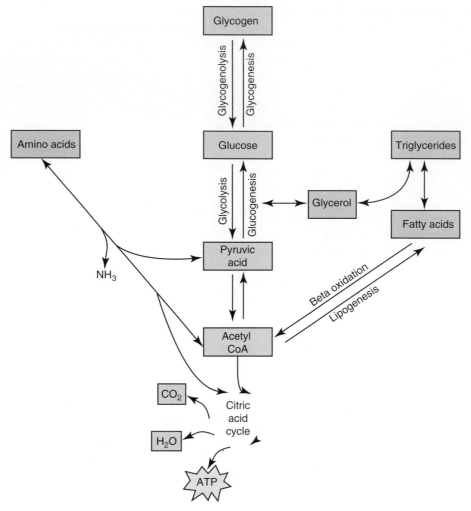

Figure 17-8 *Summary of metabolic pathways. Note that pyruvic acid and acetyl Co-A are key molecules in the metabolic pathways of carbohydrates, proteins, and lipids.*

water 1° C, for example, from 14° C to 15° C. This energy is used by the body in three different ways: basal metabolism, physical activity, and thermogenesis or assimilation of food. Together, these three factors constitute the total metabolic rate (Figure 17-9). If caloric intake exceeds the needs for metabolism, then there is weight gain.

Basal Metabolism

The **basal metabolic rate** (BMR) is the amount of energy that is necessary to maintain life and to keep the body functioning at a minimal level. It includes the functioning of the heart, lungs, nervous system, kidneys, and liver at a maintenance level in resting conditions. BMR is calculated by measuring the amount of oxygen that is used when an individual is awake but in a relaxed and resting state. The amount of oxygen used is proportional to the energy expended.

Several factors influence basal metabolism (see Figure 17-9). Males tend to have a higher metabolic rate than females of the same age and size. Individuals with greater muscle mass usually have higher rates. Fever increases the metabolic rate, and age decreases it. Hormones such as thyroxine, growth hormone, and epinephrine increase the

metabolic rate. On average, 60% to 75% of the energy used each day is for basal metabolism.

Physical Activity

Because muscle contraction requires energy, physical activity contributes to the total energy used each day. Muscular activity accounts for less than 25% of the total daily expenditure for an individual with a sedentary lifestyle; however, this increases with physical activity. Of the three ways in which the body uses energy (basal metabolism, physical activity, and thermogenesis), the only one a person can reasonably control is physical activity. Assuming that calorie (energy) intake remains constant, increasing the amount of physical activity tends to encourage weight loss.

Thermogenesis

Thermogenesis (thur-moh-JEN-eh-sis) is the production of heat in response to food intake. When food is digested, transported, absorbed, and metabolized, heat is produced. In other words, it takes energy to process the food. Less than 10% of the caloric intake is used in the assimilation of food, or thermogenesis.

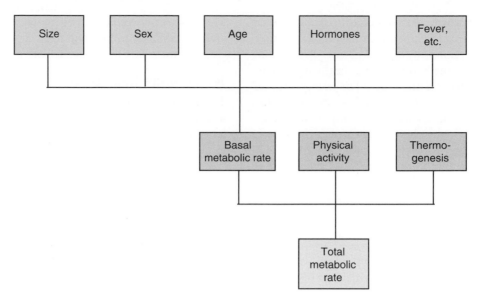

Figure 17-9 *Factors that affect basal and total metabolic rates. Total metabolic rate depends on the basal metabolic rate, physical activity, and thermogenesis.*

 FROM THE PHARMACY

Under normal conditions, adequate nutrition can be achieved through the ingestion of a balanced diet. There are times when this is not possible for a variety of reasons, for example, gastrointestinal (GI) diseases, coma, major burns, certain postsurgical conditions, and anorexia from radiation therapy and chemotherapy for cancer. In individuals with specific disease states and illnesses, nutritional therapy is necessary to prevent severe malnutrition that will impair the healing process. Clinical nutrition to promote the health and well-being of patients is an important aspect of the health care setting.

There are two main routes of administration in nutrition therapy: enteral nutrition and parenteral nutrition. In **enteral nutrition**, the nutrients directly enter some region of the gastrointestinal tract. Theoretically, normal ingestion of food by mouth is one type of enteral nutrition. More generally, it refers to the tube feeding of an individual through a nasogastric, nasoduodenal, nasojejunal, esophagostomy, gastrostomy, or jejunostomy tube. In the first three instances, a feeding tube is inserted through the nose into the stomach, duodenum, or jejunum as indicated. In an esophagostomy, gastrostomy, or jejunostomy, a tube is inserted through a surgically created opening (ostomy) into the esophagus, stomach, or jejunum as indicated. These procedures are more invasive and are used less frequently, especially for short-term nutritional therapy.

The nasogastric, esophagostomy, and gastrostomy routes allow for more natural digestion processes in the stomach. Aspiration is a risk with the nasogastric tubes because the lower esophageal sphincter is not completely closed, which may result in gastric reflux. Enteral formulations are broadly divided into four groups: oligomeric, polymeric, modular, and specialized. The selection of enteral formulations depends on the individual's nutritional needs, disease states, lactose intolerance, and GI competence. Oligomeric formulations are chemically defined preparations that require minimal digestion and produce minimal residue in the colon. Polymeric formulations contain complex nutrients that require digestion and are preferred for persons with fully functional GI tracts. Modular formulations are single-nutrient formulas. Specialized formulations are indicated for individuals with specific disease states.

Parenteral nutrition is the intravenous approach and is used for individuals who are unable to tolerate and maintain adequate enteral intake. It is often called total parenteral nutrition (TNP) or hyperalimentation. This procedure can supply all the nutrients that are needed for health-sustaining functions such as growth, weight gain, wound healing, and convalescence. Total parenteral nutrition may be infused through a peripheral vein or through a central vein. The subclavian vein is the most commonly used central vein.

 QUICK CHECK

17.11 What three factors comprise the total metabolic rate?
17.12 What type of energy expenditure can be controlled voluntarily?

Body Temperature

Humans are **warm-blooded** animals, or **homeotherms** (HO-mee-oh-therms). This means that humans have the ability to maintain a constant internal body temperature even when external environmental temperatures change. Maintenance of the body's **core temperature**, the temperature of the internal organs, is essential for enzyme function. Enzymes are temperature sensitive and function only within narrow temperature ranges. Most enzymes in the body function optimally at temperatures equal to the core temperature. The average normal rectal temperature is 37.6° C (99.7° F) and is considered close to the true core body temperature. The temperature at the body's surface, or skin, is the **shell temperature**. This is the heat loss surface and has a lower value than the core temperature. Temperatures taken orally are considered close to the shell temperature. The average normal oral temperature is 37° C (98.6° F).

Heat Production

Heat is produced by the catabolism of nutrients. Approximately 40% of the total energy that is released by catabolism is used for biological activities in the body. The remaining 60% is heat energy, which is used to maintain body temperature.

When environmental temperatures are too cold and additional heat is needed to maintain core temperature, the hypothalamus initiates additional heat-conserving and heat-promoting activities. These include constriction of the cutaneous blood vessels, increase in the metabolic rate, and shivering. Constriction of the cutaneous vessels restricts the flow of blood to the skin and conserves heat by reducing heat loss through the surface. The blood maintains warmth in the core of the body. Cold stimulates the release of epinephrine and norepinephrine. These hormones increase the metabolic rate and promote heat production. Another response to cold is involuntary contraction of skeletal muscles, which results in shivering. The muscle contractions in shivering produce heat, which helps maintain core temperature.

Heat Loss

Heat loss mechanisms protect the body from excessive heat accumulation. Mechanisms of heat loss include **radiation, conduction, convection**, and **evaporation**. Radiation is the loss of heat as infrared energy. Any object that is warmer than its environment will transfer heat to the environment by radiation. This accounts for about 60% of the body's heat loss. Conduction is the transfer of heat from one object to another by direct contact between the two objects. Convection is the transfer of heat from an object (the body) to the air around it. Conduction and convection account for 15% to 20% of the body's heat loss. Evaporation is the loss of heat from the body through water.

This is most noticeable on a warm day when perspiration evaporates on the skin surface and has a cooling effect. Evaporation accounts for 20% to 25% of the body's heat loss.

Temperature Regulation

Temperature regulation is an example of a negative feedback mechanism that helps maintain homeostasis in the body. Review Figure 5-5 for an example of how the skin functions in this negative feedback mechanism. Temperature homeostasis requires a balance between heat production and heat loss. A small region in the hypothalamus acts as the body's thermostat with a "set point" at which it maintains body temperature. Other regions in the hypothalamus detect changes in body temperature. When the body temperature increases, the hypothalamus inhibits the mechanisms that produce heat and stimulates the mechanisms that promote heat loss. When the body temperature decreases, the hypothalamus initiates responses that produce heat and reduce heat loss. These reflex responses, along with voluntary activities such as putting on or taking off clothing, help maintain homeostasis of core body temperature.

QUICK CHECK

17.13 Which is greater—shell temperature or core temperature?
17.14 By what mechanism is most heat lost from the body?
17.15 How is internal body temperature regulated and maintained?

Concepts of Nutrition

Nutrition is the science that studies the relationship of food to the functioning of the living organism. The intake of food and the digestive processes described in Chapter 16 are part of the broad subject of nutrition. The release of energy from metabolic processes, described in the first part of this chapter, is also a part of nutrition. This section primarily deals with nutrients, the substances in food that are necessary to maintain the normal functions of human organisms. Nutrients include carbohydrates, proteins, lipids, vitamins, and minerals. The functions of the nutrients are summarized in Table 17-1.

TABLE 17-1 **Nutrient Functions**

Carbohydrates	Proteins	Lipids	Vitamins	Minerals
Energy	Provide structure	Energy	Function with enzymes	Component of body structures
Bulk	Regulate body processes	Essential fatty acids		Part of enzyme molecules
Make other compounds	Enzymes	Transport for fat-soluble vitamins		Part of organic molecules
	Hormones	Structural components		(hemoglobin)
	Carrier molecules	Insulation		Fluid and electrolyte balance
	Fluid and electrolyte balance	Cushions organs		
	Energy			

Carbohydrates

Carbohydrates have three major functions in the body. Of these, the most important is to provide energy. They are the most abundant and least expensive food source of energy. On average, 1 gram of pure carbohydrate yields 4 Calories of energy, and 50% to 60% of the daily caloric intake is from this source. Carbohydrates also add bulk to the diet and are used in the synthesis of other compounds in the body, such as lipids, amino acids, and nucleic acids.

Dietary carbohydrates are divided into two groups: simple sugars and complex polysaccharides (Table 17-2). The simple sugars include the monosaccharides and disaccharides, which dissolve in water, form crystals, and taste sweet. Three monosaccharides (**glucose, fructose**, and **galactose**) are important in nutrition. Glucose is the form in which carbohydrates are carried in the blood, and it is easily broken down to provide energy. Fructose is found in honey and in many fruits. Galactose does not exist as a molecule in any food, but it is a part of the disaccharide lactose molecule. Fructose and galactose are converted to glucose in the liver. When two monosaccharide molecules combine, they form a disaccharide. Three disaccharides (**sucrose, maltose**, and **lactose**) are commonly found in food. Sucrose, the familiar table sugar, is formed from glucose and fructose. It is obtained by refining sugar cane and sugar beets. It is also found in many fruits and vegetables. Maltose, which is formed from two glucose molecules, is found in sprouting grains, corn syrup, and maple syrup. It is commonly called malt sugar. Lactose, which is a combination of glucose and galactose, is found in milk and is commonly called milk sugar.

Complex polysaccharides are formed from long chains of several thousand monosaccharide units, usually glucose. In contrast to the simple sugars, these large molecules do not dissolve in water, do not form crystals, and do not taste particularly sweet. Three types of complex polysaccharides, **starch, glycogen**, and **fiber**, are important in nutrition.

Starch is the storage form of energy in plants, where it is found in the seeds and roots. Cereal grains (wheat, oats, corn, barley), starchy roots (potatoes and turnips), and legumes (peas and beans) are important sources of calories in many countries.

Glycogen is the storage form of energy in animals. Liver and muscle tissues have the ability to combine glucose molecules into highly branched long chains to form glycogen.

TABLE 17-2 Summary of Dietary Carbohydrates

Simple Sugars	Complex Polysaccharides
Monosaccharides Glucose (blood sugar)	Starch (glucose storage in plants) Glycogen (glucose storage in animals)
Fructose (fruit sugar, honey)	Fiber (nondigestible plant polysaccharide)
Disaccharides Sucrose (table sugar)	
Maltose (malt sugar)	
Lactose (milk sugar)	

The process of converting glucose to glycogen is called glycogenesis.

Fiber, which comes from plant sources, represents numerous polysaccharides that cannot be digested by enzymes in the human digestive system. Some of the fibers, however, are broken down by bacteria in the intestine. Fiber adds bulk to the diet, which enhances the absorption of nutrients and helps move the contents of the large intestine faster and with less effort. In unprocessed foods, starch and fiber are usually found together, but processing or refining removes the fiber. The practice of eating refined or processed carbohydrates results in too little fiber in the diet. Numerous diseases are related to decreased intake of dietary fiber, including constipation, hemorrhoids, colon cancer, cardiovascular disease, obesity, and diabetes mellitus. Dietitians recommend diets that include 25 to 35 grams of fiber per day.

QUICK CHECK

17.16 Carbohydrates are an important food source of energy. On the average, how much energy is derived from 1 gram of carbohydrate?

17.17 What are the three types of complex polysaccharides that are important in nutrition?

17.18 What is the nutritional problem with eating refined or processed carbohydrate foods?

Proteins

Proteins have three major functions in the body. One function is to provide structure. The matrix of bone, muscles, teeth, tendons, and other structures is composed of protein. The cell membrane and membranous organelles within the cell contain protein to provide structure. Another function of proteins is to regulate body processes. Hormones and enzymes are proteins that regulate body functions and control chemical reactions. Plasma proteins have an important role in maintaining fluid and electrolyte balance. Other proteins act as carrier molecules to transport substances across the cell membrane or to transport substances in the blood and in this way regulate body transport processes. A third function of proteins is to supply energy for the body. If adequate carbohydrate and lipid are not available to meet the energy needs, the amino acids in proteins are metabolized to provide the necessary calories. One gram of pure protein provides 4 Calories of energy. Guidelines for protein consumption recommend that 10% to 12% of a normal healthy adult's daily caloric intake be in the form of proteins. Most Americans eat more than this amount.

The purpose of proteins in the diet is to provide amino acids. The body cannot store amino acids for later use, so a daily supply of protein is necessary. When there are excess amino acids, they are broken down and used to make glycogen or fat. If adequate amounts of the required amino acids are available, the body makes the various proteins that it needs. About half of the different amino acids that the body uses must be consumed because they cannot be synthesized in the body. These are called **essential amino acids**. The other amino acids are called **nonessential** because they can be made in the body and do not have to be supplied in the diet (see Table 17-3).

TABLE 17-3 Essential and Nonessential Amino Acids

Name	ABBREVIATIONS Three-Letter	ABBREVIATIONS One-Letter	Name	ABBREVIATIONS Three-Letter	ABBREVIATIONS One-Letter
Alanine	Ala	A	Asparagine	Asn	N
Cysteine	Cys	C	Glutamine	Gln	Q
Glycine	Gly	G	**Isoleucine**	Ile	I
Leucine*	Leu	L	**Methionine**	Met	M
Phenylalanine	Phe	F	Proline	Pro	P
Serine	Ser	S	**Threonine**	Thr	T
Tryptophan	Trp	W	Tyrosine	Tyr	Y
Valine	Val	V	Aspartic acid	Asp	D
Glutamic acid	Glu	E	**Arginine**	Arg	R
Histidine	His	H	**Lysine**	Lys	K

*Essential amino acids are indicated in bold.

A **complete protein** contains all of the essential amino acids. Animal proteins such as meat, eggs, and dairy products are complete proteins. Vegetable proteins such as nuts, grains, and legumes are **incomplete proteins** because they do not contain all of the essential amino acids. Vegetable proteins, eaten in some combinations, often complement each other to provide all of the essential amino acids and are equivalent to eating a complete protein. Another point to consider is the fact that animal proteins often contain high amounts of fat and are often more expensive than vegetable proteins. Some common combinations that provide all the essential amino acids are wheat bread with peanut butter, corn and lima beans, rice with black beans, and corn bread with split pea soup.

Lipids

Most **lipids** in the diet are triglycerides, or neutral fats, which occur in both animal and plant food. These molecules are composed of three fatty acids attached to a glycerol molecule. Other lipids are phospholipids and steroids.

Lipids have six important functions in the body. (1) They are stored in the body as adipose tissue, or fat, and represent a concentrated source of energy. Each gram of lipid provides 9 Calories, more than twice as many Calories as carbohydrates or proteins. (2) Lipids are necessary in the diet to provide essential fatty acids that cannot be synthesized in the body. (3) Fat-soluble vitamins (A, D, E, K) require lipids for their absorption and transport. (4) Numerous structural components in the body are lipid in nature. The major component of the cell membranes is phospholipid, and myelin, the covering around nerve fibers, is a fatty substance. (5) Adipose tissue under the skin helps maintain body temperature by providing insulation. (6) Adipose tissue also forms a cushion around body organs to protect them from bumps, jolts, and damage.

The preceding paragraph indicates that fats are essential components of the diet. Less than 3% of the total caloric intake in the form of fats is considered to be inadequate lipid consumption to maintain health. This most often occurs in

infants who are fed nonfat milk. These infants may not get sufficient amounts of the essential fatty acids that are important for growth and thus their development will be impaired.

Although fats are essential components of a healthy diet, excessive amounts are not desirable. Most Americans need to consciously strive to reduce their intake of dietary fat. For a healthy diet, it is recommended that no more than 30% of the daily calorie intake be in the form of fats. This means that a person who eats 2200 Calories per day should eat no more than 660 Calories, or about 75 grams, of fat. In general, the American diet has 40% or more of its calories coming from fats. This overconsumption of fats is unhealthy and is related to cardiovascular disease, obesity, diabetes, and some cancers.

The American diet not only contains too much fat but also has the wrong kinds of fat. Scientific evidence shows that the dietary intake of saturated fat, trans fat, and cholesterol raises the low-density lipoprotein (LDL) level in the blood, and this increases the risk of coronary heart disease. It is recommended that less than 10% of the daily caloric intake, or 25 grams, be in the form of saturated fats. The current average diet in the United States contains two times this amount. Most foods that contain fats have a combination of saturated and unsaturated forms. In general, if a fat is solid at room temperature, it has more saturated than unsaturated fat. Animal fats also tend to have more saturated fat. Foods high in saturated fats include beef, pork, lamb, eggs, butter, and whole-milk products. Coconut oil and palm oil are vegetable oils that are high in saturated fat.

Trans fat is produced when unsaturated liquid vegetable oils are hydrogenated to make them solid. Hydrogenation tends to increase the shelf life and enhance the flavor of foods. Many foods that are high in saturated fats are also high in cholesterol. The dietary intake of cholesterol should be limited to less than 250 milligrams per day, which is the amount in one egg yolk. Recent recommendations suggest that the dietary intake of saturated fat, cholesterol, and trans fat be as low as possible while maintaining an adequate nutritionally balanced diet. Table 17-4 compares some common sources of fats and cholesterol.

TABLE 17-4 Common Sources of Fat and Cholesterol

Food	Amount	Weight (grams)	Total Calories	Total Fat (grams)	Saturated Fat (grams)	Cholesterol (milligrams)
Milk, nonfat	1 cup	244	86	0.4 (4%)*	0.3 (3%)*	4
Milk, 2%	1 cup	244	121	4.8 (36%)	2.9 (2%)	22
Milk, whole	1 cup	244	151	8.2 (49%)	5.0 (30%)	33
Egg, hard boiled	1 whole	50	79	5.6 (64%)	1.7 (19%)	274
Cream cheese	1 ounce	28.3	100	10.0 (90%)	6.3 (57%)	31.4
Mayonnaise	1 tablespoon	13.8	99	10.9 (99%)	1.7 (15%)	8.1
Butter	1 tablespoon	14.2	102	11.5 (100%)	7.2 (63%)	31.1
Cheese, American	1 ounce	28.3	107	9.0 (76%)	5.7 (48%)	27.3
Cheese, cheddar	1 ounce	28.3	115	9.2 (72%)	5.8 (45%)	27.3
Sirloin steak	3 ounces	85	240	15.0 (66%)	6.4 (24%)	74
Ground beef	3 ounces	85	230	16.0 (63%)	6.2 (24%)	74
Chicken breast (no skin)	3 ounces	86	142	3.0 (19%)	0.9 (6%)	73
Fried chicken	3 ounces	85	187	7.8 (38%)	2.1 (10%)	79.6
Frankfurter	2 ounces	57	183	16.6 (82%)	6.1 (30%)	29
Turkey, roasted	3 ounces	85	145	4.2 (26%)	1.4 (9%)	65
Cod, broiled	3 ounces	85	97	0.9 (8%)	0.8 (7%)	51
Tuna salad	½ cup	102	188	9.5 (45%)	1.6 (8%)	40
Avocado	1 medium	201	324	30.8 (86%)	4.9 (14%)	0

*Percentage of the total calories represented by the fat.

When evaluating a person's diet, the proportion of fat is a very important aspect to consider in addition to the number of calories. The conversion of dietary fat to body fat requires less energy than the conversion of carbohydrates to fat. If two people routinely eat the same number of calories but one person's calories contain a higher proportion of fat, then that person is more likely to gain more weight because fewer calories are used to convert the dietary fat to body fat than are used to convert the dietary carbohydrates to body fat.

QUICK CHECK

17.19 What is the difference between a complete protein and an incomplete protein?

17.20 What is the American Heart Association recommendation regarding the amount of fat in the diet?

17.21 On average, how many calories are in a gram of fat?

QUICK APPLICATION

Good cholesterol. Bad cholesterol. What is the difference? Cholesterol and fats are insoluble in water, and to become soluble they combine with a blood protein. This combination of cholesterol, fat, and protein is called a lipoprotein, which functions to transport the cholesterol to and from the tissues. Small, dense lipoproteins are called high-density lipoproteins (HDL) and function to remove cholesterol from your arteries and carry it to the liver. HDL is known as "good cholesterol." Large lipoproteins are called low-density lipoproteins (LDL), and they carry cholesterol to the tissues. This "bad cholesterol" lodges in the walls of arteries; consequently, the higher the LDL level, the greater the risk of coronary artery disease. Simple ways to decrease LDL levels are through diet and exercise.

Vitamins

Vitamins are organic compounds that are needed in minute amounts to maintain growth and good health. They do not supply energy, but they are essential to release energy from the carbohydrates, lipids, and proteins. Many of the vitamins function with enzymes in the chemical reactions throughout the body. Vitamins are important in the clotting mechanism and in nucleic acid synthesis. Presently 13 vitamins have been discovered. Of these, four are **fat soluble** and nine are **water soluble**. Table 17-5 summarizes the functions, sources, and recommended daily allowances (RDA) of these vitamins.

Some people have the opinion that if vitamins are good for you, then more must be better. But this is not always the case. Excess fat-soluble vitamins, particularly vitamins A, D, and K, may accumulate to toxic levels in the fatty tissues of the body. Symptoms of vitamin A toxicity include

TABLE 17-5 Functions and Sources of Vitamins

Vitamin	Function	Food Source	Adult RDA
Fat Soluble			
Vitamin A	Healthy amucous membranes, skin, hair; essential for bone development and growth; component for pigments in retina needed for night vision	Milk and cheese; yellow, orange, green vegetables	800 to 1000 mcg
Vitamin D	Formation and development of bones and teeth; assists in absorption of calcium	Fortified milk, fish oils; made in skin when exposed to sunlight	5 to 10 mcg
Vitamin E	Conserves certain fatty acids; aids in protection against cell membrane damage	Whole grains, wheat germ, vegetable oils, nuts, green leafy vegetables	8 to 10 mcg
Vitamin K	Needed for synthesis of factors essential in blood clotting	Green leafy vegetables, cabbage; synthesized by bacteria in intestine	65 to 80 mcg
Water Soluble			
Thiamine (B_1)	Release of energy from carbohydrates and amino acids; growth; proper functioning of nervous system	Whole grains, legumes, nuts	1.5 mg
Riboflavin (B_2)	Helps transform nutrients into energy; involved in citric acid cycle	Whole grains, milk, green vegetables, nuts	7.7 mg
Niacin (B_3)	Helps transform nutrients into energy; involved in glycolysis and citric acid cycle	Whole grains, nuts legumes, fish, liver	20 mg
Pyridoxine (B_6)	Involved in amino acid metabolism	Legumes, poultry, nuts, dried fruit, green vegetables	2 mg
Cyanocobalamin (B_{12})	Aids in formation of red blood cells; helps in nervous system function	Dairy products, eggs, fish, poultry	2 mcg
Pantothenic acid	Part of coenzyme A; functions in steroid synthesis; helps in nutrient metabolism	Legumes, nuts, green vegetables, milk poultry	7 mg
Folic acid	Aids in formation of hemoglobin and nucleic acids	Green vegetables, legumes, nuts, fruit juices, whole grains	200 mcg
Biotin	Fatty acid synthesis; movement of pyruvic acid into citric acid cycle	Eggs; made by intestinal bacteria	0.3 mg
Ascorbic acid (C)	Important in collagen synthesis; helps maintain capillaries; aids in absorption of iron	Citrus fruits, tomatoes, green vegetables, berries	60 mg

anorexia, headache, irritability, enlarged liver and spleen, hair loss, scaly dermatitis, and bone thickening. Excess vitamin D is characterized by loss of weight, calcification of soft tissues, and kidney failure. Anemia, jaundice, and problems of the gastrointestinal tract may indicate vitamin K toxicity. Excessive amounts of some water-soluble vitamins also may be toxic. For example, large doses of vitamin B_6 can cause peripheral nerve damage.

QUICK APPLICATION

Vitamin C is needed for the synthesis of collagen fibers in connective tissue. A vitamin C deficiency leads to **scurvy**, a condition in which the body is unable to produce and maintain healthy connective tissues, which are important in binding the body together. A condition called **rickets** is due to a deficiency of vitamin D. This vitamin is necessary for the absorption of calcium, and a lack of it causes weak bones in children. Milk is often fortified with vitamin D to enhance the absorption of the calcium in the intestines.

Minerals

Minerals are inorganic substances that plants absorb from the soil and animals obtain by eating plants. They are necessary in small amounts to maintain good health in humans. Like vitamins, they do not supply energy, but they work with other nutrients to keep the body functioning properly. Some minerals, like calcium, are incorporated into body structures. Others are a part of enzymes that facilitate chemical reactions. Some help control fluid levels in the body, and others become part of organic molecules such as hemoglobin. Table 17-6 summarizes the functions and sources of some important minerals.

Water

An average adult requires about 2.5 liters of water every day. A regular supply of water is more vital than food. An individual can live for several weeks without food, but only a few days without water. About two thirds of the daily dietary intake is ingested in the form of water or other liquids. Solid food provides the remainder. Optimal health is dependent on the right balance of fluid intake and output, and abnormalities in this balance can be life threatening.

TABLE 17-6 Functions and Sources of Selected Minerals

Mineral	Function	Food Source	Adult RDA
Calcium	Component of bones and teeth; muscle contraction; blood clotting	Dairy products, green vegetables, legumes, nuts	800 to 1000 mcg
Chloride	Acid-base balance of blood; component of hydrochloric acid in stomach	Table salt, milk, eggs, meat	750 mg
Phosphorus	Component of bones and teeth; component of ATP and nucleic acids; component of cell membranes	Legumes, dairy products, nuts, poultry, lean meats	800 mg
Sodium	Regulates body fluid volume; nerve impulse conduction	Table salt is biggest source of sodium in diet	500 mg
Potassium	Body fluid balance; muscle contraction; nerve impulse conduction	Fruits, legumes, nuts, vegetables; widely distributed	2000 mg
Magnesium	Component of some active enzymes; releases energy from nutrients	Whole grains, legumes, green vegetables, nuts	280 to 350 mg
Iron	Component of hemoglobin and myoglobin; releases energy from nutrients	Whole grains, nuts, legumes, poultry, fish, lean meats	10 to 15 mg
Iodine	Component of thyroid hormones	Iodized table salt, dairy products, fish	150 mcg
Zinc	Component of several enzymes; formation of proteins; wound healing	Legumes, poultry, nuts, whole grains, fish, lean meats	12 to 15 mg
Flouride	Healthy bones and teeth	Fluoridated water is best source	1.5 to 4.0 mg

Water has numerous functions in the body. It is the principal component of the body, which is about 60% water by weight. It is an integral part of the body's cells and the fluid that surrounds the cells. Water provides an appropriate medium for the chemical reactions of the body and provides a transport medium for nutrients and waste products. It is the primary component of the lubricants for joints and muscles. The water contained in the body also helps maintain body temperature.

QUICK CHECK

17.22 Small amounts of vitamins and minerals are needed in the diet. What is the fundamental difference between vitamins and minerals?

17.23 Which vitamins are the fat-soluble vitamins?

17.24 About how much water does the average adult require per day?

 FOCUS ON AGING

As a person grows older, there is a steady decline in energy requirements from loss of body tissue, reduced physical activity, and lowered basal metabolic rate. Although energy requirements are altered with aging, there are no major changes in nutritional needs. Older people still need a balanced diet of carbohydrates, fats, proteins, vitamins, minerals, and water. What does change is the amount of these nutrients that should be ingested.

Although the energy reduction associated with aging varies from individual to individual, it is estimated that the average decline in basal metabolic rate is 0.5% every year between the ages of 55 and 75. This means that a person needs 5% fewer calories at age 65 than at age 55, assuming other factors remain equal. Although the older person may need fewer carbohydrates and fats because of reduced energy requirements, the need for proteins, vitamins, and minerals continues. Nutrition in the elderly is complicated further because age-related changes in the digestive tract may affect the digestion and absorption of nutrients, production of enzymes, and kidney function.

As a person ages, the body often becomes less able to metabolize glucose efficiently and the individual is prone to fluctuations in blood sugar levels. Eating complex carbohydrates instead of refined carbohydrates helps control this problem. Reducing fat intake is a good way to reduce the calories in a diet; however, fats should not be eliminated from the diet. They are necessary for the absorption of the fat-soluble vitamins. The need for proteins does not appear to decline with aging; this is an area of concern because the elderly tend to eliminate proteins from their diet because they often are expensive, require extra preparation time, and are sometimes difficult to chew. Because vitamins and minerals are essential in the diet, it is common practice to supplement dietary intake with over-the-counter tablets. This may help the deficiency problems but may lead to other difficulties because it is possible to produce toxic effects from excessive intake. Moderation is the key.

Because nutritional requirements do not change significantly with aging, there is no need for radical changes in dietary habits. Balanced meals provide good nutrition and contribute to a healthy lifestyle.

CHAPTER SUMMARY

Metabolism of Absorbed Nutrients

■ **Define the terms *metabolism* and *nutrition*.**
- Metabolism is the sum of all the chemical reactions in the body. Within a cell, the reactions are called cellular metabolism. Enzymes speed up the reactions.
- Nutrition is the acquisition, assimilation, and utilization of the nutrients is a part of metabolism.

■ **Distinguish between anabolism and catabolism.**
- Anabolism uses energy to build large molecules from smaller ones. Dehydration synthesis is a type of anabolic reaction in which a water molecule is removed when a larger molecule is synthesized from two smaller ones.
- Catabolism releases energy when large molecules break down into smaller ones. Catabolic reactions within the cell that release energy for use by the cell are termed cellular respiration.

■ **Describe the basic steps in glycolysis, the citric acid cycle, and electron transport.**
- Cellular respiration utilizes the absorbed end products of digestion and stores the energy in the high energy bonds of adenosine triphosphate (ATP).
- The first step in the catabolism of glucose is glycolysis, which takes place in the cytoplasm and is anaerobic. Two pyruvic acid molecules are produced form one glucose molecule and there is a net gain of two ATP.
- If oxygen is present, pyruvic acid enters the mitochondria for the aerobic phase of cellular respiration. The pyruvic acid is incorporated into acetyl-CoA, which enters the citric acid cycle. Hydrogen ions produced in the citric acid cycle combine with oxygen in the electron transport chain to produce water and ATP. The complete breakdown of glucose produces 36 to 38 ATP.

■ **Define *glycogen, glycogenesis, glycogenolysis,* and *gluconeogenesis*.**
- Glycogen is the storage form of glucose. Glycogen is synthesized from glucose by glycogenesis. Glycogenolysis is the breakdown of glycogen into glucose and gluconeogenesis is the production of glucose from noncarbohydrate sources.

■ **Describe the pathway by which proteins are utilized in the body.**
- The end products of protein digestion are amino acids. Amino acids are used to synthesize proteins to build new tissues and replace damaged tissues. They are also used to synthesize hemoglobin, hormones, enzymes, and plasma proteins.
- Amino acids may be used as an energy source by removing the amino group (deamination). The resulting keto acid enters the citric acid cycle to produce energy.

■ **Describe the pathway by which fatty acids are broken down to produce ATP.**
- The end products of lipid digestion are monoglycerides and fatty acids, which are an important source of energy.
- Fatty acids are catabolized by beta oxidation, which removes two carbon segments from fatty acid chains and converts the segments to acetyl-CoA. The acetyl-CoA enters the citric acid cycle to produce ATP.

■ **Name two key molecules in the metabolism and interconversion of carbohydrates, proteins, and fats.**
- The end products of carbohydrate, protein, and lipid metabolism can be interconverted when necessary. Pyruvic acid and acetyl-CoA are key molecules in the metabolism and interconversion of carbohydrates, proteins, and fats.

■ **Explain how energy from food is measured.**
- Energy from food is measured in kilocalories, sometimes called a large Calorie (with a Capital C). One Calorie is the amount of energy required to raise the temperature of 1 kilogram of water from 14° Celsius to 15° Celsius.

■ **Define basal metabolism and state four factors that influence it.**
- Basal metabolism is the energy that is necessary to keep the body functioning at a minimal level. It is influenced by sex, muscle mass, age, hormones, and body temperature.
- Increasing physical activity increases energy expenditure in the body. Physical activity is the only means of voluntarily controlling energy expenditure.
- Thermogenesis is the production of heat through the assimilation of food. Digestion, absorption, and metabolism use chemical energy, and heat energy is produced.

Body Temperature

■ **Distinguish between core temperature and shell temperature.**
- Core temperature is the temperature of the internal organs; shell temperature is the temperature at the body surface.

■ **Explain three ways in which core temperature is maintained when the environmental temperature is cold.**
- Heat is produced by the catabolism of nutrients. The body produces additional heat by muscular contraction (shivering) and increasing metabolic rate (hormones). Heat is conserved by constriction of cutaneous blood vessels, which keeps the blood warmer by keeping it away from the cold surface of the body.

■ **State four ways in which heat is lost from the body.**
- Heat is lost from the body through radiation, conduction, convection, and evaporation.

■ **Describe the general mechanism by which the body maintains core temperature and identify the region of the brain that integrates this mechanism.**
- Body temperature is regulated by a homeostatic negative feedback mechanism that is integrated by the hypothalamus of the brain.

Basic Elements of Nutrition

■ **List the functions of carbohydrates, proteins, and fats in the body.**
- Carbohydrates provide energy, add bulk to the diet, and are used to synthesize other compounds.
- Proteins provide structure, regulate body processes, and provide energy.

- Lipids are an important source of energy, provide essential fatty acids, transport vitamins, are components of certain structural elements, provide heat insulation, and form protective cushions.

■ **State the caloric value of carbohydrates, proteins, and fats.**
- One gram pure carbohydrate yields 4 Calories of energy.
- One gram of protein yields 4 Calories of energy.
- One gram of fat yields 9 Calories of energy.

■ **Explain the importance of fiber in the diet.**
- Fiber adds bulk to the diet, enhances absorption of nutrients, and helps move the feces in the large intestine. Many diseases are related to decreased dietary fiber.

■ **Distinguish between essential and nonessential amino acids and between complete and incomplete proteins.**
- Essential amino acids must be supplied in the diet; they cannot be synthesized in the body. Nonessential amino acids can be synthesized from other amino acids that are available in the body.
- Complete proteins contain all the essential amino acids. Incomplete proteins lack one or more of the essential amino acids.

Incomplete proteins should be eaten in combinations that provide all the essential amino acids.

■ **Discuss the functions of vitamins, minerals, and water in the body.**
- Vitamins are organic molecules that are necessary for good health. Many vitamins are part of enzyme molecules, which are incomplete without the vitamin portion. These enzymes are necessary for the chemical reactions throughout the body.
- Minerals are inorganic substances that are necessary in small amounts to maintain good health. They are obtained from plants since the plants absorb them from the soil. Some minerals are components of body structures; some are parts of enzyme molecules; others help control fluid levels; and others become part of larger organic molecules.
- Water is an essential component of the diet. The average adult requires about 2.5 liters of water every day. Water is an integral part of body cells, provides a medium for chemical reactions, is a transport medium, is a lubricant, and helps maintain body temperature.

CHAPTER QUIZ

Recall

Match the definitions on the left with the appropriate term on the right.

_____ 1. Anaerobic catabolism of glucose
_____ 2. Acquisition, assimilation, and use of nutrients
_____ 3. Smaller molecules react to produce larger molecules and water
_____ 4. Production of glycogen from glucose
_____ 5. Process in catabolism of fatty acids
_____ 6. Temperature of internal organs
_____ 7. A, D, E, K
_____ 8. Glucose, fructose, and galactose
_____ 9. Produces ammonia and a keto acid from an amino acid
_____ 10. Production of heat in response to food intake

A. Beta oxidation
B. Core temperature
C. Deamination
D. Dehydration synthesis
E. Fat-soluble vitamins
F. Glycogenesis
G. Glycolysis
H. Monosaccharides
I. Nutrition
J. Thermogenesis

Thought

1. Glycerol combines with three fatty acids to form a triglyceride and three molecules of water. This is an example of:
   A. Hydrolysis   **C.** Dehydration synthesis
   B. Catabolism   **D.** Glycolysis

2. When there is oxygen deficiency (anaerobic conditions), pyruvic acid:
   A. Enters the citric acid cycle
   B. Undergoes oxidative phosphorylation
   C. Is converted to acetyl-CoA
   D. Is converted to lactic acid

3. The principal reaction that prepares amino acids for use as an energy source is:
   A. Deamination   **C.** Glycolysis
   B. Beta oxidation   **D.** Hydrolysis

4. The total body metabolic rate includes three factors. The only one of these that can be controlled voluntarily is:
   A. Basal metabolic rate   **C.** Thermogenesis
   B. Caloric intake   **D.** Physical activity

5. The mechanism that accounts for most of the heat loss from the body is:
   A. Radiation   **C.** Convection
   B. Conduction   **D.** Evaporation

Application

. . . .

A single serving of pudding contains 4 grams of protein, 31 grams of carbohydrate, and 5 grams of fat. How many Calories (kilocalories) are in this serving?

What would happen to the core body temperature if the peripheral cutaneous blood vessels constricted on a hot day?

BUILDING YOUR MEDICAL VOCABULARY

Building a vocabulary is a cumulative process. As you progress through this book, you will use word parts, abbreviations, and clinical terms from previous chapters. Each chapter will present new word parts, abbreviations, and terms to add to your expanding vocabulary.

Word Parts and Combining Form with Definition and Examples

PART/COMBINING FORM	DEFINITION	EXAMPLE
ana-	up	anabolism: build up or creating large complex molecules from smaller ones
cata-	down	catabolism: breaking down complex molecules into smaller units
-gen-	producing	glycogenesis: producing glycogen
lys-	to take apart	glycolysis: taking apart glucose
mal-	bad, poor	malnutrition: poor nutrition
neo-	new	gluconeogenesis: forming new glucose form noncarbohydrate sources
nutri-	to nourish	nutrition: nourishment of the body; study of the acquisition, assimilation, and use of nutrients contained in food
-pepsia	digestion	dyspepsia: impairment of the function of digestion; poor digestion
prote/o	protein	proteolysis: to take apart proteins by hydrolysis of the peptide bonds
pyr/o	fever, fire	pyrogen: agent that causes fever
therm/o	temperature	thermanalgesia: lack of sensitivity to heat or temperature
-tion	process of	nutrition: process of nourishing the body
vita-	life	vitamin: essential for life: originally they were thought to be amines essential for life, but it was later discovered they were not all amines

Clinical Abbreviations

ABBREVIATION	MEANING
ASAP	as soon as possible
BBT	basal body temperature
BMI	body mass index
BMR	basal metabolic rate
D/W	dextrose with water
FFA	free fatty acids
FUO	fever of unknown origin
HDL	high-density lipoproteins
LDL	low-density lipoproteins
MNT	medical nutrition therapy
PEG	percutaneous endoscopic gastrostomy (feeding tube placed in the stomach)
PPBS	postprandial blood sugar
temp	temperature
TPN	total parenteral nutrition
TPR	temperature, pulse, respiration

Clinical Terms

Achlorhydria (a-klor-HYE-dree-ah) Absence of hydrochloric acid in gastric secretions

Alimentation (al-ih-men-TAY-shun) Process of providing nutrients or nutrition for the body

Celiac disease (SEE-lee-ak dih-ZEEZ) Condition in which the ingestion of gluten destroys the villi of the small intestine, resulting in a malabsorption of nutrients

Heat exhaustion (HEET eks-AWST-shun) Condition characterized by fluid and electrolyte loss because of profuse perspiration, but body temperature remains normal; symptoms include muscle cramps, dizziness, vomiting, low blood pressure, and fainting; also called heat prostration

Heat stroke (HEET STROAK) Condition that occurs when core body temperature rises above 40.5° C (105° F). The excessive heat interferes with the temperature-regulating mechanism in the hypothalamus and the sweat glands cease to function. The skin is hot and dry. Immediate measures must be taken to lower body temperature because prolonged increases in core temperature can result in cardiac failure and death

Hypervitaminosis (hye-per-vye-tah-min-OH-sis) Excess of one or more vitamins, usually from the consumption of vitamin supplements; may become toxic and deadly

Hypothermia (hye-poh-THER-mee-ah) Refers to body temperature of 35° C (95° F) or below

Jaundice (JAWN-dis) Yellow discoloration of the skin and other tissues caused by excessive bilirubin in the blood

Kwashiorkor (kwosh-ee-OR-kor) Condition in which protein intake is deficient despite normal or nearly normal calorie intake

Malabsorption syndrome (mal-ab-SOARP-shun SIN-drohm) Group of disorders in which there is subnormal absorption of dietary constituents and thus excessive loss of nonabsorbed substances in the bowel

Malnutrition (mal-noo-TRIH-shun) State of poor nutrition

Marasmus (mar-AZ-mus) Condition in which there is deficiency in both protein and calorie intake

Pica (PYE-kah) Craving for substances not normally considered nutrients, such as dirt

Rickets (RICK-ehts) Softening of bone, occurring in childhood, because of inadequate amounts of calcium and phosphorus; bones bend easily and become deformed

Scurvy (SKUR-vee) Condition caused by a deficiency of vitamin C in the diet, which results in abnormal collagen synthesis

Undernutrition (un-der-noo-TRIH-shun) Inadequate food intake

VOCABULARY QUIZ

Use word parts you have learned to form words that have the following definitions.

1. Producing fever _____

2. Taking apart glycogen _____

3. Process of nourishing _____

4. Producing new glucose _____

5. Excessive temperature _____

Using the definitions of word parts you have learned, define the following words.

6. Hydrolysis _____

7. Hypothermia _____

8. Lipogenesis _____

9. Antipyrogenic _____

10. Thermogenesis _____

Match each of the following definitions with the correct word.

_____ **11.** A substance required for life

_____ **12.** Poor nourishment

_____ **13.** Throwing down or taking apart

_____ **14.** Taking apart fats

_____ **15.** Process of removing an amino group

A. Catabolism
B. Deamination
C. Lipolysis
D. Malnutrition
E. Vitamin

16. What term refers to a "body temperature of 35° C (95° F)"? _____

17. Name the condition in which there is a deficiency in protein and calorie intake. _____

18. What is the condition in which there is an absence of hydrochloric acid in gastric secretions? _____

19. What is the meaning of the abbreviation BMR? _____

20. What is the meaning of the abbreviation HDL? _____

Urinary System and Body Fluids

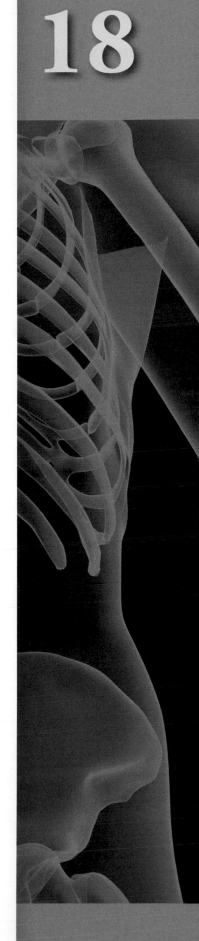

CHAPTER OBJECTIVES

Functions of the Urinary System
- State six functions of the urinary system.

Components of the Urinary System
- Describe the location and structural features of the kidneys.
- Draw and label a diagrammatic representation of a nephron.
- Name the two parts of the juxtaglomerular apparatus and state where they are located.
- Trace the pathway of blood flow through the kidney from the renal artery to the renal vein.
- Describe the location, structure, and function of the ureter, urinary bladder, and urethra.

Urine Formation
- Describe each of the three basic steps in urine formation.
- Identify three different types of pressure that affect the rate of glomerular filtration and describe how these interact.
- Explain why some substances, such as glucose, have limited reabsorption and describe what happens when concentration exceeds this limit.
- Explain how kidney function has a role in maintaining blood concentration, blood volume, and blood pressure.
- Name three hormones that affect kidney function and explain the effect of each one.
- Name the enzyme that stimulates the production of angiotensin II and is produced by the kidneys.

- Describe two mechanisms by which angiotensin II increases blood pressure.
- Define *micturition*.

Characteristics of Urine
- Describe the physical characteristics and chemical composition of urine.
- List five abnormal constituents of urine.

Body Fluids
- State the percentage of body weight that is composed of water.
- Identify the major fluid compartments in the body and state the relative amount of fluid in each compartment.
- State the sources of fluid intake and avenues of fluid output and explain how these are regulated to maintain fluid balance.
- Identify the major intracellular and extracellular ions and explain how electrolyte balance is regulated.
- State the normal pH of the blood and define the terms *acidosis* and *alkalosis*.
- Describe the three primary mechanisms by which blood pH is regulated.
- State the normal values for P_{CO_2} and HCO_3^- in the blood.
- Describe the causes and indicators of respiratory acidosis, respiratory alkalosis, metabolic acidosis, and metabolic alkalosis.

KEY TERMS

Acidosis (AS-id-oh-sis) Condition in which the blood has a lower pH than normal

Alkalosis (AL-kah-loh-sis) Condition in which the blood has a higher pH than normal

Extracellular fluid (eks-trah-SELL-yoo-lar FLOO-id) Fluid in the body that is not inside cells; includes plasma and interstitial fluid

Glomerular capsule (gloh-MER-yoo-lar KAP-sool) Double-layered epithelial cup that surrounds the glomerulus in a nephron; also called Bowman's capsule

Glomerular filtration (gloh-MER-yoo-lar fil-TRAY-shun) Movement of blood plasma across the filtration membrane in the renal corpuscle.

Glomerulus (gloh-MER-yoo-lus) Cluster of capillaries in the nephron through which blood is filtered.

Interstitial fluid (in-ter-STISH-al FLOO-id) Portion of the extracellular fluid that is found in the microscopic spaces between cells

Intracellular fluid (in-trah-SELL-yoo-lar FLOO-id) Fluid inside body cells

Intravascular fluid (in-trah-VAS-kyoo-lar FLOO-id) Portion of extracellular fluid that is in the blood; plasma

Juxtaglomerular apparatus (juks-tah-gloh-MER-yoo-lar ap-pah-RAT-us) Complex of modified cells in the afferent arteriole and the ascending limb/distal tubule in the kidney; helps regulate blood pressure by secreting renin; consists of the macula densa and juxtaglomerular cells

Micturition (mick-too-RISH-un) Act of expelling urine from the bladder; also called urination or voiding.

Nephron (NEFF-rahn) Functional unit of the kidney consisting of a renal corpuscle and a renal tubule

Renal tubule (REE-nal TOOB-yool) Tubular portion of the nephron that carries the filtrate away from the glomerular capsule; site where tubular reabsorption and secretion occur

Tubular reabsorption (TOOB-yoo-lar ree-ab-SORP-shun) Movement of filtrate from the renal tubules back into the blood during urine formation.

Tubular secretion (TOOB-yoo-lar see-KREE-shun) Movement of substances from the blood into the renal tubules during urine formation.

Functional Relationships of the
Urinary System

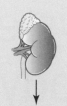

Eliminates metabolic wastes; helps maintain
pH and ion concentration of body fluids.

Reproductive

Prostate surrounds urethra in males and may
compress it to cause urine retention; pregnant
uterus pushes on bladder and causes
frequent urination.

Maintains pH and electrolyte composition for
effective gonadal hormone function; maternal
urinary system excretes metabolic wastes from developing fetus.

Digestive

Liver synthesizes urea, which is excreted
by kidneys; liver metabolizes hormones,
toxins, and drugs to forms that can be
excreted in urine; absorbs nutrients for tissues
of urinary system.

Excretes toxins absorbed by the digestive system;
excretes metabolic wastes produced by the liver.

Respiratory

Assists in regulation of
pH by removing carbon dioxide.

Helps maintain pH and electrolyte
composition of body fluids for effective
respiratory function; eliminates metabolic
waste products generated by
respiratory organs.

Lymphatic/Immune

Immune cells provide specific defense
mechanisms against pathogens that enter
the urinary tract.

Helps maintain pH and electrolyte composition
of body fluids for effective immune cell function; acid
pH and urine flow provide a barrier against urinary
tract pathogens.

Cardiovascular

Delivers wastes to be excreted by the kidneys;
adjusts blood flow to maintain kidney function; transports
hormones that regulate urine formation in the kidneys.

Helps maintain blood pH and electrolyte composition; adjusts
blood volume and pressureby altering urine composition;
initiates the renin-angiotensin-aldosterone mechanism.

Integument

Alternative excretory route for some salts and
nitrogenous wastes; limits fluid loss.

Maintains fluid and electrolyte balance which
is necessary for production of sweat.

Skeletal

Supports and protects organs of the
urinary system.

Conserves calcium and phosphate ions
for bone growth, maintenance, and repair.

Muscular

Supports and protects organs of the urinary
system; sphincters control voluntary urination.

Conserves calcium ions necessary for
muscle contraction; eliminates wastes
produced by muscle metabolism.

Nervous

Autonomic nervous system controls renal
blood pressure and blood flow, which
affect rate of urine formation; regulates
bladder emptying.

Helps maintain pH and electrolyte balance
necessary for neural function; eliminates
metabolic wastes harmful to nerve function.

Endocrine

Hormones, such as aldosterone, ADH, and atrial natriuretic
hormone, regulate urine formation in the kidneys.

Helps maintain pH and electrolyte balance necessary for
hormone function; eliminates inactivated hormones and
other metabolic wastes; releases renin and erythropoietin.

← Gives to Urinary System
→ Receives from Urinary System

Functions of the Urinary System

The overall function of the **urinary system** is to maintain the volume and composition of body fluids within normal limits. One aspect of this function is to rid the body of waste products that accumulate as a result of cellular metabolism, and because of this the urinary system is sometimes referred to as the excretory system. Although the urinary system has a major role in excretion, other organs contribute to the excretory function. Some waste products, such as carbon dioxide and water, are excreted by the lungs in the respiratory system. The skin is another excretory organ that rids the body of wastes through the sweat glands. The liver and intestines excrete bile pigments that result from the destruction of hemoglobin. The major task of excretion still belongs to the urinary system, and if the system fails the other organs cannot adequately compensate and toxic levels of waste products accumulate rapidly. In addition to ridding the body of waste materials, the urinary system maintains an appropriate fluid volume by regulating the amount of water that is excreted in the urine. Other aspects of its function include regulating the concentrations of various electrolytes in the body fluids and maintaining normal pH of the blood.

In addition to maintaining fluid homeostasis in the body, the urinary system controls red blood cell production by secreting the hormone **erythropoietin** (ee-rith-roh-poy-EE-tin). The urinary system also plays a role in maintaining normal blood pressure by secreting the enzyme **renin**, which activates angiotensin II to increase blood pressure.

Components of the Urinary System

The **urinary system** consists of the kidneys, ureters, urinary bladder, and urethra. The kidneys produce the urine and account for the other functions attributed to the urinary system. The ureters convey the urine away from the kidneys to the urinary bladder, which is a temporary reservoir for the urine. The urethra is a tubular structure that carries the urine from the urinary bladder to outside of the body. The components of the urinary system are illustrated in Figure 18-1.

QUICK **CHECK**
18.1 List six functions of the urinary system.
18.2 List the components of the urinary system.

Kidneys

The **kidneys** are the primary organs of the urinary system. They are the organs that perform the functions of the urinary system. They filter the blood, remove the wastes, and excrete the wastes in the urine. The other components are accessory structures to help eliminate the urine from the body. When the kidneys fail to remove the waste products from the blood, toxic levels of urea may accumulate, resulting in **uremia**. The body attempts to compensate by excreting urea through the sweat glands. After the perspiration evaporates, tiny crystals of urea remain on the skin. This is called **uremic frost**.

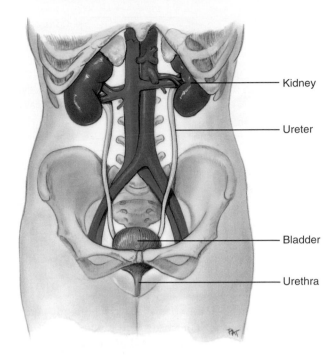

Figure 18-1 *Components of the urinary system.*

Location

The paired kidneys are located between the twelfth thoracic and third lumbar vertebrae, one on each side of the vertebral column. The right kidney usually is slightly lower than the left because the liver displaces it downward. The kidneys, partially protected by the lower ribs, lie in shallow depressions against the posterior abdominal wall and behind the parietal peritoneum. This means they are retroperitoneal. Each kidney is held in place by connective tissue, called **renal fascia**, and is surrounded by a thick layer of adipose tissue, called **perirenal fat**, that helps to protect it. A tough, fibrous connective tissue renal capsule closely envelopes each kidney and provides support for the soft tissue that is inside. **Nephroptosis**, commonly referred to as a floating kidney, occurs when the kidney is no longer held in place by the renal fascia and it drops out of its normal position. This makes the kidney more vulnerable to injury because it is no longer protected by the ribs.

Macroscopic Structure

In the adult, each kidney is approximately 3 centimeters (cm) thick, 6 cm wide, and 12 cm long. It is generally bean shaped with an indentation, called the **hilum**, on the medial side. The hilum leads to a large cavity, called the **renal sinus**, within the kidney. The **ureter** and **renal vein** leave the kidney, and the **renal artery** enters the kidney at the hilum.

A frontal (coronal) section through the kidney illustrates the macroscopic internal structure (Figure 18-2). The outer, reddish region, next to the capsule, is the **renal cortex**. This surrounds a darker reddish brown region called the **renal medulla**. The renal medulla consists of a series of **renal pyramids**, which appear striated because they contain straight tubular structures and blood vessels. The wide bases of the pyramids are adjacent to the cortex and the pointed ends, called **renal papillae**, are directed toward the center of the kidney. Portions of the

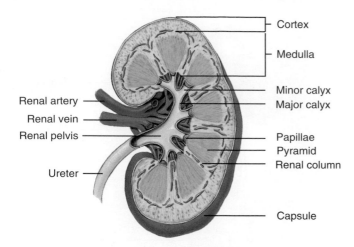

Figure 18-2 *Coronal (frontal) section through the kidney.*

renal cortex extend into the spaces between adjacent pyramids to form **renal columns**. The cortex and medulla make up the parenchyma, or functional tissue, of the kidney.

The central region of the kidney contains the **renal pelvis**, which is located in the renal sinus and is continuous with the ureter. The renal pelvis is a large cavity that collects the urine as it is produced. The periphery of the renal pelvis is interrupted by cuplike projections called **calyces**. A **minor calyx** surrounds the renal papillae of each pyramid and collects urine from that pyramid. Several minor calyces converge to form a **major calyx**. From the major calyces the urine flows into the renal pelvis and from there into the ureter.

Nephrons

Each kidney contains over a million functional units, called **nephrons**, in the parenchyma (cortex and medulla). All of the nephrons in an adult kidney are present at birth. Growth of the kidney is due to enlargement of individual nephrons. When nephrons are damaged, they are not replaced. A nephron has two parts (Figure 18-3): the renal corpuscle and the renal tubule.

The renal corpuscle consists of a cluster of capillaries, called the **glomerulus** (gloh-MER-yoo-lus), surrounded by a double-layered epithelial cup, called the **glomerular capsule (Bowman's capsule)**. Blood enters the glomerulus through an **afferent arteriole**, is filtered in the glomerulus, and leaves through an **efferent arteriole**. As the blood is filtered, the filtrate enters the glomerular capsule, which is continuous with the renal tubule. Renal corpuscles are located in the cortex of the kidney and give it a granular appearance.

The renal tubule, which carries fluid away from the glomerular capsule, consists of three regions: the proximal convoluted tubule, the nephron loop (Henle's loop), and the distal convoluted tubule. The first portion of the tubule, the proximal convoluted tubule, is highly coiled. Next the tubule straightens and dips into the medulla, makes a U-turn, and ascends back toward the cortex. This forms the nephron loop. The portion of the loop that descends from the proximal convoluted tubule into the medulla is the **descending limb**, and the part that ascends back toward the cortex is the **ascending limb**. The final region of the tubule, also coiled and found in the cortex, is the distal convoluted tubule. Figure 18-3 illustrates the regions of the renal tubule.

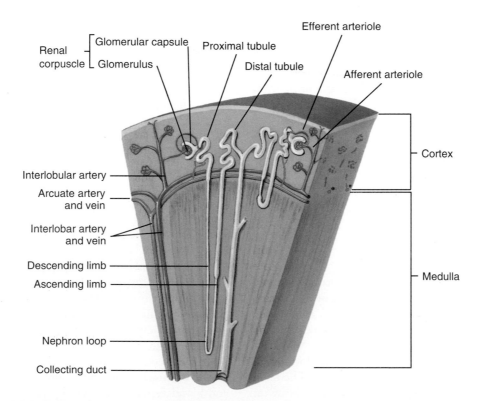

Figure 18-3 *A section of a kidney showing the structures in the cortex and those in the medulla. The renal pyramids in the medulla contain the nephron loops and collecting ducts.*

QUICK CHECK

18.3 What portion of the kidney collects urine as it is produced?

18.4 What are the two parts of a renal corpuscle?

18.5 What are the three portions of the renal tubule?

Collecting Ducts

Urine passes from the nephrons into **collecting ducts**. The distal convoluted tubules from several nephrons join with each collecting duct. The collecting ducts extend from the base of the pyramids to the renal papillae. These straight tubules, with the nephron loops and blood vessels, give the medulla its striated appearance. Fluid flows from the collecting ducts into the minor calyces that surround the renal papillae.

Juxtaglomerular Apparatus

The ascending limb of the nephron loop, in the region where it continues into the distal convoluted tubule, comes into contact with the glomerular afferent arteriole of the same nephron (Figure 18-4). In the region of contact, the cells of the ascending limb are modified to form the **macula densa**, and those in the afferent arteriole are modified to form the **juxtaglomerular** (juks-tah-gloh-MER-you-lar) **cells**. The macula densa monitors sodium chloride concentration in the urine and also influences the juxtaglomerular cells. In the afferent arteriole, the juxtaglomerular cells produce the enzyme **renin**, which has a role in the regulation of blood pressure. Together, the macula densa and juxtaglomerular cells make up the **juxtaglomerular apparatus**.

Blood Flow Through the Kidney

Blood flows through the kidneys at an approximate rate of 1200 milliliter per minute. This is about one fourth of the

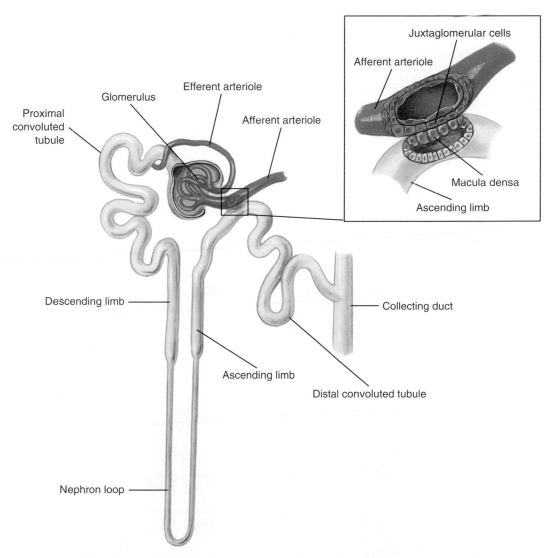

Figure 18-4 *Juxtaglomerular apparatus and its relationship to the nephron. The juxtaglomerular apparatus is in the boxes. In the region of contact, the cells of the ascending limb are modified to form the macula densa and the cells of the afferent arteriole are modified to form the juxtaglomerular cells. Together, these modified regions are the juxtaglomerular apparatus.*

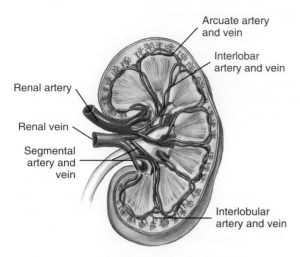

Figure 18-5 *Blood supply to the kidneys. Blood is brought to the kidney by the renal arteries. At the hilum, the renal arteries divide into segmental arteries. Blood is returned to the inferior vena cava by the renal veins.*

total cardiac output. Blood is brought to the kidneys by the renal arteries, which are branches from the abdominal aorta (Figure 18-5).

At the hilum, the renal arteries divide into **segmental arteries** that pass through the renal sinus. **Interlobar arteries** branch from the segmental arteries, pass through the renal columns, and then divide to form **arcuate arteries**, which pass over the base of the pyramids. The arcuate arteries have branches, called **interlobular arteries**, which extend into the cortex. The **afferent arterioles** are branches of the interlobular arteries. From the afferent arteriole, the blood passes through the capillaries in the glomerulus of the renal corpuscle, and then into the efferent arteriole. Each efferent arteriole divides to form an extensive capillary network, called **peritubular capillaries**, around the tubular portion of the nephron.

The peritubular capillaries eventually reunite to form **interlobular veins**. From there, the blood flows through **arcuate veins, interlobar veins, segmental veins**, and into the **renal veins**, which return the blood to the inferior vena cava (Figure 18-6).

QUICK CHECK

18.6 What are the two regions of the juxtaglomerular apparatus?
18.7 The afferent arterioles of the glomerulus branch from what arteries?
18.8 Where are the arcuate veins located in the kidney?

Ureters

Each **ureter** is a small tube, about 25 cm long, that carries urine from the renal pelvis to the urinary bladder. It descends from the renal pelvis, along the posterior abdominal wall, behind the parietal peritoneum, and enters the urinary bladder on the posterior inferior surface (Figure 18-7, *A*).

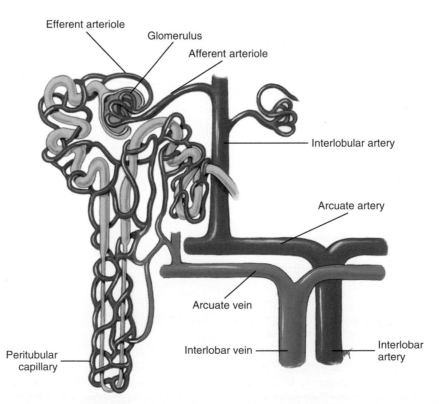

Figure 18-6 *Branching of the interlobar arteries in the kidneys. Blood flows from the interlobar arteries into the arcuate arteries, then into the interlobular arteries, then into the afferent arterioles to the glomerulus. Efferent arterioles from the glomerulus branch into peritubular capillaries.*

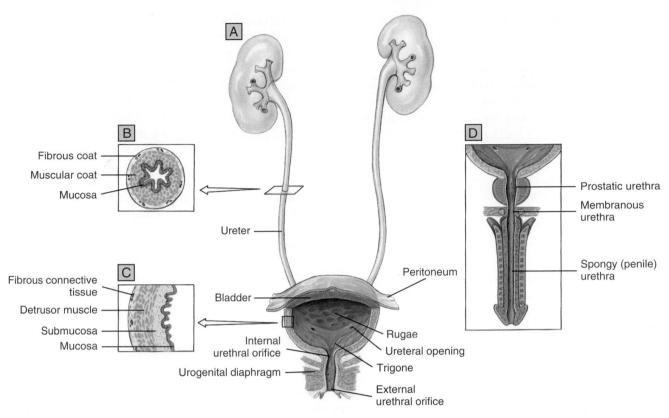

Figure 18-7 *Ureter, urinary bladder, and urethra. **A**, Urinary tract. **B**, Cross section through the ureter. **C**, Cross section of the bladder wall. **D**, Regions of the male urethra.*

The wall of the ureter consists of three layers (Figure 18-7, *B*). The outer layer, the **fibrous coat**, is a supporting layer of fibrous connective tissue. The middle layer, the **muscular coat**, consists of inner circular and outer longitudinal smooth muscle. The main function of this layer is peristalsis to propel the urine. The inner layer, the **mucosa**, is transitional epithelium that is continuous with the lining of the renal pelvis and the urinary bladder. This layer secretes mucus, which coats and protects the surface of the cells.

Urinary Bladder

The **urinary bladder** is a temporary storage reservoir for urine (see Figure 18-7, *A*). It is located in the pelvic cavity, posterior to the symphysis pubis, and below the parietal peritoneum. The size and shape of the urinary bladder vary with the amount of urine it contains and with the pressure from surrounding organs.

The inner lining of the urinary bladder is a **mucous membrane** of transitional epithelium that is continuous with that in the ureters (see Figure 18-7, *C*). When the bladder is empty, the mucosa has numerous folds called **rugae**. The rugae and transitional epithelium allow the bladder to expand as it fills. The second layer in the wall is the **submucosa** that supports the mucous membrane. It is composed of connective tissue with elastic fibers. The next layer is the **muscularis**, which is composed of smooth muscle. The smooth muscle fibers are interwoven in all directions, and collectively these are called the **detrusor** (dee-TROO-sor)

muscle. Contraction of this muscle expels urine from the bladder. On the superior surface, the outer layer of the bladder wall is parietal peritoneum. In all other regions, the outer layer is fibrous connective tissue.

There is a triangular area, called the **trigone**, formed by three openings in the floor of the urinary bladder. Two of the openings are from the ureters and form the base of the trigone. Small flaps of mucosa cover these openings and act as valves that allow urine to enter the bladder but prevent it from backing up from the bladder into the ureters. The third opening, at the apex of the trigone, is the opening into the urethra. A band of the detrusor muscle encircles this opening to form the **internal urethral sphincter**. Infections of the urinary bladder tend to persist in the region of the trigone.

Urethra

The final passageway for the flow of urine is the **urethra**, a thin-walled tube that conveys urine from the floor of the urinary bladder to outside of the body (see Figure 18-7). The opening to the outside is the **external urethral orifice**. The mucosal lining of the urethra is transitional epithelium. The urethral wall also contains smooth muscle fibers and is supported by connective tissue.

The beginning of the urethra, where it leaves the urinary bladder, is surrounded by the **internal urethral sphincter**. This sphincter is smooth (involuntary) muscle. Another sphincter, the **external urethral sphincter**, is skeletal (voluntary) muscle and encircles the urethra where it passes through the pelvic floor. These two sphincters control the flow of urine through the urethra.

In females the urethra is short, only 3 to 4 cm (about 1.5 inches) long. The external urethral orifice opens to the outside just anterior to the opening for the vagina.

In males the urethra is much longer, about 20 cm (7 to 8 inches) in length, and transports both urine and semen (see Figure 18-7, *D*). The first part, next to the urinary bladder, passes through the prostate gland and is called the **prostatic urethra**. The second part, a short region that penetrates the pelvic floor and enters the penis, is the **membranous urethra**. The third part, the **spongy urethra**, is the longest region. This portion of the urethra extends the entire length of the penis, and the external urethral orifice opens to the outside at the tip of the penis.

QUICK CHECK

18.9 Obstruction of the ureter will interfere with the flow of urine between what two structures?

18.10 Where is the detrusor muscle located?

18.11 What is the difference between the muscle in the internal and external urethral sphincters?

18.12 What are the three regions of the male urethra?

QUICK APPLICATION

Urinary tract infections (UTIs) occur more frequently in women than in men because of differences in the urethra. In females, the urethral opening is in close proximity to the anal opening, which gives intestinal bacteria easier access to the urethra. The female urethra is short, which allows any infection to spread to the urinary bladder. An infection of the urethra is called **urethritis**, and one of the urinary bladder is called **cystitis**.

Urine Formation

The work of the kidneys, performed by the nephrons, is accomplished through the formation of urine. As urine is excreted to the outside of the body, it carries with it the wastes, excess water, and excess electrolytes. At the same time, the kidneys conserve other electrolytes to maintain the appropriate balance. The formation of urine involves three basic steps: glomerular filtration, tubular reabsorption, and tubular secretion.

Glomerular Filtration

The first step in the formation of urine is **glomerular filtration**. During this process, blood plasma moves across the **filtration membrane** in the renal corpuscle and enters the glomerular capsule. The filtration membrane consists of the capillary endothelium of the glomerulus and the endothelium of the capsule. The force that moves the fluid across the membrane is **filtration pressure**, and the fluid that enters the capsule is the **filtrate**.

Blood flows through the kidneys at an average rate of 1200 milliliters per minute. As the blood passes through the glomeruli, about 19% of the plasma enters the glomerular capsule as filtrate. This is equivalent to forming filtrate at a rate of 125 milliliters per minute, or 180 liters per day. This is the total value for all the nephrons in both kidneys. The filtration membrane acts as a barrier that prevents blood cells and protein molecules from entering the capsule; therefore, they are absent from the filtrate. Normally, the filtrate in the glomerular capsule is similar in composition to blood plasma except that the filtrate lacks plasma proteins. In a diseased kidney, the filtration membrane may become too porous and allow blood cells and proteins to pass through. This alters the filtration rate, and blood cells and proteins appear in the urine.

The filtration rate is directly related to the filtration pressure. When filtration pressure increases, filtration rate increases, more filtrate is formed, and more urine is produced. If the filtration pressure decreases, filtration rate decreases, less filtrate is formed, and less urine is produced. Filtration pressure is influenced by the blood pressure in the glomerulus (glomerular hydrostatic pressure), the hydrostatic pressure of the fluid in the glomerular capsule (capsular hydrostatic pressure), and the osmotic pressure created by the plasma proteins (glomerular osmotic pressure). Figure 18-8 illustrates the factors that influence filtration pressure. In a healthy kidney, there are no proteins in the glomerular capsule so there is no capsular osmotic pressure. Net filtration pressure is the glomerular hydrostatic pressure, which moves substances into the capsule, minus the other two pressures that move substances into the glomerulus. In general, 1 mm Hg filtration pressure is equivalent to 12.5 milliliter of filtrate formed per minute.

Tubular Reabsorption

If the volume and composition of the filtrate in the glomerular capsule are compared with the volume and composition of urine, it is obvious that changes occur after filtration. First, about 180 liters (45 gallons) of filtrate are formed in a 24-hour period. This volume is reduced to 1 to 2 liters of urine. Glucose is present in the filtrate but normally absent in the urine. Urea and uric acid are present in higher concentrations in the urine than in the filtrate.

Tubular reabsorption is the first process that changes the volume and composition of the filtrate. Tubular reabsorption is the movement of substances from the filtrate in the kidney tubules into the blood in the peritubular capillaries. Only about 1% of the filtrate remains in the tubules and becomes urine. In general, water and other substances that are useful to the body are reabsorbed. Wastes remain in the filtrate and are excreted in the urine.

About 65% of the reabsorption takes place in the proximal convoluted tubule, 15% in the nephron loop (loop of Henle), and 19% in the distal convoluted tubule and collecting duct. This leaves about 1% of the filtrate to be excreted as urine. Most of the solutes (molecules and ions) are reabsorbed by active transport mechanisms. Some of the negative ions passively follow the positive ions to maintain electrical

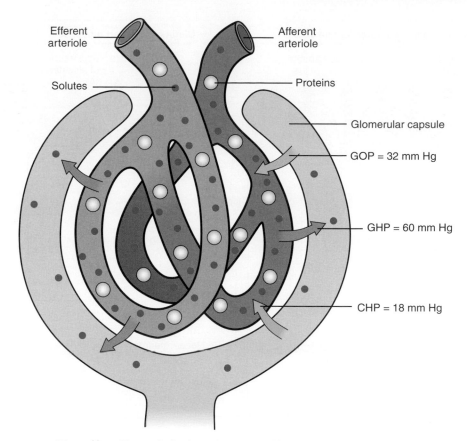

Efferent arteriole

Afferent arteriole

Solutes

Proteins

Glomerular capsule

GOP = 32 mm Hg

GHP = 60 mm Hg

CHP = 18 mm Hg

60 mm Hg = Glomerular hydrostatic pressure (GHP)
−32 mm Hg = Glomerular osmotic pressure (GOP)
−18 mm Hg = Capsular hydrostatic pressure (CHP)

10 mm Hg = Net filtration pressure

Figure 18-8 *Factors that influence filtration pressure. Glomerular hydrostatic pressure (GHP) moves water and solutes from the glomerulus into the capsule. Glomerular osmotic pressure (GOP) and capsular hydrostatic pressure (CHP) move substances from the capsule into the glomerulus.*

neutrality. Because active transport uses carrier molecules, reabsorption of some solutes is limited by the activity of the carriers. Glucose is a good example of this limitation. Under normal conditions, glucose freely passes through the filtration membrane so that the concentration in the filtrate is the same as that in the blood. Glucose is then totally reabsorbed by active transport so there is no glucose present in the urine. Under certain conditions, such as with untreated diabetes mellitus, the glucose concentration in the blood and filtrate may exceed the **renal threshold**, or transport maximum. When this happens the excess glucose remains in the filtrate and appears in the urine.

QUICK APPLICATION

Renal diabetes is a condition in which there are not enough functional carrier molecules to reabsorb normal amounts of glucose. Even though blood glucose levels are normal, glucose still appears in the urine because there is inadequate reabsorption.

Water is reabsorbed by osmosis in all parts of the tubule, except in the ascending limb of the loop. Reabsorption of the molecules and ions creates concentration differences and water follows by osmosis. The exception is in the ascending limb of the nephron loop, which is impermeable to water. Here, solutes, but very little water, are removed from the filtrate.

In the cortex of the kidney, where the proximal convoluted tubules are located, the filtrate is isotonic (same concentration) to the interstitial fluid and the plasma in the peritubular capillaries. When molecules and ions are reabsorbed by active transport, water follows by osmosis and volume is reduced.

There is a concentration gradient in the interstitial fluid of the medulla in which the nephron loops are located. Near the base, next to the cortex, the concentration is nearly the same as that in the cortex, but in the papillae the interstitial fluid is highly concentrated. As the filtrate moves down the descending limb, water leaves the tubule by osmosis, which creates a highly concentrated filtrate. The ascending limb is impermeable

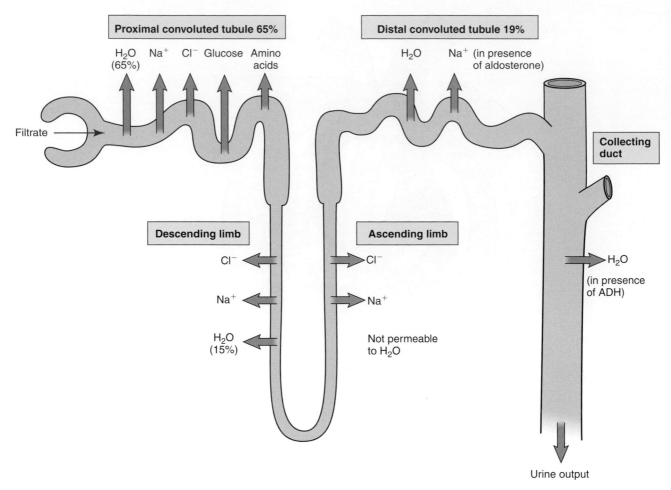

Figure 18-9 *Tubular reabsorption. Water and other useful substances move from the filtrate in the tubules into the interstitial fluid and peritubular capillaries.*

to water, but solutes leave the filtrate by active transport. This creates a dilute filtrate that enters the distal convoluted tubule. Water then leaves the distal convoluted tubule and collecting duct by osmosis (Figure 18-9).

Tubular Secretion

The final process in the formation of urine is the transport of molecules and ions into the filtrate. This is called **tubular secretion**. Most of these substances are waste products of cellular metabolism that become toxic if allowed to accumulate in the body. Tubular secretion is the method by which some drugs, such as penicillin, are removed from the body. The tubular secretion of hydrogen ions plays an important role in regulating the pH of the blood. Other molecules and ions that may enter the filtrate by tubular secretion include potassium ions, creatinine, and histamine.

The final product, urine, produced by the nephrons consists of the substances that are filtered in the renal corpuscle, minus the substances that are reabsorbed in the tubules, plus the substances that are added by tubular secretion (Figure 18-10).

QUICK CHECK

18.13 How does the glomerular filtrate differ from blood plasma?

18.14 Where does most tubular reabsorption occur?

18.15 Why does glucose appear in the urine of persons with untreated diabetes mellitus?

18.16 What region of the renal tubule is impermeable to water?

Regulation of Urine Concentration and Volume

The concentration and volume of urine depend on conditions in the internal environment of the body. Cells in the hypothalamus are sensitive to changes in the composition of the blood and initiate appropriate responses that affect the kidneys. If the concentration of solutes in the blood increases above normal, the kidneys excrete a small volume of concentrated urine. This conserves water in the body and gets rid of solutes to restore the blood to normal. If the blood solute concentration decreases below normal, the kidneys conserve solutes and get rid of water by producing large quantities of dilute urine. Urine production

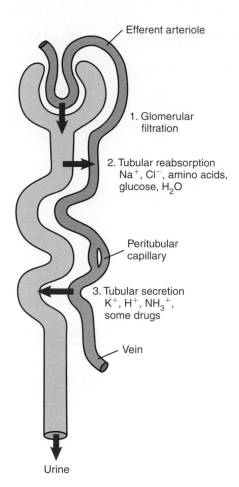

Figure 18-10 *Steps in urine formation. Urine consists of the substances that enter the tubules in glomerular filtration minus substances that are reabsorbed in the tubules plus substances that are secreted into the tubules.*

TABLE 18-1 Hormones That Influence Urine Concentration and Volume

Hormone	Source	Effect
Aldosterone	Adrenal cortex	Promotes reabsorption of sodium and water in kidney tubules; reduces urine output.
Antidiuretic hormone (ADH)	Posterior pituitary	Promotes reabsorption of water in distal tubules and collecting duct; reduces urine output.
Atrial natriuretic hormone	Heart	Promotes excretion of sodium and water in kidney; inhibits aldosterone, ADH, and rennin; increases urine output

of sodium and water by acting directly on the kidney tubules and by inhibiting the secretion of ADH, renin, and aldosterone. The result of atrial natriuretic hormone is a decrease in both blood volume and blood pressure. Table 18-1 summarizes the hormones that influence urine concentration and volume.

Renin is an enzyme that is produced by the juxtaglomerular cells in the kidney in response to low blood pressure or decreased blood sodium concentration. Renin promotes the production of **angiotensin II** in the blood. Angiotensin II is a powerful vasoconstrictor, which increases the blood pressure. Angiotensin II also stimulates the adrenal gland to secrete aldosterone, which acts on the kidney tubules to conserve sodium and water. This increases blood volume and, consequently, increases blood pressure. Figure 18-11 illustrates the role of renin in the regulation of blood volume and blood pressure.

Micturition

Micturition (mik-too-RISH-un), commonly called *urination* or *voiding*, is the act of expelling urine from the bladder. This is a reflex action in infants and very young children. By the age of three, most children learn to urinate voluntarily and to inhibit the reflex when necessary. The bladder can hold up to a liter of urine, but normally when it contains 200 to 400 milliliters, stretch receptors in the bladder wall trigger impulses that initiate the **micturition reflex**. This is an automatic and involuntary response that is coordinated in the spinal cord. Impulses are transmitted along parasympathetic nerves that cause contraction of the detrusor muscle and relaxation of the internal sphincter. Urine then enters the urethra. If the voluntarily controlled sphincter is relaxed, micturition occurs. Voluntary contraction of the external sphincter suppresses micturition. Even though the micturition reflex is involuntary, it can be inhibited or stimulated by higher brain centers.

Urinary incontinence is the inability to control urination and to retain urine in the bladder. Temporary incontinence may result when the muscles around the bladder and urethra become

plays an important role in maintaining homeostasis of blood concentration and volume. By regulating blood volume, the kidneys also play a role in regulating blood pressure because volume is directly related to pressure.

Three hormones—**aldosterone, antidiuretic hormone**, and **atrial natriuretic hormone**—influence urine concentration and volume. Aldosterone, secreted by cells of the adrenal cortex, acts on the kidney tubules to increase the reabsorption of sodium. When sodium is reabsorbed, water follows by osmosis. This reduces urine output.

Antidiuretic hormone (ADH) is produced by cells in the hypothalamus and is released from the posterior lobe of the pituitary gland. ADH makes the distal convoluted tubule and collecting duct more permeable to water. When ADH is present, more water is reabsorbed, which reduces the volume of urine and makes it more concentrated. Water is conserved in the body. In the absence of ADH, the tubules are less permeable to water and there is less reabsorption. This results in large quantities of dilute urine and water is lost from the body. Alcohol inhibits the secretion of ADH. When people drink alcohol, they experience dieresis, or excessive urination.

Special cells in the heart produce a hormone called atrial natriuretic hormone, or atriopeptin, which is secreted when the atrial cells are stretched. This hormone promotes the excretion

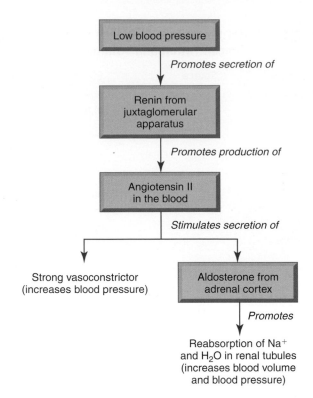

Figure 18-11 *Role of renin in the regulation of blood volume and blood pressure.*

weakened and lose muscle tone. This sometimes is caused by stretching of the muscles during childbirth. Because these muscles help restrict the outlet of the bladder, their weakness contributes to a leakage of urine. A cough or sneeze may increase pressure within the bladder sufficiently to force urine to escape. Permanent incontinence usually is caused by damage to the central nervous system or by extensive damage to the bladder or urethra.

Characteristics of Urine

The volume, physical characteristics, and chemical composition of urine change with diet, physical activity, and state of health. A chemical and microscopic examination of urine, called **urinalysis** (yoo-rin-AL-ih-sis), reveals a great deal about the physiologic condition of a person.

Physical Characteristics

A normal, healthy individual excretes 1 to 2 liters of urine per day; however, the volume varies considerably. Even under conditions of water deprivation, the body needs to excrete a certain volume of urine to free itself of toxic metabolic wastes. Freshly voided urine is normally clear and has a yellowish color. The color is due to the presence of **urochrome** (YOO-roh-krohm), a substance produced by the breakdown of bile pigments. The pH ranges between 4.6 and 8.0 with an average of about 6.0. High-protein diets increase the acidity and lower the pH. Vegetarian diets generally make the urine more alkaline and increase the pH. The specific gravity of urine varies from 1.001 to 1.035, which means that it is slightly heavier than water. This is due to the presence of solutes in the urine. The higher the concentration of solutes,

TABLE 18-2 Physical Characteristics of Urine

Characteristic	Description
Volume	1 to 2 liters per 24-hour period; average 1500 milliliters per day
pH	Varies between 4.6 and 8; average about 6; more acidic with high-protein diets and more alkaline with vegetarian diets
Specific gravity	Varies between 1.001 and 1.035; slightly heavier than water
Color	Yellow because of the presence of urochrome
Composition	95% water and 5% solutes; predominant solute is urea

the higher is the specific gravity. These characteristics are summarized in Table 18-2.

Chemical Composition

Urine is about 95% water by volume. The remaining 5% consists of various solutes. Most of the solutes are metabolic waste products that need to be eliminated from the body. Others, such as drugs, come from outside sources.

QUICK APPLICATION

Kidney stones develop when uric acid or calcium salts precipitate instead of remaining dissolved in the urine. The stones usually form in the renal pelvis, but they may also develop in the urinary bladder. If small enough, they may pass naturally with urine flow but usually cause a lot of discomfort. If kidney stones cause a serious obstruction, they may need to be surgically removed. A newer method of treatment called lithotripsy uses high-frequency sound waves to break the stone into small pieces so that it may pass naturally. The formation of stones in the urine is called **urolithiasis**.

Abnormal Constituents

Certain substances, although normally not present in the urine, may appear from time to time. Their presence in the urine may suggest some pathologic condition. Examples of abnormal constituents include albumin, glucose, blood cells, ketone bodies, and microbes (see Table 18-3).

QUICK CHECK

18.17 Name three hormones that regulate urine concentration and volume.

18.18 What is the function of renin from the juxtaglomerular cells?

18.19 What is the yellow pigment in urine?

18.20 What volume of urine is excreted by a normal healthy individual in 1 day?

FROM THE PHARMACY

Gout is a disorder associated with defective uric acid metabolism. The consistent feature of gout is hyperuricemia (elevated levels of uric acid in the blood) either from increased production or from decreased excretion of uric acid. The disease causes attacks of acute pain, swelling, and tenderness in the joints (gouty arthritis), particularly the big toe, ankle, knee, and elbow. The onset of gout is usually during middle age and predominantly affects males. In addition to the gouty arthritis caused by uric acid salts in the joints, uric acid stones may form in the kidneys, leading to nephritis. Risk factors for developing gout include obesity, hypertension, and alcohol consumption. Treatment is aimed at (1) ending the attack as soon as possible, (2) preventing recurrent attacks, (3) preventing the formation of stones in the kidney, and (4) reducing or preventing complications.

Nonsteroidal antiinflammatory drugs (NSAIDs) and corticosteroids may be used to treat the inflammation of an acute attack, but they do not address the underlying metabolic problem. The primary medications used to minimize the deposits in the joints and stones in the kidneys are colchicine, allopurinol, probenecid, and sulfinpyrazone. **Colchicine** reduces the number of urate deposits, but has no effect on the uric acid levels in the blood. **Allopurinol** decreases the production of uric acid, which decreases uric acid levels in serum and urine. **Probenecid** and **sulfinpyrazone** lower serum levels of uric acid by increasing the excretion of uric acid in the urine. The reduction of serum uric acid decreases or prevents urate deposits in the joints, and the decreased urinary urate level reduces the risk of calculi in the kidneys.

TABLE 18-3 Abnormal Constituents of Urine

Albumin	Albuminuria	Indicates increased permeability of filtration membrane because of disease, injury, or high blood pressure
Glucose	Glucosuria	Usually indicates diabetes mellitus
Erythrocytes (red blood cells)	Hematuria	Usually indicates kidney inflammation, trauma, or disease
Leukocytes (white blood cells)	Pyuria	Indicates infection in kidney or urinary tract
Ketone bodies	Ketosis	Usually indicates diabetes mellitus, but occurs in any condition in which large quantities of fatty acids are metabolized
Bilirubin	Bilirubinuria	Usually indicates excessive destruction of red blood cells
Microbes		Indicates infection in urinary tract

Body Fluids

Fluids make up 60% of the body weight in a healthy, nonobese, young adult male. The average is slightly less in females. The percentage is less in individuals with more body fat because adipose tissue contains less water than other tissues. Age also influences the amount of water in the body. About 80% of a newborn infant's body weight is fluid. This value decreases during childhood and by adolescence the adult value is reached. The percentage of body weight that is fluid decreases in the elderly because there is usually a loss of muscle mass and an increased amount of adipose.

The body fluid is not evenly distributed but is separated into compartments. Even though separated by cell membranes, fluids can move back and forth between compartments. Fluid balance suggests that there is equilibrium or homeostasis between fluid intake and fluid output and in the movement of fluid between the compartments. The urinary system has an important role in maintaining homeostasis of body fluids by adjusting total fluid volume, by modifying electrolyte concentration, and by secreting varying quantities of hydrogen ions to regulate the pH of the blood.

Fluid Compartments

About two thirds of the total body fluid is found inside the cells of the body and is called **intracellular** (in-trah-SELL-yoo-lar) **fluid** (ICF). The remaining one third is outside the cells and includes all other body fluids. It is called **extracellular** (eks-trah-SELL-yoo-lar) **fluid** (ECF). About one fourth of the ECF is **intravascular** (in-trah-VAS-kyoo-lar) **fluid (blood plasma)**, and the remaining three fourths is the **interstitial** (in-ter-STISH-al) **fluid** in the tissue spaces. Figure 18-12 illustrates the fluid compartments. A small amount of the ECF is localized as cerebrospinal fluid, aqueous and vitreous humors of the eyes, endolymph and perilymph in the ears, the serous fluid between the layers of serous membranes, and the synovial fluid in joints.

 QUICK APPLICATION

Edema is a condition in which fluid accumulates in the interstitial compartment. This is sometimes caused by a blockage of lymphatic vessels, which reduces the return of the fluid to the blood; it may also be caused by a lack of plasma proteins or sodium retention.

Intake and Output of Fluid

Normally fluid intake equals fluid output so that the total amount of fluid in the body remains constant. Average fluid intake is 2500 milliliters per day and comes from three sources

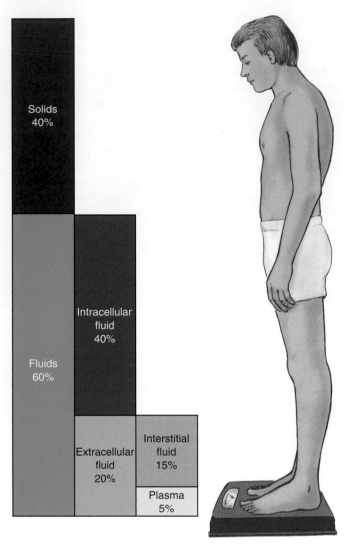

Figure 18-12 *Fluid compartments. Fluids comprise about 60% of the body weight. Two-thirds of the fluid is intracellular and one-third is extracellular, which includes interstitial fluid and plasma.*

(Figure 18-13). Beverages account for about 1600 milliliters and ingested foods provide another 700 milliliters. Water is produced as a byproduct of metabolism, and this metabolic water accounts for another 200 milliliters. The greatest regulator of fluid intake is the thirst mechanism. When body fluids become too concentrated, the increased osmotic pressure stimulates the thirst center in the hypothalamus. This results in a desire to drink fluids.

Under normal conditions, fluid output equals fluid intake to maintain fluid balance. There are four avenues of fluid loss (see Figure 18-13). On average, the kidneys excrete about 1500 milliliters of water per day. The skin loses about 400 milliliters per day through evaporation and 150 milliliters per day through perspiration for a total of 550 milliliters per day. Water vapor is exhaled with each breath, and this loss through the lungs accounts for 300 milliliters per day. Finally, about 150 milliliters is lost through the gastrointestinal tract each day. Added together, these four avenues account for 2500 milliliters of fluid output per day. The kidneys, under the influence of ADH, are the main regulators of fluid loss. If fluid intake remains constant and excess fluid is lost through diarrhea, vomiting, or perspiration, the kidneys excrete less urine to maintain fluid balance. The kidneys also adjust urine output in response to changes in electrolyte concentrations (aldosterone) and blood volume (atrial natriuretic hormone).

Electrolyte Balance

Electrolytes are in balance when the concentrations of individual electrolytes in the body fluid compartments are normal and remain relatively constant. This implies that the total electrolyte concentration is also normal and constant. Because electrolytes are dissolved in the body fluids, electrolyte balance and fluid balance are interrelated. When fluid volume changes, the concentration of the electrolytes also changes.

Sodium is the predominant cation (positive ion), and chloride is the predominant anion (negative ion) in the ECF. Bicarbonate ions are also extracellular ions. These three ions

TOTAL INTAKE—2500 mL

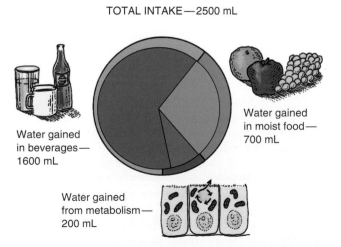

TOTAL OUTPUT—2500 mL

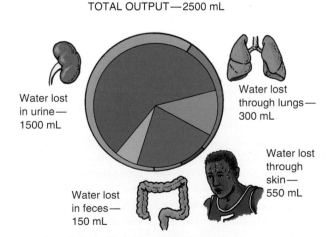

Figure 18-13 *Routes of fluid intake and output. Fluid intake comes from ingested food and beverages and from metabolism. Fluid is lost through the urine, lungs, skin, and feces.*

account for over 90% of the extracellular electrolytes. In the ICF compartment, potassium is the most abundant cation and phosphates are the major anions.

The primary regulation of electrolyte balance is through active reabsorption of positive ions. The negative ions follow by electrochemical attraction. Because sodium and potassium are the predominant cations, they are the most important ones to be regulated. Aldosterone, acting on the kidney tubules, regulates sodium and potassium levels by stimulating the reabsorption of sodium ions from the filtrate into the blood and the excretion of potassium ions in the urine. This adds sodium to the ECF and deceases the potassium level.

Because sodium (Na^+) is the predominant cation in the interstitial fluid and potassium (K^+) is the predominant cation in the ICF, changes in either of their concentrations within their respective compartments will cause a movement of fluid between the two compartments. For example, if a person has a high Na^+ concentration in the ECF (plasma and interstitial fluid), the increased osmotic pressure causes water to move from the intracellular compartment into the extracellular compartment. This movement results in edema and a reduction of fluid volume inside the cells. This illustrates that fluid balance depends on electrolyte balance. However, the opposite is also true. If the amount of water in a compartment increases, the concentration of the electrolyte will decrease. This shows that electrolyte balance depends on fluid balance. Fluid balance and electrolyte balance are interdependent.

QUICK CHECK

18.21 Approximately what percentage of an adult male's body weight is water?

18.22 In what fluid compartment is most of the body fluid located?

18.23 What is the predominant cation in extracellular fluid?

Acid-Base Balance

Blood, the intravascular fluid, has a normal pH range of 7.35 to 7.45. Deviations below this range are called **acidosis** (AS-id-oh-sis). The physiologic effect of acidosis is depression of synaptic transmission in the central nervous system. If untreated, the depression may become so severe that the person becomes disoriented and lapses into a coma. Death soon follows. Deviations of pH above 7.45 are called **alkalosis** (AL-kah-loh-sis). The principal effect of alkalosis is hyperexcitability in both the central nervous system and the peripheral nerves. This leads to extreme nervousness and muscle spasms. If untreated, alkalosis may lead to convulsions and death.

Cellular metabolism produces substances that tend to uspet the pH balance. Lactic acid is produced in the anaerobic breakdown of glucose. When carbon dioxide from aerobic metabolism combines with water, it produces carbonic acid. The metabolism of fatty acids produces acidic ketone bodies. All of these products tend to make the blood more acidic and to lower the pH. The body has three mechanisms by which it

attempts to maintain a normal blood pH: buffers, removal of carbon dioxide by the lungs, and removal of hydrogen ions by the kidneys.

Buffers (see Chapter 2) are substances that prevent significant changes in pH. If the hydrogen ion concentration is too high (acid), the buffer combines with some of the hydrogen ions to bring the pH back to normal. If the hydrogen ion concentration is too low (alkaline), the buffer releases hydrogen ions to lower the pH. Buffers are important for adjusting small changes in hydrogen ion concentration. One of the buffer pairs in the blood is sodium bicarbonate ($NaHCO_3$) and carbonic acid (H_2CO_3). When metabolic acids are produced, such as lactic acid in anaerobic metabolism, they are neutralized by the sodium bicarbonate. The carbonic acid neutralizes alkaline substances that may enter the system.

Carbon dioxide is a byproduct of cellular metabolism. When carbon dioxide combines with water, it produces carbonic acid, which releases hydrogen ions and lowers the pH of the blood. As the blood flows through the lungs, carbon dioxide is removed and exhaled. This reduces the amount of carbonic acid that is present. The lungs are important in regulating pH because they remove one of the sources of acids (CO_2). Without the work of the lungs in removing carbon dioxide, the buffers are quickly overloaded and the pH decreases below normal.

The kidneys function in acid-base balance by the tubular secretion of hydrogen ions and excreting them in the urine. If the blood is too acidic (excess hydrogen ions), the kidney tubules actively secrete hydrogen ions from the peritubular capillaries into the filtrate, and the ions are removed from the body. If the blood is too alkaline, the kidney tubules conserve hydrogen ions.

Acidosis and alkalosis can be classified according to cause as either **respiratory** or **metabolic**. Respiratory acidosis or alkalosis represents an underlying problem with respiratory mechanisms, and the most important indicator of these conditions is the partial pressure of carbon dioxide (Pco_2). The average normal Pco_2 in arterial blood is 40 mm Hg. All other acid-base imbalances are grouped together as metabolic. In other words, if it is not a respiratory problem, then it is metabolic. The most important indicator of metabolic acid-base imbalances is the concentration of bicarbonate ions (HCO_3^-) in the blood. The average normal HCO_3^- concentration in the blood is 24 mEq/L.

Changes in blood pH that lead to acidosis or alkalosis can be returned to normal by physiologic responses called **compensation**. If a person has a pH imbalance caused by some problem in the respiratory system, the kidneys attempt to compensate by changing the amount of hydrogen and bicarbonate ions that are excreted. If the imbalance is caused by something other than the respiratory system (metabolic imbalances), the respiratory system attempts to bring the pH back to normal by changing the rate and depth of breathing to either conserve or eliminate carbon dioxide, which increases or decreases acidity, respectively. If the compensation is successful, the pH returns to normal ranges.

Given that acid-base imbalances may be either respiratory or metabolic and may be either acidosis or alkalosis leads to four types of disturbances; respiratory acidosis, respiratory alkalosis, metabolic acidosis, and metabolic alkalosis.

Respiratory acidosis is a result of a lack of movement of CO_2 from the blood to the alveoli and into the atmosphere. A decreased pH and elevated P_{CO_2} are indications of this condition. The bicarbonate level is either normal (uncompensated) or increased (compensated). Conditions that may lead to respiratory acidosis include emphysema, airway obstruction, depression of the medullary respiratory center by anesthetics and sedatives, weakness of the muscles used in breathing, and pulmonary edema. The kidneys compensate by excreting hydrogen ions in the urine and by reabsorbing bicarbonate ions. Indications of respiratory acidosis are a decreased pH and increased P_{CO_2}. The bicarbonate level is either normal (uncompensated) or increased (compensated).

Respiratory alkalosis is caused by a loss of CO_2 from the lungs through hyperventilation. Conditions that may lead to respiratory alkalosis are those that stimulate the respiratory center in the medulla, such as severe anxiety, hysterical hyperventilation, oxygen deficiency because of high altitude, and the early stages of aspirin overdose, prolonged fever, caffeine overdose, and certain CNS disorders such as meningitis. The kidneys compensate for this condition by conserving hydrogen ions and increasing bicarbonate ion excretion. Indications of respiratory alkalosis are an increased pH and decreased P_{CO_2}. The bicarbonate level is either normal (uncompensated) or decreased (compensated).

Metabolic acidosis is often caused by an accumulation of metabolic acids such as occurs in uncontrolled diabetes mellitus and starvation. Severe diarrhea and renal dysfunction may result in a loss of bicarbonate ions, which leads to acidosis. Poisoning from toxic chemicals and chronic elevated levels of iron in the blood may also lead to metabolic acidosis. The respiratory system attempts to compensate for metabolic acidosis by hyperventilation. Indications of metabolic acidosis are decreased pH and bicarbonate levels. The P_{CO_2} may be normal (uncompensated) or decreased (compensated).

Metabolic alkalosis occurs when there is a nonrespiratory loss of acids such as occurs by repeated vomiting of gastric contents, by an excessive intake of alkaline substances such as sodium bicarbonate and other antacids, or from prolonged diuretic therapy. The respiratory system attempts to compensate for metabolic alkalosis by hypoventilation, which retains carbon dioxide in the body and lowers the pH. Indications of metabolic alkalosis are elevated pH and bicarbonate levels. The P_{CO_2} may be normal (uncompensated) or elevated (compensated).

Table 18-4 summarizes the four different types of acid-base imbalances, and Table 18-5 indicates the deviation in pH, P_{CO_2} and HCO_3^- for each type of imbalance

QUICK CHECK

18.24 What is the normal range for pH of the blood?

18.25 What substances in the blood provide immediate, but limited, adjustments to pH?

18.26 Abnormal hyperventilation is likely to cause what specific type of acid-base imbalance?

18.27 If a patient has a pH = 7.30, P_{CO_2} = 60 mm Hg, and HCO_3^- = 24 mEq/L, what type of acid-base imbalance is indicated?

TABLE 18-4 Acid-Base Imbalances

Imbalance	Causes	Compensation
Respiratory acidosis	Emphysema, airway obstruction, depression of respiratory center, respiratory muscle weakness, pulmonary edema	Kidneys excrete hydrogen ions and reabsorb bicarbonate ions
Respiratory alkalosis	Hyperventilation, sever anxiety, hysteria, high altitude, early aspirin overdose	Kidneys conserve hydrogen ions and excrete bicarbonate ions
Metabolic acidosis	Loss of bicarbonate ions through severe diarrhea or renal dysfunction; increased metabolic acids (diabetes mellitus)	Increased respiratory rate
Metabolic alkalosis	Loss of acids, such as by repeated vomiting; excessive intake of alkaline substances, such as antacids	Decreased respiratory rate

TABLE 18-5 Deviations in pH, Pco_2, and HCO$_3^-$ in Acid-Base Imbalances

	pH	Pco_2	HCO$_3^-$
Normal Level	7.35 to 7.45	40 mm Hg	24 mEq/L
Respiratory acidosis			
Uncompensated	<7.35	↑	Normal
Compensated	7.35 to 7.40	↑	↑
Respiratory alkalosis			
Uncompensated	>7.45	↓	Normal
Compensated	7.40 to 7.45	↓	↓
Metabolic acidosis			
Uncompensated	<7.35	Normal	↓
Compensated	7.35 to 7.40	↓	↓
Metabolic alkalosis			
Uncompensated	>7.45	Normal	↑
Compensated	7.40 to 7.45	↑	↑

FOCUS ON AGING

Some of the more obvious and familiar aging changes occur in the urinary bladder and urethra. Muscles in the walls of these structures tend to weaken and become less elastic with age. As a person ages, the bladder is unable to expand or contract as much as in younger people. This reduces the capacity of the bladder and makes it more difficult to completely empty it during urination. Awareness of the need to urinate, which usually occurs when the bladder is half full in younger people, may be delayed in the elderly until the bladder is nearly full. Thus urgency accompanies awareness. The external urethral sphincter also weakens, which adds to the problems.

Several anatomical changes occur in the kidneys as a person ages, and these changes are reflected in their related functions. There is a general atrophy of nephrons so that by the age of 80, the kidney is about 80% of its young, but mature, size. Some of the remaining glomeruli are modified, and this, along with the decrease in number, results in a decreased glomerular filtration rate so that the blood is not filtered as quickly as before.

The tubules also undergo changes as a person ages. In general, the tubule walls thicken, which makes them less able to reabsorb water to form a concentrated urine. The collecting ducts are less responsive to ADH and this, along with a diminished thirst mechanism, may result in dehydration. The ability to reabsorb glucose and sodium is also diminished. The tubules become less efficient in the secretion of ions and drugs. They have diminished ability to compensate for drastic changes in acid-base balance. Drugs that are normally eliminated from the body by tubular secretion may accumulate to toxic levels because they are not cleared from the blood as quickly as they are in younger people.

Amazingly, even with the changes caused by aging, the kidneys of elderly persons are capable of maintaining relatively stable balances in the blood and body fluids under normal conditions. However, their ability to compensate for drastic changes and abnormal conditions is diminished.

Representative Disorders of the
Urinary System and Body Fluids

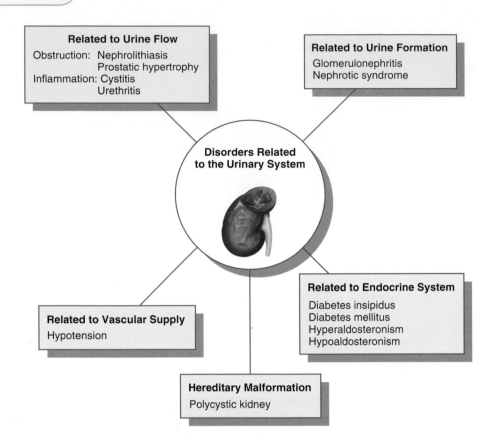

Related to Urine Flow
Obstruction: Nephrolithiasis
Prostatic hypertrophy
Inflammation: Cystitis
Urethritis

Related to Urine Formation
Glomerulonephritis
Nephrotic syndrome

Disorders Related to the Urinary System

Related to Vascular Supply
Hypotension

Related to Endocrine System
Diabetes insipidus
Diabetes mellitus
Hyperaldosteronism
Hypoaldosteronism

Hereditary Malformation
Polycystic kidney

Cystitis (sis-TYE-tis) Inflammation of the urinary bladder; may be caused by an infection descending from the kidney or one ascending from the urethra

Diabetes insipidus (di-ah-BEE-teez in-SIP-ih-dus) Metabolic disorder caused by a deficiency of antidiuretic hormone, resulting in a large quantity of urine (polyuria) and great thirst (polydipsia)

Diabetes mellitus (di-ah-BEE-teez mell-EYE-tus) Disorder caused by a deficiency of insulin from the beta cells of the pancreatic islets; characterized by a disturbance in the utilization of glucose and manifested by polyuria, polyphagia, and polydipsia

Glomerulonephritis (glo-mair-yoo-low-nef-FRY-tis) Inflammation of the capillary loops in the glomeruli of the kidney; usually secondary to an infection

Hyperaldosteronism (hye-per-al-DAHS-ter-ohn-izm) Condition in which there is excessive secretion of aldosterone from the adrenal cortex, causing sodium and fluid retention by the kidneys and accompanied by a loss of potassium

Hypoaldosteronism (hye-poh-al-DAHS-ter-ohn-izm) Condition in which there is reduced secretion of aldosterone from the adrenal cortex, causing excessive sodium and fluid loss by the kidneys and accompanied by potassium retention

Hypotension (hye-poh-TEN-shun) Reduced blood pressure; sudden and/or extremely low pressures may interfere with blood flow to the kidneys and impaired renal function

Nephrolithiasis (nef-roh-lith-EYE-ah-sis) Condition marked by presence of renal calculi, or kidney stones; common cause of urinary tract obstruction in adults

Nephrotic syndrome (nef-RAH-tick SIN-drohm) Condition marked by excessive protein in the urine; characteristic of glomerular injury; accompanied by massive edema and reduced albumin in the blood

Polycystic kidney (pah-lee-SIS-tick KID-nee) Hereditary disease in which there is massive enlargement of the kidney accompanied by the formation of cysts, which interfere with kidney function; results in renal failure; renal dialysis and kidney transplant during end-stage renal failure may prolong life

Prostatic hypertrophy (prah-STAT-ick high-PER-troh-fee) Enlargement of the prostate gland, common in men older than age 50; because the prostate gland surrounds the urethra, it may obstruct the flow of urine form the bladder

Urethritis (yoo-reth-RYE-tis) Inflammation of the urethra, usually caused by infectious organisms; urethra may swell and impede the flow of urine

CHAPTER SUMMARY

Functions of the Urinary System

■ **State six functions of the urinary system.**
- The urinary system rids the body of waste materials, regulates fluid volume, maintains electrolyte concentrations in body fluids, controls blood pH, secretes erythropoietin, and secretes renin.

Components of the Urinary System

■ **Describe the location and structural features of the kidneys.**
- The components of the urinary system are the kidneys, ureters, urinary bladder, and urethra.
- The kidneys are retroperitoneal against the posterior abdominal wall, between the levels of the twelfth thoracic and third lumbar vertebrae.
- An indentation called the hilus leads to the renal sinus. The renal artery, renal vein, and ureter penetrate the kidney at the hilus.
- The kidney is enclosed by a capsule and surrounded by perirenal fat.
- Internally, the renal cortex, renal medulla, pyramids, renal columns, papillae, calyces, and pelvis are visible.

■ **Draw and label a diagrammatic representation of a nephron.**
- The nephron is the functional unit of the kidney and consists of a renal corpuscle and a renal tubule.
- The renal corpuscle consists of a glomerulus and a glomerular capsule.
- The renal tubule consists of a proximal convoluted tubule, the nephron loop with descending and ascending limbs, and a distal convoluted tubule.
- The nephron loop is the only part of a nephron located in the pyramids. The other portions are in the cortex.
- Urine passes from the nephrons into collecting ducts, then into the minor calyces.
- Collecting ducts are located in the pyramids.

■ **Name the two parts of the juxtaglomerular apparatus and state where they are located.**
- The juxtaglomerular apparatus consists of modified cells where the ascending limb of the nephron loop (macula densa) comes in contact with the afferent arteriole (juxtaglomerular cells).
- The juxtaglomerular cells secrete renin.

■ **Trace the pathway of blood flow through the kidney from the renal artery to the renal vein.**
- Blood flows through the kidney in the following sequence: renal artery, segmental arteries, interlobar arteries, arcuate arteries, interlobular arteries, afferent arteriole, glomerulus, efferent arteriole, peritubular capillaries.
- From the capillaries back to the renal vein, the sequence is peritubular capillaries, interlobular veins, arcuate veins, interlobar veins, segmental veins, and renal vein.

■ **Describe the location, structure, and function of the ureter, urinary bladder, and urethra.**
- The ureter transports urine from the kidney to the urinary bladder and is continuous with the renal pelvis.
- As it leaves the kidney, the ureter descends along the posterior abdominal wall and is retroperitoneal.
- The ureters enter the urinary bladder on the posteroinferior surface.
- The urinary bladder is posterior to the symphysis pubis and below the parietal peritoneum in the pelvic cavity.
- The lining of the urinary bladder is a mucous membrane with folds called rugae. The smooth muscle in the wall is the detrusor muscle.
- The trigone, in the floor of the urinary bladder, is outlined by the two ureters and the internal urethral orifice.
- The urethra transports urine from the bladder to the exterior of the body.
- The flow is controlled by two sphincters. The internal sphincter is located where the urethra leaves the bladder and is smooth muscle. The external sphincter is located where the urethra penetrates the pelvic floor and is skeletal muscle.
- In females the urethra is short. In males it is longer and extends the length of the penis. The male urethra is divided into the prostatic urethra, membranous urethra, and the spongy urethra.
- The urethra opens to the exterior through the external urethral orifice.

Urine Formation

■ **Describe each of the three basic steps in urine formation.**
- The work of the kidneys is accomplished through the formation of urine, which involves glomerular filtration, tubular reabsorption, and tubular secretion.
- In glomerular filtration, plasma components cross the filtration membrane from the glomerulus into the glomerular capsule.
- Tubular reabsorption moves substances from the filtrate into the blood in the capillaries and reduces the volume of urine. Most of the solutes are reabsorbed by active transport mechanisms. Water is reabsorbed by osmosis in all parts of the tubule except the ascending limb of the nephron loop, which is impermeable to water.
- Tubular secretion adds substances to the urine. Hydrogen ions, potassium ions, creatinine, histamine, and penicillin are examples of substances that are added to the urine by tubular secretion.

■ **Identify three different types of pressure that affect the rate of glomerular filtration and describe how these interact.**
- The rate of glomerular filtration depends on the net filtration pressure. This pressure is the result of the interaction of the blood pressure in the glomerulus, hydrostatic pressure in the capsule, and osmotic pressure in the blood.

■ **Explain why some substances, such as glucose, have limited reabsorption and what happens when concentration exceeds this limit.**
- Reabsorption of some solutes is limited by carrier molecules. When the concentration of these solutes exceeds renal threshold, the excess appears in the urine.

- **Explain how kidney function has a role in maintaining blood concentration, blood volume, and blood pressure.**
- By altering the concentration of volume of urine, the kidneys have a major role in maintaining blood concentration, volume, and pressure.

- **Name three hormones that affect kidney function and explain the effect of each one.**
- Aldosterone increases sodium reabsorption in the kidney tubules. This causes sodium retention and secondarily, water retention. This will also increase blood volume and blood pressure.
- Antidiuretic hormone increases water reabsorption by the kidney tubules. This decreases urine volume and adds to the fluid in the body. This will also increase blood volume and blood pressure.
- Atrial natriuretic hormone promotes the excretion of sodium and water by acting directly on the kidney tubules and by inhibiting ADH, renin, and aldosterone. This decreases blood volume and blood pressure.

- **Name the enzyme that stimulates the production of angiotensin II and is produced by the kidneys.**
- Renin, which promotes the production of angiotensin II, is produced by the juxtaglomerular cells.

- **Describe two mechanisms by which angiotensin II increases blood pressure.**
- Angiotensin II is a vasoconstrictor, which increases blood pressure. It also stimulates the release of aldosterone, which acts on the kidney tubules to conserve sodium and water. This also increases blood volume and blood pressure.

- **Define *micturition*.**
- Micturition is the act of expelling urine from the bladder.

Characteristics of Urine

- **Describe the physical characteristics and chemical composition of urine.**
- Freshly voided urine has a clear yellow color, a specific gravity of 1.001 to 1.035, and a pH between 4.6 and 8.0.
- Urine is about 95% water and 5% solutes.

- **List five abnormal constituents of urine.**
- Abnormal constituents of urine include albumin, glucose, blood cells, ketone bodies, and microbes.

Body Fluids

- **State the percentage of body weight that is composed of water.**
- Fluids make up 60% of the adult body weight and 80% of the infant weight.

- **Identify the major fluid compartments in the body and state the relative amount of fluid in each compartment.**
- Intracellular fluid is the fluid inside the body cells and it accounts for two-thirds of the total body fluid.
- Extracellular fluid is the fluid outside the body cells and it accounts for one-third of the total body fluid.

- Extracellular fluid is further divided into intravascular fluid (one-fifth) and interstitial fluid (four-fifths).

- **State the sources of fluid intake and avenues of fluid output and explain how these are regulated to maintain fluid balance.**
- Normally fluid intake equals output. This should be about 2500 milliliters (2.5 L) per day.
- Sources of fluid intake are beverages, food, and metabolic water.
- Avenues of fluid loss are through the kidneys, skin, lungs, and gastrointestinal tract.
- Fluid intake and fluid loss are regulated by the thirst mechanism and by the water output by the kidneys.

- **Identify the major intracellular and extracellular ions and explain how electrolyte balance is regulated.**
- Electrolyte concentrations in the different fluid compartments are different, but they remain relatively constant within each compartment.
- Sodium (Na^+), chloride (Cl^-), and bicarbonate (HCO_3^-) ions are the predominant ions in the extracellular fluid.
- Potassium (K^+) and phosphates ($H_2PO_4^-$ and HPO_4^{-2}) are the predominant ions in the intracellular fluid.
- Aldosterone is the primary regulator of electrolyte concentration through reabsorption of sodium and potassium.

- **State the normal pH range of the blood and define the terms *acidosis* and *alkalosis*.**
- The normal pH of the blood ranges between 7.35 and 7.45.
- Deviations below normal are called acidosis and deviations above normal are called alkalosis.

- **Describe the three primary mechanisms by which blood pH is regulated.**
- Acid-base balance is maintained through the action of buffers, the lungs, and the kidneys.

- **State the normal values for P_{CO_2} and HCO_3^- in the blood.**
- The normal value for P_{CO_2} in the blood is 40 mm Hg and the normal value for HCO_3^- is 24 mEq/Liter.

- **Describe the causes and indicators of respiratory acidosis, respiratory alkalosis, metabolic acidosis, and metabolic alkalosis.**
- Respiratory acidosis results when CO_2 is not expelled from the body and accumulates in the blood. The indicators are a decreased pH and elevated P_{CO_2}.
- Respiratory alkalosis is caused by a loss of CO_2 through hyperventilation. The indicators are an increased pH and decreased P_{CO_2}.
- Metabolic acidosis is caused by an increase in production of metabolic acids or a decrease in bicarbonate. The indicators are a decreased pH and a decrease in HCO_3^-.
- Metabolic alkalosis results when there is a loss of acids or increase in bicarbonate. The indicators are an increase in both pH and HCO_3^-.

CHAPTER QUIZ

Recall

Match the definitions on the left with the appropriate term on the right.

_____ **1.** Functional unit of kidney
_____ **2.** Cluster of capillaries in the renal corpuscle
_____ **3.** Cortical substance between adjacent renal pyramids fluid
_____ **4.** Modified cells in the ascending limb of the nephron loop
_____ **5.** Enzyme produced by juxtaglomerular cells
_____ **6.** Folds in the mucosa of the urinary bladder
_____ **7.** Urination or voiding
_____ **8.** Responsible for yellow color of urine
_____ **9.** Accounts for about two thirds of the body fluid
_____ **10.** Increase in pH above normal

A. Alkalosis
B. Glomerulus
C. Intracellular fluid
D. Macula densa
E. Micturition
F. Nephron
G. Renal columns
H. Renin
I. Rugae
J. Urochrome

Thought

1. When blood is filtered in the kidney, the filtrate passes from the:
   A. Efferent arteriole into the glomerulus
   B. Glomerulus into the glomerular capsule
   C. Afferent arteriole into the glomerulus
   D. Glomerulus into the proximal convoluted tubule

2. Which of the following is *false* about the juxtaglomerular apparatus?
   A. It produces an enzyme called rennin.
   B. It secretes a substance that functions in the regulation of blood pressure.
   C. It occurs where the efferent arteriole contacts the afferent arteriole.
   D. It contains a region called the macula densa.

3. The greatest amount of fluid is reabsorbed from the filtrate in the:
   A. Proximal convoluted tubule
   B. Nephron loop
   C. Distal convoluted tubule
   D. Collecting duct

4. Which of the following statements about body fluids is *false*?
   A. The volume of plasma is greater than the volume of interstitial fluid.
   B. The volume of intracellular fluid is greater than the volume of extracellular fluid.
   C. The interstitial fluid is a part of the extracellular fluid.
   D. Fluids account for about 60% of the total body weight.

5. Which of the following blood pH values represent alkalosis?
   A. 7.3 and 7.6
   B. 6.7 and 6.9
   C. 7.1 and 7.3
   D. 7.5 and 7.6

Application

Name three hormones that influence urine volume, state the source of each hormone, and describe how it affects the volume.

When a child learns to control micturition, he or she controls what muscle?

BUILDING YOUR MEDICAL VOCABULARY

Building a vocabulary is a cumulative process. As you progress through this book, you will use word parts, abbreviations, and clinical terms from previous chapters. Each chapter will present new word parts, abbreviations, and terms to add to your expanding vocabulary.

Word Parts and Combining Form with Definition and Examples

PART/COMBINING FORM	DEFINITION	EXAMPLE
-atresia	without an opening	urethratresia: lack of an opening in the urethra
azot/o	urea, nitrogen	azotemia: presence of nitrogen-containing compounds in the blood
caly-	small cup	calyx: small cup-shaped structure that receives filtrate from the collecting duct at the tip of a renal pyramid
-chrom/o	color, pigment	chromaturia: abnormal coloration of the urine
-continence	to hold	incontinence: inability to hold, or control, excretory functions
cyst/o	bladder	cystitis: inflammation of the bladder usually due to a bacterial infection
-ectasia	dilation	cystectasia: dilation of the urinary bladder
juxta-	near to	juxtaglomerular: near to the glomerulus
lith/o	stone	nephrolithiasis: condition marked by presence of kidney stones
mict-	to pass	micturition: to pass urine, urination
nephr/o	kidney	nephropathy: any disease of the kidneys
noct-	night	nocturia: frequent, excessive urination at night
olig-	few, little	oliguria: little urine, diminished urine production
peri-	around	peritubular: around the tubules; peritubular capillaries are around the renal tubules

PART/COMBINING FORM	DEFINITION	EXAMPLE
-pexy	fixation	cystopexy: surgical fixation of the bladder to the abdominal wall
-phraxis	to obstruct	urethrophraxis: obstruction of the urethra
pyel/o	renal pelvis	pyelocystitis: inflammation of the renal pelvis and urinary bladder
ren/o	kidney	renal: pertaining to the kidney
-rrhaphy	suture	cystorrhaphy: suture of the bladder
-ur-	urine	anuria: suppression of urine formation by the kidney

Clinical Abbreviations

ABBREVIATION	MEANING
ADH	antidiuretic hormone
ARF	acute renal failure
BUN	blood urea nitrogen
cath	catheter, catheterize
CKD	chronic kidney disease
creat	creatinine
CRF	chronic renal failure
cysto	cystoscopic examination
EPO	erythropoietin
ERPF	effective renal plasma flow
ESRD	end-stage renal disease
GFR	glomerular filtration rate
IVP	intravenous pyelogram
KUB	kidney, ureter, bladder
mEq	milliequivalent
RPF	renal plasma flow
TUR	transurethral resection
UA	urinalysis
UTI	urinary tract infection

Clinical Terms

Anuria (an-YOO-rih-ah) Condition in which there is no formation of urine

Azotemia (az-oh-TEE-mee-ah) Presence of increased amounts of nitrogen waste products in the blood

Blood urea nitrogen (BUN) (BLUHD yoo-REE-ah NYE-troh-jen) Blood test to determine the amount of urea that is excreted by the kidneys; abnormal results indicate urinary tract disease

Catheterization (kath-eh-ter-ih-ZAY-shun) Insertion of a sterile catheter through the urethra into the urinary bladder

Cystoscopy (sis-TAHS-koh-pee) Visual examination of the urinary bladder using a cystoscope

Dialysis (dye-AL-ih-sis) Procedure to separate waste material from the blood and to maintain fluid, electrolyte, and acid-base balance when kidney function is impaired

Diuresis (dye-yoo-REE-sis) Condition of increased or excessive flow of urine

Diuretic (dye-yoo-RET-ik) Substance that increases the production of urine

Dysuria (dis-YOO-rih-ah) Difficult or painful urination

Enuresis (en-yoo-REE-sis) Involuntary emission of urine; bed-wetting

Intravenous pyelogram (in-trah-VAYN-us PYLE-oh-gram) Radiographic procedure in which a radiopaque dye is injected into a vein and its path through the kidneys, ureters, and urinary bladder is followed to visualize abnormalities in the renal vessels and urinary tract

Lithotripsy (lith-oh-TRIP-see) Crushing of a calculus (stone) in the bladder, urethra, ureter, or kidney

Nephrectomy (nef-REK-toh-mee) Surgical removal of a kidney

Nephritis (nef-RYE-tis) Inflammation of the kidney

Nocturia (nahk-TOO-rih-ah) Excessive urination at night

Oliguria (ahl-ig-YOO-rih-ah) Very little, or scanty, urination

Polyuria (pahl-ee-YOO-rih-ah) Excessive urination

VOCABULARY QUIZ

Use word parts you have learned to form words that have the following definitions.

1. Urination at night _____

2. Surgical fixation of the bladder _____

3. Kidney stone _____

4. Process of passing urine _____

5. Near the glomerulus _____

Using the definitions of word parts you have learned, define the following words.

6. Cystorrhaphy _____

7. Hematuria _____

8. Incontinence _____

9. Urethratresia _____

10. Urethrophraxis _____

Match each of the following definitions with the correct word.

_____ 11. Inflammation around the ureter

_____ 12. Glandular tumor of the kidney

_____ 13. Surgical removal of a stone from the renal pelvis

_____ 14. Painful urination

_____ 15. Dilation of the urinary bladder

A. Cystectasia
B. Dysuria
C. Nephroadenoma
D. Periureteritis
E. Pyelolithotomy

16. What is the clinical term for the "condition in which there is no formation of urine"? _____

17. What term denotes a "visual examination of the urinary bladder"? _____

18. What term denotes the "crushing of a calculus, or stone, in the urinary tract"? _____

19. What is the meaning of the abbreviation BUN? _____

20. What is the meaning of the abbreviation UTI? _____

Reproductive System

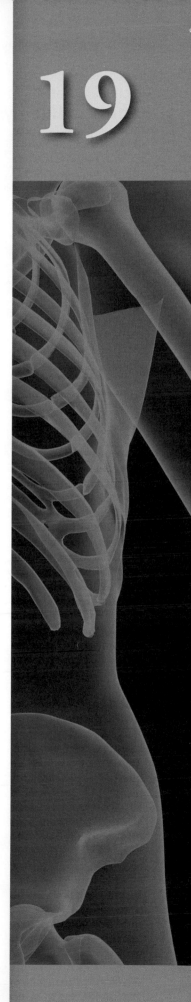

CHAPTER OBJECTIVES

Overview of the Reproductive System
- State four functions of the reproductive system.
- Distinguish between primary and secondary reproductive organs.

Male Reproductive System
- Describe the location and structure of each component of the male reproductive system.
- Draw and label a diagram or flow chart that illustrates spermatogenesis and describe the process by which spermatids become mature sperm.
- Trace the pathway of sperm from the testes to the outside of the body.
- Outline the physiologic events in the male sexual response.
- Describe the roles of GnRH, FSH, LH, and testosterone in male reproductive functions.

Female Reproductive System
- Describe the location and structure of each component of the female reproductive system.

- Draw and label a diagram or flow chart that illustrates oogenesis.
- Describe the development of ovarian follicles as they progress from primordial follicles to primary follicles, secondary follicles, vesicular follicles, corpus luteum, and, finally, the corpus albicans.
- Outline the physiologic events in the female sexual response.
- Describe the roles of GnRH, FSH, LH, estrogen, and progesterone in female reproductive functions.
- Describe what happens in each phase of the ovarian and uterine cycles, when each phase occurs, and how the cycles interact.
- Describe the structure of mammary glands.
- Discuss the hormonal control of mammary glands.

KEY TERMS

Ejaculation (ee-jak-yoo-LAY-shun) Forceful expulsion of seminal fluid from the urethra

Emission (ee-MISH-un) Discharge of seminal fluid into the urethra

Gametes (GAM-eets) Sex cells: sperm and ova

Gonads (GO-nads) Primary reproductive organs; organs that produce the gametes: testes in the male and ovaries in the female

Oogenesis (oh-oh-JEN-eh-sis) Process of meiosis in the female in which one ovum and three polar bodies are produced from one primary oocyte

Ovarian cycle (oh-VAIR-ee-an SYE-kul) Monthly cycle of events that occur in the ovary from puberty to menopause; occurs concurrently with the uterine cycle

Ovarian follicle (oh-VAIR-ee-an FAHL-ih-kul) Oocyte surrounded by one or more layers of cells within the ovaries

Pudendum (pyoo-DEN-dum) Collective term for the external accessory structures of the female reproductive system; also called the vulva

Spermatogenesis (spur-mat-oh-JEN-eh-sis) Process of meiosis in the male in which four spermatids are produced from one primary spermatocyte

Spermiogenesis (spur-mee-oh-JEN-eh-sis) Morphologic changes that transform a spermatid into a mature sperm

Uterine cycle (YOO-ter-in SYE-kul) Monthly cycle of events that occur in the uterus from puberty to menopause; also called the menstrual cycle; occurs concurrently with the ovarian cycle

Functional Relationships of the
Reproductive System

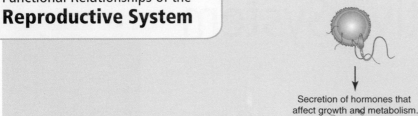

Secretion of hormones that affect growth and metabolism.

Urinary

Maintains pH and electrolyte composition for effective gonadal hormone function; maternal urinary system excretes metabolic wastes from developing fetus.

Prostate surrounds urethra and may compress it to cause urine retention; pregnant uterus pushes on urinary bladder to cause frequent urination.

Digestive

Provides nutrients for growth, maintenance, and repair of reproductive tissues; provides nutrientsfor repair of endometrium following menstruation and for developing fetus during pregnancy.

During pregnancy, expanding uterus may crowd intestines causing constipation and heartburn; gonadal steroids have an effect on metabolic rate.

Respiratory

Supplies oxygen and removes carbon dioxide to maintain metabolism in tissues of reproductive system; helps maintain pH for gonadal hormone function.

Sexual arousal stimulates changes in rate and depth of breathing; during pregnancy, the expanding uterus reduces depth of breathing, but rate of breathing increases.

Lymphatic/Immune

Provides specific defense against pathogens that enter the body through the reproductive tract; surveillance against cancer cells.

Acids and enzymes along the reproductive tract provide nonspecific defense against pathogens; blood-testis barrier prevents immune destruction of sperm.

Integument

Skin forms scrotum that protects testes; tactile receptors in skin provide sensations associated with sexual behaviors.

Gonads provide hormones that promote growth, and maintenance of skin; gonadal hormones affect growth and distribution of body hair and activity of sebaceous glands.

Skeletal

Provides protection for reproductive organs in pelvis; pelvis may hinder vaginal delivery of fetus.

Gonads produce hormones that influence bone growth, maintenance, and closure of epiphyseal plates.

Muscular

Provides support for reproductive organs in the pelvis; muscle contractions contribute to orgasm in both sexes; aid in childbirth.

Gonads produce hormones that influence muscle development and size.

Nervous

Regulates sex drive, arousal, and orgasm; stimulates the release of numerous hormones involved in sperm production, menstrual cycle, pregnancy, and parturition.

Gonads produce hormones that affect CNS development and sexual behavior; menstrual hormones affect the activity of the hypothalamus.

Cardiovascular

Transports reproductive hormones to target tissues; provides nutrients and oxygen to developing fetus, and removes wastes; vasodilation responsible for erection.

Gonads produce hormones that help maintain healthy blood vessels; testosterone stimulates erythropoiesis; some evidence that estrogens may decrease cholesterol levels.

Endocrine

Hormones have a major role in differentiation and development of reproductive organs, sexual development, sex drive, gamete production, menstrual cycle, pregnancy, parturition, and lactation.

Gonads produce hormones that feedback to influence pituitary function.

 Gives to Reproductive System
 Receives from Reproductive System

Overview of the Reproductive System

The **major function** of the reproductive system is to produce offspring. Other systems in the body, such as the endocrine and urinary systems, work continuously to maintain homeostasis for survival of the individual. The reproductive system, on the other hand, functions for the survival of the species. An individual may live a long, healthy, and happy life without producing offspring, but if the species is to continue, at least some individuals must produce offspring. Within the context of producing offspring, the reproductive system has four functions: to produce egg and sperm cells, to transport and sustain these cells, to nurture the developing offspring, and to produce hormones.

These functions of the reproductive system are divided between the **primary reproductive organs** and the **secondary**, or **accessory, reproductive organs**. The primary reproductive organs, also called **gonads**, are the ovaries and testes. These organs are responsible for producing the **gametes** or egg and sperm cells, and for producing hormones. These hormones function in the maturation of the reproductive system and the development of sexual characteristics and have important roles in regulating the normal physiology of the reproductive system. All other organs, ducts, and glands in the reproductive system are considered secondary, or accessory, reproductive organs. These structures transport and sustain the gametes and nurture the developing offspring.

Male Reproductive System

The male reproductive system produces, sustains, and transports sperm; introduces the sperm into the female vagina; and produces hormones. Figure 19-1 shows the organs of the male reproductive system.

Testes

The **testes**, or **testicles**, are the male gonads; they begin their development high in the abdominal cavity, near the kidneys. During the last 2 months before birth, or shortly after birth, they descend through the inguinal canal into the **scrotum**, a pouch that extends below the abdomen, posterior to the penis. Occasionally, the testes do not descend. This condition is called **cryptorchidism**, or hidden testes. Although this location of the testes, outside the abdominal cavity, may seem to make them vulnerable to injury, it provides a temperature about 3° C below normal body temperature. This lower temperature is necessary for the production of viable sperm. Cryptorchidism results in sterility if it is not surgically corrected before puberty. The scrotum consists of skin and subcutaneous tissue. A vertical septum, or partition, of subcutaneous tissue in the center of the scrotum divides it into two parts, each containing one testis. Smooth muscle fibers, called the **dartos muscle**, in the subcutaneous tissue contract to give the scrotum its wrinkled appearance. When these fibers are relaxed, the scrotum is smooth. Another muscle, the **cremaster muscle** in the spermatic cord, consists of skeletal muscle fibers and controls the position of the scrotum and testes. When it is cold or a man is sexually aroused, this muscle contracts to pull the testes closer to the body for warmth.

> ### QUICK APPLICATION
> The inguinal canal is a weak area in the abdominal wall that may rip open, resulting in an inguinal hernia. A portion of the intestine may pass through the opening into the scrotum. This is painful and potentially dangerous if the blood supply to the intestine is constricted. This condition is more common in men than in women. Inguinal hernias are frequently repaired by surgery.

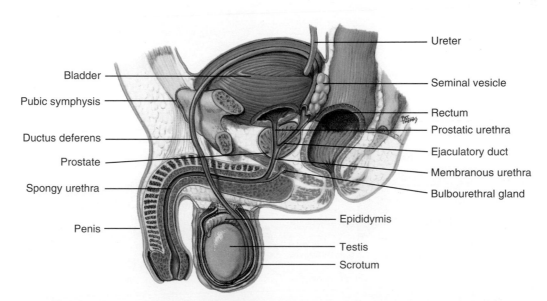

Figure 19-1 *Structures in the male reproductive system. The testes are the primary reproductive organs in the male. The ducts and glands are accessory organs.*

Structure

Each testis is an oval structure about 5 centimeters (cm) long and 3 cm in diameter (Figure 19-2). A tough, white fibrous connective tissue capsule, the **tunica albuginea** (TOO-nik-ah al-byoo-JIN-ee-ah), surrounds each testis and extends inward to form **septa** that partition the organ into **lobules**. There are about 250 lobules in each testis. Each lobule contains one to four highly coiled **seminiferous** (seh-mye-NIFF-er-us) **tubules** that converge to form a single **straight tubule**. The straight tubule leads into the **rete testis** (REE-tee TEST-is), a tubular network on one side of the testis. Short efferent ducts exit the testes. **Interstitial cells** (cells of Leydig), which produce male sex hormones, are located between the seminiferous tubules within a lobule.

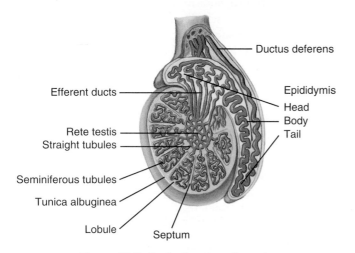

Figure 19-2 *Sagittal section of a testis.*

Spermatogenesis

Sperm are produced by **spermatogenesis** (spur-mat-oh-JEN-eh-sis) within the seminiferous tubules. A transverse section of a seminiferous tubule shows that it is packed with cells in various stages of spermatogenesis (Figure 19-3). Interspersed with these cells, there are large cells that extend from the periphery of the tubule to the lumen. These large cells are the **supporting**, or **sustentacular, cells** (Sertoli cells), which support and nourish the other cells.

Early in embryonic development, **primordial germ cells** enter the testes and differentiate into **spermatogonia** (spur-mat-oh-GOH-nee-ah), immature cells that remain dormant until puberty. Spermatogonia are diploid cells, each with 46 chromosomes (23 pairs), that are located around the periphery of the seminiferous tubules. At puberty, hormones stimulate these cells to begin dividing by mitosis. Some of the daughter cells produced by mitosis remain at the periphery as spermatogonia. Others are pushed toward the lumen, undergo some changes, and become **primary spermatocytes**. Because they are produced by mitosis, primary spermatocytes, like spermatogonia, are diploid and have 46 chromosomes.

Each primary spermatocyte goes through the first meiotic division, meiosis I, to produce two **secondary spermatocytes**, each with 23 chromosomes (haploid). Just before this division, the genetic material is replicated so that each chromosome consists of two strands, called **chromatids**, that are joined by a centromere. During meiosis I, one chromosome, consisting of two chromatids, goes to each secondary spermatocyte. In the second meiotic division, meiosis II, each secondary spermatocyte divides to produce two **spermatids**. There is no replication of genetic material in this division, but the centromere divides so that a single-stranded chromatid goes to each cell. As a result of

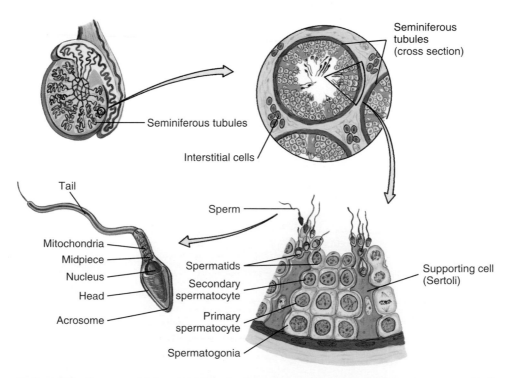

Figure 19-3 *Cross section of a seminiferous tubule showing the different cell types. Interstitial cells that produce testosterone are between the seminiferous tubules. Spermatids in the lumen become sperm by a process called spermiogenesis.*

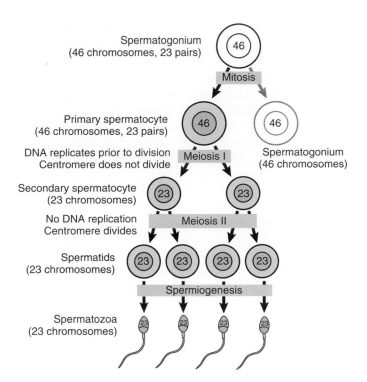

Figure 19-4 *Spermatogenesis. Each primary spermatocyte yields four spermatids by meiosis.*

Sperm production begins at puberty and continues throughout the life of a male. The entire process, beginning with a primary spermatocyte, takes about 74 days. After ejaculation, the sperm can live for about 48 hours in the female reproductive tract.

QUICK CHECK

19.1 Why is it necessary for the testes to descend into the scrotum?

19.2 Specifically, where in the testes does spermatogenesis occur?

19.3 How many spermatids are produced from one primary spermatocyte?

19.4 What is the process that transforms spermatids into spermatozoa (sperm)?

Duct System

Sperm cells pass through a series of ducts to reach the outside of the body. After they leave the testes through the efferent ducts, the sperm pass through the epididymis, ductus deferens, ejaculatory duct, and urethra.

Epididymis

Sperm leave the testes through a series of efferent ducts that enter the **epididymis** (ep-ih-DID-ih-mis) (see Figure 19-2). The epididymis is a long (about 6 meters) tube that is tightly coiled to form a comma-shaped organ located along the superior and posterior margins of the testes. Each epididymis has a region called the head, a body, and a tail. When the sperm leave the testes, they are immature and incapable of fertilizing ova. They complete their maturation process and become fertile as they move through the epididymis. Mature sperm are stored in the lower portion, or tail, of the epididymis.

Ductus Deferens

The **ductus deferens**, also called **vas deferens**, is a fibromuscular tube that is continuous with the epididymis. It begins at the bottom (tail) of the epididymis and then turns sharply upward along the posterior margin of the testes (see Figure 19-2). A **vasectomy** is a surgical procedure that severs the vas deferens and interrupts the pathway of sperm to provide male sterility. The ductus deferens enters the abdominopelvic cavity through the inguinal canal and passes along the lateral pelvic wall. It crosses over the ureter and posterior portion of the urinary bladder, and then descends along the posterior wall of the bladder toward the prostate gland (see Figure 19-1). Just before it reaches the prostate gland, each ductus deferens enlarges to form an **ampulla**. Sperm are stored in the proximal portion of the ductus deferens, near the epididymis, and peristaltic movements propel the sperm through the tube.

The proximal portion of the ductus deferens is a component of the **spermatic cord**, which contains vascular and neural structures that supply the testes. The spermatic cord contains the ductus deferens, testicular artery and veins, lymph vessels,

the two meiotic divisions, each primary spermatocyte produces four spermatids (Figure 19-4). During spermatogenesis there are two cellular divisions, but only one replication of DNA so that each spermatid has 23 chromosomes (haploid), one from each pair in the original primary spermatocyte. Each successive stage in spermatogenesis is pushed toward the center of the tubule so that the more immature cells are at the periphery and the more differentiated cells are nearer the center (see Figure 19-3).

Spermatogenesis (and oogenesis in the female) differs from mitosis (review Chapter 3) because the resulting cells have only half the number of chromosomes as the original cell. When the sperm cell nucleus unites with an egg cell nucleus, the full number of chromosomes is restored. If sperm and egg cells were produced by mitosis, then each successive generation would have twice the number of chromosomes as the preceding one.

The final step in the development of sperm is called **spermiogenesis** (spur-mee-oh-JEN-eh-sis). In this process, the spermatids formed from spermatogenesis become mature spermatozoa, or sperm. The mature sperm cell has a **head, midpiece**, and **tail** (see Figure 19-3). The head, also called the nuclear region, contains the 23 chromosomes surrounded by a nuclear membrane. The tip of the head is covered by an **acrosome** (AK-roh-sohm), which contains enzymes that help the sperm penetrate the female gamete. The midpiece, also called the metabolic region, contains mitochondria that provide adenosine triphosphate (ATP). The tail, also called the locomotor region, is a typical flagellum for locomotion. The sperm are released into the lumen of the seminiferous tubule, leave the testes, and enter the epididymis, where they undergo their final maturation and become capable of fertilizing a female gamete.

testicular nerve, cremaster muscle (which elevates the testes for warmth and at times of sexual stimulation), and a connective tissue covering.

Ejaculatory Duct

Each ductus deferens, at the ampulla, joins the duct from the adjacent seminal vesicle (one of the accessory glands) to form a short **ejaculatory** (ee-JAK-yoo-lah-to-ree) **duct** (see Figure 19-1). Each ejaculatory duct passes through the prostate gland and empties into the urethra.

Urethra

The **urethra** (yoo-REE-thrah) extends from the urinary bladder to the external urethral orifice at the tip of the penis. It is a passageway for sperm and fluids from the reproductive system and for urine from the urinary system. While reproductive fluids are passing through the urethra, sphincters contract tightly to keep urine from entering the urethra.

The male urethra is divided into three regions (see Figure 19-1). The **prostatic urethra** is the proximal portion that passes through the prostate gland. It receives the ejaculatory duct, which contains spermatozoa and secretions from the seminal vesicles, and numerous ducts from the prostate gland. The next portion, the **membranous urethra**, is a short region that passes through the pelvic floor. The longest portion is the **penile urethra** (also called spongy urethra or cavernous urethra), which extends the length of the penis and opens to the outside at the external urethral orifice. The ducts from the bulbourethral glands open into the penile urethra.

QUICK CHECK

19.5 In what structure do sperm complete the maturation process?

19.6 What duct is ligated or severed in a vasectomy?

19.7 What region of the male urethra passes through the pelvic floor?

Accessory Glands

The accessory glands of the male reproductive system are the seminal vesicles, prostate gland, and the bulbourethral glands. These glands secrete fluids that enter the urethra.

Seminal Vesicles

The paired **seminal vesicles** are saccular glands posterior to the urinary bladder (see Figure 19-1). Each gland has a short duct that joins with the ductus deferens at the ampulla to form an ejaculatory duct that empties into the urethra. The fluid from the seminal vesicles is viscous and contains fructose, which provides an energy source for the spermatozoa; prostaglandins, which contribute to the motility and viability of the sperm; and proteins, which cause slight coagulation reactions in the semen after ejaculation.

Prostate

The **prostate gland** is a firm, dense structure that is located just inferior to the urinary bladder (see Figure 19-1). It is about the size of a walnut and encircles the urethra as it leaves the urinary bladder. Numerous short ducts from the substance of the prostate gland empty into the prostatic urethra. The secretions of the prostate are thin, milky colored, and alkaline. They function to enhance the motility of the sperm.

Cancer of the prostate is a common cancer in men. It usually starts in one of the secretory glands, and as it continues, it produces a lump on the surface of the prostate. In many cases, however, by the time the lump can be palpated through the wall of the rectum, the cancer has metastasized to other areas of the body. It is hoped that blood screening techniques (for prostate-specific antigen [PSA]) in addition to rectal palpation will result in earlier detection of the tumor so that treatment can begin before metastasis occurs.

Bulbourethral Glands

The paired **bulbourethral (Cowper's) glands** (see Figure 19-1) are small, about the size of a pea, and are located near the base of the penis. A short duct from each gland enters the proximal end of the penile urethra. In response to sexual stimulation, the bulbourethral glands secrete an alkaline, mucus-like fluid. This fluid neutralizes the acidity of the urine residue in the urethra, helps to neutralize the acidity of the vagina, and provides some lubrication for the tip of the penis during intercourse.

Seminal Fluid

Seminal fluid, or **semen**, is the slightly alkaline (pH 7.5) mixture of sperm cells and secretions from the accessory glands. Secretions from the seminal vesicles make up about 60% of the volume of the **semen**, with most of the remainder coming from the prostate gland. The spermatozoa and secretions from the bulbourethral glands contribute only a small volume.

The volume of semen in a single ejaculation may vary from 1.5 to 6.0 milliliters. There are typically between 50 and 150 million sperm per milliliter of semen. Sperm counts below 10 to 20 million per milliliter usually present fertility problems. Although only one spermatozoon actually penetrates and fertilizes the ovum, it takes several million spermatozoa in an ejaculation to ensure that fertilization occurs.

QUICK CHECK

19.8 Which of the male accessory glands contributes the greatest volume to the seminal fluid?

19.9 Which male accessory gland surrounds the urethra?

19.10 The duct from which male accessory gland opens into the proximal penile urethra?

Penis

The **penis**, the male copulatory organ, is a cylindrical pendant organ located anterior to the scrotum. It functions to transfer sperm to the vagina. Figure 19-5 illustrates that the penis consists of three columns of erectile tissue that are wrapped in connective tissue and covered with skin. The two dorsal columns are the **corpora cavernosa** (KOR-por-ah kav-er-NOH-sah). The single, midline ventral column surrounds the urethra and is called the **corpus spongiosum** (KOR-pus spun-jee-OH-sum).

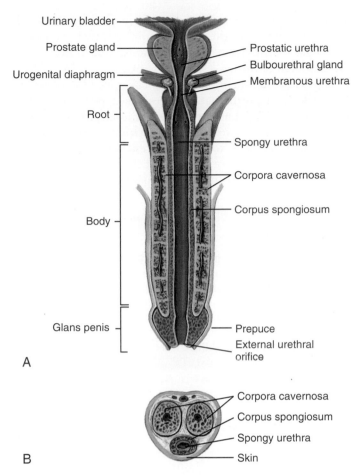

Urinary bladder

Prostate gland

Urogenital diaphragm

Root

Body

Glans penis

Prostatic urethra

Bulbourethral gland

Membranous urethra

Spongy urethra

Corpora cavernosa

Corpus spongiosum

Prepuce

External urethral orifice

A

Corpora cavernosa

Corpus spongiosum

Spongy urethra

Skin

B

Figure 19-5 *Structure of the penis.* **A,** *Longitudinal section.* **B,** *Cross section. Note that the penis has three regions: root, body, and glans penis. There are two dorsal columns of corpora cavernosa and a ventral column of corpus spongiosum.*

The penis has a **root, body** (or **shaft**), and **glans penis**. The root of the penis attaches it to the pubic arch, and the body is the visible, pendant portion. The corpus spongiosum expands at the distal end to form the glans penis. The urethra, which extends throughout the length of the corpus spongiosum, opens through the external urethral orifice at the tip of the glans penis. A loose fold of skin, called the **prepuce** (PREE-pyoos), or **foreskin**, covers the glans penis.

Circumcision is the surgical removal of the prepuce of the penis. Sometimes this is done to correct **phimosis**, a condition in which the prepuce is too tight and obstructs urine flow. In certain cultures, circumcision is performed as a religious rite or an ethnic custom. For others, it is a matter of family preference. The medical benefits of circumcision are a subject of debate in the medical community.

Male Sexual Response

In the absence of sexual arousal, the vascular sinusoids in the erectile tissue of the penis contain only a small volume of blood and the penis is flaccid. During sexual excitement, parasympathetic impulses dilate the arterioles that supply blood to the erectile tissue and constrict the veins that remove the blood. As a result, the spaces in the erectile tissue become engorged with

blood, causing the penis to enlarge and become rigid. This is called **erection** and is necessary to allow the penis to enter the female vagina. The erection reflex may be initiated by stimuli such as anticipation, memory, and visual sensations, or it may be the result of stimulation of touch receptors on the glans penis and skin of the genital area.

Impotence is the inability to achieve an erection. Psychological stresses are often blamed for impotence, but it may be caused by other factors. Impotence may result from an abnormality of the erectile tissue or failure of the parasympathetic reflexes that produce an erection. Drugs and alcohol may cause temporary impotence because they interfere with the nerve and blood vessel actions that are necessary for an erection.

Continued sexual stimulation causes the parasympathetic reflexes that promote an erection to become more and more intense until they reach a level that prompts a surge of sympathetic impulses to the genital organs. These sympathetic impulses stimulate rhythmic contractions of the epididymides, vasa deferentia, and ejaculatory ducts, along with contractions of the accessory glands. This results in **emission**—the forceful discharge of semen into the urethra. **Ejaculation**, which immediately follows emission, is the forceful expulsion of semen from the urethra to the exterior. Contraction of smooth muscle in the wall of the urethra and of skeletal muscle at the base of the penis forcefully ejects the semen from the urethra. Concurrently with emission and ejaculation, the sphincters of the urinary bladder constrict to prevent semen from entering the bladder and to inhibit the flow of urine from the bladder.

The rhythmic muscle contractions of ejaculation are accompanied by feelings of intense pleasure, increased heart rate, elevated blood pressure, and increased respiration. Together, these physiologic activities are referred to as **climax**, or **orgasm**. This is quickly followed by relaxation, and blood leaves the penis so it becomes flaccid. After orgasm, there is a latent period, lasting from several minutes to several hours, during which another erection is impossible.

Hormonal Control

The hypothalamus, anterior pituitary, and testes have significant roles in the hormonal control of male reproductive functions. The relationship between these areas is sometimes referred to as the **brain testicular axis**.

Puberty in males usually begins between the ages of 10 and 12 and continues until ages 16 to 18. During this period the male reproductive organs become sexually mature. The sequence of events that triggers the onset of puberty is unknown. It begins when certain unknown stimuli cause the hypothalamus to start secreting **gonadotropin-releasing hormone** (GnRH), which enters the blood and goes to the anterior pituitary gland.

In response to GnRH, the anterior pituitary secretes **luteinizing hormone** (LH) and **follicle-stimulating hormone** (FSH). Luteinizing hormone is often referred to as **interstitial cell-stimulating hormone** (ICSH) because it promotes the growth of the interstitial cells (cells of Leydig) in the testes and stimulates the cells to secrete **testosterone**. FSH binds with receptor sites on the sustentacular cells (Sertoli cells) in the seminiferous tubules. This action makes the spermatogenic cells respond to stimulation by testosterone. Testosterone and

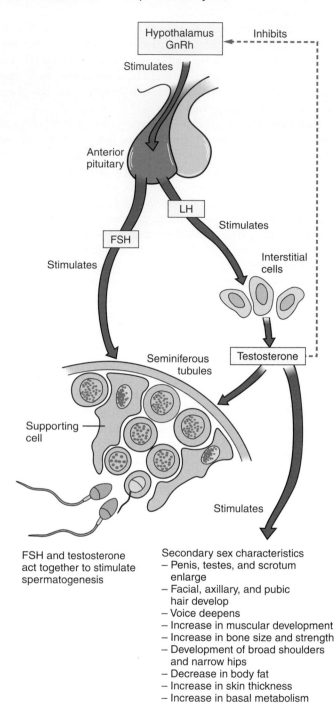

FSH and testosterone act together to stimulate spermatogenesis

Secondary sex characteristics
– Penis, testes, and scrotum enlarge
– Facial, axillary, and pubic hair develop
– Voice deepens
– Increase in muscular development
– Increase in bone size and strength
– Development of broad shoulders and narrow hips
– Decrease in body fat
– Increase in skin thickness
– Increase in basal metabolism

Figure 19-6 *Hormonal regulation of testicular function.*

FSH, acting together, stimulate spermatogenesis in the seminiferous tubules. Figure 19-6 summarizes the hormonal control of testicular functions.

Male sex hormones are collectively called **androgens**. The most abundant androgen is testosterone. Before birth and for a brief period following birth, testosterone from the adrenal cortex stimulates the development of the male reproductive organs. Between birth and puberty, testosterone levels are low. Then at puberty, under the influence of LH, the interstitial cells begin secreting high levels of testosterone. The adrenal cortex continues to secrete small

amounts of androgens. The increase in testosterone levels at puberty promotes the maturation of the male reproductive organs, stimulates spermatogenesis, and promotes the development of the male secondary sex characteristics. After puberty, testosterone production is controlled by a negative feedback mechanism that involves the hypothalamus (see Figure 19-6). High blood testosterone levels inhibit GnRH, which removes the stimulus for LH, which reduces the testosterone level back to normal. Testosterone production continues from puberty throughout the rest of a man's life, although there is some decline in quantity in old age.

QUICK CHECK

19.11 What occurs when the arteries to the penis dilate in response to parasympathetic impulses?
19.12 What is the difference between emission and ejaculation?
19.13 At puberty, what is the effect of luteinizing hormone in the male?
19.14 What will occur if there is a deficiency of testosterone before birth?

Female Reproductive System

The organs of the female reproductive system produce and sustain the female sex cells (egg cells, or ova), transport these cells to a site where they may be fertilized by sperm, provide a favorable environment for the developing offspring, move the offspring to the outside at the end of the development period, and produce the female sex hormones. The system includes the ovaries, uterine tubes, uterus, vagina, accessory glands, and external genital organs (Figure 19-7).

Ovaries

The primary reproductive organs, or gonads, in the female are the paired **ovaries**. Each ovary is a solid, ovoid structure about the size and shape of an almond, about 3.5 cm in length, 2 cm wide, and 1 cm thick. The ovaries are located in shallow depressions, called **ovarian fossae**, one on each side of the uterus, in the lateral wall of the pelvic cavity. They are held loosely in place by peritoneal ligaments.

Structure

The ovaries are covered on the outside by a layer of simple cuboidal epithelium called **germinal (ovarian) epithelium** (Figure 19-8). This is actually the visceral peritoneum that envelops the ovaries. Underneath this layer there is a dense connective tissue capsule, the **tunica albuginea**. The substance of the ovaries is indistinctly divided into an outer **cortex** and an inner **medulla**. The cortex appears more dense and granular because of the presence of numerous **ovarian follicles** in various stages of development. Each of the follicles contains an **oocyte**, a female germ cell. The medulla is loose connective tissue with abundant blood vessels, lymphatic vessels, and nerve fibers.

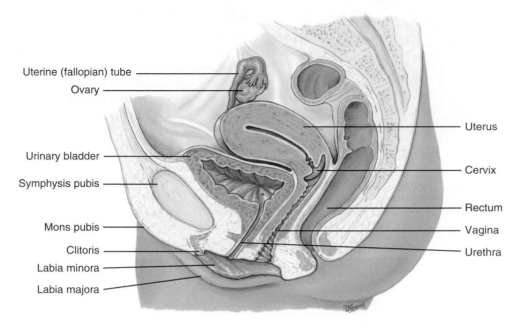

Figure 19-7 *Organs of the female reproductive system.*

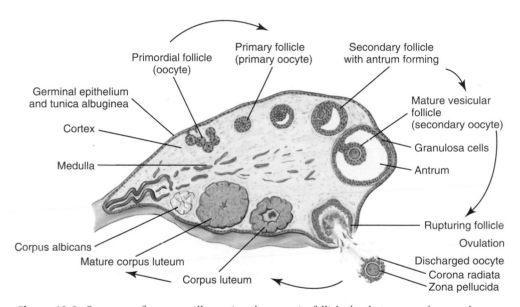

Figure 19-8 *Structure of an ovary illustrating the stages in follicle development and corpus luteum formation.*

Oogenesis

Female sex cells, or gametes, develop in the ovaries by a form of meiosis called **oogenesis** (oh-oh-JEN-eh-sis). The sequence of events in oogenesis is similar to the sequence in spermatogenesis, but the timing and final result are different (Figure 19-9). Early in fetal development, primitive germ cells in the ovaries differentiate into **oogonia** (oh-oh-GO-nee-ah). These divide rapidly to form thousands of cells, still called oogonia, which have a full complement of 46 (23 pairs) chromosomes. Oogonia then enter a growth phase, enlarge, and become **primary oocytes**. The diploid (46 chromosomes) primary oocytes replicate their DNA and begin the first meiotic division, but the process stops in prophase and the cells remain in this suspended state until puberty. Many of the primary oocytes degenerate before birth, but even with this decline, the 2 ovaries together contain approximately 700,000 oocytes at birth. This is the lifetime supply, and no more will develop. This is quite different than the male in whom spermatogonia and primary spermatocytes continue to be produced throughout the reproductive lifetime. By puberty the number of primary oocytes has further declined to about 400,000.

Beginning at puberty, under the influence of FSH, several primary oocytes start to grow again each month. One of the primary oocytes seems to outgrow the others, and it resumes meiosis I. The other cells degenerate. The large cell undergoes an unequal division so that nearly all the cytoplasm,

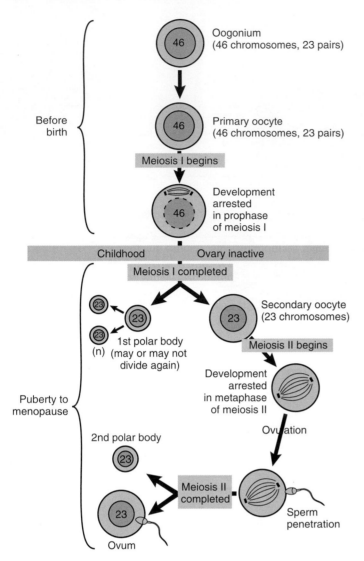

Before birth

Childhood — Ovary inactive

Puberty to menopause

Oogonium
(46 chromosomes, 23 pairs)

Primary oocyte
(46 chromosomes, 23 pairs)

Meiosis I begins

Development arrested in prophase of meiosis I

Meiosis I completed

1st polar body (n) (may or may not divide again)

Secondary oocyte (23 chromosomes)

Meiosis II begins

Development arrested in metaphase of meiosis II

Ovulation

2nd polar body

Meiosis II completed

Sperm penetration

Ovum

Figure 19-9 *Oogenesis. The first meiotic division is interrupted in prophase before birth and does not resume until after puberty. The second meiotic division is interrupted in metaphase and does not resume unless a sperm penetrates the cell.*

organelles, and half the chromosomes go to one cell, which becomes a **secondary oocyte**. The remaining half of the chromosomes go to a smaller cell called the **first polar body**. The secondary oocyte begins the second meiotic division, but the process stops in metaphase. At this point, ovulation occurs. If fertilization occurs, meiosis II continues. Again this is an unequal division with all of the cytoplasm going to the ovum, which has 23 single-stranded chromosomes. The smaller cell from this division is a **second polar body**. The first polar body also usually divides in meiosis II to produce two polar bodies. If fertilization does not occur, the second meiotic division is never completed and the secondary oocyte degenerates. Here again there are obvious differences between the male and female. In spermatogenesis, four functional spermatozoa develop from each primary spermatocyte. In oogenesis, only one functional fertilizable cell develops from a primary oocyte. The other three cells are polar bodies and they degenerate.

Ovarian Follicle Development

An ovarian follicle consists of a developing oocyte surrounded by one or more layers of cells called **follicular cells**. At the same time the oocyte is progressing through meiosis, corresponding changes are taking place in the follicular cells (see Figure 19-8). **Primordial follicles**, which consist of a primary oocyte surrounded by a single layer of flattened cells, develop in the fetus and are the stage that is present in the ovaries at birth and throughout childhood.

Beginning at puberty FSH stimulates changes in the primordial follicles. The follicular cells become cuboidal, the primary oocyte enlarges, and it is now a **primary follicle**. The follicles continue to grow under the influence of FSH, and the follicular cells proliferate to form several layers of **granulosa cells** around the primary oocyte. Most of these primary follicles degenerate along with the primary oocytes within them, but usually one continues to develop each month. The granulosa cells start secreting estrogen, and a cavity, or antrum, forms within the follicle. When the antrum starts to develop, the follicle becomes a **secondary follicle**. The granulosa cells also secrete a glycoprotein substance that forms a clear membrane, the **zona pellucida** (ZOH-nah peh-LOO-sih-dah), around the oocyte. After about 10 days of growth the follicle is a mature vesicular (graafian) follicle, which forms a "blister" on the surface of the ovary and contains a secondary oocyte ready for ovulation.

Ovulation

Ovulation, prompted by LH from the anterior pituitary, occurs when the mature follicle at the surface of the ovary ruptures and releases the secondary oocyte into the peritoneal cavity. The ovulated secondary oocyte, ready for fertilization, is still surrounded by the zona pellucida and a few layers of cells called the **corona radiata** (koh-ROH-nah ray-dee-AH-tah). If it is not fertilized, the secondary oocyte degenerates in a couple of days. If a spermatozoon passes through the corona radiata and zona pellucida and enters the cytoplasm of the secondary oocyte, the second meiotic division resumes to form a polar body and a mature ovum.

After ovulation and in response to LH, the portion of the follicle that remains in the ovary enlarges and is transformed into a **corpus luteum** (see Figure 19-8). The corpus luteum is a glandular structure that secretes progesterone and some estrogens. Its fate depends on whether fertilization occurs. If fertilization does not take place, the corpus luteum remains functional for about 10 days and then begins to degenerate into a **corpus albicans**, which is primarily scar tissue, and its hormone output ceases. If fertilization occurs, the corpus luteum persists and continues its hormone functions until the placenta develops sufficiently to secrete the necessary hormones. Again, the corpus luteum ultimately degenerates into a corpus albicans; it just remains functional for a longer period of time.

QUICK CHECK

19.15 Within what region of the ovary are the ovarian follicles located?

19.16 In what stage of oogenesis does ovulation occur?

19.17 What happens to the follicle cells after ovulation?

Genital Tract

Uterine Tubes

There are two **uterine tubes**, also called **fallopian** (fah-LOH-pee-an) **tubes** or **oviducts**. Each tube is about 4 cm long and about 1 cm in diameter and extends laterally from the upper portion of the uterus to the region of the ovary on that side (Figure 19-10). There is one tube associated with each ovary. The end of the tube near the ovary expands to form a funnel-shaped **infundibulum**, which is surrounded by fingerlike extensions called **fimbriae**. Because there is no direct connection between the infundibulum and the ovary, the oocyte enters the peritoneal cavity before it enters the uterine tube. At the time of ovulation, the fimbriae increase their activity and create currents in the peritoneal fluid that help propel the oocyte into the uterine tube. Once inside the uterine tube, the oocyte is moved along by the rhythmic beating of cilia on the epithelial lining and by the peristaltic action of the smooth muscle in the wall of the tube. The journey through the uterine tube takes about 7 days. Because the oocyte is fertile for only 24 to 48 hours, fertilization usually occurs in the uterine tube. Tubal ligation is a surgical procedure in which the uterine tubes are severed and tied off. This is a permanent method of birth control because sperm are unable to reach the egg for fertilization.

Uterus

The **uterus** is a muscular organ that receives the fertilized oocyte for implantation and provides an appropriate environment for the developing offspring. An ectopic pregnancy occurs when a fertilized egg implants in some site other than the uterus. Since other locations do not provide a suitable environment for development, an ectopic pregnancy often ends in a miscarriage. The uterus is located in the pelvic cavity, between the rectum and urinary bladder (see Figure 19-7). Before the first pregnancy, the uterus is about the size and shape of a pear, with the narrow portion directed inferiorly. After childbirth, the uterus is usually larger, and then regresses after menopause.

The upper, bulging surface of the uterus, above the entrance of the uterine tubes, is the **fundus** (see Figure 19-10). The large main portion is the **body**, and the narrow region that is directed inferiorly into the vagina is the **cervix**. The opening between the body and cervix is the **internal os**, and the opening from the cervix into the vagina is the **external os**. Normally the uterus is bent forward between the body and cervix so that the body projects anteriorly over the superior surface of the urinary bladder. In this position the uterus is said to be **anteflexed** (see Figure 19-7). Several ligaments hold the uterus in place. The largest of these is the **broad ligament**, which drapes over the uterus like a sheet and extends laterally to the lateral pelvic wall. The broad ligament also encloses the uterine tubes.

The wall of the uterus consists of perimetrium, myometrium, and endometrium. The outer serous layer, the **perimetrium**, is visceral peritoneum. The thick middle layer, the **myometrium**, is smooth muscle and makes up the bulk of the uterine wall. The inner layer, the **endometrium**, is a mucous membrane and is subdivided into two regions. The **stratum functionale** of the endometrium is the portion that is sloughed off during menstruation. The deeper, thinner **stratum basale** is more constant and provides the materials to rebuild the stratum functionale after menstruation.

Vagina

The **vagina** is a fibromuscular tube, about 10 cm long, that extends from the cervix of the uterus to the outside. It is located between the rectum and the urinary bladder. Because the vagina is tilted posteriorly as it ascends and the cervix is tilted anteriorly, the cervix projects into the vagina at nearly a right angle. The vagina provides a passageway for menstrual flow to reach the outside, receives the penis and semen during sexual intercourse (coitus), and serves as the birth canal during the birth of a baby. The smooth muscle and mucosal lining of the vaginal wall are capable of stretching to accommodate the erect penis and to permit passage of a baby. The opening of the vagina to the outside, the **vaginal orifice**, may be incompletely covered by a thin fold of mucous membrane called the **hymen**.

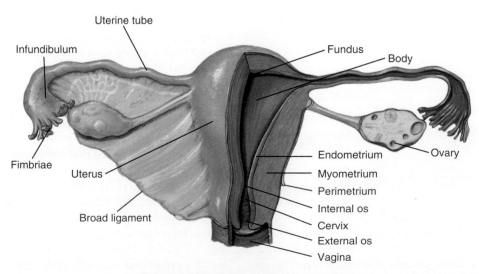

Figure 19-10 *Uterus and uterine tubes.*

QUICK APPLICATION

The **Papanicolaou smear (Pap test)** is a frequently used screening method for cervical and vaginal cancer. In this technique, the physician uses a swab or spatula to remove cells from the cervix. These cells are microscopically examined for abnormalities. This is an important tool in the early detection of cervical cancer, and every woman should take advantage of its availability.

QUICK CHECK

19.18 In what part of the genital tract does fertilization typically occur?

19.19 Which layer of the uterus is sloughed off during menstruation?

19.20 The vagina is anterior to what portion of the digestive tract?

External Genitalia

The external genitalia are accessory structures of the female reproductive system that are outside the vagina. They are also referred to as the **vulva** or **pudendum** (pyoo-DEN-dum). The external genitalia include the labia majora, mons pubis, labia minora, clitoris, and glands within the vestibule (Figure 19-11).

The **labia majora** *(labium majus)* are two large fat-filled folds of skin that enclose the other external genitalia. Anteriorly the labia majora merge to form the **mons pubis**, a rounded elevation of fat that overlies the pubic symphysis. After puberty, the mons pubis and lateral surfaces of the labia majora are covered with coarse pubic hair. The skin on the medial surfaces of the labia majora is thinner than that on the lateral surfaces and contains numerous sebaceous and sweat glands. The **labia minora** *(labium minus)* are two smaller folds of skin medial to the labia majora. The skin on the labia minora contains sebaceous glands but does not have hair, sweat glands, or adipose tissue.

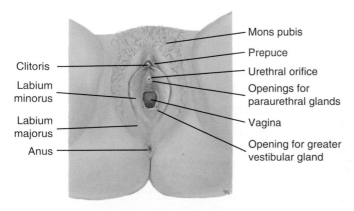

Clitoris

Labium minorus

Labium majorus

Anus

Mons pubis

Prepuce

Urethral orifice

Openings for paraurethral glands

Vagina

Opening for greater vestibular gland

Figure 19-11 *Female external genitalia. The area between the two labia minora is the vestibule.*

The area between the two labia minora is called the **vestibule**. At the anterior end of the vestibule, where the two labia minora meet, there is a small mass of erectile tissue called the **clitoris** (KLY-toh-ris). The clitoris is homologous to the male penis and becomes erect in response to sexual stimulation. The labia minora merge and form a hood, or **prepuce**, over the clitoris. Posterior to the clitoris, the urethra and vagina open into the vestibule. **Paraurethral glands** (Skene's glands) open into the vestibule on each side of the urethral orifice. These glands secrete mucus. Adjacent to the vaginal orifice, between the vagina and labia minora, the **greater vestibular glands** (Bartholin's glands) open into the vestibule. These glands produce a mucus-like secretion for lubrication during sexual intercourse.

Female Sexual Response

The female sexual response is similar to that of the male and consists of erection and orgasm. Parasympathetic responses to sexual stimuli produce increased blood flow to the erectile tissue in the clitoris, the vaginal mucosa, breasts, and nipples. The clitoris and nipples become rigid and erect. The breasts and vaginal mucosa enlarge. Glands in the cervix and the vestibular glands secrete fluids that lubricate the vaginal mucosa and aid the entry of the penis. These responses correspond to the erection phase of the male response.

With continued stimulation, the female response culminates in orgasm, but this is not accompanied by ejaculation. Sympathetic responses produce rhythmic contractions of the uterus and muscles of the pelvic floor. This helps the movement of sperm through the uterus toward the uterine tubes. The rhythmic muscle contractions are accompanied by feelings of intense pleasure, increased heart rate, elevated blood pressure, and increased respiration rate. This is followed by a general relaxation and feeling of warmth throughout the body. It is not necessary for a woman to experience orgasm to become pregnant.

QUICK CHECK

19.21 Which component of the vulva is homologous to the male penis and becomes erect during sexual arousal?

19.22 Which is more posterior, the urethra or the vagina?

19.23 What glands open into the vestibule adjacent to the vagina?

Hormonal Control

As in the male, the hypothalamus, anterior pituitary, and gonads secrete hormones that have significant roles in the control of reproductive functions (Figure 19-12). The hypothalamus secretes GnRH, the anterior pituitary secretes FSH and LH, and the ovaries secrete the sex hormones **estrogen** and **progesterone**. Unlike the male, the secretion of these hormones follows monthly cyclic patterns that affect the ovaries and the uterus. These cycles, referred to as the **ovarian cycle** and the **menstrual** (uterine) **cycle**, begin at puberty and continue for about 40 years.

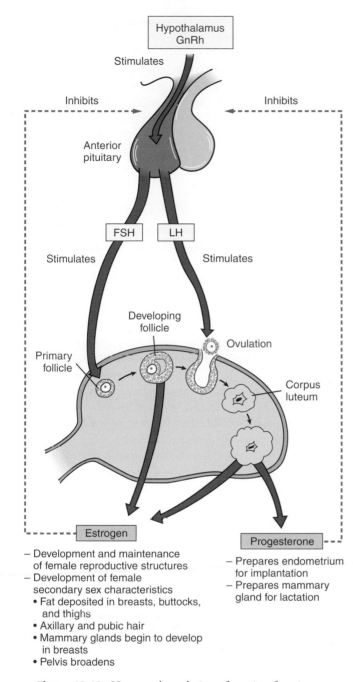

Figure 19-12 *Hormonal regulation of ovarian functions.*

Estrogen

– Development and maintenance of female reproductive structures
– Development of female secondary sex characteristics
  • Fat deposited in breasts, buttocks, and thighs
  • Axillary and pubic hair
  • Mammary glands begin to develop in breasts
  • Pelvis broadens

Progesterone

– Prepares endometrium for implantation
– Prepares mammary gland for lactation

At puberty, when the ovaries and uterus are mature enough to respond to hormonal stimulation, certain stimuli cause the hypothalamus to start secreting GnRH. This hormone enters the blood and goes to the anterior pituitary gland, where it stimulates the secretion of FSH and LH. These hormones, in turn, affect the ovaries and uterus, and the monthly cycles begin. In females, the beginning of puberty is marked by the first period of menstrual bleeding, called **menarche** (meh-NAHR-kee). After this the cycles continue, more or less regularly, until the late 40s or early 50s. At this time, the cycles become increasingly irregular until they finally stop. **Menopause** is the cessation of reproductive cycles.

Ovarian Cycle

The ovarian cycle reflects the changes that occur within the ovaries as the follicles develop (follicular phase), ovulation occurs (ovulatory phase), and the corpus luteum develops (luteal phase) (Figure 19-13).

The **follicular phase** of the cycle begins when GnRH from the hypothalamus stimulates increased secretion of FSH (and small amounts of LH) from the anterior pituitary. FSH stimulates growth of the ovarian follicles. As the follicles enlarge, estrogen secretion increases. Low levels of estrogen exert a negative feedback effect on the hypothalamus and anterior pituitary, but also increase the effect of FSH on the follicle. The follicle continues to grow and mature until the middle of the cycle.

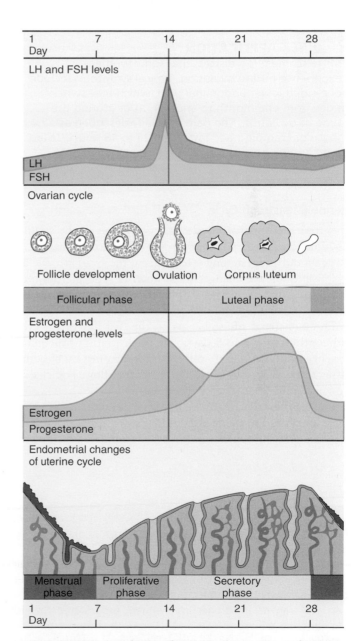

Figure 19-13 *Correlation of events in the ovarian and uterine cycles. The menstrual and proliferative phases of the uterine cycle correspond to the follicular phase of the ovarian cycle. The secretory phase of the uterine cycle corresponds to the luteal phase of the ovarian cycle. Note the relative hormone levels in each phase.*

The **ovulatory phase** is the result of high levels of estrogen from the mature follicles. Low levels of estrogen have a negative feedback effect on the pituitary, but high levels have a positive feedback effect. This results in a surge of LH and a smaller increase in FSH. The surge of LH stimulates resumption of meiosis in the oocyte and causes the rupture of the follicle, and estrogen levels decline.

The **luteal phase** occurs when the surge of LH stimulates the development of the corpus luteum from the ruptured follicle. LH also stimulates the corpus luteum to secrete progesterone and some estrogen. These have a negative feedback effect on the hypothalamus and anterior pituitary so that FSH and LH levels decline. As LH level declines, corpus luteum activity declines and the inhibitory effect is removed. The cycle starts over.

QUICK APPLICATION

Mittelschmerz is a distress that some women experience during ovulation. Typical symptoms include one-sided lower abdominal pain, which may switch sides from one month to another, at or around the time of ovulation. The pain is usually described as sharp or cramping and lasts from 24 to 48 hours. There is no known prevention, and treatment consists of analgesics.

Uterine (Menstrual) Cycle

The uterine (menstrual) cycle reflects changes in the stratum functionale of the endometrium of the uterus. Changes in estrogen and progesterone levels during the ovarian cycle are responsible for the changes in the uterus. The uterine cycle is divided into the menstrual phase, proliferative phase, and secretory phase (see Figure 19-13).

The **menstrual phase** begins on the first day of the cycle and continues for 3 to 5 days. The thick stratum functionale detaches from the uterine wall and, accompanied by bleeding, passes through the vagina as the menstrual flow. Follicles are growing in the ovary during this time.

The **proliferative phase** begins with the end of the menstrual phase and lasts for about 8 days. Increasing levels of estrogen from the growing follicles in the ovary stimulate repair of the endometrium in the uterus. The endometrium thickens, glands develop, and blood vessels grow in the new tissue. Ovulation in the ovarian cycle occurs at the end of this uterine phase.

The **secretory phase** corresponds to the luteal phase of the ovarian cycle. Progesterone from the corpus luteum stimulates continued growth and thickening of the endometrium. Arteries and glands proliferate and enlarge. The glands secrete glycogen, which will nourish a developing embryo if fertilization occurs. If fertilization does not occur, the corpus luteum in the ovary begins to degenerate. This leads to menstruation, and the cycle starts over.

Menopause

Menopause is the cessation of the female reproductive cycles. Even though menopause is marked by the lack of menstrual cycles, the first changes are in the ovary. By the age of 45 or 50, ovarian follicles cease responding to FSH and LH from the pituitary gland. As a result, the follicle cells do not produce estrogen and there is no ovulation, no corpus luteum, and no progesterone. Without estrogen and progesterone, the cyclic changes in the uterus stop and menstruation ceases. This is the visible evidence of menopause. As estrogen and progesterone levels decline, FSH and LH increase because of the lack of ovarian hormone feedback. These high levels of pituitary hormones with the low levels of ovarian hormones are believed to be responsible for a variety of symptoms associated with the onset of menopause. Some women experience hot flashes, sweating, depression, headaches, irritability, and insomnia. However, many women experience few, if any, of these symptoms.

QUICK CHECK

19.24 What are two effects of luteinizing hormone in the ovarian cycle?

19.25 What is the primary source of progesterone in the nonpregnant remale?

19.26 What happens in the uterus in response to increased estrogen levels during the second week of the menstrual cycle?

 FROM THE PHARMACY

Oral contraceptives, or birth control pills, are used to prevent pregnancy when a highly effective method is needed and heterosexual activity is regular. They act by increasing serum levels of estrogens and progestins (synthetic progesterone), which inhibits the release of FSH and LH from the anterior pituitary gland. The decrease in the pituitary hormones inhibits ovulation, causes changes in the endometrium of the uterus that impair implantation, and alters the amount and viscosity of the mucus in the cervix, which inhibits the motility of the sperm. Oral contraceptives are metabolized in the liver and excreted in the kidneys.

The use of oral contraceptives involves some risk. They should be used with caution in patients with a strong family history of breast cancer, benign breast disease, diabetes mellitus, epilepsy, gallbladder disease, hypertension, and migraine; patients who have had recent major surgery or extended immobilization are

FROM THE PHARMACY—CONT'D

also at risk. Oral contraceptives are contraindicated in persons who have had estrogen-dependent tumors in the breast or endometrium, any form of cardiovascular disease including thrombosis and phlebitis, cerebrovascular disease, and liver disease.

Combination estrogen and progestin contraceptives are divided into three types:

Monophasic—This type of contraceptive has a fixed ratio of estrogen and progestin that is taken for 21 days of the normal menstrual cycle.

Biphasic—This type of contraceptive regimen supplies a constant dose of estrogen with low levels of progestin during the

follicular phase of the cycle and higher doses of progestin during the luteal phase. The final portion of the 28-day biphasic cycle uses a placebo.

Triphasic—This contraceptive most closely resembles the natural levels of estrogen and progesterone during a normal uterine cycle. Estrogen levels are kept fairly low and constant, while the progestin level is progressively increased in three increments to simulate the natural release of hormones in the female. Because this more closely resembles the natural cycle, there are fewer adverse effects with this regimen than with the monophasic or biphasic methods.

Mammary Glands

Functionally, the mammary glands are the organs of milk production; structurally, they are modified sweat glands. Mammary glands, which are located in the breast overlying the pectoralis major muscles, are present in both sexes, but usually are functional only in the female.

Externally, each breast has a raised **nipple**, which is surrounded by a circular pigmented area called the **areola** (ah-REE-oh-lah). The nipples are sensitive to touch, and they contain smooth muscle that contracts and causes them to become erect in response to stimulation.

Internally, the adult female breast contains 15 to 20 lobes of glandular tissue that radiate around the nipple (Figure 19-14). The **lobes** are separated by connective tissue and adipose. The connective tissue helps support the breast. Some bands of connective tissue, called **suspensory** (Cooper's) **ligaments**, extend through the breast from the skin to the underlying muscles. The amount and distribution of the adipose determines the size and shape of the breast. Each lobe consists of **lobules** that contain the glandular units. A **lactiferous** (lak-TIFF-er-us) **duct** collects the milk from the lobules within each lobe and carries it to the nipple. Just before the nipple, the lactiferous duct enlarges to form a **lactiferous sinus (ampulla)**, which serves as a reservoir for milk. After the sinus, the duct again narrows and each duct opens independently on the surface of the nipple.

Mammary gland function is regulated by hormones. At puberty, increasing levels of estrogen stimulate the development of glandular tissue in the female breast. Estrogen also causes the breasts to increase in size through the accumulation of adipose tissue. Progesterone stimulates the development of the duct system. During pregnancy these hormones further enhance development of the mammary glands. Prolactin from the anterior pituitary stimulates the production of milk within the glandular tissue, and oxytocin causes the ejection of milk from the glands.

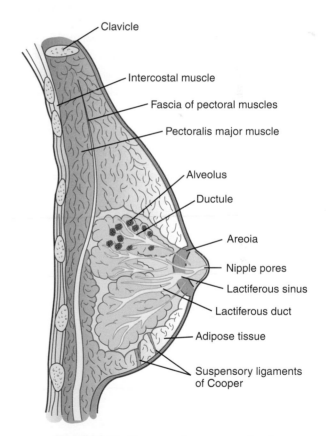

Figure 19-14 *Breasts and mammary glands.*

 QUICK CHECK

19.27 What ducts transport milk from the glandular tissue to the nipple of the breast?
19.28 What hormone stimulates the ejection of milk from the mammary glands?
19.29 What is the difference in effect on the breast at puberty between estrogen and progesterone?

FOCUS ON AGING

Men normally do not experience a sudden decline in reproductive function comparable to menopause in women. Instead, they experience a gradual and subtle decline over many years. After age 50, men have some testicular atrophy, partially caused by a decrease in the size of the seminiferous tubules and partially a result of a reduction in the number of interstitial cells. These changes are accompanied by a decline in sperm and testosterone production. Both the seminal vesicles and the prostate show a decrease in secretory activity, which results in a reduction in the volume of semen. The portion of the prostate gland that surrounds the urethra often enlarges and may constrict the urethra, making urination difficult. The penis may undergo some atrophy and become smaller with age. The blood vessels and erectile tissue in the penis become less elastic, which hinders the ability to attain an erection. Although there is a general decline in the aging male reproductive system, many men are capable of achieving erection and ejaculation into old age.

After menopause, there is a gradual decline in the female reproductive system. Most of the changes are believed to be due to the reduction in estrogen. The ovaries undergo progressive atrophy. The uterus becomes smaller, and fibrous connective tissue replaces much of the myometrium. The vagina becomes narrower and shorter, and its walls become thin and less elastic. Glands that lubricate the vagina reduce their secretory activity, and the vagina becomes dry. The vaginal secretions that remain are less acidic, which makes older women more susceptible to vaginal infections. The external genitalia and mammary glands undergo atrophic changes. The lack of estrogen also affects nonreproductive organs. This is particularly true in the case of bone metabolism, which is indicated by the increased prevalence of osteoporosis in postmenopausal women. There is also increased cardiovascular disease.

Estrogen replacement therapy is prescribed for many women to combat osteoporosis and other symptoms of menopause. However, there is controversy about the risks involved with this treatment, particularly the risks of uterine and breast cancer. Consideration should be given to the risks and benefits before beginning estrogen replacement therapy. Current practice often prescribes progesterone in conjunction with estrogen, which seems to reduce some of the risks.

There is a growing awareness that elderly people have sexual needs and enjoy sexual relations. Although age-related physical and hormonal changes that take place in the reproductive system may alter these needs and sexual functioning, studies demonstrate that sexuality remains important to many older people.

Representative Disorders of the
Reproductive System

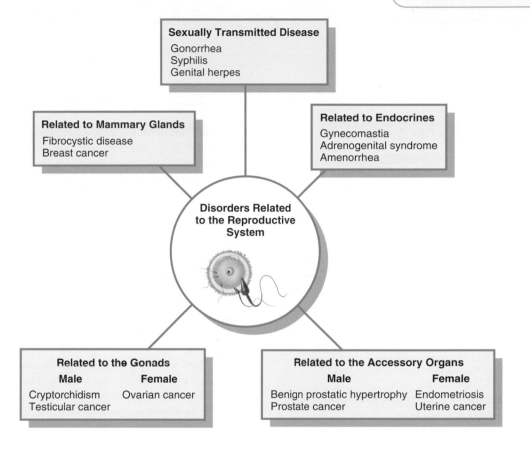

Sexually Transmitted Disease
Gonorrhea
Syphilis
Genital herpes

Related to Mammary Glands
Fibrocystic disease
Breast cancer

Related to Endocrines
Gynecomastia
Adrenogenital syndrome
Amenorrhea

Disorders Related to the Reproductive System

Related to the Gonads	
Male	**Female**
Cryptorchidism	Ovarian cancer
Testicular cancer	

Related to the Accessory Organs	
Male	**Female**
Benign prostatic hypertrophy	Endometriosis
Prostate cancer	Uterine cancer

Adrenogenital syndrome (ah-DREE-noh-jen-ih-tal SIN-drohm) Group of symptoms associated with alterations in secondary sex characteristics as a result of increased sex hormones from the adrenal cortex; increased amounts of androgens in the female lead to masculinization, and increased amounts of estrogens in the male lead to breast enlargement (gynecomastia)

Amenorrhea (ah-men-oh-REE-ah) Absence of menstruation; may be due to abnormal development or malformation of the reproductive organs, endocrine disturbances, excessive weight loss, emotional shock, or pregnancy

Benign prostatic hypertrophy (bee-NYNE prah-STAT-ick high-PER-troh-fee) Nonmalignant enlargement of the prostate gland, common in men older than age 50; because the prostate surrounds the urethra, it may obstruct the flow of urine from the bladder

Breast cancer (BREST KAN-ser) Malignancy of the breast arising from milk glands and ducts; the most common type is invasive ductal carcinoma. Other types are medullary carcinoma and lobular carcinoma; early detection by breast self examination and annual mammography are keys to early detection

Cryptorchidism (krip-TOR-kid-izm) Failure of one or both of the testes to descend into the scrotum; if both testes are involved and untreated, sterility results

Endometriosis (en-doh-mee-trih-OH-sis) A condition in which endometrial tissue occurs in various abnormal sites in the abdominal or pelvic cavity; may be caused when pieces of

menstrual endometrium pass backward through the uterine tubes into the peritoneal cavity

Fibrocystic disease (fi-broh-SIS-tick dih-ZEEZ) Condition characterized by the development of benign cysts in the breasts; cysts may be fluid-filled sacs or solid growths containing connective tissue

Genital herpes (JEN-ih-tal HER-peez) Highly contagious disease of the reproductive organs caused by the herpes simplex virus and transmitted by direct person-to-person contact; characterized by periods of dormancy and recurrent outbreaks; second most prevalent sexually transmitted disease in the United States

Gonorrhea (gahn-oh-REE-ah) Highly contagious bacterial *(Neisseria gonorrhoeae)* infection of the genitourinary system; most prevalent sexually transmitted disease in the United States

Gynecomastia (jin-eh-koh-MASS-tee-ah **or** gy-neh-koh-MASS-tee-ah) Excessive development of the mammary glands in the male as a result of hormonal disorders, especially the estrogen/testosterone ratio

Ovarian cancer (oh-VAIR-ee-an KAN-ser) Malignant tumor of the ovary; usually arising from epithelial cells; most dangerous reproductive cancer among women because it is seldom detected early

Prostate cancer (PRAH-state KAN-ser) Malignant tumor of the prostate, usually originating in a secretory portion; early detection is important because metastasis soon involves the lymphatic system, bones, and lungs

Syphilis (SIFF-ih-lis) Contagious sexually transmitted disease caused by the spirochete *Treponema pallidum*; it leads to many structural and cutaneous lesions; if untreated, it may lead to serious cardiovascular and neurologic disturbances years after infection and may be spread to the fetus during pregnancy

Testicular cancer (tess-TICK-yoo-lar KAN-ser) Malignant tumor of the testes, usually arising from the germ cells; relatively rare and curable if treated early in tumor development

Uterine cancer (YOO-ter-in KAN-ser) Malignant tumor of the uterus, usually a carcinoma that arises from the basal cells of the epithelial lining; may be detected by dysplasia of the epithelial cells in a Papanicolaou (PAP) smear; progresses to carcinoma-in-situ and finally to an invasive malignant carcinoma

CHAPTER SUMMARY

Overview of the Reproductive System

■ **State four functions of the reproductive system.**

- The reproductive system produces egg and sperm cells, transports and sustains these cells, nurtures the developing offspring, and produces hormones.

■ **Distinguish between the primary and secondary reproductive organs.**

- Primary, or essential, reproductive organs are the gonads. In the male, the gonads are the testes and in the female they are the ovaries.
- Secondary reproductive organs include all other components of the reproductive system, including the ducts, glands, and accessory organs.

Male Reproductive System

■ **Describe the location and structure of each component of the male reproductive system.**

- The testes, located in the scrotum, are surrounded by the tunica albuginea, which extends inward to divide the organ into lobules. Each lobule contains seminiferous tubules and interstitial cells. Sperm are produced in the seminiferous tubules and the interstitial cells produce testosterone.
- The epididymis is a convoluted tube on the margin of the testis. Sperm mature and become fertile in the epididymis.
- The ductus deferens begins at the epididymis and extends to the ejaculatory duct posterior to the urinary bladder.
- The ejaculatory duct is a short passage way that is formed when the ductus deferens and the duct from the seminal vesicles join. It penetrates the prostate gland and empties into the urethra.
- The male urethra is a passageway for sperm, fluids from the reproductive system, and urine. It is divided into the prostatic urethra, membranous urethra, and penile urethra.
- Seminal vesicles are paired, saccular glands posterior to the urinary bladder. Secretion from these glands contributes over 60% of the volume of seminal fluid and contains fructose, prostaglandins, and coagulation proteins.
- The prostate is a firm dense gland located inferior to the urinary bladder; it encircles the proximal part of the urethra. Prostatic secretions are alkaline and enhance motility of sperm.
- The bulbourethral glands are located near the base of the penis and secrete a viscous alkaline fluid, that provides lubrication during intercourse and neutralizes the acidity of vagina

- Seminal fluid or semen is a mixture of sperm cells and the secretions of the accessory glands. Emission is the discharge of semen into the urethra; ejaculation is the forceful expulsion of semen from the urethra.
- The penis consists of three columns of erectile tissue. The two dorsal columns are the corpora cavernosa and the ventral column is the corpus spongiosum. The root of the penis attaches it to the pubic arch, the body is the visible portion, and the glans penis is the expanded tip of corpus spongiosum. The urethra extends through the entire length of the penis.

■ **Draw and label a diagram or flow chart that illustrates spermatogenesis and describe the process by which spermatids become mature sperm.**

- Spermatogenesis, which begins at puberty, is the process by which spermatids are formed. Spermatogonia become primary spermatocytes and each spermatocyte produces four spermatids by meiosis.
- Spermiogenesis changes spermatids into mature spermatozoa, each with a head, midpiece, and tail. The head contains the nucleus, the midpiece contains mitochondria, and the tail is a flagellum for movement.

■ **Trace the pathway of sperm from the testes to the outside of the body.**

- Spermatozoa are produced in the seminiferous tubules of the testes then pass through the duct system to reach the outside.
- From the seminiferous tubules the sperm enter the straight tubules and rete testis. From there they pass through the efferent ducts into the epididymis, then the ductus deferens to the ejaculatory duct and into the urethra.

■ **Outline the physiologic events in the male sexual response.**

- The male sexual response includes erection and orgasm accompanied by ejaculation of semen.
- Orgasm is followed by a variable time period during which it is not possible to achieve another erection.

■ **Describe the roles of GnRH, FSH, LH, and testosterone in male reproductive functions.**

- At puberty, the hypothalamus secretes GnRH, which stimulates the anterior pituitary to secrete FSH and LH.
- FSH stimulates the seminiferous tubules and spermatogenesis.
- LH stimulates the interstitial cells and the production of testosterone.
- Testosterone from the interstitial cells stimulates the development of the secondary sex characteristics and spermatogenesis.

Female Reproductive System

■ **Describe the location and structure of each component of the female reproductive system.**

- The female gonads are the ovaries, which are located on each side of the uterus in the pelvic cavity. They are covered with simple cuboidal epithelium around the tunica albuginea. Numerous ovarian follicles make the cortex appear granular; the medulla is connective tissue with vessels and nerves.
- The uterine tubes, also called fallopian tubes or oviducts, extend laterally from each side of the uterus. The end of the uterine tube near the ovary expands to form the infundibulum. Fingerlike projections, called fimbriae, extend from the infundibulum.
- The uterus consists of a fundus, body, and cervix. The broad ligament is a fold of peritoneum that extends laterally from the uterus to the pelvic wall. The fundus and body of the uterus are normally anteflexed over the superior surface of the urinary bladder. The internal os is the opening from the body into the cervix; the external os is the opening from the cervix into the vagina. Visceral peritoneum forms the perimetrium, the outer layer of the uterine wall; smooth muscle makes up the thick myometrium; and the endometrium is mucous membrane. The endometrium is separated into a deeper stratum basale and superficial stratum functionale.
- The vagina extends from the cervix to the exterior. It serves as a passageway for menstrual flow, receives the erect penis during intercourse, and is the birth canal during the birth of a baby.
- Collectively, the female external genitalia are referred to as the vulva or pudendum. This includes the libia majora, mons pubis, libia minora, clitoris, and accessory glands. The area between the two libia minora is the vestibule. The clitoris (erectile tissue) is at the anterior end of the vestibule. Paraurethral and greater vestibular glands, accessory glands of the female reproductive tract, open into the vestibule. The urethra and vagina also open into the vestibule.

■ **Draw and label a diagram or flow chart that illustrates oogenesis.**

- Oogenesis begins in prenatal development with the formation of the primary oocyte. Division ceases in this stage and the oocytes remain dormant until puberty.
- Beginning at puberty, each month, a primary oocyte resumes meiosis and produces a secondary oocyte and a polar body. Division again halts.
- If a sperm penetrates the oocyte, meiosis resumes and a mature egg and another polar body are produced.
- Oogenesis differs from spermatogenesis in that each primary oocyte produces one ovum and three nonfunctional polar bodies; each primary spermatocyte produces four spermatids. Beginning at puberty, spermatogenesis is a continuous process; oogenesis occurs in monthly cycles. Spermatogenesis does not begin until puberty; oogenesis begins in prenatal development.

■ **Describe the development of ovarian follicles as they progress from primordial follicles to primary follicles, secondary follicles, vesicular follicles, corpus luteum, and, finally, the corpus albicans.**

- An ovarian follicle consists of an oocyte with one or more layers of cells surrounding it.
- Primordial follicles, each with a primary oocyte surrounded by a single layer of cells, are the follicles present at birth.
- At puberty, the primordial follicles begin to grow and become secondary follicles, and some mature and become vesicular follicles.
- Vesicular follicles rupture at ovulation and release their secondary oocyte.
- Follicles develop under the influence of FSH; follicle cells produce estrogen.
- After ovulation, the follicle cells are transformed into a corpus luteum that produces progesterone. The corpus luteum degenerates into a corpus albicans.

■ **Outline the physiologic events in the female sexual response.**

- The female sexual response includes erection and orgasm, but there is no ejaculation.
- A woman may become pregnant without having an orgasm.

■ **Describe the roles of GnRH, FSH, LH, estrogen, and progesterone in female reproductive functions.**

- GnRH, FSH, LH, estrogen, and progesterone interact to create the ovarian and uterine cycles.

■ **Describe what happens in each phase of the ovarian and uterine cycles, when each phase occurs, and how the cycles interact.**

- The monthly ovarian cycle begins with the follicle development during the follicular phase, continues with ovulation during the ovulatory phase, and concludes with the development and regression of the corpus luteum during the luteal phase.
- The follicle develops under the influence of FSH. As the follicle matures, it secretes increasing amounts of estrogen. At ovulation, the estrogen level falls, then the corpus luteum, under the influence of LH, begins secreting progesterone.
- The uterine cycle takes place simultaneously with the ovarian cycle.
- The uterine cycle begins with menstruation during the menstrual phase, continues with repair of the endometrium during the proliferative phase, and ends with growth of glands and blood vessels during the secretory phase.
- The menstrual phase is the result of decreased amounts of progesterone and estrogen from the corpus luteum; estrogen from developing follicles is responsible for the proliferative phase, and progesterone from the corpus luteum is responsible for the secretory phase.
- Menarche is the first menstrual flow; menopause is the time when the monthly ovarian and uterine cycles cease.

■ **Describe the structure of mammary glands.**

- The mammary glands, located within the breast, consist of lobules of glandular units that produce milk. Lactiferous ducts transport the milk to the nipple. Cords of connective tissue, called suspensory ligaments, help support the breast.

■ **Discuss the hormonal control of mammary glands.**

- Estrogen and progesterone stimulate the development of glandular tissue and ducts in the breast. Prolactin stimulates the production of milk. Oxytocin causes the ejection of milk from the breast.

CHAPTER QUIZ

Recall

Match the definitions on the left with the appropriate term on the right.

_____ **1.** Ovaries and testes
_____ **2.** Egg and sperm cells
_____ **3.** Smooth muscle in the subcutaneous tissue of the scrotum
_____ **4.** Skeletal muscle in the spermatic cord
_____ **5.** Produce testosterone
_____ **6.** Occurs within seminiferous tubules
_____ **7.** Clear membrane around the oocyte
_____ **8.** Produces progesterone
_____ **9.** Normal site of fertilization
_____ **10.** Smooth muscle in the uterus

A. Corpus luteum
B. Cremaster
C. Dartos
D. Gametes
E. Gonads
F. Interstitial cells
G. Myometrium
H. Spermatogenesis
I. Uterine tube
J. Zona pellucida

Thought

1. The correct pathway for passage of sperm is:
  A. Seminiferous tubules, ductus deferens, epididymis, rete testis, urethra
  B. Rete testis, seminiferous tubules, ductus deferens, epididymis, urethra
  C. Ductus deferens, seminiferous tubules, rete testis, epididymis, urethra
  D. Seminiferous tubules, rete testis, epididymis, ductus deferens, urethra

2. Spermiogenesis is:
  A. A part of meiosis
  B. The formation of mature spermatozoa from spermatids
  C. The formation of spermatids
  D. The formation of testosterone by spermatozoa

3. More than half of the volume of seminal fluid comes from the:
  A. Prostate gland
  B. Seminal vesicles
  C. Bulbourethral glands
  D. Sperm

4. The term *vulva* refers to the:
  A. Ovary, uterus, and vagina
  B. Cervix, vagina, and clitoris
  C. Labia majora and minora, mons pubis, and clitoris
  D. Penis, scrotum, and testes

5. The luteal phase of the ovarian cycle:
  A. Is characterized by high levels of progesterone
  B. Occurs at the same time as the proliferative phase of the uterine cycle
  C. Occurs just before ovulation
  D. Is characterized by high levels of luteinizing hormone

Application

A 36-year-old mother of four is considering tubal ligation (constricting the uterine tubes) as a means of ensuring that her family gets no larger. She asks her gynecologist if she will become "menopausal" after the surgery. Answer her question and explain.

Sexually transmitted diseases such as gonorrhea sometimes cause peritonitis (inflammation of the peritoneum) in females. However, peritonitis from this cause does not develop in males. Explain.

BUILDING YOUR MEDICAL VOCABULARY

Building a vocabulary is a cumulative process. As you progress through this book, you will use word parts, abbreviations, and clinical terms from previous chapters. Each chapter will present new word parts, abbreviations, and terms to add to your expanding vocabulary.

Word Parts and Combining Form with Definition and Examples

PART/ COMBINING FORM	DEFINITION	EXAMPLE
andr/o	male	<u>andr</u>ogens: hormones that produce maleness
balan/o	glans	penis <u>balan</u>itis: inflammation of the glan penis caused by an overgrowth of bacteria and yeast
colp/o	vagina	<u>colp</u>oscopy: visual examination of the vagina and cervix using a colposcope
crypt-	hidden	<u>crypt</u>orchidism: condition of hidden testes; testes fail to descend into the scrotum
ejacul-	to shoot forth	<u>ejacul</u>ation: process of forceful expulsion of seminal fluid from the urethra
episi/o	Vulva	<u>episi</u>otomy: incision through the skin of the perineum to enlarge the vaginal orifice for delivery
fimb-	fringe	<u>fimb</u>riae: fringe-like or finger-like projections at the end of the uterine tube
genit/o	organs of reproduction	<u>genit</u>ourinary: pertaining to the reproductive and urinary organs
gynec/o	female	<u>gynec</u>ogenic: producing female characteristics
hyster/o	womb or uterus	<u>hyster</u>ectomy: surgical removal of the uterus
lapar/o	abdominal wall	<u>lapar</u>orrhaphy: suture or repair of the abdominal wall
mamm/o	breast	<u>mamm</u>ogram: radiograph of the breast
men/o	menstruation	oligo<u>men</u>orrhea: infrequent or scanty menstruation
metr/o	uterus	myo<u>metr</u>ium: muscle layer of the uterine wall

PART COMBINING FORM	DEFINITION	EXAMPLE
oo-	egg, ovum	<u>oo</u>genesis: production of egg cells
oophor/o	Ovary	<u>oophor</u>ectomy: surgical removal of an ovary
orchi-	testicle	<u>orchi</u>opathy: any disease of the testes
prostat/o	prostate	<u>prostat</u>omegaly: hypertrophy of the prostate
-rrhagia	to burst forth	mono<u>rrhagia</u>: excessive menstruation
salping/o	Tube	<u>salping</u>olithiasis: presence of calcium deposits in the wall of the uterine tube
-spadias	an opening	hypo<u>spadias</u>: congenital abnormality in which the male urethral opening is on the undersurface of the penis

Clinical Abbreviations

ABBREVIATION	MEANING
BPH	benign prostatic hypertrophy or hyperplasia
BSE	breast self-examination
Cx	cervix
D&C	dilation and curettage
IUD	intrauterine device
IVF	in vitro fertilization
LMP	last menstrual period
Pap	Papanicolaou smear, stain, or test
PID	pelvic inflammatory disease
PMS	premenstrual syndrome
PSA	prostate-specific antigen
STD	sexually transmitted disease
TAH	total abdominal hysterectomy
TURP	transurethral resection of the prostate
VD	venereal disease

Clinical Terms

Anorchism (an-OAR-kizm) Congenital absence of one or both testes

Azoospermia (ah-zoh-oh-SPER-mee-ah) Absence of sperm in the semen

Circumcision (ser-kum-SIZH-un) Surgical removal of part or all of the foreskin of the penis

Coitus (KOH-ih-tus) Sexual intercourse between a man and a woman

Curettage (kyoo-reh-TAZH) Removal of a growth or other matter from a cavity by scraping or by using suction

Dysmenorrhea (dis-men-oh-REE-ah) Difficult or painful menstruation

Episiotomy (eh-peez-ee-AHT-oh-mee) Incision of the perineum to prevent tearing of the perineum and to facilitate delivery during childbirth

Hysterectomy (his-ter-ECK-toh-mee) Surgical removal of the uterus

Mittelschmerz (MIT-el-shmairts) Abdominal pain that occurs midway between the menstrual periods, at the time of ovulation

Oligospermia (ahl-ih-goh-SPER-mee-ah) Condition in which there are few sperm in the semen; low sperm count

Oophorectomy (oh-ahf-oh-RECK-toh-mee) Surgical removal of an ovary

Orchidectomy (or-kih-DECK-toh-mee) Surgical excision of a testis

Phimosis (fye-MOH-sis) Condition in which the opening of the prepuce is narrow and cannot be drawn back over the glans penis

Prostatitis (prahs-tah-TYE-tis) Inflammation of the prostate

Salpingectomy (sall-pin-JECK-toh-mee) Surgical removal of a uterine tube

Spermicide (SPER-mih-syde) Agent that kills sperm

Vaginoplasty (vah-JIH-noh-plas-tee) Surgical repair of the vagina

Vasectomy (vah-SECK-toh-mee) Excision of the vas deferens or a portion of it

VOCABULARY QUIZ

Use word parts you have learned to form words that have the following definitions.

1. Study of the female _____

2. Inflammation of the glans penis _____

3. Producing maleness _____

4. Muscular layer of the uterus _____

5. Plastic surgery of the breast _____

Using the definitions of word parts you have learned, define the following words.

6. Hysterectomy _____

7. Salpingitis _____

8. Prostatomegaly _____

9. Oophorectomy _____

10. Anorchism _____

Match each of the following definitions with the correct word.

_____ 11. Urethra opens on dorsum of penis

_____ 12. Surgical repair of the vagina and perineum

_____ 13. Excessive monthly flow

_____ 14. Suture of the vagina

_____ 15. Difficult or painful monthly flow

A. Colpoperineoplasty
B. Colporrhaphy
C. Dysmenorrhea
D. Epispadias
E. Menorrhagia

16. What is the clinical term for "the absence of sperm in the semen"? _____

17. What clinical term denotes "the surgical removal of a uterine tube"? _____

18. What is the clinical term for "the congenital absence of one or both testes"? _____

19. What is the meaning of the abbreviation PID? _____

20. What is the meaning of the abbreviation BPH? _____

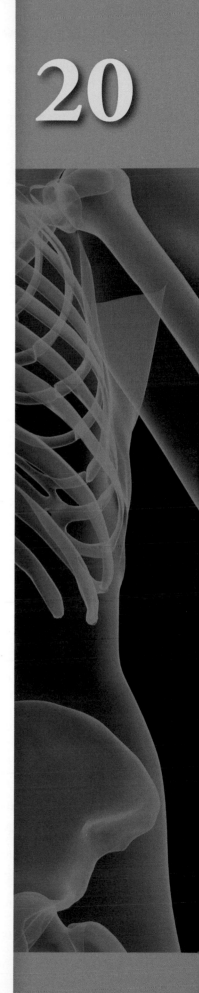

Development and Heredity

CHAPTER OBJECTIVES

Fertilization
- Describe the events in the process of fertilization, state where it normally occurs, and name the cell that is formed as a result of fertilization.
- Name the three divisions of prenatal development and state the period of time for each.

Preembryonic Period (2 Weeks)
- Describe three significant developments that take place during the preembryonic period.

Embryonic Development (6 Weeks)
- Describe three significant developments that take place during the embryonic period.
- List five derivatives from each of the primary germ layers.

Fetal Development (30 Weeks)
- State the two fundamental processes that take place during fetal development.
- Name, describe the location, and state the function of five structures that are unique in the circulatory pattern of the fetus.

Parturition and Lactation
- Describe the roles of the hypothalamus, estrogen, progesterone, oxytocin, and prostaglandins in promoting labor.

- Describe the three stages of labor.
- Describe the changes that take place in the infant's respiratory system and circulatory pathway at birth or soon after birth.
- Describe the hormonal and neural control of lactation.

Postnatal Development
- Name and define six periods in postnatal development.

Heredity
- Define the term *gene*.
- Explain the principle of segregation and the principle of independent assortment.
- Define the terms *homozygous* and *heterozygous*, *genotype* and *phenotype*, and *dominant* and *recessive genes*.
- Explain the purpose of Punnett squares.
- Explain the difference between autosomal single gene inheritance and six-linked single gene inheritance.
- Define a genetic mutation and explain how it may occur.

KEY TERMS

Allele (ah-LEEL) One of two or more alternative forms of a gene at the same site in a chromosome

Amnion (AM-nee-on) Innermost fetal membrane; transparent sac that holds the developing fetus suspended in fluid

Autosomes (AW-toh-sohm) Any of the 22 pairs of chromosomes in humans not concerned with determination of sex

Chorion (KOR-ee-on) Outermost extraembryonic, or fetal, membrane; contributes to the formation of the placenta

Cleavage (KLEE-vayj) Series of mitotic cell divisions after fertilization; resulting cells are called blastomeres

Embryonic period (em-bree-ON-ik PEER-ee-ud) Stage of development that lasts from the beginning of the third week until the end of the eighth week after fertilization; period during which the organ systems develop in the body

Fetal period (FEE-tal PEER-ee-ud) Stage of development that starts at the beginning of the ninth week after fertilization and lasts until birth

Genotype (JEE-noh-typ) Specific alleles that are present for a given trait

Heterozygous (het-er-oh-ZYE-gus) Having different alleles for a given trait

Homozygous (hoh-moh-ZYE-gus) Having identical alleles for a given trait

Implantation (im-plan-TAY-shun) Process by which the developing embryo becomes embedded in the uterine wall; usually takes about 1 week and is completed by the fourteenth day after fertilization

Parturition (par-too-RIH-shun) Act of giving birth to an infant

Phenotype (FEE-noh-typ) Visible expression of the genes for a given trait

Preembryonic period (pree-em-bree-AHN-ik PEER-ee-ud) First 2 weeks after fertilization; period of cleavage, implantation, and formation of primary germ layers

Zygote (ZYE-goht) The single diploid cell that is a fertilized ovum

The previous chapter focused on the male and female reproductive systems and the formation of the spermatozoa and oocytes. This chapter continues with the events that result in the union of the gametes and the subsequent development of a new individual. Development is a continuous process that starts with fertilization (conception) and ends with death. Birth is an awesome event that divides the total span of development into two portions. It is the culmination of 38 weeks of **prenatal development** within the uterus. Prenatal development is divided into preembryonic, embryonic, and fetal periods. The period of **postnatal development** begins with birth and lasts until death.

Embryologists describe the timing of events in development by using the term **developmental age**, which begins at fertilization. The medical community uses **clinical age**, which begins at the last menstrual period (LMP). Developmental age is 2 weeks less than clinical age. Pregnancy is also divided into three equal periods called trimesters. Three calendar months constitute a trimester.

Fertilization

Fertilization, or conception, is the union of the sperm cell nucleus with an egg cell nucleus. The product of fertilization, a single cell, is called a **zygote**.

During ovulation, a secondary oocyte, surrounded by the zona pellucida and corona radiata, is released from an ovary and enters the uterine tube. Sperm cells, deposited in the vagina during sexual intercourse, travel through the cervix and body of the uterus to the uterine tube, where they encounter the secondary oocyte. The upward movement through the female reproductive tract is accomplished by the movements of the sperm tail (flagellum) and by contractions of the uterus. Sperm have a high mortality rate, and many never reach the uterine tube. While moving through the female reproductive tract, which takes about an hour, the sperm undergo a process called **capacitation** (kah-pass-ih-TAY-shun). This weakens the membrane around the acrosome so that the enzymes can be released. When the sperm reach the egg (secondary oocyte) in the uterine tube, the acrosomal enzymes from thousands of sperm break down the corona radiata and zona pellucida to create a tiny opening. As soon as one sperm enters the cell membrane of the oocyte, changes occur in the membrane that prevent other sperm from entering.

Sperm penetration of the oocyte membrane is the stimulus for the second meiotic division to resume. This division produces a small polar body, which is pushed to the side, and a large ovum with all the cytoplasm and 23 single-stranded chromosomes. When this division is finished, the sperm nucleus moves toward the center of the egg cell, the membranes of the two nuclei (sperm and egg) disintegrate, and their chromosomes combine. This completes the process of fertilization. The resulting cell, a **zygote**, has a full complement, or diploid number (46), of chromosomes; 23 came from the sperm and 23 from the egg. The zygote is the first cell of the future offspring. Figure 20-1 depicts the events in fertilization.

The zygote occasionally divides, resulting in **monozygotic twins**. Because these twins have the same genetic makeup, they are called "identical" twins. Sometimes a woman may ovulate two or more oocytes at the same time and both oocytes are subsequently fertilized. This results in **dizygotic twins** because more than one zygote is formed. These are also called "fraternal" twins.

Ovulated secondary oocytes are viable and fertile for only about 24 hours. Most sperm are fertile in the female reproductive tract for only about 48 hours, but some may retain fertility for 3 days. This means that for conception to occur, sexual intercourse must take place between 3 days before to 1 day after ovulation. Again, because the eggs are fertile for only 24 hours, fertilization usually occurs in the uterine tube near the infundibulum.

Most of the generally accepted methods of birth control prevent fertilization of the oocyte after intercourse. These are called contraceptives because they prevent conception. Contraceptives include behavioral, barrier, chemical, and surgical methods. Some of these are described and summarized in Table 20-1.

QUICK CHECK

20.1 Why is capacitation of sperm necessary and where does it occur?

20.2 When, in the ovarian cycle, is fertilization most likely to occur?

20.3 What is the term used for a fertilized egg, the first cell of the future offspring?

Preembryonic Period

The **preembryonic period** lasts for about 2 weeks after fertilization. During this time the zygote and subsequent stages move through the uterine tube into the cavity of the uterus. Significant developments in this period are **cleavage** (KLEE-vayj), **implantation** in the uterine wall, and formation of the **primary germ layers**.

Cleavage

After fertilization, the zygote undergoes a series of mitotic cell divisions that produce increasing numbers of cells. These early cell divisions are called **cleavage** (Figure 20-2). The first cleavage division is completed about 36 hours after fertilization and results in two cells, called **blastomeres** (BLAS-toh-meerz). Subsequent divisions of the blastomeres occur at approximately 12-hour intervals. By the end of the third day after fertilization, there are 16 cells in a solid ball, called a **morula** (MOR-yoo-lah). Still enclosed within the zona pellucida, each division produces cells that are smaller and smaller so that the total volume stays about the same.

The morula moves into the uterine cavity, floats freely in the cavity, and continues cell division. By the fifth day, the zona pellucida breaks down and a cavity forms in the ball of cells. The resulting hollow sphere of cells is called a **blastocyst**

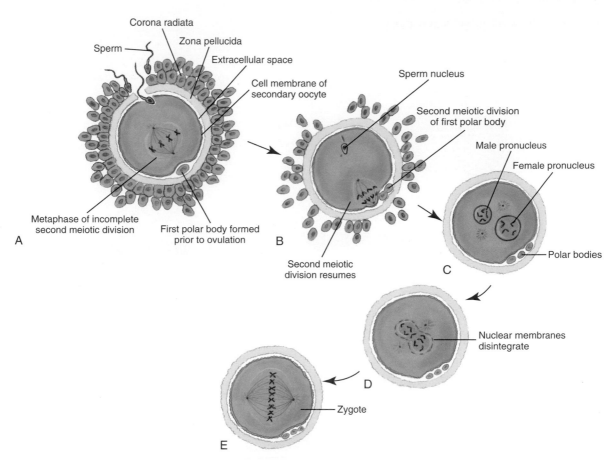

Figure 20-1 *Events in fertilization.* ***A****, Sperm penetrates secondary oocyte.* ***B****, Second meiotic division resumes.* ***C****, Male and female pronuclei come together.* ***D****, Nuclear membranes disintegrate.* ***E****, Fertilization completed with formation of zygote.*

TABLE 20-1 Methods of Birth Control

Temporary	Comments
Behavioral Methods	
Rhythm method	This method involves abstinence from intercourse for a few days before and a few days after ovulation.
Coitus interruptus	The penis is withdrawn from the vagina before ejaculation.
Barrier Methods	
Mechanical barriers	
Condom	This prevents semen from entering the female reproductive tract.
Diaphragm	This prevents sperm from entering the uterus.
Chemical barriers	These include creams, foams, and jellies with spermicidal properties that create an unfavorable environment for sperm within the vagina.
Chemical Methods	
Oral contraceptives	These are birth control pills that contain estrogen and progestin to inhibit ovulation, impair implantation, and inhibit motility of sperm.
Contraceptive patch	The patch contains estrogen and progestin, which are absorbed through the skin with the same effects as oral contraceptives.
Contraceptive injections	Depo-Provera is an injectable progestin that inhibits development of the ovum and impairs motility of sperm. One injection provides 12 weeks of protection.
Generally Considered Permanent Surgical Methods	
Vasectomy	This is a surgical procedure in which the ductus deferentia, within the scrotum, are cut and tied to prevent sperm cells from becoming part of the semen.
Tubal ligation	This is a surgical procedure in which the uterine tubes are cut and tied to prevent the sperm cells from coming in contact with the oocyte.

Two-cell stage

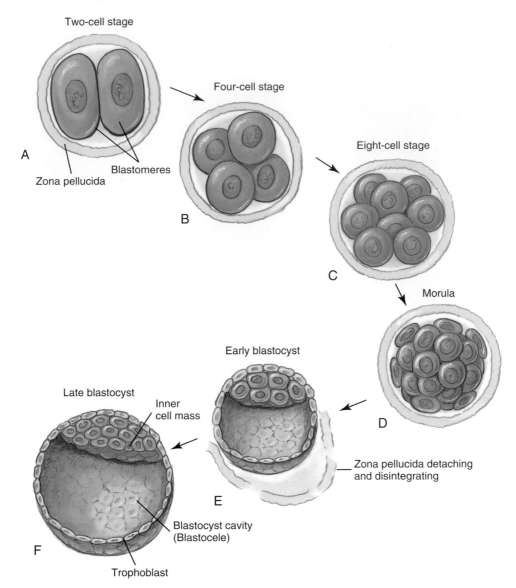

Four-cell stage

Eight-cell stage

Morula

Early blastocyst

Late blastocyst

A

Zona pellucida

Blastomeres

B

C

D

E

F

Inner cell mass

Zona pellucida detaching and disintegrating

Blastocyst cavity (Blastocele)

Trophoblast

Figure 20-2 *Cleavage and formation of blastocyst. **A**, Two-cell stage. **B**, Four-cell stage. **C**, Eight-cell stage. **D**, Morula; **E**, Early blastocyst. **F**, Late blastocyst.*

(BLAS-toh-syst). It receives nourishment from glycogen that is secreted by the endometrial glands in response to high levels of progesterone. The blastocyst consists of a single layer of flattened cells around a cavity and a cluster of cells at one side. The cavity is the **blastocele** (BLAS-toh-seel), the cells around the cavity make up the **trophoblast** (TROH-foh-blast), and the cluster of cells is the **inner cell mass**. The trophoblast functions in the formation of the chorion, which forms the fetal portion of the placenta, and the inner cell mass becomes the embryo. The blastocyst is now ready for implantation to begin. Figure 20-3 summarizes the events that occur during the first week of prenatal development.

Implantation

By the seventh day after ovulation (twenty-first day of the menstrual cycle), the endometrium of the uterus is ready to receive the blastocyst. The blastocyst approaches the endometrium, usually high in the uterus, and if the endometrium is ready,

the blastocyst attaches to it. This begins the process of **implantation** (Figure 20-4).

The blastocyst is oriented so that the inner cell mass is toward the endometrium. The trophoblast cells in this region secrete enzymes that erode the endometrium to form a hole. The blastocyst "burrows" into the thick endometrial tissue. Cells of the endometrium grow over the blastocyst until it is fully implanted. Implantation is usually completed by the fourteenth day after ovulation. If fertilization had not taken place, the corpus luteum would regress and the endometrium would slough off as menstrual flow at this time. The blastocyst saves itself from being aborted by secreting **human chorionic gonadotropin** (HCG), a hormone that acts like luteinizing hormone (LH). Pregnancy tests are based on the presence of this hormone in the blood or urine because it is not produced in a woman unless she is pregnant. HCG travels in the blood to the ovary where it causes the corpus luteum to remain functional and secrete progesterone to maintain the

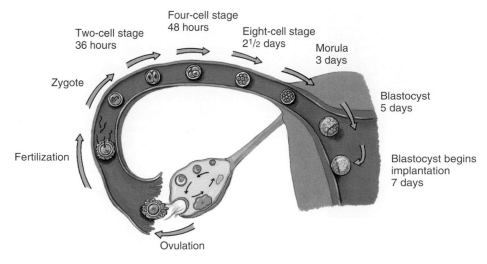

Figure 20-3 *Summary of events during the first week of prenatal development.*

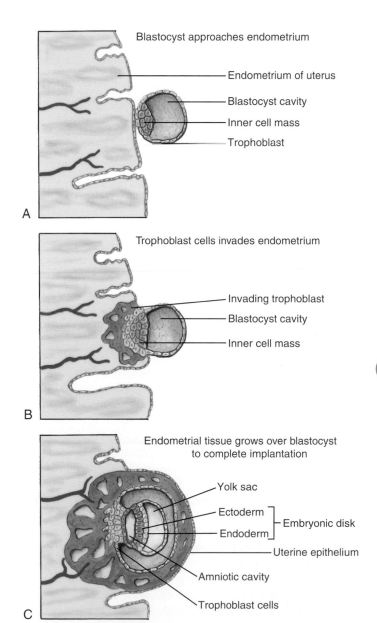

Figure 20-4 *Process of implantation.*

endometrium. The secretion of HCG begins by the eighth day after fertilization and reaches a peak about 8 weeks after fertilization, and then declines as the placenta develops and is able to secrete sufficient quantities of progesterone to maintain pregnancy.

Formation of Primary Germ Layers

While the blastocyst is implanting itself in the endometrium, changes are taking place in the inner cell mass that result in the formation of the primary germ layers. A cavity, called the **amniotic** (am-nee-AH-tik) **cavity**, develops in the inner cell mass. The portion of the inner cell mass that is adjacent to the blastocele flattens into a two-layered **embryonic disk** (see Figure 20-4). The upper layer, next to the amniotic cavity, becomes **ectoderm**, and the lower layer becomes **endoderm**. A short time later a third layer, the **mesoderm**, appears between the ectoderm and endoderm. All of the tissues and organs in the body come from these three primary germ layers.

QUICK CHECK

20.4 What develops from the inner cell mass?
20.5 About how many days after ovulation does implantation begin?
20.6 What cavity develops within the inner cell mass?
20.7 What hormone is secreted by the blastocyst? What is the function of this hormone?

Embryonic Development

Implantation and the formation of the three primary germ layers mark the end of the preembryonic period and the beginning of the embryonic period. The period of embryonic development lasts from the beginning of the third week to the end of the eighth week. The developing offspring is called an **embryo** during this time. Figure 20-5 is an ultrasound image of an

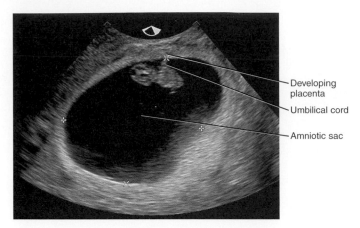

Figure 20-5 *Ultrasound image of an embryo at approximately seven weeks of development. (Adapted from Applegate E: The Sectional Anatomy Learning System, Concepts, ed. 3. St. Louis, Elsevier/Saunders, 2010.)*

embryo at approximately 7 weeks of development. Significant changes during this 6-week period include the formation of the **extraembryonic membranes, placenta,** and all of the **organ systems** in the body. This is the most critical time of development because there is a great deal of tissue differentiation and the organs are forming.

Formation of Extraembryonic Membranes

The **extraembryonic membranes** form outside the developing embryo and are responsible for its protection, nutrition, and excretion (Figure 20-6). At birth these membranes are expelled, along with the placenta, as the **afterbirth**. The extraembryonic membranes are the amnion, chorion, yolk sac, and allantois.

The **amnion** (AM-nee-on) is a thin membrane that forms as the embryonic disk separates from the outer layer of the inner cell mass. The membrane and enclosed space enlarge to form a sac that completely surrounds the developing embryo. The **amniotic sac** is filled with **amniotic fluid**, which is formed initially by absorption from the maternal blood. The amni-

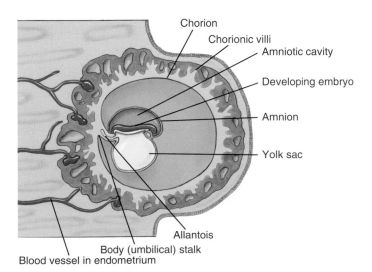

Figure 20-6 *Extraembryonic membranes. These membranes are the amnion, chorion, yolk sac, and allantois.*

otic fluid cushions and protects the developing offspring from bumps and jolts, helps maintain a constant temperature and pressure around the embryo, provides a medium for symmetrical development, and allows freedom of movement, which is necessary for the development of muscles and skeleton and for blood flow. Usually performed at the end of the embryonic period or beginning of the fetal period, **Amniocentesis** is a procedure in which a sample of amniotic fluid is aspirated from the amniotic sac. The fluid contains cells from the fetus. The cells and fluid are analyzed to detect chromosomal and biochemical abnormalities in the fetus. Before delivery, the amnion ruptures, either naturally or surgically, and the fluid is released.

The **chorion** (KOR-ee-on), which develops from the trophoblast, is the outermost extraembryonic membrane. It develops numerous fingerlike projections called chorionic villi. The villi on the side of the uterine wall enlarge, penetrate the maternal tissue, and become highly vascular. This region contributes to the formation of the placenta. The villi on the side next to the uterine cavity degenerate, and the surface becomes smooth. As the fetus enlarges, the chorion fuses with the amnion.

The **yolk sac** develops from the endoderm side of the embryonic disk. The yolk sac produces the primordial germ cells, which then migrate to the developing gonads during the fourth week. It also produces blood until the sixth week; by this time the liver is developed sufficiently to take over this task. During the sixth week, the yolk sac detaches, shrinks, and has no further purpose.

The **allantois** (ah-LAN-toys) develops as a small outgrowth of the yolk sac. It contributes to the development of the urinary bladder and the umbilical arteries and vein. During the second month, it degenerates and becomes part of the umbilical cord.

Formation of the Placenta

The **placenta** is a highly vascular disk, 15 to 20 centimeters in diameter and 2.5 centimeters thick, that develops from both embryonic and maternal tissue. It is usually formed and fully functioning by the end of the embryonic period. After the infant is born, the placenta is expelled from the uterus as part of the afterbirth.

The placenta develops as chorionic villi from the embryo penetrate the endometrium of the uterus (Figure 20-7). As this occurs, the villi become highly vascular and these vessels extend to the umbilical arteries and umbilical vein. The spaces in the endometrium surrounding the villi are filled with maternal blood. Oxygen and nutrients diffuse from the mother's blood into the fetal blood, and metabolic wastes, including carbon dioxide, diffuse from the fetal blood into the maternal blood. The membranes of the fetal capillaries and chorionic villi normally keep the fetal and maternal blood from actually mixing.

By the diffusion of substances across the membranes, the placenta functions as a nutritive, respiratory, and excretory organ. It also secretes hormones and thus functions as a temporary endocrine gland. Table 20-2 lists some of the hormones secreted by the placenta and their effects.

Normally, implantation occurs and the placenta develops in the upper portion of the uterus. In **placenta previa**, the placenta forms in the lower portion of the uterus and grows over the internal os of the cervix. This condition may result

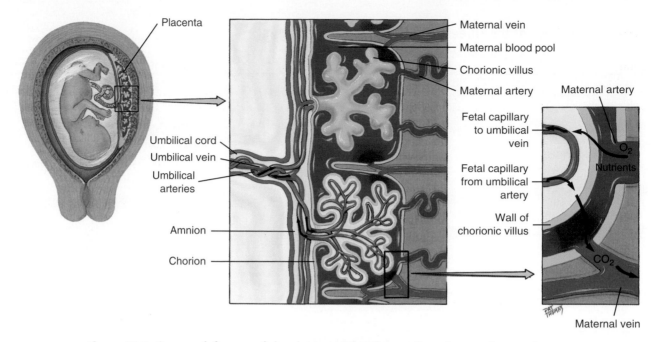

Figure 20-7 *Structural features of the placenta and exchange of nutrients and water between maternal and fetal blood.*

TABLE 20-2 Hormones Secreted by the Placenta

Hormone	Effects
Human chorionic gonadotropin	Similar to luteinizing hormone from anterior pituitary; maintains mother's corpus luteum for first 2 months of pregnancy
Estrogen	Helps maintain endometrium; stimulates mammary gland development; inhibits follicle-stimulating hormone; increases uterine sensitivity to oxytocin
Progesterone	Helps maintain endometrium; stimulates mammary gland development; inhibits prolactin; inhibits follicle-stimulating hormone

QUICK CHECK

20.8 What is the purpose of the amniotic fluid? List four functions.

20.9 What are the two primary functions of the placenta?

20.10 What are the three primary germ layers that develop into the organs of the body? When does organogenesis occur?

in reduced oxygen supply to the fetus and increases the risk of hemorrhage and infection for the mother. Cesarean delivery usually is recommended when placenta previa is diagnosed.

Organogenesis

Organogenesis (or-gan-oh-JEN-eh-sis) is the formation of body organs and organ systems. The formation of the primary germ layers sets the stage for this process because all body organs develop from them. Table 20-3 lists some examples of the derivatives of the primary germ layers. The skin, one of the earliest organs to develop, forms during the third week. By the end of the fourth week, the heart is pumping blood to all parts of the embryo. By the end of the eighth week, all the main internal body organs are established and the embryo has a humanlike appearance even though it is only about 25 millimeters (1 inch) long and weighs about 1 gram.

Fetal Development

The **fetal stage** of development starts at the beginning of the ninth week and lasts until birth. The developing offspring is called a **fetus** at this time. Because all the organ systems are formed during the embryonic period, the fetus is less vulnerable than the embryo to malformations caused by radiation, viruses, and drugs. The fetal stage is a period of growth and maturation. Table 20-4 describes some developments in each month of development.

By the end of the twentieth week, the mother commonly feels the fetus moving within the uterus. This is called **quickening**. At this time in development, the fetal skin is coated with **vernix caseosa**, a cheesy mixture of sebum and dead epidermal cells, which protects the skin from the amniotic fluid that surrounds it. Fine **lanugo hair** covers the body and helps keep the vernix caseosa intact. By the end of 28 weeks, the fetus may survive outside the uterus with medical support. The mortality rate is high with infants born at this time because their temperature control mechanisms are not mature enough to maintain constant body temperature, and their respiratory system is not ready to maintain regular respiration. By 38 weeks, the fetus is considered full term and ready for life outside the uterus.

TABLE 20-3 Derivatives of the Three Primary Germ Layers

Ectoderm	Mesoderm	Endoderm
Epidermis of skin	Dermis of skin	Epithelial lining of digestive tract
Hair, nails, skin glands	Skeletal, smooth, and cardiac muscle	Epithelium of liver and pancreas
Lens of eye	Connective tissue including cartilage and bone	Epithelium of urinary bladder and urethra
Enamel of teeth	Epithelium of serous membranes	Epithelium of respiratory tract
All nervous tissue	Epithelium of join cavities	Thyroid, parathyroid, and thymus glands
Adrenal medulla	Epithelium of blood vessels	
Sense organ receptor cells	Kidneys and ureters	
Linings of oral and nasal cavities, vagina, and anal canal	Adrenal cortex Epithelium of gonads and reproductive ducts	

TABLE 20-4 Monthly Changes During Prenatal Development

End of Month*	Size	Developments During Month
1	6 mm	Arm and leg buds form; heart forms and starts beating; body systems begin to form
2	25 to 30 mm, 1 g	Head nearly as large as body; major brain regions present; ossification begins; arms and legs distinct; blood vessels form and cardiovascular system fully functional; liver large
3	75 mm, 10 to 45 g	Facial features present; nails develop on fingers and toes; can swallow and digest amniotic fluid; urine starts to form; fetus starts to move; heartbeat detected; external genitalia develop
4	140 mm, 60 to 200 g	Facial features well formed; hair appears on head; joints begin to form
5	190 mm, 250 to 450 g	Mother feels fetal movement; fetus covered with fine hair called lanugo hair; eyebrows visible; skin coated with vernix caseosa, a cheesy mixture of sebum and dead epidermal cells
6	220 mm, 500 to 800 g	Skin reddish because blood in capillaries is visible; skin wrinkled because it lacks adipose in subcutaneous tissue
7	260 mm, 900 to 1300 g	Eyes open; capable of survival but mortality rate is high; scrotum develops; testes begin their descent
8	280 to 300 mm, 1400 to 2100 g	Testes descend into scrotum; sense of taste is present
9	310 to 340 mm, 2200 to 2900 g	Reddish skin fades to pink; nails reach tips of fingers and toes or beyond
10	350 to 360 mm, 3000 to 3400 g	Skin smooth and plump because of adipose in subcutaneous tissue; lanugo hair shed; fetus usually turns to a head-down position; full term

*These are 4-week (28-day) months.

The fetus is dependent on its mother for oxygen and nutrients, which are supplied by diffusion through membranes in the placenta. Carbon dioxide and other wastes diffuse from the fetus into the maternal blood in the lacunae of the placenta. Gases, nutrients, and wastes are carried to and from the fetus in the umbilical vessels. This arrangement necessitates that the pattern of blood flow in the fetus be different from the pattern after birth. These differences, which are discussed in Chapter 12, are reviewed in Table 20-5.

> **QUICK CHECK**
>
> 20.11 What are the fundamental changes that occur during fetal development?
> 20.12 What are the first movements of the fetus that are felt by the mother called?

Parturition And Lactation

The time of prenatal development, or pregnancy, is referred to as the **gestation** (jes-TAY-shun) period. In humans, the normal gestation period is 266 days from fertilization. Because it is difficult to determine the actual time of fertilization, "due dates" are calculated as 280 days after the beginning of the last menstrual period. **Parturition** (par-too-RIH-shun) refers to the birth of an infant, and **labor** is the process by which forceful contractions expel the fetus from the uterus.

Labor and Delivery

The onset of **true labor** is marked by rhythmic contractions, dilation of the cervix, and "show," which is a discharge of bloody mucus from the cervix and vagina. In **false labor** the contractions are weak and irregular and there is no cervical dilation and no "show."

TABLE 20-5 Summary of Special Features in Fetal Circulation

Feature	Location	Before Birth	After Birth
Umbilical arteries (2)	Umbilical cord	Transport blood from fetus to placenta	Degenerate to become lateral umbilical ligaments
Umbilical vein (1)	Umbilical cord	Transports blood from placenta to fetus	Becomes ligamentum teres (round ligament) of liver
Ductus venosus	Between umbilical vein and inferior vena cava	Carries blood directly from umbilical vein to inferior vena cava; bypasses liver	Becomes ligamentum venosum of liver
Foramen ovale	Interatrial septum	Allows blood to go directly from right atrium into left atrium to bypass pulmonary circulation	Closes after birth to become fossa ovalis
Ductus arteriosus	Between pulmonary trunk and aorta	Permits blood in pulmonary trunk to go directly into descending aorta and bypass pulmonary circulation	Becomes a fibrous cord; ligamentum arteriosum

Although the physiologic mechanisms are unclear, labor appears to be the result of hormone interactions. Near the end of pregnancy, progesterone levels decline and estrogen levels increase. This removes progesterone's inhibitory effects on uterine contractions, and at the same time estrogen sensitizes the uterus to the effects of oxytocin. Pressure of the fetal head on the cervix signals the hypothalamus to secrete oxytocin from the posterior pituitary. Oxytocin stimulates the production of prostaglandins, and together oxytocin and prostaglandins cause contractions of the myometrium. As soon as the hypothalamus becomes involved, a positive feedback cycle is set up in which the uterine contractions stimulate more oxytocin and the oxytocin stimulates uterine contractions.

Labor is divided into three stages (Figure 20-8): dilation stage, expulsion stage, and placental stage. The **dilation stage** begins with the onset of true labor and lasts until the cervix is fully dilated. This is the longest part of labor and lasts from 6 to 24 hours, or even longer. It is characterized by rhythmic and forceful uterine contractions. The amniotic sac ruptures, and the cervix dilates to a diameter of 10 centimeters.

The **expulsion stage** lasts from full cervical dilation until delivery of the fetus. This stage usually lasts less than an hour. In normal position, the head appears first (cephalic presentation) and helps dilate the cervix. This makes it easier for the rest of the body to pass through the birth canal. The head-first position also allows mucus to be suctioned from the respiratory passages so that the newborn can breathe before the rest of the body is completely through the birth canal. A breech presentation occurs in about 5% of all deliveries. In a breech presentation, the buttocks appear first rather than the head, which increases the time and difficulty of labor. A cesarean section may be necessary for a successful delivery in some cases. At the end of the expulsion stage, the umbilical cord is clamped and cut.

The final phase is the **placental stage**. Within 10 to 15 minutes after parturition, the placenta separates from the uterine wall. Forceful uterine contractions expel the placenta and the attached membranes as the afterbirth. The forceful contractions constrict torn blood vessels to prevent hemorrhage. Normally less than half a liter of blood is lost during delivery.

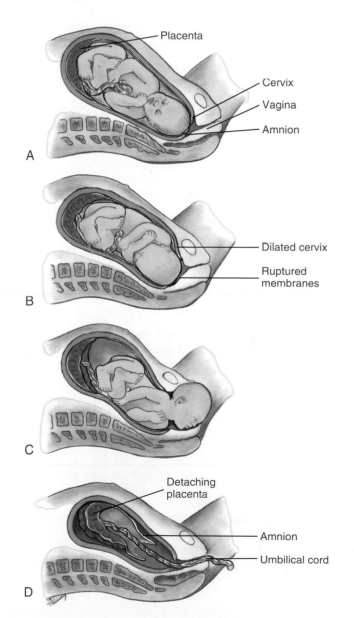

Figure 20-8 *Three stages of labor. **A**, Before labor begins. **B**, Dilation stage begins with onset of true labor and lasts until cervix is fully dilated. **C**, Expulsion stage is from full cervical dilation until delivery of the fetus. **D**, Placental stage lasts from delivery of the fetus until the placenta and extraembryonic membranes are expelled.*

FROM THE PHARMACY

The uterus is a highly muscular organ that is also highly vascular; however, when the smooth muscle of the myometrium contracts, the blood vessels are squeezed or constricted and blood flow is diminished. During parturition, strong smooth muscle contractions expel the fetus and, at the same time, constrict the blood vessels to reduce the risk of hemorrhage. There are times when it is necessary, for the safety of the mother and/or the fetus, to use preparations that stimulate uterine contractions. Oxytocics are used for this purpose. In other cases, it may be necessary to inhibit the contractions to prevent premature labor or miscarriage. Tocolytics are preparations that are used to reduce or inhibit contractions.

Oxytocin is one of the hormones released from the posterior pituitary gland. It has little effect on the nonpregnant uterus, but as pregnancy progresses, its effect normally increases until the end of gestation when it stimulates delivery. Oxytocin acts directly on the smooth muscle of the myometrium and on the mammary glands (refer to Chapters 10 and 19). Synthetic preparations of oxytocin, namely, Pitocin and Syntocinon, may be used to induce more rapid deliveries, to control hemorrhage, to help expel the placenta, and to stimulate the flow of milk from the mammary glands after parturition.

In contrast to oxytocics, tocolytics reduce or inhibit uterine contractions. Most act on the beta2-adrenergic receptors to relax the smooth muscle in the uterus and thus inhibit contractions. The most commonly used tocolytic is ritodrine hydrochloride (Yutopar). The use of smooth muscle relaxants is indicated in the management of uncomplicated premature labor and threatened miscarriage.

Adjustments of the Infant at Birth

The fetal lungs are collapsed or partially filled with amniotic fluid and are nonfunctional. As soon as the umbilical cord is cut, the oxygen supply from the mother ceases. Blood continues to circulate through the fetus, however, and the increasing carbon dioxide levels, decreasing pH (acidosis), and decreasing oxygen levels stimulate the respiratory center in the medulla. The respiratory muscles contract and the infant draws its first breath. This first breath is normally strong and deep as the infant inflates the collapsed alveoli.

Numerous changes take place in the circulatory pathway at birth or soon after (Figure 20-9 and Table 20-5). The fora-

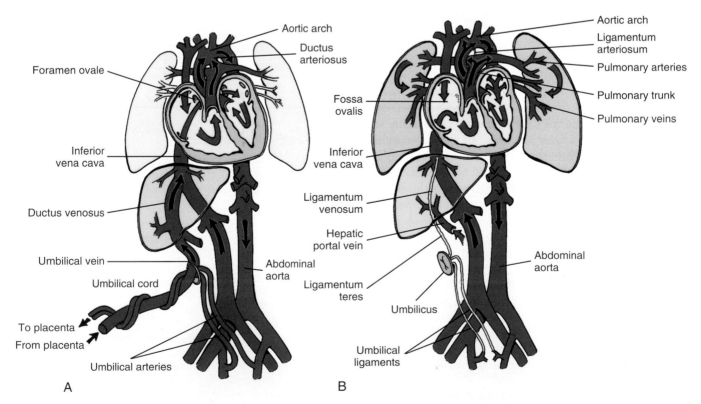

Figure 20-9 *Circulation patterns before and after birth. **A**, Fetal circulation. **B**, Circulation after birth. Before birth, the two umbilical arteries transport blood to the placenta and a single umbilical vein returns blood from the placenta to the fetus. The ductus venosus bypasses the nonfunctional fetal liver. The ductus arteriosus and foramen ovale permit blood to bypass the lungs, which are nonfunctional. After birth, the circulatory patterns change to include the liver and lungs.*

men ovale, between the right and left atria, functionally closes at the moment of birth so that the blood will flow through the pulmonary arteries to pick up oxygen in the lungs. Eventually, two flaps of tissue fuse across the opening to permanently seal it off, and it becomes the fossa ovalis. As soon as the lungs begin to function, smooth muscle in the ductus arteriosus contracts to close that vessel, again to encourage blood flow to the lungs. The remnant of the ductus arteriosus is the ligamentum arteriosum. The ductus venosus takes blood from the umbilical vein to the inferior vena cava in the fetus. As soon as the umbilical cord is severed, this shunt is no longer needed and the ductus venosus degenerates into the ligamentum venosum.

Physiology of Lactation

Lactation (lak-TAY-shun) refers to the production of milk by the mammary glands and the ejection of milk from the breasts. The most important hormone that stimulates milk production is **prolactin**. During pregnancy, the increasing levels of estrogen and progesterone from the placenta stimulate the enlargement of the mammary glands. Prolactin levels also increase during this time, but the prolactin stimulus for milk production is inhibited by the estrogen and progesterone. During parturition, when the placenta is expelled from the uterus, estrogen and progesterone levels dramatically decrease and remove the inhibition for milk production. The infant's sucking action triggers impulses to the posterior pituitary by means of the hypothalamus. These impulses stimulate the release of **oxytocin**, which causes the ejection (let-down) of milk from the breasts (Figure 20-10).

Normally, there is a 2- to 3-day delay in the start of milk production after birth. During this time a cloudy yellowish fluid, called **colostrum** (koh-LAHS-trum), is secreted by the mammary glands. Colostrum has less lactose than milk, has almost no fat, but has more protein, vitamin A, and minerals than milk. Colostrum and maternal milk contain antibodies that help provide immunity for the infant.

Prolactin levels soon return to normal after birth, and milk is not produced unless prolactin is stimulated by the infant's suckling. Each time a mother nurses her infant, impulses from the nipple to the hypothalamus stimulate the release of **prolactin-releasing hormone** (PRH). This causes a temporary surge in prolactin levels, which stimulates milk production for the next nursing period. If a mother stops nursing her infant, milk production ceases within a few days.

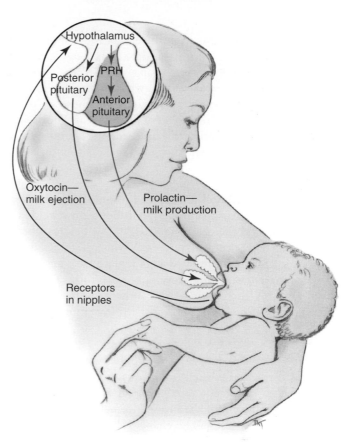

PRH = Prolactin releasing hormone

Figure 20-10 *Stimulus for lactation. Prolactin stimulates milk production and oxytocin stimulates the ejection of milk from the breast. An infant's suckling stimulates receptors in the nipple that send impulses to the hypothalamus. This stimulates the release of oxytocin and prolactin.*

Postnatal Development

Development after birth is termed **postnatal development**. It lasts from parturition until death and is divided into the following stages: neonatal period, infancy, childhood, adolescence, adulthood, and senescence.

Neonatal Period

The **neonatal** (nee-oh-NAY-tal) **period** begins at the moment of birth and lasts until the end of the first 4 weeks. During this time, the infant is called a newborn or **neonate** (NEE-oh-nayt). These 4 weeks are characterized by dramatic changes, which occur at a rapid rate. In addition to the respiratory and circulatory changes mentioned earlier, the neonate faces other adjustments to life outside the uterus. The neonate needs vitamins C and D to help harden the skeleton and iron to aid the liver in the production of red blood cells. This period is critical also because the temperature-regulating mechanism and immune system are not fully developed, which means that the neonate is vulnerable to environmental temperature changes and infections.

QUICK CHECK

20.13 What are the three manifestations of true labor?
20.14 What is the longest stage of labor?
20.15 Describe three changes that occur in the circulatory pattern of the infant after birth.
20.16 What are the functions of prolactin and oxytocin in lactation?

Infancy

The period of **infancy** lasts from the end of the first month to the end of the first year. During this time body weight generally triples and many developmental changes take place. There is a myelinization of the nervous system, which is manifest in more coordination of motor activities. The infant learns to sit, crawl, stand, and perhaps walk by the end of the first year. The deciduous incisor teeth erupt during this time, and the infant learns to communicate by smiling, laughing, and making sounds.

Childhood

The period of **childhood** lasts from the end of the first year until puberty. Bone ossification is rapid. Bone growth slows in late childhood and then accelerates again just before puberty. Bladder and bowel controls are established. The deciduous teeth erupt and then are shed and replaced by permanent teeth. Motor coordination develops more fully. Language, reading, writing, reasoning, and other intellectual skills become more refined. The child is maturing emotionally.

Adolescence

The period of **adolescence** lasts from puberty until adulthood. Puberty is the period when an individual becomes physiologically capable of reproduction. In females, the first sign of puberty is usually development of the breasts. In males, the first sign is enlargement of the testes. The changes that occur during this period are largely controlled by hormones. The secondary sex characteristics appear, and there is a rapid growth in the muscular and skeletal systems. The adolescent shows increasing levels of motor skills, intellectual ability, and emotional maturity.

Adulthood

Adulthood is the period from adolescence to senescence. The progression from adolescence to adulthood is vague and has physical, emotional, and behavioral implications. Generally, adulthood is characterized by a maintenance of existing body tissues so that the body remains unchanged anatomically and physiologically for many years. Sometime after the age of 30, degenerative changes start to occur. These changes may not be noticeable at first, but later become more significant as the individual progresses into older adulthood.

Senescence

Senescence (seh-NESS-ens) is the period of older adulthood that ends in death. As the degenerative changes that begin in adulthood continue throughout this period, the body becomes less and less capable of coping with the demands placed on it. Changes related to aging take place in all body systems; however, the rate at which they occur varies from individual to individual and from system to system. The central nervous system may become less efficient so that memory fails and motor skills are impaired. Homeostatic mechanisms may fail so the individual is more vulnerable to fluid and electrolyte changes, acid-base imbalances, and temperature variations. Sensory functions often decline and are manifest in hearing loss, poor vision, and reduced senses of taste, smell, and touch. The aging section that is presented with each body system in this book describes some of the changes in more detail. Even though there are degenerative

changes in all body systems, death usually results from cardiovascular disturbances, failure of the immune system, or disease processes that affect vital organs.

QUICK CHECK

20.17 What two mechanisms are not yet fully developed in the neonate that make this a critical period for the newborn?
20.18 What is the main characterization of adulthood?
20.19 What period is marked by degenerative changes and eventually ends in death?

Heredity

Introduction

The patterns of inheritance that are passed from generation to generation is **heredity** and **genetics** is the scientific study of heredity. In the mid 1800s, a Moravian monk, Gregor Mendel, studied the patterns of inheritance in pea plants in the monastery garden. He showed that biological traits are passed from parents to their offspring by discreet and independent units that are now called genes.

One hundred years later, in the middle of the twentieth century, two scientists, James Watson and Francis Crick, discovered the structure of deoxyribonucleic acid (DNA) found in the chromosomes in the nucleus of a cell. It was hypothesized that the genetic information that is carried by chromosomes resides in the structure of DNA. More recently, it has become possible to isolate the DNA from a cell, cut it into pieces, and insert the pieces into bacterial cells where they grow and multiply. This process now makes it possible to study the structure and function of individual genes.

Chromosomes and Genes

Every cell in the human body has 23 pairs of chromosomes in the nucleus. One member of each pair comes from the mother and the other comes from the father. For 22 pairs, the maternal chromosome and the paternal chromosome are similar in size, and shape. They have the same appearance in both sexes. These are the **autosomes**. The twenty-third pair is different and these are the **sex chromosomes**. In the female, the two members of the pair are similar and are designated XX. In the male, one of the chromosomes in the pair is smaller and has a different appearance. The male pair of chromosomes is designated as XY.

Mechanism of Gene Function

Each chromosome consists of a double helix of DNA. Recall from chapter 2 that DNA is a long chain of nucleotides and that a sequence of three nucleotide bases contains the code for one amino acid. The sequence of nucleotide bases determines the sequence of amino acids in a protein molecule. A **gene** is the specific portion of a DNA strand that contains the code for a single protein. Chromosomes contain DNA and DNA contains genes. The genes determine the proteins that are synthesized by the body through transcription and translation. Review protein synthesis in Chapter 2.

Human Genome

The entire collection of genetic material in a cell is called the genome (JEE-nohm). The human genome is all the genetic material in each typical cell of the human body. Some other organisms such as bacteria and fruit flies have also been studied. In 1991, an international scientific research project was established as the **Human Genome Project**. The primary goal of this publicly funded worldwide collaboration was to determine the sequence of nucleotide bases that make up DNA in each of the 23 pairs of chromosomes in the human body and to identify the genes from both a physical and a functional perspective. The project was completed in 2003, the fiftieth anniversary of the discovery of DNA by Watson and Crick. The human genome contains about 20,000 to 25,000 genes. Surprisingly, this is roughly the same number as in a rat or mouse. Less than 2% of the DNA in chromosomes carries genes that code for proteins. Most of the rest is "filler" that is either not used or is filtered out by mRNA. The genome of any given individual, except for identical twins and clones, is unique.

Transmission of Chromosomes to Offspring

A review of meiosis reveals that the gametes have only 23 chromosomes. This is called the haploid number. During meiosis, the members of each pair of chromosomes separate or segregate, resulting in the haploid number. This is known as Mendel's **principle of segregation**. When the haploid ovum is fertilized by a haploid sperm, the diploid number, 46 chromosomes, is restored. Another of Mendel's principles is the **principle of independent assortment**. This principle states that each chromosome assorts itself independently during meiosis. Each sperm is likely to have a different set of 23 chromosomes. Since ova are formed in the same way, it follows that each ovum will have a different set of 23 chromosomes. Combining the principle of segregation and the principle of independent assortment means that two parents can provide 2^{23} x 2^{23} (more than 65 trillion) different chromosome combinations. The variation provided by this enormous number is further enlarged by a mechanism called crossing over in which the two chromosomes of a pair may exchange segments during meiosis. These principles account for the tremendous variation in the human population.

QUICK CHECK

20.20 What is a gene?
20.21 What two Mendelian principles account for the variations seen in the human population?

Gene Expression

Dominant and Recessive Traits

Each inherited trait is controlled by a pair of genes, one from each parent. These gene versions, or alleles, are situated at the same location on the respective chromosomes. Consider eye color, for example. Each individual has two genes, or alleles, for eye color, one from each parent. Both alleles may be for brown eyes, both may be for blue eyes, or one may be for brown eyes and the other for blue eyes. Mendel discovered that some genes are **dominant** and some are **recessive**. A dominant gene always expresses itself when it is present. A recessive gene may be present but not expressed because it is masked by the dominant gene. By convention, dominant genes are represented by uppercase letters and recessive genes are represented by lower case letters. In the example of eye color, the gene for brown eyes is dominant (B) and the gene for blue eyes is recessive (b). If both alleles are identical (BB or bb), the trait is said to be **homozygous**. If the alleles are different (Bb), the trait is said to be **heterozygous**. The codes BB, Bb, and bb are called the **genotype** because they indicate the types of alleles that are present. BB and Bb genotypes will both have brown eyes because brown (B) is dominant and masks or covers the recessive (b) blue. The expression of the genes is called the **phenotype**. Both the BB and Bb genotypes have the brown eye phenotype. Only the homozygous bb will exhibit the blue eye phenotype.

Punnett Squares

A Punnett square is a chart that shows all possible gene combinations in the offspring of two parents whose genotypes are known. Figure 20-11 uses Punnett squares to show possible combinations for the brown/blue eye trait. The following steps are used to create a Punnett square.

1. Designate letters to represent the genes.
2. Write the genotype of each parent.
3. List the genes that each parent can contribute.
4. Draw a Punnett square (four small squares in the shape of a window).
5. Write the possible genes of one parent across the top and the genes of the other parent down the side.
6. Fill in each box of the Punnett square by transferring the letter above and in front of each box into the appropriate box. By convention, the upper case letter goes first and the lower case letter follows.
7. List the possible genotypes and phenotypes of the offspring.

Albinism, the total lack of melanin pigment in the skin and eyes, is another example of simple Mendelian dominant/recessive inheritance. The genes that cause albinism are recessive and the genes for normal pigmentation are dominant. The dominant gene for normal pigmentation may be represented by A and the recessive gene for albinism as a. The homozygous genotype AA will exhibit normal pigmentation phenotype. The heterozygous genotype Aa will also exhibit normal pigmentation. Only the homozygous genotype aa will exhibit the albinism phenotype. The heterozygous genotype Aa is a **genetic carrier** of albinism because that individual can transmit the albinism gene, a, to offspring. To be an albino, an offspring has to receive a recessive allele from each parent.

QUICK CHECK

20.22 What is the difference between genotype and phenotype?
20.23 Assuming that the gene for brown eyes is dominant over blue eyes, what are the possible eye colors of children with two brown-eyed parents?

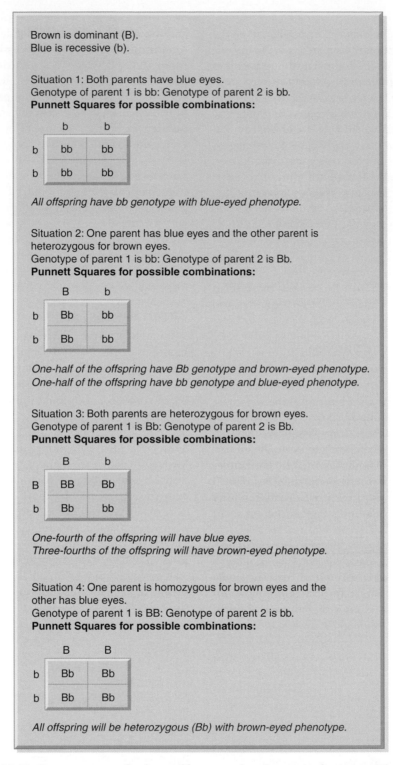

Brown is dominant (B).
Blue is recessive (b).

Situation 1: Both parents have blue eyes.
Genotype of parent 1 is bb: Genotype of parent 2 is bb.
Punnett Squares for possible combinations:

	b	b
b	bb	bb
b	bb	bb

All offspring have bb genotype with blue-eyed phenotype.

Situation 2: One parent has blue eyes and the other parent is
heterozygous for brown eyes.
Genotype of parent 1 is bb: Genotype of parent 2 is Bb.
Punnett Squares for possible combinations:

	B	b
b	Bb	bb
b	Bb	bb

One-half of the offspring have Bb genotype and brown-eyed phenotype.
One-half of the offspring have bb genotype and blue-eyed phenotype.

Situation 3: Both parents are heterozygous for brown eyes.
Genotype of parent 1 is Bb: Genotype of parent 2 is Bb.
Punnett Squares for possible combinations:

	B	b
B	BB	Bb
b	Bb	bb

One-fourth of the offspring will have blue eyes.
Three-fourths of the offspring will have brown-eyed phenotype.

Situation 4: One parent is homozygous for brown eyes and the
other has blue eyes.
Genotype of parent 1 is BB: Genotype of parent 2 is bb.
Punnett Squares for possible combinations:

	B	B
b	Bb	Bb
b	Bb	Bb

All offspring will be heterozygous (Bb) with brown-eyed phenotype.

Figure 20-11 *Punnett square for brown/blue eye color. Brown is dominant (B) and blue is recessive (b).*

Variations of Single Gene Inheritance

It is important to note that the inheritance of eye color and melanin pigmentation are not quite as simple as indicated here, but they serve to illustrate the principles of inheritance. Some genes may modify the expression of other genes and some traits are governed by several different pairs of genes. These are called **polygenic** traits to distinguish them from single gene traits, which are **monogenic**. Eye color was originally considered to be a monogenic trait, but research shows there are at least three genes for eye color and probably more. This makes it a polygenic trait. Some traits exhibit **codominance** where the two genes are equally effective, but exhibit different phenotypes. In this case the heterozygous genotype exhibits a phenotype that is intermediate between the two homozygous genotypes. Sickle cell anemia is an example of codominance. A person with two sickle cell genes will have sickle cell anemia, a person with one normal gene and one sickle cell gene will have a less severe form called sickle cell trait.

Another variation of simple single gene inheritance is that there may be more than two alleles of a given gene. There may be three alleles or more in existence in the general population, although a given individual has only two of them. An example of this is the inheritance of blood types where there are alleles for A, B, and O. The A and B alleles are dominant to O. A person with type A blood may have both alleles for A (AA) or one A allele and one O allele (AO). The same is true for type B blood. The person may be BB or BO. Type AB blood has one allele for A and one for B (AB), while type O has only O alleles (OO). Figure 20-12 uses Punnett squares to illustrate the inheritance of blood types. Table 20-6 gives possible blood types for the children of parents with given blood types. You can verify these by using Punnett squares.

Sex-Linked Traits

The previous paragraphs discussed the inheritance of traits when the alleles are located on the autosomes. The pattern of inheritance changes when the alleles are carried on the sex chromosomes. Recall that the female has two X chromosomes and the male has an X and a much smaller Y chromosome. In addition to genes that determine female sexual characteristics, the X chromosomes carry many other genes. The tiny Y chromosome carries very few genes other than those that determine male sexual characteristics. Traits carries on the X and Y chromosomes are called **sex-linked traits**, either X-linked traits if they are carried on the X chromosome or Y-linked traits if they are carried on the Y chromosome. Genes for sex-linked traits may be either dominant or recessive. In the female, a recessive trait will be expressed only if it is homozygous. In the male, a recessive gene on the X chromosome will be expressed because there is no corresponding allele on the Y chromosome to cover the effect of the recessive gene. For this reason, recessive X-linked traits occur more frequently in males.

An example of a recessive X-linked trait is red-green color blindness in which there is a deficiency in the photopigments of the retina. For a female to be color blind, she must have the recessive allele on both X chromosomes. She can, however, be a heterozygous carrier and pass the recessive allele on to her offspring. If a male has the dominant allele on his X chromosome, he will have normal color vision. If he has the recessive allele, which he received from his mother, he will exhibit red-green color blindness. Figure 20-13 illustrates the inheritance of recessive X-linked traits.

> **QUICK CHECK**
>
> 20.24 Two parents have type AB blood. What are the possible blood types for their children?
> 20.25 A man and his wife have normal color vision. They are not color blind. Is it possible for them to have a color blind child?

Genetic Mutations

A **genetic mutation** is a change in an individual's genetic code. Some mutations occur within a single gene. A mutation called a **deletion** occurs when one or more nucleotides are missing from the normal sequence. **Insertions**

occur when nucleotides are added to the sequence. Other mutations may involve damage to a portion of the chromosome instead of a single gene. For example, a portion of the chromosome may break away or a piece of another chromosome may attach itself to a different chromosome. In all of these situations, the cell is unable to read the code and translate it into the proteins that are normally encoded there. If the mutation occurs in a reproductive cell, it is passed on to the next generation.

Mutations may occur spontaneously without influences from outside the cell. More commonly, however, outside agents cause disruptions in the genetic code. These agents are called **mutagens**. Certain chemicals, radiation, and viruses are examples of mutagens.

Many mutations are harmful and inhibit survival. They usually kill the organism in which they occur or prevent their reproduction. For this reason, these mutations are not passed on to the offspring and they are not likely to be widely spread throughout a population. Relatively minor recessive mutations may persist in a population simply because they do not cause problems in the people who inherit them. The expression of the mutated recessive gene is masked by its corresponding normal dominant allele.

Hereditary Disorders

Distinction should be made between **congenital disorders**, which are present at the time of birth, and **hereditary disorders**, which are genetically transmitted. Congenital includes defects that are genetically transmitted, but also includes those that are not. Hemophilia is present at birth, therefore it is congenital. It is genetically transmitted so it is hereditary. It is congenital and hereditary. On the other hand, consider an infant born of a mother who was exposed to the rubella virus (German measles) during her first trimester of pregnancy. The infant may be born with cardiovascular defects, disorders of the eye, and learning disabilities. These are congenital defects because they are present at birth, but they are not hereditary because they are not genetically transmitted.

Hereditary disorders may be due to single genes or they may be due to aberrations in the chromosomes. Phenylketonuria, cystic fibrosis, and albinism are recessive autosomal single gene defects. Hemophilia, red-green color blindness, and some forms of cleft palate are recessive X-linked single gene defects. Some genetic disorders are not inherited in the usual sense, but are due to the abnormal presence or absence of entire chromosomes. During meiosis, the two members of a pair of chromosomes normally separate. Occasionally, the pair fails to separate. This is called **nondisjunction** and it produces a gamete with two chromosomes that are stuck together instead of the usual one. When this gamete joins with a normal gamete to form a zygote, the zygote has three chromosomes instead of the normal two. This is called **trisomy**. **Monosomy**, the presence of only one autosome instead of a pair, may also be the result of nondisjunction in meiosis. Monosomy and trisomy are usually fatal, but when they involve certain chromosomes, the person may survive for a period of time but will have developmental abnormalities. A well-known chromosomal disorder is trisomy 21, which produces a group of symptoms called Down syndrome. In this condition there is a triplet of chromosomes 21 rather than the usual pair.

Type A blood has genotype AA or AO   Type B blood has genotype BB or BO
Type AB blood has genotype AB   Type O blood has genotype OO

Situation 1: Both parents are type A.
Possible genotypes for parents are AA or AO.
Punnett Squares for possible combinations:

	A	A
A	AA	AA
A	AA	AA

	A	A
A	AA	AA
O	AO	AO

	A	O
A	AA	AO
O	AO	OO

There is a possibility of type O offspring if both parents are heterozygous.

Situation 2: One parent is type A and one parent is type B.
Possible genotypes for type A are AA and AO: Possible genotypes for typeB are BB and BO.
Punnett Squares for possible combinations:

	A	A
B	AB	AB
B	AB	AB

	A	O
B	AB	BO
B	AB	BO

	A	A
B	AB	BO
O	AO	AO

	A	O
B	AB	BO
O	AO	OO

Blood types A, B, AB, and O are all possible depending on whether the parents are heterozygous or homozygous.

Situation 3: One parent is type A and one parent is type AB.
Possible genotypes for type A are AA and AO: Possible genotypes for type AB is AB.
Punnett Squares for possible combinations:

	A	B
A	AA	AB
A	AA	AB

	A	B
A	AA	AB
O	AO	BO

Blood type A, B, and AB are possible in the offspring depending on whether the type A parent is homozygous or heterozygous.

Situation 4: Both parents are type AB.
Both parents will have A and B alleles.
Punnett Squares for possible combinations:

	A	B
A	AA	AB
B	AB	BB

Offspring may be type A, B, or AB.

Situation 5: One parent is type O and one parent is type AB.
Genotype for type O is OO: Genotype for type AB is AB.
Punnett Squares for possible combinations:

	O	O
A	AO	AO
B	BO	BO

Offspring may be type A or B.

Figure 20-12 *Inheritance of the ABO blood types.*

TABLE 20-6 Inheritance Patterns for Blood Types

Parents Blood Types	Possible Blood Types in Children
Both A	A or O
Both B	B or O
Both AB	A, B, or AB
Both O	O
One A and one B	A, B, AB, or O
One A and one AB	A, B, or AB
One A and one O	A or O
One B and one AB	A, B, or AB
One B and one O	B or O
One AB and one O	A or B

Gene Therapy

Until recently, the only way to treat a hereditary disorder was to treat the symptoms. Persons with phenylketonuria (PKU), for example, had to avoid foods that contained phenylalanine. Surgery can alleviate some of the problems inherent with cleft palate. Gene therapy now offers some hope of treating hereditary disorders.

In a procedure known as **gene replacement**, genes that code for proteins that cause disease are replaced by normal genes. Researchers use viruses as carriers to introduce the normal genes into the body. Another therapy known as **gene augmentation** introduces normal genes with the hope that they will add to the production of the needed protein. The use of gene therapy began in 1990. There are currently hundreds of ongoing gene therapy trials for diverse genetic disorders. Thousands of laboratory experiments in anticipation of human trials are also currently under way. There are still many hurdles to overcome, including the high cost and the risks involved. Who knows? Maybe you will be on the team that makes a breakthrough so that genetic diseases may be treated or cured with gene therapy!

QUICK CHECK

20.26 What are deletion mutations?
20.27 What is the definition of a congenital disorder?

X = normal X chromosome
X̶ = X chromosome with recessive gent for color blindness
Y = Y chromosome

Situation 1: Normal mother and normal father.
Mother's genotype is XX: Father's genotype is XY

Punnett Squares for possible combinations:

	X	Y
X	XX	XY
X	XX	XY

All offspring have normal color vision.

Situation 2: Heterozygous carrier mother with normal father.
Mother's genotype is XX̶: Father's genotype is XY

Punnett Squares for possible combinations:

	X	Y
X	XX	XY
X̶	XX̶	X̶Y

Half the sons will be color blind.
Half the daughters will be heterozygous carriers.

Situation 3: Normal mother and color blind father.
Mother's genotype is XX: Father's genotype is X̶Y

Punnett Squares for possible combinations:

	X̶	Y
X	XX̶	XY
X	XX̶	XY

The sons will have normal color vision.
The daughters will be heterozygous carriers.

Situation 4: Heterozygous carrier mother and color blind father.
Mother's genotype is XX̶: Father's genotype is X̶Y

Punnett Squares for possible combinations:

	X̶	Y
X	XX̶	XY
X̶	XX̶	X̶Y

Half the sons will be color blind.
Half of the daughters will be heterozygous carriers and the other half will be color blind.
This is the only combination that produces a color blind daughter.

Figure 20-13 *Inheritance of recessive X-linked color blindness.*

CHAPTER SUMMARY

Fertilization

■ **Describe the events in the process of fertilization, state where it normally occurs, and name the cell that is formed as a result of fertilization.**

- As the sperm move through the female reproductive tract, the acrosome membrane weakens. This is called capacitation.
- The process of fertilization begins when a single sperm penetrates the cell membrane of a secondary oocyte. This stimulates completion of the second meiotic division, which produces a second polar body and ovum. The nuclear membranes of the male and female pronuclei degenerate and the two nuclei fuse.
- The fertilized egg, which has a full complement of 46 chromosomes, is called a zygote.
- Fertilization normally takes place in the uterine tube.

■ **Name the three divisions of prenatal development and state the period of time for each one.**

- Prenatal development is the period from fertilization to birth. It consists of a preembryonic period (2 weeks), embryonic period (6 weeks), and fetal period (30 weeks).

Preembryonic Period (2 weeks)

■ **Describe three significant developments that take place during the preembryonic period.**

- Cleavage, implantation, and the formation of the primary germ layers occur in the preembryonic period.
- Cleavage is a rapid series of mitotic cell divisions after fertilization. The cells that result are blastomeres. A solid ball of blastomeres is a morula. When a cavity forms inside the morula it becomes a blastocyst. The cavity inside the blastocyst is the blastocele and the cells around the outside are the trophoblast. A cluster of cells on one side is the inner cell mass and represents the future embryo.
- Implantation occurs as endometrial tissue grows around the blastocyst. The entire process takes about 7 days. The trophoblast cells secrete human chorionic gonadotropin, which acts like LH to maintain the corpus luteum. Since the corpus luteum continues to secrete progesterone, the uterine lining is maintained and menstruation is inhibited.
- The three primary germ layers develop while implantation is taking place. The primary germ layers are ectoderm, mesoderm, and endoderm.

Embryonic Development (6 weeks)

■ **Describe three significant developments that take place during the embryonic period.**

- The period of embryonic development encompasses the formation of the extraembryonic membranes, formation of the placenta, and development of the organ systems.
- The amnion, chorion, yolk sac, and allantois are membranes that form outside the embryo and are called extraembryonic membranes.
- The extraembryonic membranes function in protection, nutrition, and excretion for the embryo.

- The amnion forms from the outer layer of the inner cell mass. It forms a fluid-filled sac that surrounds the growing embryo. It helps maintain constant temperature and pressure around the embryo, and provides for symmetrical development and movement.
- The chorion develops from the trophoblast and contributes to the formation of the placenta.
- The yolk sac develops from the endoderm side of the embryonic disk and produces the primordial germ cells.
- The allantois develops from the yolk sac and contributes to the formation of the umbilical arteries and vein.
- The placenta develops from the endometrium of the uterus and the chorion of the embryo.
- Nutrients and oxygen diffuse from the mother's blood in the lacunae into the blood vessels in the chorionic villi. Waste materials diffuse in the opposite direction.
- The placenta functions as a temporary endocrine gland for the mother and produces estrogen and progesterone.
- Body organs develop from the ectoderm, mesoderm, and endoderm that are formed during the preembryonic period. All organ systems are formed by the end of the embryonic period.

■ **List five derivatives from each of the primary germ layers.**

- **Ectoderm**: epidermis of the skin, hair, nails, skin glands, lens of the eye, enamel of the teeth, all nervous tissue, adrenal medulla, sense organ receptor cells, linings of the oral and nasal cavities, vagina, and anal canal.
- **Mesoderm**: dermis of the skin; skeletal, smooth, cardiac muscle; connective tissue including cartilage and bone; epithelium of serous membranes; epithelium of joint cavities; epithelium of blood vessels, kidneys and ureters; adrenal cortex; epithelium of gonads and reproductive ducts.
- **Endoderm**: epithelial lining of digestive tract; epithelium of the liver and pancreas; epithelium of urinary bladder and urethra; epithelium of the respiratory tract; and thyroid, parathyroid, and thymus glands.

Fetal Development (30 weeks)

■ **State the two fundamental processes that take place during fetal development.**

- The fetal period is one of growth and maturation of the organ systems that form during the embryonic period.

■ **Name, describe the location, and state the function of five structures that are unique in the circulatory pattern of the fetus.**

- Since the lungs and liver are nonfunctional during the fetal period, special structures in the circulatory pathway allow blood to bypass these organs. Other vessels take blood to and from the placenta for gaseous exchange.
- Two umbilical arteries carry fetal blood to the placenta and one umbilical vein returns the oxygenated blood to fetal circulation. The ductus venosus carries blood from the umbilical vein to the inferior vena cava and bypasses the liver. The foramen ovale, in the interatrial septum, and the ductus arteriosus, between the pulmonary trunk and descending aorta, allow blood to bypass the lungs.

Parturition and Lactation

■ **Describe the roles of the hypothalamus, estrogen, progesterone, oxytocin, and prostaglandins in promoting labor.**

- Gestation period is the time from fertilization to birth; it is the time of pregnancy. It normally lasts for 266 days from fertilization, or 280 days from the beginning of the last menstrual period.
- Parturition refers to the birth of a baby and labor is the process by which forceful contractions expel the fetus from the uterus.
- Near the end of gestation, estrogen levels increase and progesterone starts to decrease. This removes progesterone's inhibitory effects on the uterus. Estrogen also sensitizes oxytocin receptors. Pressure of the baby's head on the cervix signals the hypothalamus to secrete oxytocin and this, with prostaglandins, stimulates uterine contractions.

■ **Describe the three stages of labor.**

- The **dilation stage** of labor begins with the onset of true labor and lasts until the cervix is fully dilated. The **expulsion stage** lasts from full cervical dilation until delivery of the fetus. The final phase is the **placental stage** when the placenta and extraembryonic membranes are expelled.

■ **Describe the changes that take place in the baby's respiratory system and circulatory pathway at birth or soon after birth.**

- When the umbilical cord is cut, the baby's oxygen supply from the mother is terminated. Changes in blood gases stimulate the respiratory center. The first breath needs to be strong and deep to inflate the lungs.
- The special features of fetal circulation cease to function after birth and degenerate or change into their postnatal state at birth or soon after.

■ **Discuss the hormonal and neural control of lactation.**

- Lactation refers to the production of milk by the mammary glands. For the first 2 or 3 days, the mammary glands secrete colostrum. After this, the glands produce milk.
- The baby's sucking at the nipple sends signals to the hypothalamus; oxytocin is released; and this ejects milk from the mammary glands.
- In response to the baby's sucking, the hypothalamus also releases prolactin-releasing hormone. This creates a surge in prolactin, which stimulates milk production for the next feeding period.

Postnatal Development

■ **Name and define six periods in postnatal development.**

- Neonatal period begins at the moment of birth and lasts until the end of the first 4 weeks.
- Infancy lasts from the end of the first month to the end of the first year.
- Childhood lasts from the end of the first year until puberty.
- Adolescence begins at puberty and lasts until adulthood.
- Adulthood is the period from adolescence to old age.
- Senescence is the period of old age and ends in death.

Heredity

■ **Define the term *gene*.**

- A gene is a nucleotide sequence on a DNA molecule that contains the code for a given protein.

■ **Explain the principle of segregation and the principle of independent assortment.**

- The principle of segregation means that the two chromosomes of a pair separate and go to different cells during meiosis.
- The principle of independent assortment states that the 23 pairs of chromosomes separate independently and randomly during meiosis.

■ **Define the terms *homozygous* and *heterozygous*, *genotype* and *phenotype*, and *dominant* and *recessive genes*.**

- Homozygous means that both alleles of a gene pair are the same; heterozygous means they are different, one dominant and one recessive.
- Genotype indicates the type of alleles that are present; phenotype is the visible expression of the genes.
- Dominant genes are expressed if only one allele is present; recessive genes are those that must be homozygous to be expressed.

■ **Explain the purpose of Punnett squares.**

- Punnett squares are used to determine all the offspring genotypes and phenotypes that are possible from parents whose genotypes are known.

■ **Explain the difference between autosomal single gene inheritance and sex-linked single gene inheritance.**

- Autosomal inheritance is transmitted by the 22 pairs of autosomes. Each parent contributes one set of genes to the offspring.
- In autosomal inheritance, recessive genes are only expressed if they are homozygous.
- Sex-linked inheritance involves genes that are transmitted on the X and Y sex chromosomes.
- In sex-linked inheritance a single recessive gene will be expressed in a male because there is no counterpart gene to cover it.

■ **Define a *genetic mutation* and explain how they may occur.**

- A genetic mutation is a change in an individual's genetic code.
- Mutations may occur by insertion of deletion of nucleotides in the DNA chain.
- Mutations may occur spontaneously, but most are caused by mutagenic agents such as certain chemicals, radiation, and viruses.
- Most mutations are harmful and inhibit survival.

Recall

Match the definitions on the left with the appropriate term on the right.

_____ 1. The single cell produced by fertilization
_____ 2. Early cell divisions after fertilization
_____ 3. Develops from the trophoblast and contributes to the formation of the placenta
_____ 4. Hollow sphere of cells during the preembryonic period
_____ 5. Develops into three primary germ layers
_____ 6. Develops into muscle
_____ 7. Develops into epidermis and its derivatives
_____ 8. Any of the 22 chromosome pairs not concerned with determination of sex
_____ 9. Longest stage of labor
_____ 10. Stimulates uterine contractions and the ejection of milk

A. Autosome
B. Blastocyst
C. Chorion
D. Cleavage
E. Dilation
F. Ectoderm
G. Inner cell mass
H. Mesoderm
I. Oxytocin
J. Zygote

Thought

1. Capacitation:
  A. Dissolves the corona radiata so a sperm can fertilize the egg
  B. Occurs in the female reproductive tract and weakens the acrosomal membrane
  C. Occurs in the epididymis and energizes the mitochondria for locomotion
  D. Occurs in the vagina and weakens the zona pellucida

2. Which of the following statements about human chorionic gonadotropin is *false*?
  A. It is secreted by the blastocyst.
  B. It causes the corpus luteum to continue secreting progesterone.
  C. Levels reach a peak about 8 weeks after fertilization.
  D. It has the same effect as follicle-stimulating hormone.

3. Implantation begins:
  A. About 3 days after ovulation
  B. About 7 days after ovulation
  C. About 14 days after ovulation
  D. About 21 days after ovulation

4. Which of the following is *false* about the placenta?
  A. It develops from both embryonic and maternal tissue.
  B. It secretes progesterone, estrogen, and human chorionic gonadotropin.
  C. It develops before the fetal period begins.
  D. It develops from the amnion and chorion.

5. Fetal blood is carried to the placenta by the:
  A. Umbilical artery
  B. Umbilical vein
  C. Ductus venosus
  D. Ductus arteriosus

Application

Mary Smith recently gave birth to her first baby, a daughter she named Sarah. One day while nursing Sarah, she felt "cramps" in her uterus. Concerned, she called her obstetrician. Explain what was happening to cause the cramps.

How do the amnion and chorion differ in function?

Assume that eye color is an autosomal single gene trait and brown (B) is dominant and blue (b) is recessive. What are the possible genotypes for the children of a blue-eyed mother and brown-eyed father? What phenotype(s) will be evident?

BUILDING YOUR MEDICAL VOCABULARY

Building a vocabulary is a cumulative process. As you progress through this book, you will use word parts, abbreviations, and clinical terms from previous chapters. Each chapter will present new word parts, abbreviations, and terms to add to your expanding vocabulary.

Word Parts and Combining Form with Definition and Examples

PART/COMBINING FORM	DEFINITION	EXAMPLE
amni/o	a fetal membrane	amniocentesis: needle puncture of the amniotic sac to withdraw amniotic fluid for analysis
-blast	embryonic or early form	trophoblast: peripheral cells of the blastocyst which attach to the uterine wall and become the placenta
cente-	to puncture surgically	amniocentesis: needle puncture of the amniotic sac to withdraw amniotic fluid for analysis
cleav-	to divide	cleavage: early divisions of the zygote into blastomeres
contra-	against	contraceptives: agents that work against fertilization of an ovum
-cyesis	pregnancy	pseudocyesis: false pregnancy
-cyst-	hollow bag, bladder	blastocyst: hollow ball of cells that results from cleavage
-gravid-	filled, pregnant	primigravida: woman during her first pregnancy
morph/o	shape, form	morphogenesis: developmental changes in shape and form due to growth and differentiation
morul-	mulberry	morula: ball of blastomeres that resembles a mulberry
nat-	birth	prenatal: before birth
nulli-	none	nulliparous: bearing no children
oxy-	sharp, quick, rapid	oxytocin: hormone to stimulate a rapid birth
-parous	to bear, bring forth	multiparous: woman who has two or more pregnancies

PART/COMBINING FORM	DEFINITION	EXAMPLE
partur-	bring forth, give birth	parturition: process of giving birth
sen-	old	senescence: period of old age
-toc-	birth	dystocia: abnormal labor or childbirth
zyg-	paired together, union	zygote: cell formed from union of an ovum and sperm

Clinical Abbreviations

ABBREVIATION	MEANING
AB	abortion
C-section	cesarean section
CVS	chorionic villus sampling
EDC	estimated date of confinement (due date)
EDD	expected delivery date
EFM	electronic fetal monitor
FAS	fetal alcohol syndrome
FHR	fetal heart rate
GYN	gynecology
LMP	last menstrual period
NB	newborn
OB	obstetrics
SIDS	sudden infant death syndrome
TAB	therapeutic abortion

Clinical Terms

Abruptio placentae (ab-RUP-tee-oh plah-SEN-tay) Complete separation of the placenta from the uterine wall after 20 weeks but before labor; results in immediate death of the fetus and severe hemorrhage in the mother

Amniocentesis (am-nee-oh-sen-TEE-sis) Surgical procedure in which a needle is passed through the abdominal and uterine walls to obtain a specimen of amniotic fluid

Apgar score (APP-gar score) System of scoring an infant's physical condition 1 minute after birth; heart rate, respiration, color, muscle tone, and response to stimuli are rated as 0, 1, 2, with a maximum total score of 10; infants with low Apgar scores require prompt medical attention

Cesarean section (seh-SAIR-ee-an SECK-shun) Removal of the fetus by abdominal incision into the uterus

Dystocia (dihs-TOH-see-ah) Difficult and painful childbirth

Eclampsia (ee-KLAMP-see-ah) Critical condition during pregnancy or shortly after, marked by high blood pressure, proteinuria, edema, uremia, convulsions, and coma

Ectopic pregnancy (eck-TOP-ick PREG-nan-see) Fertilized egg is implanted and begins development outside the uterus

Episiotomy (eh-piz-ee-AHT-oh-mee) Surgical incision of the perineum and vagina to facilitate delivery and prevent tearing of the tissues

Eutocia (yoo-TOH-see-ah) Good, normal childbirth

Lochia (LOH-kee-ah) Vaginal discharge during the first 1 or 2 weeks after childbirth; consists of blood, mucus, and tissue

Miscarriage (miss-KAIR-ayj) Loss of an embryo or fetus before the twentieth week; most common cause is a structural or functional defect in the developing offspring; technically known as spontaneous abortion

Multipara (mull-TIP-ah-rah) Woman who has borne more than one child

Neonate (NEE-oh-nayt) Newborn infant during the first 4 weeks after birth

Nullipara (null-IP-ah-rah) Woman who has borne no offspring

Pelvimetry (pel-VIM-eh-tree) Measurement of the dimensions of the mother's pelvis to determine its capacity to allow passage of the fetus through the birth canal

Placenta previa (plah-SEN-tah PREE-vee-ah) Abnormal implantation of the placenta in the lower portion of the uterus

Preeclampsia (pree-ee-KLAMP-see-ah) Condition during pregnancy or shortly after, marked by acute hypertension, proteinuria, and edema; may progress to the more severe form, eclampsia; also called toxemia of pregnancy

Stillbirth (STILL-berth) Delivery of a lifeless infant after the twentieth week

Teratogen (TUR-ah-toh-jen) Agent or influence that causes physical defects in the developing embryo

VOCABULARY QUIZ

Use word parts you have learned to form words that have the following definitions.

1. Development of shape or form _____

2. Study of the newborn _____

3. Surgical puncture of the amniotic sac _____

4. Quick childbirth; rapid labor _____

5. Ball of cells resembling a mulberry _____

Using the definitions of word parts you have learned, define the following words.

6. Antenatal _____

7. Pseudocyesis _____

8. Dystocia _____

9. Parturition _____

10. Senescence _____

Match each of the following definitions with the correct word.

_____ 11. First pregnancy

_____ 12. Excessive vomiting during pregnancy

_____ 13. Cessation of milk secretion

_____ 14. Excessive quantity of amniotic fluid

_____ 15. Joining together of two cells

A. Galactostasis
B. Hyperemesis gravidarum
C. Polyhydramnios
D. Prima gravida
E. Zygote

16. What is the clinical term for "a newborn infant for the first month after birth"? _____

17. What clinical term denotes "a surgical incision of the perineum to facilitate delivery"? _____

18. What is the clinical term for "the abnormal implantation of the placenta in the lower portion of the uterus"? _____

19. What is the meaning of the abbreviation EDD? _____

20. What is the meaning of the abbreviation SIDS? _____

Answers for QuickCheck Questions

Chapter 1

1.1) physiologist, because immunology is a part of the study of functions in the human body; 1.2) structure, because anatomy is the study of structure and structural relationships; 1.3) organs are more complex than tissues; 1.4) respiratory system; 1.5) cardiovascular; 1.6) water, oxygen, nutrients, heat, and pressure; 1.7) negative feedback; 1.8) sagittal plane; 1.9) in the area behind the knee

Chapter 2

2.1) oxygen; 2.2) there are 17 protons, 17 electrons, and 18 neutrons; 2.3) three energy levels with 2 electrons in the first level, 8 in the second level, and 7 in the third level; 2.4) an isotope has the same number of protons and electrons but a different number of neutrons, so it has a different atomic weight than other atoms of the same element; 2.5) covalent bonds; 2.6) intermolecular bonds; 2.7) sodium, hydrogen, carbon, and oxygen; 2.8) a molecule; 2.9) 4 atoms of oxygen in the reactants and 4 atoms of oxygen in the products; 2.10) decomposition reaction; 2.11) the coal dust has a much greater surface area than the briquette; therefore it has a much faster reaction rate; 2.12) the pH is less than 7.0; 2.13) the acids are neutralized by buffers in the body; 2.14) carbon, hydrogen, and oxygen; 2.15) glucose, fructose, and galactose; 2.16) sucrose, maltose, and lactose are disaccharides; glucose and fructose form sucrose, two glucose molecules form maltose, and glucose and galactose form lactose; 2.17) functions of proteins include forming structural components of cells and tissues, acting as antibodies in the fight against disease, assisting in muscle contraction, acting as both antigens and receptor sites on cell membranes, forming enzymes, hormones, and hemoglobin, and providing a source of energy; 2.18) amino acids are the building blocks of proteins and contain carbon, hydrogen, oxygen, and nitrogen, usually sulfur, and often phosphorus; 2.19) lipids have a lower oxygen content than carbohydrates; 2.20) the bonds between the carbon atoms in a saturated fatty acid are all single covalent bonds; in an unsaturated fatty acid, one or more of the bonds are double covalent bonds; 2.21) RNA because DNA does not contain uracil but RNA does; 2.22) a lipid because carbohydrates have a higher oxygen content, proteins contain nitrogen, and nucleic acids contain phosphates and nitrogenous bases

Chapter 3

3.1) phospholipids and proteins; 3.2) intracellular fluid; 3.3) the nuclear membrane is more permeable because of the presence of nuclear pores; 3.4) RNA; 3.5) golgi apparatus; 3.6) mitochondria; 3.7) ribosomes; 3.8) cilia; 3.9) aid in the distribution of chromosomes in cell division; 3.10) cell membrane; 3.11) passive transport moves substances down a concentration gradient and does not require energy, active transport moves substances against a concentration gradient

and requires energy in the form of ATP; 3.12) simple diffusion; 3.13) facilitated diffusion require a carrier molecule and simple diffusion does not; 3.14) selectively permeable membrane; 3.15) the cells shrink or crenate; 3.16) pressure gradients; 3.17) the size of the pores in the membrane; 3.18) ATP molecules and protein carrier molecules; 3.19) endocytosis; 3.20) phagocytosis involves solid particles, pinocytosis involves fluid droplets; 3.21) telophase; 3.22) during interphase; 3.23) there are two nuclear divisions but only one replication of chromosomes so each resulting cell has ½ of the number of chromosomes; 3.24) mitosis results in 2 cells, meiosis results in 4 cells; 3.25) a gene is the portion of a DNA molecule that contains the instructions for making one particular protein molecule; 3.26) adenine; 3.27) uracil; 3.28) a codon is a sequence of 3 bases on mRNA that codes for a specific amino acid; 3.29) messenger RNA (mRNA), transfer RNA (tRNA), and ribosomal RNA (rRNA); 3.30) GUA on the codon and CAU on the anticodon

Chapter 4

4.1) stratified; 4.2) simple squamous epithelium; 4.3) simple columnar epithelium; 4.4) psendostratified ciliated columnar epithelium; 4.5) stratified squamous epithelium; 4.6) goblet cell; 4.7) apocrine glands ; 4.8) collagenous fibers; 4.9) macrophage; 4.10) loose (alveolar) connective tissue; 4.11) adipose; 4.12) dense fibrous connective tissue; 4.13) chondrocyte; 4.14) hyaline cartilage; 4.15) blood; 4.16) lamellae; 4.17) heart; 4.18) skeletal muscle; 4.19) axon; 4.20) neuroglia or glial cells; 4.21) redness, swelling, heat, and pain; 4.22) regeneration; 4.23) serous membrane; 4.24) synovial membrane

Chapter 5

5.1) stratum basale; 5.2) stratum lucidum; 5.3) dermis; 5.4) stratum basale of the epidermis; 5.5) shaft; 5.6) arrector pili muscle; 5.7) stratum basale; 5.8) because of the rich vascular supply in the underlying dermis; 5.9) merocrine sweat glands; 5.10) protection, sensory reception, regulation of body temperature, synthesis of vitamin D; 5.11) it absorbs light and helps protect underlying tissues from the harmful effects of UV light; 5.12) evaporation of perspiration cools the body and cutaneous blood vessels dilate to bring warm blood to the surface where heat is lost to the surrounding air; 5.13) second-degree burn; 5.14) 4.5% has been burned

Chapter 6

6.1) support, protection, movement, storage, and blood cell formation; 6.2) osteon or haversian system; 6.3) diploë; 6.4) diaphysis; 6.5) articular cartilage; 6.6) intramembranous ossification; 6.7) diaphysis; 6.8) growth in diameter; 6.9) 8; 6.10) sutures; 6.11) occipital bone; 6.12) 14; 6.13) hyoid; 6.14) malleus, incus, stapes; 6.15) cervical; 6.16) articulation points for the ribs; 6.17) coccyx; 6.18) thoracic vertebrae; 6.19) manubrium, body, xiphoid; 6.20) clavicle; 6.21) manubrium of sternum; 6.22) ulna; 6.23) metacarpals; 6.24) trochlea 6.25) hold the head of the femur; 6.26) ilium; 6.27) calcaneus; 6.28) fibula; 6.29) femur; 6.30) open fracture; 6.31) fracture hematoma; 6.32) knee joint; 6.33) diarthrosis or synovial joint; 6.34) hinge joint

Chapter 7

7.1) excitability, contractility, extensibility, and elasticity; 7.2) movement, posture, joint stability, heat production; 7.3) epimysium; 7.4) sarcoplasmic reticulum is the modified smooth endoplasmic reticulum of muscle tissue and its function is to store calcium ions; 7.5) I band; 7.6) it will prolong contraction because acetylcholine will remain in the synaptic cleft; 7.7) in the synaptic vessels of the axon terminal; 7.8) calcium ions are necessary to alter the configuration of troponin on the actin molecules to expose the myosin binding sites; 7.9) to energize the myosin heads and for the active transport of calcium from the sarcoplasm back into the sarcoplasmic reticulum so myosin can detach from actin; 7.10) threshold or liminal stimulus 7.11) whole muscles show a graded response due to motor unit summation in which more or less motor units respond to the stimulus depending on the weight of the load; 7.12) lag phase, contraction phase, and relaxation phase; 7.13) tetany; 7.14) ATP; 7.15) creatine phosphate; 7.16) myoglobin; 7.17) aerobic respiration; 7.18) synergist; 7.19) antagonist; 7.20) supination; 7.21) zygomaticus; 7.22) mandible; 7.23) sternocleidomastoid; 7.24) external and internal intercostals muscles; 7.25) rectus abdominis muscle; 7.26) levator ani muscle; 7.27) deltoid muscle; 7.28) flex the wrist and fingers; 7.29) posterior thigh region; 7.30) dorsiflex the foot

Chapter 8

8.1) sensory, integrative, and motor functions; 8.2) cranial nerves, spinal nerves, ganglia; 8.3) association neurons; 8.4) myelin; 8.5) sensory functions because afferent neurons are sensory neurons that conduct impulses to the CNS; 8.6) Na^+; 8.7) action potential; 8.8) myelin sheath; 8.9) relative refractory period; 8.10) potassium channels open, potassium diffuses out of the cell, membrane hyperpolarizes which makes it more difficult to generate an action potential; 8.11) divergence circuit; 8.12) receptor, sensory neuron, center, motor neuron, effector; 8.13) dura mater; 8.14) parietal lobe; 8.15) neuron cell bodies and unmyelinated nerve fibers; 8.16) a large band of white commissural fibers that connects the two cerebral hemispheres; 8.17) thalamus, hypothalamus, epithalamus; 8.18) hypothalamus; 8.19) cardiac center, vasomotor center, and respiratory center; 8.20) mediates contractions of skeletal muscle fibers for coordination of movements, posture and balance; 8.21) choroid plexus; 8.22) 31; 8.23) sensory impulses; 8.24) olfactory (I), optic (II), and vestibulocochlear (VIII); 8.25) 31 pairs; 8.26) motor fibers; 8.27) two; 8.28) sympathetic division; 8.29) parasympathetic division; 8.30) parasympathetic division

Chapter 9

9.1) free nerve endings, Meissner's corpuscles, and Pacinian corpuscles; 9.2) nociceptors; 9.3 Facial (VII) and glossopharyngeal (IX); 9.4) sense of smell; 9.5) produce tears; 9.6) orbicularis oculi and levator palpebrae superioris; 9.7) this is the area of the retina that produces the sharpest vision; 9.8) there are no receptor cells (rods and cones) in the optic disk; 9.9) pupil; 9.10) lens; 9.11) rhodopsin; 9.12) visual cortex of the occipital lobe; 9.13) tympanic membrane; 9.14) auditory or Eustachian tube; 9.15) basilar membrane; 9.16) endolymph; 9.17) movement of the stapes in the oval window; 9.18) the hair cells move against the tectorial membrane and bend; 9.19; by the portion of the basilar membrane that vibrates in response to the sound; 9.20) vestibule; 9.21) crista ampullaris is the organ of dynamic equilibrium and there is one in the ampulla at the base of each semicircular canal; 9.22) vestibular branch

Chapter 10

10.1) exocrine glands have ducts that carry their product to a surface or into a cavity, but endocrine glands secrete their product directly into the blood; 10.2) proteins and steroids; 10.3 inside the cell; 10.4) the anterior pituitary gland is regulated by releasing and inhibiting hormones from the hypothalamus but the posterior pituitary gland is regulated directly by neurons from the hypothalamus; 10.5) cortex of the adrenal gland; 10.6) the secretion of ADH will increase to conserve as much fluid in the body as possible; 10.7) regulate metabolism; 10.8) calcitonin; 10.9) regulate blood calcium levels; 10.10) aldosterone reduces urine output because it causes the kidneys to retain or conserve sodium and water; 10.11) cortisol; 10.12) epinephrine and norepinephrine, also called adrenaline and noradrenaline; 10.13) insulin; 10.14) glucagon; 10.15) the gonadotropins will be inhibited and secretion reduced; 10.16) pineal gland; 10.17) thymus gland; 10.18) heart

Chapter 11

11.1) left side; 11.2) epicardium; 11.3) left ventricle; 11.4) left atrium and left ventricle; 11.5) ascending aorta; 11.6) papillary muscles and chordae tendineae; 11.7) circumflex artery and anterior interventricular artery; 11.8) the heart rate will be slower than normal; 11.9) to allow time for the atria to finish contraction for final ventricular filling before ventricular systole begins; 11.10) P wave; 11.11) 0.3 sec; 11.12) vibrations in the blood caused by closing of valves; 11.13) the volume of blood ejected from a ventricle during one systole; 11.14) end diastolic volume and contraction strength

Chapter 12

12.1) veins; 12.2) to prevent backflow of blood; 12.3) single layer of endothelium; 12.4) blood flows directly from the arteriole into the venule and bypasses the capillary bed; 12.5) hydrostatic pressure and osmotic pressure; 12.6) protein molecules; 12.7) protein molecules are too large to pass through the capillary wall so they are not present in the interstitial fluid; 12.8) capillaries; 12.9) skeletal muscle action, respiratory movement, and constriction of veins; 12.10) vessel diameter; 12.11) muscle contraction in exercise increases venous return, which increases end-diastolic volume, which is directly related to cardiac output; 12.12) increase heart rate; 12.13) cardiac output, blood volume, peripheral resistance, and viscosity; 12.14) renin; 12.15) spleen, pancreas, stomach, liver; 12.16) subclavian arteries; 12.17) internal carotid arteries; 12.18) celiac trunk, superior mesenteric artery, inferior mesenteric artery; 12.19) hepatic portal vein; 12.20) small pulmonary artery or capillary; 12.21) right and left common iliac veins; 12.22) hepatic portal vein; 12.23) placenta; 12.24) ductus venosus

Chapter 13

13.1) transportation, regulation, and protection; 13.2) 55%; 13.3) albumins, globulins, and fibrinogen; 13.4) decrease in albumins produced by the liver would reduce osmotic pressure of the blood and affect fluid/electrolyte balance, decrease in fibrinogen would impair blood clotting process, decrease in globulins would interfere with transport of fats and fat-soluble vitamins; 13.5) reticulocytes; 13.6) low blood oxygen; 13.7) 120 days; 13.8) decrease active erythropoietin; 13.9) neutrophil; 13.10) lymphocytes and monocytes; 13.11) move through the capillary walls by diapedesis; 13.12) vascular spasm, platelet plug formation, and formation of a blood clot; 13.13) it will prevent or reduce clotting because of the lack of thrombocytes that are derived from megakaryocytes; 13.14) because most of the clotting factors are produced in the liver; 13.15) agglutinogen; 13.16) type O; 13.17) yes, because the Rh factors match and type AB blood has no agglutinins to react with the A agglutinogens on the donor blood; 13.18) there is no reaction, but the Rh- individual produces anti Rh agglutinins so there will be a reaction if there are subsequent transfusions of Rh+ blood

Chapter 14

14.1) respiratory movements, skeletal muscle contractions, smooth muscle contractions in the wall of the vessels; 14.2) as the infectious agents and cellular debris are filtered/trapped by the lymph nodes, they accumulate and cause enlargement; 14.3) there is massive hemorrhage because the spleen is a reservoir for blood; 14.4) palatine tonsils; 14.5) chemical action of complement and interferon, phagocytosis, and inflammation; 14.6) viral infections; 14.7) redness, warmth, swelling, and pain; 14.8) bring neutrophils and macrophages to the area, increased blood flow through vasodilation, increased capillary permeability; 14.9) macrophages ingest and process antigens then present them to the appropriate T cells and B cells; 14.10) helper T cells, killer T cells, suppressor T cells, and memory T cells; 14.11) secrete substances that cause B cells to produce clones of plasma cells and memory B cells; 14.12) produce antibodies; 14.13) IgG; 14.14) passive natural immunity; 14.15) active artificial immunity

Chapter 15

15.1) ventilation, external respiration, transport, internal respiration, and cellular respiration; 15.2) nasopharynx, oropharynx, laryngopharynx; 15.3) nose, pharynx, and larynx; 15.4) three; 15.5) pleura; 15.6) external respiration is the exchange of gases between the lungs and the blood; it occurs between the alveoli and the surrounding capillaries; 15.7) it increases intrapleural pressure so it is no longer less that interpulmonary pressure; 15.8) it reduces the surface tension of the fluid in the alveoli so they expand more readily; 15.9) her vital capacity was reduced because fluid displaces some of the air; 15.10) intrapulmonary volume increases and pressure decreases; 15.11) it reduces the

effectiveness of external respiration because the surface area for gas exchange is reduced; 15.12) 137 mm Hg; 15.13) surfactant, alveolar epithelium, basement membrane of epithelium, interstitial space, capillary basement membrane, and capillary endothelium; 15.14) dissolved in plasma and combined with hemoglobin as oxyhemoglobin; 15.15) dissolved in plasma, combined with hemoglobin as carbaminohemoglobin, and as bicarbonate ions; 15.16) this will lower the pH and make the blood more acidic; 15.17) it will stimulate the medullary respiratory centers to increase the rate and depth of breathing; 15.18) phrenic nerve

Chapter 16

16.1) submucosa and muscle layer; 16.2) 20 teeth; 16.3) crown; 16.4) parotid glands; 16.5) the esophagus has to expand into the soft part of the trachea to permit passage of the bolus; 16.6) lower esophageal, or cardiac sphincter; 16.7) pyloric sphincter; 16.8) parasympathetic impulses from the vagus nerve stimulate gastric secretions; 16.9) plicae circulares, villi, and microvilli; 16.10) secretin and cholecystokinin; 16.11) hepatic flexure; 16.12) teniae coli; 16.13) branches of hepatic duct, hepatic artery, and portal vein; 16.14) portal vein and hepatic artery; 16.15) bilirubin; 16.16) cholecystokinin; 16.17) stimulates the secretion of a fluid that has a high concentration of bicarbonate ions; 16.18) hydrolysis; 16.19) it activates trypsinogen into trypsin which acts on proteins; 16.20) it stimulates secretion of digestive enzymes from the pancreas; 16.21) it will interfere with fat digestion because bile will be unable to reach the duodenum to emulsify the fats; 16.22) active transport

Chapter 17

17.1) catabolism; 17.2) dehydration synthesis; 17.3) ATP or adenosine triphosphate; 17.4) pyruvic acid and ATP; 17.5) mitochondria; 17.6) glycogenesis; 17.7) aerobic reactions in the mitochondria; 17.8) ammonia and keto acids; 17.9) acetyl-CoA; 17.10) pyruvic acid and acetyl-CoA; 17.11) basal metabolic rate, physical activity, and thermogenesis; 17.12) physical activity; 17.13) core temperature; 17.14) radiation; 17.15) by negative feedback mechanisms mediated by the hypothalamus; 17.16) 4 Calories/gram; 17.17) starch, glycogen, and fiber; 17.18 includes too little dietary fiber; 17.19) a complete protein contains all of the essential amino acids, but incomplete proteins lack one or more essential amino acids; 17.20) no more than 30% of the caloric intake should be in the form of fats and less that 10% of caloric intake should be in the form of saturated fat; 17.21) 9 Calories; 17.22) vitamins are organic compounds and minerals are inorganic; 17.23) vitamins A, D, E, and K; 17.24) 2.5 liters

Chapter 18

18.1) excrete waste materials, maintain appropriate fluid volumes, maintain electrolyte balance, maintain normal blood pH, regulate red blood cell production, and maintain blood pressure; 18.2) kidneys, ureters, urinary bladder, and urethra; 18.3) renal pelvis; 18.4) glomerulus and glomerular capsule; 18.5) proximal convoluted tubule, nephron loop (of Henle), and distal convoluted tubule; 18.6) macula densa and juxtaglomerular cells; 18.7) interlobular arteries; 18.8) over the bases of the pyramids; 18.9) pelvis of kidney and urinary bladder; 18.10) in the wall of the bladder; 18.11) internal sphincter is smooth (involuntary) muscle and the external sphincter is skeletal (voluntary) muscle; 18.12) prostatic, membranous, and spongy or penile; 18.13) normally there are no proteins in the filtrate; 18.14) proximal convoluted tubule; 18.15) the amount of glucose in the filtrate exceeds the threshold for reabsorption; 18.16) ascending limb of the nephron loop; 18.17) aldosterone, antidiuretic hormone, and atrial natriuretic hormone; 18.18) promotes production of angiotensin II; 18.19) urochrome; 18.20) 1500 ml or 1-2 liters; 18.21) 60%; 18.22) intracellular fluid; 18.23) sodium (Na⁺); 18.24) 7.35 to 7.45; 18.25) buffers; 18.26) respiratory alkalosis; 18.27) respiratory acidosis

Chapter 19

19.1) because the abdominal temperature is too high to produce viable sperm; 19.2) seminiferous tubules; 19.3) four; 19.4) spermiogenesis; 19.5) epididymis; 19.6) ductus (vas) deferens; 19.7) membranous urethra; 19.8) seminal vesicles; 19.9) prostate gland; 19.10) bulbourethral glands; 19.11) erection; 19.12) emission is the discharge of semen into the urethra, ejaculation is the expulsion of semen from the urethra to the exterior; 19.13) stimulates the interstitial cells to produce testosterone; 19.14) male reproductive organs will not develop properly; 19.15) cortex; 19.16) metaphase of the second meiotic division; 19.17) they are transformed into the corpus luteum; 19.18) uterine tube; 19.19) stratum functionale of the endometrium; 19.20) rectum; 19.21) clitoris; 19.22) vagina; 19.23) greater vestibular glands (Bartholin's glands); 19.24) stimulates ovulation and the development of corpus luteum; 19.25) corpus luteum; 19.26) the cells of the endometrium proliferate to restore the uterine lining; 19.27) lactiferous ducts; 19.28) oxytocin; 19.29) estrogen stimulates development of the glandular tissue and deposition of adipose, progesterone stimulates development of the duct system

Chapter 20

20.1) capacitation weakens the membrane around the acrosome so the enzymes necessary for fertilization can be released; 20.2) 3 days before ovulation to 1 day after ovulation; 20.3) zygote; 20.4) embryo; 20.5) 7 days; 20.6) amniotic cavity; 20.7) human chorionic gonadotropin, which causes the corpus luteum to remain functional and secrete progesterone to maintain the endometrium; 20.8) cushions and protects developing embryo/fetus, maintains constant temperature and pressure, provides medium for symmetrical development, permits freedom of movement for musculoskeletal development; 20.9) placenta functions as an endocrine organ for the mother and in the nutrition, excretion, and respiration of the fetus; 20.10) ectoderm, mesoderm, and endoderm are the three primary germ layers; organogenesis occurs during the period between the beginning of the third week and the end of the eighth week following conception; 20.11) growth and maturation; 20.12) quickening; 20.13) rhythmic contractions, dilation of the cervix, and discharge of bloody mucus from the cervix and vagina: 20.14) dilation stage; 20.15) foramen ovale functionally closes so blood flows to the lungs, ductus arteriosus closes so blood flows to the lungs, ductus venosus closes so blood flows to the liver; 20.16) prolactin stimulates the production of milk and oxytocin causes contractions that eject the milk; 20.17) immune system and temperature regulation; 20.18) maintenance of existing body tissues so that the body remains unchanged anatomically and physiologically; 20.19) senescence; 20.20) specific portion of a DNA strand that codes for a given protein; 20.21) principle of segregation and principle of independent assortment; 20.22) genotype gives the specific alleles that are present, phenotype is the visible expression of the genes; 20.23) brown eyes or blue eyes, if both parents are heterozygous there is a 25% chance that an offspring will have blue eyes; 20.24) A, B, and AB; 20.25) yes, if the mother is a heterozygous carrier, one-half of the male children are likely to be color-blind; 20.26) one or more nucleotides are missing in the DNA sequence; 20.27) a disorder that is present at birth

General Glossary

A

Abdomen (ab-DOH-men or AB-doh-men) Portion of the trunk below the diaphragm; between the thorax and pelvis

Abdominal cavity (ab-DAHM-ih-nal KAV-ih-tee) Superior portion of the abdominopelvic cavity; contains the stomach, spleen, liver, pancreas, gallbladder, small intestine, and most of the large intestine

Abdominopelvic cavity (ab-dahm-ih-no-PEL-vik KAV-ih-tee) Inferior part of the ventral body cavity; subdivided into the abdominal cavity and pelvic cavity

Abducens nerve (ab-DOO-sens NERV) Cranial nerve VI; motor nerve; responsible for eye movements

Abduction (ab-DUCK-shun) Movement away from the midline or axis of the body; opposite of adduction

Absolute refractory period (AB-soh-loot ree-FRACK-toar-ee PEE-ree-od) Time during which an excitable cell cannot respond to a second stimulus regardless of the strength of the second stimulus

Absorption (ab-SOARP-shun) The passage of digestive end products from the gastrointestinal tract into the blood or lymph

Accessory nerve (ak-SES-oar-ee NERV) Cranial nerve XI; motor nerve; responsible for contraction of the trapezius and sternocleidomastoid muscles

Accommodation (ah-kahm-o-DAY-shun) Mechanism that allows the eye to focus at various distances, primarily achieved by changing the curvature of the lens

Acetabulum (as-sah-TAB-yoo-lum) Large depression in the hip bone for articulation with the femur

Acetyl-CoA (as-SEE-till-CO-A) A molecule formed from pyruvic acid in the mitochondria when oxygen is present

Acetylcholine (ah-see-till-KOH-leen) A chemical substance that is released at the axon terminals of many neurons to carry the impulse across a synaptic cleft; one of the neurotransmitters

Acetylcholinesterase (ah-see-till-koh-lin-ES-ter-ase) An enzyme that causes the decomposition of acetylcholine; also called cholinesterase

Acid (AS-id) A substance that ionizes in water to release hydrogen ions; a proton donor; a substance with a pH less than 7.0

Acidosis (AS-id-oh-sis) Condition in which the blood has a lower pH than normal

Acinar (AS-ih-nar) Shaped like a small sac

Acinar cells (AS-ih-nar SELZ) Cells in the pancreas that secrete digestive enzymes

Acromion (ah-KRO-mee-on) A process on the scapula that forms the tip of the shoulder

Acrosome (AK-roh-sohm) Structure on the head of a sperm that contains enzymes neccessary to break down the coverings around an ovum to facilitate fertilization

Actin (AK-tin) Contractile protein in the thin filaments of skeletal muscle cells

Action potential (AK-shun po-TEN-shall) A nerve impulse; a rapid change in membrane potential that involves depolarization and repolarization

Active immunity (AK-tiv ih-MYOO-nih-tee) Immunity that is produced as the result of an encounter with an antigen, with subsequent production of memory cells

Active transport (AK-tiv TRANS-port) Membrane transport process that requires cellular energy (ATP)

Addison's disease (ADD-ih-sons dys-EEZ) Condition caused by hyposecretion of glucocorticoids from the adrenal cortex

Adduction (ad-DUCK-shun) Movement toward the midline or axis of the body; opposite of abduction

Adenohypophysis (add-eh-noe-hye-PAH-fih-sis) Anterior portion of the pituitary gland

Adenoids (AD-eh-noyds) Paired masses of lymphoid tissue in the nasopharynx; pharyngeal tonsils

Adenosine diphosphate (ADP) (ah-DEN-oh-sin di-FOS-fate) A molecule that is formed when the terminal phosphate group is removed from adenosine triphosphate

Adenosine triphosphate (ATP) (ah-DEN-oh-sin try-FOS-fate) Compound that stores chemical energy within the cell and provides energy for use by body cells

Adipose (ADD-ih-pose) Fat tissue

Adolescence (add-oh-LESS-ens) Period from puberty until adulthood

Adrenal cortex (ah-DREE-null KOR-teks) Outer portion of the adrenal gland that secretes hormones called corticoids

Adrenal gland (ah-DREE-null GLAND) Endocrine gland that is located on the superior pole of each kidney; divided into cortex and medulla regions; also called suprarenal gland

Adrenal medulla (ah-DREE-null meh-DOO-lah) Inner portion of the adrenal gland; secretes epinephrine and norepinephrine

Adrenaline (ah-DREN-ah-lihn) Neurotransmitter released by some neurons in the sympathetic nervous system; also called epinephrine; also secreted by the adrenal medulla

Adrenergic fiber (add-rih-NER jik FYE-ber) A nerve fiber that releases epinephrine (adrenaline) or norepinephrine (noradrenaline) at a synapse

Adrenocorticotropic hormone (ACTH) (ah-dree-noh-kor-tih-koh-TROH-pik HOAR-moan) A hormone secreted by the anterior pituitary gland that stimulates the cortex of the adrenal gland

Adulthood (ah-DULT-hood) Period from the end of adolescence to old age

Aerobic (air-ROE-bik) Requiring molecular oxygen

Aerobic respiration (air-ROE-bik res-pih-RAY-shun) Catabolic process in cells that requires oxygen

Afferent nervous system (AF-fur-ent NER-vus SIS-tem) The portion of the peripheral nervous system that consists of all the incoming sensory nerves

Afferent neuron (AF-fur-ent NOO-ron) Nerve cell that carries impulses toward the central nervous system from the periphery; sensory neuron

Agglutination (ah-GLOO-tih-nay-shun) Clumping of red blood cells or microorganisms; typically an antigen-antibody reaction

Agglutinin (ah-GLOO-tih-nin) A specific substance in plasma that is capable of causing a clumping of red blood cells; an antibody

Agglutinogen (ah-gloo-TIN-oh-jen) A genetically determined antigen on the cell membrane of a red blood cell that determines blood types

Agranulocyte (a-GRAN-yoo-loh-syte) A white blood cell that lacks granules in the cytoplasm

Albinism (AL-bih-nizm) Lack of pigment in the skin, hair, and eyes

Albumin (al-BYOO-min) The most abundant plasma protein, which is primarily responsible for regulating the osmotic pressure of the blood

Aldosterone (al-DAHS-ter-own) Primary mineralocorticoid secreted by the adrenal cortex; helps maintain sodium homeostasis by promoting the reabsorption of sodium in the renal tubules

Alimentary canal (al-ih-MEN-tah-ree kah-NAL) Long, continuous tube that extends from the mouth to the anus; the digestive tract or gastrointestinal tract

Alkaline (AL-kuh-lihn) Pertaining to a base; having a pH greater than 7.0

Alkalosis (AL-kah-lo-sis) Condition in which the blood has a higher pH than normal

Allantois (ah-LAN-toys) Small extraembryonic membrane that develops between the amnion and chorion and contributes to the formation of the urinary bladder; degenerates and becomes part of the umbilical cord during the second month

Alleles (ah-LEEL) One of two or more alternative forms of a gene at the same site in a chromosome

All-or-none principle (ALL OR NUN PRIN-sih-pull) A property of skeletal muscle fiber contraction; when a muscle fiber receives a sufficient stimulus to contract, all sarcomeres shorten; with insufficient stimulus, none of the sarcomeres contract

Alpha cell (AL-fah SELL) Type of cell that produces glucagon in the pancreatic islets

Alveolar processes (al-VEE-oh-lar PRAH-sess-es) Bony sockets for the teeth

Alveolus (al-VEE-oh-lus) Small sac-shaped structure; most often used to denote the microscopic dilations of terminal bronchioles in the lungs where diffusion of gases occurs; air sacs in the lungs

Amino acid (ah-MEEN-oh AS-id) The structural unit of a protein molecule; an organic compound that contains an amino group (—NH$_2$) and a carboxyl group (—COOH)

Amnion (AM-nee-on) The innermost fetal membrane; transparent sac that holds the developing fetus suspended in fluid

Amniotic cavity (am-nee-AH-tik KAV-ih-tee) Fluid-filled cavity within the amnion

Amniotic fluid (am-nee-AH-tik FLOO-id) Fluid surrounding the fetus within the amnion

Amphiarthrosis (am-fee-ahr-THROH-sis) A slightly movable joint; plural, amphiarthroses

Amylase (AM-ih-lays) An enzyme that breaks down starches into disaccharides

Anabolism (ah-NAB-oh-lizm) Building up, or synthesis, reactions that require energy and make complex molecules out of two or more smaller ones; opposite of catabolism

Anaerobic (an-air-OH-bik) Not requiring molecular oxygen

Anaerobic respiration (an-air-ROE-bik res-pih-RAY-shun) Catabolic process in cells that does not require oxygen

Anaphase (AN-ah-faze) Third stage of mitosis; stage in which spindle fibers shorten and duplicate chromosomes move to opposite ends of the cell

Anatomical position (an-ah-TOM-ih-kull poh-ZIH-shun) Standard reference position for the body; body is erect, facing the observer; upper extremities are at the sides; palms and toes are directed forward

Anatomy (ah-NAT-o-mee) Study of body structure and the relationships of its parts

Androgens (AN-droh-jenz) Male sex hormones; primary androgen is testosterone; produced by the interstitial cells of the testes and a small amount by the adrenal cortex

Anemia (ah-NEE-mee-ah) Condition of the blood in which the number of red blood cells or their hemoglobin content is below normal

Angiotensin (an-jee-oh-TEN-sin) A substance formed in the blood that helps regulate blood pressure; a vasoconstrictor

Anion (AN-eye-on) A negatively charged ion

Anisotropic (an-eye-soh-TROH-pik) Dark bands in skeletal and cardiac muscle fibers; do not allow light to pass through

Antagonist (an-TAG-oh-nist) A muscle that has an action opposite to the prime mover

Antebrachial region (an-te-BRAY-kee-al REE-jun) Region between the elbow and the wrist; forearm; cubital region

Antecubital (an-te-KYOO-bih-tal) Space in front of the elbow

Anterior (an-TEER-ee-or) Toward the front or ventral; opposite of posterior or dorsal

Antibody (AN-tih-bahd-ee) Substance produced by the body that inactivates or destroys another substance that is introduced into the body; immunoglobulin

Antibody-mediated immunity (AN-tih-bahd-ee MEE-dee-ate-ed ih-MYOO-nih-tee) Immunity that is the result of B-cell action and the production of antibodies; also called humoral immunity

Anticoagulant (an-tih-koh-AG-yoo-lant) A substance that delays, suppresses, or prevents the clotting of blood

Anticodon (an-tih-KO-don) A sequence of three nucleotide bases on transfer RNA that is complementary to a codon on messenger RNA; represents a single amino acid

Antidiuretic hormone (ADH) (an-tih-dye-yoo-RET-ik HOAR-moan) Hormone produced by the hypothalamus and secreted from the posterior pituitary gland that promotes water reabsorption from the kidney tubules

Antigen (AN-tih-jen) A substance that triggers an immune response when it is introduced into the body

Antiserum (AN-tih-see-rum) A preparation that contains antibodies from a source other than the recipient and that provides short-term protection against a specific antigen

Aorta (ay-OR-ta) The main systemic vessel that emerges from the left ventricle and carries blood to systemic arteries

Aortic semilunar valve (ay-OR-tik seh-mee-LOO-nar VALVE) The valve between the left ventricle and the aorta that keeps blood from flowing back into the ventricle

Apical foramen (A-pik-al foh-RAY-men) Opening in the root of a tooth where nerves and blood vessels enter the root canal

Apneustic center (ap-NOO-stick SEN-ter) A group of neurons in the pons that affects the rate of respiration by stimulating inspiration

Apocrine gland (AP-oh-krin GLAND) Gland in which the secretory product accumulates in the apex of the cells; then that portion pinches off and is discharged with the secretion

Aponeurosis (ah-pah-noo-ROE-sis) Broad, flat sheet of connective tissue that connects one muscle to another or a muscle to bone

Appendicular (ap-pen-DIK-yoo-lar) Pertaining to the upper and lower extremities, or arms and legs

Appendicular skeleton (ap-pen-DIK-yoo-lar SKEL-eh-ton) Bones of the upper and lower extremities of the body

Appositional growth (ap-poh-ZISH-un-al GROHTH) Growth resulting from material deposited on the surface, such as the growth in diameter of long bones

Aqueous humor (AY-kwee-us HYOO-mer) Fluid that fills the anterior cavity of the eye, in front of the lens

Arachnoid (ah-RAK-noyd) The middle layer of the coverings around the brain and spinal cord

Arachnoid granulations (ah-RACK-noyd gran-yoo-LAY-shuns) Small projections of arachnoid that protrude into the superior sagittal sinus and function in returning cerebrospinal fluid into the blood; also called arachnoid villi

Arachnoid villi (ah-RACK-noyd VILL-eye) Small projections of arachnoid that protrude into the superior sagittal sinus and function in returning cerebrospinal fluid into the blood; also called arachnoid granulations

Arbor vitae (AR-bor VYE-tay) Arrangement of the internal white matter of the cerebellum

Areola (ah-REE-oh-lah) Circular pigmented area that surrounds the nipple

Areolar (ah-REE-oh-lar) Type of connective tissue that contains fibers and a variety of cells in a soft, loose matrix; loose connective tissue

Arrector pili (ah-REK-tor PY-lee) Muscle associated with hair follicles

Arteriole (ar-TEER-ee-ohl) A small branch of an artery that delivers blood to a capillary

Arteriovenous anastomosis (ar-teer-ee-oh-VAY-nus ah-NAS-toh-moh-sis) Direct connection between a small arteriole and a small venule without the intervening capillary

Artery (AR-ter-ee) A blood vessel that carries blood away from the heart

Articular cartilage (ahr-TIK-yoo-lar KAR-tih-layj) Thin layer of hyaline cartilage that covers the ends of long bones

Articulation (are-TIK-yoo-lay-shun) A joint; a point of contact between bones

Artificial immunity (art-ih-FISH-al ih-MYOO-nih-tee) Immunity that requires some deliberate action, such as a vaccination, to achieve exposure to the potentially harmful antigen

Ascending tract (ah-SEND-ing TRACT) Spinal cord nerve tract that conducts sensory impulses up to the brain

Association neuron (ah-soh-see-AY-shun NOO-ron) Nerve cell, totally within the central nervous system, that carries impulses from a sensory neuron to a motor neuron; also called an interneuron

Astrocyte (AS-troh-syte) A star-shaped neuroglial cell that supports neurons in the brain and spinal cord

Atom (AT-tum) The smallest unit of a chemical element that retains the properties of that element

Atomic number (ah-TOM-ik NUM-ber) The number of protons in the nucleus of an atom

Atrioventricular bundle (ay-tree-oh-ven-TRIK-yoo-lar BUN-dul) Portion of the conduction system of the heart that begins at the atrioventricular node, continues for a short distance down the interventricular septum, and then divides into the right and left bundle branches; involved in coordination of heart muscle contraction; also called the bundle of His

Atrioventricular node (ay-tree-oh-ven-TRIK-yoo-lar NODE) Mass of specialized cardiac muscle cells that form a part of the conduction system of the heart and that are located in the right atrium near the opening for the coronary sinus

Atrioventricular valve (ay-tree-oh-ven-TRIK-yoo-lar VALVE) Valve between an atrium and a ventricle in the heart

Atrium (A-tree-um) Thin-walled chamber of the heart that receives blood from veins

Auditory tube (AW-dih-toar-ee TOOB) Passageway between the throat and middle ear that functions to equalize pressure between the middle ear and the exterior; also called the eustachian tube

Auditory ossicles (AW-dih-toar-ee OS-sih-kulls) Three tiny bones in the middle ear that function to amplify sound waves: malleus, incus, stapes

Auerbach's plexus (OUR-bahks PLEK-sus) A network of autonomic nerves between the circular and longitudinal muscle layers of the gastrointestinal tract; also called myenteric plexus

Auricle (AW-rih-kull) Visible portion of the external ear; also called the pinna; also used to denote earlike projections of the atria of the heart

Autoimmune disease (aw-toh-im-YOON dih-ZEEZ) Tissue destruction that results when an individual produces antibodies that attack his or her own tissues

Autonomic ganglion (aw-toe-NOM-ic GANG-lee-on) A cluster of cell bodies of neurons from the autonomic nervous system that are outside the central nervous system; the region of synapse between the two neurons in an autonomic pathway

Autonomic nervous system (aw-toe-NOM-ic NER-vus SIS-tem) The portion of the peripheral efferent nervous system consisting of motor neurons that control involuntary actions

Autosomes (AW-toh-sohm) Any of the 22 pairs of chromosomes in humans not concerned with determination of sex

Avascular (ay-VAS-kyoo-lar) Without blood vessels

Axial (AK-see-al) Pertaining to the head, neck, and trunk

Axial skeleton (AK-see-al SKEL-eh-ton) Bones of the head, neck, and torso

Axillary (AK-sih-lair-ee) Pertaining to the armpit region

Axon (AKS-on) The single efferent process of a neuron that carries impulses away from the cell body

Axon collateral (AKS-on koh-LAT-er-al) One or more side branches of an axon

B

B lymphocytes (B LIM-foh-sytes) A cell of the immune system that develops into a plasma cell and produces antibodies; B cell

Basal ganglia (BAY-sal GANG-lee-ah) Paired regions of gray matter located within the white matter of the cerebrum

Basal metabolic rate (BAY-sal met-ah-BAHL-ik RAYT) Amount of energy that is necessary to maintain life and keep the body functioning at a minimal level

Base (BASE) A substance that ionizes in water to release hydroxyl (OH-) ions or other ions that combine with hydrogen ions; a proton acceptor; a substance with a pH greater than 7.0; alkaline

Basilar membrane (BAYS-ih-lar MEM-brayn) Floor of the cochlear duct; separates cochlear duct from scala tympani

Basophil (BAY-soh-fill) A white blood cell with granules in the cytoplasm that stain readily with basic dyes

Beta cell (BAY tah SELL) Type of cell that produces insulin in the pancreatic islets

Beta oxidation (BAY-tah ox-ih-DAY-shun) Catabolic reaction in which a two-carbon segment is removed from a fatty acid and is converted to acetyl-CoA

Bicuspid valve (bye-KUS-pid VALVE) Valve between the left atrium and left ventricle; also called the mitral valve

Bile (BYLE) Yellowish green fluid that is produced by the liver, stored in the gallbladder, and functions to emulsify fats

Bilirubin (bill-ih-ROO-bin) A pigment that is produced in the breakdown of hemoglobin and excreted in bile

Blastocele (BLAS-toh-seel) Cavity within a blastocyst

Blastocyst (BLAS-toh-syst) Hollow sphere of cells that forms when a cavity develops in a morula; usually present by the fifth day after conception

Blastomeres (BLAS-toh-meerz) Cells formed by early mitotic divisions of a zygote

Blood pressure (BLUHD PRESH-ur) Pressure exerted by the blood against the vessel walls; usually refers to arterial blood pressure

Bony labyrinth (BOH-nee LAB-ih-rinth) A series of interconnecting chambers in the petrous portion of the temporal bone that includes the cochlea, vestibule, and semicircular canals of the inner ear

Bowman's capsule (BOE-mans KAP-sool) Double-layered epithelial cup that surrounds the glomerulus in a nephron; also called glomerular capsule

Brachial (BRAY-kee-al) Pertaining to the arm; proximal portion of the upper limb

Brain stem (BRAYN STEM) The portion of the brain, between the diencephalon and spinal cord, that contains the midbrain, pons, and medulla oblongata

Broad ligament (BRAWD LIG-ah-ment) Largest peritoneal ligament that supports the uterus; double fold of peritoneum that extends laterally from the uterus to the pelvic wall

Bronchial tree (BRONG-kee-al TREE) The bronchi and all their branches that function as passageways between the trachea and the alveoli

Bronchopulmonary segment (brong-koh-PUL-moh-nair-ee SEG-ment) Portion of a lung surrounding a tertiary, or segmental, bronchus; lobule of the lung

Brunner's glands (BROO-nerz GLANDS) Mucous glands in the submucosa of the duodenum; also called duodenal glands

Buccal (BUK-al) Pertaining to the region of the mouth and cheek

Buffer (BUFF-fur) A substance that prevents, or reduces, changes in pH when either an acid or a base is added

Bulbourethral glands (bul-boh-yoo-REE-thral GLANDS) Small accessory glands near the base of the penis in the male; also called Cowper's glands

Bundle branches (BUN-dul BRAN-chez) Specialized cardiac muscle cells that form a part of the conduction system of the heart, are located on either side of the interventricular septum, and carry impulses from the atrioventricular node to the conduction myofibers

Bursae (BUR-see) Fluid-filled sacs that act as cushions at friction points in certain freely movable joints; singular, bursa (BUR-sah)

C

Calcaneus (kal-KAY-nee-us) Heel bone; one of the tarsal bones

Calcitonin (cal-sih-TOH-nin) A hormone produced by the thyroid gland that reduces calcium levels in the blood

Calorie (c) (KAL-or-ee) Unit of heat energy; amount of energy required to raise the temperature of 1 gram of water 1°C (for example, from 14°C to 15°C)

Calorie (C) (KAL-or-ee) Unit of heat energy used in metabolic and nutritional studies; also called a kilocalorie, equivalent to 1000 calories; amount of energy required to raise

the temperature of 1 kilogram of water 1°C (from 14°C to 15°C)

Calyx (KAY-liks) Cuplike extensions of the renal pelvis that collect the urine

Canaliculi (kan-ah-LIK-yoo-lye) Small tubular channels or passages that connect lacunae with other lacunae or the haversian canal in compact bone; singular, canaliculus

Capacitation (kah-pass-ih-TAY-shun) Process that enables sperm to penetrate an egg; membrane around the acrosome weakens so the enzymes can be released

Capillary (KAP-ih-lair-ee) Microscopic blood vessel between an arteriole and a venule, where gaseous exchange takes place

Carbaminohemoglobin (kar-bah-meen-oh-HEE-moh-gloh-bin) Compound that is formed when carbon dioxide combines with the protein portion of a hemoglobin molecule; accounts for approximately 23% of carbon dioxide transport in the blood

Carbohydrate (kar-boh-HYE-drayt) An organic compound that contains carbon, hydrogen, and oxygen with the hydrogen and oxygen present in a 2:1 ratio; examples include sugar, starch, and cellulose

Cardiac cycle (KAR-dee-ak SYE-kul) A complete heartbeat consisting of contraction and relaxation of both atria and both ventricles

Cardiac muscle (KAR-dee-ak MUS-el) Muscle tissue that is found only in the heart; involuntary striated muscle

Cardiac output (KAR-dee-ak OUT-put) The volume pumped from one ventricle in 1 minute; usually measured from the left ventricle

Cardiovascular (car-dee-oh-VAS-kyoo-lar) Relating to the heart and blood vessels

Carina (kah-RYE-nah) Ridge of hyaline cartilage in the region where the trachea divides into the right and left bronchi

Carotene (KAIR-oh-teen) Yellowish pigment that contributes to skin color

Carpal (KAR-pul) Pertaining to the wrist

Carpals (KAR-puls) Small bones in the wrist

Carpus (KAR-pus) Wrist, which consists of eight small carpal bones

Cartilage (KAR-tih-layj) A type of connective tissue in which cells and fibers are embedded in a semisolid gel matrix

Catabolism (kah-TAB-oh-lizm) Reactions that break down complex molecules into two or more smaller ones with the release of energy; opposite of anabolism

Catalyst (KAT-ah-list) A substance that speeds up chemical reactions without being changed itself

Cation (KAT-eye-on) A positively charged ion

Cauda equina (KAW-dah ee-KWYNE-ah) Collection of spinal nerve roots at the distal end of the spinal cord; literally means "horse's tail"

Cecum (SEE-kum) Proximal portion of the large intestine, below the ileocecal valve

Celiac (SEE-lee-ak) Pertaining to the abdomen

Cell (SELL) Basic unit of life; structural and functional unit of the body

Cell-mediated immunity (SELL MEE-dee-ate-ed ih-MYOO-nih-tee) Immunity that is the result of T-cell action; also called cellular immunity

Cell membrane (SELL MEM-brane) Phospholipid membrane that separates the contents of the cell from the material outside the cell

Cellular metabolism (SELL-yoo-lar meh-TAB-oh-lizm) Chemical reactions that take place inside cells

Cellular respiration (SELL-yoo-lar res-per-RAY-shun) Use of oxygen by the tissue cells

Central nervous system (CNS) (SEN-tral NER-vus SIS-tem) The portion of the nervous system that consists of the brain and spinal cord

Central sulcus (SEN-tral SULL-kus) The groove between the frontal and parietal lobes of the cerebrum; also called the fissure of Rolando

Central venous pressure (SEN-tral VAYN-us PRESH-ur) Blood pressure in the right atrium

Centrioles (SEN-tree-ohlz) Collection of microtubules that function in cell division

Centromere (SEN-tro-meer) Portion of the chromosome where the two chromatids are joined; serves as point of attachment for spindle fibers

Centrosome (SEN-tro-sohm) Dense area near the nucleus that contains the centrioles

Cephalic (seh-FAL-ik) Pertaining to the head

Cerebellar hemisphere (sair-eh-BELL-ar HEM-ih-sfeer) Either of the two halves of the cerebellum

Cerebellar peduncles (sair-eh-BELL-ar pee-DUNK-als) Bundles of nerve fibers that connect the cerebellum with other parts of the central nervous system

Cerebellum (sair-eh-BELL-um) Second largest part of the human brain, located posterior to the pons and medulla oblongata, and involved in the coordination of muscular movements

Cerebral aqueduct (seh-REE-brull AH-kweh-dukt) A narrow channel through the midbrain, between the third and fourth ventricles, that contains cerebrospinal fluid

Cerebral cortex (seh-REE-brull KOR-teks) Thin layer of gray matter, composed of neuron cell bodies and dendrites, on the surface of the brain

Cerebral hemisphere (seh-REE-brull HEM-ih-sfeer) Either of the two halves of the cerebrum

Cerebral peduncles (seh-REE-brull pee-DUNK-als) Two bands of connecting fibers on the ventral aspect of the midbrain that contain voluntary motor tracts descending from the cerebral cortex

Cerebrospinal fluid (seh-ree-broh-SPY-null FLOO-id) A fluid, similar to plasma, that fills the subarachnoid space around the brain and spinal cord and is in the ventricles of the brain

Cerebrum (seh-REE-brum) The largest and uppermost part of the human brain; site of brain associated with consciousness, learning, memory, sensations, and voluntary movements

Cerumen (see-ROOM-men) Earwax

Ceruminous gland (see-ROOM-in-us GLAND) A gland in the ear canal that produces cerumen or earwax

Cervical (SIR-vih-kal) Pertaining to the neck region

Cervix (SIR-viks) The lower, narrow portion of the uterus that projects into the vagina

Chemical bond (KEM-ih-kal BOND) Force that holds atoms together; involved in the sharing or exchange of electrons

Chemoreceptor (kee-moh-ree-SEP-tor) A sensory receptor that detects the presence of chemicals; responsible for taste, smell, and monitoring the concentration of certain chemicals in body fluids

Chief cells (CHEEF SELZ) Cells in the gastric mucosa that secrete pepsinogen

Childhood (CHYLD-hood) Period from age 1 until puberty

Cholecystokinin (koh-lee-sis-toe-KYE-nin) A hormone that is produced in the small intestine that stimulates the release of bile from the gallbladder

Cholinergic fiber (koh-lih-NER-jik FYE-ber) A nerve fiber that releases acetylcholine at a synapse

Cholinesterase (koh-lin-ES-ter-ase) An enzyme that causes the decomposition of acetylcholine; also called acetylcholinesterase

Chondrocyte (KON-droh-syte) Cartilage cell

Chordae tendineae (KOR-dee ten-DIN-ee) Fibrous, stringlike structures that attach the atrioventricular valves to the papillary muscles of the heart wall

Chorion (KOR-ee-on) Outermost extraembryonic, or fetal, membrane; contributes to the formation of the placenta

Chorionic villi (kor-ee-ON-ik VILL-eye) Fingerlike projections of the chorion that grow into the endometrium of the uterus and contain fetal blood vessels

Choroid (KOR-oyd) Dark pigmented portion of the vascular tunic of the eye; prevents scattering of light rays

Choroid plexus (KOR-oyd PLEKS-us) Specialized capillary network within the ventricles of the brain that secretes cerebrospinal fluid

Chromatid (KRO-mah-tid) One member of a duplicate pair of chromosomes

Chromatin (KRO-ma-tin) Long, slender threads of DNA in the nucleus of a cell; develops into chromosomes during mitosis

Chromosomes (KRO-mo-sohmz) Dark-staining structures that appear in the nucleus when chromatin condenses during mitosis

Chylomicrons (kye-loh-MY-krons) Small fat droplets that are covered with a protein coat in the epithelial cells of the mucosa of the small intestine

Chyme (KYME) The semifluid mixture of food and gastric juice that leaves the stomach through the pyloric sphincter

Cilia (SILL-ee-ah) Hairlike processes that project outward from the cell membrane and move substances across the surface of the cell

Ciliary body (SILL-ee-air-ee BAH-dee) Part of the vascular tunic of the eye; includes the ciliary muscle and ciliary processes

Ciliary muscle (SILL-ee-air-ee MUSS-el) Smooth muscle in the ciliary body that functions in accommodation for near vision

Ciliary processes (SILL-ee-air-ee PRAH-sess-es) Fingerlike structures in the ciliary body that secrete the aqueous humor

Circumcision (SIR-kum-sih-shun) Surgical removal of the prepuce

Circumduction (sir-kum-DUCK-shun) A conelike movement of a body part; the distal end of the part outlines a circle but the proximal end remains relatively stationary

Cisterna chyli (sis-TER-nah KY-lee) A dilation that forms the beginning of the thoracic duct

Citric acid cycle (SIT-rik AS-id SYE-kul) Aerobic series of reactions that follows glycolysis in glucose metabolism to release energy and carbon dioxide; also called Krebs cycle

Clavicle (KLAV-ih-kul) Collarbone

Cleavage (KLEE-vayj) Series of mitotic cell divisions after fertilization; resulting cells are called blastomeres

Clitoris (KLY-toe-ris) Small mass of erectile tissue at the anterior end of the vestibule in females; homologous to the penis in males

Coagulant (koh-AG-yoo-lant) A substance that promotes the clotting of blood

Coagulation (koh-ag-yoo-LAY-shun) The process of blood clotting

Coccyx (KOK-siks) Lowest part of the vertebral column, composed of four vertebrae fused together

Cochlea (KOK-lee-ah) The spiral, or coiled, portion of the bony labyrinth

Cochlear duct (KOK-lee-ar DUKT) Portion of the membranous labyrinth that is inside the cochlea; contains endolymph and the organ of hearing

Codon (KO-don) A set of three nucleotides on a messenger RNA molecule; contains the code for a single amino acid

Collagenous fibers (koh-LAJ-eh-nuss FYE-burs) Strong and flexible connective tissue fibers that contain the protein collagen

Collateral ganglion (koh-LAT-er-al GANG-lee-on) An autonomic ganglion outside the sympathetic chain and located near selected abdominal blood vessels; part of the sympathetic pathway; also called prevertebral ganglion

Colostrum (koh-LAHS-trum) Fluid secreted by the mammary glands before the beginning of true milk production

Columnar (ko-LUM-nar) Shaped like a column; vertical dimension is greater than horizontal dimension

Columns (KOLL-ums) Regions of white matter in the spinal cord; also called funiculi

Complement (KOM-pleh-ment) A group of proteins normally found in serum that provides nonspecific resistance by promoting phagocytosis and inflammation

Complete protein (kum-PLEET PRO-teen) A protein that contains all of the essential amino acids

Compound (KAHM-pownd) A substance formed from two or more elements joined by chemical bonds in a definite, or fixed, ratio; smallest unit of a compound is a molecule

Concentration gradient (kon-sen-TRAY-shun GRAY-dee-ent) Difference in concentration of substances between two different areas

Conception (kon-SEP-shun) Union of a sperm cell nucleus with an egg cell nucleus; fertilization

Conchae (KONG-kee) Scroll-like bones or projections of bones; also called turbinates; singular, concha (KONG-kah)

Conduction myofibers (kon-DUK-shun my-o-FYE-bers) Cardiac muscle cells specialized for conducting action potentials to the myocardium; part of the conduction system of the heart; also called Purkinje fibers

Conduction system (kon-DUCK-shun SIS-tem) Specialized cardiac muscle cells that coordinate the contraction of the heart chambers

Conductivity (kon-duck-TIV-ih-tee) The ability of neurons to transmit an impulse from one point in the body to another

Condyle (KON-dial) Smooth, rounded articular surface on a bone

Cone (KONE) A photoreceptor in the retina that is specialized for color vision

Connective tissue (ko-NEK-tiv TISH-yoo) Most abundant and widespread tissue in the body; includes bone, cartilage, adipose, blood, and various fibrous tissues

Contractility (kon-track-TILL-ih-tee) The ability of muscle cells to shorten to produce movement

Conus medullaris (KOE-nus med-yoo-LAIR-is) Tapered distal portion of the spinal cord below the lumbar enlargement

Convergence circuit (kon-VER-jens SIR-cut) A conduction pathway in which several presynaptic neurons synapse with a single postsynaptic neuron within a neuronal pool

Coracoid (KOR-ah-koyd) A process on the scapula that projects forward and downward below the clavicle and provides attachments for arm and chest muscles

Core temperature (KOR TEM-per-ah-chur) The temperature deep in the body; temperature of the internal organs

Cornea (KOR-nee-ah) Transparent anterior portion of the outer layer of the eyeball; anterior portion of the fibrous tunic

Corona radiata (koh-ROH-nah ray-dee-AH-tah) Several layers of cells that surround a secondary oocyte when it is released from the ovary

Coronal plane (ko-ROH-nal PLANE) A vertical plane that extends from side to side and divides the body or part into anterior and posterior portions; also called a frontal plane

Coronary artery (KOR-oh-nair-ee AHR-tur-ee) Vessel that carries oxygenated blood to the myocardium

Coronary sinus (KOR-oh-nair-ee SYE-nus) A large venous channel on the posterior surface of the heart that collects deoxygenated blood from coronary circulation and returns it to the right atrium

Corpora cavernosa (KOR-por-ah kav-er-NOH-sah) Two dorsal columns of erectile tissue found in the penis

Corpora quadrigemina (KOR-por-ah kwad-rih-JEM-ih-nah) Four rounded structures composed of the superior and inferior colliculi on the dorsal surface of the midbrain

Corpus albicans (KOR-pus AL-bih-kans) Scar tissue in the ovary that forms when the corpus luteum degenerates

Corpus callosum (KOR-pus kah-LOH-sum) Large band of myelinated nerve fibers that connects the two halves of the cerebrum

Corpus luteum (KOR-pus LOO-tee-um) Under the influence of luteinizing hormone, the structure that develops from the mature follicle after ovulation

Corpus spongiosum (KOR-pus spun-jee-OH-sum) Ventral column of erectile tissue found in the penis

Cortex (KOR-tex) The outer portion of an organ

Corticospinal tract (kor-tih-koh-SPY-null TRACT) Descending, or motor, tract that begins in the cerebral cortex and ends in the spinal cord; also called pyramidal tract

Cortisol (KOR-tih-sahl) Primary glucocorticoid secreted by the adrenal cortex; functions in the regulation of carbohydrate metabolism and also has anti-inflammatory effects

Costal (KAHS-tal) Pertaining to the ribs

Covalent bond (koe-VAY-lent BOND) Chemical bond formed by two atoms sharing one or more pairs of electrons

Cowper's glands (KOW-pers GLANDS) Small accessory glands near the base of the penis in the male; also called bulbourethral glands

Cranial (KRAY-nee-al) Pertaining to the skull

Cranial cavity (KRAY-nee-al KAV-ih-tee) Part of the dorsal body cavity that contains the brain

Craniosacral division (kray-nee-oh-SAY-kral dih-VIH-shun) One of two divisions of the autonomic nervous system; primarily concerned with processes that involve the conservation and restoration of energy; sometimes referred to as the "rest and repose" division; also called the parasympathetic division

Cranium (KRAY-nee-um) Bones of the skull that surround the brain; includes the frontal, parietal, occipital, temporal, ethmoid, and sphenoid bones

Creatine phosphate (KREE-ah-tin FOS-fate) High-energy molecule in muscle cells that is used to rapidly regenerate ATP

Crenation (kree-NAY-shun) Shrinkage of red blood cells when they are placed in a hypertonic solution

Crest (KREST) Narrow ridge of bone

Cretinism (KREE-tin-izm) Dwarfism caused by insufficient thyroid hormone production in children

Cribriform plate (KRIB-rih-form PLATE) Portion of the ethmoid bone that contains olfactory foramina

Cricoid cartilage (KRY-koyd KAR-tih-layj) Most inferior cartilage of the larynx

Crista ampullaris (KRIS-tah amp-yoo-LAIR-is) Receptor organ located within the ampulla of the semicircular canals; functions in dynamic equilibrium

Crista galli (KRIS-tah GAL-lee) Upward projecting process on the ethmoid bone

Cristae (KRIS-tee) Shelflike ridges formed by folds of the inner membrane of the mitochondria

Cubital (KYOO-bih-tal) Pertaining to the forearm; region between the elbow and wrist; antebrachial

Cuboidal (kyoo-BOYD-al) Shaped like a cube; horizontal and vertical dimensions approximately equal

Cupula (KEW-pew-lah) A gelatinous mass over the crista ampullaris in the ampulla of the semicircular canals; functions in dynamic equilibrium

Cushing's syndrome (KOOSH-ings SIN-drohm) Condition caused by hypersecretion of glucocorticoids from the adrenal cortex

Cutaneous (kyoo-TAY-nee-us) Pertaining to the skin

Cutaneous membrane (kyoo-TAY-nee-us MEM-brayn) One of the types of epithelial membranes; primary organ of the integumentary system; also called the skin

Cuticle (KEW-tih-kuhl) A fold of stratum corneum at the proximal border of the visible portion of a nail; also called eponychium

Cystic duct (SIS-tik DUKT) Duct from the gallbladder

Cytokinesis (sye-toe-kih-NEE-sis) Division of the cytoplasm at the end of mitosis to form two separate daughter cells

Cytoplasm (SYE-toe-plazm) Gel-like fluid inside the cell, exclusive of the organelles

Cytoskeleton (sye-toe-SKEL-eh-ton) Complex network of filaments and tubules that function in support and movement of the cell and organelles

D

Deamination (dee-am-ih-NAY-shun) Catabolic reaction in which an amino group is removed from an amino acid to form ammonia and a keto acid; occurs in the liver as part of protein catabolism

Deciduous teeth (dee-SID-yoo-us TEETH) The teeth that appear first and then are shed and replaced by permanent teeth; also called primary teeth or baby teeth

Decussation (dee-kuh-SAY-shun) A crossing over; usually refers to motor fibers that cross over to the opposite side in the medulla oblongata

Deep (DEEP) Away from the surface; opposite of superficial

Defecation (def-eh-KAY-shun) The expulsion of indigestible wastes, or feces, through the anus

Deglutition (dee-gloo-TISH-un) The process of swallowing

Dehydration synthesis (dee-hye-DRAY-shun SIN-the-sis) A reaction in which a larger molecule is made from two or more smaller molecules by removing a water molecule

Dendrite (DEN-dryte) The branching afferent process of a neuron that receives impulses from other neurons and transmits them toward the cell body

Deoxyhemoglobin (dee-ok-see-hee-moh-GLOH-bin) The reduced form of hemoglobin; hemoglobin that is not combined with a full load of oxygen

Deoxyribonucleic acid (DNA) (dee-ahk-see-rye-boh-noo-KLEE-ik AS-id) A nucleic acid that contains deoxyribose sugar; genetic material of the cell

Depolarization (dee-poh-lar-ih-ZAY-shun) A reduction in membrane potential; the interior side of the cell membrane becomes less negative (more positive) relative to the exterior

Dermis (DER-mis) Inner layer of the skin that contains the blood vessels, nerves, glands, and hair follicles; also called stratum corium

Descending tract (dee-SEND-ing TRACT) Spinal cord nerve tract that conducts motor impulses down the cord from the brain

Detrusor muscle (dee-TROO-sor MUSS-el) The smooth muscle in the wall of the urinary bladder

Diabetes mellitus (dye-ah-BEE-teez MEL-ih-tus) Condition caused by insufficient insulin production and characterized by high blood glucose levels

Diapedesis (dye-ah-peh-DEE-sis) The process by which white blood cells squeeze between the cells in a vessel wall to enter the tissue spaces outside the blood vessel

Diaphysis (dye-AF-ih-sis) The long, straight shaft of a long bone

Diarthrosis (dye-ahr-THROW-sis) Freely movable joint characterized by a joint cavity; also called a synovial joint; plural, diarthroses (dye-ahr-THROW-sees)

Diastole (dye-AS-toh-lee) Relaxation phase of the cardiac cycle; opposite of systole

Diastolic pressure (dye-ah-STAHL-ik PRESH-ur) Blood pressure in the arteries during relaxation of the ventricles

Diencephalon (dye-en-SEF-ah-lon) Part of the brain between the cerebral hemispheres and the midbrain; includes the thalamus, hypothalamus, and epithalamus

Differentiation (dif-er-en-she-AY-shun) Process by which cells become structurally and functionally specialized

Diffusion (dif-YOO-zhun) Movement of atoms, ions, or molecules from a region of high concentration to a region of low concentration

Digestion (dye-JEST-shun) Process of converting food into chemical substances that can be utilized by the body

Digestive (dye-JES-tiv) Relating to digestion

Diploë (DIP-loh-ee) Layer of spongy bone between the inner and outer tables of compact bone in the flat bones of the skull

Disaccharide (die-SAK-ah-ride) A sugar formed from two monosaccharide molecules; a double sugar such as sucrose, maltose, and lactose

Distal (DIS-tal) Farther from a point of attachment or origin; opposite of proximal

Divergence circuit (dye-VER-jens SIR-cut) A conduction pathway in which a single neuron synapses with multiple neurons within a neuronal pool

Dopamine (DOH-pah-meen) One of several neurotransmitters

Dorsal (DOR-sal) Toward the back or posterior; opposite of ventral or anterior

Dorsal root (DOR-sal ROOT) Sensory branch of a spinal nerve by which the nerve is attached to the spinal cord

Dorsal root ganglion (DOR-sal ROOT GANG-lee-on) Collection of sensory neuron cell bodies in the dorsal root of a spinal nerve

Dorsiflexion (dor-sih-FLEK-shun) Bending backward of a body part; usually refers to the movement in which the top of the foot is lifted upward to decrease the angle between the foot and leg; opposite of plantar flexion

Ductus arteriosus (DUCK-tus ar-teer-ee-OH-sus) A small vessel between the pulmonary trunk and the aorta in the fetus; it permits blood to bypass the lungs in the fetal circulatory pathway

Ductus deferens (DUCK-tus DEFF-er-enz) Tubular structure that is continuous with the epididymis; it ascends through the inguinal canal and transports sperm to the ejaculatory duct

Ductus venosus (DUCK-tus veh-NOH-sus) A small vessel that connects the umbilical vein to the inferior vena cava in the fetus and permits blood to bypass the liver in the fetal circulatory pathway

Duodenal glands (doo-oh-DEE-nal GLANDS) Mucous glands in the submucosa of the duodenum; also called Brunner's glands

Duodenum (doo-oh-DEE-num) The first part of the small intestine, about 25 cm long

Dura mater (DOO-rah MAY-ter) Tough, outermost layer of the coverings around the brain and spinal cord; one of the meninges; literally means "tough mother"

Dural sinus (DOO-ral SYE-nus) A channel, or large vein, within the dura mater of the cranial cavity

Dynamic equilibrium (dye-NAM-ik ee-kwi-LIB-ree-um) Equilibrium of motion; maintaining balance when the head or body is moving

E

Ectoderm (EK-toh-derm) The outermost of the three primary germ layers; develops into the nervous system and the epidermis of the skin

Efferent ducts (EF-fur-ent DUCTS) Tubules that carry sperm from the rete testis to the epididymis

Efferent nervous system (EF-fur-ent NER-vus SIS-tem) The portion of the peripheral nervous system that consists of all the outgoing motor nerves

Efferent neuron (EF-fur-ent NOO-ron) Nerve cell that carries impulses away from the central nervous system toward the periphery; motor neuron

Ejaculation (ee-jak-yoo-LAY-shun) Forceful expulsion of seminal fluid from the urethra

Ejaculatory duct (ee-JAK-yoo-lah-tor-ee DUKT) Short duct that penetrates the prostate gland and empties sperm and fluid from the seminal vesicles into the prostatic urethra

Elastic fibers (ee-LAS-tick FYE-burs) Yellow connective tissue fibers that are not particularly strong, but can be stretched and will return to their normal shape when released

Elasticity (ee-lass-TIS-ih-tee) The ability of tissue to return to its original shape after contraction or extension

Electrocardiogram (ECG) (ee-lek-troh-KAR-dee-oh-gram) A graphic recording of the electrical changes that occur during a cardiac cycle

Electroencephalogram (EEG) (ee-lek-troh-en-SEF-al-oh-gram) A graphic recording of the electrical activity associated with the function of neural tissue

Electrolyte (ee-LEK-troh-lite) A substance that forms positive and negative ions in a solution, which makes it capable of conducting an electric current

Electron (ee-LEK-tron) A negatively charged particle found in the nucleus of an atom

Element (EL-eh-ment) Simplest form of matter that cannot be broken down by ordinary chemical means

Embryo (EM-bree-oh) Stage of development that lasts from the beginning of the third week to the end of the eighth week after fertilization; period during which the organ systems develop in the body

Embryonic disk (em-bree-ON-ik DISK) Cells of the early embryo that develop into the three primary germ layers

Embryonic period (em-bree-ON-ik PEER-ee-ud) Stage of development that lasts from the beginning to the third week until the end of the eighth week after fertilization; period during which the organ systems develop in the body

Emission (ee-MISH-un) Discharge of seminal fluid into the urethra

Endergonic reaction (en-der-GAHN-ik ree-AK-shun) A chemical reaction that uses energy

Endocardium (en-doh-KAR-dee-um) The thin, smooth inner lining of each chamber of the heart

Endochondral ossification (en-doh-KON-dral ah-sih-fih-KAY-shun) Method of bone formation in which cartilage is replaced by bone

Endocrine (EN-doh-krin) Relating to glands that secrete their product directly into the blood; opposite of exocrine

Endocytosis (en-doh-sye-TOH-sis) Formation of vesicles to transfer substances from outside the cell to inside the cell

Endoderm (EN-doh-derm) The innermost of the three primary germ layers; it develops into the digestive tract, urinary organs, respiratory structures, and other organs and glands

Endolymph (EN-doh-lymf) Fluid that fills the membranous labyrinth of the inner ear

Endometrium (end-oh-MEE-tree-um) Innermost layer of the uterus, which is a mucous membrane

Endomysium (end-oh-MY-see-um) Connective tissue that surrounds individual muscle fibers (cells)

Endoneurium (end-oh-NOO-ree-um) Connective tissue that surrounds individual nerve fibers

Endoplasmic reticulum (end-oh-PLAZ-mik ree-TIK-yoo-lum) Membrane-enclosed channels within the cytoplasm

Endorphin (en-DOR-fin) A chemical in the central nervous system that influences pain perception and acts as a natural painkiller

Endosteum (end-AH-stee-um) The membrane that lines the medullary cavity of bones

Enkephalin (en-KEF-ah-lin) A chemical in the central nervous system that influences pain perception and acts as a natural painkiller

Enterokinase (en-ter-oh-KYE-nays) An enzyme in the small intestine that activates trypsinogen from the pancreas

Enzyme (EN-zime) An organic catalyst; a substance that affects the rates of biochemical reactions; usually a protein

Eosinophil (ee-oh-SIN-oh-fill) A white blood cell with granules in the cytoplasm that stain readily with acidic (eosin) dyes

Ependymal cell (ee-PEN-dih-mal SELL) Neuroglial cell that lines the ventricles of the brain and the central canal of the spinal cord

Epicardium (eh-pih-KAR-dee-um) The outer layer of the heart wall; the visceral pericardium

Epicondyle (ep-ih-KON-dial) Bony bulge adjacent to, or above, a condyle on a bone

Epidermis (ep-ih-DER-mis) Outermost layer of the skin

Epididymis (ep-ih-DID-ih-mis) Tightly coiled tubule along the posterior margin of each testis; functions in the maturation and storage of sperm

Epidural space (ep-ih-DOO-ral SPACE) The space between the dura mater and surrounding bone, especially between the spinal dura mater and the vertebrae

Epigastric region (ep-ih-GAS-trik REE-jun) The upper middle portion of the abdomen

Epiglottis (eh-pih-GLAHT-is) Long, leaf-shaped, movable cartilage of the larynx that covers the opening of the larynx and prevents food from entering the respiratory tract during swallowing

Epimysium (ep-ih-MY-see-um) Fibrous connective tissue that surrounds a whole muscle

Epinephrine (ep-ih-NEFF-rihn) Hormone secreted by the adrenal medulla that produces effects similar to the sympathetic nervous system; neurotransmitter released by some neurons in the sympathetic nervous system; also called adrenaline

Epineurium (ep-ih-NOO-re-um) Fibrous connective tissue that surrounds a whole nerve

Epiphyseal plate (ep-ih-FIZ-ee-al PLATE) The cartilaginous plate between the epiphysis and diaphysis of a bone; responsible for the lengthwise growth of a long bone

Epiphyseal line (ep-ih-FIZ-ee-al LINE) The remnant of the epiphyseal plate after the cartilage calcifies and growth ceases

Epiphysis (ee-PIF-ih-sis) The end of a long bone

Epiploic appendages (ep-ih-PLOH-ik ah-PEN-day-jez) Pieces of fat-filled connective tissue attached to the outer surface of the colon

Epithalamus (ep-ih-THAL-ah-mus) A tiny region, located superior to the thalamus and hypothalamus in the diencephalon, that includes the pineal body

Epithelial tissue (ep-ih-THEE-lee-al TISH-yoo) Tissue that covers the body and its parts; lines parts of the body; classified according to shape and arrangement

Eponychium (eh-poh-NICK-ee-um) A fold of stratum corneum at the proximal border of the visible portion of a nail; also called cuticle

Equilibrium (ee-kwi-LIB-ree-um) A state of balance between opposing forces

Erection (ee-REK-shun) Condition when the erectile tissue of the penis is filled with blood

Erythrocyte (ee-RITH-roh-syte) Red blood cell

Erythropoiesis (ee-rith-roh-poy-EE-sis) The process of red blood cell formation

Erythropoietin (ee-rith-roh-POY-ee-tin) A hormone released by the kidneys that stimulates red blood cell production

Esophagus (ee-SAHF-ah-gus) Collapsible muscular tube that serves as a passageway for food between the pharynx and stomach

Essential amino acids (ee-SEN-chul ah-MEEN-oh AS-ids) Amino acids that cannot be synthesized in the human body and must be supplied in the diet

Estrogen (ESS-troh-jen) Hormone secreted by the ovarian follicles that stimulates the development and maintenance of female secondary sex characteristics and the cyclic changes in the uterine lining

Ethmoid bone (ETH-moyd BONE) Cranial bone that occupies most of the space between the nasal cavity and the orbits of the eyes

Eustachian tube (yoo-STAY-shee-an TOOB) Tubular passageway that connects the middle ear cavity and the nasopharynx and functions to equalize pressure between the middle ear cavity and the exterior; also called auditory tube

Eversion (ee-VER-zhun) Act of turning a body part outward; usually refers to movement of the sole of the foot outward or laterally; opposite of inversion

Excitability (eks-eye-tah-BILL-ih-tee) The ability of muscle and nerve tissue to receive and respond to stimuli; also called irritability

Excitatory transmission (eks-EYE-tah-toar-ee trans-MIH-shun) Nerve impulse conduction in which the neurotransmitter causes a depolarization (excitation) of the postsynaptic membrane

Exergonic reaction (eks-er-GAHN-ik ree-AK-shun) A chemical reaction that releases energy

Exhalation (eks-hah-LAY-shun) Process of letting air out of the lungs during the breathing cycle; also called expiration

Exocrine gland (EKS-oh-krihn GLAND) A gland that secretes its product to a surface or cavity through ducts

Exocytosis (eks-oh-sye-TOH-sis) Formation of vesicles to transfer substances from inside the cell to outside the cell

Exophthalmos (eks-off-THAL-mus) An abnormal bulging or protruding eyeball

Expiration (eks-per-RAY-shun) Process of letting air out of the lungs during the breathing cycle; also called exhalation

Expiratory reserve volume (ERV) (eks-PYE-rah-tor-ee ree-ZERV VOL-yoom) Maximum amount of air that can be forcefully exhaled after a tidal expiration

Extensibility (eks-ten-sih-BILL-ih-tee) The ability of muscle tissue to stretch when pulled

Extension (ek-STEN-shun) A movement that increases the angle between two parts; opposite of flexion

External auditory meatus (eks-TER-nal AW-dih-toar-ee mee-AY-tus) The curved tube that extends from the auricle of the ear into the temporal bone, and ends at the tympanic membrane; also called the ear canal or external auditory canal

External ear (eks-TER-nal EER) The outer portion of the ear; includes the pinna and external auditory canal, and terminates at the tympanic membrane

External nares (eks-TER-nal NAY-reez) Openings through which air enters the nasal cavity; nostrils

External respiration (eks-TER-nal res-per-RAY-shun) Exchange of gases between the lungs and the blood

Extracellular (eks-trah-SELL-yoo-lar) Outside the cell

Extracellular fluid (ECF) (eks-trah-SELL-yoo-lar FLOO-id) Fluid in the body that is not inside cells; includes plasma and interstitial fluid

Extrapyramidal tract (eks-trah-pih-RAM-ih-dal TRACT) Any descending, or motor, tract in the spinal cord that is not a pyramidal tract

Extrinsic eye muscles (eks-TRIN-sik EYE MUSS-els) Six skeletal muscles external to the eyeball that control movements of the eye

F

Facet (FASS-et) Smooth, nearly flat articular surface on a bone

Facial nerve (FAY-shall NERV) Cranial nerve VII; mixed nerve; responsible for taste sensations and for stimulating the muscles of facial expression

Facilitated diffusion (fuh-SIL-ih-tay-ted dif-YOO-zhun) A type of passive membrane transport that requires a carrier molecule

Falciform ligament (FALL-sih-form LIG-ah-ment) Fold of peritoneum that attaches the anterior surface of the liver to the anterior abdominal wall

Fallopian tubes (fah-LOH-pee-an TOOBS) The tubes that extend laterally from the upper portion of the uterus to the region of the ovaries; also called uterine tubes or oviducts

Falx cerebelli (FALKS sair-ee-BELL-eye) Small extension of dura mater between the cerebellar hemispheres

Falx cerebri (FALKS SAYR-ee-brye) A fold of dura mater that extends into the longitudinal fissure between cerebral hemispheres

Fasciculus (fah-SICK-yoo-lus) A small bundle or cluster of muscle or nerve fibers (cells); also called fascicle (FAS-ih-kull); plural, fasciculi (fah-SICK-yoo-lye)

Fatty acid (FAT-tee AS-id) Building blocks of fat molecules; consists of a long chain of carbon and hydrogen atoms

Fauces (FAW-seez) Opening from the oral cavity into the oropharynx

Feces (FEE-seez) Material discharged from the rectum, consisting of bacteria, indigestible food residue, and secretions

Femoral (FEM-or-al) Pertaining to the thigh; the part of the lower extremity between the hip and the knee

Femur (FEE-mer) Large, long bone in the thigh

Fertilization (fir-tih-lih-ZAY-shun) Union of a sperm cell nucleus with an egg cell nucleus; conception

Fetal period (FEE-tal PEER-ee-od) Stage of development that starts at the beginning of the ninth week after fertilization and lasts until birth

Fetus (FEE-tus) Term used for the developing offspring from the beginning of the ninth week after fertilization until birth

Fiber (FYE-bur) Complex polysaccharides that cannot be digested by enzymes in the human digestive system but add bulk to the diet

Fibrin (FYE-brin) An insoluble, fibrous protein that is formed by the action of thrombin on fibrinogen during the process of blood clotting

Fibrinogen (fye-BRIN-oh-jen) A soluble plasma protein that is converted to insoluble fibrin by the action of thrombin during the process of blood clotting

Fibrinolysis (fye-brin-AHL-ih-sis) Mechanism by which a blood clot dissolves

Fibroblast (FYE-broh-blast) Connective tissue cell that produces fibers

Fibrosis (fye-BROH-sis) Replacement of damaged tissue with fibrous connective tissue

Fibrous pericardium (FYE-bruss pair-ih-KAR-dee-um) The outer, tough, white fibrous connective tissue layer of the pericardial sac

Fibrous tunic (FYE-bruss TOO-nik) Outermost layer of the eyeball

Fibula (FIB-yoo-lah) Small bone on the lateral side of the leg

Filtrate (FILL-trayt) The fluid that enters the glomerular capsule when blood is filtered by the glomerulus in a nephron of the kidney

Filtration (fil-TRAY-shun) The movement of a fluid through a membrane in response to hydrostatic pressure

Filum terminale (FYE-lum term-ih-NAL-ee) Slender thread of pia mater that extends inferiorly from the conus medullaris of the spinal cord to the coccyx

Fissure (FISH-ur) Narrow cleft or slit between bones or separating body parts

Fissure of Rolando (FISH-ur of roh-LAN-doh) The groove between the frontal and parietal lobes of the cerebrum; also called the central sulcus

Fissure of Sylvius (FISH-ur of SYL-vee-us) The deep groove between the temporal lobe below and the frontal and parietal lobes above; also called the lateral fissure or lateral sulcus

Flagellum (fluh-JELL-um) Long projection of the cell membrane that functions in the motility of the cell; plural, flagella

Flexion (FLEK-shun) A movement that decreases the angle between two parts; opposite of extension

Follicle-stimulating hormone (FSH) (FAHL-ik-yool STIM-yoo-lay-ting HOAR-moan) Hormone secreted by the anterior pituitary gland in both males and females; one of the gonadotropins

Foramen (foh-RAY-men) A hole or opening

Foramen of Monro (foh-RAY-men of mun-ROH) A small opening between each lateral ventricle and the third ventricle of the brain for the passage of cerebrospinal fluid; also called the interventricular foramen

Foramen ovale (foh-RAY-men oh-VAL-ee) An opening in the interatrial septum of the fetal heart that permits blood to flow directly from the right atrium into the left atrium and to bypass the lungs in the fetal circulatory pathway; also an opening in the sphenoid bone that transmits the mandibular branch of the trigeminal nerve

Formed elements (formed EL-eh-ments) Red blood cells, white blood cells, and platelets in the blood

Fossa (FAW-sah) A smooth, shallow depression

Fossa ovalis (FAW-sah oh-VAL-is) A region in the interatrial septum of the heart that represents the region of the foramen ovale in the fetus

Fovea (FOE-vee-ah) A small pit or depression

Fovea centralis (FOE-vee-ah sen-TRAL-is) Depression in the center of the macula lutea where vision is sharpest because the cones are most numerous at this location

Free nerve ending (FREE NERV END-ing) A nerve ending that has no connective tissue covering; responds to pain

Frenulum (FREN-yoo-lum) Membranous fold of tissue that loosely attaches the tongue to the floor of the mouth

Frontal (FRUN-tal) Pertaining to the forehead region

Frontal bone (FRUN-tal BONE) Bone of the cranium that forms the forehead

Frontal plane (FRUN-tal PLANE) A vertical plane that extends from side to side and divides the body or part into anterior and posterior portions; also called a coronal plane

Functional residual capacity (FRC) (FUNK-shun-al ree-ZID-yoo-al kah-PASS-ih-tee) Amount of air remaining in the lungs after a tidal expiration; equals residual volume plus expiratory reserve volume

Funiculi (fuh-NIK-yoo-lee) Regions of white matter in the spinal cord; also called columns

G

Gametes (GAM-eets) Sex cells; sperm and ova

Gamma-aminobutyric acid (GABA) (gam-mah-ah-meen-oh-byoo-TEER-ik AS-id) One of several neurotransmitters

Ganglion (GANG-lee-on) A group of nerve cell bodies that lie outside the central nervous system; plural, ganglia

Genotype (JEE-noh-typ) The specific alleles that are present for a given trait

Gaster (GAS-ter) The fleshy part of a muscle; also called the belly

Gastrointestinal tract (gas-troh-in-TEST-ih-nal TRACT) Long, continuous tube that extends from the mouth to the anus; the digestive tract or alimentary canal

Gene (JEEN) Portion of a DNA molecule that contains the genetic information for making one particular protein molecule

Gestation (jes-TAY-shun) Time of prenatal development or pregnancy

Gingiva (JIN-jih-vah) The soft tissue that covers the alveolar processes of the mandible and maxillae; also called gums

Glans penis (glanz PEE-nis) Distal end of the penis through which the urethra opens to the exterior

Glenoid cavity (GLEN-oyd KAV-ih-tee) A large depression on the lateral side of the scapula that articulates with the head of the humerus to form the shoulder joint

Globulin (GLOB-yoo-lin) One type of protein in the blood plasma

Glomerular capsule (gloh-MER-yoo-lar KAP-sool) Double-layered epithelial cup that surrounds the glomerulus in a nephron; also called Bowman's capsule

Glomerular filtration (gloh-MER-yoo-lar fil-TRAY-shun) The movement of blood plasma across the filtration membrane in the renal corpuscle

Glomerulus (gloh-MER-yoo-lus) Cluster of capillaries in the nephron through which blood is filtered

Glossopharyngeal nerve (glos-so-fah-RIN-jee-al NERV) Cranial nerve IX; mixed nerve; responsible for taste sensations and for stimulating the muscles used in swallowing

Glucagon (GLOO-kah-gahn) A hormone secreted by the alpha cells of the pancreatic islets; increases the glucose level in the blood

Glucocorticoids (gloo-koh-KOR-tih-koyds) A group of hormones secreted by the adrenal cortex that regulate carbohydrate and fat metabolism; primary glucocorticoid is cortisol

Gluconeogenesis (gloo-koh-nee-oh-JEN-eh-sis) Process of forming glucose from noncarbohydrate nutrient sources such as proteins and lipids

Gluteal (GLOO-tee-al) Pertaining to the buttock region

Glycogen (GLY-koh-jen) Complex polysaccharide that is the storage form of energy in animals

Glycogenesis (gly-koh-JEN-eh-sis) Series of reactions that convert glucose or other monosaccharides into glycogen for storage

Glycogenolysis (gly-koh-jen-AHL-ih-sis) Series of reactions that convert glycogen into glucose

Glycolysis (gly-KAHL-ih-sis) Anaerobic series of reactions that produces two molecules of pyruvic acid from one molecule of glucose; first series of reactions in the catabolism of glucose

Goiter (GOY-ter) An enlargement of the thyroid gland

Golgi apparatus (GOAL-jee ap-ah-RA-tus) Membranous sacs within the cytoplasm that process and package cellular products

Golgi tendon organ (GOAL-jee TEN-don OAR-gan) A receptor, usually found near the junction of a muscle and tendon, that is stimulated by changes in muscle length or tension; a type of proprioceptor

Gonadocorticoids (go-nad-oh-KOR-tih-koyds) Sex hormones secreted by the adrenal cortex

Gonadotropic hormones (go-nad-oh-TROH-pik HOAR-moans) Hormones secreted by the anterior pituitary gland that stimulate the ovaries or testes; also called gonadotropins

Gonads (GO-nads) Primary reproductive organs; organs that produce the gametes: testes in the male and ovaries in the female

Granulocyte (GRAN-yoo-loh-syte) White blood cell that has granules in the cytoplasm

Granulosa cells (gran-yoo-LOH-sah SELZ) Cells that surround an oocyte during the development of ovarian follicles

Gray commissure (GRAY KOM-ih-shur) Narrow strip of gray matter that connects the two larger regions of gray matter in the spinal cord

Greater vestibular glands (GRAY-ter ves-TIB-yoo-lar GLANDS) Female accessory glands located adjacent to the vaginal orifice; also called Bartholin's glands

Growth hormone (GH) (GROWTH HOAR-moan) A hormone secreted by the anterior pituitary gland that influences the rate of skeletal growth; also called somatotropic hormone

Gustatory (GUS-tah-toar-ee) Refers to taste

Gustatory cell (GUS-tah-toar-ee SELL) Specialized chemoreceptor in the tongue for the sense of taste

Gyrus (JYE-rus) One of the raised folds on the surface of the cerebrum; also called convolution; plural, gyri (JYE-rye)

H

Hair follicle (HAIR FAH-lih-kal) Cells that surround the root of the hair

Hair root (HAIR ROOT) The part of the hair that is below the surface of the skin

Hair shaft (HAIR SHAFT) The visible part of hair

Hard palate (HARD PAL-at) Anterior portion of the floor of the nasal cavity and roof of the mouth; is supported by bone

Haustra (HAWS-trah) A series of pouches along the length of the colon

Haversian canal (hah-VER-shun kah-NAL) Central canal in a haversian system, contains blood vessels and nerves; also called osteonic canal

Haversian system (hah-VER-shun SIS-tem) Structural unit of bone consisting of concentric rings of cells and matrix around a central canal; also called osteon

Head (HED) Enlarged, often rounded, end of a bone

Hematocrit (hee-MAT-oh-krit) The percentage of red blood cells in a given volume of blood

Hematopoiesis (hee-mat-oh-poy-EE-sis) Blood cell production; occurs in the red bone marrow; also called hemopoiesis

Hemocytoblast (hee-moh-SYTE-oh-blast) A stem cell in the bone marrow from which the blood cells arise

Hemoglobin (hee-moh-GLOH-bin) The iron-containing protein in red blood cells that is responsible for the transport of oxygen

Hemolysis (hee-MAHL-ih-sis) The escape of hemoglobin from a red blood cell into the surrounding medium, usually caused by rupture of the cell

Hemopoiesis (hee-moh-poy-EE-sis) Blood cell production; occurs in the red bone marrow; also called hematopoiesis

Hemostasis (hee-moh-STAY-sis) The stoppage of bleeding

Heparin (HEP-ah-rihn) A substance that inhibits blood clotting

Hepatic flexure (heh-PAT-ik FLEK-shur) Curve between the ascending colon and transverse colon; also called the right colonic flexure

Hepatocytes (heh-PAT-oh-sytes) Liver cells

Hering-Breuer reflex (HER-ing BREW-er REE-fleks) Stretch reflex in the lungs that prevents overinflation of the lungs

Heterozygous (het-er-oh-ZYE-gus) Having different alleles for a given trait

Histamine (HISS-tah-meen) A substance that promotes inflammation

Histology (hiss-TAHL-oh-jee) Branch of microscopic anatomy that studies tissues

Holocrine gland (HOH-loh-krin GLAND) Gland in which the cells are discharged with the secretory product

Homeostasis (hoh-mee-oh-STAY-sis) A normal stable condition in which the body's internal environment remains the same; constant internal environment

Homeotherms (HOH-mee-oh-therms) Warm-blooded animals; have the ability to maintain a constant internal temperature

Homozygous (hom-moh-ZYE-gus) Having identical alleles for a given trait

Homunculus (hoh-MUNK-yoo-lus) An imaginary figure that represents the distribution of body regions in the primary sensory and motor areas of the cerebral cortex

Hormone (HOAR-moan) A substance secreted by an endocrine gland

Horn (HORN) Regions of gray matter in the spinal cord

Human chorionic gonadotropin (HCG) (HYOO-man kor-ee-ON-ik goh-nad-oh-TROH-pin) Hormone secreted by the trophoblast; it has an effect similar to luteinizing hormone and causes the corpus luteum to remain functional to maintain pregnancy

Humerus (HYOO-mur-us) Bone in the arm, or brachium

Humoral immunity (HYOO-mur-al ih-MYOO-nih-tee) Immunity that is the result of B-cell action and the production of antibodies; also called antibody-mediated immunity

Hyaline cartilage (HYE-ah-lihn KAR-tih-layj) Most abundant type of cartilage; appears glossy

Hydrogen bond (HYE-droh-jen BOND) Weak chemical bond that is formed between the partial positive charge on a covalently bound hydrogen atom and the partial negative charge on another covalent molecule

Hydrolysis (hye-DRAHL-ih-sis) Chemical breakdown of complex molecules by the addition of water

Hydrophilic (hye-droh-FILL-ik) "Water loving"; attracts water

Hydrophobic (hye-droh-FOE-bik) "Water fearing"; will not mix with water

Hydrostatic pressure (hye-droh-STAT-ik PRESH-ur) Pressure or force caused by a fluid

Hymen (HYE-men) Thin fold of mucous membrane that may cover the vaginal orifice

Hyoid (HYE-oyd) Bone in the neck, between the mandible and the larynx, that supports the tongue

Hyperextension (hye-per-eks-TEN-shun) A movement in which a part of the body is extended beyond the anatomical position

Hyperpolarization (hye-per-poh-lar-ih-ZAY-shun) An increase in the difference of electrical charges between the inside and outside of the cell membrane, which makes it more difficult to generate an action potential

Hypertonic (hye-per-TAHN-ik) A solution that has a greater concentration of solutes than another solution

Hypochondriac region (hye-poh-KAHN-dree-ak REE-jun) The upper lateral portions of the abdomen, on either side of the epigastric region

Hypodermis (hye-poh-DER-miss) Below the skin; a sheet of areolar connective tissue and adipose beneath the dermis of the skin; also called the subcutaneous layer or superficial fascia

Hypogastric region (hye-poh-GAS-trik REE-jun) The central portion of the abdomen inferior to the umbilical region

Hypoglossal nerve (hye-poh-GLAHS-al NERV) Cranial nerve XII; motor nerve; responsible for tongue movements

Hypophysis (hye-PAH-fih-sis) See Pituitary gland

Hypothalamus (HYE-poh-thal-ah-mus) A small region, located just below the thalamus in the diencephalon, that is important in autonomic and neuroendocrine functions

Hypotonic (hye-poh-TAHN-ik) A solution that has a lesser concentration of solutes than another solution

I

Ileocecal junction (ill-ee-oh-SEE-kul JUNK-shun) Connection between the small intestine and large intestine

Ileum (ILL-ee-um) Terminal portion of the small intestine

Iliac region (ILL-ee-ak REE-jun) The lateral inferior portions of the abdomen, on either side of the hypogastric region; also called the inguinal region

Iliosacral (ill-ee-oh-SAY-kral) Pertaining to the ilium and sacrum

Ilium (ILL-ee-um) One of the parts of the os coxa or hip bone

Immunity (ih-MYOO-nih-tee) Specific defense mechanisms that provide resistance to invading pathogens

Immunoglobulins (ih-myoo-noh-GLAHB-yoo-lins) Substances produced by the body that inactivate or destroy another substance that is introduced into the body; antibodies

Implantation (im-plan-TAY-shun) Process by which the developing embryo becomes embedded in the uterine wall; usually takes about 1 week and is completed by the fourteenth day after fertilization

Incomplete protein (IN-kum-pleet PRO-teen) A protein that does not contain all of the essential amino acids

Incus (INK-us) One of the tiny bones in the middle ear, between the malleus and the stapes

Infancy (IN-fan-see) Period from the end of the first month to the end of the first year after birth

Inferior (in-FEER-ee-or) Lower; one part is below another; opposite of superior

Inferior colliculi (in-FEER-ee-or koh-LIK-yoo-lye) Two lower bodies of the corpora quadrigemina on the dorsal surface of the midbrain; function in auditory reflexes

Inferior vena cava (in-FEER-ee-or VEE-nah KAY-vah) Large vein that collects blood from all parts of the body inferior to the heart and returns it to the right atrium

Inflammation (in-flah-MAY-shun) Nonspecific defense mechanism that involves a group of responses to tissue irritants and is characterized by redness, heat, swelling, and pain

Ingestion (in-JEST-chun) The process of taking in food

Inguinal (IN-gwih-nal) Pertaining to the depressed region between the abdomen and the thigh; groin

Inguinal canal (IN-gwih-nal kah-NAL) A passageway in the abdominal wall that transmits the spermatic cord in the male and the round ligament of the uterus in the female

Inguinal region (IN-gwih-nal REE-jun) The lateral inferior portions of the abdomen, on either side of the hypogastric region; also called the iliac region

Inhalation (in-hah-LAY-shun) Process of taking air into the lungs; also called inspiration

Inhibitory transmission (in-HIB-ih-toar-ee trans-MIH-shun) Nerve impulse conduction in which the neurotransmitter causes hyperpolarization of the postsynaptic membrane and reduces the chances of generating an action potential

Inorganic compound (in-or-GAN-ik KAHM-pownd) Chemical components that do not contain both carbon and hydrogen

Inner cell mass (IN-ner SELL MASS) Cluster of cells at one side of a blastocyst; cells that develop into the embryo

Inner ear (IN-er EER) The internal ear or labyrinth; located in the temporal bone; contains the organs of hearing and balance

Insertion (in-SIR-shun) The end of a muscle that is attached to a relatively movable part; the end opposite the origin

Inspiration (in-spur-RAY-shun) Process of taking air into the lungs; also called inhalation

Inspiratory capacity (IC) (in-SPY-rah-tor-ee kah-PASS-ih-tee) Maximum amount of air that can be inhaled; equals tidal volume plus inspiratory reserve volume

Inspiratory reserve volume (IRV) (in-SPY-rah-tor-ee ree-ZERV VOL-yoom) Maximum amount of air that can be forcefully inhaled after a tidal inspiration

Insula (IN-sull-ah) A triangular area of the cerebral cortex that lies deep within the lateral sulcus, beneath the frontal, parietal, and temporal lobes and hidden from view; also called the island of Reil (RY-al)

Insulin (IN-suh-lin) A hormone secreted by the beta cells of the pancreatic islets; decreases the glucose level in the blood

Integumentary (in-teg-yoo-MEN-tar-ee) Pertaining to the skin and related structures

Interatrial septum (in-ter-AY-tree-al SEP-tum) The partition between the right atrium and left atrium

Intercalated disk (in-TER-kuh-lay-ted DISK) Specialized intercellular connection that appears as a dark band in cardiac muscle

Intercellular matrix (in-ter-SELL-yoo-lar MAY-triks) Nonliving material that fills the spaces between cells in a tissue

Interferon (in-ter-FEER-on) A substance that is produced by the body in response to the presence of a virus, which, in turn, offers some protection against that virus by inhibiting its multiplication

Internal nares (in-TER-nal NAY-reez) Openings from the nasal cavity into the pharynx

Internal respiration (in-TER-nal res-per-RAY-shun) Exchange of gases between the blood and tissue cells

Interneuron (in-ter-NOO-ron) Nerve cell, totally within the central nervous system, that carries impulses from a sensory neuron to a motor neuron; also called association neuron

Interphase (IN-ter-faze) The period of the cell cycle between active cell divisions

Interstitial cells (in-ter-STISH-al SELZ) Cells between the seminiferous tubules in the testes; produce testosterone; also called cells of Leydig

Interstitial cell-stimulating hormone (ICSH) (in-ter-STISH-al SELL STIM-yoo-lay-ting HOAR-moan) Hormone secreted by the anterior pituitary gland in males; is the same as luteinizing hormone in females; one of the gonadotropins

Interstitial fluid (in-ter-STISH-al FLOO-id) Portion of the extracellular fluid that is found in the microscopic spaces between cells

Interventricular foramen (in-ter-ven-TRICK-yoo-lar foh-RAY-men) Opening between the lateral and third ventricles in the brain; also called the foramen of Monro

Interventricular septum (in-ter-ven-TRIK-yoo-lar SEP-tum) The partition between the right ventricle and left ventricle

Intraalveolar pressure (in-trah-al-VEE-oh-lar PRESH-ur) Pressure inside the alveoli of the lungs; also called intrapulmonary pressure

Intracellular (in-trah-SELL-yoo-lar) Inside the cell

Intracellular fluid (ICF) (in-trah-SELL-yoo-lar FLOO-id) The fluid inside body cells

Intramembranous ossification (in-trah-MEM-bran-us ah-sih-fih-KAY-shun) Method of bone formation in which the bone is formed directly in a membrane

Intrapleural pressure (in-trah-PLOO-ral PRESH-ur) Pressure within the pleural cavity, between the visceral and parietal layers of the pleura

Intrapulmonary pressure (in-trah-PUL-mon-air-ee PRESH-ur) Pressure inside the alveoli of the lungs; also called intraalveolar pressure

Intravascular fluid (in-trah-VAS-kyoo-lar FLOO-id) Portion of extracellular fluid that is in the blood; plasma

Intrinsic factor (in-TRIN-sik FAK-tor) A substance produced by cells in the stomach lining that facilitates the absorption of vitamin B_{12}

Inversion (in-VER-zhun) Act of turning a body part inward; usually refers to movement of the sole of the foot inward or medially; opposite of eversion

Ion (EYE-on) Electrically charged atom or group of atoms; an atom that has gained or lost one or more electrons

Ionic bond (eye-ON-ik BOND) Chemical bond that is formed when one or more electrons are transferred from one atom to another

Ionic compound (eye-ON-ik KAHM-pownd) Chemical compounds that are formed by the force of attraction between cations and anions

Iris (EYE-rihs) The colored portion of the eye that is seen through the cornea; contains smooth muscle that regulates the size of the pupil; part of the middle tunic of the eye

Ischium (IS-kee-um) One of the parts of the os coxa or hip bone

Island of Reil (EYE-land of RY-al) A triangular area of cerebral cortex that lies deep within the lateral sulcus, beneath the frontal, parietal, and temporal lobes and hidden from view; also called the insula

Islets of Langerhans (EYE-lets of LAHNG-er-hanz) See Pancreatic islets

Isometric contraction (eye-so-MET-rik kon-TRACK-shun) Type of muscle contraction in which the muscle tension increases but no movement is produced

Isotonic (eye-soh-TAHN-ik) A solution that has the same concentration of solutes as another solution

Isotonic contraction (eye-soh-TAHN-ik kon-TRACK-shun) Type of muscle contraction in which the muscle maintains the same tension and there is movement between two parts

Isotope (EYE-so-tohp) Atoms of a given element that have different numbers of neutrons and consequently different atomic weights

Isotropic (eye-soh-TROH-pik) Bands in skeletal and cardiac muscle fibers that appear light; light passes through

J

Jejunum (jeh-JOO-num) Middle portion of the small intestine

Jugular notch (JUG-yoo-lar NOTCH) Depression on the superior margin of the manubrium of the sternum

Juxtaglomerular apparatus (juks-tah-gloh-MER-yoo-lar ap-pah-RAT-us) Complex of modified cells in the afferent arteriole and the ascending limb/distal tubule in the kidney; helps regulate blood pressure by secreting renin; consists of the macula densa and juxtaglomerular cells

K

Keratin (KER-ah-tin) Hard, fibrous protein found in the epidermis, hair, and nails

Keratinization (ker-ah-tin-ih-ZAY-shun) Process by which the cells of the epidermis become filled with keratin and move to the surface, where they are sloughed off

Kidney (KID-nee) Organ of the urinary system that filters the blood and functions to maintain the homeostasis of body fluids

Kilocalorie (KILL-oh-kal-or-ee) Unit of heat energy used in metabolic and nutritional studies; equivalent to 1000 standard calories; also called Calorie

Korotkoff sounds (koh-ROT-kof SOUNDS) The sounds heard in the stethoscope while taking blood pressure

Krebs cycle (KREBZ SYE-kul) Aerobic series of reactions that follows glycolysis in glucose metabolism to release energy and carbon dioxide; also called citric acid cycle

L

Labia majora (LAY-bee-ah mah-JOR-ah) Two large fat-filled folds of skin that enclose the other female external genitalia; homologous to the scrotum in the male

Labia minora (LAY-bee-ah mye-NOR-ah) Two small folds of skin medial to the labia majora

Labor (LAY-bor) Process by which forceful contractions expel the fetus from the uterus

Lacrimal (LAK-rih-mal) Small bone anterior to the ethmoid in the orbit of the eye

Lacrimal apparatus (LAK-rih-mal ap-pah-RAT-us) The structures that produce and convey tears

Lacrimal gland (LAK-rih-mal GLAND) A glandular structure in the superior and lateral region of the orbit that produces tears

Lactase (LAK-tays) An enzyme that acts on the disaccharide lactose and breaks it into a molecule of glucose and a molecule of galactose

Lactation (lak-TAY-shun) Milk production and ejection from the mammary glands

Lacteal (lak-TEEL) Lymph capillary found in the villi of the small intestine

Lactic acid (LAK-tik AS-id) A molecule that is formed from pyruvic acid when oxygen is lacking; product of anaerobic respiration in muscles during exercise

Lactiferous duct (lak-TIFF-er-us DUKT) Duct that collects milk from the lobules of a mammary gland and carries it to the nipple

Lacuna (lah-KOO-nah) Space or cavity; space that contains bone or cartilage cells; pleural, lacunae

Lamellae (lah-MEL-ee) Concentric rings of hard calcified matrix in bone

Lamellated corpuscle (lam-el-LAY-ted KOAR-pus-al) See Pacinian corpuscle

Laryngopharynx (lah-ring-go-FAIR-inks) Portion of the pharynx that is posterior to the larynx and extends from the level of the hyoid bone to the lower margin of the larynx

Larynx (LAIR-inks) Passageway for air between the pharynx and trachea; commonly called the voice box

Lateral (LAT-er-al) Toward the side, away from the midline; opposite of medial

Lateral sulcus (LAT-er-al SULL-kus) The deep groove between the temporal lobe below and the frontal and parietal lobes above; also called the lateral fissure or fissure of Sylvius

Left atrium (LEFT AY-tree-um) Chamber of the heart that receives oxygenated blood from the lungs through the pulmonary veins

Left ventricle (LEFT VEN-trih-kul) Chamber of the heart that pumps oxygenated blood through the aorta to the systemic circulation

Leg (LEG) Portion of the lower extremity between the knee and the foot; also called the crural region

Lens (LENZ) Transparent biconvex structure that is posterior to the iris in the eye and functions in the refraction of light rays

Leukocyte (LOO-koh-syte) White blood cell

Ligament (LIG-ah-ment) Band of dense fibrous connective tissue that connects one bone to another

Ligamentum arteriosum (lig-ah-MEN-tum ahr-teer-ee-OH-sum) The fibrous remnant that results from the atrophy of the ductus arteriosus

Ligamentum venosum (lig-ah-MEN-tum veh-NO-sum) The fibrous remnant that results from the atrophy of the ductus venosus

Liminal stimulus (LIM-ih-null STIM-yoo-lus) Minimum level of stimulation that is required to start a nerve impulse or muscle contraction; also called threshold stimulus

Lipid (LIP-id) A class of organic compounds that includes oils, fats, and related substances

Lipogenesis (lip-oh-JEN-eh-sis) Series of reactions in which lipids are formed from other nutrients

Longitudinal fissure (lonj-ih-TOO-dih-null FISH-ur) Deep groove that divides the cerebrum into two halves

Lower respiratory tract (LOW-er res-PYE-rah-tor-ee TRAKT) Portion of the respiratory tract that is below the larynx, including the trachea, bronchial tree, and lungs

Lumbar region (LUM-bar REE-jun) The middle lateral portions of the abdomen, on either side of the umbilical region

Lunula (LOO-nyoo-lah) The small curved white area at the base of a nail

Luteinizing hormone (LH) (LOO-ten-eye-zing HOAR-moan) Hormone secreted by the anterior pituitary gland in both males and females; one of the gonadotropins; in males, it may be called interstitial cell-stimulating hormone

Lymph (LIMF) Fluid that is derived from interstitial fluid and found in the lymphatic vessels

Lymph capillary (LIMF KAP-ih-lair-ee) Smallest lymphatic vessel that picks up lymph from the interstitial fluid for its return to the circulating blood

Lymph node (LIMF NODE) A small, bean-shaped aggregate of lymphoid tissue along a lymphatic vessel that filters the lymph before it is returned to the blood circulation

Lymphatic (lim-FAT-ik) Relating to lymph

Lymphocyte (LIM-foh-syte) A type of white blood cell that lacks granules in the cytoplasm and has an important role in immunity

Lysosome (LYE-so-sohm) Membrane-enclosed sac of digestive enzymes within the cytoplasm

Lysozyme (LYE-soh-zyme) An enzyme found in tears, saliva, and perspiration that inhibits bacteria

M

Macrophage (MAK-roh-fahj) Large phagocytic connective tissue cell that functions in immune responses; name given to a monocyte after it leaves the blood and enters the tissues

Macula (MAK-yoo-lah) The structure in the utricle and saccule that detects a change in position of the head; functions in static equilibrium

Macula lutea (MAK-yoo-lah LOO-tee-ah) Yellowish spot near the center of the retina where the cones are concentrated

Malleus (MAL-lee-us) One of the tiny bones in the middle ear, adjacent to the tympanic membrane

Maltase (MAWL-tays) An enzyme that acts on the disaccharide maltose and breaks it into two molecules of glucose

Mammary (MAM-ah-ree) Pertaining to the breast

Mammary glands (MAM-ah-ree GLANDS) Organs of milk production located within the breast

Mandible (MAN-dih-bul) Bone of the lower jaw

Mandibular (man-DIB-yoo-lar) Relating to the mandible

Manubrium (mah-NOO-bree-um) Upper portion of the sternum

Mass number (MASS NUM-bur) The total number of protons and neutrons in the nucleus of an atom of an element

Mast cell (MAST SELL) A connective tissue cell that produces heparin and histamine; the name given to a basophil after it leaves the blood and enters the tissues

Mastication (mas-tih-KAY-shun) The process of chewing

Maxillae (maks-ILL-ee) Bones of the upper jaw

Meatus (mee-ATE-us) A tubelike passageway through a bone; a tunnel or canal

Mechanoreceptor (mek-ah-noh-ree-SEP-tor) A sensory receptor that responds to a bending or deformation of the cell; examples include receptors for touch, pressure, hearing, and equilibrium

Medial (MEE-dee-al) Toward the middle; opposite of lateral

Medulla (meh-DOO-lah or meh-DULL-ah) The inner portion of an organ

Medulla oblongata (meh-DOO-lah ahb-long-GAH-tah) Lowest part of the brain stem; it contains the vital cardiac center, vasomotor center, and respiratory center

Medullary cavity (MED-yoo-lair-ee KAV-ih-tee) Space in the shaft of a long bone that contains yellow marrow

Megakaryocyte (meg-ah-KAIR-ee-oh-syte) A large cell that contributes to the formation of platelets

Meiosis (my-OH-sis) Type of nuclear division in which the number of chromosomes is reduced to one half the number found in a body cell; results in the formation of an egg or sperm

Meissner's corpuscle (MYS-nerz KOAR-pus-al) A sensory receptor near the surface of the skin that detects light touch; also called corpuscle of touch

Meissner's plexus (MYS-nerz PLEK-sus) A network of autonomic nerves in the submucosa of the gastrointestinal tract; also called submucosal plexus

Melanin (MEL-ah-nin) A dark brown or black pigment found in parts of the body, especially the skin and hair

Melanocyte (meh-LAN-oh-syte) Specialized cell that produces melanin; found in the stratum basale of the epidermis

Melatonin (mell-ah-TOH-nihn) Hormone produced by the pineal body; regulates the body's internal clock and daily rhythms; responds to varying light levels; regulates onset of puberty and menstrual cycle

Membrane potential (MEM-brayn po-TEN-shall) The difference in electrical charge between inside and outside a cell membrane

Membranous labyrinth (MEM-brah-nus LAB-ih-rinth) A series of membranes located within the bony labyrinth and separated from it by perilymph; includes the cochlear duct, the utricle and saccule, and the membranous semicircular canals (ducts)

Membranous urethra (MEM-brah-nus yoo-REE-thrah) Portion of the male urethra that passes through the membranous pelvic floor

Menarche (meh-NAHR-kee) First period of menstrual bleeding at puberty

Meninges (meh-NIN-jeez) Connective tissue membranes that cover the brain and spinal cord

Meningitis (meh-nin-JYE-tis) Inflammation of the meninges

Meniscus (meh-NIS-kus) Fibrocartilaginous pads found in certain freely movable joints

Menopause (MEN-oh-pawz) Cessation of menstrual bleeding; termination of uterine cycles

Menses (MEN-seez) Periodic shedding of the stratum functionale of the uterine lining; menstruation

Menstrual cycle (MEN-stroo-al SYE-kul) Monthly cycle of events that occur in the uterus from puberty to menopause; also called the uterine cycle; occurs concurrently with the ovarian cycle

Menstruation (men-stroo-AY-shun) Periodic shedding of the stratum functionale of the uterine lining; menses

Merocrine gland (MER-oh-krin GLAND) Gland that discharges its secretions directly through the cell membrane

Mesentery (MEZ-en-tair-ee) Extensions of peritoneum that are associated with the intestine

Mesoderm (MEZ-oh-derm) The middle of the three primary germ layers; it develops into connective tissues, muscles, bones, and blood

Messenger RNA (mRNA) (MES-en-jer RNA) A molecule of RNA that transfers information for protein synthesis from the DNA in the nucleus to the cytoplasm

Metabolism (meh-TAB-oh-lizm) The total of all biochemical reactions that take place in the body; includes anabolism and catabolism

Metacarpals (meh-tah-KAR-pulls) Five bones that form the palm of the hand

Metacarpus (meh-tah-KAR-pus) Palm of the hand, which consists of five metacarpal bones

Metaphase (MET-ah-faze) Second stage of mitosis; stage in which visible chromosomes become aligned along the center of the cell

Metarteriole (met-ahr-TEER-ee-ohl) A microscopic vessel that directly connects an arteriole to a venule without an intervening capillary network; an arteriovenous shunt

Metatarsals (meh-tah-TAHR-sahls) Five bones that form the instep of the foot

Metatarsus (meh-tah-TAHR-sis) Instep of the foot, which consists of five metatarsal bones

Micelles (my-SELZ) Tiny droplets of monoglycerides and free fatty acids that are coated with bile salts

Microfilaments (my-kroh-FIL-ah-ments) Long, slender rods of protein within a cell

Microglia (my-kroh-GLEE-ah) Neuroglial cell that is capable of phagocytosis

Microtubules (my-kroh-TOOB-yools) Thin cylinders of protein within a cell; composed of the protein tubulin (TOOB-yoo-lin)

Microvillus (my-kroh-VIL-us) Small projection of the cell membrane that is supported by microfilaments; plural, microvilli

Micturition (mik-too-RISH-un) Act of expelling urine from the bladder; also called urination or voiding

Midbrain (MID-brayn) Region of the brain stem between the diencephalon and the pons

Middle ear (MID-dull EER) Small epithelial-lined cavity in the temporal bone that contains the three auditory ossicles; also called the tympanic cavity

Midsagittal plane (mid-SAJ-ih-tal PLANE) A vertical plane that divides the body or organ into equal right and left parts; a sagittal plane that is in the midline; also called the median plane

Mineralocorticoids (min-er-al-oh-KOR-tih-koyds) A group of hormones secreted by the adrenal cortex that regulates electrolyte balance in the body; primary mineralocorticoid is aldosterone

Minerals (MIN-er-als) Inorganic substances that are needed in minute amounts in the diet to maintain growth and good health but do not supply energy

Mitochondria (my-tohe-KAHN-dree-ah) Organelles that contain the enzymes essential for producing ATP; singular, mitochondrion

Mitosis (my-TOH-sis) Process by which the nucleus of a body cell divides to form two new cells, each identical to the parent cell

Mitral valve (MY-tral VALVE) Valve between the left atrium and left ventricle; also called the bicuspid valve

Mixed nerve (MIK-st NERV) A nerve that contains both sensory and motor fibers

Mixture (MIX-chur) A combination of two or more substances that can be separated by ordinary physical means

Molecule (MAHL-eh-kyool) A particle composed of two or more atoms that are chemically bound together; smallest unit of a compound

Monocyte (MAHN-oh-syte) A type of white blood cell that lacks granules in the cytoplasm and is capable of phagocytosis

Monosaccharide (mahn-oh-SAK-ah-ride) Building block of carbohydrates; a simple sugar such as glucose, fructose, and galactose

Mons pubis (MAHNZ PYOO-bis) Rounded elevation of fat that overlies the pubic symphysis in females

Morula (MOR-yoo-lah) Solid ball of cells formed by early mitotic divisions of a zygote, usually present by the end of the third day after fertilization

Motor nerve (MOH-toar NERV) A nerve that contains primarily motor, or efferent, fibers

Motor neuron (MOH-toar NOO-ron) Nerve cell that carries impulses away from the central nervous system toward the periphery; efferent neuron

Motor unit (MOH-toar YOO-nit) A single neuron and all the muscle fibers it stimulates

Mucosa (MYOO-koh-sah) Epithelial membranes that secrete mucus and line body cavities that open directly to the exterior; also called mucous membranes

Mucous cell (MYOO-cus SELL) Cell that secretes a thick fluid called mucus

Mucous membrane (MYOO-cus MEM-brayn) Epithelial membrane that secretes mucus and lines body cavities that open directly to the exterior; also called mucosa

Multicellular (muhl-tih-SELL-yoo-lar) Consisting of many cells

Multiple motor unit summation (MUHL-tih-pul MOH-toar YOO-nit sum-MAY-shun) Type of response in which numerous motor units are stimulated simultaneously, which increases contraction strength

Multiple wave summation (MUHL-tih-pul WAYV sum-MAY-shun) Type of response in which stimuli are so rapid that the muscle is not able to relax completely between successive stimuli and sustained or more forceful contractions result

Muscle spindle (MUSS-el SPIN-dull) A receptor in skeletal muscles that is stimulated by changes in muscle length or tension; a type of proprioceptor

Muscle tone (MUSS-el TOAN) Sustained partial muscle contraction; produces a constant tension in the muscles

Muscular (MUSS-kyoo-lar) Relating to the muscles

Myelin (MY-eh-lin) White, fatty substance that surrounds many nerve fibers

Myenteric plexus (my-en-TAIR-ik PLEK-sus) A network of autonomic nerves between the circular and longitudinal muscle layers of the gastrointestinal tract; also called Auerbach's plexus

Myocardium (my-oh-KAR-dee-um) Middle layer of the heart wall; composed of cardiac muscle tissue

Myofibrils (my-oh-FYE-brills) Threadlike structures that run longitudinally through muscle cells and are composed of actin and myosin myofilaments

Myofilaments (my-oh-FILL-ah-ments) Ultramicroscopic, threadlike structures in the myofibrils of muscle cells; composed of the contractile proteins actin and myosin

Myoglobin (my-oh-GLOH-bin) The iron-containing protein in the sarcoplasm of muscle cells that binds with oxygen and stores it; gives the red color to muscle

Myometrium (my-oh-MEE-tree-um) Thick middle layer of the uterus; it is composed of smooth muscle

Myoneural junction (my-oh-NOO-ral JUNK-shun) The area of communication between the axon terminal of a motor neruon and the sarcolemma of a muscle fiber; also called a neuromuscular junction

Myosin (MY-oh-sin) Contractile protein in the thick filaments of skeletal muscle cells

Myxedema (mik-seh-DEE-mah) Condition caused by insufficient thyroid hormone in adults

N

Nares (NAY-reez) Openings of the nasal cavity

Nasal conchae (NAY-zal KONG-kee) Bony ridges that project medially into the nasal cavity from the lateral walls of the nasal cavity

Nasopharynx (nay-zo-FAIR-inks) Portion of the pharynx that is posterior to the nasal cavity and extends from the base of the skull to the uvula

Natural immunity (NAT-yoor-al ih-MYOO-nih-tee) Immunity acquired through normal processes of daily living

Negative feedback (NEG-ah-tiv FEED-bak) A mechanism of response in which a stimulus initiates reactions that reduce the stimulus

Neonatal period (nee-oh-NAY-tal PEER-ee-ud) The first month after birth

Neonate (NEE-oh-nayt) Term for a baby during the neonatal period, or the first month after birth

Nephron (NEFF-rahn) Functional unit of the kidney consisting of a renal corpuscle and a renal tubule

Nerve tracts (NERV TRAKTS) Bundles of myelinated nerve fibers in the spinal cord

Nervous (NER-vus) Relating to the nerves and brain

Nervous tissue (NER-vus TISH-yoo) Specialized tissue found in the nerves, brain, and spinal cord

Nervous tunic (NERV-us TOO-nik) The innermost layer of the eyeball; also called the retina

Neurilemma (noo-rih-LEM-mah) The layer of Schwann cells that surrounds a nerve fiber in the peripheral nervous system and, in some cases, produces myelin; also called Schwann's sheath

Neuroglia (noo-ROG-lee-ah) Supporting cells of nervous tissue; cells in nervous tissue that do not conduct impulses

Neurohypophysis (noo-roh-hye-PAH-fih-sis) Posterior portion of the pituitary gland

Neuromuscular junction (noo-roe-MUSK-yoo-lar JUNK-shun) The area of communication between the axon terminal of a motor neuron and the sarcolemma of a muscle fiber; also called a myoneural junction

Neuron (NOO-ron) Nerve cell, including its processes; conducting cell of nervous tissue

Neuronal pool (noo-ROH-nal POOL) Functional group of neurons within the central nervous system that receives information, processes and integrates that information, and then transmits it to some other destination

Neurotransmitter (noo-roh-TRANS-mit-ter) A chemical substance that is released at the axon terminals to stimulate muscle fiber contraction or an impulse in another neuron

Neutron (NOO-tron) An electrically neutral particle found in the nucleus of an atom

Neutrophil (NOO-troh-fill) A type of white blood cell that has granules in the cytoplasm that stain with acidic and basic dyes and is capable of phagocytosis

Nociceptor (noh-see-SEP-tor) A sensory receptor that responds to tissue damage; pain receptor

Node of Ranvier (NODE OF rahn-vee-AY) Short space between two segments of myelin in a myelinated nerve fiber

Nonessential amino acids (NON-ee-sen-chul ah-MEEN-oh AS-ids) Amino acids that can be synthesized in the human body

Nonspecific resistance (non-speh-SIF-ik ree-SIS-tans) Body's ability to counteract all types of harmful agents

Noradrenalin (nor-ah-DREN-ah-lihn) A hormone secreted by the adrenal medulla that produces effects similar to epinephrine and the sympathetic nervous system; neurotransmitter released by some neurons in the sympathetic nervous system; also called norepinephrine

Norepinephrine (nor-ep-ih-NEFF-rihn) A hormone secreted by the adrenal medulla that produces effects similar to epinephrine and the sympathetic nervous system; neurotransmitter released by some neurons in the sympathetic nervous system; also called noradrenaline

Nostrils (NAHS-trils) Openings through which air enters the nasal cavity; external nares

Nucleolus (noo-KLEE-oh-lus) A dense, dark-staining body within the nucleus; contains a high concentration of RNA

Nucleoplasm (NOO-klee-oh-plazm) The fluid inside the nucleus of a cell

Nucleotide (NOO-klee-oh-tide) Building block of nucleic acids; consists of a pentose sugar, an organic nitrogenous base, and a phosphate group

Nucleus (NOO-klee-us) Largest structure within the cell; contains the DNA

Nutrient (NOO-tree-ent) A chemical substance that provides energy, forms new body components, or assists in the metabolic processes of the body

Nutrient foramen (NOO-tree-ent for-A-men) Small opening in the diaphysis of bone for passage of blood vessels

Nutrition (noo-TRIH-shun) Science that studies the relationship of food to the functioning of the living organism; acquisition, assimilation, and use of nutrients contained in food

O

Occipital (ahk-SIP-ih-tal) Pertaining to the lower portion of the back of the head

Occipital bone (ahk-SIP-ih-tal BONE) Bone that forms the back of the skull and the base of the cranium

Oculomotor nerve (ahk-yoo-loh-MOH-tor NERV) Cranial nerve III; motor nerve; controls eye movements

Odontoid (oh-DON-toyd) Tooth-shaped projection on the second cervical vertebra; also called dens

Olfaction (ohl-FAK-shun) Sense of smell

Olfactory (ohl-FAK-toar-ee) Relating to the sense of smell

Olfactory bulb (ohl-FAK-toar-ee BULB) Mass of gray matter on either side of the crista galli of the ethmoid bone where olfactory neurons synapse

Olfactory cortex (ohl-FAK-toar-ee KOR-teks) Region in the temporal lobe where the sense of smell is interpreted

Olfactory epithelium (ohl-FAK-toar-ee ep-ih-THEE-lee-um) Tissue in the mucous membrane of the upper portion of the nasal cavity; contains bipolar neurons and supporting cells

Olfactory nerve (ohl-FAK-toar-ee NERV) Cranial nerve I; sensory nerve; responsible for the sense of smell

Olfactory neuron (ohl-FAK-toar-ee NOO-ron) Bipolar neuron in the upper portion of the nasal cavity that converts odors into neural signals

Olfactory tract (ohl-FAK-toar-ee TRACT) A bundle of axons that extends from the olfactory bulb to the cortex in the temporal lobe where the sense of smell is interpreted

Oligodendrocyte (ah-lee-go-DEN-droh-site) A neuroglial cell that produces myelin within the central nervous system

Oogenesis (oh-oh-JEN-eh-sis) Process of meiosis in the female in which one ovum and three polar bodies are produced from one primary oocyte

Oogonia (oh-oh-GO-nee-ah) Stem cells that give rise to ova or egg cells

Ophthalmic (off-THAL-mik) Pertaining to the eyes

Opsin (OP-sin) The protein component of the pigment in rods

Optic chiasma (OP-tik kye-AZ-mah) Region, just anterior to the pituitary gland, where the right and left optic nerves meet and some fibers cross over to the opposite side

Optic disk (OP-tik DISK) Area in the retina where the optic nerve fibers leave the eye and there are no rods and cones; also called the blind spot

Optic nerve (OP-tik NERV) Cranial nerve II; sensory nerve; conducts visual information to the brain

Optic radiations (OP-tik ray-dee-AY-shuns) Nerve fibers in the visual pathway between the thalamus and the visual cortex in the occipital lobe

Optic tract (OP-tik TRAKT) Bundles of fibers in the visual pathway between the optic chiasma and the thalamus

Oral (OH-ral or AW-ral) Pertaining to the mouth

Organ (OR-gan) Group of tissues that work together to perform a specific function

Organ of Corti (OR-gan OV KOAR-tee) The organ of hearing, consisting of supporting cells and hair cells that rest on the basilar membrane and project into the endolymph of the cochlear duct; also called spiral organ

Organelles (or-guh-NELZ) Little organs; highly organized structures suspended in the cytoplasm that are specialized to perform specific cellular activities

Organic compound (or-GAN-ik KAHM-pownd) Chemical components that contain carbon and hydrogen atoms covalently bonded together

Organism (OR-gan-izm) A living entity

Organogenesis (or-gan-oh-JEN-eh-sis) The process of organ formation in the embryo

Origin (OR-ih-jin) The end of a muscle that is attached to a relatively immovable part; the end opposite the insertion

Oropharynx (or-oh-FAIR-inks) Portion of the pharynx that is posterior to the oral cavity and extends from the uvula to the level of the hyoid bone

Osmosis (os-MOH-sis) Diffusion of water through a selectively permeable membrane

Osseous tissue (AS-see-us TISH-yoo) Bone tissue; rigid connective tissue

Ossification (ah-sih-fih-KAY-shun) Formation of bone; also called osteogenesis

Osteoblast (AH-stee-oh-blast) Bone-forming cell

Osteoclast (AH-stee-oh-clast) Cell that destroys or resorbs bone tissue

Osteocyte (AH-stee-oh-syte) Mature bone cell

Osteogenesis (AH-stee-oh-jen-eh-sis) Formation of bone; also called ossification

Osteon (AH-stee-ahn) Structural unit of bone; haversian system

Osteonic canal (AH-stee-ahn-ik kah-NAL) Central canal in an osteon, contains blood vessels and nerves; also called haversian canal

Otic (OH-tik) Pertaining to the ears

Otoliths (OH-toh-liths) Calcium carbonate particles associated with the macula in the utricle and the saccule in the inner ear; involved with static equilibrium

Oval window (OH-val WIN-dow) Small opening between the middle ear and the inner ear where the stapes fits; also called the fenestra vestibuli

Ovarian follicle (oh-VAIR-ee-an FAHL-ih-kul) An oocyte surrounded by one or more layers of cells within the ovaries

Ovarian cycle (oh-VAIR-ee-an SYE-kul) Monthly cycle of events that occur in the ovary from puberty to menopause; occurs concurrently with the uterine cycle

Ovaries (OH-vah-reez) Primary reproductive organs in the female; produce the ova or eggs

Oviducts (OH-vih-dukts) The tubes that extend laterally from the upper portion of the uterus to the region of the ovaries; also called uterine tubes or fallopian tubes

Ovulation (ah-vyoo-LAY-shun) The release of a secondary oocyte from a mature follicle at the surface of an ovary

Oxygen debt (AHKS-ee-jen DET) The amount of oxygen that must be supplied after physical exercise to convert the accumulated lactic acid into glucose

Oxyhemoglobin (ahk-see-HEE-moh-gloh-bin) Compound that is formed when oxygen binds with hemoglobin; form in which most of the oxygen is transported in the blood

Oxytocin (ahk-see-TOH-sin) Hormone produced by the hypothalamus and secreted from the posterior pituitary gland that causes uterine muscle contraction and ejection of milk from the lactating breast

P

P wave (P WAYV) Deflection on an electrocardiogram that corresponds to atrial depolarization

Pacinian corpuscle (pah-SIN-ee-an KOAR-pus-al) A sensory receptor deep in the dermis of the skin that detects pressure on the surface; also called a lamellated corpuscle

Palate (PAL-at) Floor of the nasal cavity; it separates the nasal cavity from the oral cavity

Palatine (PAL-ah-tyne) Bone that forms a portion of the roof of the mouth

Palmar (PAWL-mar) Pertaining to the palm of the hand

Pancreas (PAN-kree-ahs) A glandular organ in the abdominal cavity that has both exocrine and endocrine functions; the exocrine portion consists of acinar cells; the endocrine portion is the islets of Langerhans

Pancreatic islets (pan-kree-AT-ik EYE-lets) Endocrine portion of the pancreas; consist of alpha cells that secrete glucagon and beta cells that secrete insulin

Papillary layer (PAP-ih-lair-ee LAY-er) Upper layer of the dermis

Papillary muscle (PAP-ih-lair-ee MUSS-el) Projections of cardiac muscle that extend inward from the heart wall into the chambers of the ventricles

Parasagittal plane (pair-ah-SAJ-ih-tal PLANE) A sagittal plane that is not in the midline

Parasympathetic division (pair-ah-sim-pah-THET-ik dih-VIH-shun) One of two divisions of the autonomic nervous system; primarily concerned with processes that involve the conservation and restoration of energy; sometimes referred to as the "rest and repose" division; also called the craniosacral division

Parathyroid glands (pair-ah-THYE-royd GLANDS) A set of small glands embedded on the posterior aspect of the thyroid gland

Parathyroid hormone (PTH) (pair-ah-THYE-royd HOAR-moan) Hormone secreted by the parathyroid glands that functions to increase blood calcium levels; also called parathormone

Paraurethral glands (pair-ah-yoo-REE-thral GLANDS) Mucus-secreting glands located on each side of the urethral orifice in the female; also called Skene's glands

Paravertebral ganglion (pair-ah-ver-TEE-brull GANG-lee-on) A chain of autonomic ganglia, the sympathetic chain, that extends longitudinally along each side of the vertebral column; a region of synapse between the two neurons in a sympathetic pathway

Parietal (pah-RYE-eh-tal) Pertains to the wall of a body cavity

Parietal bone (pah-RYE-eh-tal BONE) Bone of the cranium immediately posterior to the frontal bone; forms the top of the head

Parietal cells (pah-RYE-eh-tal SELZ) Cells in the gastric mucosa that secrete hydrochloric acid and intrinsic factor

Parietal pericardium (pah-RYE-eh-tal pair-ih-KAR-dee-um) Layer of serous membrane that lines the fibrous sac around the heart

Parturition (par-too-RIH-shun) Act of giving birth to an infant

Passive immunity (PASS-iv ih-MYOO-nih-tee) Immunity that results when an individual receives the immune agents from some source other than his or her own body

Passive transport (PASS-iv TRANS-port) Membrane transport process that does not require cellular energy

Patella (pah-TELL-ah) Kneecap

Pectoral (PEK-toh-ral) Pertaining to the chest region

Pedal (PED-al) Pertaining to the foot

Pelvic (PEL-vik) Pertaining to the inferior region of the abdominopelvic cavity; lower portion of the trunk

Pelvic cavity (PEL-vik KAV-ih-tee) Inferior portion of the abdominopelvic cavity; contains the urinary bladder, part of the large intestine, and internal reproductive organs

Penile urethra (PEE-nye-al yoo-REE-thrah) Portion of the male urethra that is surrounded by corpus spongiosum and passes through the length of the penis; also called spongy urethra

Peptidase (PEP-tih-days) An enzyme that acts on protein segments called peptides

Peptide bond (PEP-tide BOND) The chemical bond that forms between two amino acids

Pericardial sac (pair-ih-KAR-dee-al SAK) Loose-fitting sac surrounding the heart that consists of fibrous connective tissue lined with a serous membrane

Pericardial cavity (pair-ih-KAR-dee-al KAV-ih-tee) Potential space between the parietal pericardium and visceral pericardium that contains a small amount of serous fluid for lubrication

Pericardium (pair-ih-KAR-dee-um) Membrane that surrounds the heart; usually refers to the pericardial sac

Perichondrium (pair-ih-KAHN-dree-um) Connective tissue covering that surrounds cartilage

Perilymph (PAIR-ih-limf) Fluid inside the bony labyrinth but outside the membranous labyrinth of the inner ear

Perimetrium (pair-ih-MEE-tree-um) Outermost layer of the uterus, which is visceral peritoneum

Perimysium (pair-ih-MY-see-um) Fibrous connective tissue that surrounds a bundle, or fasciculus, of muscle fibers (cells)

Perineal (pair-ih-NEE-al) Pertaining to the region between the anus and pubic symphysis; includes the region of the external reproductive organs

Perineurium (pair-ih-NOO-ree-um) Fibrous connective tissue that surrounds a bundle, or fasciculus, of nerve fibers

Periosteum (pair-ee-AH-stee-um) The tough, white outer membrane that covers a bone and is essential for bone growth, repair, and nutrition

Peripheral nervous system (PNS) (per-IF-er-al NER-vus SIS-tem) The portion of the nervous system that is outside of the brain and spinal cord; consists of the nerves and ganglia

Peripheral resistance (per-IF-er-al ree-SIS-tans) Opposition to blood flow caused by friction of the blood vessel walls

Perirenal fat (pair-ih-REE-nal FAT) Capsule of adipose tissue that surrounds and protects the kidney

Peristalsis (pair-ih-STALL-sis) Rhythmic contractions of the intestines that move food along the digestive tract

Peritoneum (pair-ih-toe-NEE-um) Serous membrane associated with the abdominopelvic cavity

Peritubular capillaries (pair-ih-TOOB-yoo-lar KAP-ih-lair-eez) Extensive capillary network around the tubular portions of the nephrons in the kidneys

Permanent teeth (PER-mah-nent TEETH) The teeth that replace the deciduous teeth; also called secondary teeth

Pernicious anemia (per-NISH-us ah-NEE-mee-ah) A type of anemia that is caused by a deficiency of intrinsic factor

Phagocytic (fag-oh-SIT-ik) Capable of phagocytosis in which solid particles are engulfed and taken in by a cell

Phagocytosis (fag-oh-sye-TOH-sis) Cell eating; a form of endocytosis in which solid particles are taken into the cell

Phalanges (fah-LAN-jeez) Bones of the fingers and toes; singular, phalanx (fah-LANKS)

Pharynx (FAIR-inks) Passageway for air and food that extends from the base of the skull to the larynx and esophagus; throat

Phenotype (FEE-noh-typ) The visible expression of the genes for a given trait

Phospholipid (fahs-foh-LIP-id) A fat molecule that contains phosphates; an important constituent of cell membranes

Photoreceptor (foh-toh-ree-SEP-tor) A sensory receptor that detects light; located in the retina of the eye

Physiology (fiz-ee-AHL-oh-jee) Study of the functions of living organisms and their parts

Pia mater (PEE-ah MAY-ter) The innermost layer of the coverings around the brain and spinal cord; literally means "soft mother"

Pineal body (PYE-nee-al BAH-dee) A region of the epithalamus in the diencephalon that is thought to be involved with regulating the "biological clock"; also called the pineal gland; secretes melatonin

Pineal gland See Pineal body

Pinealocytes (PYE-nee-al-oh-cytes) Secretory cells of the pineal body; secrete melatonin

Pinna (PIN-nah) Visible portion of the external ear; also called the auricle

Pinocytosis (pin-oh-sye-TOH-sis) Cell drinking; a form of endocytosis in which fluid droplets are taken into the cell

Pituitary gland (pih-TOO-ih-tair-ee GLAND) Endocrine gland located in the sella turcica of the sphenoid bone, near the base of the brain; also called the hypophysis

Placenta (plah-SEN-tah) Structure that anchors the developing fetus to the uterus and provides for the exchange of gases, nutrients, and waste products between the maternal and fetal circulations

Plantar (PLAN-tar) Pertaining to the sole of the foot

Plantar flexion (PLAN-tar FLEK-shun) A movement at the ankle that increases the angle between the foot and leg; opposite of dorsiflexion

Plasma (PLAZ-mah) Liquid portion of blood

Plasma cell (PLAZ-mah SELL) A cell that develops from an activated B lymphocyte and produces antibodies

Platelet (PLATE-let) A formed element in the blood that functions in blood clotting; also called a thrombocyte

Platelet plug (PLATE-let PLUG) Accumulation of platelets at the site of blood vessel damage to prevent blood loss

Pleura (PLOO-rah) Serous membrane that surrounds the lungs; consists of a parietal layer and a visceral layer

Pleural cavity (PLOO-ral KAV-ih-tee) The small space between the parietal and visceral layers of the pleura

Plexus (PLEK-sus) A complex network of blood vessels or nerves

Plicae circulares (PLY-kee sir-kyoo-LAIR-eez) Circular folds in the mucosa and submucosa of the small intestine

Pneumotaxic center (noo-moh-TACK-sik SEN-ter) A group of neurons in the pons that affects the rate of respiration by inhibiting inspiration

Polar body (POH-lar BAH-dee) A small cell resulting from the unequal division of cytoplasm during the meiotic division of an oocyte

Polar covalent bond (POH-lar koh-VAY-lent BOND) An unequal sharing of electrons in a covalent bond

Polysaccharide (pahl-ee-SAK-ah-ride) A substance that consists of long chains of monosaccharides linked together; examples include starch, cellulose, and glycogen

Pons (PONZ) Middle portion of the brain stem, between the midbrain and the medulla oblongata

Popliteal (pop-LIT-ee-al or pop-lih-TEE-al) Pertaining to the area behind the knee

Positive feedback (POS-ih-tiv FEED-back) A mechanism of response in which a stimulus initiates reactions that increase the stimulus and the reaction keeps building until a culminating event occurs that halts the process

Postcentral gyrus (post-SEN-trull JYE-rus) The convolution of the brain surface immediately posterior to the central sulcus; this is the primary sensory area of the brain

Posterior (pos-TEER-ee-or) Toward the back or dorsal surface; opposite of anterior or ventral

Postganglionic fiber (post-gang-lee-AHN-ik FYE-ber) The axon that transmits impulses from a cell body in an autonomic ganglion to an effector organ

Postganglionic neuron (post-gang-lee-AHN-ik NOO-ron) A neuron that conducts impulses from an autonomic ganglion to an effector organ

Postnatal development (POST-nay-tal dee-VELL-op-ment) Development that begins with birth and lasts until death

Postsynaptic membrane (post-sih-NAP-tik MEM-brayn) The membrane that receives an impulse at a synapse

Postsynaptic neuron (post-sih-NAP-tik NOO-ron) The neuron that receives an impulse from an adjacent neuron in neuron-to-neuron communication; the neuron after the synapse

Precentral gyrus (pree-SEN-trull JYE-rus) The convolution of the brain surface immediately anterior to the central sulcus; this is the primary motor area of the brain

Preembryonic period (pree-em-bree-AHN-ik PEER-ee-ud) First 2 weeks after fertilization; period of cleavage, implantation, and formation of primary germ layers

Preganglionic fiber (pree-gang-lee-AHN-ik FYE-ber) The axon that transmits impulses from a cell body in the central nervous system to an autonomic ganglion, where it synapses with a second neuron

Preganglionic neuron (pree-gang-lee-AHN-ik NOO-ron) A neuron that conducts impulses from the central nervous system to an autonomic ganglion

Pregnancy (PREG-nan-see) Presence of a developing offspring in the uterus

Prenatal development (PREE-nay-tal dee-VELL-op-ment) Development within the uterus

Prepuce (PREE-pyoos) Fold of skin that covers the distal end of the penis; also called foreskin; fold of skin that covers the clitoris in females

Presynaptic neuron (pree-sih-NAP-tik NOO-ron) The neuron that transmits an impulse to an adjacent neuron in neuron-to-neuron communication; the neuron before the synapse

Primary response (PRY-mair-ee ree-SPONS) The initial reaction of the immune system to a specific antigen

Primary teeth (PRY-mair-ee TEETH) The teeth that appear first and then are shed and replaced by permanent teeth; also called deciduous teeth

Prime mover (PRYM MOO-ver) The muscle that is mainly responsible for a particular body movement; also called agonist

Process (PRAH-sess) Any projection on a bone, often pointed and sharp

Procoagulant (pro-koh-AG-yoo-lant) A factor in the blood that promotes blood clotting

Progesterone (proh-JESS-ter-ohn) Hormone secreted by the corpus luteum of the ovaries; prepares the uterine lining for implantation, maintains pregnancy, and prepares mammary glands for lactation

Prolactin (pro-LAK-tin) Hormone secreted by the anterior pituitary gland during pregnancy to stimulate mammary gland development for lactation

Pronation (pro-NAY-shun) Act of assuming the prone position; when applied to the hand, it is the movement of the forearm that turns the palm of the hand toward the back or downward; opposite of supination

Prophase (PRO-faze) First stage of mitosis; stage of mitosis during which the chromosomes become visible

Proprioception (proh-pree-oh-SEP-shun) Sense of position, or orientation, and movement

Proprioceptor (proh-pree-oh-SEP-tor) A type of mechanoreceptor located in muscles, tendons, and joints that provides information about body position and movements

Prostaglandins (prahss-tih-GLAN-dins) A group of substances, derived from fatty acids, that are produced in small amounts and have an immediate, short-term, localized effect; sometimes called local hormones

Prostate gland (PRAHS-tayt GLAND) Accessory gland that is located below the urinary bladder in males and surrounds the proximal portion of the urethra; it produces part of the seminal fluid

Prostatic urethra (prah-STAT-ik yoo-REE-thrah) Portion of the male urethra that passes through the prostate gland

Protein (PRO-teen) An organic compound that contains nitrogen and consists of chains of amino acids linked together by peptide bonds

Prothrombin (pro-THROM-bin) A protein that is produced by the liver and released into the blood, where it is converted to thrombin during the process of blood clotting

Prothrombin activator (pro-THROM-bin AK-tih-vay-tor) A substance that is produced in the process of blood clotting and functions to change prothrombin into thrombin

Proton (PRO-ton) A positively charged particle found in the nucleus of an atom

Proximal (PRAHK-sih-mal) Next or nearest; closer to a point of attachment; opposite of distal

Puberty (PYOO-ber-tee) Period during which secondary sex characteristics begin to appear and capability for sexual reproduction becomes possible

Pubis (PYOO-biss) One of the parts of the hip bone

Pudendum (pyoo-DEN-dum) Collective term for the external accessory structures of the female reproductive system; also called the vulva

Pulmonary semilunar valve (PULL-mon-air-ee seh-mee-LOO-nar VALVE) Valve between the right ventricle and pulmonary trunk that keeps blood from flowing back into the ventricle

Pulmonary trunk (PULL-mon-air-ee TRUNK) Large vessel that receives deoxygenated blood from the right ventricle

Pulmonary vessels (PULL-mon-air-ee VES-els) Blood vessels that transport blood from the heart to the lungs and then return it to the left atrium

Pulse (PULS) Expansion and recoil of arteries caused by contraction and relaxation of the heart

Pulse points (PULS POYNTZ) Sites where the pulse can be palpated; where an artery passes over a bone or other firm base near the surface

Pulse pressure (PULS PRESH-ur) Difference between systolic and diastolic pressures

Pupil (PYOO-pill) The hole in the center of the iris through which light enters the posterior part of the eye

Purkinje fibers (per-KIN-jee FYE-bers) Cardiac muscle cells specialized for conducting action potentials to the myocardium; part of the conduction system of the heart; also called conduction myofibers

Pyramidal tract (pih-RAM-ih-dal TRAKT) Descending, or motor, tract in the spinal cord; also called the corticospinal tract

Pyrogen (PYE-roh-jen) A chemical agent in inflammation that causes an increase in temperature

Pyruvic acid (pye-ROO-vik AS-id) A molecule with three carbons that is produced in glycolysis

Q

QRS complex (QRS KOHM-plex) Deflection on an electrocardiogram that reflects ventricular depolarization

R

Radioactive isotope (ray-dee-oh-ACK-tiv EYE-so-tohp) An isotope with an unstable atomic nucleus that decomposes, releasing energy or atomic particles

Radius (RAY-dee-us) Bone on the lateral side of the forearm

Ramus (RAY-mus) Vertical portion of the mandible

Reactant (ree-AK-tant) Initial substance that is changed during a chemical reaction

Reflex arc (REE-fleks ARK) Smallest unit of the nervous system that can receive a stimulus and generate a response; functional unit of the nervous system

Refraction (ree-FRAK-shun) The bending of light as it passes from one medium to another

Refractory period (ree-FRAK-toar-ee PEE-ree-od) Time during which an excitable cell cannot respond to a stimulus that is usually adequate to start an action potential

Regeneration (ree-jen-er-A-shun) Replacement of damaged tissue cells with cells that are identical to the original ones

Relative refractory period (RELL-ah-tiv ree-FRAK-toar-ee PEE-ree-od) Time during which an excitable cell can respond to a second stimulus only if the second stimulus is stronger than that normally required to start an action potential

Renal capsule (REE-nal KAP-sool) Fibrous connective tissue covering the kidney

Renal corpuscle (REE-nal KOAR-pu-sel) Portion of the nephron where filtration occurs; consists of a glomerulus and glomerular capsule

Renal cortex (REE-nal KOAR-teks) Outer portion of the kidney that appears granular

Renal erythropoietin factor (REE-nal ee-rith-roh-poy-EE-tin FAK-tor) A substance produced by the kidneys that activates erythropoietin to stimulate the production of erythrocytes

Renal medulla (REE-nal meh-DOO-lah) Inner portion of the kidney consisting of renal pyramids

Renal papillae (REE-nal pah-PILL-ee) Pointed ends of the renal pyramids that are directed toward the center of the kidney

Renal pelvis (REE-nal PELL-vis) Large cavity in the central region of a kidney that collects the urine as it is produced

Renal pyramids (REE-nal PEER-ah-mids) Triangular-shaped regions in the kidney that appear striated

Renal sinus (REE-nal SYE-nus) Cavity within the kidney that contains the renal pelvis and branches of the renal vessels

Renal tubule (REE-nal TOOB-yool) Tubular portion of the nephron that carries the filtrate away from the glomerular capsule; portion of nephron where tubular reabsorption and secretion occur

Renin (REE-nin) An enzyme secreted by the kidneys that functions in blood pressure regulation by stimulating the formation of angiotensin

Reproductive (ree-pro-DUK-tiv) Relating to reproduction

Residual volume (RV) (ree-ZID-yoo-al VAHL-yoom) Amount of air that remains in the lungs after a maximum expiration

Resistance (ree-SIS-tans) Body's ability to counteract the effects of pathogens and other harmful agents

Respiration (res-per-Ray-shun) Exchange of oxygen and carbon dioxide between the atmosphere and the body cells

Respiratory (reh-SPY-rah-tor-ee or res-per-ah-TOR-ee) Relating to respiration

Respiratory membrane (reh-SPY-rah-tor-ee MEM-brayn) Surfaces in the lungs where diffusion occurs; consists of the layers that the gases must pass through to get into or out of the alveoli

Rete testis (REE-tee TEST-is) Network of tubules on one side of the testis

Reticular layer (ree-TIK-yoo-lur LAY-er) Lower layer of the dermis; collagenous fibers in this region provide strength to the skin

Retina (RET-ih-nah) The innermost layer of the eyeball; also called the nervous tunic; contains the photoreceptor cells for vision

Retinal (RET-ih-nal) A derivative of vitamin A that is a component of rhodopsin and is involved in reactions that trigger nerve impulses that result in vision

Rhodopsin (roh-DAHP-sin) Photosensitive pigment in the rods; also called visual purple

Ribonucleic acid (RNA) (rye-boh-noo-KLEE-ik AS-id) A nucleic acid that contains ribose sugar; functions in protein synthesis

Ribosomal RNA (rRNA) (RYE-boh-soh-mal RNA) RNA in the ribosomes in the cytoplasm; functions in protein synthesis

Ribosome (RYE-boh-sohm) Granules of RNA in the cytoplasm that function in protein synthesis

Right ventricle (RYTE VEN-trih-kul) Chamber of the heart that pumps deoxygenated blood through the pulmonary trunk to the lungs

Right lymphatic duct (RYTE lim-FAT-ik DUKT) The collecting duct of the lymphatic system that collects lymph from the upper right quadrant of the body

Right atrium (RYTE AY-tree-um) Chamber of the heart that receives deoxygenated blood from coronary circulation through the coronary sinus and from systemic circulation through the superior vena cava and inferior vena cava

Rod (RAHD) A photoreceptor in the retina that is specialized for vision in dim light

Rotation (roh-TAY-shun) Movement of a part around its own axis in a pivot joint

Rough endoplasmic reticulum (RUFF end-oh-PLAZ-mik ree-TIK-yoo-lum) Endoplasmic reticulum that has ribosomes attached to it

Round window (ROWND WIN-dow) Small, membrane-covered opening between the middle ear and the inner ear, just below the oval window; also called the fenestra cochleae

Rugae (ROO-jee) Longitudinal folds in the mucosa of the stomach

S

Saccule (SAK-yool) One of the divisions of the membranous labyrinth located in the vestibule of the inner ear; involved with static equilibrium

Sacral (SAY-kral) Pertaining to the posterior region between the hip bones

Sacrum (SAY-krum) Triangular structure, composed of five vertebrae fused together, that forms the base of the vertebral column

Sagittal plane (SAJ-ih-tal PLANE) A vertical plane that divides the body or organ into right and left portions

Saltatory conduction (SAL-tah-toar-ee kon-DUCK-shun) Process in which a nerve impulse travels along a myelinated nerve fiber by jumping from one node of Ranvier to the next

Sarcolemma (sar-koh-LEM-mah) The cell membrane of a muscle fiber (cell)

Sarcomere (SAR-koh-meer) A functional contractile unit in a skeletal muscle fiber

Sarcoplasm (SAR-koh-plazm) Cytoplasm of muscle fibers (cells)

Sarcoplasmic reticulum (sar-koh-PLAZ-mik ree-TICK-yoo-lum) Network of tubules and sacs in muscle cells; similar to endoplasmic reticulum in other cells

Satellite cell (sat-eh-LYTE SELL) Cell that binds neuron cell bodies together in peripheral ganglia

Scala tympani (SKAY-lah TIM-pah-nee) Lower portion of the cochlea, inferior to the basilar membrane; it contains perilymph and extends from the apex to the round window

Scala vestibuli (SKAY-lah ves-TIB-yoo-lee) Portion of the cochlea that is superior to the vestibular membrane; it contains perilymph and extends from the oval window to the apex

Scapula (SKAP-yoo-lah) Shoulder blade

Schwann cells (SHVON SELZ) Large cells that wrap around nerve fibers in the peripheral nervous system and, in some cases, produce myelin

Sclera (SKLEE-rah) White outer coat of the posterior part of the eyeball; posterior portion of the fibrous tunic

Scrotum (SKROH-tum) A pouch of skin and subcutaneous tissue that extends below the abdomen and contains the testes

Sebaceous gland (see-BAY-shus GLAND) An oil gland of the skin that produces sebum or body oil

Secondary response (SEK-on-dair-ee ree-SPONS) Rapid and intense reaction to antigens on second and subsequent exposures attributable to memory cells

Secondary teeth (SEK-on-dair-ee TEETH) The permanent teeth that replace the deciduous teeth; also called permanent teeth

Secretin (see-KREE-tin) A hormone that is produced in the small intestine and stimulates the pancreas to secrete a fluid with a high bicarbonate ion concentration

Selectively permeable (sel-EK-tiv-lee PER-me-ah-bul) Restricts the passage of some substances but permits the passage of other substances

Sella turcica (SELL-ah TUR-sih-kah) Depression on the superior surface of the sphenoid bone that houses the pituitary gland

Semen (SEE-men) Mixture of sperm cells and secretions from the accessory glands in the male; also called seminal fluid

Semicircular canals (seh-mee-SIR-kew-lar kah-NALS) Three curved passageways in the bony labyrinth of the inner ear; are filled with perilymph and contain the membranous semicircular ducts

Semicircular ducts (seh-mee-SIR-kew-lar DUKTS) Three curved membranous channels located within the bony labyrinth of the inner ear; are filled with endolymph and surrounded by perilymph; function in dynamic equilibrium

Semilunar valve (seh-mee-LOO-nar VALV) Valve between a ventricle of the heart and the vessel that carries blood away from the ventricle; also pertains to the valves in veins

Seminal vesicles (SEM-ih-nal VES-ih-kulz) Accessory glands located posterior to the urinary bladder in the male; secretion accounts for 60% of the semen volume

Seminal fluid (SEM-ih-nal FLOO-id) Mixture of sperm cells and secretions from the accessory glands in the male; also called semen

Seminiferous tubules (seh-mih-NIFF-er-us TOOB-yools) Tightly coiled structures within which sperm are produced in the testes

Senescence (seh-NESS-ens) Period of old age

Sensory adaptation (SEN-soh-ree add-dap-TAY-shun) Phenomenon in which some receptors respond when a stimulus is first applied but decrease their response if the stimulus is maintained; receptor sensitivity decreases with prolonged stimulation

Sensory nerve (SEN-soar-ee NERV) A nerve that contains primarily sensory, or afferent, fibers

Sensory neuron (SEN-soar-ee NOO-ron) Nerve cell that carries impulses toward the central nervous system from the periphery; afferent neuron

Serosa (see-ROS-ah) Epithelial membranes that secrete a serous fluid, line the closed body cavities, and cover the organs within those cavities; also called serous membranes

Serotonin (sair-oh-TONE-in) One of several neurotransmitters

Serous cell (SEER-us SELL) Glandular cell that secretes a watery fluid, usually with a high enzyme content

Serous membrane (SEER-us MEM-brayn) Epithelial membrane that secretes a serous fluid, lines the closed body cavities, and covers the organs within those cavities; also called serosa

Sertoli cells (sir-TOH-lee SELZ) Cells within the seminiferous tubules that do not produce gametes but support and nourish the gamete-producing cells; also called supporting cells

Serum (SEE-rum) The fluid that remains after a blood clot has formed; plasma minus the clotting factors

Sesamoid bone (SEE-sah-moyd BONE) A small bone, usually found in a tendon

Shell temperature (SHELL TEM-per-ah-chur) The temperature at or near the body surface

Simple epithelium (SIM-pull ep-ih-THEE-lee-um) Epithelial tissue that is only one layer thick

Simple series circuit (SIM-pull SEER-ees SIR-cut) The simplest conduction pathway in which a single neuron synapses with another neuron

Sinoatrial node (sye-noh-AY-tree-al NODE) Mass of specialized cardiac muscle cells that form a part of the conduction system of the heart and that are located in the right atrium near the opening for the superior vena cava; often referred to as the pacemaker of the heart

Sinus (SYE-nus) A cavity or hollow space in a bone or other body part

Skeletal muscle (SKEL-eh-tal MUS-el) Muscle that is under voluntary or willed control; also called voluntary striated muscle

Skeletal (SKEL-eh-tal) Relating to the bones of the body

Smooth muscle (SMOOTH MUS-el) Muscle tissue that is neither striated nor controlled voluntarily; also called visceral muscle

Smooth endoplasmic reticulum (SMOOTH end-o-PLAZ-mik ree-TICK-yoo-lum) Endoplasmic reticulum that does not have ribosomes attached to it and appears smooth

Soft palate (SOFT PAL-at) Posterior portion of the floor of the nasal cavity and roof of the mouth that consists of soft tissue and has no bony support

Solute (SOL-yoot) A substance that is dissolved in a solution

Solvent (SOL-vent) Fluid in which substances dissolve

Somatic nervous system (soh-MAT-ik NER-vus SIS-tem) The portion of the peripheral efferent nervous system consisting of motor neurons that control voluntary actions of skeletal muscles

Somatic sense (soh-MAT-ik SENS) A general sense that is not localized but is found throughout the body; includes touch, pressure, temperature, and pain

Somatomotor cortex (soh-mat-oh-MOH-ter KOR-teks) The primary motor area of the brain, which is located in the precentral gyrus

Somatosensory cortex (soh-mat-oh-SEN-soar-ee KOR-teks) The primary sensory area of the brain, which is located in the postcentral gyrus

Somatotropic hormone (STH) (soh-mat-oh-TROH-pik HOAR-moan) A hormone secreted by the anterior pituitary gland that influences the rate of skeletal growth; also called growth hormone

Specific resistance (speh-SIF-ik ree-SIS-tans) Body's ability to counteract certain types of harmful agents; immunity

Spermatic cord (spur-MAT-ik KORD) Composite structure that contains the ductus deferens, blood vessels, nerves, lymphatic vessels, and the cremaster muscle

Spermatids (SPUR-mah-tids) Haploid cells that are the product of meiosis in the male (spermatogenesis)

Spermatogenesis (spur-mat-oh-JEN-eh-sis) Process of meiosis in the male in which four spermatids are produced from one primary spermatocyte

Spermatogonia (spur-mat-oh-GOH-nee-ah) Stem cells that give rise to sperm cells

Spermiogenesis (spur-mee-oh-JEN-eh-sis) Morphologic changes that transform a spermatid into a mature sperm

Sphenoid bone (SFEE-noyd BONE) Bone that forms a portion of the cranial floor

Sphygmomanometer (sfig-moh-mah-NAHM-eh-ter) Device for measuring blood pressure

Spinal cavity (SPY-nal KAV-ih-tee) Part of the dorsal body cavity that contains the spinal cord

Spindle fibers (SPIN-dul FYE-burs) Microtubules that extend from the centromeres to the centrioles during cell division

Spinothalamic tract (spy-noh-thah-LAM-ik TRACT) Ascending, or sensory, tract that begins in the spinal cord and conducts impulses to the thalamus

Spirometer (spy-RAHM-eh-ter) An instrument used to measure the volume of air that moves into and out of the lungs

Spleen (SPLEEN) Large lymphoid organ that filters blood and acts as a reservoir for blood

Splenic flexure (SPLEH-nik FLEK-shur) Curve between the transverse colon and the descending colon; also called the left colonic flexure

Spongy urethra (SPUN-jee yoo-REE-thrah) Portion of the male urethra that is surrounded by corpus spongiosum and passes through the length of the penis; also called penile urethra

Squamous (SKWAY-mus) Flat, platelike, scalelike; horizontal dimension is greater than vertical dimension

Stapes (STAY-peez) One of the tiny bones in the middle ear, adjacent to the oval window

Starch (STARCH) Complex polysaccharide that is the storage form of energy in plants

Starling's law of the heart (STAR-lings LAW OV THE HART) Principle that the more cardiac muscle fibers are stretched, the greater the contraction strength of the heart

Static equilibrium (STAT-ik ee-kwi-LIB-ree-um) Sensing and evaluating the position of the head relative to gravity

Sternal (STIR-nal) Pertaining to the anterior midline of the thorax

Sternum (STIR-num) Breastbone

Stimulus (STIM-yoo-lus) Any agent that produces a reaction in a receptor or excitable (irritable) tissue

Straight tubule (STRAYT TOOB-yool) Single tube from each lobule of the testes that leads into the rete testis

Stratified epithelium (STRAT-ih-fyed ep-ih-THEE-lee-um) Epithelial tissue that has multiple layers of cells

Stratum basale (STRAY-tum BAY-sah-lee) Deepest layer of the epidermis; basal layer where the cells are actively mitotic; also deep layer of the endometrium that is constant and responsible for rebuilding the stratum functionale after menstruation

Stratum corium (STRAY-tum KOR-ee-um) Another name for the dermis

Stratum corneum (STRAY-tum KOR-nee-um) Outermost layer of the epidermis; consists of flattened, dead keratinized cells

Stratum functionale (STRAY-tum FUNK-shun-al-ee) Portion of the endometrium that is sloughed off during menstruation

Stratum germinativum (STRAY-tum JER-mih-nah-tiv-um) Term used for combined stratum basale and stratum spinosum in the epidermis

Stratum granulosum (STRAY-tum gran-yoo-LOH-sum) Epidermal layer in which keratinization begins and cells appear granular; "granular layer"

Stratum lucidum (STRAY-tum LOO-sih-dum) Epidermal layer in thick skin between the stratum corneum and stratum granulosum; "clear layer"

Stratum spinosum (STRAY-tum spy-NOH-sum) Epidermal layer directly above the stratum basale; cells appear "spiny"

Striae (STRY-ee) Streaks or bands; tiny white scars that appear when elastic fibers in the dermis are stretched too much; stretch marks

Striated muscle (STRY-ate-ed MUS-el) Muscle tissue that appears to have cross-bars; voluntary striated muscle is skeletal muscle and involuntary striated muscle is cardiac muscle

Stroke volume (STROAK VAHL-yoom) The volume of blood ejected from one ventricle during one contraction; normally about 70 ml

Subarachnoid space (sub-ah-RAK-noyd SPACE) The space between the arachnoid and pia mater layers of the meninges; it contains cerebrospinal fluid

Subcutaneous layer (sub-kyoo-TAY-nee-us LAY-er) Below the skin; a sheet of areolar connective tissue and adipose beneath the dermis of the skin; also called hypodermis or superficial fascia

Subdural space (sub-DOO-ral SPACE) The space between the dura mater and arachnoid layers of the meninges

Subliminal stimulus (sub-LIM-ih-null STIM-yoo-lus) A weak stimulus that is of insufficient intensity to start a nerve impulse or muscle contraction; also called subthreshold stimulus

Submucosal plexus (sub-myoo-KOH-sal PLEK-sus) A network of autonomic nerves in the submucosa of the gastrointestinal tract; also called Meissner's plexus

Subthreshold stimulus (sub-THRESH-hold STIM-yoo-lus) A weak stimulus that is of insufficient intensity to start a nerve impulse or muscle contraction; also called subliminal stimulus

Sucrase (SOO-krays) An enzyme that acts on the disaccharide sucrose and breaks it into a molecule of glucose and a molecule of fructose

Sudoriferous gland (soo-door-IF-er-us GLAND) A gland in the skin that produces perspiration; also called sweat gland

Sulcus (SULL-kus) A groove or furrow between parts; often refers to the grooves between the convolutions on the surface of the brain; plural, sulci (SULL-see)

Superficial (soo-per-FISH-al) On or near the body surface; opposite of deep

Superior (soo-PEER-ee-or) Higher; one part is above another; opposite of inferior

Superior colliculi (soo-PEER-ee-or koh-LIK-yoo-lye) Two upper bodies of the corpora quadrigemina on the dorsal surface of the midbrain; function in visual reflexes

Superior vena cava (soo-PEER-ee-or VEE-nah KAY-vah) Large vein that collects blood from all parts of the body superior to the heart and returns it to the right atrium

Supination (soo-pih-NAY-shun) Act of assuming the supine position; when applied to the hand, it is the movement of the forearm that turns the palm of the hand toward the front or upward; opposite of pronation

Supporting cells (suh-POR-ting SELZ) Cells within the seminiferous tubules that do not produce gametes, but support and nourish the gamete-producing cells; also called Sertoli cells

Suprarenal gland (soo-prah-REE-null GLAND) Endocrine gland that is located on the superior pole of each kidney; divided into cortex and medulla regions; also called adrenal gland

Surfactant (sir-FAK-tant) A substance produced by certain cells in lung tissue that reduces surface tension between fluid molecules that line the respiratory membrane and helps keep the alveolus from collapsing

Susceptibility (sus-sep-tih-BILL-ih-tee) Lack of resistance to disease

Suspensory ligament (sus-PEN-soar-ee LIG-ah-ment) Stringlike structures attached to the lens of the eye that hold the lens in place and function in accommodation for near vision

Sutural bone (SOO-cher-ahl BONE) A small bone located within a suture between certain cranial bones; also called a wormian bone

Suture (SOO-cher) An immovable fibrous joint between the flat bones in the skull

Sympathetic division (sim-pah-THET-ik dih-VIH-shun) One of the two divisions of the autonomic nervous system; primarily concerned with processes that involve the expenditure of energy; sometimes referred to as the "fight or flight" division; also called the thoracolumbar division

Synapse (SIN-aps) The region of communication between two neurons

Synaptic cleft (sih-NAP-tik KLEFT) Small space between the synaptic knob of a neuron and the cell membrane of an adjacent neuron or muscle cell

Synaptic knob (sih-NAP-tik NAHB) An enlargement at the distal end of telodendria; it contains vesicles of neurotransmitters

Synarthrosis (sin-ahr-THROH-sis) An immovable joint; plural, synarthroses

Synergist (SIN-er-jist) A muscle that assists a prime mover but is not capable of producing the movement by itself

Synovial fluid (sih-NOH-vee-al FLOO-id) Fluid secreted by the synovial membrane for lubrication of freely movable joints

Synovial membrane (sih-NOH-vee-al MEM-brayn) Membrane that lines the cavity of freely movable joints and secretes synovial fluid for lubrication

System (SIS-tem) A group of organs that work together to perform complex functions

Systemic vessels (sis-TEM-ik VES-els) Blood vessels that transport blood from the heart to all parts of the body and back to the right atrium; excludes pulmonary circulation

Systole (SIS-toh-lee) Contraction phase of the cardiac cycle; opposite of diastole

Systolic pressure (sis-TAHL-ik PRESH-ur) Blood pressure in the arteries during contraction of the ventricles

T

T wave (T WAYV) Deflection on an electrocardiogram that reflects ventricular repolarization

T tubules (T TOOB-yools) Invaginations of the sarcolemma that form transverse tubules in a muscle cell and permit electrical impulses to travel deeper into the cell

T lymphocytes (T LIM-foh-sytes) Cells of the immune system that differentiate in the thymus gland and are responsible for cell-mediated immunity; T cells

Talus (TAL-us) Tarsal bone that articulates with the tibia

Target tissue (TAR-get TISH-yoo) A tissue (cells) that responds to a particular hormone because it has receptor sites for that hormone

Tarsal (TAHR-sal) Pertaining to the ankle and instep of the foot

Tarsus (TAHR-sis) Ankle

Taste hair (TAYST HAIR) Specialized projection of the gustatory cell that extends through the taste pore and functions in the sense of taste; also called gustatory hair

Tectorial membrane (tek-TOH-ree-al MEM-brayn) A gelatinous membrane over the hair cells of the organ of Corti in the cochlear duct

Telodendria (tell-oh-DEN-dree-ah) Short branches at the distal end of an axon or axon collateral

Telophase (TELL-oh-faze) Final stage of mitosis; membrane forms around the genetic material at each end of the cell to establish two separate nuclei

Temporal bone (TEM-por-al BONE) Bone that forms the side of the cranium and surrounds the ear

Temporomandibular (tem-por-oh-man-DIB-yoo-lar) Pertaining to the temporal bone and the mandible

Tendon (TEN-dun) Band of dense fibrous connective tissue that attaches muscle to bone or another structure

Teniae coli (TEE-nee-ah KOH-lye) Bands of longitudinal muscle fibers in the large intestine

Tentorium cerebelli (ten-TOAR-ee-um sair-eh-BELL-eye) Extension of dura mater in the transverse fissure between the cerebrum and the cerebellum

Terminal ganglion (TER-mih-null GANG-lee-on) An autonomic ganglion located near, or within, the wall of the effector organ; part of the parasympathetic pathway

Testes (TEST-eez) Primary reproductive organs in the male; produce the sperm; also called testicles

Testosterone (tess-TAHS-ter-ohn) Male sex hormone produced by the interstitial cells of the testes; see Androgens

Tetanus (TET-ah-nus) A smooth, sustained contraction produced by a series of very rapid stimuli to a muscle

Thalamus (THAL-ah-mus) Region of gray matter located in the diencephalon that channels sensory impulses to the appropriate regions of the cortex for discrimination, localization, and interpretation

Thermogenesis (thur-moh-JEN-eh-sis) Production of heat in response to food intake

Thermoreceptor (ther-moh-ree-SEP-tor) A sensory receptor that detects changes in temperature

Thoracic (tho-RAS-ik) Pertaining to the chest; part of the trunk inferior to the neck and superior to the diaphragm

Thoracic cavity (tho-RAS-ik KAV-ih-tee) Superior part of the ventral body cavity; contains the heart, lungs, esophagus, and trachea

Thoracic duct (tho-RAS-ik DUKT) The primary collecting duct of the lymphatic system that collects lymph from all regions of the body except the upper right quadrant

Thoracolumbar division (thoar-ah-koh-LUM-bar dih-VIH-shun) See Sympathetic division

Threshold stimulus (THRESH-hold STIM-yoo-lus) Minimum level of stimulation that is required to start a nerve impulse or muscle contraction; also called liminal stimulus

Thrombin (THROM-bin) The active substance formed from prothrombin that functions to convert fibrinogen into fibrin

Thrombocyte (THROM-boh-syte) One of the formed elements of the blood; functions in blood clotting; also called platelet

Thrombocytopenia (throm-boh-syte-oh-PEE-nee-ah) An abnormal decrease in the number of platelets in the blood

Thymosin (THYE-moh-sin) Hormone produced by the thymus; important in the development of the body's immune system

Thymus (THYE-mus) Endocrine gland and lymphoid organ located in the mediastinum; secretes thymosin; plays an important role in the body's immune system

Thyroid cartilage (THYE-royd KAR-tih-layj) Large, anterior cartilage of the larynx; commonly called Adam's apple

Thyroid gland (THYE-royd GLAND) Endocrine gland that is located anterior to the trachea at the base of the neck

Thyroid hormone (THYE-royd HOAR-moan) Hormone produced by the thyroid gland that accelerates metabolism; includes thyroxine and triiodothyronine

Thyroid-stimulating hormone (TSH) (THYE-royd STIM-yoo-lay-ting HOAR-moan) A hormone secreted by the anterior pituitary gland that stimulates the thyroid gland to produce thyroid hormone; also called thyrotropin

Thyrotropin (thye-roh-TROH-pin) See Thyroid-stimulating hormone

Thyroxine (T₄) (thye-RAHK-sin) One of the thyroid hormones that stimulates cellular metabolism

Tibia (TIB-ee-ah) Large bone on the medial side of the leg

Tidal volume (TV) (TYE-dal VAHL-yoom) Amount of air that is inhaled and exhaled in a normal quiet breathing cycle

Tincture (TINK-chur) A solution that uses alcohol as the solvent

Tissue (TISH-yoo) Group of similar cells specialized to perform a certain function

Tonsil (TAHN-sil) Aggregate of lymphoid tissue embedded in mucous membranes that line the nose, mouth, and throat

Total lung capacity (TLC) (TOH-tal LUNG kah-PASS-ih-tee) Amount of air in the lungs after a maximum inspiration; equals residual volume plus expiratory reserve volume plus tidal volume plus inspiratory reserve volume

Trabeculae (trah-BEK-yoo-lee) Thin plates of bone tissue arranged in an irregular latticework; found in spongy bone

Trabeculae carneae (trah-BEK-yoo-lee KAR-nee-ee) Ridges of myocardium in the ventricles of the heart

Trachea (TRAY-kee-ah) Passageway for air that extends inferiorly from the larynx to the carina; commonly called the windpipe

Transcription (trans-KRIP-shun) Process in which a single strand of DNA acts as a template for the formation of an RNA molecule that transfers information from the nucleus to the cytoplasm

Transfer RNA (tRNA) (TRANS-fur RNA) A molecule of RNA that carries an amino acid to a ribosome during protein synthesis

Translation (trans-LAY-shun) The process of creating a new protein on the ribosome of a cell in response to messenger RNA codons

Transverse fissure (TRANS-vers FISH-ur) The deep fissure that separates the cerebrum from the cerebellum

Transverse plane (TRANS-vers PLANE) A horizontal plane that divides the body or part into superior and inferior portions

Treppe (TREP-peh) The gradual increase in the strength of muscle contraction caused by rapid, repeated stimuli of the same intensity

Tricuspid valve (trye-KUS-pid VALVE) Valve between the right atrium and right ventricle

Trigeminal nerve (trye-JEM-ih-nal NERV) Cranial nerve V; mixed nerve; responsible for chewing movements and facial sensations

Triglyceride (trye-GLIS-ser-ide) A lipid composed of glycerol and three fatty acid molecules

Trigone (TRYE-goan) Triangular area in the floor of the urinary bladder and formed by the openings for the urethra and the two ureters

Triiodothyronine (T₃) (trye-eye-oh-doh-THYE-roh-neen) One of the thyroid hormones that stimulates cellular metabolism

Trochanter (troh-KAN-tur) Large, blunt, irregularly shaped projection on a bone

Trochlear nerve (TROH-klee-ar NERV) Cranial nerve IV; motor nerve; responsible for eye movements

Trophoblast (TROH-foh-blast) Layer of cells that form the outer surface of a blastocyst; functions in the formation of the placenta

Tubercle (TOO-burr-kul) Small, rounded, knoblike projection on a bone

Tuberosity (too-burr-AHS-ih-tee) An elevation or protuberance on a bone, similar to a tubercle

Tubular reabsorption (TOOB-yoo-lar ree-ab-SORP-shun) The movement of filtrate from the renal tubules back into the blood in response to the body's needs during urine formation

Tubular secretion (TOOB-yoo-lar see-KREE-shun) The movement of substances from the blood into the renal tubules in response to the body's needs during urine formation

Tunica adventitia (TOO-nih-kah ad-ven-TISH-ah) Outermost layer of the blood vessel wall, composed of tough, fibrous connective tissue; also called tunica externa

Tunica albuginea (TOO-nih-kah al-byoo-JIN-ee-ah) Fibrous connective tissue capsule that surrounds each testis in the male; in the female, it is the layer directly under the simple cuboidal epithelium

Tunica externa (TOO-nih-kah ex-TER-nah) Outermost layer of the blood vessel wall; composed of tough, fibrous connective tissue; also called tunica adventitia

Tunica interna (TOO-nih-kah in-TER-nah) Endothelium that lines the blood vessels; also called tunica intima

Tunica intima (TOO-nih-kah IN-tih-mah) Endothelium that lines the blood vessels; also called tunica interna

Tunica media (TOO-nih-kah MEE-dee-ah) Middle layer of the blood vessel wall; primarily composed of smooth muscle

Turbinate (TURB-ih-nayte) Scroll-like bone or projection of a bone; also called concha

Tympanic cavity (tim-PAN-ik KAV-ih-tee) See Middle ear

Tympanic membrane (tim-PAN-ik MEM-brayn) Membranous partition between the external ear and the middle ear; also called the eardrum

U

Ulna (UHL-nah) Bone on the medial side of the forearm

Umbilical (um-BIL-ih-kal) Pertaining to the navel; middle region of the abdomen

Umbilical region (um-BIL-ih-kal REE-jun) The central portion of the abdomen, where the belly button or navel is located

Unicellular (yoo-nih-SELL-yoo-lar) Consisting of one cell

Universal recipient (yoo-nih-VER-sal ree-SIP-ee-ent) An individual who has no blood type agglutinins in the plasma and thus may receive transfusions of any blood type; type AB positive

Universal donor (yoo-nih-VER-sal DOH-nor) An individual who has no blood type agglutinogens on the red blood cells and thus may donate blood to all other types; type O negative

Upper respiratory tract (UP-per res-PYE-rah-tor-ee TRACT) Portion of the respiratory tract that includes the nose, pharynx, and larynx

Ureter (yoo-REE-ter) Tubular structure that carries urine from the renal pelvis to the urinary bladder

Urethra (yoo-REE-thrah) Passageway that conveys urine from the urinary bladder to the exterior

Urinalysis (yoo-rin-AL-ih-sis) A chemical and microscopic examination of urine

Urinary (YOO-rin-air-ee) Relating to the system responsible for eliminating most fluid wastes from the body

Urinary bladder (YOO-rin-air-ee BLAD-der) Storage reservoir for urine; located in the pelvic cavity

Urochrome (YOO-roh-kroam) Yellow pigment, produced by the decomposition of bile pigments, that appears in the urine

Uterine tubes (YOO-ter-in TOOBS) The tubes that extend laterally from the upper portion of the uterus to the region of the ovaries; also called fallopian tubes or oviducts

Uterine cycle (YOO-ter-in SYE-kul) Monthly cycle of events that occur in the uterus from puberty to menopause; also called the menstrual cycle; occurs concurrently with the ovarian cycle

Uterus (YOO-ter-us) Hollow, muscular organ in females that is the site of menstruation, receives the developing embryo, and supports the fetus until birth

Utricle (YOO-trih-kull) One of the divisions of the membranous labyrinth located in the vestibule of the inner ear; involved with static equilibrium

Uvula (YOO-vyoo-lah) Posterior projection of the soft palate

V

Vaccine (vak-SEEN) A preparation of weakened or destroyed antigens introduced into an individual to stimulate the development of immune agents against that specific antigen

Vagina (vah-JYE-nah) Fibromuscular tube that extends from the cervix to the exterior; also called the birth canal

Vagus nerve (VAY-gus NERV) Cranial nerve X; mixed nerve; responsible for sensations and movements of the internal organs

Vasa vasorum (VAS-ah vah-SOR-um) Small blood vessels that supply nutrients to the tissues in the walls of the large blood vessels

Vascular tunic (VAS-kyoo-lar TOO-nik) Middle layer of the eyeball; also called the uvea

Vasoconstriction (vaz-oh-kon-STRIK-shun) A narrowing of blood vessels; decrease in the size of the lumen of blood vessels

Vasodilation (vaz-oh-dye-LAY-shun) An enlarging of blood vessels; increase in the size of the lumen of blood vessels

Vein (VAYN) A blood vessel that carries blood toward the heart

Ventilation (ven-tih-LAY-shun) Movement of air into and out of the lungs; breathing

Ventral (VEN-tral) Toward the front or anterior; opposite of dorsal or posterior

Ventral root (VEN-tral ROOT) Motor branch of a spinal nerve by which the nerve is attached to the spinal cord

Ventricle (VEN-trih-kull) A cavity, such as the fluid-filled cavities in the brain or in the heart

Venule (VAYN-yool) A small vessel that receives blood from a capillary and delivers it to a vein for return to the heart

Vertebrae (VER-teh-bray) Bones that make up the spinal column

Vertebral (ver-TEE-bral or VER-teh-bral) Pertaining to the spinal column; backbone

Vertebrochondral (ver-TEE-broh-kahn-dral) Pertaining to the vertebral column and the costal cartilage of the ribs

Vertebrosternal (ver-TEE-broh-stirnal) Pertaining to the vertebral column and the sternum

Vestibular membrane (ves-TIB-yoo-lar MEM-brayn) Roof of the cochlear duct; separates the cochlear duct from the scala vestibuli; also called Reissner's membrane

Vestibule (ves-TIB-yool) A small space at the entrance to a passage; in the inner ear, the vestibule is located adjacent to the oval window, between the cochlea and the semicircular canals; in the female, it is the space between the two labia minora

Vestibulocochlear nerve (ves-TIB-yoo-loh-koh-klee-ar NERV) Cranial nerve VIII; sensory nerve; responsible for hearing and equilibrium

Visceral (VIS-er-al) Pertains to internal organs or the covering of the organs

Visceral muscle (VIS-er-al MUSS-el) Muscle tissue that is found in the walls of internal organs; smooth muscle

Visceral pericardium (VIS-er-al pair-ih-KAR-dee-um) Layer of serous mem brane that forms the outermost layer of the heart wall; also called the epicardium

Vital capacity (VC) (VYE-tal kah-PASS-ih-tee) Maximum amount of air that can be exhaled after a maximum inspiration; equals tidal volume plus inspiratory reserve volume plus expiratory reserve volume

Vitamins (VYE-tah-mins) Organic compounds that are needed in minute amounts to maintain growth and good health; do not supply energy but are necessary to release energy from carbohydrates, proteins, and lipids

Vitreous humor (VIT-ree-us HYOO-mer) Jellylike substance that fills the posterior cavity of the eye, between the lens and the retina; also called vitreous body

Vomer (VOH-mer) Bone that forms the inferior part of the nasal septum

Vulva (VUL-vah) Collective term for the external accessory structures of the female reproductive system; also called the pudendum

W

Wormian bone (WER-mee-an BONE) A small bone located within a suture between certain cranial bones; also called sutural bone

X

Xiphoid (ZYE-foyd) Most inferior portion of the sternum

Y

Yolk sac (YOHK SAK) An extraembryonic membrane that, in humans, produces the primordial germ cells and is an early source of blood cells

Z

Zona pellucida (ZOH-nah peh-LOO-sih-dah) Glycoprotein membrane that surrounds a secondary oocyte

Zygomatic (zye-goh-MAT-ik) Triangular bone that forms the prominence of the cheek; cheekbone

Zygote (ZYE-goht) The single diploid cell that is a fertilized ovu

Index

Note: Page numbers followed by *f* indicate figures; *t*, tables; and *b*, boxes

A

A (anisotropic) bands, of muscle fiber, 142, 143
abdominal aorta, 277
abdominal cavity, 10
abdominal viscera
  arteries of, 278
  veins of, 280
abdominal wall
  muscles, 155–156
  transversus abdominis, 155*f*
  veins of, 280
abdominopelvic cavity
  body cavities, 10
  quadrants of, 11*f*
  regions of, 11*f*
abduction, types of movement, 150*f*
ABO blood groups, 297–298
  inheritance of, 456*f*, 457*t*
abruptio placentae, 461
absolute refractory period, in conduction along a neuron, 175
absorption
  defined, 351
  digestive processes, 353
  illustration of, 369*f*
  of nutrients, 368–369, 370*t*
accelerated heart rate, endocrine disorder, 239
accessory (secondary) reproductive organs, 421
accessory glands, male reproductive system
  bulbourethral (Cowper's) glands, 424
  overview of, 424
  prostate, 424
  seminal fluid, 424
  seminal vesicles, 424
accessory organs of digestion, 363–367
  gallbladder, 365
  liver, 363–365
  overview of, 353
  pancreas (islets of Langerhans), 365–367
accommodation
  defined, 201
  visual, 208, 208*f*
acetabulum, of pelvic girdle, 121
acetaminophen (Tylenol, Anacin-3, Liquiprin), 126*b*
acetylcholine (ACh)
  in contraction of skeletal muscles, 143
  as neurotransmitter, 176
  release of, 190
acetylcholinesterase (AChE), 143–144
acetyl-CoA
  carbohydrate digestion and, 378–380
  defined, 377
  ketone bodies due to excess of, 381*b*
acetylsalicylic acid (ASA), 126*b*
ACh. *See* acetylcholine (ACh)
AChE (acetylcholinesterase), 143–144
Achilles tendon, 161–162
achlorhydria, 392
acid-base balance
  causes and compensations, 410*t*
  deviations in pH, Pco_2, and HCO_3, 411*t*
  overview of, 409–410

acidosis
  acid-base balance and, 409
  defined, 395
acids
  defined, 21
  neutralization reactions, 33
  overview of, 31–32
  pH scale for measuring, 32–33
acinar glands, 73, 73*f*
acne vulgaris
  disorders of integumentary system, 96
  overview of, 92*b*
acquired immunity, 319
  active, 319
  illustration of, 320*f*
  passive, 319
acquired immunodeficiency syndrome (AIDS), 322
acromegaly, endocrine disorder, 239
acromion process, in pectoral girdle, 118
acrosome, covering head of sperm, 423
ACTH (adrenocorticotropic hormone), hormones of anterior lobe, 228
actin
  contractile proteins in muscle tissue, 77
  defined, 139
  thin muscle filaments formed by, 142
action potential
  propagation of, 175, 175*f*
  recording, 174*f*
  stimulation of neurons and, 174
active immunity
  artificial immunity, 319
  defined, 307
  natural immunity, 319
active transport
  in movement of substances across plasma membrane, 51*f*
  pumps, 45, 54–55
acute diseases, 18
adapalene (Differin), for treating acne, 93
Addison's disease, endocrine disorder, 239
adduction, types of movement, 150*f*
adductor brevis, thigh muscles, 159
adductor longus, thigh muscles, 159
adductor magnus, thigh muscles, 159
adenocarcinoma, 74
adenohypophysis, 223, 226–229
adenoids, 312, 330
adenoma, 74, 243
adenosine diphosphate (ADP), 378
adenosine triphosphate (ATP)
  in anabolism, 378
  components of, 39*f*
  energy exchange in, 379*f*
  overview of, 38–39
  sources of energy in muscle contraction, 146–147, 147*f*
ADH (antidiuretic hormone), 229, 405
adhesion, 85
adipose (fat) tissue
  illustration of, 75*f*
  overview of, 75
  suction lipectomy for removal of, 75*b*

adolescence, 452
ADP (adenosine diphosphate), 378
adrenal cortex, 232–233
adrenal glands
  hormones of, 232–234
  hormones of adrenal cortex, 232–233
  hormones of adrenal medulla, 234
  hypothalamic control of, 234*f*
adrenal medulla, 234
adrenergic antagonist, 271
adrenergic fibers, release of norepinephrine by, 190
adrenocorticotropic hormone (ACTH), hormones of anterior lobe, 228
adrenogenital syndrome
  disorders of reproductive system, 435
  endocrine disorder, 239
adulthood, 452
adventitia, layers of digestive tract, 354
aerobic, 377
aerobic exercise, 162
aerobic respiration, in muscular contraction, 147, 148*f*
afferent (sensory)
  division of peripheral nervous system, 171
  neurons, 173
afferent arteriole, 398, 400
afferent lymphatic vessels, 311
afterbirth, 446
agglutination, 289, 297
agglutinins
  defined, 289
  illustration of, 298*f*
  overview of, 297
agglutinogens
  defined, 289
  illustration of, 298*f*
  overview of, 297
aging
  blood and, 300*b*
  blood vessels and, 284*b*
  body fluids, 411*b*
  cardiovascular system and, 256*b*
  cellular effects of, 60*b*
  digestive system and, 370*b*
  endocrine system and, 238*b*
  muscle mass loss in, 162
  nervous system and, 193*b*
  reproductive system and, 434*b*
  respiratory system and, 343*b*
  skeletal system and, 129*b*
  skin and, 95*b*
  special senses and, 215*b*
  tissues and, 77–78
  urinary system, 411*b*
agonists (prime movers), in movement produced by skeletal muscles, 149
agranulocytes, types of leukocytes, 295
AIDS (acquired immunodeficiency syndrome), 322
airway, clearing blocked, 331*b*
ala, of coxal bones, 122
albinism, 91, 99
albumins, plasma proteins, 291
alcoholism, cirrhosis of the liver due to, 381*b*

aldosterone
  arterial blood pressure and, 271
  hormones influencing concentration and volume
    of urine, 405
  regulation of, 234*f*
alendronate (Fosamax), 126*b*
alimentary canal. *See* digestive tract
alimentation, 392
alkaline solutions, measuring, 32–33
alkalosis
  acid-base balance and, 409
  defined, 395
allantois, formation of, 446
alleles, 441
allergens, 325
allergies, 320*b*
allopurinol, for treating gout, 407
all-or-none principle
  defined, 139
  muscular contraction and, 145
  nerve fibers and, 176
alopecia, integumentary disorder, 96
alprazolam (Xanax), 187
altered metabolic rate, endocrine disorder, 239
alveolar ducts, 332
alveolar gland, 73
alveolar process, 113
alveoli, 332
alveolus, 327
Alzheimer's disease, 194
amenorrhea, 435
amino acids
  as building block of proteins, 35, 225
  catabolism of, 380*f*
  essential and nonessential, 385, 386*t*
  list of common, 35*t*
  overview of, 380
  structural formula for leucine and
    valine, 35*f*
Amniocentesis, 461
  overview of, 446
amnion
  defined, 441
  formation of extraembryonic membranes, 446
amniotic cavity, 445
amniotic fluid, 446
amniotic sac, 446
amphiarthroses, 101, 127
ampulla, 423
ampulla, in dynamic equilibrium, 214
amylase
  enzyme of salivary glands, 357
  pancreatic amylase, 366
amyotrophic lateral sclerosis, 199
anabolic steroids, 243
anabolism, 1, 6–7, 378
  defined, 377
anaerobic reactions
  breakdown of pyruvic acid, 380*f*
  catabolism of glucose, 378
  defined, 377
anaerobic respiration, in muscular
  contraction, 147, 148*f*
anal canal, 362
anal columns, 362
analgesic drugs, for pain relief, 126*b*
anaphase, of mitosis, 56
anaphylaxis, lymphatic disorder, 322
anaplasia, 66
anatomical position
  defined, 1
  illustration of, 10*f*
  overview of, 9

anatomical terms
  anatomical position, 9
  body cavities, 10–11
  body regions, 11
  directional terms, 9–10
  overview of, 9–11
  planes and sections of body, 10
anatomy, 1, 2
androgens
  hormones of adrenal cortex, 233
  hormones of testes, 236
  male sex hormones, 426
anemia
  blood disorder, 301
  pernicious anemia, 293
anencephaly, nervous system disorder, 194
aneurysm, blood disorder, 301
angina pectoris, blood disorder, 301
angiography, 288
angioplasty, 288
angiotensin II, 405
angiotensin, in regulation of arterial blood
  pressure, 271
anions, negatively charged ions, 24–25
anisotropic (A) bands, of muscle fiber, 142, 143
ankle muscles, 161–162
ankylosing spondylitis, skeletal system disorder, 131
anomaly, 66
anorchism, 440
anosmia, sensory disorder, 216
ANS. *See* autonomic nervous system (ANS)
antacids, 367
antagonists, 139
antagonists, in movement produced by skeletal
  muscles, 149, 149*f*
antebrachium (forearm)
  arteriogram of vessels of, 278*f*
  muscles, 159
  overview of, 119–120
  radius and ulna, anterior view, 119*f*
  radius and ulna, markings on, 120*t*
anteflexed position of uterus, 429
anterior
  body directions, 9
  view of skull, 110*f*
anterior crest, 124
anterior inferior iliac spine, 122
anterior interventricular artery, 251–252
anterior lobe, 228–229
anterior superior iliac spine, 122
anterior tibial arteries, 278
anterior tibial vein, 282
antiarrhythmic, 288
antibodies
  classes of, 319*t*
  creation of, 317
  defined, 307
antibody-mediated immunity
  defined, 307
  illustration of, 318*f*
  overview of, 317
anticholinergic drugs, for GI disorders, 367
anticoagulants
  overview of, 296
  pharmacology for, 297*b*
anticodons, in protein synthesis, 59
antidiarrheal drugs, for GI disorders, 367
antidiuretic hormone (ADH), 229, 405
antigens
  ability of immune system to distinguish self and
    nonself, 315–316
  defined, 307
  vaccine as, 319

antihypertensive agents, 288
antiserum, in passive artificial immunity, 319
antitussives, 342
anucleate, mature RBCs, 293
anuria, 417
anvil (incus)
  auditory ossicles, 113
  of middle ear, 211
aorta
  aortic arch, 276
  ascending aorta, 272
  branches of, 272, 276*f*
  descending aorta, 277
aortic arch, 276, 276*f*
aortic semilunar valve
  pathway of blood through the heart, 251
  valves of heart, 250
apex, of heart, 247
apgar score, 461
apical foramen, 356
apneustic area, of midbrain, 182
apocrine glands
  classification of glands by secretion, 73*f*
  sweat glands, 73–74, 93
aponeurosis, muscle attachment and, 142
appendicitis, 371
appendicular region, of body, 11
appendicular skeleton
  lower extremity, 123–125
  names of bones in, 109*t*
  overview of, 107, 117–125
  pectoral girdle, 117–118
  pelvic girdle, 121–123
  upper extremity, 118–120
appositional growth
  of bones, 107
  defined, 101
aqueous humor, in structure of eyeball, 207–208
arachnoid granulations, 184
arachnoid layer, of meninges, 81, 179
arbor vitae, structures of cerebellum, 183
arcuate arteries, blood flow through kidneys, 400
arcuate veins, blood flow through kidneys, 400
areola, structures of breasts and mammary glands, 433
areolar (loose connective tissue), 74
areolar (loose) connective tissue, 74, 75*f*
arm. *See* brachium (arm)
arrector pili
  defined, 87
  muscle associated with hair follicles, 92
arrhythmias, 253*b*, 301
arterectomy, 288
arteries
  of abdominal viscera, 278
  aortic arch, 276
  descending aorta, 277
  of head and neck, 277
  of lower extremity, 278
  major systemic, 272–278
  overview of, 264–265
  pulmonary, 272
  supply to brain, 277*f*
  umbilical, 283
  of upper extremity, 277
arteriosclerosis, 288
arteriovenous anastomoses, 265
arthritis, skeletal system disorder, 131
arthrocentesis, 136
arthroscopy, 136
articular cartilage
  diarthroses and, 128
  features of long bones, 105
  ossification and, 107

articulations
  amphiarthroses, 127
  diarthroses, 128–129
  overview of, 127–129
  projections for, 106t
  synarthroses, 127
artificial immunity, 307
artificial pacemaker, 260
arytenoid, larynx cartilage, 330, 331
ASA (acetylsalicylic acid), 126b
ascending aorta, pathway of blood through the
  heart, 251
ascending colon, 362
ascending limb, of nephrons, 398
ascending tracts, pathways of spinal cord, 185
aspiration, 349
aspirin, 126b
association areas, of cerebral cortex, 181
association fibers, in transmission within cerebral
  hemispheres, 180
association neurons, 173
asthma, disorder of respiratory system, 344
astigmatism, disorder of the senses, 216
atelectasis, 349
atherectomy, 288
atheroma, 288
atherosclerosis, disorder of blood, 301
atlas, cervical vertebrae, 115
atmospheric pressure, in pulmonary
  ventilation, 333
atomic number, 23
atomic radiation, 24
atoms
  defined, 21
  diagrams of atomic structure of biologically
    important elements, 24f
  isotopes, 23–24
  principal particles of matter, 28t
  structure of, 23–24
ATP. See adenosine triphosphate (ATP)
atrial diastole, in cardiac cycle, 254
atrial natriuretic hormone, 237, 405
atrial systole, in cardiac cycle, 254
atriopeptin, 237
atrioventricular (AV) node, in conduction
  system, 253
atrioventricular (AV) valve
  defined, 245
  valves of heart, 249–250
  ventricular relaxation and contraction, 250f
atrioventricular bundle, in conduction system, 253
atrophy, 66
atrophy, disorder of muscular system, 163
attachment, skeletal muscle, 142
auditory ossicles
  of face, 113
  of middle ear, 211
auditory pathway, 213
auditory senses
  overview of, 210–213
  physiology of hearing, 212–213
  pitch and loudness, 213
  structure of ear, 210–212
auditory tubes
  of middle ear, 211
  opening into nasopharynx, 330
auricles
  chambers of heart, 249
  external ear and, 210
auricular surface, 122
auscultation, 260
autoimmune diseases, 315–316, 325
autonomic ganglion, 188

autonomic nervous system (ANS)
  comparing sympathetic and parasympathetic
    actions, 191t
  comparing sympathetic and parasympathetic
    division, 192t
  divisions of PNS, 187
  general features of, 188
  parasympathetic division of, 190–193
  structure and function of, 192f
  subdivisions of peripheral nervous system, 171
  sympathetic division of, 188–190
autosomes
  chromosomes, 452
  defined, 441
avascular epithelial tissue, 70
axial region, of body, 11
axial skeleton
  cranium, 108–109
  names of bones in, 109t
  overview of, 107–117
  skull, 107–113
axillary artery, 277
axillary vein, 278
axis, cervical vertebrae, 115
axon collaterals, 173
axons, 171–173
azoospermia, 440
azotemia, 417
azygos vein, of thoracic and abdominal wall, 280

B

B cells (B lymphocytes)
  in antibody-mediated immunity, 317
  development of, 316
B lymphocytes (B cells)
  in antibody-mediated immunity, 317
  development of, 316
baclofen (Lioresal), 160
bacterial endocarditis, disorder of blood, 301
balance, functions of cerebellum, 183
ball and socket joint, types of freely movable
  joints, 130f
barriers, mechanical barriers to disease, 312–313
basal cell carcinoma, disorder of integumentary
  system, 96
basal ganglia
  functions of, 181
  overview of, 180
basal metabolic rate (BMR)
  defined, 377
  factors affecting, 383f
  overview of, 382
base, of heart, 247
basement membrane, of epithelial tissue, 70, 72f
bases
  components of nucleotides, 38f
  defined, 21
  neutralization reactions, 33
  overview of, 32
  pH scale for measuring, 32–33
basilar artery, of head and neck, 277
basilar membrane, of inner ear, 211
basilic vein, of shoulders and arms, 278–280
basophils
  granulocytes, 295
  illustration of, 292f
bedsores, 91b
bedsores (decubitus ulcers), 91b
Bell's palsy, 199
belly, of muscle, 142
benign, 66
benign prostatic hypertrophy, 435

benzoyl peroxide-erythromycin (Benzamycin), 93
beta oxidation
  defined, 377
  of fatty acids, 381
bicarbonate ions, in carbon dioxide transport, 339
biceps brachii, forearm and hand muscles, 159
biceps femoris, leg muscles, 161
biconcave disks, 293
bicuspid valve
  AV valves of heart, 250
  pathway of blood through the heart, 251
bicuspids (premolars), 356
bifid, of cervical vertebrae, 115
bile
  cholecystokinin regulating flow of, 365
  production by liver, 365
  secretion of, 364
bile canaliculi, structures of liver, 363–364
bile salts
  as emulsifying agents, 365
  synthesis of, 364
bilirubin, in bile pigmentation, 365
biopsy, 85, 167
biphasic contraceptives, 432b
birth control, 443t. See also oral contraceptives
bisoprolol fumarate (Zebeta), 271
bitter taste, 204b
bladder. See urinary bladder
blastocele, 442–444
blastocyst
  in cleavage phase, 442–444
  formation of, 444f
blastomeres, 442
blisters, between dermis and epidermis, 90b
blood
  ABO blood groups, 297–298
  agglutinogens and agglutinins, 297
  aging and, 300b
  coagulation, 296
  composition of, 290–295, 290f
  connective tissue, 76–77
  disorders of, 301
  erythrocytes, 293–294
  formed elements, 291–295, 292t
  functions and characteristics of, 290
  hemostasis, 295–296
  leukocytes, 295
  light micrograph of, 293f
  pathway of blood through the heart, 251
  pharmacology for, 297b
  plasma, 290–291
  platelet plug formation, 296
  Rh blood groups, 299–300
  summary, 301–302
  thrombocytes, 295
  typing and transfusions, 297–300
  vascular constriction, 296
blood cell formation (hematopoiesis), 103, 291–293
blood flow, 267–268
  fetal. See fetal circulation
  fetal circulation compared with circulation after
    birth, 450f
  Korotkoff sounds and, 269–270
  relationship to blood pressure, 267
  relationship to resistance, 267–268
  through kidneys, 399–400
  velocity of, 267
  venous blood flow, 268
blood pressure
  factors affecting, 270
  measuring, 270f
  overview of, 269–270
  pulse and, 269–271

blood pressure *(Continued)*
    regulation of arterial blood pressure,
        270–271
    relationship of blood flow to, 267
    role of renin in regulation of, 397
    in systemic vessels, 267*f*
blood sugar, maintaining levels of, 34
blood supply
    to liver, 364
    to myocardium, 251–252, 252*f*
    for skeletal muscles, 143
blood types
    ABO blood groups, 297–298
    preferred and permissible for
        transfusion, 299*t*
    Rh blood groups, 299–300
    transfusions and, 297–300
blood urea nitrogen (BUN), 417
blood vessels
    aging and, 284*b*
    aortic arch, 276
    arteries, 264–265
    arteries of abdominal viscera, 278
    arteries of head and neck, 277
    arteries of lower extremity, 278
    arteries of upper extremity, 277
    blood flow, 267–268
    capillaries, 265
    classification and structure of, 264–266
    descending aorta, 277
    fetal circulation, 283, 284*t*
    major systemic arteries, 272–278
    major systemic veins, 278–283
    pharmacology for, 271*b*
    pulmonary circuit, 272
    pulse and blood pressure, 269–271
    role of capillaries in circulation, 266–267
    structure of, 264*f*
    summary, 285
    systemic circuit, 272–283
    veins, 265–266
    veins of abdominal and pelvic organs, 280
    veins of head and neck, 278
    veins of lower extremity, 282–283
    veins of shoulders and arms, 278–280
    veins of thoracic and abdominal walls, 280
blood volume, blood pressure effected by, 270
blood-brain barrier, 183*b*
BMR. *See* basal metabolic rate (BMR)
body
    of mandible, 113
    of sternum, 116–117
    of vertebrae, 114–115
body (shaft), of penis, 425
body cavities, 10–11, 11*f*
body fluids
    acid-base balance, 410*t*
    acid-base balance and, 409–410
    aging and, 411*b*
    disorders of, 412
    electrolyte balance and, 408–409
    fluid compartments, 407, 408*f*
    intake and output of, 408*f*
    intake and output of fluids, 407–408
    overview of, 407–410
    summary, 414
body membranes. *See* membranes
body region, of stomach structure, 357–358
body regions
    overview of, 11
    terms, 12*t*, 13*f*
body systems, levels of organization in human
    body, 2–3

body temperature
    heat loss, 384
    heat production, 384
    overview of, 383–384
    regulation of, 384
    summary, 390
body temperature, in regulation of respiration, 342
body tissues. *See* tissues
body, of uterus, 429
boils, disorder of integumentary system, 96
bone marrow, 105
bones. *See also* skeletal system
    classification of, 104
    connective tissue, 76
    development (ossification) of, 105–107
    of foot, 126*f*
    fractures and fracture repair, 126–127
    growth, 107
    long bones, 104–105
    names by category, 109*t*
    overview of, 104–105
    structure of bone tissue, 103
    terms related to bone markings, 106*t*
bony callus, fracture repair and, 127
bony labyrinth, of inner ear, 211, 211*f*
Bowman's capsule (glomerular capsule), 395, 398
brachial artery, 277
brachial vein, 278
brachialis, forearm and hand muscles, 159
brachiocephalic artery, 276
brachiocephalic veins, 278
brachioradialis, forearm and hand muscles, 159
brachium (arm)
    muscles, 156–159
    overview of, 118–119
    veins of, 278–280
bradycardia, 260
brain
    arterial supply to, 277*f*
    brain stem, 182–183
    cerebellum, 183
    cerebrum, 179–181
    diencephalon, 182
    impulses from brain overriding respiratory system, 342
    midsagittal section of, 180*f*
    organs of central nervous system, 171
    overview of, 179–183
brain stem
    overview of, 182–183
    respiratory center in, 340*f*
brain testicular axis, 425
breast cancer, 435
breasts, 433*f. See also* mammary glands
breathing. *See* respiratory system
broad ligament, of uterus, 429
bronchi, 332
bronchial tree
    CT image of, 333*f*
    defined, 327
    overview of, 332
    terminal branching of, 332*f*
bronchioles, 332
bronchitis, disorder of respiratory system, 344
bronchodilators, 342
bronchogenic carcinoma, 349
bronchopulmonary segment, 327
bronchoscopy, 349
Brunner's glands (duodenal glands), 361
brush border, microvilli and, 359–361
buccinator muscle, 354
buccinator, muscles of facial expression, 152
buffers, 21, 33

bulbourethral (Cowper's) glands, 424
bulbus oculi (eyeball), 206–208, 206*f*
BUN (blood urea nitrogen), 417
bunion, 136
burns
    classification of, 95*f*
    disorder of integumentary system, 96
    overview of, 94–95
bursae, of knee, 128–129
bursitis, 128–129, 131

C

C (carbon), 24*f*
Ca$_{++}$ imbalance, disorder of endocrine system, 239
CABG (coronary artery bypass grafting), 260
calcaneal tendon, foot and ankle muscles, 161–162
calcaneus (heel) bone, 124–125
calcitonin
    effects on blood calcium levels, 232*f*
    hormones of thyroid gland, 231
calcium channel blockers, 271
callus, 99
calories, measuring energy derived from food, 381–382
calyces, macroscopic structures of kidneys, 398
canaliculi, bone tissue and, 76
cancellous (spongy) bone
    diploë, 104*f*
    overview of, 103
    structure of, 104*f*
cancer cells, spreading via lymphatic system, 311*b*
canine teeth, 356
capacitation, in fertilization, 442
capillaries
    circulatory role of, 266–267
    of lung, 272
    lymph capillaries, 310, 310*f*
    microcirculation, 266*f*
    organization of capillary network, 265*f*
    overview of, 265
capitulum, of arm, 118–119
caprylic acid, 36*f*
capsular hydrostatic pressure (CHP), 403*f*
capsule, surrounding lymph node, 311
captopril (Capoten), 271
carbaminohemoglobin
    carbon dioxide transport, 339
    defined, 327
carbidopa-levodopa (Sinemet), 187
carbohydrates
    chemical digestion, 367
    defined, 21
    dietary, 385*t*
    digestion of, 368*f*
    interconversion of carbohydrates, proteins, and
        lipids, 381
    list of important, 35*t*
    metabolism by liver, 365
    metabolism of, 378–380
    nutrition and, 385
    overview of, 33–34
carbon (C), 24*f*
carbon dioxide
    diffusion of oxygen and carbon dioxide in lungs, 52*f*
    transport, 339–340, 339*f*
carbonic acid, in carbon dioxide transport, 339
carbonic anhydrase, in carbon dioxide transport, 339
carcinogens, 66
carcinomas
    adenocarcinoma, 74
    basal cell, 96
    bronchogenic carcinoma, 349
    overview of, 85
    squamous cell, 97

cardiac arrest, 260
cardiac catheterization, 260
cardiac center
 mediating heart rate, 256
 of midbrain, 182–183
cardiac cycle
 defined, 245
 illustration of, 254*f*
 overview of, 254–255
cardiac muscle tissue, 78, 78*f*
cardiac notch, 332
cardiac output
 blood pressure effected by, 270
 defined, 245
 heart rate, 255–256
 overview of, 255–256
 stroke volume, 255
cardiac region, of stomach structure, 357–358
cardiac veins, 252
cardiomegaly, 260
cardiomyopathy, 260
cardiovascular system
 aging and, 256*b*
 blood. *See* blood
 blood supply to myocardium, 251–252, 252*f*
 blood vessels. *See* blood vessels
 cardiac cycle, 254–255
 cardiac output, 255–256
 chambers of heart, 248–249
 conduction system, 252–254
 coverings of heart, 247
 CT image of heart and vessels, 248*f*
 form, size, and location of heart, 247
 frontal view of mediastinum, 247*f*
 functional relationships of, 246*f*
 heart sounds, 255
 illustration of, 4*f*
 internal view of heart, 249*f*
 layers of heart wall, 248
 overview of, 6
 pathway of blood through the heart,
  251, 251*f*
 pharmacology for, 253*b*
 relationship with lymphatic system, 309*f*
 summary, 257
 valves of heart, 249–250
caries (dental cavities), 356*b*
carina, 332
carisoprodol (Soma, Muslax, Rotalin), 160
carotene, skin color and, 91
carotid sinus, 277
carpal bones, 120
carpal tunnel syndrome, 136
carpus (wrist)
 CT image of, 121*f*
 overview of, 120
cartilage
 articular cartilage, 105, 107, 128
 connective tissue, 76
 costal cartilage of ribs, 117
 hyaline cartilage, 76, 76*f*
 larynx cartilage, 330, 331
 semilunar cartilage of knee, 128–129
 spaces in cartilage matrix, 76
catabolism, 1, 6–7, 378
 of amino acids, 380*f*
 defined, 377
 heat production, 384
catalysts, in chemical reactions, 30
cataracts, vision disorder, 216
catheterization, 417
cations, positively charged ions, 24
cauda equina, of spinal cord, 185

caudate lobe, of liver, 363
cavities, bones, 106*t*
cecum region, of large intestine, 362
celiac artery
 arteries of abdominal viscera, 278
 branches of abdominal aorta, 277
celiac disease, 392
cell body
 nerve cells, 78
 of neuron, 171–173
cell division
 meiosis, 57
 mitosis, 56–57
 overview of, 56–57
cell functions
 active transport pumps, 54–55
 cell division, 56–57
 DNA replication, 58–59
 endocytosis, 55–56
 exocytosis, 56
 facilitated diffusion, 52–53
 filtration, 54–56
 medical vocabulary, 65–67
 meiosis, 57
 mitosis, 56–57
 movement of substances across cell membrane, 51–54
 osmosis, 53–54
 overview of, 51
 protein synthesis, 59–60
 quiz, 64
 simple diffusion, 52
 summary, 63–64
cell membrane. *See* plasma membrane
cell structure
 cytoplasm, 48
 cytoplasmic organelles, 49–50
 filamentous protein organelles, 50
 functions of cellular components, 47*t*
 illustration of, 46*f*
 medical vocabulary, 65–67
 nucleus, 48–49
 overview of, 46–50
 plasma membrane, 46–48
 summary, 63
cell-mediated immunity
 defined, 307
 illustration of, 317*f*
 overview of, 316
cells of Leydig (interstitial cells), 422
cellular levels, in human bod, 2–3
cellular metabolism, 378
cellular physiology, 2
cellular respiration, 329, 378
cellulitis, disorder of integumentary system, 96
cementum, 356
central canal, of spinal cord, 185
central nervous system (CNS)
 brain, 179–183
 brain stem, 182–183
 cerebellum, 183
 cerebrum, 179–181
 diencephalon, 182
 meninges, 178–179
 overview of, 178–186
 spinal cord, 184–186
 subdivisions of nervous system, 171
 ventricles and cerebrospinal fluid, 184
central sulcus, of cerebrum, 179–180
central vein
 blood supply to liver, 364
 of liver, 363–364
central venous pressure, 263, 267
centrioles, 47*t*, 50

centromere, 56
centrum, of vertebrae, 114–115
cephalic phase, of gastric secretion, 359
cephalic vein, of shoulders and arms, 278–280
cerebellar cortex, 183
cerebellar hemispheres, 183
cerebellar peduncles, 183
cerebellum
 functions of, 183
 overview of, 183
cerebral aqueduct, of midbrain, 182, 184
cerebral concussion, 194
cerebral contusion, 194
cerebral cortex
 functional regions of, 181*t*
 overview of, 180
cerebral hemispheres, 179
cerebral palsy, 199
cerebral peduncles, of midbrain, 182
cerebrospinal fluid (CSF)
 lumbar punctures and, 186*b*
 overview of, 184
cerebrovascular accident (CVA), 199
cerebrum
 functions of hemispheres of, 180*b*
 lobes and landmarks of, 179*f*
 overview of, 179–181
cerumen, 93, 210
ceruminous gland
 defined, 87
 external ear and, 210
 overview of, 93
cervical curvature, of vertebral column, 113
cervical enlargement, of spinal cor, 185
cervical nerves (C1 to C8), 188
cervical vertebrae, 115
cervix, 429
cesarean section, 461
chambers of heart, 248–249, 249*f*
cheeks, mouth components in digestion, 354
chemical actions, in disease resistance, 313
chemical bonds
 covalent bonds, 25–26
 hydrogen bonds, 27
 ionic bonds, 24–25
 overview of, 24–27
chemical digestion, 367–368
 carbohydrate digestion, 367
 digestive processes, 353
 lipid digestion, 368
 protein digestion, 367–368
chemical equations, 28–29
chemical level, of organization in human body, 2–3
chemical names, vs. generic names of drugs, 30*b*
chemical reactions
 chemical equations, 28–29
 neutralization reactions, 33
 overview of, 28–30
 reaction rates, 29–30
 reversible, 30
 types of, 29
chemistry
 atomic structure, 23–24
 chemical bonds, 24–27
 chemical reactions, 28–30
 compounds and molecules, 27–28
 electrolytes, acids, bases, and buffers, 31–33
 elements, 22
 medical vocabulary, 42–44
 mixtures, solutions, and suspension, 31
 organic compounds, 33–39
 overview of, 22
 summary, 39

chemoreceptors
    affecting breathing, 340–341
    defined, 201
    in gustatory sense, 203
    in olfactory sense, 204
    types of sense receptors, 202
chief cells, in stomach structure, 358
childhood, 452
chiropractor, 136
chlorine (Cl)
    diagrams of atomic structure of, 24f
    formation of ionic bonds, 25f
chloroform (trichloromethane), use as anesthetic, 46b
chloroform, use as anesthetic, 46b
chlorzoxazone (Paraflex, Parafon Forte
    DSC), 160
cholecystitis, 371
cholecystokinin, 237
    flow of bile and, 365
    hormones of digestive system, 361
    pancreas and, 366–367
cholelithiasis, 371
cholesterol
    good vs. bad, 387b
    sources of fat and cholesterol, 387t
cholinergic fibers, 190–193
cholinesterase, 176
chondrin, cartilage protein, 76
chondrocytes
    cartilage cells, 76
    defined, 69
chordae tendineae, of AV valve, 249–250
chorion
    defined, 441
    formation of extraembryonic
        membranes, 446
choroid plexus, 184
choroid, in structure of eyeball, 207
CHP (capsular hydrostatic pressure), 403f
chromatids, in spermatogenesis, 422–423
chromatids, of chromosomes, 56
chromatin
    overview of, 48
    structure and function of cellular components, 47t
chromosomes
    chromatids of, 56
    hereditary disorders, 455
    overview of, 48
    transmission to offspring, 453
chronic diseases, 18
chronic obstructive pulmonary disease (COPD), 349
chyle, 309
    in absorption of nutrients, 369
chylomicrons
    in absorption of nutrients, 369
    defined, 351
chyme
    defined, 351
    production in stomach, 359
cicatrix, 99
cilia
    aiding in movement of secretions along cell
        surface, 71
    overview of, 50
    structure and function of cellular components, 47t
ciliary body, structure of eyeball, 207
ciliary muscle, structure of eyeball, 207
ciliary processes, structure of eyeball, 207
circle of Willis, 277, 277f
circumcision, 425, 440
circumduction, types of movement, 150f
circumflex artery, blood supply to myocardium, 251–252
cirrhosis, 371, 381b

cisterna chyli, 310
citric acid cycle (Krebs cycle)
    carbohydrate digestion and, 378–380
    defined, 377
CK (creatine kinase), enzyme indicators of myocardial
    infarction, 252b
Cl (chlorine)
    diagrams of atomic structure of, 24f
    formation of ionic bonds, 25f
clavicle (collarbone), 117, 118, 118f
cleavage
    defined, 441
    formation of blastocyst, 444f
    in preembryonic period, 442–444
cleavage lines, in dermis, 90
climax, in male sexual response, 425
clinical age, 442
clitoris, 430
clone, T cell, 316
closed reduction, fracture repair, 126
CNS. See central nervous system (CNS)
Co (coccygeal nerve), 188
coagulation
    defined, 289
    hemostasis and, 296
coagulopathy, 305
coccygeal nerve (Co), 188
coccyx, 116
cochlea
    of inner ear, 211
    pathway of pressure waves in, 213f
    photomicrograph of, 212f
    section of, 211f
cochlear branch, of inner ear, 211–212
cochlear duct, of inner ear, 211
cochlear nerve, auditory pathway and, 213
codominance, of genetic traits, 454
codons, in protein synthesis, 59
coitus, 440
colchicine, for treating gout, 407
cold receptors, in proprioception, 202–203
collagenous fibers, 69, 74
collarbone (clavicle), 117, 118, 118f
collateral ganglia, of ANS, 190
collecting ducts, 399
colloidal suspension, 31
colon, 362
color blindness, 209b, 216, 457f
colostrum, 451
columnar cells
    classification of epithelia, 70
    pseudostratified columnar epithelium, 71
    simple columnar epithelium, 71
combining vowels, linking word parts, 16
commissural fibers, transmission within cerebral
    hemispheres, 180
common bile duct, 365
common hepatic artery, 278, 364
common iliac veins, 282
compact bone
    overview of, 103
    structure of, 104f
compensation, for acidosis or alkalosis, 409
complement proteins, as barrier to disease, 313
complete proteins
    defined, 377
    overview of, 386
complex sugars (polysaccharides), 34, 35f, 385
composition, of blood, 290–295, 290f
compound glands, 73, 73f
compounds
    defined, 21
    nature of, 27–28

computed tomography (CT), 199
concentration gradients, in movement of substances
    across plasma membrane, 52
conducting passages, of respiratory system
    bronchi and bronchial tree, 332
    illustration of, 344f
    larynx, 330–331
    lungs, 332–333
    nose and nasal cavities, 329–330
    overview of, 329–333
    pharynx, 330
    trachea, 331
conduction deafness, hearing disorder, 216
conduction myofiber, 245, 253
conduction system
    atrioventricular bundle, bundle branches, and
        conduction myofiber, 253
    atrioventricular node, 253
    defined, 245
    electrocardiogram of, 254
    illustration of, 253f
    overview of, 252–254
    sinoatrial node, 252–253
conduction, sources of heat loss, 384
conductivity, of neurons, 174
condyles
    lateral and medial of leg, 124
    lateral and medial of thigh, 123
condyloid (ellipsoidal) joint, types of freely movable
    joints, 130f
cones
    comparing with rods, 209t
    as photoreceptors, 208
    of retina (nervous tunic), 207f
    in structure of eyeball, 207
congenital disorder, 66, 455
congestive heart failure, 260, 344
conjunctiva, 205
connective tissue
    adipose, 75
    blood, 76–77
    bone, 76
    cartilage, 76
    dense fibrous connective tissue, 75–76
    elastic connective tissue, 76
    loose connective tissue (areolar), 74
    overview of, 74–77
    summary of, 75t
contact dermatitis
    disorder of integumentary system, 96
    disorder of lymphatic system, 322
contraception
    methods of birth control, 443t
    oral contraceptives, 432b
    pharmacology for, 432b
contraction phase, in contraction of whole muscle, 145
contraction, of skeletal muscles
    energy sources and oxygen debt in, 146–149
    functions of, 141
    sarcomere contraction, 144–145
    sliding filament theory of, 144–145, 144f
    stimulus for, 143–144
    summary of, 145t
    of whole muscle, 145–146
conus medullaris, of spinal cor, 184
convection, sources of heat loss, 384
convergence circuit, nerve impulses, 177
Cooper's (suspensory) ligaments, of
    breast, 433
coordination, functions of cerebellum, 183
COPD (chronic obstructive pulmonary disease), 349
cor pulmonale, 260
coracoid process, in pectoral girdle, 118

core temperature
    of body, 383–384
    defined, 377
cornea, of eye, 206
corniculate, larynx cartilage, 330
corona radiata, in ovulation, 428
coronal plane, of body, 10
coronal suture, parietal bones and, 108
coronary arteries
    blood supply to myocardium, 251–252
    CT image of, 251*f*
coronary artery bypass grafting (CABG), 260
coronary artery disease, 301
coronoid fossa, of arm, 118–119
coronoid process, of antebrachium (forearm), 119–120
corpora cavernosa, 424
corpora quadrigemina, of midbrain, 182
corpus albicans, in ovulation, 428
corpus callosum, 179, 180
corpus luteum, in ovulation, 428
corpus spongiosum, 424
cortex, 91–92, 426
corticospinal tracts, pathways of spinal cord, 185
cortisol
    hormones of adrenal cortex, 233
    regulation of, 234*f*
coryza, 349
costal cartilage, of ribs, 117
cough, antitussives for, 342
covalent bonds
    defined, 21
    double covalent bond in carbon dioxide, 26*f*
    double covalent bond in oxygen gas, 26*f*
    overview of, 25–26
    polar covalent bond, 26
    polar covalent bond between oxygen and
        hydrogen, 27*f*
    single covalent bond in hydrogen gas, 25*f*
    single covalent bond in methane gas, 26*f*
    triple covalent bond in nitrogen gas, 27*f*
coverings of heart, 247
Cowper's (bulbourethral) glands, 424
coxal bones
    lateral view, 121*f*
    markings on, 122*t*
    of pelvic girdle, 121
cramp, 167
cranial cavity, 10
cranial nerves
    illustration of, 189*f*
    peripheral nervous system, 187–188
    summary of, 189*t*
craniosacral division, of ANS, 190–193
cranium
    cranial floor viewed from above, 111*f*
    ethmoid bone, 109
    frontal bone of, 108
    occipital bone of, 108
    overview of, 108–109
    parietal bones of, 108
    sphenoid bone, 109
    temporal bones, 108
creatine kinase (CK), enzyme indicators of myocardial
    infarction, 252*b*
creatine phosphate, sources of energy in muscle
    contraction, 147, 147*f*
cremaster muscle, 421
crenate, osmosis and, 53
cretinism, disorder of endocrine system, 239
cribrifrom plate, of ethmoid, 109
cricoid, larynx cartilage, 330, 331
crista ampullaris, in dynamic equilibrium, 214, 214*f*
crista galli, cranial cavity, 109

Crohn's disease, 371
croup, 349
crown, of teeth, 356
cryptorchidism, 421, 435
CSF (cerebrospinal fluid)
    lumbar punctures and, 186*b*
    overview of, 184
CT (computed tomography), 199
cuboidal cells
    classification of epithelia, 70
    simple cuboidal epithelium, 71
cuneiform cartilage, larynx, 330
cupula, dynamic equilibrium and, 214
curettage, 440
Cushing's syndrome, disorder of endocrine system, 239
cuspids (canines), 356
cusps, of AV valve, 249–250
cutaneous membrane, 80
    epidermis, 89–90
    glands of, 92–93
    layers of, 89
cuticle (eponychium), of nails, 91–92
CVA (cerebrovascular accident), 199
cyclobenzaprine (Flexeril), 160
cyclosporine, interleukin-2 inhibitor, 316*b*
cystic duct, attaching gallbladder to liver, 365
cystic fibrosis, 48*b*, 344
cystitis, 412
cystoscopy, 417
cytokinesis
    defined, 45
    division of cytoplasm, 56, 57
cytology, 66
cytoplasm
    cytokinesis (division), 57
    overview of, 48
    structure and function of cellular components, 47*t*
cytoplasmic organelles
    endoplasmic reticulum, 49
    Golgi apparatus, 49–50
    lysosomes, 50
    mitochondria, 49
    overview of, 49–50
    ribosomes, 49
cytoskeleton
    filamentous protein organelles, 50
    structure and function of cellular components, 47*t*

D

dalteparin, as anticoagulant, 297
Dalton's law of partial pressures, 335
dartos muscle, 421
deamination
    of amino acids, 380
    defined, 377
deciduous (primary) teeth, 355
deciduous teeth, 355*f*
decomposition reactions, types of chemical reactions,
    29
decubitus ulcers (bedsores), 91*b*
decussation, 182–183
deep back muscles, 153–154
deep, directions of body, 10
defecation, 353
defense mechanisms, illustration of, 313*f*
defense mechanisms, nonspecific
    chemical actions in disease resistance,
      313
    components of, 315*f*
    inflammation in disease resistance, 314
    mechanical barriers to disease, 312–313
    overview of, 312–314
    phagocytosis in disease resistance, 314

defense mechanisms, specific
    acquired immunity, 319
    antibody-mediated immunity, 317
    cell-mediated immunity, 316
    lymphocytes developed by immune system, 316
    overview of, 315–317
    recognition of self and nonself by immune system,
      315–316
defibrillation, 260
deglutition, 353
dehydration synthesis, 378
deletions, genetic mutations, 455
delivery, 448–449
deltoid tuberosity, of arm, 118–119
deltoid, shoulder and arm muscles, 159
dendrites
    nerve cells, 78
    neurons and, 171–173
dens, cervical vertebrae, 115
dense fibrous connective tissue, 75–76
dental cavities (caries), 356*b*
dentin, 356
deoxyhemoglobin, 293
deoxyribonucleic acid (DNA), 37
    compared with RNA, 31*t*
    DNA molecule before replication, 59*f*
    DNA replication, 58–59, 60*f*
    in protein synthesis, 59–60
    structure of, 38*f*
deoxyribose, pentose monosaccharides, 34
depolarization, stimulation of neurons and, 174
depressions, bone, 106*t*
dermatitis
    contact dermatitis, 96
    defined, 99
dermatomes, in nerve enervation, 188
dermis
    defined, 87
    overview of, 90
    in structure of skin, 89*f*
descending aorta, 277
descending colon, 362
descending limb, of nephrons, 398
descending tracts, pathways of spinal cord, 185
desmopressin (DDAVP), 229–230
detoxification, functions of liver, 365
detrusor muscle, of urinary bladder, 401
deuterium, atomic structure of, 24*f*
development
    embryonic period, 445–447
    fertilization, 442
    fetal stage, 447–448
    implantation, 444–445
    overview of, 442
    parturition and lactation, 448–451
    postnatal, 451–452
    preembryonic period, 442–445
developmental age, 442
diabetes insipidus, 239, 412
diabetes mellitus, 240, 371, 412
diagnosis, 18
dialysis, 417
diapedesis, 289, 295
diaphragm
    inspiration and, 334
    thoracic wall muscles, 154
diaphysis
    defined, 101
    of long bones, 105
diarthroses
    defined, 101
    overview of, 128–129
    types of freely movable joints, 130*f*

diastole
    in cardiac cycle, 254
    defined, 245
diastolic pressure
    blood pressure, 269
    defined, 263
diazoxide (Hyperstat), 271
diencephalon, 182
differentiation
    of cells, 46
    defined, 1
    life processes, 7
diffusion
    defined, 45
    facilitated diffusion, 52–53, 53*f*
    of oxygen and carbon dioxide in lungs, 52*f*
    simple diffusion, 52, 52*f*
digestion, 7
    defined, 351
digestive enzymes
    digestive processes, 353
    overview of, 368*t*
    secretions of small intestines, 361–362
digestive system
    absorption, 368–369
    accessory organs of digestion, 363–367
    aging and, 370*b*
    carbohydrate digestion, 367
    chemical digestion, 367–368
    disorders of, 371–372
    enzymes and hormones of, 368*t*
    esophagus, 357
    functional relationships of, 353*f*
    gallbladder, 365
    general structure of digestive tract, 353–354
    illustration of, 4*f*
    large intestine, 362–363
    lipid digestion, 368
    liver, 363–365
    mouth (oral cavity), 354–357
    organs of, 353*f*
    overview of, 6, 353
    pancreas (islets of Langerhans), 365–367
    pharmacology for GI disorders, 367*b*
    pharynx, 357
    protein digestion, 367–368
    small intestine, 359–362
    stomach, 357–359
    summary, 372–373
digestive tract, 354
    esophagus, 357
    general structure of, 353–354
    large intestine, 362–363
    layers of, 354*f*
    mouth (oral cavity), 354–357
    overview of, 353
    pharynx, 357
    small intestine, 359–362
    stomach, 357–359
digoxin (Lanoxin, Cardoxin), 253*b*
dihydroergotamine mesylate (DHE45,
    Migranal), 187
dilation stage, of labor, 449
diltiazem hydrochloride (Cardizem,
    Tiazac), 271
diploë, middle layer of spongy bone, 104
direct attachment, muscles, 142
directions, in body, 9–10
disaccharide
    pentose monosaccharides, 34
    representation of, 35*f*
disease resistance. *See* resistance to disease
dislocation, disorder of skeletal system, 131

disorders, 322
    of blood, 301
    digestive system, 371–372
    of endocrine system, 239*f*
    hereditary, 455
    of integumentary system, 96–97
    of muscular system, 163*f*
    of nervous system, 194*f*
    of reproductive system, 435–436
    of respiratory system, 344–345
    of skeletal system, 131*f*
    of special senses, 216*f*
    of urinary system and body fluids, 412
dissociation
    of electrolytes in solution, 31
    of sodium chloride in water, 32*f*
distal, directions of body, 9–10
disuse atrophy, disorder of muscular system, 163
diuresis, 417
diuretics, 229*b*, 271, 417
divergence circuit, nerve impulses, 177
dizygotic twins, 442
DNA. *See* deoxyribonucleic acid (DNA)
dominant genetic traits, 453
donors, blood, 298
Doral funiculi, of white matter in spinal cord, 185
Doral horns, of gray matter in spinal cord, 185
dorsal (posterior) median sulcus, of spinal cor, 185
dorsal cavity, 10
dorsal root ganglion, 188
dorsal root, of spinal nerves, 188
dorsalis pedis artery, arteries of lower extremity, 278
dorsiflexion, types of movement, 150*f*
double replacement, chemical reactions, 29
dual innervation, ANS and, 188
duct system, male reproduction, 423–424
    ductus deferens, 423–424
    ejaculatory duct, 424
    epididymis, 423
    overview of, 423
    urethra, 424
ductus arteriosus, fetal circulation and, 283
ductus deferens, 423–424
duodenal glands (Brunner's glands), 361
duodenum, 361
dura matter, of meninges, 81, 178–179
dural sinuses, in meninges, 178–179
dwarfism, disorder of endocrine system, 240
dynamic equilibrium
    defined, 201
    in sense of equilibrium, 213
dysmenorrhea, 440
dysphagia, 371
dysplasia, 66
dystocia, 462
dysuria, 417

E
ears
    external ear, 210
    inner ear, 211–212
    middle ear, 210–211
    overview of, 210
    pharmacology for, 206*b*
    structure of, 210–212
ecchymosis, 99, 305
ECF. *See* extracellular fluid (ECF)
ECG (electrocardiogram), 254, 254*f*
echocardiography, 260
eclampsia, 462
ectoderm
    derivatives of, 448*t*
    formation of primary germ layers, 445

ectopic focus, 253*b*
ectopic pregnancy, 462
eczema, 96
edema, 407*b*
EEG (electroencephalography), 199
effectors, muscles and glands as, 171
efferent (motor)
    division of peripheral nervous system, 171
    neurons, 173
efferent arteriole, 398
efferent lymphatic vessels, 311
ejaculation
    defined, 419
    of semen, 425
ejaculatory duct, 424
EKG (electrocardiogram), 254, 254*f*
elastic cartilage, 76
elastic connective tissue, 76
elastic fibers, 69, 74
elbow, relationship of radius, ulna, and
    humerus, 120*f*
electrocardiogram (ECG or EKG),
    254, 254*f*
electroencephalography (EEG), 199
electrolytes
    balance of, 408–409
    overview of, 31
    in plasma, 291
electromyography, 167
electrons
    in atomic structure, 23
    principal particles of matter, 28*t*
electron-transport chain, 378–380
elements
    abbreviated periodic table, 23*f*
    defined, 21
    diagrams of atomic structure of biologically
        important elements, 24*f*
    essential for human life, 22*t*
    overview of, 22
elimination, digestive processes, 353
ellipsoidal (condyloid) joint, types of freely movable
    joints, 130*f*
embolus, 305
embryo
    defined, 445–446
    ultrasound of, 446*f*
embryonic disk, 445
embryonic period
    defined, 441
    formation of extraembryonic membranes, 446
    formation of the placenta, 446–447
    organogenesis, 447
    overview of, 445–446
emission
    defined, 419
    of semen into urethra, 425
emphysema, disorder of respiratory system, 344
emulsification, of fats
    bile salts for, 365
    overview of, 368
enamel, covering of teeth, 356
encephalitis, disorder of nervous system, 194
endemic, 43
endergonic, chemical reactions, 29
endochondral bones, 105–107
endochondral ossification
    defined, 101
    events in, 106*f*
endocrine cells
    of pancreas, 366*f*
    in stomach structure, 358

endocrine glands
    defined, 223
    exocrine gland compared with, 225
    glandular epithelium, 73
    hormones of, 227*t*
    illustration of, 226*f*
    lesser glands, 237
    list of glands and hormones, 227*t*
endocrine system
    adrenal glands, 232–234
    aging and, 238*b*
    anterior lobe hormones, 228–229
    characteristics of hormones, 225–226
    comparing exocrine and endocrine glands, 225
    comparing with nervous system, 225
    disorders of, 239*f*
    endocrine glands and hormones, 227*t*
    functional relationships of, 224*f*
    gonads, 236
    illustration of, 4*f*
    lesser endocrine glands, 237
    overview of, 6
    pancreas (islets of Langerhans), 235–236
    parathyroid glands, 231–232
    pharmacology for, 229*b*
    pineal gland, 237
    pituitary gland hormones, 226–230
    posterior lobe hormones, 229–230
    prostaglandins, 237–238
    thyroid gland, 230–231
endocrinology, 243
endocrinopathy, 243
endocytosis, 45, 55–56
endoderm
    derivatives of, 448*t*
    formation of primary germ layers, 445
endolymph
    defined, 201
    of inner ear, 211
endometriosis, 435
endometrium layer, of uterus, 429
endomysium, in muscle structure, 141–142
endoneurium, in nerve structure, 187
endoplasmic reticulum
    cytoplasmic organelles, 49
    structure and function of cellular components, 47*t*
endosteum
    defined, 101
    features of long bones, 105
energy from foods
    aging and, 389*b*
    basal metabolic rate (BMR), 382
    carbohydrates, 378–380
    lipids, 381
    overview of, 378–381
    physical activity, 382
    proteins, 380–381
    thermogenesis, 382
    uses of, 381–382
energy levels (shells), electrons located in, 23
energy sources, for muscular contraction, 146–149
enoxaparin, as anticoagulant, 297
enteral drug administration, 54*b*
enteral nutrition, 383
enterokinase, digestive enzyme, 361
enuresis, 417
environmental requirements, for life
    heat, 7
    nutrients, 7
    overview of, 7
    oxygen, 7
    pressure, 7
    water, 7

enzymes
    digestive. *See* digestive enzymes
    of pancreas, 366
enzymes, as organic catalyst, 30
eosinophils
    granulocytes, 295
    illustration of, 292*f*
epicardium, of heart, 247, 248
epicondyles
    of arm, 118–119
    of thigh, 123
epidemic, 43
epidermal derivatives
    glands, 92–93
    hair and hair follicles, 91–92
    nails, 92
    overview of, 91–93
epidermis
    defined, 87
    overview of, 89–90
    in structure of skin, 89*f*
epididymis, in duct system, 423
epidural space, of spinal cord, 184
epigastric region, of abdominopelvic cavity, 11
epiglottis cartilage, larynx, 330, 331
epimysium, in muscle structure, 141–142
epinephrine
    effects and control of, 235*f*
    hormones of adrenal medulla, 234
epineurium, structure of nerves and, 187
epiphyseal line
    bone growth and, 107
    of long bones, 105
epiphyseal plate
    defined, 101
    of long bones, 105
epiphysis
    defined, 101
    of long bones, 105
epiphysis cerebri. *See* pineal gland
epiploic appendages, structures of large intestine, 362
episiotomy, 440, 462
epithalamus, in diencephalon, 182
epithelial tissues
    classification by shape and layers, 70*f*
    glandular epithelium, 73–74
    overview of, 70–74
    pseudostratified columnar epithelium, 71
    simple columnar epithelium, 71
    simple cuboidal epithelium, 71
    simple squamous epithelium, 70
    stratified squamous epithelium, 71–72
    summary, 72*t*
    transitional epithelium, 72
eponychium (cuticle), of nails, 91–92
eponym, 18
equilibrium
    chemical reactions and, 30
    in movement of substances across plasma
      membrane, 52
equilibrium, sense of, 213–214
erection, in male sexual response, 425
erector spinae, vertebral column muscles, 153–154
erythrocytes
    in blood composition, 290
    characteristics and functions of, 293
    defined, 69, 289
    destruction of, 293–294
    formed elements of blood, 291
    hemolysis, 53–54
    illustration of, 292*f*
    life cycle of, 294*f*
    overview of, 293–294

erythrocytes *(Continued)*
    production of, 293
    red blood cells, 76–77, 77*f*
    regulation of production of, 294*f*
    role of erythropoietin in production of, 397
erythrocytosis, 305
erythropoiesis, 289
erythropoietin
    defined, 289
    in production of erythrocytes, 293
    in red blood cell production, 397
eschar, 99
esophagus, 357
essential amino acids, 385, 386*t*
estrogens
    hormones of adrenal cortex, 233
    hormones of female sexuality, 430
    hormones of ovaries, 236
    produced by placenta during
      pregnancy, 237
    secretion by placenta, 447*t*
ethmoid bone, of cranium, 109
ethmoidal sinuses, 109
etiology, 43
eutocia, 462
evaporation, sources of heat loss, 384
eversion, types of movement, 150*f*
excitability, of neurons, 174
excitatory transmission, 176–177
excretion, as life process, 7
excretion, functions of liver, 365
exercise, benefits of and types of, 162
exergonic, chemical reactions, 29
exhalation. *See* expiration
exocrine cells, of pancreas, 366*f*
exocrine gland
    defined, 223
    endocrine gland compared with, 225
    glandular epithelium, 73
exocytosis, 45, 56, 56*f*
exophthalmic, 243
exostosis, 136
expiration
    expiratory area of respiratory center, 340
    overview of, 334
expiratory reserve volume, lungs, 335
expulsion stage, of labor, 449
extension, types of movement, 150*f*
external anal sphincter, 362
external auditory meatus
    external ear and, 210
    temporal bones and, 108
external carotid artery, of head and neck, 277
external ear, 210
external genitalia
    illustration of, 430*f*
    overview of, 430
external iliac artery, 278
external iliac veins, 282
external intercostal muscles
    inspiration and, 334
    thoracic wall muscles, 154
external jugular veins, 278
external nares, 329
external oblique, abdominal wall
    muscles, 155–156
external os, structures of uterus, 429
external respiration, 327, 329, 337, 338*f*
external urethral orifice, 401
external urethral sphincter, 401
extracellular fluid (ECF)
    defined, 395
    overview of, 407

extraembryonic membranes
    formation in embryonic period, 446
    illustration of, 446f
extrapyramidal tracts, pathways of spinal cord, 185
eyeball (bulbus oculi), 206–208, 206f
eyebrows, 205
eyelashes, 205
eyelids, 205
eyes
    muscles of, 205t
    pathway of light and refraction, 208
    pharmacology for, 206b
    protective features and accessory structures of, 205
    structure of eyeball, 206–208, 206f

F

facial bones
    auditory ossicles, 113
    inferior nasal conchae, 113
    lacrimal bones, 113
    mandible, 113
    maxillary bones, 113
    nasal bones, 113
    overview of, 113
    palatine bones, 113
    vomer, 113
    zygomatic bones, 113
facial expression, muscles of, 152
facial nerve, in gustatory sense, 204
facilitated diffusion, 52–53, 53f
falciform ligament, structures of liver, 363
fallopian tubes. See uterine tubes
false labor, 448
false pelvis, 122–123
false ribs, 117
false vocal cords, 331
falx cerebelli, of cerebellum, 183
falx cerebri, of cerebrum, 179
fasciculus
    defined, 139
    in muscle structure, 141–142
    structure of nerves and, 187
fasting blood sugar, 243
fat soluble vitamins, 387
fat tissue (adipose). See adipose (fat) tissue
fats. See triglycerides
fatty acids
    beta oxidation of, 381
    in formation of triglycerides, 37f
    free fatty acids, 368
    list of, 37t
    overview of, 36
    sources of energy in muscle contraction, 147
fauces, 330, 357
female reproductive system
    external genitalia, 430
    female sexual response, 430
    genital tract, 429
    hormonal control, 430–432
    mammary glands, 433
    organs of, 427f
    ovaries, 426–428
    overview of, 426–433
female sexual response, 430
femoral arteries, 278
femoral veins, 282
femur, 123
    anterior and posterior views, 123f
    head of, 123
    markings on, 124t
    neck of, 123
    radiograph showing relationship of femur, tibia, and fibula, 125f

fertilization
    events in, 443f
    overview of, 442
fetal circulation
    blood vessels, 283, 284t
    compared with circulation after birth, 450f
    illustration of, 283f
    special features of, 449t
    summary of structures in, 284t
fetal period
    defined, 441
    overview of, 447–448
fetus, 447
fever, in systemic inflammation, 314
fiber, types of polysaccharides, 385
fibers
    collagenous and elastic, 74
    nerve, 171–173
    oxygen stored in muscle fibers, 147
    skeletal muscle, 142–143
fibrillation, 260
fibrinogen, plasma proteins, 291
fibrinolysis, in coagulation and healing
    process, 296
fibroblast cells
    defined, 69
    types of cells in connective tissue, 74
fibrocartilage, 76
fibrocartilaginous callus, fracture repair
    and, 127
fibrocystic disease, 435
fibrosis, tissue repair, 79
fibrous coat, of ureters, 401
fibrous pericardium, coverings of
    heart, 247
fibrous tunic, structure of eyeball, 206
fibula
    anterior view, 124f
    head of, 124
    of leg, 124
    markings on, 125t
    radiograph showing relationship of femur, tibia,
        and fibula, 125f
fight-or-flight, 188–190
filamentous protein organelles, 50
filtering blood, functions of liver, 365
filtrates, glomerular filtration, 402
filtration
    active transport pumps, 54–55
    endocytosis, 55–56
    exocytosis, 56
    overview of, 54–56
filtration membrane, glomerular filtration, 402
filtration pressure
    factors in, 403f
    glomerular filtration, 402
filum terminale, of spinal cord, 184
fimbriae, of uterine tubes, 429
fingers (phalanges), 120
first heart sound (lubb), 255
first messenger, mechanisms of hormone actions, 226
first polar body, 427–428
flagella, 47t, 50
flat bones, 104
flexion, types of movement, 150f
floating ribs, 117
fluid compartments, body fluids, 407, 408f
fluoxetine (Prozac), 187
follicles, ovarian. See ovarian follicles
follicle-stimulating hormone (FSH), 425–426
follicular cells, 428
follicular phase, of ovarian cycle, 431
fontanels, of skull, 109

foot
    bones of, 126f
    muscles of, 161–162
    overview of, 124–125
foramen magnum, occipital bone and, 108
foramen ovale, fetal circulation and, 283
foramina (openings)
    bones, 106t
    interventricular foramina, 184
    obturator foramen, 121
    olfactory foramina, 109
    optic foramina, 109
    of skull, 107, 112t
    transverse foramina, 115
    vertebral foramen, 114–115
forearm. See antebrachium (forearm)
foreskin. See prepuce
formed elements, of blood
    in blood composition, 290
    defined, 290
    erythrocytes, 293–294
    leukocytes, 295
    overview of, 291–295
    thrombocytes, 295
formulas, molecular, 28
fossa (glenoid cavity), in pectoral girdle, 118
fossa ovalis, chambers of heart, 249
fourth ventricle, 184
fovea capitis, of femur, 123
fovea centralis, in structure of eyeball, 207
fracture hematoma, 127
fractures
    classification and description of, 128f
    disorder of skeletal system, 131
    overview of, 126–127
free edge, of nails, 92
free fatty acids, lipid digestion and, 368
free nerve endings, receptors for touch and
    pressure, 202
freely moveable joints. See diarthroses
frenulum, of tongue, 355
frontal bone, of cranium, 108
frontal lobe, of cerebrum, 179–180
frontal plane, of body, 10, 10f
frontal sinuses, 108
frontalis, muscles of facial expression, 152
fructose, 34, 34f
    carbohydrate digestion and, 367
    types of monosaccharides, 385
FSH (follicle-stimulating hormone), 425–426
functional disorder, 43
functional residual capacity, lung
    volumes, 335
fundus
    stomach structure and, 357–358
    uterus and, 429
furosemide (Lasix), 271

G

galactose, 34, 34f
    carbohydrate digestion and, 367
    types of monosaccharides, 385
gallbladder
    overview of, 365
    sonogram showing gallstones, 365f
    structures of liver, 363
gallstones
    formation of, 365
    sonogram showing, 365f
gametes, 419
ganglia
    of cranial nerves, 187
    organs of peripheral nervous system, 171

gangrene, 99
gases, in plasma, 291
gases, in respiration
 carbon dioxide transport, 339–340
 Dalton's law of partial pressures, 335–336
 external respiration, 337
 Henry's law of gases, 336
 internal respiration, 337
 oxygen transport, 338–340
 partial pressure (PP) of gases in atmosphere and alveolar air, 336t
 properties of, 335–336
gaster, of muscle, 142
gastric glands
 overview of, 358
 regulation of gastric secretions, 360f
 secretions of, 358t
gastric mucosa, 237
gastric phase, of gastric secretion, 359
gastric pits, 358
gastric secretions
 regulation of, 359
 types of, 358–359
gastrin, 237, 358
gastritis, 371
gastrocnemius, foot and ankle muscles, 161–162
gastroesophageal reflux, 371
gastrointestinal (GI) tract. See digestive tract
gene augmentation, 457
gene expression, 453–455
gene replacement, 457
gene therapy, 457
general senses, 202
 overview of, 202–203
 pain, 203
 proprioception, 202
 receptors for touch, pressure, and proprioception, 203f
 temperature, 202–203
 touch and pressure, 202
generic names, vs. chemical names of drugs, 30b
genes
 dominant and recessive traits, 453
 human genome, 453
 mechanism of gene function, 452
 Punnett squares showing possible combinations, 453
 sex-linked traits, 455
 variations of single gene inheritance, 454–455
genetic carrier, 453
genetic disorder, 66
genetic mutations, 455
genetics, 452
genital herpes, 435
genital tract, 429
 uterine tubes, 429
 uterus, 429
 vagina, 429
genotypes, 441, 453
germ layers, 447
 derivatives of, 448t
 formation in preembryonic period, 445
germinal (ovarian) epithelium, 426
gestation, 448
GH (growth hormone), 228
GHP (glomerular hydrostatic pressure), 403f
gigantism, 240
gingiva (gums), 356
glands
 acinar glands, 73
 alveolar gland, 73
 apocrine glands, 73–74
 ceruminous gland, 93
 classification by secretion, 73f

glands (Continued)
 classification by structure, 73f
 as effectors, 171
 endocrine glands. See endocrine glands
 exocrine glands. See exocrine gland
 holocrine glands, 73–74
 lacrimal glands, 205
 merocrine glands, 73–74
 multicellular glands, 73
 overview of, 92–93
 pineal gland, 182
 pituitary gland, 226–230, 231f
 sebaceous glands, 92, 205
 simple, compound, and tubular, 73
 sudoriferous gland, 93
 unicellular and multicellular, 73
glandular epithelium
 classification of glands by secretion, 73f
 classification of glands by structure, 73f
 summary of modes of secretion, 74t
glans penis, 425
glaucoma, vision disorder, 216
glenoid cavity (fossa), in pectoral girdle, 118
glial (neuroglia) cells
 defined, 69
 nerve cells, 78
 overview of, 173
 types of, 171, 173t
gliding joint, types of freely movable joints, 130f
globulins
 in antibody-mediated immunity, 317
 plasma proteins, 291
glomerular capsule (Bowman's capsule), 395, 398
glomerular filtration, 395, 402
glomerular hydrostatic pressure (GHP), 403f
glomerular osmotic pressure (GOP), 403f
glomerulonephritis, 412
glomerulus
 defined, 395
 renal corpuscle and, 398
glossopharyngeal nerve, in gustatory sense, 204
glottis, 331
glucagon, pancreatic hormones, 235
glucocorticoids, hormones of adrenal cortex, 233
gluconeogenesis, 380
 defined, 377
glucose
 carbohydrate digestion and, 367
 homeostasis of blood glucose, 380f
 overview of, 34
 renal threshold of, 402–403
 sources of energy in muscle contraction, 147
 storage by liver, 365
 structural formula for, 34f
 types of monosaccharides, 385
glucose tolerance test (GTT), 243
gluteal tuberosity, 123
gluteus maximus, thigh muscles, 159
gluteus medius, thigh muscles, 159
gluteus minimus, thigh muscles, 159
glycerol
 in formation of triglycerides, 37f
 overview of, 36
 structural formula for, 36f
glycogen
 glucose stored as, 380
 types of polysaccharides, 385
glycogenesis, 377, 380
glycogenolysis, 380
 defined, 377
glycolysis
 in catabolism of glucose, 378, 379f
 defined, 377

GnRH (gonadotropin-releasing hormone), 425
goblet cells, 71
Golgi apparatus
 overview of, 49–50
 structure and function of cellular components, 47t
Golgi tendon organs, 202
gonadal arteries, 277, 278
gonadocorticoids, hormones of adrenal cortex, 233
gonadotropic hormones, hormones of anterior lobe, 228–229
gonadotropin-releasing hormone (GnRH), 425
gonads
 defined, 419
 ovaries, 236
 overview of, 236
 primary reproductive organs, 421
 testes, 236
gonorrhea, 435
GOP (glomerular osmotic pressure), 403f
gout, 407
gout, disorder of skeletal system, 131
gracilis, thigh muscles, 159
granulation tissue, fibrosis and, 79
granulocytes, types of leukocytes, 295
granulosa cells, 428
gray commissure, 185
gray matter
 in cerebral hemispheres, 180
 in spinal cord, 185
great saphenous veins, of lower extremity, 282–283
greater curvature, of stomach, 357–358
greater sciatic notch, 122
greater trochanters, of thigh, 123
greater tubercles, of arm, 118–119
greater vestibular glands, 430
greater wings, of sphenoid bone, 109
growth hormone (GH), 228
growth, life processes, 7
GTT (glucose tolerance test), 243
Guillan-Barré syndrome, 344
gums (gingiva), 356
gustatory cells, 203
gustatory senses, 203–204
gynecomastia, 435
gyri, of cerebrum, 179

H

H (hydrogen)
 breathing rate as means of adjusting hydrogen ion concentration in blood, 30b
 diagrams of atomic structure of, 24f
H zone (H band), muscle bands and, 143
hair
 overview of, 91–92
 structure of, 91f
hair bulb, 92
hair cells, of inner ear, 211–212
hair follicles
 overview of, 91–92
 structure of, 91f
hammer (malleus)
 auditory ossicles, 113
 of middle ear, 211
hamstrings, leg muscles, 161
hand
 illustration of, 121f
 muscles, 159
 overview of, 120
hard palate, 354–355
hard palate, in nasal cavity, 329
haustra, structures of large intestine, 362
haversian (osteonic) canal, in bone tissue, 76

haversian systems, in bone tissue, 76
HCG. *See* human chorionic gonadotropin (HCG)
hCG (human chorionic gonadotropin), 237
head
    of femur, 123
    of fibula, 124
    of humerus, 118–119
    of pancreas, 365
    of radius, 119–120
    of sperm, 423
head and neck arteries, 277
head and neck muscles, 152–153
    illustration of, 152*f*
    muscles of facial expression, 152
    muscles of mastication, 153
    neck muscles, 153
    summary of, 153*t*
head and neck veins, 278
hearing. *See* auditory senses
heart. *See* cardiovascular system
heart block, disorder of blood, 301
heart rate, 255–256
heart sounds, 255
heart valves. *See* valves of heart
heat exhaustion, 392
heat loss, 384
heat production, 384
heat receptors, proprioception, 202–203
heat stroke, 392
heat, environmental requirements for life, 7
heavy metal poisoning, disorder of nervous
    system, 194
helper T cells, 316
hemangioma, 288
hematocrit
    in blood composition, 290
    defined, 305
hematopoieses, 101
hematopoiesis (blood cell formation), 103, 291–293
hemocytoblast
    defined, 289
    formed elements of blood, 291–293
hemoglobin, in RBCs, 293
hemolysis, cell rupture and, 53–54
hemolytic disease, of newborn, 299*f*
hemophilia, disorder of blood, 301
hemopoiesis, 289
hemoptysis, 349
hemorrhage, 194
hemorrhoids, 288
hemostasis
    coagulation, 296
    illustration of, 296*f*
    overview of, 295–296
    platelet plug formation, 296
    vascular constriction, 296
Henry's law of gases, 336
heparin, as anticoagulant, 297
hepatic (right colonic) flexure, regions of large
    intestine, 362
hepatic artery, 363
hepatic ducts, 363
hepatic portal, 363
hepatic portal circulation, 282*f*
hepatic portal system, 280–282
hepatic portal vein, 280–282, 364
hepatic veins, 280, 363–364
hepatitis, 371
hepatocytes (liver cells), 363–364
heredity
    chromosomes and genes and, 452–453
    gene expression, 453–455
    gene therapy, 457

heredity (*Continued*)
    hereditary disorders, 455
    inheritance of ABO blood types, 456*f*, 457*t*
    inheritance of recessive X-linked
        color blindness, 457*f*
    overview of, 452
Hering-Breuer reflex, 327, 341–342
herniated disks, 136
herniated intervertebral disk, 194
herpes simplex, disorder of integumentary
    system, 96
herpes zoster, disorder of integumentary system, 96
heterozygous
    defined, 441
    genetic traits, 453
hexoses, monosaccharides, 34
hiatal hernia, 358*b*, 371
hilum, 311, 332, 397
hinge joint, types of freely movable joints, 130*f*
hip girdle. *See* pelvic girdle
hirsutism, disorder of integumentary system, 96
histology
    defined, 69
    overview of, 85
    study of tissues, 70
HIV (human immunodeficiency virus) disease, 322
holocrine glands
    classification of glands by secretion, 73*f*
    overview of, 73–74
homeostasis
    of blood glucose, 380*f*
    defined, 1
    negative feedback mechanisms, 8, 8*f*
    overview of, 7–9
    positive feedback mechanisms, 8–9
homeotherms, humans as, 383–384
homozygous
    defined, 441
    genetic traits, 364, 453
horizontal plates, forming hard palate, 113
hormonal control
    female reproductive system, 430–432
    male reproductive system, 425–426
    menopause, 432
    ovarian cycle, 431–432
    uterine (menstrual) cycle, 432
hormonal immunity, 317
hormones
    adrenal cortex, 232–233
    adrenal medulla, 234
    anterior lobe, 228–229
    chemical nature of, 225
    controlling action of, 226
    defined, 223
    of digestive system, 368*t*
    endocrine gland, 225, 227*t*
    gonads, 236
    influencing concentration and volume of urine, 405
    mechanism of action, 225–226
    pancreas (islets of Langerhans), 235–236
    parathyroid glands, 231–232
    pineal gland, 237
    pituitary gland, 226–230
    posterior lobe, 229–230
    receptor sites for, 225*f*
    in red blood cell production, 397
    secreted by placenta, 447*t*
    thyroid gland, 230–231
human body
    anatomical terms, 9–11
    anatomy and physiology of, 2
    environmental requirements for life, 7
    homeostasis, 7–9

human body (*Continued*)
    levels of organization, 2–3
    life processes, 6–7
    medical vocabulary, 16
    organ systems, 3–6
    overview of, 2
    summary, 14
human chorionic gonadotropin (HCG)
    hormones produced by placenta during
        pregnancy, 237
    implantation and, 444–445
    secretion by placenta, 447*t*
Human Genome Project, 453
human immunodeficiency virus (HIV) disease, 322
humerus
    of arm, 118–119
    markings on, 119*t*
    posterior and anterior views, 105–107
    relationship to radius and ulna, 120*f*
hyaline cartilage, 76, 76*f*
hydrocephalus, 184, 195
hydrochloric acid, production in
    stomach, 358
hydrocortisone (Hycort, Cortril), 93
hydrogen (H)
    breathing rate as means of adjusting hydrogen
        ion concentration in blood, 30*b*
    diagrams of atomic structure of, 24*f*
hydrogen bonds
    intermolecular in water, 27*f*
    overview of, 27
hydrogenated fats, 36
hydrolysis
    catabolic reactions, 378
    in chemical digestion, 367
    digestive processes, 353
hydrophilic layer, of plasma membrane, 46
hydrophobic layer, of plasma membrane, 46
hydrostatic pressure, in movement of substances
    across plasma membrane, 54
hydroxyamphetamine, 206*b*
hymen, 429
hyoid bone
    overview of, 113
    position and shape of, 115*f*
hyperaldosteronism, 412
hyperextension, types of movement,
    150*f*
hyperopia, disorder of the senses, 216
hyperparathyroidism, disorder of skeletal
    system, 131
hyperplasia, 66
hyperpolarized, 176–177
hypertension
    disorders of endocrine system, 240
    pharmacology for, 271
hypertonic solutions, in osmosis, 53
hypertrophy, 66
hypervitaminosis, 393
hypoaldosteronism, 412
hypochondriac region, of abdominopelvic
    cavity, 11
hypodermis. *See also* subcutaneous layer
    overview of, 91
    in structure of skin, 89*f*
hypogeusia, disorder of the senses, 216
hypogonadism, disorder of endocrine
    system, 240
hypoparathyroidism, disorder of skeletal
    system, 131
hypoperfusion, 288
hypophysis. *See* pituitary gland
hyposmia, disorder of the senses, 216

hypotension, 412
hypothalamus
    interaction of pituitary, hypothalamus, and thyroid
        glands, 231*f*
    regions of diencephalon, 182
hypothermia, 393
hypotonic solutions, in osmosis, 53–54
hypovolemia, disorder of blood, 301
hysterectomy, 440

I

I (isotropic) band, of muscle fiber, 142, 143
ibuprofen (Motrin, Advil), 126*b*
ICF. *See* intracellular fluid (ICF)
ICSH (interstitial cell-stimulating hormone), 425–426
idiopathic disorder, 43
ileocecal junction, 362
ileocecal sphincter, 362
ileocecal valve, 362
ileum, 361
iliac crest, 122
iliac fossa, 122
iliac region, of abdominopelvic cavity, 11
iliopectineal line, 122
iliopsoas, thigh muscles, 159
iliosacral joints, of pelvic girdle, 121
ilium, of coxal bones, 121
immune system. *See also* lymphatic system
    acquired immunity, 319
    antibody-mediated immunity, 317
    cell-mediated immunity, 316
    illustration of, 88*f*
    recognition of self and nonself by, 315–316
immunity
    defined, 307
    pharmacology for, 320*b*
    specific resistance, 312, 315
immunoelectrophoresis, 325
immunoglobulins (Ig)
    in antibody-mediated immunity, 317
    defined, 307
immunologist, 325
immunosuppression
    disorders of lymphatic system, 322
    pharmacology for, 320*b*
immunotherapy, 325
impetigo, disorder of integumentary system, 96
implantation, 444–445
    defined, 441
    in preembryonic period, 444–445
    process of, 445*f*
impotence, in male sexual response, 425
incisors, 356
inclusions, in cytoplasm, 48
incomplete proteins, 386
incus (anvil)
    auditory ossicles, 113
    of middle ear, 211
indirect attachment, muscle attachment, 142
infants
    adjustments of infant at birth, 450–451
    postnatal development, 452
infectious disease, 43
inferior cerebellar peduncles, 183
inferior colliculi, of midbrain, 182
inferior mesenteric artery
    of abdominal viscera, 278
    branches of abdominal aorta, 277
inferior nasal conchae, of face, 113
inferior vena cava, 278, 280
    structures of liver, 363
inferior view, of skull, 111*f*
inferior, directions of body, 9

inflammation
    in disease resistance, 314
    of meninges (meningitis), 81
    of serous membranes, 69*b*
    steps in, 314*f*
    tissues, 79
infraspinatus, shoulder and arm muscles, 159
infundibulum, 182, 429
ingestion, 353
inguinal canal, 421*b*
inguinal hernia, 421*b*
inguinal region, of abdominopelvic cavity, 11
inhalation. *See* inspiration
inheritance. *See* heredity
inhibitory transmission, 176–177
inner cell mass, 442–444
inner circular layer, of digestive trac, 354
inner ear, 211–212
inner tables, of flat bones, 104
innominate bones, of pelvic girdle, 121
inorganic compounds, 21, 28–30
insect bites, disorder of integumentary
        system, 96
insertion
    defined, 139
    of muscle, 142
insertions, genetic mutations, 455
inspiration
    inspiratory area of respiratory center, 340
    overview of, 334
inspiratory capacity, of lungs, 335
inspiratory reserve volume, lungs, 335
insula (island of Reil), of cerebrum, 179–180
insulin, pancreatic hormone, 236
integration, functions of nervous system, 171
integumentary system
    burns, 94–95
    dermis, 90
    disorders of, 96–97
    epidermal derivatives, 91–93
    epidermis, 89–90
    functional relationships of, 88*f*
    glands, 92–93
    hair and hair follicles, 91–92
    illustration of, 4*f*
    medical vocabulary, 99–100
    nails, 92
    overview of, 3–6, 89
    protective function of skin, 93
    sensory function of skin, 93
    skin color, 91
    structure of skin, 89–91
    subcutaneous layer of skin, 91
    summary, 97–98
    temperature regulation by skin, 94
    vitamin D synthesis, 94
interatrial septum, chambers of heart, 249
intercalated disks, in myocardium, 248
intercellular matrix, 70
intercondylar notch, 123
intercostal arteries, 277
intercostal nerves
    in nerve enervation, 188
    in respiration, 340
interferon
    defined, 325
    viral protection from, 313
interleukins
    defined, 326
    secreted by T cells, 316*b*
interlobar arteries, blood flow through kidneys, 400
interlobar veins, blood flow through
    kidneys, 400

intermolecular bonds
    hydrogen bonds, 27
    hydrogen bonds in water, 27*f*
internal anal sphincter, 362
internal carotid artery, of head and neck, 277
internal iliac artery, 278
internal iliac veins, 282
internal intercostal muscles
    expiration and, 334
    thoracic wall muscles, 154
internal jugular veins, 278
internal nares, 329
internal oblique, abdominal wall muscles, 155–156
internal os, 429
internal respiration, 338*f*
    defined, 327
    overview of, 329, 337
internal urethral sphincter, 401
interneurons, 173
interphase, of mitosis, 56
interstitial cells (cells of Leydig), 422
interstitial cell-stimulating hormone
    (ICSH), 425–426
interstitial fluid
    capillaries and, 266
    defined, 395
    fluid compartments, 407
intertrochanteric crest, 123
intertrochanteric line, 123
intertubercular groove, of arm, 118–119
interventricular foramina, 184
interventricular septum, chambers of
    heart, 249
intervertebral disks, 113
intestinal glands, 359–361
intestinal lipase, digestive enzymes, 361
intestinal phase, of gastric secretion, 359
intracellular fluid (ICF)
    defined, 395
    overview of, 407
intracellular fluid, in cytoplasm, 48
intramembranous ossification, 101, 105
intramolecular bonds, hydrogen, 27
intrapleural pressure, in pulmonary ventilation, 333
intrapulmonary (intraalveolar) pressure, in pulmonary
    ventilation, 333
intravascular fluid (blood plasma), 395, 407
intravenous (IV), parenteral nutrition via, 383
intravenous pyelogram, 417
intrinsic factor
    production in stomach, 358
    in production of erythrocytes, 293
inversion, types of movement, 150*f*
involuntary muscles, 77
ionic bonds
    defined, 21
    formation of, 25*f*
    important ions in body, 25*t*
    overview of, 24–25
ionic compounds, 25
ions
    breathing rate as means of adjusting hydrogen ion
        concentration in blood, 30*b*
    principal particles of matter, 28*t*
iris, in structure of eyeball, 207
ischemia, 288
ischial spine, 122
ischial tuberosity, 122
ischium, of coxal bones, 121
island of Reil (insula), of cerebrum, 179–180
islets of Langerhans. *See also* pancreas
    hormones, 365–367
    structures of pancreas, 365

isometric contraction, muscular, 146
isopropyl alcohol in glycerin (Swim-Ear drops), 206*b*
isotonic contraction, in muscular
    contraction, 146
isotonic solutions, in osmosis, 53, 54*f*
isotopes, 23–24
    defined, 21
    diagram of atomic structure of
        deuterium, 24*f*
isotropic (I) band, of muscle fiber, 142, 143

J
jaundice, 393
jejunum, 361
joint capsule, diarthroses and, 128
joint cavity, diarthroses and, 128
joints
    amphiarthroses, 101
    freely moveable. *See* diarthroses
    immovable. *See* synarthroses
    slightly movable. *See* amphiarthroses
jugular (suprasternal) notch, 116–117
juxtaglomerular apparatus, 399*f*
    defined, 395
    overview of, 399
juxtaglomerular cells, 399

K
Kaposi's sarcoma, 326
keloid, 100
keratin, 89
keratinization
    defined, 87
    of epidermis, 89
ketone bodies, with excess of acetyl-CoA, 381*b*
kidney stones, 406*b*
kidneys, 397
    aging and, 411
    blood flow through, 399–400, 400*f*
    collecting ducts, 399
    coronal (frontal) section, 398*f*
    juxtaglomerular appartus, 399
    location of, 397
    macroscopic structure of, 397–398
    nephrons, 398, 398*f*
    overview of, 397
    renal arteries, 277, 278
    renal erythropoietic factor (REF) in, 293
    renal veins, 280
killer T cells, 316
kilocalories, measuring energy derived from food,
    381–382
knee
    radiograph of, 125*f*
    sagittal section of, 129*f*
    semilunar cartilage of, 128–129
kneecap. *See* patella
Korotkoff sounds
    blood flow and, 269–270
    defined, 263
Krebs cycle. *See* citric acid cycle (Krebs cycle)
Kupffer cells, 365
Kwashiorkor, 393

L
labia majora, 430
labia minora, 430
labor
    overview of, 448–449
    stages of, 449*f*
labyrinthitis, disorder of the senses, 216
lacrimal apparatus, 205

lacrimal bones, of face, 113
lacrimal canals, 205
lacrimal ducts, 205
lacrimal gland, 205
lacrimal groove, 113
lacrimal sac, 205
lactase, 361
lactase dehydrogenase (LDH), enzyme indicators of
    myocardial infarction, 252*b*
lactation, 448–451
    physiology of, 451
    stimulus for, 451*f*
lacteals
    structures of small intestines, 359–361
    villi and, 309
lactic acid, in carbohydrate digestion, 378
lactiferous duct, 433
lactiferous sinus (ampulla), 433
lactose
    carbohydrate digestion and, 367
    types of monosaccharides, 385
lactose intolerance, 371
lacunae, spaces in cartilage matrix, 76
lag phase, in contraction of whole muscle, 145
lambdoid suture, occipital bone and, 108
lamellae, osteonic canal and, 76
lanugo hair, covering fetus, 447
large intestine, 362–363
    functions of, 362–363
    overview of, 362
    regions of, 362
    structure and features of, 362, 362*f*
laryngitis, 349
laryngopharynx, 330
    regions of pharynx, 357
larynx
    clearing blocked airway, 331*b*
    conducting passages for respiration, 330–331
lateral condyles, of leg, 123, 124
lateral epicondyles, of arm, 118–119
lateral funiculi, of white matter in spinal cord, 185
lateral horns, of gray matter in spinal cord, 185
lateral malleolus, 124
lateral meniscus, of knee, 128–129
lateral sulcus, of cerebrum, 179–180
lateral ventricles, of brain, 184
lateral view, of skull, 110*f*
lateral, directions of body, 9
latissimus dorsi, shoulder and arm muscles, 159
latrogenic illness, 43
laxatives, 367
layers of heart wall, 248, 248*f*
LDH (lactase dehydrogenase), enzyme indicators of
    myocardial infarction, 252*b*
left atrioventricular bundle branch, in conduction
    system, 253
left atrium
    chambers of heart, 249
    pathway of blood through the heart, 251
    pulmonary circuit, 272
left colonic splenic flexure, 362
left common carotid artery, of aortic arch, 276
left common iliac artery
    of abdominal aorta, 277
    of lower extremity, 278
left coronary arteries
    ascending aorta, 272
    blood supply to myocardium, 251–252
left gastric artery, 278
left gastric veins, of abdominal and pelvic
    organs, 280–282
left hepatic duct, 363–364
left lobe, of liver, 363

left lung, 332
left primary bronchi, 332
left subclavian artery, of aortic arch, 276
left ventricle
    chambers of heart, 249
    pathway of blood through the heart, 251
leg
    lower extremity, 124
    muscles, 161
lens, structure of eyeball, 207
lesser curvature, of stomach, 357–358
lesser sciatic notch, 122
lesser trochanters, of thigh, 123
lesser tubercles, of arm, 118–119
leucine, structural formula for, 35*f*
leukemia, 301
leukocytes
    characteristics and functions of, 295
    defined, 69, 289
    formed elements of blood, 291
    illustration of, 292*f*
    types of, 295
    while blood cells, 77*f*
leukocytosis, 305
leukopenia, 305
leuteinizing hormone (LH), 425–426
levator ani, pelvic floor muscles, 156
levator palpebrae superioris, 205
levothyroxine sodium (Levoxyl, Levothroid,
    Synthroid), 230
LH (leuteinizing hormone), 425–426
life processes, 6–7
    differentiation, 7
    digestion, 7
    excretion, 7
    growth, 7
    metabolism, 6–7
    movement, 7
    organizational scheme, 6–7
    reproduction, 7
    respiration, 7
    responsiveness, 7
life, environmental requirements. *See* environmental
    requirements, for life
ligaments
    dense fibrous connective tissue, 75–76
    suspensory ligaments, in structure of eyeball, 207
ligamentum teres, structures of liver, 363
ligamentum venosum, structures of liver, 363
light, pathway of and refraction, 208
liminal (threshold) stimulus, in muscular
    contraction, 145
linea alba, abdominal wall muscles, 155–156
linea aspera, 123
lingual tonsils, 312, 330, 355
linolenic acid, structural formula for, 37*f*
lipase, 366
lipids
    chemical digestion, 368
    defined, 21
    digestion of, 368*f*
    important lipid groups, 36*t*
    interconversion of carbohydrates, proteins, and
        lipids, 381
    metabolism by liver, 365
    metabolism of, 381
    nutrition and, 386–387
    overview of, 36–37
lipogenesis, 377, 380
lipoma, 85
lips, mouth components of digestive tract, 354
lithium carbonate (Eskalith, Lithane), 187
lithotripsy, 417

liver
accessory organs of digestion, 363–365
bile production by, 365
blood supply to, 364, 364f
functions of, 364–365
overview of, 363
structure and features of, 363–364
structure of liver lobules, 364f
surface features, 363f
liver cells (hepatocytes), 363–364
liver lobules
overview of, 363–364
structure of, 364f
loading, of oxygen in lungs, 338
lobes, structures of breasts and mammary glands, 433
lobules
structures of breasts and mammary glands, 433
of testes, 422
localized inflammation, 314
lochia, 462
long bones
examples of, 104
features of, 104–105, 105f
longitudinal fissure, of cerebrum, 179
loose connective tissue (areolar), 74
loudness, auditory senses, 213
lower esophageal sphincter, 357
lower extremity
arteries of, 278
foot, 124–125
leg, 124
overview of, 123–125
patella, 125
thigh, 123
veins of, 282–283
lower extremity muscles
ankle and foot, 161–162
leg, 161
overview of, 159–162
summary of, 160t
thigh, 159
lower respiratory tract
conducting passages for respiration, 329
defined, 327
illustration of, 331f
lumbar arteries
of abdominal aorta, 277
of abdominal viscera, 278
lumbar curvature, of vertebral column, 113
lumbar enlargement, structures of spinal cord, 185
lumbar region, of abdominopelvic cavity, 11
lumbar vertebrae, 115–116
lumber nerves (L1 to L5), 188
lungs
carbon dioxide transport, 339f
conducting passages for respiration, 332–333
CT image of, 333f
diffusion of oxygen and carbon dioxide, 52f
features of, 333f
loading of oxygen in lungs, 338f
pathway of blood through the heart, 251
lunula, of nails, 92
luteal phase, of ovarian cycle, 432
lyme disease, 136
lymph, 310
lymph capillaries, 310
lymph nodes, 311
location of clusters of, 311f
overview of, 310
structure of, 311f
lymph nodules, 311
lymph sinuses, 311
lymphadenitis, 326

lymphadenopathy, disorder of lymphatic system, 322
lymphangiogram, 326
lymphatic ducts, 310
lymphatic organs, 310–312
lymphatic system
acquired immunity, 319
aging and, 321b
antibody-mediated immunity, 317
cell-mediated immunity, 316
chemical actions in disease resistance, 313
components of, 309–312
disorders of, 322
functional relationships of, 308f
functions of, 309
illustration of, 4f
inflammation in disease resistance, 314
lymph, 310
lymph nodes, 311
lymphatic organs, 310–312
lymphatic vessels, 310
lymphocytes developed by immune system, 316
mechanical barriers to disease, 312–313
nonspecific defense mechanisms, 312–314
overview of, 6
phagocytosis in disease resistance, 314
pharmacology for, 320b
recognition of self and nonself by immune system, 315–316
relationship with cardiovascular system, 309f
resistance to disease, 312–319
specific defense mechanisms, 315–317
spleen, 312
summary, 323–324
thymus, 312
tonsils, 312
lymphatic trunks, 310
lymphatic vessels, 310
lymph capillaries in tissue space, 310f
regions drained by lymphatic ducts, 310f
lymphedema, 311, 322
lymphocytes
agranulocytes, 295
developed by immune system, 316
development of, 316, 316f
illustration of, 292f
in immunity, 315
lymphatic organs and, 310–311
T lymphocytes (T cells), 312
lymphoma, disorder of lymphatic system, 322
lysis, cell rupture and, 53–54
lysosomes
overview of, 50
structure and function of cellular components, 47t

M

M line, muscle bands and, 143
macrophages
defined, 69
in immunity, 315
monocytes becoming, 295
as phagocytic cell, 314
types of cells in connective tissue, 74
macula
static equilibrium and, 214
structure of, 214f
macula densa, 399
macula lutea, structure of eyeball, 207
macular degeneration, vision disorder, 216
magnetic resonance imaging (MRI), 199
major calyx, 398
malabsorption syndrome, 393

malar bones, 113
male reproductive system
accessory glands, 424
duct system, 423–424
hormonal control, 425–426
male sexual response, 425
overview of, 421–426
penis, 424–425
structures of, 421f
testes, 421–423
male sexual response, 425
malignant, 66
malignant melanoma, disorder of integumentary system, 96
malleus (hammer)
auditory ossicles, 113
of middle ear, 211
malnutrition, 393
maltase, 361
maltose
carbohydrate digestion and, 367
types of monosaccharides, 385
mamillary bodies, of hypothalamus, 182
mammary glands, 433, 433f
mandible, of face, 113
mandibular condyle, 113
mandibular fossa, 108
manubrium, of sternum, 116–117
marasmus, 393
Marfan syndrome, 85
mass number, of an atom, 23
masseter, muscles of mastication, 153
mast cells
defined, 69
types of cells in connective tissue, 74
mastectomy, 311
mastication
digestive processes, 353
muscles of, 153
mastoid process, temporal bones and, 108
matter
defined, 22
principal particles of, 28t
maxillary bones, of face, 113
maxillary sinus, 113
mechanical digestion, 353
mechanoreceptors
for auditory sense, 210
defined, 201
for proprioception, 202
for touch and pressure, 202
types of sense receptors, 202
medial condyles, of leg, 123, 124
medial epicondyles, of arm, 118–119
medial malleolus, 124
medial meniscus, of knee, 128–129
medial, directions of body, 9
median cubital vein, of shoulders and arms, 278–280
medications, routes of administration, 54b
medulla, 182–183
auditory pathway and, 213
of hair, 91–92
in ovarian structure, 426
medulla oblongata, 182–183
medullary cavity, of long bones, 105
megakaryocytes, thrombocytes as fragments of, 295
meiosis, 56
compared with mitosis, 58f, 58t
defined, 45
overview of, 57
Meissner's corpuscles, 202

melanin
    defined, 87
    skin color and, 90, 91
melanocytes, in stratum basale, 90
melanomas, malignant, 96
melatonin, hormones of pineal gland, 237
membranes
    illustration of, 81f
    meninges, 81
    mucous membranes, 80
    overview of, 80–81
    serous membranes, 80–81
    summary, 83
    synovial membranes, 81
membranous labyrinth, of inner ear, 211
membranous urethra, 402, 424
memory B cells, 317
memory T cells, 316
memory, in specific defense mechanism, 315
menarche, 431
Ménière's disease, 216
meninges
    of central nervous system, 178f
    as membrane, 81
    overview of, 178–179
meningitis, 81, 195
meniscus, of knee, 128–129
menopause, 431, 432
menstrual cycle. See uterine (menstrual) cycle
menstrual phase, of uterine cycle, 432
merocrine glands
    classification of glands by secretion, 73f
    secretions of, 73–74
    sweat glands, 93
mesentery
    defined, 351
    structures of small intestines, 361
mesoderm
    derivatives of, 448t
    formation of primary germ layers, 445
mesothelium, 80–81
messenger RNA, 59
metabolic
    acidosis and alkalosis, 410t
    acidosis or alkalosis, 409, 410
metabolism
    aging and, 389b
    altered metabolic rate, endocrine
        disorder, 239
    anabolism, 378
    basal metabolism, 382
    carbohydrates, 378–380
    catabolism, 378
    defined, 1
    energy from foods, 378–381
    factors affecting, 383f
    life processes and, 6–7
    lipids, 381
    overview of, 378
    proteins, 380–381
    summary, 390
    uses of energy, 381–382
metacarpal bones, 120
metacarpus, of hand, 120
metaphase, of mitosis, 56
metaplasia, 66
metarteriole, 263
metarterioles, 265
metastasis
    defined, 66, 326
    of neoplasm, 57b
metatarsal bones, 124–125
metatarsus, of foot, 124–125

MI (myocardial infarction)
    disorder of blood, 301
    enzyme indicators of myocardial infarction, 252b
micelles, in absorption of nutrients, 369
microfilaments, cytoskeleton composed on, 50
microtubules
    cytoskeleton composed on, 50
    spindle fibers, 56
microvilli, 50, 71
    structures of small intestines, 359–361
micturition
    defined, 395
    overview of, 405–406
micturition reflex, 405
midbrain
    apneustic area of, 182
    auditory pathway and, 213
    brain stem and, 182
    cerebral aqueduct of, 182, 184
    cerebral peduncles of, 182
    corpora quadrigemina of, 182
middle cerebellar peduncles, 183
middle ear, 210–211
middle nasal conchae, 109
midpiece, of sperm, 423
midsagittal plane, of body, 10
midsagittal section, of skull, 112f
mineral storage, functions of skeletal system, 103
mineralocorticoids
    aldosterone, 232–233
    hormones of adrenal cortex, 232–233
minerals
    in nutrition, 388
    sources and functions of, 389t
minor calyx, 398
miscarriage, 462
mitochondria
    overview of, 49
    structure and function of cellular components, 47t
mitosis
    compared with meiosis, 58f, 58t
    defined, 45
    overview of, 56–57
    summary of mitotic events, 57t
mitral valve
    AV valves of heart, 250
    prolapse, 260
Mittelschmerz, 432b
mittelschmerz, 440
mixed nerves, 187
mixing movements, in digestion, 353
mixtures, of substances, 31
molars, 356
molecules
    defined, 21
    molecular formulas, 28
    nature of compounds and, 27–28
    principal particles of matter, 28t
monoclonal antibody, 326
monocytes
    agranulocytes, 295
    illustration of, 292f
monogenic traits, 454
monoglycerides, 368
mononucleosis, 326
monophasic contraceptives, 432b
monosaccharides (simple sugars)
    overview of, 34
    representation of, 35f
monosaccharides (simple), 385
monosomy, hereditary disorders, 455
monozygotic twins, 442
mons pubis, 430

morula, 442
motor (efferent)
    division of peripheral nervous system, 171
    neurons, 173
motor areas, of cerebrum, 180–181
motor nerves, 187
motor neurons, in contraction of skeletal muscles, 143
motor output (function), of nervous system, 171
motor unit summation, 145
motor units
    in contraction of skeletal muscles, 143, 145
    defined, 139
mouth
    alveolar process of, 113
    palatine process of, 113
mouth (oral cavity), 354–357
    features of, 354f
    lips and cheeks, 354
    overview of, 354
    palate, 354–355
    salivary glands, 357
    teeth, 355–356
    tongue, 355
movement
    functions of skeletal system, 103, 149
    life processes, 7
movements, digestive processes, 353
MRI (magnetic resonance imaging), 199
mucolytics, 342
mucosa
    drugs for restoring/maintaining, 367
    layers of digestive tract, 353–354
    layers of ureters, 401
mucosae, 80
mucous cells, 73–74, 358
mucous membranes, 80, 401
mucous production, in stomach, 358
multicellular glands, 73
multipara, 462
multiple motor unit summation, in contraction of
    whole muscle, 145
multiple myeloma, 305
multiple sclerosis (MS)
    disorder of muscular system, 163
    disorder of nervous system, 195
multiple wave summation, 145
murmurs, listening to heart sounds, 255
muscle attachments, projections for, 106t
muscle biopsy, 167
muscle fibers, 77
muscle spindles, receptors for proprioception, 202
muscle tissue
    cardiac muscle, 78
    movement produced by, 78b
    overview of, 77–78
    skeletal, 77
    smooth muscle, 77
    summary of, 77t, 141t
muscle tone, 146
muscular bruise, disorder of muscular system, 163
muscular coat, of ureter, 401
muscular dystrophy, disorder of muscular system, 163
muscular layer (muscularis), of digestive tract, 354
muscular system
    anterior view, 157f
    characteristics and functions of, 141
    contraction of whole muscle, 145–146
    contraction stimulus, 143–144
    disorders of, 163f
    energy sources and oxygen debt in muscle
        contraction, 146–149
    eye muscles, 205t
    functional relationships of, 140f

muscular system *(Continued)*
  illustration of, 4*f*
  movements produced by skeletal muscle, 149
  muscles of head and neck, 152–153
  muscles of lower extremity, 159–162
  muscles of trunk, 153–156
  muscles of upper extremity, 156–159
  naming skeletal muscles, 150–152
  neck muscles, 153
  nerves and blood supply for skeletal muscles, 143
  overview of, 6
  pharmacology for, 160*b*
  posterior view, 158*f*
  sarcomere contraction, 144–145
  skeletal muscle, 141–142
  skeletal muscle attachments, 142
  skeletal muscle fibers, 142–143
  skeletal muscle groups, 150–162
  summary, 164–165
muscular tears, disorder of muscular system, 163
muscularis, of urinary bladder, 401
mutagens, 455
myasthenia gravis
  disorder of muscular system, 163
  disorder of respiratory system, 344
myelin (sheath), 173
myenteric plexus, in digestion, 354
myocardial infarction (MI)
  disorder of blood, 301
  enzyme indicators of myocardial infarction, 252*b*
myocardium
  blood supply to, 251–252, 252*f*
  layers of heart wall, 248
myofiber, conduction, 253
myofibrils
  muscle fibers and, 142
  sarcomere and, 143
myofilaments
  arrangements of, 143*f*
  in muscle structure, 142
myoglobin
  aerobic respiration and, 147
  defined, 139
myoma, 85
myometrium, of uterus, 429
myoneural junction. *See* neuromuscular (myoneural) junction
myoparesis, 167
myopathy, 167
myopia, disorders of the senses, 216
myorrhexis, 167
myosin
  contractile proteins in muscle tissue, 77
  defined, 139
  thick muscle filaments formed by, 142
myositis, disorders of muscular system, 163
myxedema, disorders of endocrine system, 240

N

N (nitrogen). *See* nitrogen (N)
Na (sodium)
  diagrams of atomic structure of, 24*f*
  formation of ionic bonds, 25*f*
Na+ /K+ imbalance, disorders of endocrine system, 240
nail bed, 92
nail body, 92
nail matrix, 92
nail root, 92
nails
  overview of, 92
  structure of, 92*f*
naproxen (Aleve), 126*b*

nasal bones, 113
nasal cavities, 329–330
nasal conchae
  conducting passages for respiration, 329
  inferior, 113
nasal septum, 113, 329
nasolacrimal duct, 205
nasopharynx, 330, 357
natural immunity, 307
neck muscles. *See* head and neck muscles
neck, of femur, 123
neck, of teeth, 356
necrosis, 66
negative feedback
  defined, 1, 223
  in homeostasis, 8, 8*f*
  physiologic example of, 9*f*
neonatal period, 451
neonates
  defined, 462
  postnatal development, 451
neoplasm, 66
nephrectomy, 417
nephritis, 417
nephrolithiasis, 412
nephrons
  defined, 395
  illustration of, 398*f*
  overview of, 398
nephroptosis, 397
nephrotic syndrome, 412
nerve cells, 78
nerve gas, 195
nerve impulses
  conduction across synapses, 176–177
  conduction along neurons, 175–176
  overview of, 174–178
  pathways, 177
  reflex arcs, 177–178
  resting membrane in nerve impulses, 174
  stimulation of neurons, 174
nerve tissue, 171–173
nerve tracts, 185
nerves
  cranial nerves, 187–188
  organs of peripheral nervous system, 171
  for skeletal muscles, 143
  spinal nerves, 188
  structure of, 187, 188*f*
  types of, 187
nervous stimulation, in regulation of hormone secretion, 226
nervous system
  aging and, 193*b*
  autonomic. *See* autonomic nervous system (ANS)
  central. *See* central nervous system (CNS)
  conduction across synapses, 176–177
  conduction along neurons, 175–176
  disorders of, 194*f*
  endocrine system compared with, 225
  functional relationships of, 170*f*
  functions of, 171–173
  illustration of, 4*f*
  nerve impulses, 174–178
  nerve tissue, 171–173
  organization of, 171, 172*f*
  overview of, 6
  peripheral. *See* peripheral nervous system (PNS)
  pharmacology for, 187*b*
  reflex arcs, 177–178
  resting membrane in nerve impulses, 174
  stimulation of neurons, 174
  summary, 195–197

nervous tissue, 78
nervous tunic. *See* retina (nervous tunic)
nervous tunic (retina). *See* retina (nervous tunic)
neurilemma, 173
neuroglia. *See* glial (neuroglia) cells
neurohypophysis, 223, 226–228
neuromuscular (myoneural) junction
  in contraction of skeletal muscles, 143
  defined, 139
  illustration of, 144*f*
neuronal pools, 177, 177*f*
neurons
  classified by function, 173*t*
  conduction across a synapse, 176–177
  conduction along, 175–176
  defined, 69
  excitability and conductivity of, 174
  functions of, 171–173
  motor neurons, 143
  nervous tissue conducting, 78, 78*f*
  olfactory neurons, 204
  stimulation of, 174
  structure of, 172*f*
  types of nerve cells, 171
neurotoxins, disorder of respiratory system, 344
neurotransmitters
  defined, 139
  list of common, 177*t*
  synaptic vesicles containing, 176
neutral solutions, pH scale for measuring, 32–33
neutrons
  in atomic structure, 23
  principal particles of matter, 28*t*
neutrophils
  granulocytes, 295
  illustration of, 292*f*
  as phagocytic cell, 314
nevus, disorder of integumentary system, 96
nipple, 433
nitrogen (N)
  atomic structure of, 24*f*
  nonprotein molecules containing, 291
  triple covalent bond in, 27*f*
nitroglycerin (Nitrostat, Nitrogard), 253*b*
nociceptors
  defined, 201
  sense of pain, 203
  types of sense receptors, 202
nocturia, 417
nodes of Ranvier, 173
nondisjunction, hereditary disorders, 455
nonessential amino acids, 385, 386*t*
nonrespiratory air movements, 342, 342*t*
nonspecific defense mechanisms. *See* defense mechanisms, nonspecific
nonspecific resistance, 312. *See also* defense mechanisms, nonspecific
nonsteroidal anti-inflammatory drugs (NSAIDs), 126*b*, 407
norepinephrine
  hormones of adrenal medulla, 234
  release of, 190
nose, conducting passages for respiration, 329–330
nosocomial infection, 43
nostrils, 329
NSAIDs (nonsteroidal anti-inflammatory drugs), 126*b*, 407
nuclear membrane
  overview of, 48
  structure and function of cellular components, 47*t*

nucleic acids
overview of, 37–38
structure of, 38f
nucleolus
overview of, 48–49
structure and function of cellular components, 47t
nucleoplasm, 48
nucleotides
components of, 38f
overview of, 37
nucleus
overview of, 48–49
structure and function of cellular components, 47t
viewing with electron microscope, 48f
nullipara, 462
nutrient foramen
defined, 101
features of long bones, 105
nutrients
environmental requirements for life, 7
in plasma, 291
nutrition
aging and, 389b
carbohydrates and, 385
functions of, 384t
lipids and, 386–387
minerals and, 388
overview of, 378, 384
pharmacology for, 383b
proteins and, 385–386
summary, 390–391
vitamins and, 387–388
water, 388–389
nystagmus, disorder of the senses, 216

O

O (oxygen). See oxygen (O)
obstruction, 371
obturator foramen, of pelvic girdle, 121
occipital bone, of cranium, 108
occipital condyles, 108
occipital lobe, of cerebrum, 179–180
odontoid process, of cervical vertebrae, 115
olecranon fossa, of arm, 118–119
olecranon process, of antebrachium (forearm), 119–120
olfaction, 204
olfactory bulb, 204
olfactory cortex, 204
olfactory epithelium, of naval cavity, 204
olfactory foramina, 109
olfactory neurons, 204
olfactory senses
overview of, 204
structure of olfactory receptors in nasal cavity, 204f
olfactory tract, 204
oligodendrocytes, 173
oligospermia, 440
oliguria, 417
oncologist, 326
oocytes
overview of, 426
primary oocytes, 427
secondary oocytes, 427–428
oogenesis
defined, 419
illustration of, 428f
overview of, 427–428
oogonia, 427
oophorectomy, 440
open reduction, fracture repair, 126
openings. See foramina (openings)
optic chiasma, visual pathway and, 209–210
optic disk, in structure of eyeball, 207

optic foramina, in cranial cavity, 109
optic nerves
structure of eyeball, 207
visual pathway and, 209–210
optic radiations, 209–210
optic tract, 209–210
oral cavity. See mouth (oral cavity)
oral contraceptives, 432b
orbicularis oculi, 205
orbicularis oris, 152
orbit (socket), of eye, 205
orchidectomy, 440
organ level, of organization in human
bod, 2–3
organ of Corti, of inner ear, 211–212, 212f
organ systems
cardiovascular system, 6
digestive system, 6
endocrine system, 6
illustrations of, 4f
integumentary system, 3–6
lymphatic system, 6
muscular system, 6
nervous system, 6
overview of, 3–6
reproductive system, 6
respiratory system, 6
skeletal system, 3–6
urinary system, 6
organelles
cytoplasmic, 49–50
overview of, 48
organic chemistry, 33. See also chemistry
organic compounds, 28–30
adenosine triphosphate (ATP), 38–39
carbohydrates, 33–34
defined, 21
lipids, 36–37
nucleic acids, 37–38
overview of, 33–39
proteins, 35–36
organic disorder, 43
organogenesis, 447
orgasm, in male sexual response, 425
origin, of muscle, 139, 142
oropharynx, 330, 357
orthopedist, 136
osmosis, 45, 53–54
ossa coxae, of pelvic girdle, 121
osseous tissue (bone), 76, 76f
ossification
endochondral ossification, 105–107
events in endochondral ossification, 106f
intramembranous ossification, 105
overview of, 105
primary ossification center, 105–107
secondary ossification center, 107
osteitis fibrosa cystica, 232b
osteoarthritis, 136
osteoblasts
defined, 101
role in bone development, 105
osteoclasts
defined, 101
role in bone development, 105
osteocytes
bone cells, 76
defined, 69, 101
role in bone development, 105
osteogenesis, 105
osteomalacia, disorder of skeletal system, 131
osteomyelitis, disorder of skeletal system, 131
osteonic (haversian) canal, in bone tissue, 76

osteons
in bone tissue, 76
defined, 101
photomicrograph of, 104f
osteoporosis, disorder of skeletal system, 131
osteosarcoma, disorder of skeletal system, 131
otomycosis, disorder of the senses, 216
otosclerosis, 212b
outer longitudinal layer, of digestive tract, 354
outer tables, of flat bones, 104
oval window, of middle ear, 210
ovarian cancer, 435
ovarian cycle
correlation of events in ovarian and uterine
cycles, 431f
cyclic patterns of secretion of sex hormones, 430
defined, 419
overview of, 431–432
ovarian follicles
defined, 419
development, 428
in ovarian structure, 426
ovarian fossae, 426
ovaries, 426–428
hormones of, 236, 431f
oogenesis, 427–428
ovarian follicle development, 428
overview of, 426
ovulation, 428
primary reproductive organs, 421
structure of, 426, 427f
oviducts. See uterine tubes
ovulation, 428
ovulatory phase, of ovarian cycle, 432
oxygen (O)
diagrams of atomic structure of, 24f
diffusion of oxygen and carbon dioxide
in lungs, 52f
environmental requirements for life, 7
oxygen debt
carbohydrate digestion and, 378
following strenuous exercise, 149
muscular contraction and, 143
oxygen transport, in respiration
loading of oxygen in lungs, 338f
overview of, 338–340
unloading of oxygen in tissues, 338f
oxyhemoglobin, 293
defined, 327
in oxygen transport, 338
oxytocin (Pitocin, Syntocinon), 229,
230, 450

P

P wave, on ECG, 254
pacemaker, of heart, 252–253
pacinian (lamellated) corpuscles, receptors for touch
and pressure, 202
packed cell volume (PCV), in blood
composition, 290
pain
analgesic drugs for, 126b
in localized inflammation, 314
sense of, 203
palate
mouth components of digestive tract,
354–355
in nasal cavity, 329
palatine bones, of face, 113
palatine process, of mouth, 113
palatine tonsils, 312, 330, 357
palmar artery, of upper extremity, 277
palmic acid, structural formula for, 36f

pancreas
  exocrine and endocrine glands of, 366f
  glucagon and, 235
  hormones of, 235–236
  insulin and, 236
  location and features of, 366f
  overview of, 365–367
pancreatic acinar cells, 365–366
pancreatic amylase, 366
pancreatic duct, 365–366
pancreatic lipase, 366
pancreatitis, 371
pandemic, 43
Papanicolaou smear (Pap test), 430b
papilla, of dermis, 92
papillae, 90, 203, 355
papillary layer, of dermis, 90
papillary muscles, chambers of heart, 249
papilloma, 85
parageusia, disorder of the senses, 217
paralysis, 168
paranasal sinuses, 329
parasympathetic nervous system
  comparing sympathetic and parasympathetic
    actions, 191t
  overview of, 190–193
  subdivisions of peripheral nervous system, 171
parathyroid glands, 231–232
parathyroid hormone (PTH), 231–232, 232f
paraurethral glands (Skene's glands), 430
paravertebral ganglia, of ANS, 190
parenteral drug administration, 54b
parenteral nutrition, 383
parietal bones, of cranium, 108
parietal cells, 358
parietal layer, of serous membrane, 80–81
parietal lobe, of cerebrum, 179–180
parietal pericardium, coverings of heart, 247
parietal pleura, of lungs, 332–333
parietal, directions of body, 10
Parkinson's disease, 195
paronychia, disorder of integumentary system, 96
parosmia, disorder of the senses, 217
parotid glands, 357
partial pressure (PP), of gases in respiration, 336
parturition, 448–451
  adjustments of infant at birth, 450–451
  defined, 441
  labor and delivery, 448–449
  overview of, 448
  pharmacology for, 450b
  physiology of lactation, 451
passive immunity
  artificial immunity, 319
  defined, 307
  natural immunity, 319
passive transport
  defined, 45
  in movement of substances across plasma
    membrane, 51f
patella
  anterior and posterior views, 123f
  as example of sesamoid bone, 107
  lower extremity, 125
patellar surface, 123
patent foramen ovale, disorder of blood, 301
pathogens, 312
pathology, 85
PCV (packed cell volume), in blood
  composition, 290
pectoral girdle
  components of, 118f
  overview of, 117–118

pectoralis major, shoulder and arm muscles, 159
pedicle, of vertebrae, 114–115
pelvic brim (pelvic inlet), 122–123
pelvic cavity, 10
pelvic diaphragm, 156
pelvic floor muscles, 156
pelvic girdle
  anterior view, 121f
  overview of, 121–123
pelvic organs, veins of, 280
pelvic outlet, 122–123
pelvimetry, 462
penile urethra, 424
penis, 424–425
pentoses, 34
peptic ulcers, 371
peptidase, 361
peptide bonds, linking amino acids, 35
percutaneous drug administration, 54b
pericardial cavity, 247
pericardiocentesis, 260
pericarditis, 247b
pericardium, 81, 247
perichondrium, connective tissue covering
    cartilage, 76
perilymph
  defined, 201
  of inner ear, 211
perimetrium layer, of uterus, 429
perimysium, in muscle structure, 141–142
perineurium, structure of nerves and, 187
periodontal ligaments, 356
periosteum
  defined, 101
  features of long bones, 105
peripheral nervous system (PNS)
  cranial nerves, 187–188
  overview of, 187–193
  spinal nerves, 188
  structure of nerves, 187
  subdivisions of nervous system, 171
peripheral resistance, 263, 270
perirenal fat, protecting kidneys, 397
peristalsis
  defined, 351
  digestive processes, 353
peritoneum, 81
peritonitis, 69b
peritubular capillaries, 400
permanent (secondary) teeth, 355
permanent teeth, 355f
pernicious anemia, 293, 371
peroneal artery, of lower extremity, 278
peroneal vein, of lower extremity, 282
peroneus, foot and ankle muscles, 161–162
perpendicular plate, of ethmoid, 109
perspiration, 93, 397
pertussis, 349
PET (positron emission tomography), 199
petechia, 100, 305
pH scale
  for measuring strength of acid or base, 32–33
  values of common substances, 32f
phagocytes, 314
phagocytic, 289
phagocytosis, 55–56
  defined, 45
  in disease resistance, 314
phalanges
  fingers, 120
  toes, 124–125
pharmaceutical, 12
pharmacist, 12

pharmacodynamics, 79b
pharmacokinetics, 79b
pharmacologist, 12
pharmacology
  chemical names vs. generic names, 30b
  for GI disorders, 367b
  overview of, 12
pharmacopeia, 12
pharmacotherapeutics, 79b
pharmacy, 12
pharyngeal tonsils, 312, 330
pharyngitis, 349
pharynx, 330, 357
phenotypes, 441, 453
pheochromocytoma, 243
phimosis, 425, 440
phlebitis, 288
phlebotomy, 288
phosphate, 38f
phospholipids, 37, 46, 47f
photoreceptors
  defined, 201
  overview of, 208–209
  types of sense receptors, 202
phrenic nerve, 340
phrenitis, 349
physical activity, uses of energy from food, 382
physiology
  defined, 1
  of hearing, 212–213
  overview of, 2
pia mater, in meninges, 81, 179
pica, 393
pineal gland
  in epithalamus, 182
  hormones, 237
pinealocytes, 237
pinocytosis, 45, 55–56
pitch, auditory, 213
pituitary gland
  hormones, 226–230
  interaction of pituitary, hypothalamus, and thyroid
    glands, 231f
pivot joint, types of freely movable
    joints, 130f
placenta
  fetal circulation and, 283
  formation in embryonic period, 446–447
  hormones produced by placenta during pregnancy, 237
  hormones secreted by, 447t
  structural features of, 447f
placenta previa, 446–447, 462
placental stage, of labor, 449
planes, of body, 10
plantar flexion, types of movement, 150f
plasma
  blood cells suspended in, 77f
  in blood composition, 290
  electrolytes, 291
  formed elements in, 290
  intravascular fluid (blood plasma), 407
  nonprotein molecules containing nitrogen, 291
  nutrients and gases, 291
  overview of, 290–291
  plasma proteins, 291
  solutes in, 291t
plasma cells, in antibody-mediated immunity, 317
plasma membrane
  movement of substances across, 51–54
  overview of, 46–48
  structure and function of cellular components, 47t
  structure of, 47f
  transport mechanisms, 51t

plasma proteins, 291, 364
plasmapheresis, 305
platelet plug formation, 296
platelets. *See* thrombocytes
pleura, 81
   defined, 327
   features of, 333*f*
   of lungs, 332–333
pleural cavity, 332–333
pleural effusion, 344
pleuritis, 333*b*
plexuses, of nerves, 188, 191*t*
plicae circulares
   defined, 351
   structures of small intestines, 359–361
plural forms, of words, 17
pneumoconiosis, disorder of respiratory system, 344
pneumonectomy, 349
pneumonia, disorder of respiratory system, 344
pneumotaxic area, of midbrain, 182
pneumothorax, 334*b*, 344
PNS. *See* peripheral nervous system (PNS)
polar covalent bond
   overview of, 26
   between oxygen and hydrogen, 27*f*
polio, 163
polycystic kidney, 412
polycythemia, 301
polydipsia, 243
polygenic traits, 454
polyphagia, 243
polysaccharides (complex sugars), 34, 35*f*, 385
polyuria, 243, 417
pons, of midbrain, 182
popliteal arteries, of lower extremity, 278
popliteal vein, of lower extremity, 282
porta, 363
portal triads, 363–364
positive feedback
   defined, 1
   in homeostasis, 8–9
positron emission tomography (PET), 199
postcentral gyrus, 180
posterior lobe, 229–230
posterior superior iliac spine, 122
posterior tibial arteries, of lower extremity, 278
posterior tibial vein, of lower extremity, 282
posterior, directions of body, 9
postganglionic fiber, ANS and, 188, 191*f*
postnatal development, 451–452
   adolescence, 452
   adulthood, 452
   childhood, 452
   defined, 442
   infancy, 452
   neonatal period, 451
   overview of, 451
   senescence, 452
postsynaptic neuron, 176
posture, functions of cerebellum, 183
PP (partial pressure), of gases in respiration, 336
precapillary sphincters, 265
precentral gyrus, 180–181
precocious puberty, 240
prednisone (Deltasone, Orasone), 230
preeclampsia, 462
preembryonic period
   cleavage, 442–444
   defined, 441
   formation of primary germ layers, 445
   implantation, 444–445
   overview of, 442
prefixes, word parts, 16–17

preganglionic fiber, ANS and, 188, 191*f*
premature heart contraction, 253*b*
premolars (bicuspids), 356
prenatal development
   defined, 442
   events in, 445*f*
   monthly changes during, 448*t*
prepuce
   covering glans penis, 425
   over clitoris, 430
presbyopia
   aging and, 215
   vision disorder, 217
pressure, environmental requirements for life, 7
presynaptic neuron, 176
PRH (prolactin-releasing hormone), 451
primary (deciduous) teeth, 355
primary follicle, 428
primary oocytes, 427
primary ossification center, 105–107
primary reproductive organs, 421
primary response
   in antibody-mediated immunity, 317
   compared with secondary, 318*f*
   defined, 307
primary spermatocytes, 422
prime movers (agonists), in movement produced by
   skeletal muscles, 149
primordial follicles, 428
primordial germ cells, 422
principle of independent assortment, in chromosome
   transmission, 453
principle of segregation (Mendel), in chromosome
   transmission, 453
PRL. *See* prolactin (PRL)
PRL (prolactin), hormones of anterior lobe, 229
probenecid, for treating gout, 407
procallus, fracture repair and, 127
procoagulant, 296
products, of chemical reactions, 28
progeria, 243
progesterone
   hormones of female sexuality, 430
   hormones of ovaries, 236
   produced by placenta during pregnancy, 237
   secretion by placenta, 447*t*
prognosis, 18
projection fibers, transmission within cerebral
   hemispheres, 180
prolactin (PRL)
   hormones of anterior lobe, 229
   physiology of lactation, 451
prolactin-releasing hormone (PRH), 451
proliferative phase, of uterine cycle, 432
pronation, types of movement, 150*f*
pronunciation, of medical terminology, 18
propagated action potential, 175. *See also* nerve
   impulses
prophase, of mitosis, 56
prophylactic agents, in pharmacology, 93
proprioception, sense of, 202
proprioceptors, 201
prostaglandins, 223, 237–238
prostate, 424
prostate cancer, 435
prostatic hypertrophy, 412
prostatic urethra, 402, 424
prostatitis, 440
protection
   activities of blood, 290
   functions of skeletal system, 103
   functions of skin, 93
protein synthesis, 59–60, 62*f*

proteins
   chemical digestion, 367–368
   classification of hormones as, 225
   defined, 21
   digestion of, 368*f*
   interconversion of carbohydrates, proteins, and
     lipids, 381
   metabolism by liver, 365
   metabolism of, 380–381
   nutrition and, 385–386
   overview of, 35–36
   structural components of plasma membrane, 46, 47*f*
prothrombin, in coagulation, 296
proton acceptors, bases as, 32
proton donors, acids as, 31–32
protons
   in atomic structure, 23
   principal particles of matter, 28*t*
proximal, directions of body, 9–10
pruritus, 100
pseudostratified columnar epithelium, in respiratory
   tract, 72*f*
psoriasis, 100
PTH (parathyroid hormone), 231–232, 232*f*
pubic arch, 122
pubic rami, 122
pubis, of coxal bones, 121
pudendum
   defined, 419
   external genitalia, 430
pulmonary arteries, 251, 272
pulmonary circuit
   illustration of, 272*f*
   overview of, 264*f*, 272
pulmonary edema, 344
pulmonary embolism, 344
pulmonary hypertension, 345
pulmonary semilunar valve
   heart valves, 250
   pathway of blood through the heart, 251
   pulmonary circuit, 272
pulmonary trunk
   pathway of blood through the heart, 251
   pulmonary circuit, 272
pulmonary veins, 251, 272
pulmonary ventilation
   overview of, 333
   pressures in, 334*f*
pulmonary vessels, 264
pulp cavity, of teeth, 356
pulp, of teeth, 356
pulse (PULS)
   blood pressure and, 269–271
   defined, 263
pulse points
   location of, 269*f*
   overview of, 269
pulse pressure, 269
Punnett squares
   for brown/blue eye color, 454*f*
   showing possible genetic combinations, 453
pupil, in structure of eyeball, 207
purpura, 100, 305
pustule, 100
pyloric region, of stomach, 357–358
pyloric sphincter, 357–358
pyramidal tracts, pathways of spinal cord, 185
pyrogens
   defined, 326
   in systemic inflammation, 314
pyruvic acid
   anaerobic breakdown of, 380*f*
   carbohydrate digestion and, 378, 379*f*

Q

QRS complex, on ECG, 254
quadrate lobe, of liver, 363
quadriceps femoris, leg muscles, 161
quickening, of fetus, 447
quinidine (Quinaglute, Cardioquin, Qunora), 253b

R

radial artery, of upper extremity, 277
radial notch, of antebrachium (forearm), 119–120
radial tuberosity, of antebrachium (forearm), 119
radial vein, of shoulders and arms, 278
radiation, sources of heat loss, 384
radioactive isotopes, 21, 24
radius
    of antebrachium (forearm), 119
    markings on, 120t
    relationship to humerus and ulna, 120f
ramus, of mandible, 113
Raynaud's phenomenon, 288
RBCs (red blood cells). See erythrocytes
reactants, chemical, 28
reaction rates, chemical, 29–30
receptor sites, for hormones, 225–226, 225f
receptors, sensory. See sense receptors
recessive traits
    in genetics, 453
    inheritance of recessive X-linked color
        blindness, 457f
recipients, blood, 298
rectum, 362
rectus abdominis, abdominal wall muscles, 155–156
rectus femoris, leg muscles, 161
red blood cells (RBCs). See erythrocytes
red fibers, aerobic respiration and, 147
red pulp, spleen tissue, 312
redness, in localized inflammation, 314
REF (renal erythropoietic factor), 293
reflex arcs
    components of, 178t
    illustration of, 178f
    overview of, 177–178
reflexes, clinically significant, 186t
refraction, of light, 201, 208
refractory period, conduction along a
    neuron, 175
regeneration, tissue repair, 79
regulation of body temperature, 384
regulation of respiration
    body temperature, 342
    chemoreceptors affecting breathing, 340–341
    impulses from brain overriding, 342
    overview of, 340–342
    respiratory center, 340
    stretch receptors and Hering-Breuer reflex, 341–342
regulation, activities of blood, 290
relative refractory period, conduction along a
    neuron, 175
relaxation phase, in contraction of whole muscle, 145
remission, 18
remodeling, fracture repair and, 127
renal arteries
    of abdominal aorta, 277
    of abdominal viscera, 278
    macroscopic structures of
        kidneys, 397
renal columns, 397–398
renal cortex, 397–398
renal diabetes, 403b
renal erythropoietic factor (REF), 293
renal fascia, 397
renal medulla, 397–398

renal papillae, 397–398
renal pelvis, 398
renal pyramids, 397–398
renal sinus, 397
renal threshold, of glucose, 402–403
renal tubule, 395
renal veins
    of abdominal and pelvic organs, 280
    blood flow through kidneys, 400
    macroscopic structures of kidneys, 397
renin, 271
    angiotensin II synthesis and, 405
    juxtaglomerular cells producing, 399
    in regulation of blood pressure, 397
    regulation of blood volume and pressure, 406f
renin-angiotensin-aldosterone mechanism
    inhibitor, 271
    in regulation of arterial blood pressure, 271
repetitive stress disorder, 168
repolarization, stimulation of neurons and, 174
reproduction, life processes, 7
reproductive system
    accessory glands of males, 424
    aging and, 434b
    disorders of, 435–436
    duct system of males, 423–424
    external genitalia of female, 430
    female, 426–433
    female sexual response, 430
    genital tract, 429
    hormonal control of female, 430–432
    hormonal control of male, 425–426
    illustration of, 4f
    male, 421–426
    male sexual response, 425
    mammary glands, 433
    ovaries, 426–428
    overview of, 6, 421
    penis, 424–425
    summary, 436–437
    testes, 421–423
residual volume, lung volumes, 335
resistance (strengthening) exercise, 162
resistance to disease
    acquired immunity, 319
    antibody-mediated immunity, 317
    cell-mediated immunity, 316
    chemical actions in disease resistance, 313
    defined, 307
    inflammation in disease resistance, 314
    lymphocytes developed by immune
        system, 316
    mechanical barriers to disease, 312–313
    nonspecific defense mechanisms, 312–314
    overview of, 312–319
    phagocytosis in disease resistance, 314
    recognition of self and nonself by immune system,
        315–316
    specific defense mechanisms, 315–317
respiration, life processes, 7
respiratory
    acidosis and alkalosis, 410t
    acidosis or alkalosis, 409, 410
respiratory (pulmonary) capacity, 335
respiratory center
    factors influencing breathing, 341f
    of midbrain, 182–183, 340f
    regulation of respiration, 340
respiratory distress syndrome, 345
respiratory membrane
    components of, 337f
    defined, 327
    external respiration and, 337

respiratory movements, venous blood flow and, 268
respiratory system
    aging and, 343b
    body temperature and, 342
    breathing rate as means of adjusting hydrogen ion
        concentration in blood, 30b
    bronchi and bronchial tree, 332
    carbon dioxide transport, 339–340, 339f
    chemoreceptors affecting breathing, 340–341
    conducting passages for, 329–333
    disorders of, 344–345, 344f
    expiration, 334
    external respiration, 337, 338f
    function and overview of respiration, 329
    functional relationships of, 328f
    Henry's law of gases, 336
    illustration of, 4f
    impulses from brain overriding, 342
    inspiration, 334
    internal respiration, 337, 338f
    larynx, 330–331
    lungs, 332–333
    mechanics of ventilation, 333
    nonrespiratory air movements, 342
    nose and nasal cavities, 329–330
    overview of, 6
    oxygen transport, 338–340
    pharmacology for, 342b
    pharynx, 330
    properties of gases, 335–336
    regulation of respiration, 340–342
    respiratory center, 340
    respiratory volumes and capacities, 335
    stretch receptors and Hering-Breuer reflex, 341–342
    summary, 345–346
    trachea, 331
    ventilation, 329–335
    volumes and capacities, 335f, 336t
responsiveness, life processes, 7
resting membrane
    depolarization and, 174
    overview of, 174
resting membrane potential, 174
reticular layer, of dermis, 90
reticulocytes
    defined, 305
    immature RBCs, 293
retina (nervous tunic)
    detached, 207b
    photomicrograph of, 208f
    rods and cones of, 207f
    in structure of eyeball, 207
reverse polarization, stimulation of neurons and, 174
reversible reactions, chemical, 30
Reye's syndrome, 199
Rh blood groups, 299–300
rheumatic fever, 131
rheumatic heart disease, 301
rheumatoid arthritis
    defined, 136
    disorders of lymphatic system, 322
rheumatologist, 136
rhinitis, 349
rhinoplasty, 349
Rho(D) immune globulin (Gamulin Rh, RhoGAM,
    HypRho-D), 297
rhodopsin
    defined, 201
    rods of eye containing, 209
ribonucleic acid (RNA)
    compared with DNA, 31t
    overview of, 38
    in protein synthesis, 59–60

ribose, pentose monosaccharides, 34
ribosomal RNA (rRNA)
    in protein synthesis, 59
    transcription, 61f
ribosomes
    cytoplasmic organelles, 49
    structure and function of cellular components, 47t
ribs, 117
rickets, 131, 388b, 393
right atrioventricular bundle branch, in conduction system, 253
right atrium
    chambers of heart, 249
    pathway of blood through the heart, 251
    pulmonary circuit, 272
right colonic hepatic flexure, 362
right common carotid artery, 276
right common iliac artery
    of abdominal aorta, 277
    of lower extremity, 278
right coronary artery
    ascending aorta, 272
    blood supply to myocardium, 251–252
right gastric veins, 280–282
right gonadal veins, 280
right hepatic duct, 363–364
right lobe, of liver, 363
right lung
    bronchopulmonary segments (lobules), 332
    point of attachment for lungs, 332
right lymphatic duct, 310
right primary bronchi, 332
right subclavian artery, 276
right suprarenal vein, 280
right ventricle
    chambers of heart, 249
    pathway of blood through the heart, 251
    pulmonary circuit, 272
ringworm, 97
RNA. See ribonucleic acid (RNA)
rods
    comparing with cones, 209t
    as photoreceptors, 208
    of retina (nervous tunic), 207f
    in structure of eyeball, 207
root
    of penis, 425
    of teeth, 356
    of tongue, 355
root canal, 356
root, of hair, 91–92
root, point of attachment for lungs, 332
roots, word parts, 16
rotation, types of movement, 150f
rotator cuff muscles, of shoulder and arm, 159
round windows, of middle ear, 210
rRNA (ribosomal RNA)
    in protein synthesis, 59
    transcription, 61f
rugae
    defined, 351
    in stomach structure, 358
    of urinary bladder, 401
rule of nines, classification of burns, 95, 95f

S

SA (sinoatrial) node, in conduction system, 252–253
saccule, static equilibrium and, 214
sacral curvature, of vertebral column, 113
sacral nerves (S1 to S5), 188
sacroiliac joint, 116
sacrum, 116

saddle joint, types of freely movable joints, 130f
sagittal plane, of body, 10, 10f
sagittal suture, parietal bones and, 108
saliva, 357
salivary glands
    components of digestive tract, 357
    location of, 357f
salpingectomy, 440
salt, neutralization reactions producing, 33
saltatory conduction, conduction along a neuron, 175, 175f
saphenous veins, of lower extremity, 282–283
sarcolemma, muscle fibers and, 142
sarcomas, 74, 85
sarcomere
    defined, 139
    myofibrils and, 143
sarcomere contraction, 144–145
sarcoplasm, 142
sarcoplasmic reticulum, 142
sartorius, leg muscles, 161
saturated fats, 36, 37f
saturated fatty acid, 36
scabies, 100
scala tympani, of inner ear, 211
scala vestibuli, of inner ear, 211
scapula (shoulder blade), in pectoral girdle, 117, 118, 118f
sclera, in structure of eyeball, 206
scrotum, 421
scurvy, 85, 388b, 393
sebaceous gland, 205
    defined, 87
    overview of, 92
sebum, 92
second heart sound (dupp), 255
second messenger, mechanisms of hormone actions, 226
second polar body, 427–428
secondary (accessory) reproductive organs, 421
secondary (lobar) bronchi, 332
secondary (permanent) teeth, 355
secondary follicle, 428
secondary oocytes, 427–428
secondary ossification center, 107
secondary response
    in antibody-mediated immunity, 317
    compared with primary, 318f
    defined, 307
secondary spermatocytes, 422–423
secretin, 237
    hormones of digestive system, 361
    pancreas and, 366–367
secretion
    classification of glands by, 73f
    of gastric glands, 358t
    of liver (bile), 364
    regulation of gastric secretions, 360f
    of small intestines, 361–362
    summary of modes of secretion, 74t
secretory phase, of uterine cycle, 432
sections, of body, 10
segmental (tertiary) bronchi, 332
segmental arteries, 400
segmental veins, 400
selective permeability
    of cells, 46
    osmosis and, 53
sella turcica, in cranial cavity, 109
semicircular canals, dynamic equilibrium and, 214
semilunar (SL) valves, 245, 250, 251
semilunar cartilage, of knee, 128–129
semimembranosus, leg muscles, 161

seminal fluid (semen), 424
seminal vesicles, 424
seminiferous tubules
    cross section of, 422f
    of testes, 422
semitendinosus, leg muscles, 161
senescence, 452
sensations, 202
sense receptors
    for auditory sense, 210
    of dermis, 90
    for gustatory sense, 203
    for olfactory sense, 204
    overview of, 202
    photoreceptors, 208–209
    for proprioception, 202
    structure of olfactory receptors in nasal cavity, 204f
    for temperature, 202–203
    for touch and pressure, 202
    for touch, pressure, and proprioception, 203f
    types of, 202, 202t
senses
    aging and, 215b
    auditory. See auditory senses
    disorders of, 216f
    equilibrium, 213–214
    general and special, 202
    gustatory, 203–204
    olfactory, 204
    pain, 203
    proprioception, 202
    receptors and sensations, 202
    temperature, 202–203
    touch and pressure, 202
    visual. See visual senses
sensorineural deafness, 217
sensory (afferent)
    division of peripheral nervous system, 171
    neurons, 173
sensory adaptation
    defined, 201
    overview of, 202
    thermoreceptors and, 202–203
sensory areas, of cerebrum, 180
sensory function, of skin, 93
sensory input, functions of nervous system, 171
sensory nerves, 187
septa, separating testes, 422
septal defects, 301
serosa, 80–81, 354
serotonin, role in vascular constriction, 296
serous cells, 73–74
serous fluid, 80–81
serous membranes
    inflammation of (peritonitis), 69b
    lungs, 332–333
    overview of, 80–81
serratus anterior, shoulder and arm muscles, 156–159
serum, 291
sesamoid bones, in cranium, 107
sex chromosomes, 452
sex-linked genetic traits, 455
shaft
    of hair, 91–92
    shape determines curl or straightness of hair, 92b
shaft, of penis, 425
sheath (myelin), 173
shell temperature, of body, 383–384
shin splint, 168
shingles, 199
short bones, 104
shoulder (pectoral) girdle, 117

shoulder blade (scapula), in pectoral girdle, 117, 118, 118*f*
shoulders
    muscles, 156–159
    veins of, 278–280
sigmoid colon, 362
signs, of disease, 18
simple columnar epithelium, in lining of stomach and
    intestines, 71*f*
simple cuboidal epithelium, in kidney tubules, 71*f*
simple diffusion, 52, 52*f*
simple epithelia, vs. stratified, 70
simple glands, 73, 73*f*
simple series circuit, nerve impulses, 177
simple squamous epithelium, 71*f*
simple sugars (monosaccharides)
    overview of, 34
    representation of, 35*f*
single replacement reactions, chemical, 29
sinoatrial (SA) node, in conduction system, 252–253
sinuses
    ethmoidal sinuses, 109
    frontal sinuses, 108
    maxillary sinus, 113
    paranasal sinuses, 329
    of skull, 107
    sphenoid sinuses, 109
sinusitis, 107*b*
sinusoids, 363–364
skeletal muscle action, in venous blood flow, 268
skeletal muscles
    attachments, 142
    fibers, 142–143
    groups, 150–162
    illustration of, 77*f*
    micrography of longitudinal section, 142*f*
    movements produced by, 149
    naming, 150–152
    nerves and blood supply for, 143
    organization and connective tissues, 142*f*
    overview of, 77, 141–142
skeletal system
    amphiarthroses, 127
    appendicular skeleton, 117–125
    articulations, 127–129
    axial skeleton, 107–117
    bone development and growth, 105–107
    classification of bones, 104
    cranium, 108–109
    diarthroses, 128–129
    disorders of, 131*f*
    divisions of, 107
    fractures and fracture repair, 126–127
    functional relationships of, 102*f*
    functions of, 103
    hyoid bone, 113
    illustration of, 4*f*
    long bones, 104–105
    lower extremity, 123–125
    medical vocabulary, 135–137
    overview of, 3–6
    pectoral girdle, 117–118
    pelvic girdle, 121–123
    skull, 107–113
    structure of bone tissue, 103
    summary, 132–133
    synarthroses, 127
    thoracic cage, 116–117
    upper extremity, 118–120
    vertebral column, 113–116
Skene's glands (paraurethral glands), 430
skin. *See* cutaneous membrane
skin cancer, exposure to sunlight and, 90*b*
skin color, 91

skull, 112*t*
    anterior view, 110*f*
    cranial floor viewed from above, 111*f*
    cranium, 108–109
    CT image of lateral surface of, 114*f*
    facial bones, 113
    fontanels of, 109
    foramina (openings) of, 112*t*
    inferior view, 111*f*
    lateral view, 110*f*
    midsagittal section, 112*f*
    overview of, 107–113
SL (semilunar) valves, 245, 250, 251
sliding filament theory, of muscular contraction,
    144–145, 144*f*
small intestine, 359–362
small intestines
    absorption of nutrients and, 369*f*
    overview of, 359
    secretions of, 361–362
    structure and features of, 359–361
    walls of, 361*f*
small intestines, hormones of, 237
small saphenous veins, of lower extremity, 282–283
smell, sense of. *See* olfactory senses
smoker's respiratory syndrome, 349
smooth muscle tissue, 77, 78*f*
SNS. *See* sympathetic nervous system (SNS)
socket (orbit), of eye, 205
sodium (Na)
    diagrams of atomic structure of, 24*f*
    formation of ionic bonds, 25*f*
soft palate, 354–355
soft palate, in nasal cavity, 329
soleus, foot and ankle muscles, 161–162
solutes
    defined, 21
    osmosis and, 53
    as part of solution, 31
    in plasma, 291*t*
solutions
    in osmosis, 53–54
    overview of, 31
solvent
    defined, 21
    osmosis and, 53
    as part of solution, 31
soma, of neuron, 171–173
somatic nervous system, 171, 187
somatomotor cortex, 180–181
somatosensory cortex, 180
spasm (vascular constriction), 296
special senses, 202
    aging and, 215*b*
    auditory sense. *See* auditory senses
    disorders of, 216*f*
    equilibrium, 213–214
    gustatory sense, 203–204
    olfactory sense, 203–204
    visual sense. *See* visual senses
specific defense mechanisms. *See* defense
    mechanisms, specific
specific resistance, 312, 315
specificity, in systemic inflammation, 315
spelling, of medical terminology, 17
sperm production. *See* spermatogenesis
spermatic cord, 423–424
spermatids, 422–423
spermatocytes, primary and secondary, 422–423
spermatogenesis
    defined, 419
    illustration of, 423*f*
    testes and, 422–423

spermatogonia, 422
spermicide, 440
spermiogenesis, 419, 423
sphenoid bone, of cranium, 109
sphenoid sinuses, 109
sphygmomanometer, for measuring blood pressure,
    269, 270*f*
spina bifida
    defined, 136
    disorders of nervous system, 195
spinal cavity, 10
spinal cord, 184–186
    cross section of, 186*f*
    organs of central nervous system, 171
spinal cord injuries, 163
spinal nerves
    peripheral nervous system, 188
    plexuses, 191*t*
spindle fibers, microtubules, 56
spine, in pectoral girdle, 118
spinothalamic tracts, 185
spinous process, of vertebrae, 114–115
spirogram, 335
spirometer, 335
spirometry, 335
spleen, 312
splenic (left colonic) flexure, 362
splenic artery, of abdominal viscera, 278
splenic vein, of abdominal and pelvic organs, 280–282
splenomegaly, disorder of lymphatic system, 322
spongy (cancellous) bone
    diploë, 104*f*
    overview of, 103
    structure of, 104*f*
spongy urethra, 402
sprains, disorder of skeletal system, 131
squamous cell carcinoma, 97
squamous cells
    classification of epithelia, 70
    simple squamous epithelium, 70
    stratified squamous epithelium, 71–72
squamous suture, temporal bones, 108
stapes (stirrup)
    auditory ossicles, 113
    of middle ear, 211
starch, 385
Starling's law of the heart, 245, 255
static equilibrium
    defined, 201
    overview of, 214
    in sense of equilibrium, 213
sternal angle, 116–117
sternocleidomastoid, muscles of neck, 153
sternum, 116–117
steroids, 37, 225
stethoscope, listening to heart sounds, 255
stillbirth, 462
stimulus
    for muscular contraction, 143–144
    threshold (liminal) and subthreshold (subliminal)
        stimuli, 145
stirrup (stapes)
    auditory ossicles, 113
    of middle ear, 211
stomach, 357–359
    emptying, 359
    features of, 358*f*
    gastric secretions, 358–359
    overview of, 357
    structure of, 357–358
stomach, hormones of, 237
storage, of glucose by liver, 365
strabismus, disorder of the senses, 217

straight tubules, of testes, 422
strains, disorder of skeletal system, 131
stratified epithelia, vs. simple, 70
stratified squamous epithelium, 72f
stratum basale, 90, 429
stratum corium. *See* dermis
stratum corneum, in structure of skin, 90
stratum functionale, 429
stratum germinativum, in structure of skin, 90
stratum granulosum, in structure of skin, 90
stratum lucidum, in structure of skin, 90
stratum spinosum, in structure of skin, 90
strengthening (resistance) exercise, 162
stressors, disrupting homeostatic balances, 8
stretch marks (striae), on skin, 90
stretch receptors, and Hering-Breuer reflex, 341–342
stretching exercises, 162
striae (stretch marks), on skin, 90
striated appearance, of skeletal muscle tissue, 77
stroke
    disorder of muscular system, 163
    disorder of nervous system, 195
stroke volume
    cardiac output, 255
    defined, 245
structural formulas
    bonding arrangements as, 28
    for caprylic acid, 36f
    for glucose, 34f
    for glucose, fructose, and galactose, 34f
    for glycerol, 36f
    for leucine and valine, 35f
    for linolenic acid, 37f
    for palmic acid, 36f
stye, 205
styloid process
    of antebrachium (forearm), 119–120
    temporal bones and, 108
subarachnoid space, in meninges, 179
subclavian arteries, of upper extremity, 277
subclavian veins
    of head and neck, 278
    of shoulders and arms, 278
subcutaneous layer
    defined, 87
    overview of, 91
subcutaneous tissue, layers of skin, 89
subliminal (subthreshold) stimulus, in muscular contraction, 145
sublingual glands, 357
subluxation, disorder of skeletal system, 131
submandibular glands, 357
submucosa
    layers of digestive tract, 354
    of urinary bladder, 401
subscapularis, shoulder and arm muscles, 159
subthreshold (subliminal) stimulus
    in muscular contraction, 145
    stimulation of neurons and, 174
sucrase, 361
sucrose
    carbohydrate digestion and, 367
    types of monosaccharides, 385
suction lipectomy, for removal of adipose tissue, 75b
sudoriferous gland
    defined, 87
    overview of, 93
suffixes, word parts, 16, 17
sugars
    components of nucleotides, 38f
    monosaccharides (simple sugars), 34, 35f
    polysaccharides (complex sugars), 34, 35f

sulci, of cerebrum, 179
sulfinpyrazone, for treating gout, 407
superficial fascia, subcutaneous layer and, 91
superficial, directions of body, 10
superior cerebellar peduncles, 183
superior colliculi, of midbrain, 182
superior mesenteric artery
    of abdominal aorta, 277
    of abdominal viscera, 278
superior mesenteric vein, of abdominal and pelvic organs, 280–282
superior nasal conchae, 109
superior sagittal sinus, in meninges, 178–179
superior vena cava, 278
superior, directions of body, 9
supination, types of movement, 150f
support function, of skeletal system, 103
supporting (sustentacular) cells, in sperm production, 422
supporting cells, of inner ear, 211–212
suppressor T cells, 316
supraorbital foramen, 108
suprarenal arteries, of abdominal viscera, 278
supraspinatus, shoulder and arm muscles, 159
suprasternal (jugular) notch, 116–117
surfactant
    defined, 327
    preventing collapse of lungs during expiration, 334
susceptibility
    defined, 307
    to disease, 312
suspension, of mixtures, 31
suspensory (Cooper's) ligaments, of breast, 433
suspensory ligaments, in structure of eyeball, 207
sustentacular (supporting) cells, in sperm production, 422
sutural bones, in cranium, 107
sutures
    immovable joints of skull, 127
    of skull, 107
sweat glands. *See* sudoriferous gland
sweat pores, 93
swelling, in localized inflammation, 314
Swim-Ear drops (isopropyl alcohol in glycerin), 206b
sympathetic nervous system (SNS)
    comparing sympathetic and parasympathetic actions, 191t
    overview of, 188–190
    subdivisions of peripheral nervous system, 171
symphysis pubis, of pelvic girdle, 121
symptoms, of disease, 18
synapses
    components of, 176f
    conduction across, 176–177
synaptic bulbs, 173
synaptic cleft, in contraction of skeletal muscles, 143
synaptic vesicles, 176
synarthroses, 101, 127
syndromes, 18
synergists
    defined, 139
    in movement produced by skeletal muscles, 149
synovial fluid, 81, 128
synovial joints
    diarthroses and, 128
    structure of, 128f
synovial membranes, 81, 128
synthesis
    of bile salts, 364
    of plasma proteins, 364

synthesis reactions, chemical, 29
syphilis, 435–436
systemic arteries
    of abdominal viscera, 278
    aortic arch, 276
    ascending aorta, 272
    descending aorta, 277
    diagram of, 274f
    of head and neck, 277
    illustration of, 273f
    of lower extremity, 278
    overview of, 272–278
    summary of arteries and regions supplied by, 275t
    of upper extremity, 277
systemic circuit, 277
    aortic arch, 276
    arteries of abdominal viscera, 278
    arteries of head and neck, 277
    arteries of lower extremity, 278
    arteries of upper extremity, 277
    blood pressure in, 267f
    descending aorta, 277
    major systemic arteries, 272–278
    major systemic veins, 278–283
    overview of, 264, 264f, 272–283
    veins of abdominal and pelvic organs, 280
    veins of head and neck, 278
    veins of lower extremity, 282–283
    veins of shoulders and arms, 278–280
    veins of thoracic and abdominal walls, 280
systemic inflammation, 314
systemic lupus erythematosus, 85, 322
systemic veins
    of abdominal and pelvic organs, 280
    diagram of, 280f
    of head and neck, 278
    illustration of, 279f
    of lower extremity, 282–283
    overview of, 278–283
    of shoulders and arms, 278–280
    summary of veins and regions supplied by, 281t
    of thoracic and abdominal walls, 280
systole
    in cardiac cycle, 254
    defined, 245
systolic pressure
    blood pressure, 269
    defined, 263

T

T (transverse) tubules, 142
T cells. *See* T lymphocytes (T cells)
T lymphocytes (T cells)
    cell-mediated immunity, 316
    development of, 316
    overview of, 312
T wave, on ECG, 254
tachycardia, 260
tail
    of pancreas, 365
    of sperm, 423
talipes, 136
talus, 124–125
target tissue
    defined, 223
    for hormones, 225–226
tarsal bones, 124–125
tarsus, 124–125
taste. *See* gustatory senses
taste buds, 203

taste cells, 203
taste hairs, 203
taste pore, 203
tectorial membrane, of inner ear, 211–212
teeth
 ages for eruption and shedding, 356*t*
 components of digestive tract, 355–356
 illustration of, 355*f*
 longitudinal section of, 356*f*
telodendria, 173
telophase, of mitosis, 57
temperature regulation
 overview of, 384
 by skin, 94, 94*f*
temperature, sense of, 202–203
temporal bones, of cranium, 108
temporal lobe
 auditory pathway and, 213
 of cerebrum, 179–180
temporal process, of zygomatic bones, 113
temporalis, muscles of mastication, 153
temporomandibular joint, 113
tendinitis, disorder of skeletal system, 131
tendons
 dense fibrous connective tissue, 75–76
 muscle attachment and, 142
teniae coli, 362
tenomyoplasty, 168
tenoplasty, 168
tenorrhaphy, 168
tensor fasciae latae, thigh muscles, 159
tentorium cerebelli, structures of
 cerebellum, 183
teratogen, 462
teres minor, shoulder and arm muscles, 159
terminal ganglia, in parasympathetic nervous
 system, 190–193
tertiary (segmental) bronchi, 332
testes (testicles)
 hormonal regulation of, 426*f*
 overview of, 421
 primary reproductive organs, 421
 sagittal section of, 422*f*
 spermatogenesis, 422–423
 structure of, 422
testes, hormones of, 236
testicular cancer, 436
testosterone
 hormones of reproduction, 425–426
 hormones of testes, 236
 production of, 236*f*
tetanus, disorder of nervous system, 195
tetany
 sustained muscular contraction, 145
 types of whole muscle contraction, 146*f*
tetralogy of Fallot, 301
thalamus
 auditory pathway and, 213
 optic tract and, 209–210
 regions of diencephalon, 182
therapeutic agents, in pharmacology, 93
thermogenesis
 defined, 377
 use of energy from food, 382
thermoreceptors
 defined, 201
 sensory adaptation and, 202–203
 types of sense receptors, 202
thigh
 lower extremity, 123
 muscles, 159
third ventricle, of brain, 184, 184*f*
thoracic aorta, 277

thoracic cage
 overview of, 116–117
 ribs, 117
 sternum, 116–117
thoracic cavity, 10
thoracic curvature, of vertebral column, 113
thoracic duct, 310
thoracic nerves (T1 to T12), 188
thoracic veins, 280
thoracic vertebrae, 115
thoracic wall muscles, 154, 155*f*
thoracocentesis, 349
thoracolumbar division, of ANS, 190. *See also*
 sympathetic nervous system (SNS)
threshold (liminal) stimulus
 in muscular contraction, 145
 stimulation of neurons and, 174
throat. *See* pharynx
thrombocytes
 blood tissues, 77*f*
 defined, 69, 289
 formed elements of blood, 291, 295
 illustration of, 292*f*
thrombocytopenia, 305
thrombus
 defined, 305
 disorders of blood, 301
thymosin, 237, 312
thymus, 312
thymus gland, 237
thyroid gland
 hormones, 230–231
 interaction of pituitary, hypothalamus, and thyroid
 glands, 231*f*
 ultrasound of, 230*f*
thyroid, larynx cartilage, 330, 331
thyroid-stimulating hormone (TSH), 228
thyroxine, hormones of thyroid gland, 230–231
TIA (transient ischemic attack), 199
tibia
 anterior view, 124*f*
 of leg, 124
 markings on, 125*t*
 radiograph showing relationship of femur, tibia, and
 fibula, 125*f*
tibial tuberosity, 124
tibialis anterior, foot and ankle muscles, 161–162
tic douloureux, 199
ticlopidine, 168
tidal volume, lung volumes, 335
timolol (Timoptic), 206*b*
tinctures, of alcohol in solution, 31
tissue repair
 fibrosis, 79
 overview of, 79
 regeneration, 79
 steps in, 80*f*
 summary, 83
tissues, 66
 adipose, 75
 blood, 76–77
 bone, 76
 bone tissue, 103
 carbon dioxide transport, 339*f*
 cardiac muscle, 78
 cartilage, 76
 connective, 74–77
 defined, 69
 dense fibrous connective tissue, 75–76
 elastic connective tissue, 76
 epithelial, 70–74
 glandular epithelium, 73–74
 inflammation, 79

tissues *(Continued)*
 levels of organization in human body, 2–3
 muscle, 77–78, 141*t*
 nerve tissue, 171–173
 nervous, 78
 overview of, 85
 pseudostratified columnar epithelium, 71
 simple columnar epithelium, 71
 simple cuboidal epithelium, 71
 simple squamous epithelium, 70
 skeletal, 77
 smooth muscle, 77
 stratified squamous epithelium, 71–72
 summary, 82–83
 transitional epithelium, 72
 unloading of oxygen in tissues, 338*f*
tocolytics, 450
toes, 124–125
tongue, mouth components of digestive tract, 355
tonsillitis, 322
tonsils, 312, 330
torso (trunk) region, of body, 11
torticollis, 168
total lung capacity, 335
total organism, levels of organization in human
 body, 2–3
touch and pressure
 receptors, 203*f*
 sense of, 202
trabeculae carneae, chambers of heart, 249
trabeculae, plates of bone, 103
trachea
 clearing blocked airway, 331*b*
 conducting passages for respiration, 331
 c-shaped cartilage ring of, 332*f*
 CT image of, 333*f*
tracheotomy, 349
trans fat, 386
transcription
 in protein synthesis, 59
 ribosomal RNA (rRNA), 61*f*
transfer RNA (tRNA), 59
transfusions
 defined, 297
 donors and recipients, 298
 preferred and permissible blood types, 299*t*
 typing and transfusions, 297–300
 universal donors and recipients, 298
transient ischemic attack (TIA), 199
transitional epithelium, 72*f*
translation, in protein synthesis, 59–60, 62*f*
transplants, preventing rejection, 316*b*
transport, of gases in respiration, 329
transportation, activities of blood, 290
transverse (T) tubules, 142
transverse colon, 362
transverse fissure, separating lobes of
 cerebellum, 183
transverse foramina, of cervical vertebrae, 115
transverse plane, of body, 10, 10*f*
transverse process
 of cervical vertebrae, 115
 of vertebrae, 114–115
transversus abdominis, abdominal wall muscles,
 155–156
trapezius, shoulder and arm muscles, 156–159
treppe
 staircase effect in muscle contraction, 146
 types of whole muscle contraction, 146*f*
triceps brachii, forearm and hand
 muscles, 159
trichloromethane (chloroform), use as
 anesthetic, 46*b*